Springer Series on

ATOMIC, OPTICAL, AND PLASMA PHYSICS 25

Springer Series on
ATOMIC, OPTICAL, AND PLASMA PHYSICS

The Springer Series on Atomic, Optical, and Plasma Physics covers in a comprehensive manner theory and experiment in the entire field of atoms and molecules and their interaction with electromagnetic radiation. Books in the series provide a rich source of new ideas and techniques with wide applications in fields such as chemistry, materials science, astrophysics, surface science, plasma technology, advanced optics, aeronomy, and engineering. Laser physics is a particular connecting theme that has provided much of the continuing impetus for new developments in the field. The purpose of the series is to cover the gap between standard undergraduate textbooks and the research literature with emphasis on the fundamental ideas, methods, techniques, and results in the field.

27 **Quantum Squeezing**
By P.D. Drumond and Z. Ficek

28 **Atom, Molecule, and Cluster Beams I**
Basic Theory, Production and Detection of Thermal Energy Beams
By H. Pauly

29 **Polarization, Alignment and Orientation in Atomic Collisions**
By N. Andersen and K. Bartschat

30 **Physics of Solid-State Laser Physics**
By R.C. Powell
(Published in the former Series on Atomic, Molecular, and Optical Physics)

31 **Plasma Kinetics in Atmospheric Gases**
By M. Capitelli, C.M. Ferreira, B.F. Gordiets, A.I. Osipov

32 **Atom, Molecule, and Cluster Beams II**
Cluster Beams, Fast and Slow Beams, Accessory Equipment and Applications
By H. Pauly

33 **Atom Optics**
By P. Meystre

34 **Laser Physics at Relativistic Intensities**
By A.V. Borovsky, A.L. Galkin, O.B. Shiryaev, T. Auguste

35 **Many-Particle Quantum Dynamics in Atomic and Molecular Fragmentation**
Editors: J. Ullrich and V.P. Shevelko

36 **Atom Tunneling Phenomena in Physics, Chemistry and Biology**
Editor: T. Miyazaki

37 **Charged Particle Traps**
Physics and Techniques of Charged Particle Field Confinement
By V.N. Gheorghe, F.G. Major, G. Werth

38 **Plasma Physics and Controlled Nuclear Fusion**
By K. Miyamoto

Vols. 1–26 of the former Springer Series on Atoms and Plasmas are listed at the end of the book

D. Kremp M. Schlanges W.-D. Kraeft

Quantum Statistics of Nonideal Plasmas

In Collaboration with T. Bornath

With 141 Figures

Springer

Professor Dr. Dietrich Kremp
PD Dr. Thomas Bornath
Universität Rostock, Institut für Physik
Universitätsplatz 3, 18051 Rostock, Germany
E-mail: dietrich.kremp@uni-rostock.de

Professor Dr. Manfred Schlanges
Professor Dr. Wolf-Dietrich Kraeft
Ernst-Moritz-Arndt-Universität Greifswald, Institut für Physik
Domstr. 10a, 17489 Greifswald, Germany
E-mail: schlanges@physik.uni-greifswald.de
wolf-dietrich.kraeft@uni-rostock.de

ISSN 1615-5653

ISBN 3-540-65284-1 Springer Berlin Heidelberg New York

Library of Congress Control Number: 2004115767

Springer is a part of Springer Science+Business Media.

springeronline.com

Printed in The Netherlands

Typesetting and prodcution: PTP-Berlin, Protago-TEX-Production GmbH, Berlin
Cover concept by eStudio Calmar Steinen
Cover design: *design & production* GmbH, Heidelberg

Printed on acid-free paper SPIN: 10481745 57/3141/YU - 5 4 3 2 1 0

Preface

During the last decade impressive development and significant advance of the physics of nonideal plasmas in astrophysics and in laboratories can be observed, creating new possibilities for experimental research. The enormous progress in laser technology, but also ion beam techniques, has opened new ways for the production and diagnosis of plasmas under extreme conditions, relevant for astrophysics and inertially confined fusion, and for the study of laser-matter interaction. In shock wave experiments, the equation of state and further properties of highly compressed plasmas can be investigated.

This experimental progress has stimulated the further development of the statistical theory of nonideal plasmas. Many new results for thermodynamic and transport properties, for ionization kinetics, dielectric behavior, for the stopping power, laser-matter interaction, and relaxation processes have been achieved in the last decade. In addition to the powerful methods of quantum statistics and the theory of liquids, numerical simulations like path integral Monte Carlo methods and molecular dynamic simulations have been applied.

This situation encourages us to present this new book on the quantum statistical theory of nonideal plasmas. The goal of this book is to present the basic theory of nonideal partially ionized plasmas from a unified point of view of quantum field theoretical methods. This book arose out of lectures given by the authors and out of their extensive experience in the field of quantum statistics and the theory of charged many-particle systems. On the one hand, an introduction is given into the quantum statistics of equilibrium and non-equilibrium systems on the basis of the methods of real-time Green's functions. On the other hand, the dynamical, the thermodynamic and the kinetic properties of strongly coupled plasmas are dealt with on a wide scale.

This book is intended as a graduate-level textbook and as a monograph on quantum statistical theory of charged many-particle systems, especially nonideal plasmas. We hope that it will be also useful to researchers in the field of plasma physics and quantum statistics.

We would like to thank all those who have encouraged and assisted us in this task. First, we are grateful to Günter Ecker for his motivation and help in the realization of this volume. We thank the team at Springer, especially Adelheid Duhm and Claus Ascheron, for their constructive collaboration.

In particular, we are grateful to Thomas Bornath. The parts about two-particle properties, kinetic equations, ionization kinetics, and laser–plasma interaction include many results worked out together with him and were written in fruitful collaboration. Our thanks go also out to Valery Bezkrovniy, Dirk Gericke, and Dirk Semkat for many helpful discussions and for critically reading parts of the manuscript.

This monograph involves many results of the long-time pleasant collaboration with our friends and colleagues Werner Ebeling, Gerd Röpke, Yury Lvovich Klimontovich (†), Klaus Kilimann, Michael Bonitz, Hubertus Stolz (†), Roland Zimmermann, Ronald Redmer, Hugh E. DeWitt, and Piet Schram. Essential results presented here have been obtained and published in cooperation with these colleagues.

Furthermore, we gratefully thank Stefan Arndt, Roman Fehr, Gordon Grubert, Paul Hilse, Hauke Juranek, Ulrike Kraeft, Sylvio Kosse, Sandra Kuhlbrodt, Renate Nareyka, Ralf Prenzel, Jörg Riemann, and Jan Vorberger for their cooperation and assistance in different stages of the genesis of this monograph while preparing the manuscript.

Finally, it is a pleasure to thank many fellow scientists of the plasma community and Green's function specialists including W. Däppen, J. Dufty, V. Filinov, V. Fortov, K. Henneberger, F. Hensel, D.H.H. Hoffmann, M. Knaup, H.S. Köhler, H.J. Kull, H.J. Kunze, N.H. Kwong, P. Lipavský, B. Militzer, K. Morawetz, P. Mulser, M. Murillo, S.V. Peletminski, F.J. Rogers, R. Sauerbrey, V.G. Špička, W. Theobald, C. Toepffer, and G. Zwicknagel for illuminating discussions and collaboration.

We gratefully acknowledge the generous support of the Deutsche Forschungsgemeinschaft.

Rostock and Greifswald
December 2004

Dietrich Kremp
Manfred Schlanges
Wolf-Dietrich Kraeft

Table of Contents

1. Introduction

In the process of the formation of matter from elementary particles up to condensed matter, the plasma state is an essential stage. In this state, matter consists of electrons and protons or ions. By the formation of bound states and by phase transitions, more complex states of matter such as liquids and solids evolve out of the plasma state. Consequently, plasmas are interesting and essential many-particle systems which are of importance for our fundamental understanding of condensed matter.

Plasmas play a fundamental role in nature. Probably more than 99 percent of visible matter in the universe exist in the plasma state. Plasmas exist, e.g., as interstellar gas, in stellar atmospheres, inside the sun, in giant planets, and in white dwarfs.

In the laboratories, plasmas were investigated first in connection with gas discharges. At the present time, plasmas are of interest in connection with magnetically confined fusion, and, at high densities, with the goal to achieve inertial fusion. Furthermore, laser produced plasmas, electron–hole plasmas in highly excited semiconductors, electrons in metals, ultracold plasmas and dusty plasmas are of importance.

Obviously, the development of a quantum statistical theory of charged many-particle systems is of great principal and practical importance. The theoretical investigation of charged many-particle systems is faced with a number of typical difficulties.

A plasma consists of freely moving charged particles, which produce charge and current densities and, therefore, an electromagnetic field; plasma particles interact with the electromagnetic field. In general, the dynamics of the plasma particles and that of the field have to be dealt with self-consistently. For many problems it is sufficient to account only for the coupling to the longitudinal field, i.e., to consider only the Coulomb interaction between the plasma particles.

The Coulomb interaction is characterized by its long range. This property of the Coulomb interaction leads to typical pecularities of charged many-particle systems, namely to the collective behavior of the plasma particles such as dynamical screening and plasma oscillations. In the theoretical description, the long range character leads to special difficulties, which turn out to be divergencies in the determination of thermodynamic or transport properties by means of cluster expansions. Binary or few-particle collision approximations are not appropriate to describe the collective interaction of plasma particles.

The collective behavior of a plasma has to be described in a self-consistent scheme of particles and fields. This was done for the first time by Debye. Debye determined the field produced by the charged system of particles in static approximation and invented the fundamental idea of a screened potential. Exploiting this idea, a number of problems of the theory of many-particle systems with Coulomb interaction were solved.

Later, in extension of the more elementary ideas of Debye, there was progress in carrying out ring summations in order to re-normalize cluster expansions as invented by Mayer for classical statistical mechanics and later by Macke in the field of quantum statistics. These techniques were further developed in the kinetic theory by Balescu, Lennard, Silin and Klimontovich where convergent collision integrals for plasmas were formulated.

Because of the complexity of plasma systems, for the complete theoretical description general methods of quantum statistics are necessary. Essential progress in the theory of nonideal plasmas was achieved by the introduction of quantum field-theoretical methods into statistical physics of equilibrium systems by Matsubara, Martin and Schwinger, and by Abrikosov, Gor'kov and Dzyaloshinski. On the basis of such techniques, the theory of the electron gas was formulated and further elaborated by Gell-Mann and Brueckner, by DuBois, Pines and Noziéres, and the equation of state for quantum plasmas was given by Montroll and Ward and by DeWitt.

Of special importance for the quantum statistics of plasmas is the generalization of the quantum field-theoretical methods to non-equilibrium systems which was achieved by Kadanoff and Baym and by Keldysh. In the papers of these authors, equations of motion for the real-time Green's functions, well known as Kadanoff–Baym equations, were derived. On the basis of the Kadanoff–Baym equations, essential progress was achieved in non-equilibrium statistical mechanics. Especially, one should remark that there now exists the possibility to deal with processes on very short time scales applying non-Markovian kinetic equations. In connection with the non-Markovian form of the collision integrals, the problem of conservation laws in nonideal plasmas was solved. On the basis of ideas of Keldysh and of Kadanoff and Baym, DuBois developed a quantum electrodynamics of plasmas and gave the foundation for a general theory of matter-radiation interaction.

As we want to stress the interrelation between theoretical methods and the physical topic, we will present the method of real-time Green's functions in statistical physics and its application to the complex problems of plasma physics.

To begin, we start with an introductory chapter on nonideal plasmas. On an elementary level, this chapter presents essential concepts of the physics of nonideal plasmas, such as static and dynamical screening and self-energy.

Self-energy and screening lead to a lowering of the ionization energy and may result in a vanishing of bound states. This latter effect is referred to as Mott effect and was for the first time investigated for plasmas by Ecker and Weizel. Furthermore, the influence of many-particle effects on thermodynamic and transport properties of plasmas is described. Especially the consequences of the Mott effect are discussed such as pressure ionization, plasma phase transition and metal insulator transition. Finally, transport processes and the ionization kinetics are considered on an elementary level.

In Chap. 3, an introduction is given to the method of real-time Green's functions in statistical physics. With real-time Green's functions, both equilibrium and non-equilibrium properties of plasmas may be described from a unique point of view. In addition to the information from the single-time distribution functions, or density matrices, respectively, the real-time Green's functions deliver dynamical information on the plasma, such as single- and two-particle energies, damping and other (quasiparticle) properties. Another advantage of such Green's functions is the existence of highly effective methods for their determination, such as Feynman diagram techniques, functional techniques and the formulation of equations of motion.

We start with the definitions and with the spectral properties of Green's functions, which are of importance for the interpretation of the apparatus and which, roughly speaking, subdivide the statistical and the dynamical information inherent in the Green's functions. Then, we outline the problem of determining these functions from the Martin–Schwinger hierarchy, and the Keldysh contour is introduced. By introduction of the self-energy, a central quantity of the approach, formally closed equations for the one-particle Green's function are derived which account for initial correlations. These equations are the Kadanoff–Baym equations which are the basic equations for an essential part of this monograph. Finally we show the connection of Green's functions with the expectation values of physical quantities, such as density, mean value of the energy, etc., i.e., with macroscopic quantities.

In Chap. 4, the theory of many-particle systems with Coulomb interaction is developed. In this chapter, the basic equations of quantum statistics of Coulomb systems, especially the equation for the self-energy, are rearranged such that the Coulomb potential is replaced by a dynamically screened potential. Here we follow the ideas of Martin and Schwinger, of Kadanoff and Baym, and of Bonch–Bruevich and Tyablikov using functional techniques. The dynamically screened potential reflects the collective behavior of the plasma and removes, at the same time, the formal difficulties of the theory of plasmas. Of course, this reformulation is much more complicated, as the dynamics of the particles and of the screened potential have to be determined self-consistently now. Finally, the dielectric function is formulated in terms of the screened potential. Furthermore, the random-phase approximation (RPA) is introduced as one of the standard approximations of many-particle theory, and on this basis, collective properties such as dynamical screening, plasma oscillations, and plasma fluctuations are extensively discussed.

However, the RPA discussed in Chap. 4 is not appropriate for the description of a number of essential plasma properties such as bound states, scattering states beyond the Born approximation, and the lowering of energy of ionization and the related Mott-effect which means pressure ionization. The simplest approximation which allows for the description of bound states is the ladder approximation, or, in plasmas, the *screened* ladder approximation, respectively. Therefore, a representation of such approximations is given in Chap. 5. The two-particle properties in a plasma are appropriately described in terms of the two-particle Green's function, which, in turn, is determined by the Bethe–Salpeter equation. We consider this equation both in statically and in dynamically screened ladder approximation, and, on this basis, we discuss the influence of the plasma on bound and scattering states, and the lowering of the ionization energy and the Mott effect.

On the basis of the approach developed so far, the thermodynamic properties of the plasma are presented in Chap. 6. We start from exact relations for the determination of thermodynamic functions of the plasma in grand canonical ensemble, and we write such functions in terms of the screened potential. Then these equations are considered on the level of the dynamically screened ladder approximation. First, the contributions of first and second order, i.e., Hartree–Fock and Montroll–Ward contributions are investigated in detail. For the aim of practical application of quantum statistical results Padé formulae are presented.

In order to account for effects of strong correlations, the contribution of higher ladder contributions to the equation of state (EOS) are investigated and represented in the form of generalized cluster coefficients. With such contributions, bound states are included in the EOS. The problem of the subdivision of the cluster coefficient into bound and scattering states at higher densities is difficult for Coulomb systems and is analyzed using higher order Levinson theorems. In connection to this, the role of the so-called Planck–Larkin sum of bound states is discussed.

In this physical picture, the plasma is a system of charged particles in bound and scattering states. The EOS in grand canonical ensemble enables us, introducing the *bound states* as *atoms*, to perform a transformation from the physical picture into a chemical description with an equilibrium of ionization. This transformation gives, in addition to the EOS in the chemical picture, a mass action law (Saha equation) for the determination of the plasma composition.

An essential part of Chap. 6 is devoted to the important problem of the EOS of the H-plasma. The hydrogen plasma is a simple but very important and interesting many-particle system. Hydrogen is the simplest and at the same time the most abundant element in the cosmos. It has been a subject of great interest due to its importance for problems such as in astrophysics, inertial fusion and our fundamental understanding of matter. In the last decade, in shock experiments, important results for the EOS of highly compressed hydrogen were obtained. Of course the screened ladder approximation presented up to now gives only asymptotic results and cannot describe the EOS over the full range of density and temperature. Here methods of liquid theory and simulations like path integral Monte-Carlo or molecular dynamic simulations are necessary.

The next chapters are devoted to the non equilibrium properties of plasmas. In Chap. 7 we develop the theory of quantum kinetic equations of nonideal plasmas. The starting point are the Kadanoff–Baym equations as the most general basis for the description of non-equilibrium processes. First we give a discussion of short-time processes and energy relaxation by numerical solutions of the Kadanoff–Baym equations. Then we derive kinetic equations relevant for plasma physics, like the quantum Landau equation, the Boltzmann equation and the Lenard–Balescu equation starting from the time diagonal Kadanoff–Baym equations. We consider the kinetic equations in their non-Markovian form and in the Markovian approximation. The properties of these equations, and especially the problem of conservation laws for nonideal systems are discussed. An essential question is the appearance of bound states in kinetic equations.

In Chap. 8, we consider relaxation processes in the hydrodynamic stage. We start with rate equations to investigate the ionization and population kinetics in nonideal plasmas. Further we treat the quantum kinetic theory of the stopping power.

Finally, in Chap. 9, we consider the behavior of dense plasmas under the influence of external electro-magnetic fields. The first part of this chapter deals with the theory of plasmas in time-dependent electric fields. Due to the recent impressive progress in laser technology, which makes femto-second laser pulses of very high intensity available, this problem is of current interest. Since the appearance of the fundamental papers of Silin we know that the description of ultra fast processes under the influence of high frequency fields needs non-Markovian kinetic equations. Therefore, we present first the generalization of the kinetic equations including laser fields. Starting from these equations a quantum kinetic theory of laser-matter interaction is developed. In weak laser fields, a generalization of the Drude formulas with frequency dependent collision frequencies is discussed, and on this basis absorption and emission of radiation in weak laser fields is described. In the case of strong laser fields, nonlinear effects are treated, and results concerning collisional absorption (inverse bremsstrahlung) are presented.

The second part of Chap. 9 gives a treatment of the theory of the statical conductivity. On the basis of the kinetic equations derived in Chap. 7, the conductivity is considered for fully and partially ionized plasmas using standard methods of transport theory. In the case of a partially ionized plasma we have to start from the kinetic equations with bound states and the conductivity is determined together with the Saha equations for the plasma composition. Taking into account the lowering of the ionization energy this consideration allows the description of the Mott transition by the behavior of the conductivity.

The bibliography at the end of the volume includes relevant monographs and original papers.

2. Introduction to the Physics of Nonideal Plasmas

2.1 The Microscopic and Statistical Description of a Fully Ionized Plasma

A plasma is a system of many charged particles. In general, a plasma consists of several components, such as electrons and different ions, with masses m_a, number densities n_a and charges e_a. In the simplest case, the ions are protons, and we have a hydrogen plasma. The H-plasma is of great importance for astrophysical problems and is a simple model to study many of the theoretical problems of plasma physics. If the formation of bound states between the particles is possible, we get a partially ionized plasma which also contains neutral particles such as atoms, molecules, and clusters.

The physical properties of plasmas are essentially determined by the fact that the motion of free charged particles is connected with current- and charge-densities, which produce an electromagnetic field. Therefore, we have electromagnetic interactions between the plasma particles.

Let us consider, for a first discussion, the plasma from the classical point of view. The most detailed description of a classical plasma is given by the location and the velocity of each plasma particle as a function of time. Following an idea invented by Klimontovich (1975), the micro state of the plasma is completely specified by the microscopic phase density in the six-dimensional phase space

$$N_a(x,t) = h^3 \sum_{i=1}^{N_a} \delta(x - x_i(t)); \qquad x = (\boldsymbol{r}, \boldsymbol{p}). \tag{2.1}$$

Here, $x_i(t)$ are the exact solutions of the equations of motion for the plasma particles

$$\frac{d}{dt}\boldsymbol{p}_i(t) = \boldsymbol{F}_i \ ; \quad \boldsymbol{p}_i(t) = m_i \frac{d}{dt}\boldsymbol{r}_i(t). \tag{2.2}$$

Let us remark that $N_a(x,t)$ is not a distribution function. It is simply a mathematical possibility of indicating, where all particles are located in the six-dimensional phase space. Clearly the total number of particles of species a N_a follows from ($h = 2\pi\hbar$)

$$\int \frac{dx}{(2\pi\hbar)^3} N_a(x,t) = N_a. \tag{2.3}$$

From the equation of motion (2.2) and from the particle conservation $dN_a/dt = 0$, the equation for the temporal evolution of $N_a(x,t)$ is easily obtained

$$\frac{\partial N_a}{\partial t} + \boldsymbol{v} \cdot \frac{\partial N_a}{\partial \boldsymbol{r}} + \boldsymbol{F}_a \cdot \frac{\partial N_a}{\partial \boldsymbol{p}} = 0\,. \tag{2.4}$$

Any microscopic observable of the plasma may be expressed by the microscopic phase densities. The densities of the electric charge and of the current of species a, $\rho_a(\boldsymbol{r},t)$ and $\boldsymbol{j}_a(\boldsymbol{r},t)$, are given by

$$\rho_a^{\mathrm{M}}(\boldsymbol{r},t) = e_a \int \frac{d\boldsymbol{p}}{(2\pi\hbar)^3} N_a(\boldsymbol{p},\boldsymbol{r},t)\,, \tag{2.5}$$

$$\boldsymbol{j}_a^{\mathrm{M}}(\boldsymbol{r},t) = e_a \int \frac{d\boldsymbol{p}}{(2\pi\hbar)^3} \frac{\boldsymbol{p}}{m_a} N_a(\boldsymbol{p},\boldsymbol{r},t)\,, \tag{2.6}$$

and are related to the particle density $n_a(\boldsymbol{r},t)$ by $\rho_a = e_a n_a$ to and the mean value of the particle velocity $\boldsymbol{u}_a(\boldsymbol{r},t)$ by $\boldsymbol{j}_a(\boldsymbol{r},t) = n_a \boldsymbol{u}_a$. Now the microscopic electromagnetic fields $\boldsymbol{E}^{\mathrm{M}}$ and $\boldsymbol{B}^{\mathrm{M}}$, which are due to the motion of the particles, and the external sources ρ^{ext} and j^{ext}, are determined directly by Maxwell's equations in terms of the charge and current densities (2.5) and (2.6). We can write

$$\begin{aligned} \nabla \cdot \boldsymbol{E}^{\mathrm{M}} = 4\pi(\rho^{\mathrm{M}} + \rho^{\mathrm{ext}}), \quad & \nabla \cdot \boldsymbol{B}^{\mathrm{M}} = 0, \\ \nabla \times \boldsymbol{E}^{\mathrm{M}} + \frac{1}{c}\frac{\partial}{\partial t}\boldsymbol{B}^{\mathrm{M}} &= \frac{4\pi}{c}(\boldsymbol{j}^{\mathrm{M}} + \boldsymbol{j}^{\mathrm{ext}}), \\ \nabla \times \boldsymbol{B}^{\mathrm{M}} - \frac{1}{c}\frac{\partial}{\partial t}\boldsymbol{E}^{\mathrm{M}} &= 0. \end{aligned} \tag{2.7}$$

Here, $\rho^{\mathrm{M}} = \sum_a \rho_a^{\mathrm{M}}$ and $\boldsymbol{j}^{\mathrm{M}} = \sum_a \boldsymbol{j}_a^{\mathrm{M}}$ are the total charge and current densities. The motion of the plasma particles and the temporal evolution of the phase density is therefore determined under the influence of the Lorentz force

$$\boldsymbol{F}^{\mathrm{Lor}} = e_a \boldsymbol{E}^{\mathrm{M}} + \frac{e_a}{c}[\boldsymbol{v} \times \boldsymbol{B}^{\mathrm{M}}]. \tag{2.8}$$

The exact relations (2.4), (2.7), and (2.8), which are referred to as Klimontovich equations, are a closed set of equations for the determination of the microscopic functions N_a, $\boldsymbol{E}^{\mathrm{M}}$, and $\boldsymbol{B}^{\mathrm{M}}$.

For a nonrelativistic plasma, i.e., if the thermal velocity is much smaller than the speed of light, it is in most cases sufficient to use the electrostatic (Coulomb) approximation. Then we may write in well-known manner

$$\Delta\phi = \sum_a e_a \int \frac{d\boldsymbol{p}}{(2\pi\hbar)^3} N_a(\boldsymbol{p},\boldsymbol{r},t)\ ;\ \boldsymbol{B}^{\mathrm{M}} = 0 \tag{2.9}$$

with the solution

$$\boldsymbol{E}^{\mathrm{M}} = -\nabla\phi = \sum_a e_a \frac{\partial}{\partial \boldsymbol{r}} \int \frac{d\boldsymbol{r}' d\boldsymbol{p}'}{(2\pi\hbar)^3} \frac{1}{|\boldsymbol{r}-\boldsymbol{r}'|} N_a\left(\boldsymbol{p}', \boldsymbol{r}', t\right) . \tag{2.10}$$

This approximation leads to a considerable simplification. With (2.10), the electric field can be completely eliminated from the Klimontovich equations, and the plasma is now a system of charged particles, which interact via the binary Coulomb potential.

$$V_{ab}(|\boldsymbol{r}_a - \boldsymbol{r}_b|) = e_a \phi_b = \frac{e_a e_b}{|\boldsymbol{r}_a - \boldsymbol{r}_b|} . \tag{2.11}$$

Sometimes such a system is called a Coulomb plasma. Let us remark that this approximation produces typical peculiarities. The long range character of the Coulomb potential leads to the fact that many-particles interact simultaneously, i.e., the plasma shows a collective behavior. As a consequence, usual approximations of statistical physics such as the binary collision approximation, or density expansions, lead to divergencies and are not applicable.

Because of the complexity of the plasma system, it is not reasonable to use a microscopic description for real plasmas. In order to describe such a complex system, methods of statistical mechanics have to be applied.

We know from textbooks of statistical mechanics that the central quantity for the statistical description of a many-particle system is the distribution function $P_N(x_1 \dots x_N, t)$ in the 6N-dimensional Γ- space. This function is the probability of finding the system at the point $\Gamma = (x_1 \dots x_N)$ in the phase space at time t. The normalization is $(1/h^{3N}) \int P_N d\Gamma = 1$. In the statistical description, macroscopic properties are given by mean values with respect to P_N. The mean value of the microscopic phase density is of special importance,

$$\begin{aligned} \langle N_a(t) \rangle &= \int \frac{dx_1 \dots dx_N}{h^{3N}} N_a(x) P_N(x_1 \dots x_N) \\ &= N_a \int \frac{dx_2 \dots dx_N}{h^{3(N-1)}} P_N(x_1 \dots x_N) = f_a(x_1) , \\ &\int \frac{dx_1}{h^3} f_a(x_1) = N_a , \end{aligned} \tag{2.12}$$

which defines the single-particle distribution function. The single-particle distribution function determines most of the properties of the plasma.

The averaged densities of the electric charge and of the current $\rho_a(r,t)$, $\boldsymbol{j}_a(\boldsymbol{r},t)$ of species a are given by

$$\rho_a(r,t) = e_a \int \frac{d\boldsymbol{p}}{(2\pi\hbar)^3} f_a\left(\boldsymbol{p}, \boldsymbol{r}, t\right) , \tag{2.13}$$

$$\boldsymbol{j}_a(\boldsymbol{r},t) = e_a \int \frac{d\boldsymbol{p}}{(2\pi\hbar)^3} \frac{\boldsymbol{p}}{m_a} f_a\left(\boldsymbol{p}, \boldsymbol{r}, t\right) \tag{2.14}$$

with the mean value of the particle velocity $\boldsymbol{u}_a(\boldsymbol{r},t)$.

Now the averaged electromagnetic field due to the motion of the particles is determined directly by Maxwell's equations in terms of the averaged densities of the charge and of the electric current (2.13) and (2.14).

In order to complete the statistical description of the plasma, we introduce the first moments describing the fluctuations of the microscopic quantities

$$\delta A = A(x,t) - \langle A(x,t)\rangle \ . \tag{2.15}$$

Here, the fluctuations of the phase densities δN and of the fields $\delta \boldsymbol{E}$ and $\delta \boldsymbol{B}$ are of special importance. Using these definitions, the equation for the evolution of the single-particle distribution function follows easily from (2.4)

$$\frac{\partial f_a}{\partial t} + \boldsymbol{v}\cdot\frac{\partial f_a}{\partial \boldsymbol{r}} + \left\langle \boldsymbol{F}_a^{\mathrm{Lor}}\right\rangle \cdot \frac{\partial f_a}{\partial \boldsymbol{p}} = -\frac{1}{n}\frac{\partial}{\partial \boldsymbol{p}}\cdot \langle \delta \boldsymbol{F}_a \delta N_a\rangle \ . \tag{2.16}$$

Equation (2.16) is usually called kinetic equation. The left hand terms describe the change of the distribution function due to the continuous drift into and out of a space element of the phase space and are called the drift term. The right hand term is the collision integral and describes the change of the distribution function by collisions. This interpretation is not fully correct for a plasma, because the force $\delta \boldsymbol{F}_a$ is produced by all plasma particles. Kinetic equations are the basis for the description of the non-equilibrium properties of plasmas. In order to determine the distribution function from (2.16), the explicit expression for the collision integral must be kown. This is a complicated problem. In the classical case, this problem is considered in the monograph by Klimontovich (1975). This problem will be later dealt with in detail from the point of view of quantum mechanics.

The scheme presented here was called by Klimontovich also the second quantization in phase space. This formulation of the classical theory is, therefore, also an appropriate frame for the quantum mechanical description of the plasma which meets the aim of this book. Especially, the quantum electrodynamics of the plasma was given in papers by DuBois (1968) and by Bezzerides and DuBois (1972).

2.2 Equilibrium Distribution Function. Degenerate and Non-degenerate Plasmas

The determination of the momentum distribution function under equilibrium and non-equilibrium conditions is one of the most important tasks of plasma theory and will be dealt with below in detail.

In the case of a classical equilibrium plasma, the distribution function is the well-known Boltzmann distribution function

$$f_a(p) = e^{-\beta(p^2/2m_a - \mu_a)} = \tilde{z}_a e^{-p^2/(2m_a k_B T)} \ . \tag{2.17}$$

Here, k_B is the Boltzmann constant, μ_a is the chemical potential, $\tilde{z}_a = \exp(\beta\mu_a)$ is called fugacity, and β is used for $\beta = 1/k_B T$.

From the Boltzmann distribution function, the thermodynamic properties of ideal classical plasmas may be obtained (Huang 1963). Using (2.17), the expressions for the fugacity and for the chemical potential follow simply as

$$\tilde{z}_a = \frac{n_a \Lambda_a^3}{2s_a + 1}, \qquad \mu_a = k_B T \ln\left(\frac{n_a \Lambda_a^3}{2s_a + 1}\right), \tag{2.18}$$

where Λ_a is the thermal wavelength defined by

$$\Lambda_a = (2\pi\hbar^2/m_a k_B T)^{1/2}.$$

Furthermore, it is easy to obtain the expressions for the internal energy and the pressure

$$U_a = \frac{3}{2} N_a k_B T, \qquad p_a = n_a k_B T. \tag{2.19}$$

The description discussed so far is realistic only for rarefied plasmas. The main goal of this book is, however, the consideration of dense plasmas. Under such conditions, we cannot neglect the quantum character of the plasma particles.

From the point of view of quantum mechanics, the state of a particle with momentum $\boldsymbol{p}$ is characterized by the Ψ-function instead of the trajectory in phase space. This is due to the Heisenberg uncertainty principle and gives a first fundamental modification of classical statistical mechanics.

The probability to find a plasma particle with momentum $\boldsymbol{p}$ at the position $\boldsymbol{r}$ is then given by

$$dw(\boldsymbol{pr}) = |\Psi_{\boldsymbol{p}}(\boldsymbol{r})|^2 \, d\boldsymbol{r}. \tag{2.20}$$

The spatial extension of this probability in thermal equilibrium is determined by the thermal wavelength Λ_a.

On the other hand, the mean particle distance d_a is

$$d_a \sim 1/\sqrt[3]{n_a}. \tag{2.21}$$

In dense plasmas, this distance may be of the same order as the thermal wavelength Λ_a, i.e., $n_a \Lambda_a^3 \approx 1$. In this case, we have an overlap of the probability clouds and have to take into account the indistinguishability of the plasma particles. This leads to a second important modification of classical statistical mechanics. The behavior of a many-particle system is now essentially determined by the spin statistic theorem. The state vectors of Bose particles are symmetric, and those of Fermi particles are antisymmetric. Therefore, we can roughly subdivide plasmas into (i) non-degenerate plasmas, if $n_a \Lambda_a^3 \ll 1$, and (ii) strongly degenerate plasmas, if $n_a \Lambda_a^3 \gg 1$.

As a consequence of the spin statistic theorem, the Boltzmann distribution function is strongly modified. For Fermi particles, i.e., for plasma particles with spin 1/2, 3/2, ..., we have now

$$f_a(\boldsymbol{p}) = \frac{1}{e^{\beta(p^2/2m_a - \mu_a)} + 1}, \tag{2.22}$$

and for Bose particles, i.e., for particles with spin $0, 1, 2, \ldots$, it follows

$$f_a(\boldsymbol{p}) = \frac{1}{e^{\beta(p^2/2m_a - \mu_a)} - 1}. \tag{2.23}$$

The expressions (2.22) and (2.23) determine the mean occupation numbers of the momentum states. They have different signs in the denominator which is of importance for low temperatures. This leads, for Fermi systems, to the existence of the Fermi energy (Pauli principle), and, for Bose systems, to the possibility of a macroscopic occupation of the ground state known as Bose–Einstein condensation.

For non-degenerate plasmas, $n_a \Lambda_a^3 \ll 1$, these functions have the common limit given by (2.17). The border between the degenerate and the non-degenerate plasmas is roughly given by the equation

$$n_a \Lambda_a^3 = n_a \left(h/\sqrt{2\pi m_a k_B T} \right)^3 = 1 .$$

The Fermi and Bose distributions (2.22) and (2.23) determine, in a well-known manner, all thermodynamic properties of an ideal quantum plasma. The chemical potential has to be determined from the normalization

$$n_a = \frac{2s_a + 1}{\Lambda_a^3} I_{1/2}(x_a), \tag{2.24}$$

and the average kinetic energy is given by

$$U_a = \langle E_{\rm kin} \rangle = \frac{3}{2} N_a k_B T \, \frac{(2s_a + 1)}{n_a \Lambda_a^3} I_{3/2}(x_a). \tag{2.25}$$

Further, one can show that the equation of state (Fermi pressure) follows from

$$p_a V = \frac{2}{3} U_a .$$

Here, we introduced the useful Fermi integrals by

$$I_\nu(x) = \frac{1}{\Gamma(\nu + 1)} \int_0^\infty dt \frac{t^\nu}{e^{t-x} + 1}, \tag{2.26}$$

where $x = \mu/k_B T$ is the dimensionless chemical potential.

Let us consider some properties of the Fermi integrals. First, $I_\nu(x)$ is a monotonically increasing function and may uniquely be inverted with respect to x. Furthermore, the Fermi integrals for different ν are related to each other by the differentiation rule

$$\frac{d}{dx} I_\nu(x) = I_{\nu-1}(x) \, . \tag{2.27}$$

For practical calculations, the behavior in limiting cases is of importance. For this purpose it is more useful to consider the Fermi integral as a function of the fugacity, i.e., $I_\nu = I_\nu(\tilde{z})$ with $\tilde{z} = \exp(\mu/k_B T)$.

Then we have for small fugacities ($\tilde{z} \ll 1$)

$$I_\nu(\tilde{z}) = \tilde{z} - \frac{1}{2^{\nu+1}} \tilde{z}^2 + \cdots \, . \tag{2.28}$$

In the opposite case $\tilde{z} \gg 1$, the well-known Sommerfeld asymptotic expansion follows, i.e.,

$$I_\nu(\tilde{z}) = \frac{1}{(\nu+1)\Gamma(\nu+1)} (ln\tilde{z})^{\nu+1} (1 + \cdots) \, . \tag{2.29}$$

We can apply these formulae to calculate the chemical potential in the limiting situations discussed above. Using (2.28), we obtain for a non-degenerate plasma (Boltzmann statistics)

$$n_a = \frac{2s_a + 1}{\Lambda_a^3} \tilde{z}_a \, , \qquad \mu_a = k_B T \ln \left(\frac{n_a \Lambda_a^3}{2s_a + 1} \right) \, . \tag{2.30}$$

In the case of a strongly degenerate plasma, we find by inversion of (2.28)

$$\mu_a = \mu_a^0 \left(1 - \frac{\pi^2}{12} \left(\frac{k_B T}{\mu_a} \right)^2 + \cdots \right) \, . \tag{2.31}$$

Here, μ_a^0 denotes the chemical potential for zero temperature which is usually called Fermi energy and denoted by ϵ^F. It is one of the most important quantities of strongly degenerate Fermi systems and is given by

$$\mu_a^0 = \epsilon_a^F = \frac{\hbar^2}{2m_a} \left(\frac{6\pi^2 n_a}{2s_a + 1} \right)^{2/3} \tag{2.32}$$

In general, the chemical potential has to be determined from (2.24) by inversion. This requires the numerical evaluation of the Fermi integrals. Results for the chemical potential as a function of the density are shown in Fig. 2.1.

For practical considerations, it is useful to have an analytical expression for the chemical potential as a function of density and temperature. Such an interpolation formula can be found by taking into account the asymptotic behavior of μ_a for weak and strong degeneracy. A good approximation is given by Zimmermann (1988)

$$\begin{aligned} \frac{\mu_a(n_a, T)}{k_B T} &= \ln y_a + 0,3536 y_a - 0,00495 y_a^2 + 0,000125 y_a^3 \, , \quad && y_a < 5.5 \, , \\ &= 1,209 y_a^{2/3} - 0,6803 y_a^{-2/3} - 0,85 y_a^{-2} \, , && y_a \geq 5.5 \, , \end{aligned} \tag{2.33}$$

where $y_a = n_a \Lambda_a^3/(2s_a + 1)$.

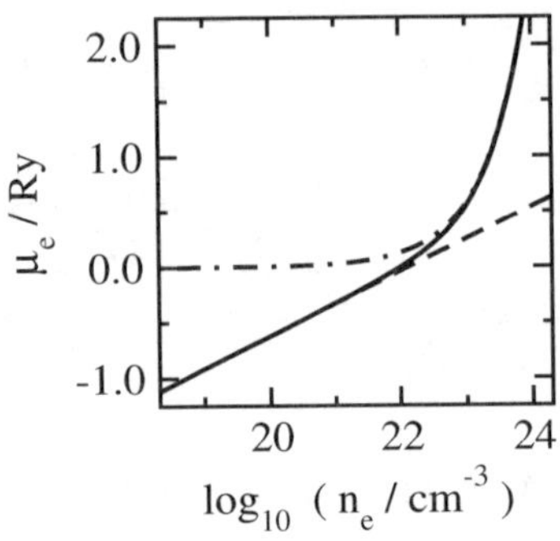

Fig. 2.1. The chemical potential of an electron gas as a function of density for a temperature $T = 12000\,\mathrm{K}$. The *solid curve* shows the full result (2.24). The *dashed line* presents the non-degenerate case according to (2.30), and the *dash-dotted* one that of the strongly degenerate case (2.31)

2.3 The Vlasov Equation

So far, we considered the equilibrium distribution function $f_a(p) = f_a^0(p)$. In non-equilibrium situations, however, we have to determine the temporal evolution of the distribution function from a kinetic equation of type (2.16). In the simplest approximation, we neglect the collision term and obtain

$$\left\{\frac{\partial}{\partial t} + \frac{\boldsymbol{p}}{m_a}\cdot\frac{\partial}{\partial \boldsymbol{r}} + \left(e_a\left\langle \boldsymbol{E}^M\right\rangle + \frac{e_a}{c}[v\times\left\langle \boldsymbol{B}^M\right\rangle]\right)\cdot\frac{\partial}{\partial \boldsymbol{p}}\right\} f_a(\boldsymbol{p},\boldsymbol{r},t) = 0\,, \tag{2.34}$$

that means, the plasma particles are considered to move independently in the average field. The averages $\left\langle \boldsymbol{E}^{\mathrm{M}}\right\rangle = \boldsymbol{E}$ and $\left\langle \boldsymbol{B}^{\mathrm{M}}\right\rangle = \boldsymbol{B}$ are produced by the charge- and current-densities of all plasma particles. Together with the Maxwell equations for the average fields

$$\begin{aligned} \nabla\cdot\boldsymbol{E} &= 4\pi(\rho+\rho^{\mathrm{ext}}), \qquad \nabla\cdot\boldsymbol{B} = 0, \\ \nabla\times\boldsymbol{E} + \frac{1}{c}\frac{\partial}{\partial t}\boldsymbol{B} &= \frac{4\pi}{c}\,(\boldsymbol{j}+\boldsymbol{j}^{\mathrm{ext}}), \\ \nabla\times\boldsymbol{B} - \frac{1}{c}\frac{\partial}{\partial t}\boldsymbol{E} &= 0\,, \end{aligned} \tag{2.35}$$

we get a self-consistent set of equations for the distribution function and for the electromagnetic field. This set of equations was for the first time investigated by Vlasov (1938), and (2.34) is therefore called *Vlasov equation.* The Vlasov equation is one of the basic equations of plasma physics. It describes most of the collective phenomena such as dynamical screening, plasma waves, instabilities, plasma turbulence, etc. (Ecker 1972). For Coulomb plasmas, the Vlasov equation simplifies to

$$\left\{\frac{\partial}{\partial t} + \frac{\boldsymbol{p}}{m_a}\cdot\frac{\partial}{\partial \boldsymbol{r}} - \frac{\partial}{\partial \boldsymbol{r}}U_a^{\mathrm{eff}}(\boldsymbol{r},t)\cdot\frac{\partial}{\partial \boldsymbol{p}}\right\} f_a(\boldsymbol{p},\boldsymbol{r},t) = 0\,. \tag{2.36}$$

Here $U_a^{\mathrm{eff}}(\boldsymbol{r},t) = e_a\Phi^{\mathrm{eff}}(\boldsymbol{r},t)$ is the effective single-particle potential given by the external and the average fields following from (2.10), i.e.,

$$U_a^{\text{eff}}(\boldsymbol{r},t) = U_a^{\text{ext}}(\boldsymbol{r},t) + \sum_b \int d\boldsymbol{r}' V_{ab}(|\boldsymbol{r}-\boldsymbol{r}'|) \int \frac{d\boldsymbol{p}}{(2\pi\hbar)^3} f_b(\boldsymbol{p},\boldsymbol{r},t) \,. \quad (2.37)$$

In an unperturbed plasma, we have $U_a^{\text{ext}}(\boldsymbol{r},t) = 0$, and the charge density vanishes as a consequence of electroneutrality.

The Vlasov equation is a complicated nonlinear equation for $f_a(\boldsymbol{p},\boldsymbol{r},t)$. In order to find a solution to (2.36), the distribution function and the effective potential have to be determined in a self-consistent manner.

Let us consider Landau's solutions (Lifschitz and Pitajewski 1983) of (2.36) as the most important and fundamental example of the Vlasov theory. The Vlasov equation is a first order differential equation with respect to the time and, therefore, initial conditions are necessary. We consider the following situation. At time $t \to -\infty$, we assume the plasma to be in equilibrium described by the homogeneous isotropic distribution function $f_a^0(p)$. Then the external potential $U_a^{\text{ext}} = e_a \Phi^{\text{ext}}$ is adiabatically switched on, i.e.,

$$\Phi^{\text{ext}}(\boldsymbol{r},t) \longrightarrow \Phi^{\text{ext}}(\boldsymbol{r},t)\, e^{\epsilon t} \,. \quad (2.38)$$

Here, ϵ is an arbitrarily small quantity, and we let $\epsilon \to 0^+$ at the end of the calculation. Clearly, for any finite time, $\Phi^{\text{ext}}(\boldsymbol{r},t)e^{\epsilon t}$ differs from $\Phi^{\text{ext}}(\boldsymbol{r},t)$ by an arbitrarily small amount, and it gives zero for $t \to -\infty$. Due to the presence of the external field, there is also a perturbation $\delta f_a(\boldsymbol{p},\boldsymbol{r},t)$ describing the deviation of the distribution function from the unperturbed one. Under this condition, the solution of the Vlasov equation should have the form

$$f_a(\boldsymbol{p},\boldsymbol{r},t) = f_a^0(p) + \delta f_a(\boldsymbol{p},\boldsymbol{r},t) \,, \quad (2.39)$$

and the corresponding effective potential $U_a^{\text{eff}} = e_a \Phi^{\text{eff}}$ is

$$\Phi^{\text{eff}}(\boldsymbol{r},t) = \Phi^{\text{ext}}(\boldsymbol{r},t) + \delta\Phi^{\text{eff}}(\boldsymbol{r},t) \,. \quad (2.40)$$

Therefore, the task is to find δf_a. Introducing (2.39) in the Vlasov equation, we obtain a nonlinear equation for δf_a. If the disturbance remains small during the time evolution, the Vlasov equation can be linearized. We get

$$\frac{\partial}{\partial t}\delta f_a + \frac{\boldsymbol{p}}{m_a}\frac{\partial}{\partial \boldsymbol{r}}\,\delta f_a + \frac{\partial}{\partial \boldsymbol{r}}\, e_a \delta\Phi^{\text{eff}} \cdot \frac{\partial f_a^0}{\partial \boldsymbol{p}} = 0 \,. \quad (2.41)$$

To solve (2.41), we apply the Fourier transformation in space and time

$$\delta f_a(\boldsymbol{p},\boldsymbol{k},\omega) = \int d\boldsymbol{r} \int_{-\infty}^{+\infty} dt\, e^{-i\boldsymbol{k}\cdot\boldsymbol{r}+i(\omega+i\epsilon)t}\, \delta f_a(\boldsymbol{p},\boldsymbol{r},t) \,. \quad (2.42)$$

Now it is easy to determine δf_a. From (2.41), we get an algebraic equation with the solution

$$\delta f_a(\boldsymbol{p},\boldsymbol{k},\omega) = \frac{e_a}{\boldsymbol{k}\cdot\frac{\boldsymbol{p}}{m_a} - \omega - i\epsilon}\,\boldsymbol{k}\,\delta\Phi^{\text{eff}}(\boldsymbol{k},\omega)\cdot\frac{\partial f_a^0(p)}{\partial\boldsymbol{p}}\,. \tag{2.43}$$

The small imaginary contribution $i\epsilon$ follows from the adiabatic switching factor $e^{\epsilon t}$ with $(\epsilon \to 0^+)$ included in (2.41). Using the inverse Fourier transformation, we find the distribution function in space and time to be

$$f_a(\boldsymbol{p},\boldsymbol{r},t) = f_a^0(p) + \int \frac{d\boldsymbol{k}}{(2\pi)^3} \int_{-\infty}^{+\infty} \frac{d\omega}{2\pi}\,\delta f_a(\boldsymbol{p},\boldsymbol{k},\omega)\,e^{i(\boldsymbol{kr}-\omega t)}\,. \tag{2.44}$$

This solution was first given by Landau (1946) and plays an important role in plasma physics. By the method of adiabatic switching, a rule to handle the pole in the propagator of (2.43) was found. The contribution $i\epsilon$ leads to the famous Landau damping (van Kampen and Felderhof 1967; Lifschitz and Pitajewski 1983).

The knowledge of δf_a allows us to find the self-consistent potential U^{eff}. After Fourier transformation, (2.37) takes the form

$$e_a\Phi^{\text{eff}}(\boldsymbol{k},\omega) = e_a\Phi^{\text{ext}}(\boldsymbol{k},\omega) + \sum_b V_{ab}(k)\int\frac{d\boldsymbol{p}}{(2\pi\hbar)^3}\,\delta f_b(\boldsymbol{p},\boldsymbol{k},\omega)\,. \tag{2.45}$$

Inserting (2.43) into this equation, we get

$$\Phi^{\text{eff}}(\boldsymbol{k},\omega) = \frac{\Phi^{\text{ext}}(\boldsymbol{k},\omega)}{1 - \sum_c V_{cc}(k)\int\frac{d\boldsymbol{p}}{(2\pi\hbar)^3}\frac{\boldsymbol{k}\cdot\partial f_c^0/\partial\boldsymbol{p}}{(\boldsymbol{k}\cdot\frac{\boldsymbol{p}}{m_c}-\omega-i\epsilon)}}\,. \tag{2.46}$$

Usually, the dynamical dielectric function is introduced by

$$\varepsilon(\boldsymbol{k},\omega+i\epsilon) = 1 + \sum_c V_{cc}(k)\int\frac{d\boldsymbol{p}}{(2\pi\hbar)^3}\frac{\boldsymbol{k}\cdot\partial f_c^0/\partial\boldsymbol{p}}{\omega - \boldsymbol{k}\cdot\boldsymbol{p}/m_c + i\epsilon}\,, \tag{2.47}$$

and we can write

$$\Phi^{\text{eff}}(\boldsymbol{k},\omega) = \Phi^{\text{ext}}(\boldsymbol{k},\omega)/\varepsilon(\boldsymbol{k},\omega)\,. \tag{2.48}$$

In this way, the change of the effective potential as the response on the external potential can be written as

$$\delta\Phi^{\text{eff}}(\boldsymbol{k},\omega) = \left(\frac{1}{\varepsilon(\boldsymbol{k},\omega)} - 1\right)\Phi^{\text{ext}}(\boldsymbol{k}\omega)\,. \tag{2.49}$$

Furthermore, from (2.43), we can find the change of the density with respect to the applied external field. If we introduce the useful quantity

$$\Pi_{aa}(\boldsymbol{k},\omega) = -\int\frac{d\boldsymbol{p}}{(2\pi\hbar)^3}\frac{\boldsymbol{k}\cdot\partial f_a^0/\partial\boldsymbol{p}}{\omega - \boldsymbol{k}\cdot\boldsymbol{p}/m_a + i\epsilon}\,, \tag{2.50}$$

the density response follows to be

$$\delta n_a(\boldsymbol{k},\omega) = \frac{\Pi_{aa}(\boldsymbol{k},\omega)}{1+\sum\limits_c V_{cc}(k)\,\Pi_{cc}(\boldsymbol{k},\omega)}\, e_a \Phi^{\rm ext}(\boldsymbol{k},\omega)\,. \tag{2.51}$$

By this formula, the physical interpretation of the quantity Π_{aa} is given. We see, if the interaction potential V_{cc} is neglected, Π_{aa} just describes the density response of a free particle system to the applied external field.

Finally, we consider the relation of the quantities discussed above to macroscopic electrodynamics in media. The electric field is described by the field strength $\boldsymbol{E}$. The sources of $\boldsymbol{E}$ are given by the total charge density. This can be expressed by

$$i\boldsymbol{k}\Phi^{\rm eff}(\boldsymbol{k},\omega) = \boldsymbol{E}(\boldsymbol{k},\omega). \tag{2.52}$$

Moreover, the dielectric displacement $\boldsymbol{D}(\boldsymbol{k},\omega)$ is introduced with sources given by the external charges only, i.e.,

$$i\boldsymbol{k}\Phi^{\rm ext}(\boldsymbol{k},\omega) = \boldsymbol{D}(\boldsymbol{k},\omega). \tag{2.53}$$

Then we find from (2.52), (2.48) and (2.53)

$$\boldsymbol{D}(\boldsymbol{k},\omega) = \varepsilon(\boldsymbol{k},\omega)\,\boldsymbol{E}(\boldsymbol{k},\omega) \tag{2.54}$$

which is a relation well-known from electrodynamics. In space-time representation, this equation gives the nonlocal spatial and temporal relation

$$\boldsymbol{D}(\boldsymbol{r},t) = \int d\boldsymbol{r}' \int\limits_{-\infty}^{t} dt' \varepsilon(\boldsymbol{r}-\boldsymbol{r}',t-t')\,\boldsymbol{E}(\boldsymbol{r}',t').$$

Therefore, the dielectric function $\varepsilon(\boldsymbol{k},\omega)$ describes the causal response of the plasma to an external perturbation. The dielectric function has the meaning of a retarded response function. It is one of the most important characteristics of the plasma.

2.4 Dynamical Screening

Let us consider a Coulomb plasma. Then the plasma properties are essentially determined by the long range Coulomb interaction.

A charged particle in a plasma does not only interact with a few neighbors but with a large number of surrounding charged particles. The result is a collective behavior of the particles in the plasma. In fact, the binary collision approximation is clearly not applicable to plasmas; it would lead formally to Coulomb divergencies.

The collective behavior which was discussed in the previous section, leads to physical effects such as dynamical screening of the Coulomb potential as expressed in (2.48) by the dielectric function ε, to plasma oscillations and plasma waves.

Furthermore, thermodynamic quantities cannot be given by virial expansions as known from the theory of gases, i.e., screening leads to a modified dependence with respect to the density.

The idea of screening was first introduced by Debye and Hückel in their famous work about the theory of electrolyte solutions (Falkenhagen 1971). It became a fundamental concept to treat many-particle systems with Coulomb interactions. In this section, we want to consider again the problem of screening from the physically obvious point of view of the Debye concept.

In order to introduce the concept of screening, we will consider a "test particle" with charge e_0 moving with the velocity $\boldsymbol{v}_0$ in the plasma. The charge density produced at $\boldsymbol{r}$ is then

$$\varrho_0(\boldsymbol{r},t) = e_0\,\delta\left(\boldsymbol{r} - \boldsymbol{v}_0 t\right) . \tag{2.55}$$

We have to expect that the charge of such a "test particle" will polarize the plasma and produce an induced charge density $\varrho^{\text{ind}}(\boldsymbol{r},t) = \sum_c \varrho_c^{\text{ind}}(\boldsymbol{r},t)$ where the summation runs over all the plasma species. Then a charged particle of species a at $\boldsymbol{r}$ responds to both the charge of the test particle and the induced charge density. The produced effective potential follows from the Poisson equation

$$\nabla^2 \Phi^{\text{eff}}\left(\boldsymbol{r},t\right) = -4\pi \left\{ e_0\,\delta\left(\boldsymbol{r} - \boldsymbol{v}_0 t\right) + \sum_c \varrho_c^{\text{ind}}\left(\boldsymbol{r},t\right) \right\} . \tag{2.56}$$

After Fourier transformation with respect to space and time, we can write the effective potential as the sum of the test particle contribution Φ^{test} and an induced part Φ^{ind}, i.e.,

$$\Phi^{\text{eff}}\left(\boldsymbol{q},\omega\right) = \Phi^{\text{test}}\left(\boldsymbol{k},\omega\right) + \Phi^{\text{ind}}\left(\boldsymbol{k},\omega\right), \tag{2.57}$$

where

$$\Phi^{\text{test}}(\boldsymbol{k},\omega) = \frac{4\pi}{k^2} e_0\,\delta(\omega - \boldsymbol{k}\cdot\boldsymbol{v}_0)\,, \tag{2.58}$$

and

$$\Phi^{\text{ind}}(\boldsymbol{k},\omega) = \frac{4\pi}{k^2} \sum_c \varrho_c^{\text{ind}}(\boldsymbol{k},\omega)\,. \tag{2.59}$$

It turns out that the determination of the effective potential reduces to find the induced charge densities $\varrho_c^{\text{ind}}(\boldsymbol{k},\omega) = e_c \delta n_c(\boldsymbol{k},\omega)$. The latter follow from the density response of the test particle to the field. The density response δn_c can be calculated from the Vlasov theory if the external field in (2.51) is replaced by the field (2.58) of the test particle. Introducing this into (2.59), the induced potential becomes

$$\Phi^{\mathrm{ind}}(\boldsymbol{k},\omega) = \left(\frac{1}{\varepsilon(\boldsymbol{k},\omega)} - 1\right)\Phi^{\mathrm{test}}(\boldsymbol{k},\omega)\,. \tag{2.60}$$

According to (2.57), the effective potential is then given by

$$\Phi^{\mathrm{eff}}(\boldsymbol{k},\omega) = \frac{\Phi^{\mathrm{test}}(\boldsymbol{k},\omega)}{\varepsilon(\boldsymbol{k},\omega)}\,. \tag{2.61}$$

Let us now consider the time behavior of Φ^{eff}. It follows by inverse Fourier transformation of (2.61)

$$\Phi^{\mathrm{eff}}(\boldsymbol{r},t) = 4\pi e_0 \int \frac{d^3k}{(2\pi)^3}\frac{1}{k^2}\frac{1}{\varepsilon(\boldsymbol{k},\boldsymbol{k}\cdot\boldsymbol{v}_0)}\, e^{i\boldsymbol{k}\cdot(\boldsymbol{r}-\boldsymbol{v}_0 t)}\,. \tag{2.62}$$

If the test particle is at rest ($\boldsymbol{v}_0 = 0$), one obtains

$$\Phi^{\mathrm{eff}}(\boldsymbol{r}) = 4\pi e_0 \int \frac{d^3k}{(2\pi)^3}\frac{1}{k^2}\frac{1}{\varepsilon(\boldsymbol{k},0)}\, e^{i\boldsymbol{k}\cdot\boldsymbol{r}}\,, \tag{2.63}$$

where $\varepsilon(\boldsymbol{k},0)$ is the static dielectric function.

We come back to the expression (2.61). Inserting (2.58), we find

$$\Phi^{\mathrm{eff}}(\boldsymbol{k},\omega) = \frac{4\pi}{k^2}\frac{1}{\varepsilon(\boldsymbol{k},\omega)}\, 2\pi e_0\,\delta(\omega - \boldsymbol{k}\cdot\boldsymbol{v}_0)\,. \tag{2.64}$$

There is a δ-function which characterizes the motion of the test particle. Furthermore, the effective potential is determined by

$$\Phi^{S}(\boldsymbol{k},\omega) = \frac{4\pi e_0}{k^2}\frac{1}{\varepsilon(\boldsymbol{k},\omega)}\,. \tag{2.65}$$

This quantity is the Fourier transform of the Coulomb potential modified by the dielectric function. Therefore, it is obvious to call $\Phi^{S}(\boldsymbol{k},\omega)$ the dynamically screened Coulomb potential.

The screened potential $\Phi^{S}(\boldsymbol{k},\omega)$ is a complicated physical quantity. However, in many cases, it is sufficient to take the important static limit. We will show that this special case gives the well-known Debye screening for a non-degenerate plasma, and the Thomas–Fermi screening for a strongly degenerate one. In the static limit, the screened potential is given by

$$\Phi^{S}(\boldsymbol{k}) = \frac{4\pi e_0}{k^2}\frac{1}{\varepsilon(\boldsymbol{k},0)}\,. \tag{2.66}$$

In order to determine the static dielectric function, we start from (2.47). Taking into account that the equilibrium distribution function has the form $f_a^0(p) = f_a^0(p^2/2m_a - \mu_a)$, we arrive at

$$\begin{aligned}\varepsilon(\boldsymbol{k},0) &= 1+\sum_c \frac{4\pi e_c^2}{k^2}\int \frac{d\boldsymbol{p}}{(2\pi\hbar)^3}\frac{\partial}{\partial\left(\frac{p^2}{2m}\right)} f_c\left(\frac{p^2}{2m}-\mu_c\right)\\ &= 1+\sum_c \frac{4\pi e_c^2}{k^2 k_B T}\frac{d}{d(\beta\mu_c)} n_c(\mu_c)\,,\end{aligned} \tag{2.67}$$

where $n_c(\mu_c, T)$ is the number density of species c. The relations (2.67) and the following ones are valid in the quantum case, too. Instead of the Boltzmann distribution, now the Fermi or Bose functions have to be used. We will show this in Chap. 4.

According to (2.67), we can write the static dielectric function in the following form

$$\varepsilon(\boldsymbol{k},0) = 1 + \frac{1}{k^2 r_0^2}\,, \tag{2.68}$$

where we introduced the screening length r_0 given by

$$r_0^{-2} = \kappa^2 = \frac{4\pi \sum_c e_c^2}{k_B T}\frac{\partial n_c}{\partial(\beta\mu_c)}\,. \tag{2.69}$$

With the dielectric function (2.68), we get the following expression for the (statically) screened potential

$$\Phi^S(\boldsymbol{k},0) = \frac{4\pi e_0}{k^2+\kappa^2}\,. \tag{2.70}$$

Fourier inversion gives then

$$\Phi^S(r) = \frac{e_0}{r}\exp(-r/r_0)\,. \tag{2.71}$$

Thus, the long range Coulomb potential of the test particle is shielded by the plasma particles, and, consequently, the screened field $\Phi^S(r)$ is of short range. The range of the potential is given by the screening radius r_0. For distances smaller than r_0, the screened potential is a Coulomb like potential whereas for larger distances the potential decreases exponentially. This screening effect is an important and fundamental property of a many-particle system with Coulomb interaction.

Now we are able to consider the effective interaction between two particles a and b in the plasma. We denote this effective interaction potential by $V_{ab}^S(|\boldsymbol{r_a}-\boldsymbol{r_b}|)$. Practically, any charged particle in the plasma behaves like a charged test particle. It polarizes the surrounding plasma, and any other particle "sees" the test particle and its screening cloud. Therefore, the effective interaction between the plasma particles is a screened one, i.e., we have in the static limit

$$V_{ab}^s(|\boldsymbol{r_a}-\boldsymbol{r_b}|) = \frac{e_a e_b}{|\boldsymbol{r_a}-\boldsymbol{r_b}|}e^{-|\boldsymbol{r_a}-\boldsymbol{r_b}|/r_0}\,. \tag{2.72}$$

Let us discuss the screening length r_0 in more detail. For a plasma of Fermi particles, this quantity is determined by the properties of Fermi integrals. From (2.69) and (2.24), we get

$$\begin{aligned} r_0^{-2} &= \kappa^2 = 4\pi \sum_c e_c^2 \frac{(2s_c+1)}{\Lambda_c^3} \frac{d}{d\mu_c} I_{1/2}(\beta\mu_c) \\ &= \sum_c \kappa_c^2 , \end{aligned} \tag{2.73}$$

where κ_c is the inverse screening length for species c. Using the limiting expressions (2.28) and (2.29) for $I_{1/2}(\beta\mu_a)$, the plasma may be discussed in the non-degenerate and highly degenerate limits. In the non-degenerate case $(n_a\Lambda_a^3 \ll 1)$, we have according to (2.28)

$$I_{1/2}(\beta\mu_a) = e^{\beta\mu_a} = \frac{n_a\Lambda_a^3}{2s_a+1} . \tag{2.74}$$

Then the screening length is given by

$$r_0 = r_D = \left(\frac{4\pi}{k_B T} \sum_c n_c e_c^2 \right)^{-1/2} . \tag{2.75}$$

This screening length is known as Debye radius, and the corresponding screened potential (2.72) is called Debye potential. The Debye potential was first introduced by Debye in connection with his work on electrolyte solutions. The expression (2.75) describes the dependence of the range r_D of the screened Coulomb potential on plasma density and temperature.

For the highly degenerate plasma $(n_a\Lambda_a^3 \gg 1)$, we find from (2.27) and (2.29)

$$\frac{d}{d\mu_a} I_{1/2}(\beta\mu_a) = \frac{1}{k_B T} \frac{(\beta\epsilon_a^F)^{1/2}}{\Gamma\left(\frac{3}{2}\right)} = \frac{3}{2\,\epsilon_a^F} \frac{n_a\Lambda_a^3}{2s_a+1} . \tag{2.76}$$

With (2.76), we get the Thomas–Fermi screening length

$$r_0 = r_{TF} = \left(\sum_c \frac{6\pi e_c^2 n_c}{\epsilon_c^F} \right)^{-1/2} = \left(\sum_c \frac{4 m_c e_c^2}{\hbar^2} \left(\frac{3 n_c}{\pi} \right)^{1/3} \right)^{-1/2} . \tag{2.77}$$

The Thomas–Fermi theory of screening was developed by Thomas and Fermi seventy years ago to give a simple model of an atom having many electrons. If we introduce r_{TF} in (2.72), we obtain the Thomas–Fermi potential. This potential decreases rapidly at large distances because the Thomas–Fermi length r_{TF} has, e.g., for electrons in metals, typical values of 1 Angström.

For arbitrary degeneracy, we have to consider the full Fermi integrals in (2.73). Because of the differentiation rule (2.27), the screening length can be written as

$$\kappa = r_0^{-1} = \left(\frac{4\pi}{k_B T} \sum_c e_c^2 \frac{2s_c + 1}{\Lambda_c^3} I_{-1/2}(\beta\mu_c) \right)^{1/2} . \tag{2.78}$$

To find the screening length as a function of density, we express the Fermi integral $I_{-1/2}(x_a)$ with $x_a = \beta\mu_a$ in terms of the dimensionless quantity $y_a = n_a \Lambda_a^3/(2s_a + 1)$. For this purpose the following relation is used

$$I_{-1/2}(x_a) = \left(\frac{dx_a}{dI_{1/2}(x_a)} \right)^{-1} = \left(\frac{dx_a}{dy_a} \right)^{-1} . \tag{2.79}$$

Here, (2.24) was taken into account, i.e., $y_a = I_{1/2}(x_a)$. The derivation dx_a/dy_a may be carried out using (2.33). Then it follows

$$I_{-1/2}(x_a) = \begin{cases} y_a/(1 + 0,353 y_a - 0,0099 y_a^2 + 0,000375 y_a^3) & , \quad y_a < 5.5, \\ y_a^{1/3}/(0,806 + 0,4535 y_a^{-4/3} + 1,7 y_a^{-8/3}) & , \quad y_a \geq 5.5 . \end{cases} \tag{2.80}$$

This interpolation formula is convenient to handle the static screening length for arbitrary degeneracy (Zimmermann 1988).

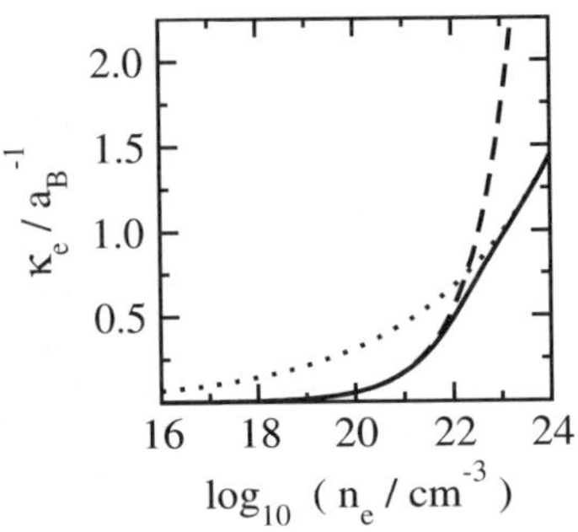

Fig. 2.2. The electron inverse screening length κ_e as a function of the density for the temperature $T = 20000\,$K. The *solid curve* represents the result for arbitrary degeneracy according to (2.73), the *dashed* one is the Debye result (2.75), and the *dotted curve* gives the inverse Thomas–Fermi screening length according to (2.77)

In order to discuss the screening length we consider the electron gas model. In Fig. 2.2, the inverse screening length is shown for different approximations as a function of density for a fixed temperature.

At the end of this section, we will deal with the question how to generalize the theory if the particles have a finite extension. This problem was considered in the classical theory of electrolytes (Falkenhagen 1971). The simplest model is to describe the plasma particles as hard spheres with a mean contact distance a. Restricting ourselves to static screening, the screened potential $\Phi^S(r)$ has the same form as for point charges. The difference is an additional constant A, i.e.,

$$\Phi^S(r) = A \frac{e_0}{r} e^{-\kappa r} , \qquad r > a . \tag{2.81}$$

In order to determine the constant, we consider the induced charge density. From the Poisson equation, it follows

$$-4\pi\varrho^{\text{ind}}(r) = A\,\kappa^2\frac{e_0}{r}\,e^{-\kappa r}\,. \tag{2.82}$$

Due to the fact that the charge of the test particle has to compensate the induced charge cloud, we have

$$4\pi\int_a^\infty drr^2\varrho^{\text{ind}}(r) = -e_0\,. \tag{2.83}$$

From this electroneutrality condition, we find

$$A = \frac{e^{\kappa a}}{1+\kappa a}\,.$$

Inserting this expression into (2.82), the induced charge density reads

$$\varrho^{\text{ind}}(r) = -\frac{1}{4\pi}\frac{\kappa^2 e^{\kappa a}}{1+\kappa a}\frac{e_0}{r}\,e^{-\kappa r}\,. \tag{2.84}$$

Finally, the statically screened interaction between two plasma particles a and b takes the form

$$V^S_{ab}(r) = \frac{e_a e_b}{r}\frac{e^{\kappa a}}{1+\kappa a}e^{-\kappa r}\,. \tag{2.85}$$

This is the Debye–Hückel potential known from the theory of electrolyte solutions. It accounts for the finite size of the interacting particles.

2.5 Self-Energy and Stopping Power

In an ideal plasma, the energy of a *probe* or *test particle* is simply $E(p) = p^2/2m$. In strongly correlated many-particle systems, however, this energy is modified by the interaction of the test particle with the surrounding plasma. The determination of the single-particle energy is a complicated problem of quantum statistical theory. This problem will be dealt with later on in greater detail.

In this section, we will give an elementary picture of plasma effects on the single-particle energy. For this purpose, let us consider a test particle with charge e_0 and velocity $\boldsymbol{v}_0$ added to the plasma. The test particle is assumed to be at the origin at time $t = 0$. We start from the mean potential energy produced by the particles of the plasma. This potential is given by

$$e_a\Phi(\boldsymbol{r},t) = \sum_b\int d\boldsymbol{r}'V_{ab}(|\boldsymbol{r}-\boldsymbol{r}'|)\int\frac{d\boldsymbol{p}}{(2\pi\hbar)^3}\left[f_b^0(p)+\delta f_b(\boldsymbol{p},\boldsymbol{r},t)\right]\,. \tag{2.86}$$

The contribution of the unperturbed distribution function $f_b^0(p)$ vanishes for electro-neutral plasmas. It follows

$$\Phi(\boldsymbol{r},t) = \Phi^{\mathrm{ind}}(\boldsymbol{r},t), \tag{2.87}$$

where the induced potential $\Phi^{\mathrm{ind}}(\boldsymbol{r},t)$ is determined by the response of the distribution function $\delta f_b(\boldsymbol{p},\boldsymbol{r},t)$ on the test particle.

In the previous section, the expression (2.60) could be derived for the induced potential. In space–time representation, it follows

$$\Phi^{\mathrm{ind}}(\boldsymbol{r},t) = 4\pi e_0 \int \frac{d\boldsymbol{k}}{(2\pi)^3}\frac{1}{k^2}\left[\frac{1}{\varepsilon(\boldsymbol{k},\boldsymbol{k}\cdot\boldsymbol{v}_0)} - 1\right] e^{i\boldsymbol{k}\cdot(\boldsymbol{r}-\boldsymbol{v}_0 t)}\,. \tag{2.88}$$

Now, we introduce the self-energy Σ_a. The self-energy is the potential energy of the test particle located at $\boldsymbol{r} = \boldsymbol{v}_0 t$ in the field of all the plasma particles, i.e., it is the interaction energy of the test particle with the surrounding plasma. Then we have

$$\Sigma_a(\boldsymbol{v}_0) = e_0\,\Phi^{\mathrm{ind}}(\boldsymbol{r},t)\Big|_{\boldsymbol{r}=\boldsymbol{v}_0 t} = 4\pi e_0^2 \int \frac{d\boldsymbol{k}}{(2\pi)^3}\frac{1}{k^2}\left[\frac{1}{\varepsilon(\boldsymbol{k},\boldsymbol{k}\cdot\boldsymbol{v}_0)} - 1\right]\,. \tag{2.89}$$

This relation can be applied to any particle in the plasma. Thus, the energy of a particle in a plasma is given by the kinetic energy term and by an additional self-energy contribution

$$E_a(\boldsymbol{p}) = \frac{p^2}{2m_a} + \mathrm{Re}\,\Sigma_{\mathrm{a}}\,. \tag{2.90}$$

This form of the one-particle energy suggests to introduce "new particles", usually called *quasiparticles* with the energy $E_a(\boldsymbol{p})$. A more rigorous foundation of the quasiparticle concept will be given in Chap. 3.

In order to get a first simple expression for the self-energy, we consider the static limit, i.e., we take $\boldsymbol{v}_0 = 0$ in (2.89). Then the expression (2.68) may be used for $\varepsilon(\boldsymbol{k},0)$. The integration is carried out easily with the result

$$\Sigma_a(0) = -\frac{e_a^2}{r_0} = -\kappa e_a^2\,. \tag{2.91}$$

This expression can also be obtained from (2.88) if the Fourier transformation with respect to $\boldsymbol{k}$ is performed. With $\boldsymbol{r} = \boldsymbol{v}_0 t$ and $\boldsymbol{v}_0 = 0$, we get

$$\Sigma_a(0) = \frac{e_a^2}{r}\left(e^{-r/r_0} - 1\right)\Big|_{r=0} = -\kappa e_a^2\,. \tag{2.92}$$

The modified behavior of the one-particle energy including the self-energy Σ_a is demonstrated in Fig. 2.3. Finally, we will give the static self-energy for particles with finite extension. For this purpose, we start from (2.86) and find for the static self-energy

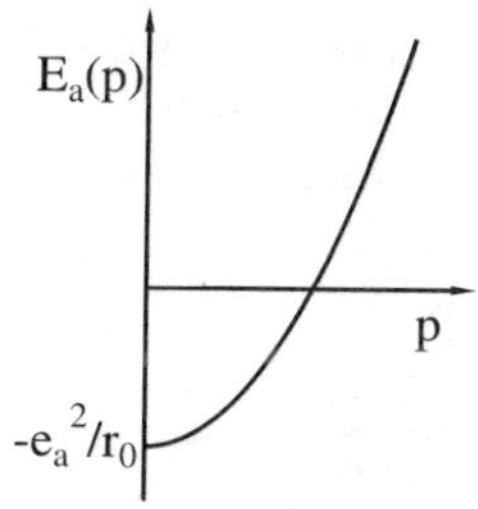

Fig. 2.3. The energy of a single-particle embedded in a plasma

$$\begin{aligned}\Sigma_a(0) &= e_a \Phi^{\mathrm{ind}}(\boldsymbol{r},t)|_{\boldsymbol{r}=0} \\ &= e_a \int d\boldsymbol{r}' \frac{\varrho^{\mathrm{ind}}(\boldsymbol{r}',t)}{|\boldsymbol{r}-\boldsymbol{r}'|}\Big|_{\boldsymbol{r}=0} . \end{aligned} \tag{2.93}$$

Now, we use the expression (2.84) for the induced charge density of particles with finite size. A simple integration gives with (2.84)

$$\Sigma_a(0) = -\frac{\kappa e_a^2}{1+\kappa a} . \tag{2.94}$$

The physical consequences of the self-energy and screening of the plasma particles are manifold and will be considered in the next sections.

An important quantity which is related to the self-energy is the stopping power. This quantity characterizes the kinetic energy loss of the test particle in the plasma. From (2.89) we can find the force acting on the moving test charge due to the surrounding plasma (Ichimaru 1992)

$$\boldsymbol{F}(\boldsymbol{v}_0) = 4\pi e_0^2 \int \frac{d\boldsymbol{k}}{(2\pi)^3} \frac{\boldsymbol{k}}{k^2} \,\mathrm{Im}\varepsilon^{-1}(\boldsymbol{k}, \boldsymbol{k}\cdot\boldsymbol{v}_0) . \tag{2.95}$$

The change of kinetic energy $E = \frac{1}{2} m_0 v_0^2$ per unit time then follows from (2.95). Using the integration variable $\omega = \boldsymbol{k}\cdot\boldsymbol{v}_0$, we get

$$\frac{dE}{dt} = -\frac{8\pi^2 e_0^2}{v_0} \int_0^{k_{\mathrm{max}}} \frac{dk}{(2\pi)^3} \frac{1}{k} \int_{-kv_0}^{+kv_0} d\omega\, \omega \,\mathrm{Im}\varepsilon^{-1}(\boldsymbol{k}, \omega) . \tag{2.96}$$

This expression represents the stopping power formula derived in the frame of dielectric theory. In other formulations, instead of the stopping power, the *stopping force* dE/dx is considered, which describes the change of the kinetic energy per unit length (in x-direction). The two quantities are related by the equation $dE/dt = v_0 dE/dx$.

In the classical approach, the cut-off parameter k_{max} is introduced to avoid divergencies. Usually, it is chosen to be the inverse of the *distance of closest approach*. For an electron to be the plasma particle, we have $k_{\mathrm{max}} = \mu_{0e} v_0^2/(e_0 e)$ with $e_0 = Z_0 e$ (Peter and Meyer-ter-Vehn 1991a; Ichimaru 1992).

At high velocities, the de Broglie wave length begins to exceed the distance of closest approach. Then one can use $k_{\max} = 2\mu_{0e}v_0/\hbar$. Furthermore, the sum rule for the dielectric function (see Chap. 4) can be applied to (2.96), and we get

$$\frac{dE}{dt} = -\frac{e_0^2\omega_{\rm pl}^2}{v_0}\int_{k_{\min}}^{k_{\max}}\frac{dk}{k} = -\frac{e_0^2\,\omega_{pl}^2}{v_0}\ln\left(\frac{2\mu_{0e}v_0^2}{\hbar\,\omega_{pl}}\right) \tag{2.97}$$

with $k_{\min} = \omega_{\rm pl}/v_0$ accounting for collective effects. Here, the contribution of the plasma electrons to the test particle stopping was included only, with μ_{0e} being the reduced mass, and $\omega_{pl} = (4\pi e^2 n_e/m_e)^{1/2}$ being the plasma frequency. With (2.97) a Bethe-type expression (Bethe 1930) is given for the energy loss due to free plasma electrons, and it shows the typical $e_0^2 = Z_0^2 e^2$ dependence on the charge of the test particle.

Stopping power calculations based on quantum kinetic equations will be discussed in Chap. 8. The energy deposition into a plasma using intense heavy ion beams is one of the possibilities to investigate strong coupling between particles (Tahir et al. 2003).

2.6 Thermodynamic Properties of Plasmas. The Plasma Phase Transition

For the description of the plasma in the equilibrium state, one needs the equation of state and other thermodynamic functions. Because of the long range Coulomb interaction, we have many peculiarities in the thermodynamics of charged particle systems. The thermodynamic behavior is essentially influenced by screening, self-energy and other collective properties. The rigorous theory of the equilibrium plasma requires the complicated formalism of quantum theory of many-particle systems and is considered later. For a first overview, we give an elementary approach due to Debye and Hückel.

For the determination of the thermodynamic functions, we start with the determination of the interaction energy $U_{\rm int}$ (Falkenhagen 1971). We know from electrodynamics that the interaction energy of a system of charged particles is

$$U_{\rm int} = \frac{1}{2}\sum_a \int \varrho_a(\boldsymbol{r}')\,\Phi(\boldsymbol{r}')\,d\boldsymbol{r}' \,. \tag{2.98}$$

Here, $\Phi(\boldsymbol{r}')$ is the potential produced by all charges at the position of the ion a, and ϱ is the charge density given by

$$\varrho_a(\boldsymbol{r}) = N_a e_a \delta(\boldsymbol{r} - \boldsymbol{r}_a)\,. \tag{2.99}$$

According to the consideration of the preceding section, we find

$$U_{\rm int} = \frac{1}{2}\sum_a N_a e_a \Phi(\boldsymbol{r}_a) = \frac{1}{2}\sum_a N_a e_a \Phi^{\rm ind}(\boldsymbol{r}_a)\,, \tag{2.100}$$

where $e_a\Phi^{\rm ind}(\boldsymbol{r}_a)$ is just the static self-energy according to (2.93). Using the simplest approximation $e_a\Phi^{\rm ind}(\boldsymbol{r}_a) = -\kappa e_a^2$, we obtain easily

$$U_{\rm int}(n,T) = -\sum_a N_a \frac{\kappa e_a^2}{2} = -V k_B T \frac{\kappa^3}{8\pi}\,. \tag{2.101}$$

We see from this expression that every particle contributes to the total interaction energy by the quantity

$$\Delta_a = -\frac{1}{2}\kappa e_a^2\,. \tag{2.102}$$

Subsequently, the expression Δ_a will be called *averaged self-energy* or *rigid shift.*

For the description of the thermodynamic properties of the plasma, it is more convenient to consider the free energy $F(T,V)$ instead of the internal energy $U(T,V)$. The latter is not a thermodynamic potential in these variables, but from thermodynamics we know the following relation between these quantities:

$$F - F_{id} = -T\int \frac{U_{\rm int}}{T^2}\,dT + C(V)\,. \tag{2.103}$$

Inserting the expression (2.101) for the interaction energy, we get

$$F - F_{id} = -\frac{1}{3}\sum_a N_a e_a^2 \kappa = -k_B T V \frac{\kappa^3}{12\pi}\,. \tag{2.104}$$

From the evident condition that $F - F_{id}$ vanishes for infinite temperatures, the integration constant C must be zero.

Of course, the most popular thermodynamic relation is the equation of state. It can be obtained from the free energy using $p = -(\partial F/\partial V)_T$. The result is

$$p - p_{id} = -k_B T \frac{\kappa^3}{24\pi} = \frac{1}{3V} U_{\rm int}\,. \tag{2.105}$$

This equation of state formula was first given by Debye and Hückel in 1923 for the osmotic pressure of ions in electrolyte solutions.

Further, the chemical potential μ_a can be obtained from the free energy by using $(\partial F/\partial N_a)_{T,V} = \mu_a$. From (2.104), we then find

$$\mu_a - \mu_a^{id} = -\frac{\kappa e_a^2}{2} = \mu_a^{\rm int}\,. \tag{2.106}$$

In general, the ideal chemical potential μ_a^{id} has to be determined by inversion of (2.24). For convenience, one can use the interpolation formula (2.33).

It is interesting to remark that the interaction part of the chemical potential in this approximation is just the averaged self-energy Δ_a, i.e.,

$$\mu_a^{\mathrm{int}}(n_a, T) = \Delta_a(n_a, T) . \tag{2.107}$$

We further write some thermodynamic functions for a plasma with ions described as charged hard spheres. Following the same line given above, the free energy reads (Falkenhagen 1971)

$$F = F_{id} - \frac{\kappa^3 k_B T V}{12\pi} \frac{3}{(\kappa a)^3} \left[\ln(1 + \kappa a) - \kappa a + \frac{1}{2}(\kappa a)^2 \right] . \tag{2.108}$$

The derivative with respect to the volume gives the pressure

$$\begin{aligned} p - p^{\mathrm{id}} &= -k_B T \frac{\kappa^3}{24\pi} \phi(\kappa a) , \\ \phi(x) &= \frac{3}{x^3} \left(1 + x - \frac{1}{1+x} - 2\log(1+x) \right) \sim 1 - \frac{3}{2}x + \cdots . \end{aligned} \tag{2.109}$$

After derivation of the free energy with respect to the particle number N_a, we have

$$\mu_a = \mu_a^{id} - \frac{\kappa e_a^2}{2} \frac{1}{1 + \kappa a} . \tag{2.110}$$

An interesting aspect of the thermodynamic relation of a charged particle system is their square root dependence on the density which is characteristic of a system with Coulomb interaction. This means that the thermodynamic functions are no analytic functions with respect to the density, and thus there is no virial expansion. This is a consequence of the long range character of the Coulomb potential. For the Coulomb potential, the second virial coefficient is divergent. These divergencies are removed by screening and lead to the square root density dependence. As we will see later, at higher orders in powers of the charge, there occurs also a logarithmic density dependence which is due to the Coulombic long range, too.

Let us now consider the chemical potential for a hydrogen plasma. The behavior of electrons and protons is determined by quantum mechanics. Therefore μ_a^{id} is given by the quantum statistical expressions (2.33), and the inverse screening length $\kappa = r_0^{-1}$ has to be taken from (2.78). Furthermore, due to the uncertainty relation, the point like plasma particles with mean thermal velocities $\langle v \rangle_a$ are localized only up to

$$\Delta r_a = \frac{\hbar}{2m_a} \frac{1}{\langle v \rangle_a} , \tag{2.111}$$

where the mean momentum $\langle p \rangle_a$ is given by

$$m_a \langle v \rangle_a = \langle p \rangle_a = 4\sqrt{\frac{m_a kT}{2\pi}} \,. \tag{2.112}$$

As a result, the uncertainty of the position is

$$\Delta r_a = \frac{1}{8}\sqrt{\frac{2\pi\hbar^2}{m_a kT}} = \frac{1}{8}\Lambda_a \,. \tag{2.113}$$

Therefore, it is obvious to consider $\Lambda_a/8$ as the *contact distance* between plasma particles, which is species-dependent. In the Debye–Hückel expression (2.110), a species-independent contact distance is used. For a rough estimation, we adopt the choice

$$a = \frac{\Lambda}{8} \sim \frac{\Lambda_e}{8} \,.$$

On the basis of this more qualitative discussion, we get for the chemical potentials of electrons and protons

$$\mu_a = \mu_a^{id} - \frac{1}{2}\frac{e_a^2 \kappa}{1 + \kappa\Lambda/8}, \qquad (a = e, p) \,. \tag{2.114}$$

Such an expression was proposed by Ebeling (a) et al. (1976) with the choice $a = \Lambda_e/8$.

In Fig. 2.4, we present the ideal and the total chemical potentials both for electrons (upper two curves) and for the electron–proton plasma $\mu_e + \mu_p$ (lower two curves) for arbitrary degeneracy. The behavior of the chemical potentials is essentially determined by the asymptotics of μ^{id} in the highly degenerate and in the non-degenerate limits. For finite densities, the ideal curves are modified by the interaction. A remarkable feature of the curves is the occurrence of Van der Waals loops for $T < T_{\text{crit}}$. These loops are produced

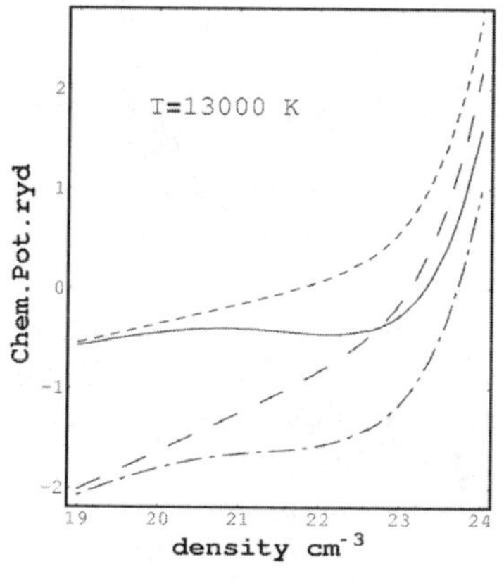

Fig. 2.4. Chemical potential of free electrons (*dashed curve*), interacting electrons (*full line*), and the plasma chemical potential (ideal electrons and protons, *long dashes*) and e–p plasma including interaction (*dash-dotted line*). The temperature is $T = 13000\,\mathrm{K}$

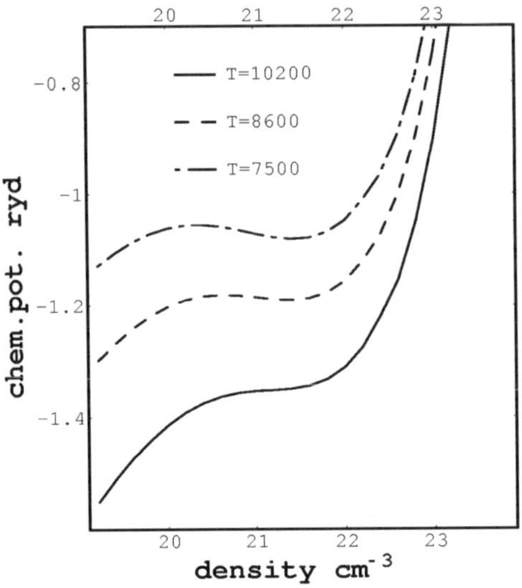

Fig. 2.5. Isotherms for the plasma chemical potential $\mu_e + \mu_p$ for various temperatures

by the interaction and lead, in certain density regions, to a violation of the stability condition

$$-\frac{\partial p}{\partial V} = \frac{n^2}{V}\frac{\partial \mu}{\partial n} \geq 0\,. \tag{2.115}$$

Such an instability may be a hint to a phase transition. The system should avoid the instability by splitting up into two coexisting phases and completely go over to the second phase on further increase of the density. The discussion of isotherms corresponding to higher temperatures leads to a critical point at

$$n_{\mathrm{crit}} = 6 \cdot 10^{21}\,\mathrm{cm}^{-3}\,, \quad T_{\mathrm{crit}} = 10200\,\mathrm{K}\,.$$

The curves in Fig. 2.5 are a hint to a special phase transition which might occur in a hydrogen plasma. This so called *plasma phase transition* was, on the basis of ideas of Landau and Zeldovich (1943), introduced by Norman and Starostin (1968) into plasma physics and discussed for ionic solutions by Ebeling (1971) and by Ebeling and Sändig (1973). This transition was considered for hydrogen and (solid state) electron–hole plasmas by Ebeling (a), Kraeft, and Kremp (1976) and by Ebeling et al. (1977). More accurate evaluations can be found in many papers (see, e.g., Ebeling and Richert (1985), Haronska et al. (1987)). In past years, this phase transition was considered very carefully in the region where the interaction with neutrals is important, see Saumon and Chabrier (1989), Schlanges et al. (1993). For an overview and new results, see Redmer (1997). Recently, this phase transition was discussed in connection with shock wave experiments and astrophysical problems (Saumon and Chabrier 1992; Redmer 1997; Ross 1996).

We will come back to this problem later and especially deal with the inclusion of bound states and the essential modifications thus produced.

2.7 Bound States in Dense Plasmas. Lowering of the Ionization Energy

In a plasma consisting of negatively and positively charged particles, the formation of bound states is possible in certain density–temperature ranges due to the attractive character of Coulomb forces. There may be reactions of the type

$$e + i_Z \rightleftharpoons i_{Z-1} \,. \tag{2.116}$$

In thermodynamic equilibrium, there is an ionization equilibrium between electrons (e), ions (i_Z) and the bound electron–ion pairs (i_{Z-1}). This equilibrium is controlled by the following relation

$$\mu_e + \mu_Z = \mu_{Z-1} \,. \tag{2.117}$$

Here, μ_Z and μ_{Z-1} denote the chemical potentials of ions with charge numbers Z and $Z-1$.

Subsequently, we consider a hydrogen plasma characterized by the total electron number density n_e^{tot} and the proton one n_p^{tot}. We assume electroneutrality and have, therefore, $n_e^{\mathrm{tot}} = n_p^{\mathrm{tot}}$. If the formation of bound states is possible, the plasma is partially ionized and consists of free electrons with density n_e, free protons with density n_p, and hydrogen atoms with density n_H. The plasma composition is described by the degree of ionization

$$\alpha = \frac{n_e}{n_e^{\mathrm{tot}}} \,, \qquad n_e^{\mathrm{tot}} = n_e + n_H \,. \tag{2.118}$$

Thus, α is an essential characteristic of the plasma.

The investigation of a partially ionized plasma has to answer three questions:

(i) How does the surrounding plasma influence the properties of an atom in dependence of density and temperature?
(ii) How do the bound states modify the thermodynamical, transport and optical properties?
(iii) How has the degree of ionization α to be determined?

In this section, we investigate the problem of an H-atom in a surrounding plasma from an elementary point of view. In quantum mechanics, the properties of the two-particle system have to be determined from the stationary Schrödinger equation

$$\mathrm{H}_{ep} \psi_{\alpha \boldsymbol{P}}(\boldsymbol{r}_1 \boldsymbol{r}_2) = E_{\alpha \boldsymbol{P}} \, \psi_{\alpha \boldsymbol{P}}(\boldsymbol{r}_1 \boldsymbol{r}_2) \,. \tag{2.119}$$

The eigen values $E_{\alpha \boldsymbol{P}}$ are the energies possible for the electron–proton pair. The eigen functions $\psi_{\alpha \boldsymbol{P}}(\boldsymbol{r}_1 \boldsymbol{r}_2)$ determine the spatial structure via $|\psi_{\alpha \boldsymbol{P}}(\boldsymbol{r}_1 \boldsymbol{r}_2)|^2$. The Hamiltonian of an isolated electron–proton pair has the form

$$\mathrm{H}_{ep} = \mathrm{H}_{ep}^0 = \frac{\mathrm{p}_1^2}{2m_e} + \frac{\mathrm{p}_2^2}{2m_p} + \mathrm{V}_{ep} \,. \tag{2.120}$$

V_{ep} is the Coulomb potential given by $V_{ep}(r) = -e^2/r$ with $r = |\boldsymbol{r}_1 - \boldsymbol{r}_2|$. Here, one has to deal with the usual theory of the hydrogen atom for $E < 0$ and that of the electron–proton scattering for $E > 0$.

If an electron–proton pair is considered to be embedded in a plasma, the Hamiltonian (2.120) has to be modified (Ecker and Weizel 1956)Ecker–Weizel potential. Now effects of the medium due to the surrounding plasma particles have to be taken into account. As shown in the preceding section, the plasma changes the single-particle energy by an additional self-energy correction, and according to such considerations, a single particle contributes to the Hamiltonian H_{ep} with the energy

$$E_a(p) = \frac{p^2}{2m_a} + \Delta_a \,, \tag{2.121}$$

where the self-energy shift Δ_a takes the value (Ebeling et al. 1977)

$$\Delta_a = -\frac{e_a^2}{2r_0} = -\frac{\kappa e_a^2}{2} \,. \tag{2.122}$$

Furthermore, the bare Coulomb potential has to be replaced by an effective potential. This is, in the simplest approximation, the statically screened Coulomb potential (Rogers et al. 1970) (2.72), i.e.,

$$V_{ep}^S(r) = -\frac{e^2}{r} \exp\left(-\kappa r\right) \,. \tag{2.123}$$

The effective Hamiltonian of an electron–proton pair, embedded in the plasma, then reads

$$\begin{aligned} \mathrm{H}_{ep} &= \frac{\mathrm{p}_1^2}{2m_e} + \frac{\mathrm{p}_2^2}{2m_p} + \Delta_{ep} + \mathrm{V}_{ep}^S \\ &= \mathrm{H}_{ep}^0 + \mathrm{H}_{ep}^{\mathrm{plasma}} \end{aligned} \tag{2.124}$$

with $\Delta_{ep} = \Delta_e + \Delta_p$ (Ebeling et al. 1977). We have introduced $\mathrm{H}_{ep}^{\mathrm{plasma}}$ to write the plasma part of the effective two-particle Hamiltonian explicitly

$$\mathrm{H}_{ep}^{\mathrm{plasma}} = \Delta_{ep} + \mathrm{V}_{ep}^S - \mathrm{V}_{ep} \,. \tag{2.125}$$

Now we use center-of-mass coordinates and such for relative motion. Taking into account that the effective Hamiltonian commutes with the square of the angular momentum operator L^2 and with the third component L_3 as well, the eigen functions of (2.124) may be written in the shape

$$\psi_{\alpha \boldsymbol{P}}\left(\boldsymbol{r}_1 \boldsymbol{r}_2\right) = \frac{u_{E,\ell}(r)}{r}\, Y_\ell^m(\vartheta, \phi)\, e^{-\frac{i}{\hbar}\boldsymbol{P}\cdot\boldsymbol{R}} \,, \tag{2.126}$$

where $\boldsymbol{R}$ is the center-of-mass coordinate, $\boldsymbol{P}$ is the total momentum, and E is the energy of relative motion. $Y_\ell^m(\vartheta, \phi)$ are the spherical harmonics with ℓ, m being the angular momentum quantum numbers. The function $u_{E,\ell}(r)$ and the possible E values have to be determined from the radial Schrödinger equation

$$\left(\frac{\hbar^2}{2\mu_{ep}} \frac{d^2}{dr^2} - \Delta_{ep} - \frac{\hbar^2}{2\mu_{ep}} \frac{\ell(\ell+1)}{r^2} - V_{ep}^S(r) + E \right) u_{E,\ell}(r) = 0 \,, \tag{2.127}$$

where $\mu_{ep} = m_e m_p/(m_e + m_p)$ is the reduced mass.

The general solution has the shape

$$u_{E,\ell}(r) = c_1\, u_{E,\ell}^{(1)}(r) + c_2\, u_{E,\ell}^{(2)}(r) \tag{2.128}$$

with the asymptotic behavior at $r \to \infty$

$$u_{E,\ell}(r) \longrightarrow c_1\, e^{-\lambda r} + c_2\, e^{\lambda r} \,. \tag{2.129}$$

Here we introduced $\lambda = \sqrt{-2\mu_{ep}(E - \Delta_{ep})/\hbar^2}$.
The eigen functions $u_{E,\ell}$ have to be finite for all r. Especially, for small r-values, the physical (regular) solution has to behave like

$$\lim_{r \to 0} = r^\ell \,. \tag{2.130}$$

From (2.128), it follows the condition

$$-\frac{c_2}{c_1} = \frac{u_{E,\ell}^{(1)}(r)}{u_{E,\ell}^{(2)}(r)} = f_\ell(E) \,. \tag{2.131}$$

There is a different behavior of the solutions depending on the value of $(E - \Delta_{ep})$. For $(E - \Delta_{ep}) < 0$, a finite solution is realized only if the constant c_2 in (2.129) is taken to be $c_2 = 0$. It follows

$$u_{E,\ell}(r) = c_1\, e^{-\lambda r} \,; \quad \lambda = \sqrt{\frac{2m|E - \Delta_{ep}|}{\hbar^2}} \,. \tag{2.132}$$

In this case, $u_{E,\ell}(r)$ is zero for large r. The electron is located at finite distance from the proton, i.e., the electron–proton pair forms a bound state. Due to the fact that $c_2 = 0$, the energy eigenvalues are restricted by the condition

$$f_\ell(E) = 0 \,, \tag{2.133}$$

The result is a discrete energy spectrum. The zeros of (2.133) determine a finite set of possible binding energies $E_{n\ell}$ classified by the principal quantum number n and the angular momentum quantum number ℓ

$$n = 1, 2, \cdots \,, \qquad \ell = 0, 1, \cdots, (n-1) \,.$$

For $(E - \Delta_{ep}) > 0$, the solutions are oscillating, and we may take finite values for both c_1 and c_2. Physically, the latter solutions have the following meaning: The probability density $|\psi_{\alpha \boldsymbol{P}}|^2$ is finite for $r \to \infty$, i.e., we have an unrestricted motion. Therefore, electron–proton scattering is described for $(E - \Delta_{ep}) > 0$.

Let us now consider the influence of the plasma medium on the discrete energy spectrum starting from (2.127). We have a radial Schrödinger equation with a statically screened Coulomb potential. This equation cannot be solved analytically. For a preliminary discussion, we expand the potential according to

$$V_{ep}^S(r) \approx -\frac{e^2}{r} + \kappa e^2 . \tag{2.134}$$

In this approximation, the first correction to the bare Coulomb potential compensates the self-energy shift Δ_{ep} in the effective Schrödinger equation (2.127). This potential was used by Ecker and Weizel (1956). It follows that the bound states are equal to that of the Coulomb potential, and the ground state energy of the H-atom is approximated by

$$E_{10} = -\frac{e^2}{2a_B} + \mathcal{O}(\kappa^2 e^2) \tag{2.135}$$

with $a_B = \hbar^2 / \left(m_e e^2\right)$ being the Bohr radius. Thus, we observe a compensation between the self-energy correction and screening. Only at higher densities, the Coulomb ground state energy is modified; see later.

The scattering energies $E = E_{\boldsymbol{p}}$ of the electron proton pair have to be determined from the asymptotic behavior of the Schrödinger equation (2.127). For $r \to \infty$, we get

$$E_{\boldsymbol{p}} = \frac{p^2}{2\mu_{ep}} + \Delta_e + \Delta_p . \tag{2.136}$$

The position of the continuum edge E_{cont} is of special importance. To determine E_{cont}, we take zero relative momentum $p = 0$. In the center of mass system, we may write

$$E_{\text{cont}} = \Delta_e + \Delta_p = -\kappa e^2 . \tag{2.137}$$

Up to now, a more qualitative discussion was given. Of course, the Schrödinger equation (2.127) can be solved numerically. This was done using the Numerov algorithm (Numerov 1923). Applying (2.130) and the boundary condition (2.132), the wave functions $u_{n\ell}$ and the binding energies $E_{n\ell}$ were calculated for different inverse screening lengths $\kappa = r_D^{-1} = (8\pi n_e e^2 / k_B T)^{1/2}$. Results for the ground state energy and the energies of excited states are shown in Fig. 2.6 as a function of density.

It turns out that the plasma modifies the hydrogen atom in the following way:

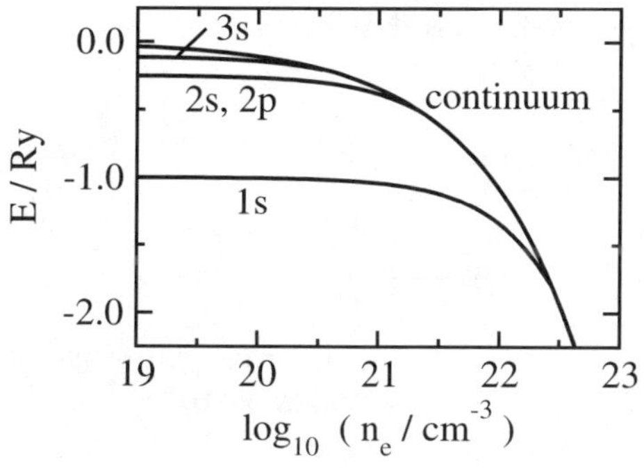

Fig. 2.6. Continuum edge and binding energies for an H-atom in the hydrogen plasma as a function of the free electron density. The temperature is $T = 20000\,\mathrm{K}$. The difference between the 2s and 2p energies cannot be resolved in this scale. Statical screening (Ecker–Weizel model); see also Sect. 5.9

(i) There is a lowering of the continuum edge given by (2.137).

(ii) The bound state energies can be written as

$$E_{n\ell} = E_n^0 + \Delta_{n\ell}\,, \tag{2.138}$$

where $\Delta_{n\ell}$ is determined by the solution of (2.127). The energies $E_{n\ell}$ are lowered, but they show a weaker density dependence as compared to the continuum edge. This follows from the compensation effect between the self-energy shift Δ_{ep} and the screened potential discussed above.

(iii) The influence of the plasma leads to a lowering of the ionization energy, i.e., we have an effective ionization energy. In our approximation it is given by

$$\begin{aligned} I_{n\ell}^{\text{eff}} &= |E_{n\ell}| + \Delta_{ep} \\ &= |E_n^0| - \Delta_{n\ell} + \Delta_e + \Delta_p\,. \end{aligned} \tag{2.139}$$

In contrast to the isolated atom, we have only a finite number of bound states.

(iv) For $I_{n\ell}^{\text{eff}} = 0$, the bound state vanishes and merges into the scattering continuum. This is referred to as the *Mott effect* (Mott 1961). The critical density (Mott density) for the break up of the atomic ground state is determined by $r_0 = 1.19a_B$ (Rogers et al. 1970).

2.8 Ionization Equilibrium and Saha Equation. The Mott-Transition

In the previous section, we considered the properties of an atom embedded in a dense plasma. Now we focus on the question how the atoms modify the properties of the plasma. Especially, the plasma composition following from the ionization equilibrium is discussed in this section. In order to present the basic concepts, we restrict ourselves to the simplest system, namely, the hydrogen plasma with the reaction

$$e + p \rightleftharpoons H(j)\,. \tag{2.140}$$

Here, $H(j)$ denotes a hydrogen atom in the state $|j\rangle$ with the set of quantum number $j = n, \ell, m$. More complex systems including molecules and highly charged ions are discussed below.

The hydrogen plasma is considered to be in thermodynamic equilibrium and to have the total electron density $n_e^{tot} = n_p^{tot}$. On account of the ionization and recombination reactions, the electrons and protons are freely moving in scattering states, or they are kept in bound states. The densities of free electrons and protons are denoted by n_e and n_p. The total density n_H of atoms is

$$n_H = \sum_j n_H^j \tag{2.141}$$

with n_H^j being the occupation number of atoms in the state $|j\rangle$.

The plasma composition can be described by the degree of ionization

$$\alpha = \frac{n_e}{n_e^{tot}}\,, \qquad n_e^{tot} = n_e + n_H\,, \tag{2.142}$$

where the second equation represents the electron density balance. It is known that the chemical equilibrium for the reaction (2.140) is controlled by the relation

$$\mu_e + \mu_p = \mu_j \tag{2.143}$$

with μ_e and μ_p being the chemical potentials of free electrons and protons, while μ_j is the chemical potential of atoms in state $|j\rangle$. For the determination of the plasma composition from (2.143), we need explicit expressions for the chemical potentials as a function of density. The chemical potentials consist of a contribution corresponding to the ideal plasma, and of one accounting for the interaction between the plasma particles. In the simplest approximation, for the charged particles, we may use the expressions derived in Sect. 2.6, i.e.,

$$\mu_a\left(n_a, T\right) = \mu_a^{id}\left(n_a, T\right) + \Delta_a\left(n_a, T\right)\,, \tag{2.144}$$

where the interaction contribution to the chemical potential is given by

$$\Delta_a = \mu_a^{\text{int}} = -\frac{\kappa e_a^2}{2}\,; \qquad a = e, p\,. \tag{2.145}$$

Accounting for a contact distance between the plasma particles, we have according to (2.114)

$$\Delta_a = \mu_a^{\text{int}} = -\frac{\kappa e_a^2}{2}\frac{1}{1+\kappa\frac{\Lambda_a}{8}}\,. \tag{2.146}$$

The first contribution on the r.h.s. of (2.144) is the ideal chemical potential. It follows from (2.24) by inversion. A useful approximation was given by the interpolation formulae (2.33).

We assume the atoms to be quasiparticles with internal degrees of freedom. The energy states are given by the effective Schrödinger equation (2.127) discussed in Sect. 2.7. The influence of the plasma medium leads to energy eigen values $E_j(n_e, T)$ depending on density and temperature.

Hydrogen atoms are Bose particles. For atoms in the state $|j\rangle$, the momentum distribution function is

$$F_j(P) = \frac{1}{e^{\beta\left(\frac{P^2}{2M}+E_j-\mu_j\right)}-1} . \tag{2.147}$$

Here, $M = m_e + m_p$ is the atomic mass, $\boldsymbol{P} = \boldsymbol{p}_e + \boldsymbol{p}_p$ is the total momentum, and $\beta = 1/k_B T$. According to the results obtained in the previous Sect. 2.7 the bound state energies can be written as $E_j(n_e, T) = E_j^0 + \Delta_j(n_e, T)$ with E_j^0 being the energy of the isolated atom, and Δ_j being the self-energy shift. The chemical potential μ_j follows (by inversion) from the formula

$$n_j\left(\mu_j, T\right) = \int \frac{d\boldsymbol{P}}{(2\pi\hbar)^3} \frac{1}{e^{\beta\left(\frac{P^2}{2M}+E_j-\mu_j\right)}-1} . \tag{2.148}$$

The relations (2.143–2.148) determine, in principle, the plasma composition and thus the degree of ionization α as a function of temperature and density. In many cases, it is sufficient to consider the non-degenerate case. For the atoms, we then get the simpler expression

$$\mu_j = k_B T \ln\left(\frac{n_j \Lambda_H^3\, e^{\beta E_j}}{\left(2s_e+1\right)\left(2s_p+1\right)}\right) \tag{2.149}$$

with Λ_H being the thermal wavelength of the atoms $\Lambda_H = (2\pi\hbar^2/Mk_BT)^{1/2}$. In the non-degenerate case, the chemical potentials of the free electrons and protons are

$$\mu_a = k_B T \ln\left(\frac{n_a \Lambda_a^3}{2s_a+1}\right) + \Delta_a , \qquad a = e, p . \tag{2.150}$$

Inserting (2.150) and (2.149) into (2.143), we get after summation over the states j

$$\frac{n_H}{n_e n_p} = \Lambda_e^3\, \sigma_H\left(n_e, T\right) e^{\beta(\Delta_e+\Delta_p)} = K_H\left(n_e, T\right) . \tag{2.151}$$

Here, we used the fact that $m_e \ll m_p$. K_H is the mass action constant, and the abbreviation σ_H denotes the atomic sum of states

$$\begin{aligned} \sigma_H\left(n_e, T\right) &= \sum_j e^{-\beta E_j(n_e,T)} \\ &= \sum_{n\ell} (2\ell+1)\, e^{-\beta(E_n^0+\Delta_{n\ell})} . \end{aligned} \tag{2.152}$$

Equation (2.151) represents a mass action law describing the ionization equilibrium between free electrons, protons, and hydrogen atoms. In plasma physics, it is referred to as the *Saha equation*.

For two distinct atomic levels with energies $E_r(n_e, T)$ and $E_s(n_e, T)$, we may write

$$\frac{n_H^r}{n_H^s} = e^{-\beta(E_r - E_s)}\,. \tag{2.153}$$

With (2.151) and (2.153), the composition of the nonideal plasma in thermodynamic equilibrium is determined, i.e., the densities of free and bound particles as well as the occupation numbers of the different levels can be calculated.

Using the degree of ionization α, we may write instead of (2.151)

$$\frac{1-\alpha}{\alpha^2} = n_e^{\text{tot}} \Lambda_e^3\, \sigma_H(n_e, T)\, e^{\beta(\Delta_e + \Delta_p)}\,. \tag{2.154}$$

If we insert the expression (2.152) for the sum over the atomic states into (2.154), the Saha equation reads

$$\frac{1-\alpha}{\alpha^2} = n_e^{\text{tot}} \Lambda_e^3 \sum_j e^{\beta I_j^{\text{eff}}(n_e,T)} = n_e^{\text{tot}} K_H(n_e, T)\,. \tag{2.155}$$

Here, we introduced the effective ionization energy

$$I_j^{\text{eff}}(n_e, T) = \left|E_j^0\right| - \Delta_j + \Delta_e + \Delta_p\,, \tag{2.156}$$

which is just the ionization energy (2.139) determined by the effective Schrödinger equation in the previous section. It should be pointed out that this follows from our simple model used for the self-energy correction. In this model, the self-energy shift is equal to the interaction part of the chemical potential. Of course, the self-energy is a dynamical quantity and, in the general case, it cannot be identified directly with the chemical potential. This will be discussed later in more detail.

In the case of an ideal plasma, the self energy shifts in (2.155) are zero, and the mass action constant on the r.h.s. is determined by the bound state energies of the isolated atom, i.e., $I_j^{\text{eff}} = |E_j^0|$. For a nonideal plasma, the ionization energies I_j^{eff} depend on density and temperature, and we observe a lowering of the ionization energy with increasing plasma density. This leads to drastic changes in the degree of ionization.

In Fig. 2.7, we see solutions α of the mass action law (2.155) for constant temperatures as a function of the total electron density. In the atomic sum of states σ_H, the ground state was taken into account only. As an approximation, the atomic self-energy shift was neglected, i.e., $E_{10} = E_1^0$. The electron and proton shifts were used in the approximation given by (2.145).

At lower densities, we find decreasing α values on account of the formation of atoms, i.e., the equilibrium is shifted towards the atomic side. On further

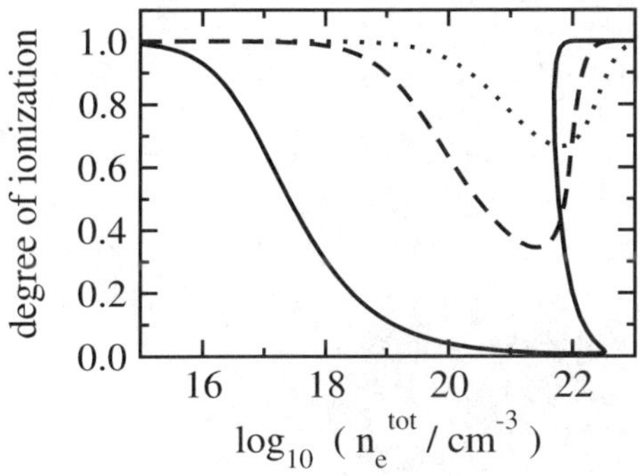

Fig. 2.7. Degree of ionization as a function of the total electron density for various temperatures: *solid*: 15000 K, *dashes*: 30000 K, *dots*: 50000 K

increase of the density, the plasma becomes more nonideal, and the ionization energy is lowered. As a consequence, the degree of ionization increases up to the value $\alpha = 1$ for a fully ionized plasma (Kraeft et al. 1975). This effect is referred to as Mott transition and represents pressure ionization. The Mott transition is one of the most important nonideality effects, and it essentially influences the properties of partially ionized plasmas at high densities.

Let us consider the mass action constant $K_H(n_e, T)$ defined by (2.151) and (2.155) in more detail. In general, it contains the bound state energies $E_j = E_j^0 + \Delta_j$ following from the effective Schrödinger equation (2.127). The sum over j contains only a finite number of terms, and $K_H(n_e, T)$ is well defined. Nevertheless, there are certain problems.

(i) For $n_e^{\rm tot} \to 0$, the mass action constant tends to be the sum of bound states for the Coulomb potential, i.e.,

$$\lim_{n_e^{\rm tot} \to 0} K_H(n_e, T) = \Lambda_e^3 \sum_{n=0}^{\infty} n^2 e^{-\beta E_n^0}$$

with the energies $E_n^0 = -\frac{e^2}{2a_B}\frac{1}{n^2}$ (n: principal quantum number). The Coulomb potential has an infinite number of bound states with a point of accumulation at $E = 0$. This means, the mass action constant $K_H(n_e, T)$ diverges for $n_e^{\rm tot} \to 0$.

(ii) $K_H(n_e, T)$ exhibits jumps at the Mott densities of the different levels.

Formally, these problems are produced by the first terms of a power series expansion of $e^{\beta I_j^{\rm eff}}$. By a subtraction of the first two terms, we get for the mass action constant

$$K_H^{PL}(n_e, T) = \Lambda_e^3 \sum_j \left(e^{\beta I_j^{\rm eff}(n_e, T)} - 1 - \beta I_j^{\rm eff}(n_e, T) \right) . \tag{2.157}$$

This expression was established by Planck and Brillouin and later by Vedenov and Larkin (1959) and by Ebeling (1967) and is now usually referred to as *Planck–Larkin sum of states.* For $n_e^{\text{tot}} \to 0$, i.e., for $I_j^{\text{eff}}(n_e, T) = |E_j^0|$, we get an expression derived by *Planck*, *Brillouin* and others which is *convergent* and represents a modification of the Coulomb sum of bound states. The introduction of the Planck–Larkin sum of bound states, however, demands a rigorous justification which will be given in Chap. 6. At this point, however, we want to stress the fact that the problems just mentioned arise from the replacement of the complete sum of two-particle states in (2.151) by the sum of bound states only, i.e., by $\sigma_H(n_e, T)$. For $n_e^{\text{tot}} \to 0$, and also for high densities near the Mott points, however, both the bound state and the scattering state contributions to the two particle sum of states are of equal importance (Ebeling (a) et al. 1976). Therefore, they have to be treated on the same level to give a correct description of the plasma properties. Indeed, the jumps of $\sigma_H(n_e, T)$ at the corresponding Mott densities are exactly compensated by (opposite) jumps of the scattering contributions, and thus the physical properties remain continuous (Kremp et al. 1971). This behavior is included partially by using the Planck–Larkin sum of states (2.157). As shown by Rogers et al. (1971), Bollé (1979), Kremp et al. (1993), the Planck–Larkin sum of states follows from the total second virial coefficient applying the first and second order Levinson theorems. With the help of Levinson's theorems, a modified subdivision into bound and scattering contributions may be found leading to the Planck–Larkin sum of states and a remainder. At sufficiently low temperatures only, the remainder may be neglected.

The Planck–Larkin sum of states K_H^{PL} has the following properties: (i) K_H^{PL} is convergent in the limit $n_e^{\text{tot}} \to 0$. (ii) It is continuous at the Mott densities. (iii) At low temperatures, K_H^{PL} is a good approximation for the full two-particle sum of states. However, the range of validity of the Planck–Larkin sum of states is restricted to low density quantum plasmas. Many-body effects of higher order have to be included at high densities.

2.9 The Density–Temperature Plane

Our considerations so far showed that the behavior and the properties of strongly correlated plasmas are essentially determined by the spin statistic theorem and by the Coulomb interaction:

(i) According to the spin statistic theorem, we have a degenerate plasma obeying Bose or Fermi statistics if $n_a \Lambda_a^3 > 1$. For $n_a \Lambda_a^3 \ll 1$, the plasma is of Boltzmann type and non-degenerate. Roughly speaking, both cases may be subdivided by the equation

$$n_a \Lambda_a^3 = 1 \,. \tag{2.158}$$

(ii) The Coulomb interaction is of special importance if the mean values of kinetic and potential energies are related as

$$\langle E_{\rm pot}\rangle \geq \langle E_{\rm kin}\rangle\ .$$

In this case, we have a strongly correlated (nonideal) plasma.

To estimate the region of density and temperature where the plasma is strongly correlated we introduce characteristic lengths. First, we determine the distance l_{ab} between particles a and b where the Coulomb interaction is of the order of the mean kinetic energy. For a non-degenerate plasma, the mean kinetic energy may be estimated as $\sim k_B T$, and the distance is defined by

$$k_B T = \left| \frac{e_a e_b}{l_{ab}} \right| . \qquad (2.159)$$

This gives the known classical expression of the so-called *Landau length* l_{ab}. The relation (2.159) can be generalized using (2.25) to include both the non-degenerate and degenerate case

$$k_B T \frac{(2s_a+1)}{n_a \Lambda_a^3} I_{3/2}(\beta\mu_a) = \left| \frac{e_a e_b}{l_{ab}} \right| . \qquad (2.160)$$

Here, l_{ab} has the meaning of a generalized Landau length.

Further, we define the mean particle distance

$$\left(\frac{3}{4\pi n_a}\right)^{1/3} = d_a\,, \qquad d_e = d_i = d\,. \qquad (2.161)$$

Now, the mean values of kinetic and potential energies can be compared introducing the nonideality parameter $\Gamma_{ab} = l_{ab}/d$. Strong correlations have to be expected if

$$\Gamma_{ab} = \frac{l_{ab}}{d} \geq 1 \qquad (2.162)$$

At high densities, nonideality can also be characterized by the *Brueckner parameter* r_S:

$$r_S = \frac{d}{a_B}\,, \qquad r_S = 0.74\,\Gamma_{ee} \qquad (2.163)$$

With the parameters thus introduced we may describe the plasma state in the density–temperature plane. To give an example, the density–temperature plane is shown in Fig. 2.8 for a hydrogen plasma.

The line given by (2.158) subdivides the plane into the non-degenerate (below) and the degenerate plasma (above), respectively.

The region of the strongly correlated plasma is determined by the inequality (2.162), and is located within the area enclosed by the lines $\Gamma_{ab} = 1$ and $r_S = 1$. This means that the plasma becomes ideal at very high densities and/or at high temperatures. In the region of strong correlations ($\Gamma_{ab} > 1$) the plasma is essentially characterized by nonideality effects such as dynamical screening, self-energy, Pauli blocking, and bound states. It is worthwhile

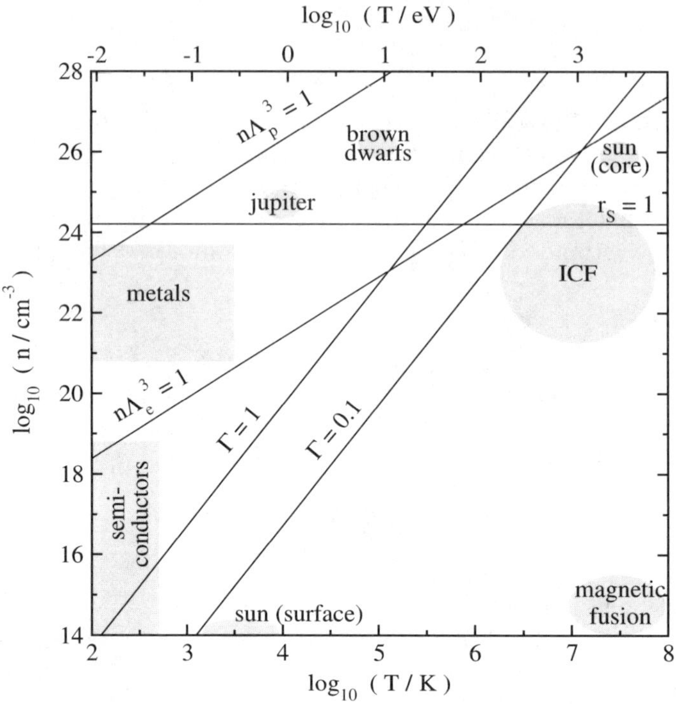

Fig. 2.8. Density–temperature plane. Explanation see text. The parameter Γ was taken for singly charged particles; its subscript was omitted. Relativistic limits: $k_B T = m_e c^2$ for $T = 4.9 \times 10^9$ K, $E_F = m_e c^2$ for $n = 2.7 \times 10^{31}\,\mathrm{cm}^{-3}$

to characterize the area where bound states are possible. On behalf of the Mott effect discussed in Sect. 2.7, the condition

$$r_0 = a_B \tag{2.164}$$

subdivides the plane in Fig. 2.8 roughly into an area where bound states are possible ($r_0 > a_B$) and where bound states cannot exist ($r_0 < a_B$). Further, $|E_1^0|/k_B T \gg 1$ gives an estimate for the fully ionized plasma region because of thermal effects.

We further introduce the ratio

$$\xi = \frac{e^2}{k_B T \lambda}; \qquad \lambda = \frac{\hbar}{\sqrt{2 m_e k_B T}} \tag{2.165}$$

which is referred to as Born parameter and which is of interest for the characterization of quantum plasmas. One can expect essential quantum effects in the domain $r_s < 1$ and and $\xi < 1$.

In Fig. 2.8, several physical systems are indicated which are of relevance for scientific and for technical reasons. Among the latter, we mention magnetically and inertially confined fusion (ICF).

2.10 Boltzmann Kinetic Equation

Up to now, we considered mainly equilibrium properties of a plasma. However, many plasma properties such as transport phenomena, relaxation processes, ionization kinetics, and stability problems demand the development of a non-equilibrium theory. Also in this case, the plasma properties are determined by the time dependent single-particle distribution function $f_a(\boldsymbol{p}, \boldsymbol{r}, t)$.

The most important problem consists, therefore, in the formulation of an equation for the time evolution of the distribution function. Such dynamic equations are called kinetic equations. In this section, we will give an elementary foundation of kinetic equations.

Already in Sect. 2.1, we saw that the kinetic equation has the general form

$$\frac{\partial f}{\partial t} + \frac{\boldsymbol{p}}{m} \cdot \frac{\partial f}{\partial \boldsymbol{r}} + \boldsymbol{F} \cdot \frac{\partial f}{\partial \boldsymbol{p}} = -\frac{1}{n}\frac{\partial}{\partial \boldsymbol{p}} \cdot \langle \delta \boldsymbol{F} \delta N \rangle = I[f] \,, \tag{2.166}$$

where $I[f]$ is called the collision integral. Now the problem is to find an explicit expression for this term.

A collision between two plasma particles with momenta $\boldsymbol{p}_1$ and $\boldsymbol{p}_2$ changes the momenta and thus the average occupation number $f(\boldsymbol{p}_1)$. The total number of collisions between particles with momenta $\boldsymbol{p}_1$ and $\boldsymbol{p}_2$ into momentum states $\bar{\boldsymbol{p}}_1$ and $\bar{\boldsymbol{p}}_2$ at fixed $\boldsymbol{p}_1$ is given by (out-scattering)

$$\begin{aligned} I^{\text{out}}(\boldsymbol{p}_1) &= \int d\boldsymbol{p}_2 d\bar{\boldsymbol{p}}_1 d\bar{\boldsymbol{p}}_2 \, W(\boldsymbol{p}_1 \boldsymbol{p}_2 \bar{\boldsymbol{p}}_1 \bar{\boldsymbol{p}}_2) \, f(\boldsymbol{p}_1) \, f(\boldsymbol{p}_2) \\ &\times \; (1 \pm f(\bar{\boldsymbol{p}}_1)) \, (1 \pm f(\bar{\boldsymbol{p}}_2)) \,. \end{aligned} \tag{2.167}$$

The quantity $I^{\text{out}}(\boldsymbol{p}_1)$ is determined by the transition probability per unit time $W(\boldsymbol{p}_1 \boldsymbol{p}_2 \bar{\boldsymbol{p}}_1 \bar{\boldsymbol{p}}_2)$ for the corresponding scattering process. Furthermore, I^{out} has to be proportional to the occupation number for the state with momenta $\boldsymbol{p}_1$ and $\boldsymbol{p}_2$. It was assumed that this quantity is approximately given by $f(\boldsymbol{p}_1)f(\boldsymbol{p}_2)$, i.e., the particles enter the collision process uncorrelated. This assumption is the famous Boltzmann "Stoßzahlansatz" (molecular chaos assumption). For Fermi particles, we additionally have to take into account that they can scatter only into states which are not occupied, while Bose particles prefer states already occupied. In (2.167), this is reflected by phase space occupation factors $(1 \pm f(\bar{\boldsymbol{p}}_1))$, etc. Here, the upper sign refers to Bose particles, and the lower sign to Fermi particles.

Correspondingly, the mean occupation number $f(\boldsymbol{p}_1)$ is changed by scattering into the state $\boldsymbol{p}_1$ (in-scattering). The collision rate is

$$\begin{aligned} I^{\text{in}}(\boldsymbol{p}_1) &= \int d\boldsymbol{p}_2 d\bar{\boldsymbol{p}}_1 d\bar{\boldsymbol{p}}_2 \, W(\bar{\boldsymbol{p}}_1 \bar{\boldsymbol{p}}_2 \boldsymbol{p}_1 \boldsymbol{p}_2) \, f(\bar{\boldsymbol{p}}_1) \, f(\bar{\boldsymbol{p}}_2) \\ &\times \; (1 \pm f(\boldsymbol{p}_1)) \, (1 \pm f(\boldsymbol{p}_2)) \,. \end{aligned} \tag{2.168}$$

As a result of the two types of collisions, we have

$$\frac{df\left(\boldsymbol{p}_1\right)}{dt} = I^{\text{in}}(\boldsymbol{p}_1) - I^{\text{out}}(\boldsymbol{p}_1) = I\left[f\right] . \tag{2.169}$$

We take into account the reversibility of quantum mechanical dynamics, i.e.,

$$W\left(\boldsymbol{p}_1\boldsymbol{p}_2\bar{\boldsymbol{p}}_1\bar{\boldsymbol{p}}_2\right) = W\left(\bar{\boldsymbol{p}}_1\bar{\boldsymbol{p}}_2\boldsymbol{p}_1\boldsymbol{p}_2\right) . \tag{2.170}$$

Then, we get the following nonlinear integro-differential equation for the single-particle distribution function

$$\begin{aligned} \frac{\partial f_1}{\partial t} + \frac{\boldsymbol{p}_1}{m_1}\cdot\frac{\partial f_1}{\partial \boldsymbol{r}_1} - \frac{\partial U_1^{\text{eff}}}{\partial \boldsymbol{r}_1}\cdot\frac{\partial f_1}{\partial \boldsymbol{p}_1} &= \int d\boldsymbol{p}_2 d\bar{\boldsymbol{p}}_1 d\bar{\boldsymbol{p}}_2 W\left(\boldsymbol{p}_1\boldsymbol{p}_2\bar{\boldsymbol{p}}_1\bar{\boldsymbol{p}}_2\right) \\ \times \left[f_1 f_2\left(1\mp\bar{f}_1\right)\left(1\mp\bar{f}_2\right) - \bar{f}_1\bar{f}_2\left(1\mp f_1\right)\left(1\mp f_2\right)\right] . \end{aligned} \tag{2.171}$$

For an abbreviation, we used the notation, e.g., $\bar{f}_1 = f(\bar{\boldsymbol{p}}_1, \boldsymbol{r}, t)$. Here U_1^{eff} is defined by (2.37). An equation of this type was given for the first time by *L. Boltzmann* in 1872. This equation is one of the most fundamental equations of statistical physics.

(i) It describes the irreversible relaxation of an arbitrary initial distribution into the thermodynamic equilibrium state.
(ii) It is the basic equation for the theory of transport processes in many-particle systems.

For further investigation, we have to find an explicit expression for the transition probabilities $W\left(\boldsymbol{p}_1\boldsymbol{p}_2\bar{\boldsymbol{p}}_1\bar{\boldsymbol{p}}_2\right)$.

If the particles interact via a weak potential $V_{12}(\boldsymbol{r}_1 - \boldsymbol{r}_2)$, we get from the time dependent perturbation theory of quantum mechanics

$$\begin{aligned} W\left(\boldsymbol{p}_1\boldsymbol{p}_2\bar{\boldsymbol{p}}_1\bar{\boldsymbol{p}}_2\right) &= \frac{1}{(2\pi\hbar)^6}\left|V_{12}\left(\boldsymbol{p}_1 - \bar{\boldsymbol{p}}_1\right)\right|^2 \delta\left(\boldsymbol{p}_1 + \boldsymbol{p}_2 - \bar{\boldsymbol{p}}_1 - \bar{\boldsymbol{p}}_2\right) \\ &\times \frac{2\pi}{\hbar}\,\delta\left(E_{12} - \bar{E}_{12}\right) , \end{aligned} \tag{2.172}$$

referred to as Fermi's "golden rule". Here, $V_{12}\left(\boldsymbol{p}_1 - \bar{\boldsymbol{p}}_1\right)$ is the Fourier transform of the two-body interaction potential defined by

$$V_{12}\left(\boldsymbol{p}\right) = \int d\boldsymbol{r}\, e^{-\frac{i}{\hbar}\boldsymbol{p}\cdot\boldsymbol{r}}\, V_{12}\left(\boldsymbol{r}\right) . \tag{2.173}$$

Equation (2.172) explicitly controls momentum and energy conservation. E_{12} and $\bar{E}_{12}$ are the energies before and after the collision, respectively, and they are given by the sum of the one-particle energies. In the simplest case of free particles we have

$$E_{12} = \frac{p_1^2}{2m_1} + \frac{p_2^2}{2m_2} .$$

Later we will derive "better" approximations for the quantity W (see Sect. 6.5). An important approximation is the binary collision approximation which reads

$$W(\boldsymbol{p}_1\boldsymbol{p}_2\bar{\boldsymbol{p}}_1\bar{\boldsymbol{p}}_2) = \frac{2\pi}{\hbar}\left|\langle\boldsymbol{p}_1\boldsymbol{p}_2|\,T_{12}(E+i\varepsilon)\,|\bar{\boldsymbol{p}}_2\bar{\boldsymbol{p}}_1\rangle\right|^2\delta\left(E_{12}-\bar{E}_{12}\right). \quad (2.174)$$

Here, $< |T_{12}| >$ is the T-matrix of scattering theory which has to be determined from the *Lippmann–Schwinger equation*

$$\mathrm{T}_{12}(\omega+i\varepsilon) = \mathrm{V}_{12} + \mathrm{V}_{12}\frac{1}{\omega-\mathrm{H}^0_{12}+i\varepsilon}\mathrm{T}_{12}(\omega+i\varepsilon). \quad (2.175)$$

For simplicity, we did not include here the exchange contribution. Equation (2.172) follows from (2.174), if the T-matrix is determined in first Born approximation, which is, according to (2.175), simply the potential V_{12}. With (2.172), the kinetic equation (2.171) reads

$$\left\{\frac{\partial}{\partial t}+\frac{\boldsymbol{p}_1}{m_1}\frac{\partial}{\partial\boldsymbol{r}_1}-\frac{\partial}{\partial\boldsymbol{r}_1}U_1^{\mathrm{eff}}\frac{\partial}{\partial\boldsymbol{p}_1}\right\}f_1 = \frac{1}{\hbar}\int\frac{d\boldsymbol{p}_2}{(2\pi\hbar)^3}\frac{d\bar{\boldsymbol{p}}_1}{(2\pi\hbar)^3}\frac{d\bar{\boldsymbol{p}}_2}{(2\pi\hbar)^3}$$
$$\times\left|V_{12}(\boldsymbol{p}_1-\bar{\boldsymbol{p}}_1)\right|^2(2\pi\hbar)^3\,\delta(\boldsymbol{p}_1+\boldsymbol{p}_2-\bar{\boldsymbol{p}}_1-\bar{\boldsymbol{p}}_2)\,2\pi\,\delta\left(E_{12}-\bar{E}_{12}\right)$$
$$\times\left[f_1f_2\left(1\pm\bar{f}_1\right)\left(1\pm\bar{f}_2\right)-\bar{f}_1\bar{f}_2\left(1\pm f_1\right)\left(1\pm f_2\right)\right]. \quad (2.176)$$

Equation (2.176) is frequently referred to as Landau equation which was written here in the quantum mechanical version. Compared to (2.172), it still contains the exchange contribution which accounts for the indistinguishability of identical particles. Here we have to remark that $f(\boldsymbol{p},\boldsymbol{r},t)$ is, in principle, the Wigner distribution function which is approximately equal to the usual momentum distribution function (see below, Chaps. 3 and 7).

The kinetic equation (2.176) has a number of properties which are proven in textbooks of statistical physics (Huang 1963; Lifschitz and Pitajewski 1983), namely:

(i) With the definition of the entropy $S = -k_B\int f\ln f\,d\boldsymbol{p}d\boldsymbol{r}$, the validity of the Boltzmann H-theorem is shown

$$\frac{dS}{dt}\geq 0\,; \quad (2.177)$$

(ii) the kinetic equation leads to conservation laws for the number density, for the momentum, and for the kinetic energy;

(iii) the equilibrium solution for the distribution function, i.e., the condition $I[f]=0$, leads to Bose or Fermi distribution functions.

Equation (2.176) may be solved numerically. We have done this for an electron gas, i.e., an electron plasma with positively charged background is considered.

For the interaction potential V_{12}, we used the statically screened Coulomb potential with the inverse screening length $\kappa = r_0^{-1} = 0.38a_B^{-1}$. Results of the numerical evaluation are shown in Fig. 2.9 for the spatially homogeneous system assuming an isotropic momentum distribution (Bonitz 1998; Kremp et al. 1996).

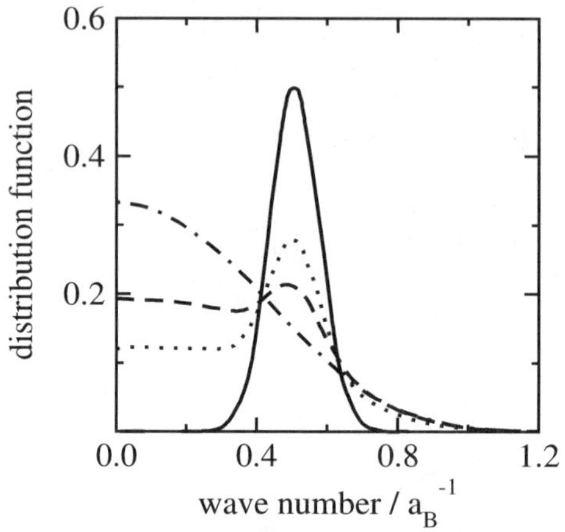

Fig. 2.9. Relaxation of the distribution function for an electron gas assuming statically screened Coulomb interaction. The inverse screening length is fixed to be $\kappa = 0.38a_B^{-1}$. *Solid line*: $t = 0.00fs$, *dots*: $0.24fs$, *dashes*: $0.38fs$, *dash-dots*: $3.34fs$

The irreversible relaxation is demonstrated, starting from a Gauss-shape initial distribution at $t = 0$, ending up with the thermodynamic equilibrium at time $t \equiv \tau = 2.43fs$. As it could be expected, after the relaxation time τ, we found the Fermi distribution.

Though (2.176) shows a number of fundamental features, it has several serious shortcomings (Klimontovich 1982). Among them we mention the following ones (Kremp et al. 1996).

(i) The kinetic equation (2.176) is valid only for times $t > \tau^{\text{corr}}$, where τ^{corr} is the relaxation time of correlations, i.e., the time for the decay of initial correlations.
(ii) The kinetic equation (2.176) only conserves the mean kinetic energy instead of the total one which includes potential energy contributions.

Both these shortcomings are overcome by a more general kinetic theory which will be outlined in Chap. 7.

2.11 Transport Properties

The Boltzmann equation is the basic equation of the transport theory in plasmas. Transport of mass, charge, momentum and energy are important phenomena. The latter are connected with a flow $\boldsymbol{j}_A$ of a physical quantity A.

From the phenomenological point of view, the flow $\boldsymbol{j}_A$, produced by the external forces $\boldsymbol{X}_B$, is determined in linear response regime (see, e.g., Ebeling et al. (1983), Groot and Mazur (1962)) by

$$\boldsymbol{j}_A = \sum_B L_{AB} \boldsymbol{X}_B \,, \tag{2.178}$$

where L_{AB} are the Onsager coefficients. Together with the local conservation laws of the densities of mass, charge, momentum, and energy, one can derive equations of motion for the macroscopic observables. It is clear that the determination of transport properties from the microscopic point of view has to start from the definition of the local density

$$n(\boldsymbol{r},t) = \int \frac{d\boldsymbol{p}}{(2\pi\hbar)^3}\, f(\boldsymbol{p},\boldsymbol{r},t) \,, \tag{2.179}$$

of the momentum flow

$$\boldsymbol{j}(\boldsymbol{r},t) = \int \frac{d\boldsymbol{p}}{(2\pi\hbar)^3} \frac{\boldsymbol{p}}{m}\, f(\boldsymbol{p},\boldsymbol{r},t) \,, \tag{2.180}$$

and of the energy flow

$$\boldsymbol{j}_E(\boldsymbol{r},t) = \int \frac{d\boldsymbol{p}}{(2\pi\hbar)^3} \frac{\boldsymbol{p}}{m} \frac{p^2}{2m}\, f(\boldsymbol{p},\boldsymbol{r},t) \,. \tag{2.181}$$

The central task of the kinetic theory of transport properties is, therefore, the determination of the distribution function from the kinetic equation.

Let us first consider a nearly homogeneous partially ionized plasma in an external electric field $\boldsymbol{E}$. Viscous flow is neglected. Further, only the contribution of electrons to the transport is considered. As known from irreversible thermodynamics, Onsager's linear response relations may be written in this case in the following form

$$\begin{aligned} \boldsymbol{j}_E &= L_{qq}\left(-\frac{\nabla T}{T}\right) + L_{qe}\left[\boldsymbol{E} - \frac{T}{e}\nabla\left(\frac{\mu_e}{T}\right)\right] , \\ \boldsymbol{j}_e &= L_{eq}\left(-\frac{\nabla T}{T}\right) + L_{ee}\left[\boldsymbol{E} - \frac{T}{e}\nabla\left(\frac{\mu_e}{T}\right)\right] . \end{aligned} \tag{2.182}$$

Here, $\boldsymbol{E}$ is the electric field, and $\boldsymbol{j}_e$ the electric current density. The Onsager coefficients L_{AB} fulfill the condition $L_{eq} = L_{qe}$. The electrical conductivity is determined by

$$\sigma = L_{ee} \,. \tag{2.183}$$

For the heat conductivity we have

$$\lambda = \left(L_{qq} - L_{qe}^2/L_{ee}\right)/T \,, \tag{2.184}$$

and the thermoelectrical coefficient has to be determined from

$$\nu = -L_{eq}/(L_{ec}T) . \tag{2.185}$$

The electron kinetic equation for the plasma consisting of electrons, singly charged ions and atoms reads

$$\frac{\partial}{\partial t} f_e + \frac{\boldsymbol{p}_e}{m_e} \cdot \frac{\partial f_e}{\partial \boldsymbol{r}} - e\boldsymbol{E} \cdot \frac{\partial f_e}{\partial \boldsymbol{p}} = I_{ei} + I_{ee} + I_{eA} . \tag{2.186}$$

According to (2.176), the collision integrals between charged particles are given in first Born approximation by ($c = e, p$)

$$I_{ec} = \frac{1}{\hbar} \int \frac{d\boldsymbol{p}_c d\bar{\boldsymbol{p}}_e d\bar{\boldsymbol{p}}_c}{(2\pi\hbar)^6} \left| V^S_{ec}(\boldsymbol{p}_e - \bar{\boldsymbol{p}}_e) \right|^2 \delta(\boldsymbol{p}_e + \boldsymbol{p}_c - \bar{\boldsymbol{p}}_e - \bar{\boldsymbol{p}}_c)$$
$$\times 2\pi\delta\left(E_{ec} - \bar{E}_{ec}\right) \left\{ \bar{f}_e (1 - f_e) \bar{f}_c (1 - f_c) - f_e \left(1 - \bar{f}_e\right) f_c \left(1 - \bar{f}_c\right) \right\} . \tag{2.187}$$

where the abbreviations $f_a = f_a(\boldsymbol{p}_a, \boldsymbol{r}, t)$, etc. are used. In order to account for plasma screening, the interaction potential between the charged particles was chosen to be the statically screened Coulomb potential

$$V^S_{ab}(\boldsymbol{q}) = \frac{V_{ab}(q)}{\varepsilon(q, 0)} . \tag{2.188}$$

$V_{ab}(q) = 4\pi\hbar^2 e_a e_b / q^2$ is the bare Coulomb potential, and $\varepsilon(q, 0) = 1 + \hbar^2\kappa^2/q^2$ is the static dielectric function as given by (2.68). In Chap. 7 we will show that such a collision integral (2.187) follows in the static limit from the quantum Lenard–Balescu kinetic equation.

The collision integral between electrons and atoms is taken to be elastic scattering on the atomic ground state only and reads ($c = A$)

$$I_{eA} = \frac{1}{\hbar} \int \frac{d\boldsymbol{p}_A d\bar{\boldsymbol{p}}_e d\bar{\boldsymbol{p}}_A}{(2\pi\hbar)^6} \left| V^S_{eA}(\boldsymbol{p}_e - \bar{\boldsymbol{p}}_e) \right|^2 \delta(\boldsymbol{p}_e + \boldsymbol{p}_A - \bar{\boldsymbol{p}}_e - \bar{\boldsymbol{p}}_A)$$
$$\times 2\pi\delta\left(E_{eA} - \bar{E}_{eA}\right) \left\{ \bar{f}_e (1 - f_e) \bar{F}_A - f_e \left(1 - \bar{f}_e\right) F_A \right\} . \tag{2.189}$$

Here, V^S_{eA} denotes the statically screened electron–atom potential. Furthermore, the atoms were considered to be non-degenerate, i.e., $(1 - f_A) \to 1$. We mention again that the spin sum in (2.187) and (2.189) is dropped for simplicity.

The Boltzmann equation (2.186) is a complicated nonlinear equation for f_e. For this reason it is in general not solvable analytically, especially for physically realistic potential models and situations. Therefore, we will discuss approximate solutions to the Boltzmann equation under the condition that the system is nearly in an equilibrium state. All time and spatial deviations are then small. On the basis of these assumptions, the Boltzmann equation may be written in the following symbolic form according to Chapman–Enskog (see, e.g., Chapman and Cowling (1952))

$$\frac{\partial f_e}{\partial t} + \frac{\boldsymbol{p}_e}{m_e} \cdot \frac{\partial f_e}{\partial \boldsymbol{r}} - e\boldsymbol{E} \cdot \frac{\partial f_e}{\partial \boldsymbol{p}} = \frac{1}{\varepsilon} \sum_c I_{ec} [f_e, f_c] \tag{2.190}$$

where ε is a small parameter. Further we expand the distribution functions f in a series of ε

$$f_a = f_a^0 + \varepsilon f_a^{(1)} + \varepsilon^2 f_a^{(2)} + \cdots . \tag{2.191}$$

When (2.191) is substituted into (2.190) and terms of equal powers in ε are equalized on both sides of the equation, we get a hierarchy of perturbation equations for f_a^0, $f_a^{(1)}, \cdots$.. The first equations are

$$\sum_c I \left[f_e^0 f_c^0\right] = 0 , \tag{2.192}$$

$$\left(\frac{\partial}{\partial t} + \frac{\boldsymbol{p}}{m_e} \frac{\partial}{\partial \boldsymbol{r}} - e\boldsymbol{E} \cdot \frac{\partial}{\partial \boldsymbol{p}} \right) f_e^0 = \sum_c \left(I_{ec} \left[f_e^0 f_c^1\right] + I_{ec} \left[f_e^1 f_c^0\right] \right) . \tag{2.193}$$

The solution of (2.192) for the distribution function f_a of species a must be of the form

$$f_a^0 (\boldsymbol{p}, \boldsymbol{r}, t) = \left[\exp \left(\frac{(\boldsymbol{p} - m_a \boldsymbol{u}(\boldsymbol{r}, t))^2}{2 m_a k_B T(\boldsymbol{r}, t)} - \frac{\mu_a(\boldsymbol{r}, t)}{k_B T(\boldsymbol{r}, t)} \right) \pm 1 \right]^{-1} . \tag{2.194}$$

This local Fermi (Bose)-distribution describes the system as being in local equilibrium. To complete this expression we have to determine the five unknown functions $T(\boldsymbol{r}, t)$, $\mu_a(\boldsymbol{r}, t)$, $\boldsymbol{u}(\boldsymbol{r}, t)$ by making use of the five local conservation laws for the particle number, the momentum, and the energy following from the Boltzmann kinetic equation. For a first discussion of the transport properties, we restrict ourselves to the stationary case $\partial f_e / \partial t = 0$ and take $\boldsymbol{u}(\boldsymbol{r}) = 0$. Introducing (2.194) in the first order perturbation (2.193), it follows

$$\begin{aligned} &\left[\left(\frac{p^2}{2m_e} - \mu_e \right) \left(- \frac{\nabla T}{T} \right) - (\nabla \mu_e + e\boldsymbol{E}) \right] \frac{\partial f_e^0(p, \boldsymbol{r})}{\partial \boldsymbol{p}} \\ &= \sum_{c=e,i,A} \left(I_{ec} \left[f_e^0 f_c^1\right] + I_{ec} \left[f_e^1 f_c^0\right] \right) . \end{aligned} \tag{2.195}$$

This is a linear integral equation for the determination of the non-equilibrium part f_e^1 of the distribution function.

Further progress in the theory of stationary transport processes is connected with the consideration of the right-hand side of (2.195). For the determination of the collision integrals, we introduce momenta for the relative motion $\boldsymbol{p}_{ec}$ and those for the center of mass $\boldsymbol{P}_{ec}$ by

$$\boldsymbol{P}_{ec} = \boldsymbol{p}_e + \boldsymbol{p}_c\,, \qquad \boldsymbol{p}_{ec} = \frac{1}{M}\left(m_c\boldsymbol{p}_e - m_e\boldsymbol{p}_c\right)\,, \tag{2.196}$$

where $M = m_e + m_c$. Introducing the reduced mass $\mu_{ec} = m_e m_c/(m_e + m_c)$, the kinetic energy reads

$$E_{ec} = \frac{p_e^2}{2m_e} + \frac{p_c^2}{2m_c} = \frac{P_{ec}^2}{2M} + \frac{p_{ec}^2}{2\mu_{ec}}\,. \tag{2.197}$$

In the following, we consider the Lorentz plasma model which already gives essential properties of transport coefficients in partially ionized plasmas. In this model, the electron–electron collision integral I_{ee} is neglected. Furthermore, on account of the large value of the masses of ions and atoms as compared to that of the electrons, we adopt the adiabatic approximation, i.e., we assume that the momentum of the heavy particle does not change during the collision procedure. For $c = i, A$, we then have

$$\boldsymbol{p}_{ec} \simeq \boldsymbol{p}_e\,; \qquad \boldsymbol{P}_{ec} \simeq \boldsymbol{p}_c\,. \tag{2.198}$$

Under this condition, the electron–ion (atom) collision integral reads

$$\begin{aligned} I_{ec}(\boldsymbol{p}_e) &= \frac{1}{\hbar}\int \frac{d\bar{\boldsymbol{p}}_e d\boldsymbol{p}_c d\bar{\boldsymbol{p}}_c}{(2\pi\hbar)^6} \left|V_{ec}^S\left(\boldsymbol{p}_e - \bar{\boldsymbol{p}}_e\right)\right|^2 \delta\left(\boldsymbol{p}_c - \bar{\boldsymbol{p}}_c\right) f_c\left(p_c\right) \\ &\times\ 2\pi\,\delta(E_{ec} - \bar{E}_{ec})\left\{f_e\left(\bar{\boldsymbol{p}}_e\right)\left[1 - f_e\left(\boldsymbol{p}_e\right)\right] - f_e\left(\boldsymbol{p}_e\right)\left[1 - f_e\left(\bar{\boldsymbol{p}}_e\right)\right]\right\}\,. \end{aligned} \tag{2.199}$$

Equation (2.195), together with (2.199), is the starting point for the determination of transport coefficients in the considered model.

As an example, we consider the calculation of the electrical conductivity of a partially ionized isothermal and homogeneous plasma. Equation (2.195) then takes the shape

$$-e\boldsymbol{E}\cdot\frac{\partial f_e}{\partial \boldsymbol{p}} = \sum_{c=e,i,A}\left(I_{ec}\left[f_e^0 f_c^1\right] + I_{ec}\left[f_e^1 f_c^0\right]\right)\,. \tag{2.200}$$

In the collision integral (2.199), we use the normalization condition for the ion distribution function, and we introduce the differential cross section in first Born approximation

$$\frac{d\sigma_{ec}(p_e,\Omega)}{d\Omega} = \frac{(2\pi)^4\hbar^2 m_e^2}{(2\pi\hbar)^6}\left|\langle\boldsymbol{p}_e|\,\mathrm{V}_{ec}^S\,|\bar{\boldsymbol{p}}_e\rangle\right|_{p=\bar{p}}^2\,, \tag{2.201}$$

with Ω being the solid angle. For simplicity, the subscripts on the momenta which label the species will be dropped in the following. Then the collision integral (2.199) reads

$$I_{ec}(\boldsymbol{p}) = n_c \int d\Omega\,\frac{p}{m_e}\,\frac{d\sigma_{ec}(p,\Omega)}{d\Omega}\left\{f_e\left(\bar{\boldsymbol{p}}\right)\left[1 - f_e\left(\boldsymbol{p}\right)\right] - f_e\left(\boldsymbol{p}\right)\left[1 - f_e\left(\bar{\boldsymbol{p}}\right)\right]\right\}\,. \tag{2.202}$$

On behalf of the axial symmetry, the distribution function has the structure

$$f_e(\boldsymbol{p}) = f_e(p, \vartheta) \; ,$$

where ϑ is the angle between the electric field and the vector $\boldsymbol{p}$, and p is the modulus of $\boldsymbol{p}$. We define a coordinate system in which the z-direction is $\hat{\boldsymbol{e}} = \frac{\boldsymbol{E}}{E}$ so that

$$p_z = \hat{\boldsymbol{e}} \cdot \boldsymbol{p} = p \cos\vartheta , \;\; \bar{p}_z = \hat{\boldsymbol{e}} \cdot \bar{\boldsymbol{p}} = \bar{p} \cos\bar{\vartheta} , \;\; \boldsymbol{p} \cdot \bar{\boldsymbol{p}} = p\bar{p} \cos\theta \, ,$$

and

$$p^2(\bar{\boldsymbol{p}} \cdot \hat{\boldsymbol{e}}) = (\boldsymbol{p} \cdot \bar{\boldsymbol{p}})(\boldsymbol{p} \cdot \hat{\boldsymbol{e}}) + |\boldsymbol{p} \times \bar{\boldsymbol{p}}||\boldsymbol{p} \times \hat{\boldsymbol{e}}| \, cos\phi \, . \tag{2.203}$$

Equation (2.203) follows from the usual vector algebra and corresponds to the cosine rule of spherical trigonometry.

Here, it is appropriate to expand the distribution function in terms of Legendre polynomials

$$f_e(\boldsymbol{p}) = \sum_n f_e^{(n)}(p) P_n(\cos\vartheta) \, . \tag{2.204}$$

We restrict ourselves to the first two terms and get

$$f_e(\boldsymbol{p}) = f_e^0(p) + f_e^1(p) \cos\vartheta = f_e^0(p) + \frac{p_z}{p} f_e^1(p) \, . \tag{2.205}$$

The functions f_e^0 and f_e^1 of the expansion (2.204) fulfill the relations

$$8\pi \int_0^\infty \frac{dp}{(2\pi\hbar)^3} \, p^2 f_e^0(p) = n_e \, ,$$

$$\frac{8\pi}{3} \int_0^\infty \frac{dp}{(2\pi\hbar)^3} \, p^3 f_e^1(p) = \langle p_z \rangle \, . \tag{2.206}$$

Here, the summation over the spins was carried out. With the expansion (2.205) the collision integral takes the shape

$$I_{ec}(\boldsymbol{p}) = n_c \, \frac{p}{m_e} \int d\Omega \, \frac{d\sigma_{ec}(p, \Omega)}{d\Omega} \left(\frac{\bar{p}_z}{\bar{p}} f_e^1(\bar{p}) - \frac{p_z}{p} f_e^1(p) \right) \Big|_{\bar{p}=p} \, . \tag{2.207}$$

Using the relations (2.203) and taking into account that the $\cos\phi$-term does not contribute when we do the integral with respect to ϕ, the remaining term may be written as

$$I_{ec}(\boldsymbol{p}) = -\frac{p_z}{p} f_e^1(p) \frac{p}{m_e} n_c \, Q_{ec}^T(p) \, , \tag{2.208}$$

where

$$Q_{ec}^T(p) = 2\pi \int_0^\pi d\theta \, \sin\theta (1 - \cos\theta) \frac{d\sigma_{ec}(p,\theta)}{d\Omega} \tag{2.209}$$

is the transport cross section, and θ is the scattering angle, i.e., the angle between the momenta $\boldsymbol{p}$ and $\bar{\boldsymbol{p}}$ of the scattered electron.

In order to stress the physical meaning of relation (2.208), it is convenient to introduce the relaxation time τ_{ec}. Then we have

$$I_{ec} = -\frac{f_e - f_e^0}{\tau_{ec}}, \qquad \tau_{ec}^{-1}(p) = n_c \frac{p}{m_e} Q_{ec}^T(p) \,. \tag{2.210}$$

If we do not introduce the cross section according to (2.201), we can derive other useful representations of the collision integral. Defining the transfer momentum by $\hbar \boldsymbol{k} = \boldsymbol{p} - \bar{\boldsymbol{p}}$, we get, e.g., the following relation for the electron–ion collision term

$$-\frac{p}{m_e} n_i \, Q_{ei}^T = n_c \frac{2\pi}{\hbar} \int \frac{d\boldsymbol{k}}{(2\pi)^3} \left(\frac{4\pi e_e e_i}{k^2 + \kappa^2} \right)^2 \delta \left(\frac{\hbar \boldsymbol{k} \cdot \boldsymbol{p}}{m_e} - \frac{\hbar^2 k^2}{2 m_e} \right) \frac{\hbar^2 k^2}{2 p^2} \,. \tag{2.211}$$

The integration over the angles can be carried out. With the definition (2.210), we find for the electron–ion relaxation time

$$\frac{1}{\tau_{ei}(p)} = \frac{n_i m_e}{2 p^3} \int_0^{2p/\hbar} \frac{dk}{2\pi} k^3 \left(\frac{4\pi e_e e_i}{k^2 + \kappa^2} \right)^2 \,. \tag{2.212}$$

Starting from (2.200), we get an equation for the determination of $f^1(p)$

$$eE \frac{\partial}{\partial p} f_e^0 = f_e^1 \frac{p}{m_e} \sum_c n_c \, Q_{ec}^T \,. \tag{2.213}$$

Thus we finally get for the electron distribution function

$$f_e(p,\vartheta) = f_e^0(p) + \frac{eE}{\sum_c n_c Q_{ec}^T(p)} \frac{m_e}{p} \frac{\partial}{\partial p} f_e^0(p) \frac{p_z}{p} \,. \tag{2.214}$$

With the distribution function we may determine the electric current from

$$\boldsymbol{j}_e = -(2s+1) e \int \frac{d\boldsymbol{p}}{(2\pi\hbar)^3} \frac{\boldsymbol{p}}{m_e} f_e(\boldsymbol{p}) = \sigma \boldsymbol{E} \,. \tag{2.215}$$

Here, the factor $(2s+1)$ results from the spin sum. The first term of (2.214) gives no contribution to $\boldsymbol{j}_e$; thus, the current is proportional to the second term. The angle integral gives contributions only for the p_z-component, that means, the current flows only in the direction of $\boldsymbol{E}$. In this way we find for the conductivity

$$\sigma^L = -(2s+1)\frac{4\pi e^2}{3}\int_0^\infty \frac{dp}{(2\pi\hbar)^3}\frac{p^2\frac{\partial}{\partial p}f_e^0(p)}{\sum_c n_c\, Q_{ec}^T(p)}\,. \tag{2.216}$$

Performing the derivation of the Fermi function f_e^0, this expression can also be written in the form

$$\sigma^L = (2s+1)\frac{4\pi e^2}{3m_e k_B T}\int_0^\infty \frac{dp}{(2\pi\hbar)^3}\frac{p^3 f_e^0(p)(1-f_e^0(p))}{\sum_c n_c\, Q_{ec}^T(p)}\,. \tag{2.217}$$

Let us first consider a fully ionized hydrogen plasma. We then need only the electron–proton transport cross section Q_{ep}^T determined by the corresponding differential cross section. Using the statically screened Coulomb potential (2.188), we get from (2.201) in first Born approximation

$$\frac{d\sigma_{ep}}{d\Omega} = \frac{4m_e^2 e^4}{[2p^2(1-\cos\theta)+\hbar^2\kappa^2]^2}\,. \tag{2.218}$$

We then get for the transport cross section $Q_{ep}^T(p)$ of the electron–proton scattering

$$Q_{ep}^T(k) = \frac{2\pi}{k^4 a_B^2}\left[\ln(1+4y) - \frac{4y}{1+4y}\right]\,, \tag{2.219}$$

where $y = k^2/\kappa^2$ with κ being the inverse screening length, and $k = p/\hbar$ is the wave number.

This scheme for the determination of the plasma conductivity is also referred to as the Lorentz or relaxation time approximation. The expression (2.217) determines the *dc* electrical conductivity for arbitrary degeneracy and can be evaluated only numerically. Analytical formulas can be found in limiting cases. First, let us consider the non-degenerate H-plasma. For an approximate evaluation, the square bracket of (2.219) may be taken out of the integral at its maximum value

$$(k^2 r_D^2)\Big|_{max} = 6m_e k_B T r_D^2/(\hbar^2)\,. \tag{2.220}$$

Then we get an expression for the conductivity which is referred to as the *Brooks–Herring formula* (Brooks 1951)

$$\sigma^{BH} = 1.0159\frac{(k_B T)^{3/2}}{e^2 m_e^{1/2}}\left[\frac{1}{2}\ln(1+\gamma_D^2) - \frac{\gamma_D^2}{2(1+\gamma_D^2)}\right]^{-1}\,. \tag{2.221}$$

Here, the abbreviation $\gamma_D = \sqrt{48\pi}\, r_D/\Lambda_e$ was used. In this expression, the divergencies of the classical conductivity theory are avoided without any arbitrary cut-off procedure. Divergencies at small wave numbers k do not occur due to the application of a screened potential, and divergencies at large k cannot exist because quantum mechanics provide for a *natural cut-off*, namely the thermal wavelength Λ_e.

A formula frequently used for the low density conductivity is due to Spitzer (Spitzer 1967) and reads

$$\sigma^{SP,ei} = 1.0159 \frac{(k_B T)^{3/2}}{e^2 m_e^{1/2}} (\ln(3/\mu))^{-1} \tag{2.222}$$

with $\mu = l/r_D$ being the plasma parameter. In contrast to the expression (2.221), Spitzer's result (2.222) is a classical one and, therefore, a cut-off at the Landau length $l = e^2/(k_B T)$ was applied avoiding thus divergencies at large wave numbers k.

Both in the Brooks–Herring formula (2.221) and in the Spitzer formula (2.222), the contribution of the electron–electron scattering to the conductivity was neglected. In the special case of the Spitzer theory (Spitzer and Härm 1953), the inclusion of the electron–electron scattering gives

$$\sigma^{SP} = 0.591 \frac{(k_B T)^{3/2}}{e^2 m_e^{1/2}} (\ln(3/\mu))^{-1} . \tag{2.223}$$

This result shows that formula (2.222) is essentially modified by electron–electron scattering in the H-plasma. Later on in the course of this book, we will incorporate this interaction, which leads to essential changes of the conductance formulae presented so far.

We now consider the fully ionized degenerate Lorentz plasma, where it is more convenient to write the conductivity (2.216) in terms of the relaxation time

$$\sigma^L = -\frac{8\pi e^2}{3} \int_0^\infty \frac{dp}{(2\pi\hbar)^3} \frac{p^3}{m_e} \tau_{ei}(p) \frac{\partial}{\partial p} f_e^0(p) \tag{2.224}$$

with $\tau_{ei}(p)$ given by (2.212).

In the zero temperature limit, $f_e^0(p)$ is given by the step function $\Theta(p_F - p)$, and we have

$$\lim_{T\to 0} \frac{\partial}{\partial p} f_e^0(p) = -\delta(p - p_F)\,; \quad p_F^3 = 3\pi^2 n_e \hbar^3 .$$

Then we get immediately

$$\sigma = \frac{n_e\, e^2}{m_e} \tau_{ei}(p_F) \tag{2.225}$$

with the relaxation time at the Fermi surface

$$\tau_{ei}(p_F) = \frac{12\pi^3\hbar^3}{m_e} \left[\int_0^{2k_F} dk k^3 \left(\frac{4\pi e^2}{k^2 + \kappa^2} \right)^2 \right]^{-1} . \tag{2.226}$$

It is important to mention that often the Ziman theory is used for the conductance of a system of ions and degenerate electrons (Ziman 1961). The Ziman expression for the conductivity generalized by Faber to finite temperatures may be written as

$$\sigma^Z = \frac{12\pi^3 e^2 n_e^2 \hbar^3}{m_e^2 n_i} \left\{ \int_0^\infty dk k^3 f_e^0(k/2) \left| \frac{V_{ei}(k)}{\varepsilon(k,0)} \right|^2 S_{ii}(k) \right\}^{-1} . \qquad (2.227)$$

Here, $\varepsilon(k,0)$ is the static dielectric function for an electron gas, $S_{ii}(k)$ is the static ion-ion structure factor, and f_e is the Fermi distribution functions of the electrons. In the case $T = 0$, the formulae (2.227) are similar to (2.225) and (2.226), up to the structure factor and the dielectric function. For finite temperatures, however, (2.227) are essentially different from (2.224), (2.212), because the approximation leading to the Ziman formula corresponds to the first step only in the polynomial expansion leading to the exact Lorentz model. This will be shown in Sect. 9.2.3.

In order to include the structure factor in the Lorentz model, we have to start from the Lenard–Balescu equation. The latter takes into account the dynamical character of the screening and will be derived in Sect. 9.2.2. Following an idea of Williams and DeWitt (1969) and Boercker, Rogers, and DeWitt (1982), it is possible to determine the modification of the electron–ion relaxation time due to dynamical screening as

$$\frac{1}{\tau_{ei}} = \frac{n_i m_e}{2p^3} \int_0^{2p/\hbar} \frac{dk}{2\pi} \left| \frac{4\pi e^2}{k^2 + \kappa_e^2} \right|^2 k^3 S_{ii}(k) . \qquad (2.228)$$

Again, $S_{ii}(k)$ is the static structure factor, and κ_e is the electronic Debye screening quantity. Then, the conductivity follows from (2.224).

Let us now consider the partially ionized hydrogen plasma. In this case, the electrical conductivity can be calculated from the expression (2.217) with $\sum_c n_c Q_{ec}^T = n_p Q_{ep}^T + n_H Q_{eH}^T$. Here, n_p is the density of the free protons, and n_H is the density of the atoms assumed to be in the ground state. Therefore, two additional problems have to be solved. One has to determine the plasma composition, i.e., the degree of ionization α, and one has to calculate the transport cross section Q_{eH}^T. For the first problem, we have solved the Saha equation (2.155) using the same level of approximation as it was done for the results shown in Fig. 2.7 (see Sect. 2.8).

As in the case of electron–proton scattering, the transport cross section Q_{eH}^T is calculated in first Born approximation. We start again from formula (2.209), and we determine the differential cross section for elastic scattering of an electron on a hydrogen atom in the ground state. In order to account for plasma effects, static screening was included in the electron–atom scattering potential. For the atoms, we used unscreened wave functions for simplicity. The differential cross section in first Born approximation then is

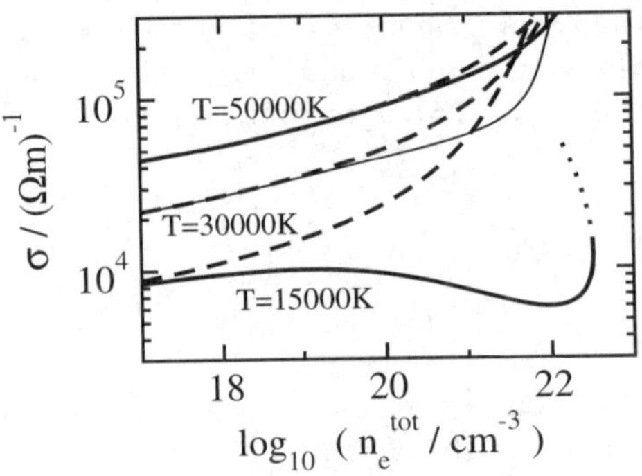

Fig. 2.10. The electrical conductivity of a partially ionized hydrogen plasma as a function of the total electron density for various temperatures (*solid lines*). The *dashed lines* show the results assuming a fully ionized hydrogen plasma

$$\frac{d\sigma_{eH}}{d\Omega} = \frac{4m_e^2 e^4}{(q^2+\hbar^2\kappa^2)^2}\left[1 - F_{11}(q)\right]^2 , \qquad (2.229)$$

with $F_{11}(q)$ being the atomic form factor given by

$$F_{11}(q) = \left[1 + \left(\frac{qa_B}{2\hbar}\right)^2\right]^{-2} .$$

Here, $\boldsymbol{q} = \boldsymbol{p}_e - \bar{\boldsymbol{p}}_e$ is the momentum transfer of the scattered electron, and κ is the inverse screening length.

With the knowledge of the degree of ionization α and the transport cross sections Q^T_{ep} and Q^T_{eH}, we are able to determine the electrical conductivity of the partially ionized hydrogen plasma from (2.217). Results for the conductivity using the cross sections in Born approximation and the simple Saha equation mentioned above are shown in Fig. 2.10. Curves are presented as a function of total electron density for different temperatures.

At high temperatures, the conductivity merges into that of a fully ionized plasma, i.e., we have a monotonous increase with density. This behavior changes drastically at lower temperatures. A characteristic feature is the minimum with a subsequent sharp increase at high densities (Ebeling et al. 1977; Kremp (b) et al. 1983; Kremp (b) et al. 1984; Höhne et al. 1984; Reinholz et al. 1995). This behavior demonstrates the strong influence of nonideality effects on the electrical conductivity of dense partially ionized plasmas. First, there is a decrease of the conductivity determined by the lowered degree of ionization due to the formation of bound states, and by the electron–atom scattering contribution. The sharp increase results from the increase of the degree of ionization caused by the lowering of the ionization energy which describes pressure ionization in dense hydrogen (Mott effect). Consequently, there is a transition between states of low and high conductivities which can be interpreted as a nonmetal–metal transition. The multi-valued behavior at lower temperatures reflects that of the degree of ionization shown in Fig. 2.7.

At the end of this section, we want to mention again that only simple approximations were used here to consider the dc electrical conductivity for nonideal plasmas. More general considerations are given in Sect. 9.2. There, many references concerning theoretical and experimental investigations of the electrical conductivity of dense plasmas can be found, too.

2.12 Ionization Kinetics

In the previous sections, we determined the plasma composition by the Saha equation (2.151). This equation follows from the condition (2.143) valid for plasmas in thermodynamic equilibrium. For non-equilibrium plasmas, the chemical equilibrium is, in general, not established, and the degree of ionization and the population densities of the different atomic states are varying functions in space and time. Therefore, one has to start from kinetic equations which account for the formation and the decay of bound particles.

In this section, we give an introduction in ionization kinetics from an elementary point of view. The aim is not to give a detailed discussion of all the manifold ionization phenomena in plasmas. For this point, we refer to some monographs on plasma physics, e.g., Biberman et al. (1987), Griem (1974), Sobel'man et al. (1988), Elton (1990). We will focus here on the question of how the plasma medium influences the ionization kinetics of strongly coupled plasmas. To obtain first results, simple cases are considered.

We start from the density balance equation, and consider the rate of change of the free electron density given by

$$\frac{\partial}{\partial t} n_e(\boldsymbol{r}, t) + \nabla_{\boldsymbol{r}} \cdot \boldsymbol{j}_e(\boldsymbol{r}, t) = W_e(\boldsymbol{r}, t) \,. \tag{2.230}$$

The current density $j_e(\boldsymbol{r}, t)$ describes the flux of electrons in a spatially inhomogeneous plasma. $W_e(\boldsymbol{r}, t)$ denotes the source function which accounts for the gain and the loss of electrons due to ionization and recombination processes. In some cases, the current density can be written in the form

$$\boldsymbol{j}_e(\boldsymbol{r}, t) = D_e(\boldsymbol{r}, t) \nabla_{\boldsymbol{r}} n_e(\boldsymbol{r}, t) \tag{2.231}$$

which represents Fick's law of diffusion.

With (2.231), the equation of evolution (2.230) is usually called reaction–diffusion equation. Such equations are frequently used in a phenomenological approach to study non-equilibrium systems. A discussion of reaction–diffusion processes on the basis of kinetic theory is given, e.g., in Ebeling et al. (1989) and Kremp (b) et al. (1993).

Let us now proceed with the elementary consideration restricting ourselves to spatially homogeneous plasmas. For simplicity, we consider a plasma consisting of free electrons and singly charged ions with number densities n_e, n_i and atoms in state $|j\rangle$ with the population density n_A^j. Here, j denotes the set of internal quantum numbers of the atomic bound state.

Writing the source function W_e in terms of rate coefficients, the ionization balance equation for the free electron density is given by

$$\frac{d}{dt}n_e = \sum_j (n_e n_A^j \alpha_j - n_e^2 n_i \beta_j + n_A^j \alpha_j^R - n_e n_i \beta_j^R) . \qquad (2.232)$$

Here, α_j and β_j are the collisional rate coefficients for the reactions

$$A_j + e \rightleftharpoons e + e + i , \qquad (2.233)$$

where α_j is the coefficient of electron impact ionization from the atomic level j, and β_j is the three-body recombination coefficient describing the inverse process. The radiative rate coefficients are denoted by α_j^R and β_j^R. They account for photo-ionization and radiative recombination according to

$$A_j + \hbar\omega \rightleftharpoons e + i . \qquad (2.234)$$

A similar rate equation can be found for the rate of change of the population densities of the ground and excited atomic states. It can be written in the form

$$\frac{d}{dt}n_A^j = \sum_{j'} (n_e n_A^{j'} K_{j'j} - n_e n_A^j K_{jj'} + n_a^{j'} K_{j'j}^R - n_A^j K_{jj'}^R)$$
$$+ (n_e^2 n_i \beta_j - n_e n_A^j \alpha_j + n_e n_i \beta_j^R - n_A^j \alpha_j^R) . \qquad (2.235)$$

Here, the temporal evolution of the population density n_A^j is determined by the ionization and recombination coefficients and also by the coefficients of excitation and de-excitation. We have included excitation and de-excitation by electron impact given by $K_{jj'}$ and $K_{j'j}$. Furthermore, radiative transitions are included by the rate coefficients $K_{jj'}^R$ of spontaneous emission and by $K_{j'j}^R$ for the inverse process. Of course, further processes have to be taken into account if more complex systems are considered (see, e.g., Biberman et al. (1987), Elton (1990)).

In order to get the free electron density and the atomic population densities from the rate equations, one has to know the rate coefficients. Since we are interested in the properties of dense plasmas, the radiative transitions included in (2.232) and (2.235) are of minor importance. The kinetics is mainly determined by the collision processes (Griem 1964; Fujimoto and McWhriter 1990).

In this section, we will determine the collisional rate coefficients in a simple manner. A more rigorous calculation of these coefficients will be given in Chap. 8 on the basis of quantum statistical theory. Let us first consider the coefficient of impact ionization.

From an elementary point of view, the probability of ionization of an atom in the state $|j\rangle$ can be written as $w_j^{\text{ion}} = n_e v \sigma_j^{\text{ion}}$, where v is the velocity of the incident electron and σ_j^{ion} is the ionization cross section. Averaging this expression with respect to the electron distribution function, we get for the ionization coefficient

$$\begin{aligned} \alpha_j &= \langle v\, \sigma_j^{\text{ion}} \rangle \\ &= \frac{1}{n_e} \int \frac{d\boldsymbol{p}}{(2\pi\hbar)^3} \frac{p}{m_e}\, \sigma_j^{\text{ion}}(\boldsymbol{p})\, f_e(\boldsymbol{p}, t)\,. \end{aligned} \tag{2.236}$$

with $p = m_e v$ being the momentum and f_e the distribution function normalized to the free electron density. We want to remember that the spin summation is not indicated explicitly. Two problems have to be solved in order to calculate α_j. The first one is the knowledge of the distribution function, and the second one is the calculation of the ionization cross section. The first problem requires, in general, the solution of the electron kinetic equation. But, in many situations, the momentum distribution can be assumed to be in equilibrium. In fact, the momentum distribution reaches local equilibrium earlier than equilibrium is reached for the plasma composition. This is due to the high frequency of electron–electron collisions in dense plasmas leading to very short relaxation times for $f_e(\boldsymbol{p}, t)$. Therefore, the electron distribution function is used in the form

$$f_e(\boldsymbol{p}, t) = \frac{n_e \Lambda_e^3}{2} e^{-p^2/(2m_e k_B T_e)} \tag{2.237}$$

where the number density $n_e = n_e(t)$ and the electron temperature $T_e = T_e(t)$ are time dependent. With (2.237), a non-degenerate plasma is considered, and we get

$$\alpha_j = \frac{8\pi m_e}{(2\pi m_e k_B T_e)^{3/2}} \int_{I_j}^{\infty} d\epsilon\, \epsilon\, \sigma_j^{\text{ion}}(\epsilon)\, e^{-\epsilon/k_B T_e} \tag{2.238}$$

where $\epsilon = p^2/2m_e$ is the electron impact energy, and $I_j = |E_j|$ is the ionization energy of the atom in the state $|j\rangle$.

The second problem determining the ionization coefficient from (2.236) is the calculation of the ionization cross section. This problem has been investigated for a long time in scattering theory. The ionization of hydrogen atoms was considered by Bethe (1930) who derived an analytic expression for the ionization cross section for high electron impact energies. In subsequent papers, a lot of work was done to get analytic expressions for excitation and ionization cross sections which can be approximately applied to the lower or the entire energy range. Thus, fit formulas and semiempirical expressions for hydrogen and for other atoms and ions were proposed (Van Regenmorter 1962; Lotz 1968; Drawin and Emard 1977). For an overview and for further references see, e.g., Biberman et al. (1987) and Sobel'man et al. (1988). But, all these formulae are based on the assumption that the elementary processes are not influenced by the surrounding plasma medium, i.e., many-particle effects are not included. This leads to a validity of the rate coefficients restricted to the ideal plasma state only. However, at high densities, there are strong correlations, and the plasma becomes nonideal. Now, one has to expect a density and temperature dependence of the ionization cross section and of the rate coefficient as a result of the many-particle effects. Of course, ionization of atomic states is essentially influenced by the lowering of the ionization energy due to the surrounding plasma. In order to include this important effect in an elementary way, one has to generalize the cross section (Schlanges et al. 1988; Bornath et al. 1988). We start from the Bethe-type expression given by Biberman et al. (1987)

$$\sigma_j^{\rm ion}(\epsilon) = 2.5\pi a_B^2\, n^3 \frac{|E_n^0|}{\epsilon} \ln \frac{\epsilon}{|E_n^0|} \, . \tag{2.239}$$

where $j = (n, l, m)$ with n being the principal quantum number. We modify this cross section assuming an effective ionization threshold given by (2.241). Then, we have

$$\sigma_j^{\rm ion}(\epsilon) = 2.5\pi a_B^2\, n^3 \frac{|E_j^0|}{\epsilon} \ln \frac{\epsilon - \Delta_e - \Delta_i + \Delta_j}{|E_j^0|} \, . \tag{2.240}$$

Here, $\Delta_e = \Delta_i = -\kappa e^2/2$ are the self-energy shifts of electrons and singly charged ions with κ being the inverse screening length discussed in Sect. 2.4. E_j^0 is the binding energy of the isolated electron-ion pair, and Δ_j is the atomic self-energy correction defined by (2.138).

The ionization cross section (2.240) includes an effective ionization threshold given by

$$|E_j^0| - \Delta_j + \Delta_e + \Delta_i = I_j^{\rm eff} \, . \tag{2.241}$$

This is the effective ionization energy of an atom in the state $|j\rangle$ as introduced in Sect. 2.7. Therefore $\sigma_j^{\rm ion}$ represents an in-medium ionization cross section accounting for the lowering of the ionization energy.

In Fig. 2.11, the ionization cross section (2.240) for hydrogen atoms in the ground state is shown as a function of electron impact energy for different screening parameters κa_B. With increasing values of κa_B (increasing plasma density), the effective threshold energy decreases to zero energy where the bound state vanishes due to the Mott effect.

Inserting the generalized formula (2.240) into (2.238), one can perform the integration with the result (Schlanges et al. 1988; Bornath et al. 1998)

$$\alpha_j = \alpha_j^{\rm ideal} \exp\left[(\Delta_j - \Delta_e - \Delta_i)/k_B T_e\right] \, , \tag{2.242}$$

where we introduced the ideal ionization coefficient

$$\alpha_j^{\rm ideal} = \frac{10\pi a_B^2 E_j}{(2\pi m_e k_B T_e)^{1/2}} j^3 Ei\left(-\frac{|E_j|}{k_B T_e}\right) \tag{2.243}$$

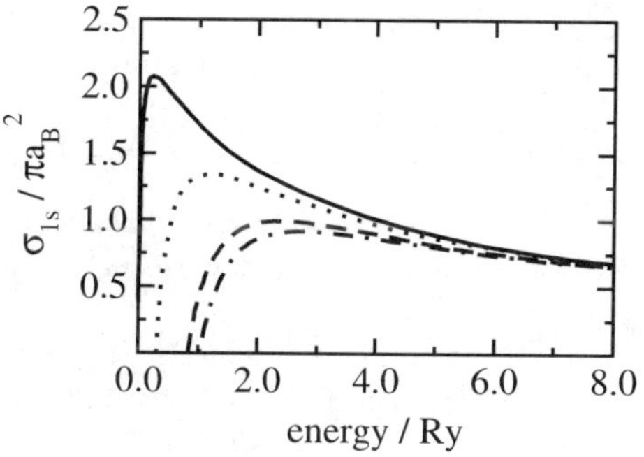

Fig. 2.11. Ionization cross section for hydrogen atoms in the ground state as a function of electron impact energy for different screening parameters: $\kappa a_B = 1.0$ (*solid line*), $\kappa a_B = 0.5$ (*dotted*), $\kappa a_B = 0.1$ (*dashed*), $\kappa a_B = 0.05$ (*dash-dotted*)

with $Ei(x)$ being the exponential integral function

$$Ei(x) = \int_{-\infty}^{x} \frac{e^t}{t} dt \,. \tag{2.244}$$

In the approximation considered, the ionization coefficient has an interesting structure. According to (2.242), it is written as a product of a density independent contribution $\alpha_j^{\mathrm{ideal}}$ and a nonideal contribution which accounts for effects of the medium by the self-energy shifts Δ_e, Δ_i, and Δ_j.

If the bound state shift Δ_j is neglected, and if we use $\Delta_e = \Delta_i = -\kappa e^2/2$, it follows that

$$\alpha_j = \alpha_j^{\mathrm{ideal}} \exp\left(\kappa e^2 / k_B T_e\right) . \tag{2.245}$$

In Fig. 2.12, the ionization coefficient (2.245) for H-atoms in the ground state is shown as a function of the free electron density for different temperatures. At low densities, the influence of many-particle effects is small, and α_1 tends to be equal to its density independent ideal contribution $\alpha_1 = \alpha_1^{\mathrm{ideal}}$. The latter follows from (2.236) using the cross section (2.239). However, the behavior of the ionization coefficient changes drastically at high density where the plasma is nonideal. We observe a strong increase of α_1 with increasing density due to the lowering of the ionization energy.

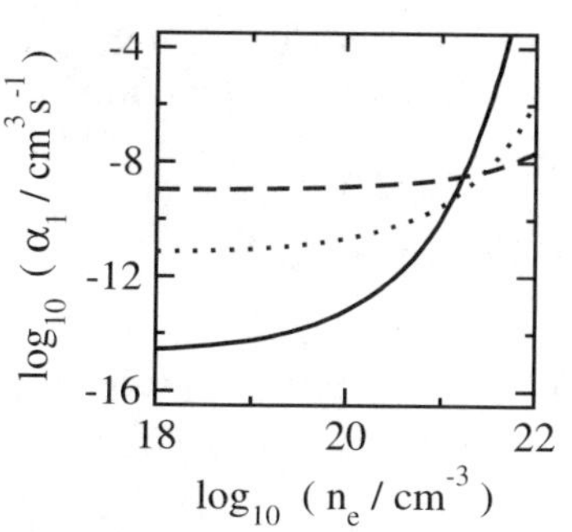

Fig. 2.12. Coefficient of impact ionization from the atomic ground state of hydrogen versus free electron density for various temperatures: $T_e = 10000\,\mathrm{K}$ (*solid line*), $T_e = 20000\,\mathrm{K}$ (*dotted*), $T_e = 50000\,\mathrm{K}$ (*dashed*)

It is interesting to consider another approximation. If we take the cross section (2.240) near the threshold and apply $\ln x \approx x-1$, we get from (2.238)

$$\alpha_j = 5\sqrt{\pi}\, n^3 a_B^2 (2k_B T_e/m_e)^{1/2} e^{-I_j^{\rm eff}/k_B T_e} \,. \tag{2.246}$$

This expression represents the well-known Arrhenius law of chemical reaction kinetics generalized to strongly coupled plasmas, where the activation energy is given by the effective ionization energy $I_j^{\rm eff}$.

Let us now consider the three-body recombination coefficient β_j. In order to calculate β_j it is useful to start from the rate equations (2.232) or (2.235). In thermodynamic equilibrium, it follows for a hydrogen plasma

$$\frac{n_A^j}{n_e n_i} = \frac{\beta_j}{\alpha_j} = \Lambda_e^3 \exp\left[-\beta(E_j - \Delta_e - \Delta_i + \Delta_j)\right] \tag{2.247}$$

where the Saha equation (2.151) is used.

From (2.247), we get

$$\beta_j = \Lambda_e^3\, \alpha_j\, \exp\left(I_j^{\rm eff}/k_B T_e\right). \tag{2.248}$$

This relation should also be valid for local equilibrium with a time dependent density determined by the rate equation (2.232). According to the relation (2.248), it is easy to calculate the recombination coefficient if the corresponding ionization coefficient is already known. Inserting (2.242) into (2.248), we arrive at (Schlanges et al. 1988)

$$\beta_j = \Lambda_e^3 \exp(-E_j/k_B T_e)\, \alpha_j^{\rm ideal} = \beta_j^{\rm ideal} \,. \tag{2.249}$$

Therefore, nonideality has an effect on the ionization coefficient only, whereas the recombination coefficient remains unaffected. This is a consequence of the simple approximation used in our model to include many-particle effects. In Sect. 8.1, we will show that there is a density dependence of β_j, too, if many-particle effects are included in a more rigorous manner starting from quantum kinetic equations (Schlanges and Bornath 1993; Bornath and Schlanges 1993). Modifications of the rate coefficients by static electric fields are considered in Kremp (a) et al. (1993).

In a similar way, one can also find simple expressions for the excitation coefficients including effects of the medium. The coefficient of electron impact excitation for the transition $j \to j'$ is

$$K_{jj'} = \frac{8\pi m_e}{2\pi m_e k_B T_e)^{3/2}} \int_{\Delta I_{jj'}^{eff}}^{\infty} d\epsilon\, \epsilon\, \sigma_{jj'}^{\rm exc}(\epsilon)\, e^{(-\beta\epsilon)} \,, \tag{2.250}$$

where $\sigma_{jj'}^{\rm exc}$ is the excitation cross section. For the de-excitation coefficients we have, like in the case of ionization and recombination,

$$K_{j'j} = K_{jj'} \exp\left(\Delta E_{jj'}^{\rm eff}/k_B T_e\right). \tag{2.251}$$

Expressions for in-medium excitation cross sections can be found similar to ionization starting from Bethe–Born-type cross section formulae. In (2.250), $\Delta I_{jj'}^{\text{eff}}$ is an effective excitation threshold energy given by

$$\Delta E_{jj'}^{\text{eff}}(n_e, T) = E_{j'}^0 - E_j^0 + \Delta_{j'} - \Delta_j \,. \tag{2.252}$$

Plasma effects are accounted for by the atomic self-energy corrections Δ_j and $\Delta_{j'}$. From the results given in Sect. 2.7 follows that the difference $(\Delta_{j'} - \Delta_j)$ leads to a small correction in (2.252). The result is a smaller influence of many-particle effects on the excitation coefficients compared to ionization where the lowering of the ionization energy drastically affects the threshold. Therefore, the excitation coefficients can be calculated from (2.250) neglecting medium effects in a first approximation, i.e., the coefficients valid for the ideal plasma state can be used (Biberman et al. 1987; Sobel'man et al. 1988).

The knowledge of the rate coefficients makes it possible to calculate the degree of ionization and the atomic level population. Neglecting the radiative processes, we get from (2.232) for a hydrogen plasma

$$\frac{d}{dt} n_e = \sum_j (n_e n_H^j \alpha_j - n_e^2 n_p \beta_j) \,. \tag{2.253}$$

In many cases, the temporal evolution of the free electron density is strongly coupled with the energy balance of the reacting plasma. Therefore, the relaxation of the densities has to be considered together with the relaxation of the temperatures (Ohde et al. 1995; Ohde et al. 1996). In order to find equations for the temperature, we start from the total energy accounting for the fact that it is constant in time. Using the results obtained in Sect. 2.6, the internal energy is in our simple model

$$\frac{U(t)}{V} = \sum_{a=e,p} \left(\frac{3}{2} n_a k_B T_a + n_a \Delta_a \right) + \sum_j \left(\frac{3}{2} n_H^j k_B T_H + n_H^j E_j \right) , \tag{2.254}$$

where $\Delta_a = -\kappa e_a^2/2$ is the self-energy contribution of the charged particles given by the Debye shift, and $E_j = E_j^0 + \Delta_j$ is the binding energy of a hydrogen atom in the state $|j\rangle$. The expression (2.254) is valid for a nonideal plasma because self-energy corrections are included.

In the following, the heavy-particle temperature $T_h = T_p = T_H$ is assumed to be constant in time. Furthermore, the atomic self energy is approximated as $\Delta_j \approx 0$. Then, from (2.254), we get

$$\frac{d}{dt} T_e(t) = \sum_j \frac{\frac{3}{2} k_B T_e - \frac{3}{2} \kappa e^2 - E_j}{\frac{3}{2} k_B n_e + \frac{1}{2} n_e \kappa e^2 \frac{1}{T_e^2} (\frac{1}{T_e} + \frac{1}{T_h})^{-1}} \frac{d}{dt} n_H^j \,. \tag{2.255}$$

If only the atomic ground state contribution is included and if the medium corrections are neglected, the temperature equation reduces to (Biberman et al. 1987)

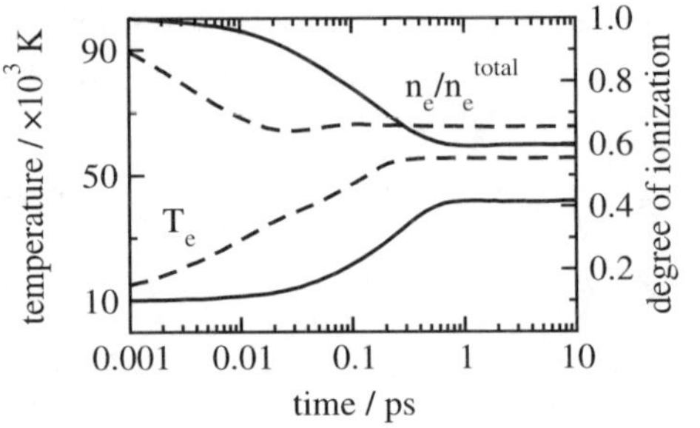

Fig. 2.13. Degree of ionization and electron temperature as a function of time. The initial state is a fully ionized hydrogen plasma with $T_e(0) = T_h(0) = 10000\,\mathrm{K}$. The total electron density is $n_e^{tot} = 10^{21}\,\mathrm{cm}^3$. For comparison results are shown neglecting nonideality corrections (*dashed lines*). Degree of ionization (*upper pair of lines*), electron temperature (*lower pair*)

$$\frac{d}{dt}T_e(t) = \frac{E_1 - \frac{3}{2}k_BT_e}{\frac{3}{2}k_Bn_e}\frac{d}{dt}n_e(t)\,. \tag{2.256}$$

Here, we used

$$\sum_j \frac{dn_H^j}{dt} = -\frac{dn_e}{dt}\,.$$

Finally, let us use the simple approach developed in this section to study some features of the ionization kinetics in a nonideal hydrogen plasma. For this purpose, we solve the coupled set of equations (2.253) and (2.255) for the number densities and the electron temperature. We determine the ionization coefficients α_j from the expression (2.245) where j labels the principal quantum number. The lowering of the continuum edge is accounted for by the shifts $\Delta_e + \Delta_p = -\kappa e^2$ of electrons and protons. The three-body recombination coefficients follow from the ionization coefficients using (2.248).

In Fig. 2.13, the temporal evolution of the degree of ionization and the electron temperature is shown for a hydrogen plasma which is fully ionized in the initial state. The total electron density is assumed to be constant in time. Results for the isothermal case are given by Ebeling and Leike (1991), Bornath et al. (1998).

In order to demonstrate the influence of the plasma medium on the density–temperature relaxation, the curves in Fig. 2.13 are compared to those obtained for the ideal plasma state (dashed curves). As expected, the degree of ionization decreases with time due to the recombination processes whereas the temperature increases. The influence of plasma nonideality reduces the effect of recombination, this means that higher ionization occurs in the nonideal plasma. For times $t > 1ps$ the stationary state is reached. Chemical equilibrium is established with a degree of ionization determined by the Saha equation (2.155). A more detailed discussion of ionization and population kinetics in strongly coupled plasmas is given in Chap. 8.

3. Quantum Statistical Theory of Charged Particle Systems

3.1 Quantum Statistical Description of Plasmas

Strongly coupled plasmas are many-particle systems the behavior of which is determined by the long-range Coulomb interaction and by the laws of quantum mechanics. In Chap. 2, we gave an elementary description of the properties of such systems. However, an exact theory may only be given on the basis of quantum theory and especially of quantum statistics.

We now intend to give a very brief introduction to quantum statistics as the appropriate tool for the description of strongly correlated many-particle systems. From the point of view of quantum mechanics, the many-particle system is characterized by its Hamiltonian, i.e., by

$$\mathrm{H} = \sum_{i=1}^{N} \frac{p_i^2}{2m} + \sum_{i<j}^{N} V(\boldsymbol{r}_i - \boldsymbol{r}_j) \,, \tag{3.1}$$

where, e.g., in the plasma case, we have to take the Coulomb interaction

$$V(\boldsymbol{r}_i - \boldsymbol{r}_j) = \frac{e_i e_j}{|\boldsymbol{r}_i - \boldsymbol{r}_j|} \tag{3.2}$$

with $1/(4\pi\varepsilon_0) = 1$. Here, $\boldsymbol{r}_i$ and $\boldsymbol{p}_i$ are the position and momentum operators of the plasma particles. The micro-state (pure state) of a plasma is determined by a measurement of a complete set of observables $b_1 \ldots b_N$ and is represented by a vector

$$|b_1 \ldots b_N\rangle \in \mathcal{H}_N \tag{3.3}$$

in the space of states $\mathcal{H}_N$ (unitary vector space) of the $N-$particle system. In (3.3), b_1 denotes a complete single-particle observable, e.g., $\boldsymbol{r}_1, s_1^z$ or $\boldsymbol{p}_1, s_1^z$ with s_1^z being the third component of the spin. A plasma consists of electrons and ions which form subsystems of identical particles. In quantum physics, these particles have to be considered as indistinguishable particles. For the description of systems of identical particles, it is necessary to introduce an additional postulate, namely the *Spin-Statistics Postulate*:

The space of states of Fermi particles (half-integer spin) is the antisymmetric subspace $\mathcal{H}_N^-$ of $\mathcal{H}_N$, while the space of the states of Bose particles (integer spin) is the symmetric subspace $\mathcal{H}_N^+$.

This postulate is of great importance for the quantum statistical description of plasmas. In fact, plasmas consist of Fermi particles, but they can also contain Bose particles which lead to different quantum effects. Therefore, the spin-statistics postulate has consequences up to the macroscopic properties of plasma systems.

The properties of a quantum system of many identical particles is most conveniently described in terms of creation and annihilation operators defined by the following relation

$$|b_1 \ldots b_N\rangle^{\pm} = \frac{1}{\sqrt{N!}} \mathrm{a}^\dagger(b_1) \ldots \mathrm{a}^\dagger(b_N) \, |0\rangle \, . \tag{3.4}$$

Here, $|0\rangle$ is the state without particles which is also referred to as vacuum state. Correspondingly, the operators $\mathrm{a}(b_1) \ldots \mathrm{a}(b_N)$ are defined by the adjoint relation. The operators a and $a^\dagger$ act in the Fock space. The completeness relation in the Fock space then reads

$$\sum_N \int db_1 \ldots db_N \; |b_1 \ldots b_N\rangle^{\pm} \, \langle b_N \ldots b_1| = 1 \, . \tag{3.5}$$

The condition that the many-particle states have to be symmetric for Bose systems and antisymmetric for Fermi ones is fulfilled if, and only if, the construction operators obey the following commutation rules

$$\begin{aligned} &[\mathrm{a}(b), \mathrm{a}(b')]_{\mp} = 0, \qquad [\mathrm{a}^\dagger(b), \mathrm{a}^\dagger(b')]_{\mp} = 0, \\ &[\mathrm{a}(b), \mathrm{a}^\dagger(b')]_{\mp} = \delta(b - b') \, . \end{aligned} \tag{3.6}$$

In the subscripts, the upper sign refers to Bose, and the lower one to Fermi particles. The brackets $[\ldots]_{\mp}$ are commutators or anti-commutators, respectively. The spin-statistics theorem is fulfilled in a simple manner merely by the commutation rules of the construction operators. This fact is one of the most important advantages of the method of second quantization in many-particle physics and quantum statistics.

After having seen how to construct states out of the vacuum state using the operators a and $\mathrm{a}^\dagger$, we now have to consider the question how to present the operators A of quantum mechanics in terms of the operators $\mathrm{a}^\dagger(b)$ and $\mathrm{a}(b)$. In most cases, N-particle operators are of additive or of binary types, and in general of s-particle types, namely

$$\mathrm{A}_N = \sum_{1 \cdots s} \mathrm{A}_{1 \ldots s} \, .$$

One can show that such operators may be expressed in terms of products of s operators a and s operators $\mathrm{a}^\dagger$

$$\mathrm{A}_N = \frac{1}{s!} \int db_1 \ldots db_s db_1' \ldots db_s' \, \langle b_1 \ldots b_s | \mathrm{A}_{1\ldots s} | b_s' \ldots b_1' \rangle^{\pm} \\ \times \mathrm{a}^\dagger(b_1) \ldots \mathrm{a}^\dagger(b_s) \mathrm{a}(b_s') \ldots \mathrm{a}(b_1') \,. \qquad (3.7)$$

In many cases, it is useful to apply the representation of the operators $\mathrm{a}(b_1)$ in the coordinate space, i.e., $b_1 = \{\boldsymbol{r}_1, s_1^z\}$. We then write

$$\mathrm{a}(b_1) = \mathrm{a}(\boldsymbol{r}_1, s_1^z) = \psi(\boldsymbol{r}_1, s_1^z) \,.$$

The operators $\psi(\boldsymbol{r}_1, s_1^z)$ are called *field operators*. It is easy to express the Hamiltonian of the plasma in terms of field operators, i.e.,

$$\begin{aligned} \mathrm{H} &= \int d\boldsymbol{r}_1 \, \psi^\dagger(\boldsymbol{r}_1) \left(-\frac{\nabla_1^2}{2m} \right) \psi(\boldsymbol{r}_1) \\ &+ \frac{1}{2} \int d\boldsymbol{r}_1 d\boldsymbol{r}_2 \, \psi^\dagger(\boldsymbol{r}_1) \psi^\dagger(\boldsymbol{r}_2) \mathrm{V}(\boldsymbol{r}_1 - \boldsymbol{r}_2) \psi(\boldsymbol{r}_2) \psi(\boldsymbol{r}_1) \,. \qquad (3.8) \end{aligned}$$

For simplicity, the spin variables were dropped.

Let us now take into consideration that a plasma is a macroscopic many-particle system with particle numbers of the order of $N \approx 10^{23}$, i.e., with a practically infinite number of particles. It is well known that such a system can no longer be described by the microstate $|b_1 \ldots b_N\rangle$. The unavoidable coupling of such a system to its surrounding leads to a permanent uncertainty concerning its microstate. Therefore, the system can be described only by its possible states $|\Psi_n\rangle$ and the corresponding probabilities P_n, according to which the states are realized. We will call a state defined in this way a *mixed* or *quantum statistical state*. It is useful to condense the information given by mixed states into the density operator

$$\rho = \sum_n P_n \, |\Psi_n\rangle \, \langle \Psi_n| \,, \qquad \mathrm{Tr}\, \rho = 1 \,. \qquad (3.9)$$

Then the quantum statistical mean value of a physical quantity follows from

$$\langle \mathrm{A} \rangle = \mathrm{Tr}\,(\rho \mathrm{A}) \,. \qquad (3.10)$$

Obviously, the mean value of an $s-$particle quantity may be represented in the following form

$$\begin{aligned} \langle \mathrm{A}_N \rangle &= \frac{1}{s!} \int db_1 \ldots db_s db_1' \ldots db_s' \, \langle b_1 \ldots b_s | \mathrm{A}_{1\ldots s} | b_s' \ldots b_1' \rangle \\ &\times \left\langle \mathrm{a}^\dagger(b_1) \ldots \mathrm{a}^\dagger(b_s) \mathrm{a}(b_s') \ldots \mathrm{a}(b_1') \right\rangle \,. \qquad (3.11) \end{aligned}$$

Mean values are therefore given by the mean value of products of creation and annihilation operators. Such quantities are referred to as *density matrices.* They are defined by

$$\begin{aligned} n^s\langle b_1 \dots b_s|\mathrm{F}_{1\dots s}|b_s' \dots b_1'\rangle &= \left\langle \mathrm{a}^\dagger(b_1') \dots \mathrm{a}^\dagger(b_s')\mathrm{a}(b_s) \dots \mathrm{a}(b_1)\right\rangle \\ &= \mathrm{Tr}\{\rho\,\mathrm{a}^\dagger(b_1) \dots \mathrm{a}^\dagger(b_s)\mathrm{a}(b_s) \dots \mathrm{a}(b_1)\}\,. \end{aligned} \quad (3.12)$$

Any quantum statistical mean value is again determined by (3.11). Therefore, the reduced density matrix contains all statistical information about the many-particle system. With the relation (3.12), we introduced the *second quantization formalism* into quantum statistics. On the basis of this formalism, it is possible to introduce the highly effective methods of quantum field theory into statistical mechanics.

Of special interest is, of course, the one-particle density matrix. In terms of field operators, we have

$$n\,\langle \boldsymbol{r}_1\,|\,\mathrm{F}_1\,|\,\boldsymbol{r}_1'\rangle = \mathrm{Tr}\{\rho\,\psi^\dagger(\boldsymbol{r}'_1)\psi(\boldsymbol{r}_1)\}\,. \quad (3.13)$$

With this quantity, mean values of observables of additive type ($A_N = \sum_i A_i$) can be calculated according to

$$\langle A_N\rangle = n\int d\boldsymbol{r}_1 d\bar{\boldsymbol{r}}_1\,\langle \boldsymbol{r}_1|\mathrm{A}_1|\bar{\boldsymbol{r}}_1\rangle\,\langle \bar{\boldsymbol{r}}_1|\mathrm{F}_1|\boldsymbol{r}_1\rangle\,. \quad (3.14)$$

If we introduce sum and difference variables

$$\boldsymbol{R} = \frac{\boldsymbol{r}_1 + \boldsymbol{r}'_1}{2}\,, \qquad \boldsymbol{r} = \boldsymbol{r}_1 - \boldsymbol{r}'_1\,,$$

we can rewrite (3.13) into

$$n\,\langle \boldsymbol{R} + \boldsymbol{r}/2\,|\,\mathrm{F}_1\,|\,\boldsymbol{R} - \boldsymbol{r}/2\rangle = \mathrm{Tr}\left\{\rho\,\psi^\dagger(\boldsymbol{R} - \boldsymbol{r}/2)\,\psi(\boldsymbol{R} + \boldsymbol{r}/2)\right\}\,. \quad (3.15)$$

Finally, we take the Fourier transform with respect to $\boldsymbol{r}$

$$f(\boldsymbol{p}, \boldsymbol{R}) = \int d\boldsymbol{r}\, e^{-i\boldsymbol{p}\cdot\boldsymbol{r}}\,\mathrm{Tr}\left\{\rho\,\psi^\dagger(\boldsymbol{R} - \boldsymbol{r}/2)\,\psi(\boldsymbol{R} + \boldsymbol{r}/2)\right\}\,. \quad (3.16)$$

$f(\boldsymbol{p}, \boldsymbol{R})$ is the Wigner representation of the one-particle density matrix. The Wigner function $f(\boldsymbol{p}, \boldsymbol{R})$ is an extremely useful function. If we consider mean values of observables of the form $A = A(\boldsymbol{p}, \boldsymbol{R})$, relation (3.11) reads in Wigner representation

$$\langle A_N\rangle = \int \frac{d\boldsymbol{p}}{(2\pi)^3} d\boldsymbol{R}\; A(\boldsymbol{p}, \boldsymbol{R})\; f(\boldsymbol{p}, \boldsymbol{R})\,, \quad (3.17)$$

where $A(\boldsymbol{p}, \boldsymbol{R})$ is the classical phase space function of the physical observable A. In a homogeneous equilibrium system, the Wigner function does not depend on $\boldsymbol{R}$ and may be interpreted as the momentum distribution function or as the mean occupation number. We know from Chap. 2 that $f(\boldsymbol{p})$ is then, in the case of ideal particles, the Bose or the Fermi distribution, respectively.

Let us now consider the dynamics of many-particle systems in the language of second quantization. In the Heisenberg picture, the time evolution of the system is determined by Heisenberg's equations of motion for the construction operators $\mathrm{a}^\dagger(b,t)$ and $\mathrm{a}(b,t)$, i.e., by the equation

$$i\frac{\partial}{\partial t}\mathrm{a}(b,t) = [\mathrm{a}(b,t), \mathrm{H}] \,. \tag{3.18}$$

We look, e.g., at the equation of motion for the field operators $\psi(\boldsymbol{r}_1, t_1) = \psi(1)$. Using the Hamiltonian (3.8), we get after simple rearrangements

$$\left(i\frac{\partial}{\partial t} + \frac{\nabla_1^2}{2m}\right)\psi(1) = \int d\bar{1}\, V(1-\bar{1})\, \psi^\dagger(\bar{1})\psi(\bar{1})\psi(1) \,. \tag{3.19}$$

Here we introduced a notation which will be used subsequently:

$$1 = \boldsymbol{r}_1, t_1, s_1^z; \qquad \int d1 = \sum_{s_1^z}\int d\boldsymbol{r}_1 dt_1;$$
$$V(1-1') = V(\boldsymbol{r}_1 - \boldsymbol{r}_1')\,\delta(t_1 - t_1')\,\delta_{s_1^z, s_{1'}^z} \,. \tag{3.20}$$

It is of some interest to remark that the equation of motion has a similar structure as the Schrödinger equation, however, now ψ is the field operator.

The equations of motion for the creation and annihilation operators completely determine the dynamics of the plasma. Especially, the time evolution of the density matrix is, in the Heisenberg picture, determined by (3.18). Besides the Heisenberg picture, there is the Schrödinger picture. In the latter one, the operators $\mathrm{a}(b)$ are time-*in*dependent, and the density operator depends on time. The temporal evolution of the density operator follows from the *von Neumann equation*

$$i\frac{\partial\rho}{\partial t} = -\,[\rho, \mathrm{H}] \,. \tag{3.21}$$

Furthermore, we want to consider the state of thermodynamic equilibrium.

Of course, the natural condition for the equilibrium is

$$i\frac{\partial\rho}{\partial t} = -\,[\rho, \mathrm{H}] = 0 \,. \tag{3.22}$$

This condition leads to the statement that the operator ρ has to be a functional of the integrals of motion. For its explicit determination, we need an additional condition. An obvious generalization of the Boltzmann H-theorem is the assumption

$$S = -k_B\,\mathrm{Tr}\,(\rho\ln\rho) = \max \,. \tag{3.23}$$

For the solution of the variational task connected with (3.23), we have to adopt additional conditions which lead to different density operators. For our investigations, only the grand canonical density operator is appropriate.

This means that we consider a system with both energy and particle exchange with the surrounding under the condition that the mean values are given (fixed). We then get the known equilibrium density operator

$$\rho = \frac{1}{Z_G}\, e^{-\beta(\mathrm{H}-\mu\mathrm{N})}; \qquad Z_G = \mathrm{Tr}\left\{e^{-\beta(\mathrm{H}-\mu\mathrm{N})}\right\} \tag{3.24}$$

with

$$\beta = \frac{1}{k_B T}, \qquad \mu - \text{chemical potential}. \tag{3.25}$$

Here, Z_G in the grand canonical sum of states. The equation of state using the relations of thermodynamics follows as

$$\begin{aligned} p(\mu,T) &= \frac{k_B T}{\Omega} \ln Z_G(\Omega,\mu,T), \\ n(\mu,T) &= \frac{\partial}{\partial\mu} p(\mu,T). \end{aligned} \tag{3.26}$$

To summarize, we can state the following:

A many-particle system can be fully described by the operators of second quantization $\mathrm{a}^\dagger(b), \mathrm{a}(b)$. The states can be constructed using these creation and annihilation operators and the spin-statistics theorem is automatically satisfied by their commutation rules. The physical quantities may be represented in terms of $\mathrm{a}^\dagger$ and a. The dynamics of the system is described by the Heisenberg equation of motion and the statistical information is condensed in the density matrix.

3.2 Method of Green's Functions

3.2.1 Correlation Functions and Green's Functions

In the previous section we showed that the statistical information concerning a many-particle system is determined by the density matrices. These quantities are given, according to (3.13), by mean values over field operators $\psi^\dagger(\boldsymbol{r}_1,t_1), \psi(\boldsymbol{r}_1,t_1)$ in the Heisenberg picture

$$n^s\left\langle \boldsymbol{r}_1\cdots\boldsymbol{r}_s|\mathrm{F}_{1\cdots s}(t)|\boldsymbol{r}'_s\cdots\boldsymbol{r}'_1\right\rangle = \left\langle \psi^\dagger(\boldsymbol{r}'_1,t)\cdots\psi^\dagger(\boldsymbol{r}'_s,t)\psi(\boldsymbol{r}_s,t)\cdots\psi(\boldsymbol{r}_1,t)\right\rangle. \tag{3.27}$$

The symbol $\langle\cdots\rangle$ denotes an averaging over the statistical operator ρ, i.e., $\langle\cdots\rangle = \mathrm{Tr}(\rho\cdots)$.

An obvious generalization of the s-particle density matrix follows if the field operators have different times. Such multi-time functions can be defined as

$$g^<(1\cdots s,1'\cdots s')=\left(\pm\frac{1}{i}\right)^s\left\langle\psi^\dagger(1')\cdots\psi^\dagger(s')\psi(s)\cdots\psi(1)\right\rangle\,. \tag{3.28}$$

The upper sign refers to Bose particles and the lower sign to Fermi particles.

For simplicity, we often use $\hbar=1$. Later, if the theory is elaborated in detail, $\hbar$ will be accounted for explicitly.

The function (3.28) is referred to as multi-time s-particle correlation function. In the special case $t_1=\ldots=t_s=\ldots=t_1'$, the latter reduces to the the density matrix. Therefore, the correlation functions contain the statistical information about the many-particle system.

Further correlation functions corresponding to different arrangements of the field operators may be introduced. In connection with (3.28), it is of importance to consider

$$g^>(1\cdots s,1'\cdots s')=\left(\frac{1}{i}\right)^s\left\langle\psi(1)\cdots\psi(s)\psi^\dagger(s')\cdots\psi^\dagger(1')\right\rangle\,. \tag{3.29}$$

According to the physical meaning of the $\psi^\dagger(1'),\psi(1)$, the correlation functions provide, in addition to the statistical information, information about the dynamics of an s-particle subsystem in an interacting many-particle system. In detail, this is shown by the following explanations.

Let us first consider the single-particle correlation function $g_1^<(1,1')$,

$$g^<(1,1')=\pm\frac{1}{i}\left\langle\psi^\dagger(1')\psi(1)\right\rangle\,. \tag{3.30}$$

In the special case $t_1=t_1'$, we get the single-particle density matrix, i.e.,

$$g^<(1,1')\Big|_{t_1=t_1'=t}=\pm\frac{1}{i}\left\langle\psi^\dagger(\boldsymbol{r}_1',t)\psi(\boldsymbol{r}_1,t)\right\rangle\,. \tag{3.31}$$

Using the cyclic invariance of the trace, one can easily show that in the case $[\mathrm{H},\rho]=0$, i.e., especially in thermodynamic equilibrium, the correlation functions depend only on the difference of the time arguments, t_1,t_1', this means $g_1^<(t_1,t_1')=g_1^<(t_1-t_1')$.

This is, however, not the case for non-equilibrium situations. Here, it is convenient to introduce the new variables

$$\begin{aligned} t &= \frac{t_1+t_1'}{2}\,,\quad \tau=t_1-t_1'\,,\\ \boldsymbol{R} &= \frac{\boldsymbol{r}_1+\boldsymbol{r}_1'}{2}\,,\quad \boldsymbol{r}=\boldsymbol{r}_1-\boldsymbol{r}_1'\,. \end{aligned} \tag{3.32}$$

In a rough interpretation, the sum variables t and $\boldsymbol{R}$ describe the processes on the macroscopic space-time scale, while τ and $\boldsymbol{r}$ are responsible for the small (microscopic) scale processes. The correlation function may then be written in the form

$$g^<(1,1') = g^<(\boldsymbol{r}\tau, \boldsymbol{R}t)\,. \tag{3.33}$$

The Fourier transform of the correlation function with respect to the difference variables is given by

$$g^<(\boldsymbol{p}\omega, \boldsymbol{R}t) = \int d\boldsymbol{r}d\tau e^{-i\boldsymbol{p}\boldsymbol{r}+i\omega\tau} g^<(\boldsymbol{r}\tau, \boldsymbol{R}t)\,. \tag{3.34}$$

This function is closely connected to the Wigner function. Accounting for the fact that the single-particle density matrix follows from $g^<(1,1')$ for $t_1 = t_1'$, the Wigner function is determined by the relation

$$f(\boldsymbol{p}, \boldsymbol{R}t) = \pm i \int \frac{d\omega}{2\pi} g^<(\boldsymbol{p}\omega, \boldsymbol{R}t)\,. \tag{3.35}$$

In the next step, we consider the multi-time two-particle correlation functions defined by (3.28) and (3.29) for s=2. In many applications, one needs the special case $t_1 = t_2 = t$, $t_1' = t_2' = t'$. We then have

$$g^>(\boldsymbol{r}_1\boldsymbol{r}_2t, \boldsymbol{r}_1'\boldsymbol{r}_2't') = \frac{1}{i^2}\left\langle \psi(\boldsymbol{r}_1, t)\psi(\boldsymbol{r}_2, t)\psi^\dagger(\boldsymbol{r}_2', t')\psi^\dagger(\boldsymbol{r}_1', t')\right\rangle \tag{3.36}$$

and

$$g^<(\boldsymbol{r}_1\boldsymbol{r}_2t, \boldsymbol{r}_1'\boldsymbol{r}_2't') = \frac{1}{i^2}\left\langle \psi^\dagger(\boldsymbol{r}_1', t')\psi^\dagger(\boldsymbol{r}_2', t')\psi(\boldsymbol{r}_2, t)\psi(\boldsymbol{r}_1, t)\right\rangle\,. \tag{3.37}$$

Obviously, these functions describe the behavior of a pair of particles in a many-particle system between the times t and t'. For $t = t'$, the two-particle density matrix follows from the correlation function given by (3.37).

Mainly for technical reasons, it is useful to introduce different types of Green's functions, in particular linear combinations of correlation functions.

Causal Green's functions are a first example. The single-particle causal Green's function is given by

$$g^c(1,1') = \Theta(t_1 - t_1')g^>(1,1') + \Theta(t_1' - t_1)g^<(1,1')\,. \tag{3.38}$$

Another representation is possible if one uses the chronological operator T, namely

$$g^c(1,1') = \frac{1}{i}\left\langle \mathrm{T}\{\psi(1)\psi^\dagger(1')\}\right\rangle\,. \tag{3.39}$$

The chronological, or time ordering operator, arranges the order of the operators such that the operators with the earlier time are positioned right from those with the later ones. Especially, we have for the single-particle Green's function

$$\begin{aligned} \mathrm{T}\{\psi(1)\psi^\dagger(1')\} &= \psi(1)\psi^\dagger(1')\,, \quad t_1 > t_1'\,, \\ &= \pm\psi^\dagger(1')\psi(1)\,, \quad t_1 < t_1'\,. \end{aligned} \tag{3.40}$$

With the representation (3.39), it is easy to have a generalization to an s-particle causal Green's function,

$$g^c(1\ldots s,1'\ldots s') = \left(\frac{1}{i}\right)^s \langle \mathrm{T}\{\psi(1)\ldots\psi(s)\psi^\dagger(s')\ldots\psi^\dagger(1')\}\rangle\,. \tag{3.41}$$

This expression represents, for the various time orders, $(2s)!$ different correlation functions. In the case s=2, there are 24 two-particle correlation functions. Those which are closely connected with the density matrix are given by (3.36) and (3.37).

Correlation functions describing density fluctuations follow from (3.41) in the special case $t_1' = t_1^+$ and $t_2' = t_2^+$. Here, t_1^+ means an infinitesimally larger time than t_1. For the description of density fluctuations it is useful to define, in addition to (3.41), the causal function

$$L^c(12,1'2') = i\Big[g^c(12,1'2') - g^c(1,1')g^c(2,2')\Big]\,. \tag{3.42}$$

For $t_1' = t_1^+$, $t_2' = t_2^+$, it can be written as

$$L_{12}^c(t_1,t_2) = \Theta(t_1-t_2)L_{12}^>(t_1,t_2) + \Theta(t_2-t_1)L_{12}^<(t_1,t_2)\,. \tag{3.43}$$

To simplify the expressions an operator notation was used. The correlation functions of density fluctuations $L^{\gtrless}$ are given by

$$\begin{aligned} iL_{12}^>(t_1,t_2) &= \langle\delta\rho\,(\boldsymbol{r}_1,\boldsymbol{r}_1',t_1)\,\delta\rho\,(\boldsymbol{r}_2,\boldsymbol{r}_2',t_2)\rangle \\ iL_{12}^<(t_1,t_2) &= \langle\delta\rho\,(\boldsymbol{r}_2,\boldsymbol{r}_2',t_2)\,\delta\rho\,(\boldsymbol{r}_1,\boldsymbol{r}_1',t_1)\rangle\,. \end{aligned} \tag{3.44}$$

Here we have introduced the operator

$$\delta\rho(\boldsymbol{r},\boldsymbol{r}',t) = \psi^\dagger(\boldsymbol{r}',t)\psi(\boldsymbol{r},t) - \big\langle\psi^\dagger(\boldsymbol{r}',t)\psi(\boldsymbol{r},t)\big\rangle \tag{3.45}$$

which is just the operator of density fluctuations in the case $\boldsymbol{r} = \boldsymbol{r}'$.

Let us return to the causal Green's functions, which are of special importance when dealing with many-particle systems using perturbation theory. In this connection, it is useful to represent Green's functions in terms of diagrams. We introduce the following diagram for the single-particle Green's function

$$g^c(1,1') = \qquad 1 \;\text{———}\; 1' \tag{3.46}$$

Correspondingly, the correlated motion of two particles is represented by the diagram

$g^c(12,1'2') =$ 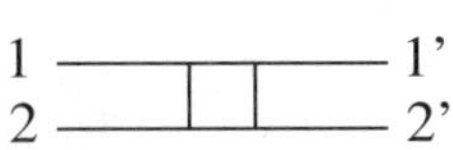

(3.47)

Such diagrams are elements of the Feynman diagram technique which is successfully applied in the perturbation theory. However, we want to remark that we do *not* use diagram rules and diagram techniques. Diagrams are used for illustrative purposes only.

To complete the apparatus of Green's functions, we still introduce retarded and advanced Green's functions. The retarded and advanced single-particle Green's functions are defined by

$$g^{R/A}(1,1') = \pm\Theta[\pm(t_1 - t_1')]\left(g^>(1,1') - g^<(1,1')\right) . \tag{3.48}$$

We will see that these functions are especially useful because:

(i) They are, in the mathematical sense, the Green's functions belonging to the correlation functions.
(ii) They have simple and useful analytical properties in the complex energy plane.

From the definitions (3.30, 3.38, and 3.48), we obtain the useful connections between the different types of Green's functions

$$\begin{aligned} g^R = g^c - g^< = g^> - g^a \ , \ g^A = g^c - g^> = g^< - g^a \ , \\ g^R - g^A = g^> - g^< \ , \ g^R + g^A = g^c - g^a \ . \end{aligned} \tag{3.49}$$

Equivalent relations can be found for any other single-particle function. Like in the case of correlation functions, it is convenient to represent the retarded and advanced Green's functions in terms of sum and difference variables. In the single-particle case we have according to (3.32)

$$g^{R/A}(1,1') = g^{R/A}(\boldsymbol{r}\tau, \boldsymbol{R}t) , \tag{3.50}$$

and one may introduce the Fourier transform via

$$g^{R/A}(\boldsymbol{p}\omega, \boldsymbol{R}t) = \int d\boldsymbol{r}d\tau e^{-i\boldsymbol{p}\boldsymbol{r}+i\omega\tau} g^{R/A}(\boldsymbol{r}\tau, \boldsymbol{R}t) . \tag{3.51}$$

In the next section, we want to deal with the analysis of the properties of $g^{\gtrless}(\boldsymbol{p}\omega, \boldsymbol{R}T)$ and $g^{R/A}(\boldsymbol{p}\omega, \boldsymbol{R}T)$, where we will pay special attention to the analytical behavior in the complex ω-plane.

The method of Green's functions introduced here is one of the most effective and general tools for the description of interacting many-particle systems. The ideas of quantum statistics, together with the mathematical techniques of quantum field theory are used to solve the complex problems of many-particle physics. The method provides powerful techniques which can be applied to elementary particle physics, nuclear physics, solid state physics, and to the physics of strongly correlated plasmas. The theory of Green's functions was developed in different versions and is outlined in a variety of papers and monographs. We mention the famous paper by Martin and Schwinger (1959), the monographs by Kadanoff and Baym (1962) and by Abrikosov et al. (1975), where a theory of imaginary-time Green's functions was given on the basis of the Kubo–Martin–Schwinger (KMS) condition. The theory of real-time Green's functions is more general and was presented in the excellent review paper by Zubarev (1960). Especially for non-equilibrium systems, the theory was developed and extended by Keldysh (1964) and DuBois (1968). Further representations of real-time non-equilibrium Green's functions are given by Kremp et al. (1986), Danielewicz (1990), Botermans and Malfliet (1990), and by Zubarev et al. (1996).

In this monograph, the theory of real-time Green's functions will be applied for the description of both equilibrium and non-equilibrium properties of strongly correlated plasmas.

3.2.2 Spectral Representations and Analytic Properties of Green's Functions

In this subsection, we want to investigate the exact properties of correlation and Green's functions, i.e., we will consider properties which follow from the general principles of quantum statistics.

Let us first derive spectral representations of the time dependent correlation functions. Spectral representations may be obtained from the definition of correlation functions. We consider

$$g^{<}(1,1') = \pm\frac{1}{i}\,\mathrm{Tr}\left\{\rho\,\psi^\dagger\left(\boldsymbol{R}-\frac{\boldsymbol{r}}{2},t-\frac{\tau}{2}\right)\psi\left(\boldsymbol{R}+\frac{\boldsymbol{r}}{2},t+\frac{\tau}{2}\right)\right\}\,. \tag{3.52}$$

It is useful to express the Wigner shifts in the field operators by space and time translations, i.e.,

$$\psi\left(\boldsymbol{R}+\frac{\boldsymbol{r}}{2},t+\frac{\tau}{2}\right) = e^{\frac{i}{2}(\mathrm{H}\tau-\boldsymbol{P}\boldsymbol{r})}\,\psi(\boldsymbol{R},t)\,e^{-\frac{i}{2}(\mathrm{H}\tau+\boldsymbol{P}\boldsymbol{r})}\,.$$

Here, H is the Hamiltonian, and $\boldsymbol{P}$ is the operator of total momentum.

In order to determine the trace, we use the eigen states $|\Psi_\lambda\rangle$ of the complete set of observables $\mathrm{H}, \boldsymbol{P}, \mathrm{N}, \cdots$. Taking into account the orthogonality and the completeness relations

$$\langle\Psi_\lambda|\Psi_{\lambda'}\rangle = \delta(\lambda,\lambda'), \qquad \sum\!\!\!\!\!\!\int d\lambda\,|\Psi_\lambda\rangle\,\langle\Psi_\lambda| = 1\,, \tag{3.53}$$

we can use the states $|\Psi_\lambda\rangle$ to be a basis in $\mathcal{H}$. Using this basis, and performing the Fourier transformation with respect to the variables $\boldsymbol{r}$ and τ, we find

$$\begin{aligned} ig^{>}(\boldsymbol{p}\omega,\boldsymbol{R}t) &= \sum\!\!\!\!\!\!\int d\lambda d\lambda' d\lambda''\,\langle\Psi_\lambda|\,\rho\,|\Psi_{\lambda'}\rangle \\ &\times\langle\Psi_{\lambda'}|\psi(\boldsymbol{R},t)|\Psi_{\lambda''}\rangle\,\langle\Psi_{\lambda''}|\psi^\dagger(\boldsymbol{R},t)|\Psi_\lambda\rangle \\ &\times(2\pi)^4\,\delta\left(\omega-\left(E''-\frac{E'+E}{2}\right)\right)\,\delta\left(\boldsymbol{p}-\left(\boldsymbol{P}''-\frac{\boldsymbol{P}'+\boldsymbol{P}}{2}\right)\right)\,, \end{aligned} \tag{3.54}$$

and correspondingly

$$\pm ig^{<}(\boldsymbol{p}\omega, \boldsymbol{R}t) = \sum\!\!\!\!\!\!\int d\lambda d\lambda' d\lambda'' \langle \Psi_\lambda | \rho | \Psi_{\lambda'} \rangle$$
$$\times \langle \Psi_{\lambda'} | \psi^\dagger(\boldsymbol{R}, t) | \Psi_{\lambda''} \rangle \langle \Psi_{\lambda''} | \psi(\boldsymbol{R}, t) | \Psi_\lambda \rangle$$
$$\times (2\pi)^4 \, \delta\left(\omega - \left(\frac{E' + E}{2} - E''\right)\right) \delta\left(\boldsymbol{p} - \left(\frac{\boldsymbol{P}' + \boldsymbol{P}}{2} - \boldsymbol{P}''\right)\right). \quad (3.55)$$

These relations are the most general forms of spectral representations. Since no assumption has been made concerning the special choice of the density operator ρ, the relations are valid both in non-equilibrium and in equilibrium.

From the meaning of the operators ψ and $\psi^\dagger$ to be annihilation and creation operators, respectively, and from the orthogonality relations in the Fock space, it follows from (3.54) that

$$E'' = E''(N+1), \quad E = E(N), \quad E' = E'(N).$$

It is clear that the values of ω for which the argument of the δ-distribution in (3.55) vanishes, are just the energy differences

$$\omega = E''(N+1) - \frac{1}{2}\left(E'(N) + E(N)\right)$$

between an $(N+1)$- and an N-particle system. Therefore, the spectral representation describes the energy change in the spectrum of a many-particle system produced by an added particle with the momentum $\boldsymbol{p}$. Similarly, the correlation function $g^{<}(\boldsymbol{p}\omega, \boldsymbol{R}t)$ given by (3.55) describes the energy change produced by a particle when removed from the system.

If the Hamiltonian H is approximately given by a sum of free or Hartree–Fock single-particle Hamiltonians, we find that the correlation functions are proportional to $\delta(\omega - E(\boldsymbol{p}))$, where $E(\boldsymbol{p})$ are the single-particle energies. In general, the energy spectrum is sufficiently complex, so that $g^{\gtrless}(\boldsymbol{p}\omega, \boldsymbol{R}t)$ appear to be continuous functions of ω.

Further simplifications are achieved in the case of thermodynamic equilibrium, i.e.,

$$\rho = \frac{1}{Z_G} e^{-\beta(\mathrm{H} - \mu \mathrm{N})}. \quad (3.56)$$

Since $|\Psi_\lambda\rangle$ are eigen vectors of H and N, the matrix elements of the density operator are diagonal, i.e.,

$$\langle \Psi_\lambda | \rho | \Psi_{\lambda'} \rangle = \frac{1}{Z_G} e^{-\beta(E - \mu N)} \, \delta(\lambda, \lambda'). \quad (3.57)$$

From these relations, we easily get the following spectral representation

$$\pm ig^{<}(\boldsymbol{p}, \omega) = I(\boldsymbol{p}, \omega), \qquad ig^{>}(\boldsymbol{p}, \omega) = I(\boldsymbol{p}, \omega)\, e^{\beta(\omega - \mu)}, \quad (3.58)$$

where we have introduced the spectral density by

$$I(\boldsymbol{p},\omega) = \int d\lambda d\lambda' \frac{1}{Z_G} e^{-\beta(E-\mu N)} \langle \Psi_\lambda | \psi^\dagger | \Psi_{\lambda'} \rangle \langle \Psi_{\lambda'} | \psi | \Psi_\lambda \rangle \times (2\pi)^4 \delta(\omega - E + E') \delta(\boldsymbol{p} - \boldsymbol{P} + \boldsymbol{P}') . \quad (3.59)$$

Consequently, in thermodynamic equilibrium, the correlation function can be determined from the following two relations, i.e.,

$$\begin{aligned} \pm i g^<(\boldsymbol{p}, t-t') &= \int \frac{d\omega}{2\pi} I(\boldsymbol{p},\omega)\, e^{-i\omega(t-t')} \\ i g^>(\boldsymbol{p}, t-t') &= \int \frac{d\omega}{2\pi} I(\boldsymbol{p},\omega)\, e^{\beta(\omega-\mu)}\, e^{-i\omega(t-t')} . \end{aligned} \quad (3.60)$$

Spectral representations of this type were introduced for the first time in quantum electrodynamics by Lehmann, Källen and others. In equilibrium quantum statistics, such relations were intensely discussed by Zubarev (1960), Zubarev (1974).

From the spectral representations (3.60), a fundamental property in thermodynamic equilibrium follows, namely a relation between the correlation functions which is usually referred to as the *Kubo–Martin–Schwinger relation* (KMS)

$$\pm g^<(\boldsymbol{p},\omega) = e^{-\beta(\omega-\mu)} g^>(\boldsymbol{p},\omega) . \quad (3.61)$$

This means that, in thermodynamic equilibrium, the two correlation functions are not independent but connected by the condition (3.61). This becomes especially obvious, if the spectral function

$$a(\boldsymbol{p},\omega) = i\Big[g^>(\boldsymbol{p},\omega) - g^<(\boldsymbol{p},\omega)\Big] \quad (3.62)$$

is introduced. From the KMS condition, the following spectral representations can be obtained

$$\begin{aligned} \pm i g^<(\boldsymbol{p},\omega) &= a(\boldsymbol{p},\omega) f(\omega) , \\ i g^>(\boldsymbol{p},\omega) &= a(\boldsymbol{p},\omega)(1 \pm f(\omega)) , \end{aligned} \quad (3.63)$$

where the function $f(\omega)$ is given by

$$f(\omega) = \frac{1}{e^{\beta(\hbar\omega-\mu)} \mp 1} . \quad (3.64)$$

For the upper sign we have a Bose-like, and for the lower one a Fermi-like distribution function. The explicit knowledge of the function $f(\omega)$ is a consequence of the KMS relation and ultimately of the knowledge of the statistical operator ρ in thermodynamic equilibrium. In particular, $f(\omega)$ describes the statistical information. In thermodynamic equilibrium, only the spectral function $a(\boldsymbol{p}, \omega)$ has to be determined, which carries the dynamic information about the many-particle system. From the definition and the equal-time commutation relations, we find the important sum rule

$$\int \frac{d\omega}{2\pi} a(\boldsymbol{p}\omega) = 1 \,. \tag{3.65}$$

Similar to the single-particle case, one can derive spectral representations for the two-time two-particle correlation functions $g_{12}^{\gtrless}(t, t')$ defined by (3.36) and (3.37). Again, a *Kubo–Martin–Schwinger relation* can be found in thermodynamic equilibrium, i.e.,

$$g_{12}^{<}(\omega) = e^{-\beta(\omega - 2\mu)} g_{12}^{>}(\omega) \,. \tag{3.66}$$

Introducing a two-particle spectral function according to

$$A_{12}(\omega) = i^2 \left[g_{12}^{>}(\omega) - g_{12}^{<}(\omega) \right] , \tag{3.67}$$

we get the following spectral representation for the two-time two-particle correlation functions

$$\begin{aligned} i^2 g_{12}^{<}(\omega) &= A_{12}(\omega) F(\omega) \,, \\ i^2 g_{12}^{>}(\omega) &= A_{12}(\omega) \left[1 + F(\omega) \right] , \end{aligned} \tag{3.68}$$

where $F(\omega)$ is a Bose-like distribution function given by

$$F(\omega) = \frac{1}{e^{\beta(\omega - 2\mu)} - 1} \,. \tag{3.69}$$

Let us return to the single-particle correlation functions. If the Hamiltonian is assumed to be a sum of single-particle operators, i.e., $\mathrm{H}_N = \sum_i \mathrm{H}_i$, the eigen value problem of H_N can be solved. In this case, the spectral function is determined by (3.62), (3.54), and (3.55). The equation that follows is

$$a(\boldsymbol{p}, \omega) = 2\pi \delta(\omega - E(\boldsymbol{p})) \,, \tag{3.70}$$

where $E(\boldsymbol{p}) = \frac{p^2}{2m} + \Delta(\boldsymbol{p})$ are the eigen values of the single-particle Hamiltonian with $\Delta(\boldsymbol{p})$ being the corresponding self-energy contribution, e.g., in Hartree–Fock approximation. In time variables, the correlation function $g^{<}$ then reads

$$\pm i g^{<}(\boldsymbol{p}, t - t') = e^{-iE(\boldsymbol{p})(t - t')} f\left(E(\boldsymbol{p}) \right) \,, \tag{3.71}$$

and the single-particle density matrix is

$$\begin{aligned} n \langle \boldsymbol{p} | F_1 | \boldsymbol{p}' \rangle &= \pm i g^{<}(\boldsymbol{p}, t - t') \Big|_{t=t'} (2\pi)^3 \delta(\boldsymbol{p} - \boldsymbol{p}') \\ &= f\left(E(\boldsymbol{p}) \right) (2\pi)^3 \delta(\boldsymbol{p} - \boldsymbol{p}') \,. \end{aligned} \tag{3.72}$$

Of course, the spectral function is drastically modified if the interaction between the particles is accounted for. In simple approximations, we have a broadening of the spectral function i.e., $a(\boldsymbol{p}\omega)$ could have the shape

$$a(\boldsymbol{p},\omega) = \frac{\Gamma(\boldsymbol{p})}{(\omega - E(\boldsymbol{p}))^2 + \left(\frac{\Gamma(\boldsymbol{p})}{2}\right)^2} \,. \tag{3.73}$$

Here, $E(\boldsymbol{p}) = \frac{p^2}{2m} + \Delta(\boldsymbol{p})$ is the single-particle energy, and $\Gamma(\boldsymbol{p})$ corresponds to the damping of single-particle state which is small for weak interaction, i.e., if $\Gamma \ll E$. In general, the damping leads to a decay of the single-particle correlation function, namely

$$\pm i g^<(t-t') = \int_{-\infty}^{+\infty} \frac{d\omega}{2\pi} e^{-i\omega(t-t')} f(\omega) \int_0^\infty d\bar{t}\, e^{-\frac{\Gamma}{2}\bar{t}} \cos\left[(\omega - E)\,\bar{t}\right] . \tag{3.74}$$

Here, we used an integral representation of (3.73). As will be shown later, the spectral function is a more complex function with respect to ω. Therefore, the correlation functions cannot be expressed in the simple form as given by (3.74).

In non-equilibrium situations, the theory is more complicated. In such a case, a Kubo–Martin–Schwinger condition does not exist, i.e., $g^>$ and $g^<$ are independent functions. Nevertheless, one can find relations which are similar to the equilibrium case.

Again, we introduce the spectral function

$$a(\boldsymbol{p}\omega, \boldsymbol{R}t) = i[g^>(\boldsymbol{p}\omega, \boldsymbol{R}t) - g^<(\boldsymbol{p}\omega, \boldsymbol{R}t)] \,. \tag{3.75}$$

In analogy to (3.63), a generalization of the function $f(\omega)$ is introduced via

$$\begin{aligned} i g^>(\boldsymbol{p}\omega, \boldsymbol{R}t) &= a(\boldsymbol{p}\omega, \boldsymbol{R}t)[1 \pm f(\boldsymbol{p}\omega, \boldsymbol{R}t)] \,, \\ \pm i g^<(\boldsymbol{p}\omega, \boldsymbol{R}t) &= a(\boldsymbol{p}\omega, \boldsymbol{R}t) f(\boldsymbol{p}\omega, \boldsymbol{R}t) \,. \end{aligned} \tag{3.76}$$

In contrast to equilibrium, we now have to consider the following facts:

(i) In non-equilibrium, every function depends on the local variables $\boldsymbol{R}$ and t.
(ii) In contrast to the equilibrium situation, the function $f(\boldsymbol{p}\omega, \boldsymbol{R}t)$ is unknown and has to be determined from a kinetic equation.

In any case, the determination of the spectral function $a(\boldsymbol{p}\omega, \boldsymbol{R}t)$ turns out to be one of the central problems of many-particle theory. For the solutions of the problems just mentioned, the application of Green's functions is extremely useful.

3.2.3 Analytical Properties, Dispersion Relations

Besides the correlation functions, discussed in Sect. 3.2, causal, retarded, and advanced Green's functions were introduced to provide further useful quantities for the description of interacting many-particle systems.

Now we want to show how these Green's functions can also be expressed by the spectral function $a(\boldsymbol{p}\omega, \boldsymbol{R}t)$. For this purpose, the analytic properties of Green's functions in the complex ω-plane will be investigated. In the definition of Green's functions, the step function $\Theta(t-t')$ plays an essential role. For this function, one may easily derive the following complex representation

$$\Theta(t-t') = \frac{i}{2\pi} \int_{-\infty}^{\infty} \frac{\exp[-ix(t-t')]}{x+i\varepsilon}\, dx\,, \qquad \epsilon > 0\,. \tag{3.77}$$

Using the convolution theorem, we get for the Fourier transform of the causal Green's function

$$g^c(\boldsymbol{p}\omega, \boldsymbol{R}t) = \int \frac{d\bar{\omega}}{2\pi} \left\{ \frac{ig^>(\boldsymbol{p}\bar{\omega}, \boldsymbol{R}t)}{\omega - \bar{\omega} + i\varepsilon} - \frac{ig^<(\boldsymbol{p}\bar{\omega}, \boldsymbol{R}t)}{\omega - \bar{\omega} - i\varepsilon} \right\} . \tag{3.78}$$

With the formula (3.76), (3.78) is known as *Landau spectral representation*, see, e.g., Landau and Lifshits (1977). According to (3.78), the causal Green's function has singularities in the lower half plane and in the upper half plane of complex energies. Thus, there exists no analytical continuation in both half planes.

In contrast to that, the analytical properties of the retarded and advanced Green's functions are more convenient. From (3.48) and (3.77), we get the following spectral representation

$$g^{R/A}(\boldsymbol{p}\omega, \boldsymbol{R}t) = \int \frac{d\bar{\omega}}{2\pi} \frac{a(\boldsymbol{p}\bar{\omega}, \boldsymbol{R}t)}{\omega - \bar{\omega} \pm i\varepsilon} . \tag{3.79}$$

From this relation one can see that $g^R(\omega)$ may be continued analytically into the upper half plane, while $g^A(\omega)$ may be continued into the lower one. Both functions may be combined to a unique function which may be represented as

$$g(\boldsymbol{p}z, \boldsymbol{R}t) = \int \frac{d\bar{\omega}}{2\pi} \frac{a(\boldsymbol{p}\bar{\omega}, \boldsymbol{R}t)}{z - \bar{\omega}} . \tag{3.80}$$

This means that both functions are combined in an integral of Cauchy's type which "carries" the properties of both the retarded and the advanced functions (carrier function).

Integrals of the Cauchy type, and therefore $g(z)$, have the following properties:

(i) $g(z)$ defines two analytic functions,

$$\begin{aligned} g_I(z) &= g^R(z) \quad \text{for} \quad \mathrm{Im}z > 0, \\ g_{II}(z) &= g^A(z) \quad \text{for} \quad \mathrm{Im}z < 0\,; \end{aligned} \tag{3.81}$$

(ii) it has a branch cut with the limiting behavior (Plemlj–Sochozki)

$$g(\omega \pm i\varepsilon) = \mathrm{P}\int \frac{d\bar{\omega}}{2\pi}\frac{a(\bar{\omega})}{\omega - \bar{\omega}} \mp \frac{i}{2}a(\omega)\,. \tag{3.82}$$

In a more compact manner, this behavior is expressed by the Dirac identity

$$\frac{1}{\omega \pm i\varepsilon} = \mathrm{P}\left(\frac{1}{\omega}\right) \mp i\pi\delta(\omega)\,. \tag{3.83}$$

(iii) From (3.82) for the discontinuity at the cut, we get

$$a(\omega) = i\Big[g(\omega + i\varepsilon) - g(\omega - i\varepsilon)\Big]\,. \tag{3.84}$$

(iv) It is possible to continue $g_I(z)$ and $g_{II}(z)$ analytically beyond the cut, and we get

$$\begin{aligned} g_I(z) &= g_{II}(z) + a(z) \quad \text{for} \quad \mathrm{Im}z < 0\,, \\ &= g_{II}(z) - a(z) \quad \text{for} \quad \mathrm{Im}z > 0\,. \end{aligned} \tag{3.85}$$

These functions are no longer analytic, they may have singularities which are those of the spectral function $a(z)$ in the complex energy plane.

Thus, it was shown that all types of Green's functions are determined by the spectral function $a(\boldsymbol{p}\omega, \boldsymbol{R}t)$ via spectral representations. On the other hand, the retarded and advanced Green's functions allow for the determination of the spectral function according to (3.84).

At the end of this subsection, we will subdivide $g^{R/A}(\omega)$ into real and imaginary parts. This decomposition is possible using the Dirac identity. The imaginary parts of the retarded and advanced Green's functions are

$$\mathrm{Im}\, g^{R/A}(\omega) = \mp\frac{1}{2}\, a(\omega)\,, \tag{3.86}$$

and for the real parts we get

$$\mathrm{Re}\, g^{R/A}(\omega) = \mathrm{P}\int \frac{d\bar{\omega}}{2\pi}\frac{a(\bar{\omega})}{\omega - \bar{\omega}}\,. \tag{3.87}$$

From these formulae, we get a relation between $\mathrm{Im}g^{R/A}(\omega)$ and $\mathrm{Re}g^{R/A}(\omega)$

$$\mathrm{Re}\, g^{R/A}(\omega) = \mp i\,\mathrm{P}\int \frac{d\bar{\omega}}{\pi}\frac{\mathrm{Im}\, g^{R/A}(\bar{\omega})}{\omega - \bar{\omega}}\,. \tag{3.88}$$

Such relations are referred to as dispersion relations. They are a consequence of the analyticity of the retarded and advanced Green's functions $g^{R/A}$.

Relations of the type (3.88) do also exist for other quantities, e.g., for the dielectric function $\varepsilon(\omega)$. In that connection they are referred to as *Kramers–Kronig relations* (see Chap. 4, where the dielectric properties of plasmas are considered).

In the subsequent considerations of this book, we furtheron will use retarded and advanced two-time quantities that have the general shape

$$F^{R/A}(t,t') = F^0\,\delta(t-t') \pm \Theta[\pm(t-t')]\Big[F^>(t,t') - F^<(t,t')\Big]\,. \tag{3.89}$$

A special realization was given by the retarded and advanced single-particle Green's function according to (3.48). It turns out that a static contribution F^0 might occur in the general case.

For any of such quantities, the spectral and analytic properties obtained for $g^{R/A}$ may be used analogously. We especially have the important relation corresponding to (3.79)

$$F^{R/A}(\boldsymbol{p}\omega,\boldsymbol{R}t) = F^0(\boldsymbol{p},\boldsymbol{R}t) \mp \int \frac{d\,\bar{\omega}}{\pi}\,\frac{\mathrm{Im}F^{R/A}(\boldsymbol{p}\bar{\omega},\boldsymbol{R}t)}{\omega-\bar{\omega}\pm i\varepsilon}\,, \tag{3.90}$$

with $2i\,\mathrm{Im}F^{R/A}(\omega) = \pm\left(F^>(\omega) - F^<(\omega)\right)$. The formula (3.90) represents a rather general dispersion relation.

Finally, let us come back to the KMS condition (3.61). By Fourier inversion, we get the KMS relation for the correlation functions as a function of time, i.e.,

$$g^<(\boldsymbol{r},t-t') = \pm e^{\beta\mu} g^>(\boldsymbol{r},t-t'-i\beta)\,. \tag{3.91}$$

In this form, the KMS condition is the basic relation for a special version of the Green's function method in statistical physics. This approach is referred to as the complex time Green's function technique and was developed systematically by Martin and Schwinger (1959), Kadanoff and Baym (1962), and by Abrikosov et al. (1975). The method is used very successfully in the theory of many-particle systems in thermodynamic equilibrium and for the determination of thermodynamic functions.

Indeed, the KMS relation points out that the imaginary axis from $-i\beta$ to $i\beta$ plays a special role for the equilibrium correlation functions. It is an obvious idea to extend $g^{\gtrless}(\boldsymbol{r},t-t')$ to complex times. Then it is possible to find the KMS condition for the causal single-particle Green's function

$$g^c(t,t')|_{t=0} = \pm e^{\beta\mu} g^c(t,t')|_{t=-i\beta}\,. \tag{3.92}$$

This means, the causal Green's function is quasi periodic along the imaginary time axis with a periodicity interval of $[0,-i\beta]$.

Because of the KMS relation (3.92), the causal single-particle Green's function may be expanded into a Fourier series in the interval $[0,-i\beta]$, namely

$$g^c(t,t') = \frac{i}{\beta}\sum_{\nu} g(z_\nu)e^{-iz_\nu(t-t')} . \tag{3.93}$$

Then the KMS condition is automatically fulfilled if we make the following choice for the frequencies z_ν

$$z_\nu = \frac{\pi\nu}{-i\beta} + \mu , \tag{3.94}$$

where

$$\begin{aligned} \nu &= \pm 1, \pm 3, \pm 5 \ldots \quad \text{for Fermions,} \\ \nu &= 0, \pm 2, \pm 4, \ldots \quad \text{for Bosons .} \end{aligned}$$

The frequencies z_ν are known as Matsubara frequencies.

Now it is easy to show that the Fourier coefficients follow from the spectral representation of the retarded Green's function (3.80) at the Matsubara frequencies z_ν

$$g^{\mathrm{R}}(\boldsymbol{p}, z_\nu) = \int_{-\infty}^{\infty} \frac{d\omega}{2\pi} \frac{a(\boldsymbol{p},\omega)}{z_\nu - \omega} . \tag{3.95}$$

The result is of fundamental importance for equilibrium systems. For the imaginary time Green's functions, there are powerful methods for their determination. Especially, one may find a perturbation expansion, and one may develop a Feynman diagram technique. By analytic continuation of the Fourier coefficients of this function, the spectral function may be determined via formula (3.84) .

3.3 Equations of Motion for Correlation Functions and Green's Functions

3.3.1 The Martin–Schwinger Hierarchy

We have shown that correlation and Green's functions contain important dynamic and statistical information about the many-particle system. Therefore, a central task of the theory is the determination of these functions. The problem may, in principle, be solved using the corresponding equations of motion, i.e., using equations similar to the Bogolyubov hierarchy for reduced density operators or density matrices.

The dynamics of the many-particle system is completely determined by the equations of motion for the field operators. If the system is considered in the presence of an external field U, we have the equation (see (3.19))

$$\left(i\frac{\partial}{\partial t_1} + \frac{\nabla_1^2}{2m} - U(1)\right)\psi(1) = \int d\bar{1}\, V(1-\bar{1})\psi^\dagger(\bar{1})\psi(\bar{1})\psi(1) \tag{3.96}$$

and the adjoint one for $\psi^\dagger(1)$. Now, we use the definitions of the correlation and the Green's functions. With (3.96), the equations of motion for these functions are generated straightforwardly. For the causal Green's function defined by (3.38), the equation reads

$$\left(i\frac{\partial}{\partial t_1}+\frac{\nabla_1^2}{2m}-U(1)\right)g^c(1,1')\mp i\int d2V(1,2)g^c(12,1'2^+)=\delta(1-1')\,. \tag{3.97}$$

Equation (3.97) is not closed. Because of the interaction $V(1,2)$, the single-particle Green's function is coupled to the two-particle Green's function, and thus, (3.97) is the first member of a chain of equations for many-particle Green's functions. We therefore get, like in the case of density operators, a hierarchy of equations coupled by the interaction between the particles. The second equation of this hierarchy follows easily and reads

$$\begin{aligned}&\left(i\frac{\partial}{\partial t_1}+\frac{\nabla_1^2}{2m}-U(1)\right)g^c(12,1'2')\mp i\int d3\,V(1,3)g^c(123,1'2'3^+)\\&=\delta(1-1')g^c(2,2')\pm\delta(1-2')g^c(2,1')\,.\end{aligned} \tag{3.98}$$

The general form of the chain of equations corresponding to (3.97) and (3.98) follows to be

$$\begin{aligned}&\left(i\frac{\partial}{\partial t_1}+\frac{\nabla_1^2}{2m}-U(1)\right)g^c(1\ldots s,1'\ldots s')\\&\mp i\int d(s+1)V(1-(s+1))g^c(1\ldots(s+1),1'\ldots(s+1)^+)\\&=\sum_{\nu=1'}^{s'}(\pm)^{\nu-1'}\delta(1-\nu)g^c(2\ldots s,1'\ldots(\nu-1)\,(\nu+1)\ldots s')\,.\end{aligned} \tag{3.99}$$

This chain of equations of motion and of the adjoint ones are referred to as the *Martin–Schwinger hierarchy* (Martin and Schwinger 1959).

A similar form is achieved for the hierarchy of equations for the correlation functions. For the single-particle correlation function $g^<(1,1')$ we get

$$\left(i\frac{\partial}{\partial t_1}+\frac{\nabla_1^2}{2m}-U(1)\right)g^<(1,1')\mp i\int d2\,V(1,2)g^<(12,1'2^+)=0\,. \tag{3.100}$$

This is just the homogeneous equation corresponding to (3.97). In addition to this equation we get an equation for the second argument $1'$. This adjoint equation follows from the equation of motion for the field operator $\psi^\dagger(1')$.

A very important problem is now to find a system of closed equations, i.e., in the simplest case, a closed equation for the single-particle Green's function.

A remark is important in connection with this problem: In order to get a closed system of equations, we have to find approximate solutions for the higher order Green's functions. Furthermore, we want to draw the attention to the fact that, in deriving the Martin–Schwinger equations, nothing was assumed about the kind of averaging, and that the equations are first order differential equations in time. Therefore, boundary or initial conditions are necessary to fix the solutions precisely. For example, in the case of thermodynamic equilibrium, using the grand canonical density operator, the solutions of the equations of motion are determined by the well-known Kubo–Martin–Schwinger (KMS) condition for the imaginary time Green's functions (see Kadanoff Baym). For the causal single-particle Green's function, this condition reads

$$g^c(1,1')|_{t_1=0} = \pm e^{\beta\mu} g^c(1,1')|_{t_1=-i\beta} \,, \tag{3.101}$$

where $\beta = 1/(k_B T)$, and μ is the chemical potential. Then, the equation of motion together with this quasi-periodic boundary condition determines the equilibrium Green's function.

In non-equilibrium situations, the condition (3.101) is not valid, and one has to work in the real-time domain. Kadanoff and Baym developed a method starting from the equations of motion for imaginary time Green's functions. The method is based on the definition of complex time Green's functions and the possibility to define an analytic continuation from complex times $t = i\tau + t_0$ to real times, namely (U being an external potential)

$$\lim_{t_0 \to -\infty} g(1,1';U;t_0) = g(1,1';U) \,. \tag{3.102}$$

This relation was derived (Kadanoff and Baym 1962) assuming the system to be initially in thermodynamic equilibrium.

The most general and natural idea to fix the solution of the hierarchy for real times is of course an initial condition. For the causal two-particle Green's function, we can write

$$g^c(12,1'2')|_{t_0} = g^c(1,1')g^c(2,2')|_{t_0} \pm g^c(1,2')g^c(2,1')|_{t_0} + c(\boldsymbol{r}_1\boldsymbol{r}_2,\boldsymbol{r}_1'\boldsymbol{r}_2';t_0) \,. \tag{3.103}$$

The notation on the l.h.s. means that all times are equal to the initial time t_0, i.e., $t_1 = t_2 = t_1' = t_2' = t_0$. A chosen infinitesimal time ordering in realizing (3.103) selects the initial condition for one of the 4! correlation functions. Usually a special case of this condition, known as Bogolyubov condition of weakening of initial correlations, is applied in non-equilibrium statistical mechanics. In terms of correlation functions it reads

$$\lim_{t_0 \to -\infty} g^<(12,1'2')\Big|_{t_0} = g^<(1,1')g^<(2,2') \pm g^<(1,2')g^<(2,1')\Big|_{t_0} \,. \tag{3.104}$$

Equation (3.104) is the correct mathematical expression for Boltzmann's assumption of molecular chaos. In order to explain the Bogolyubov condition, we introduce the correlation time τ_{corr} and the time between two collisions τ_{coll}. Under the condition

$$\tau_{\text{corr}} \ll \tau_{\text{coll}} \tag{3.105}$$

any group of particles, e.g., any pair of particles, enters the collision process statistically independent. For times long before the collision, i.e., if we choose $t_0 \to -\infty$, the particles are uncorrelated, and we get the relation (3.104). This condition may be used in situations characterized by (3.105) as a boundary condition at the time boundary $t \to -\infty$; it is a kind of asymptotic condition.

In many situations, we must replace (3.104) with a more flexible condition. Such situations are given, e.g., in systems with bound states. Bound states are long living correlations, and we have to replace (3.104) with a condition of partial weakening of initial correlations. For the two-particle correlation function $g^<(12, 1'2')$ we have in the case of distinct particles

$$\lim_{t_0 \to -\infty} g^<(12, 1'2')\Big|_{t_0} = g^<(1, 1')g^<(2, 2') + g^{\text{bound}}(12, 1'2')\Big|_{t_0} . \tag{3.106}$$

Here, g^{bound} is the bound state part of the two-particle density matrix in binary collision approximation. In (3.106), the weakening of initial correlations is assumed only for the scattering part. As a consequence, it is not possible to get a closed equation for the single-particle Green's function, which is of special relevance for the description of partially ionized plasmas.

3.3.2 The Hartree–Fock Approximation

In the previous subsection, we derived the equations of motion for the causal single-particle Green's function given by (3.97). From this latter equation, the single-particle Green's function may be determined in terms of the two-particle Green's function. This means that, for the determination of $g^c(1, 1')$, the knowledge of $g^c(12, 1'2')$ is necessary.

Of course, an exact knowledge of the two-particle Green's function cannot be achieved under general conditions. Therefore, we have to perform an approximate decoupling, or, in other words, a truncation of the hierarchy.

The simplest decoupling of the hierarchy is achieved by the application of the Hartree–Fock approximation

$$g^c(12, 1'2') = g^c(1, 1')g^c(2, 2') \pm g^c(1, 2')g^c(2, 1') . \tag{3.107}$$

This approximation leads to the quantum mechanical version of the Vlasov theory, which was considered from the classical point of view already in Chap. 2. In the Hartree–Fock approximation, the particles are not only uncorrelated for $t = -\infty$ (like in Bogolyubov's weakening of initial correlations) but for any finite time. Using (3.107), we get the following closed equation of motion (corresponding to a complete truncation of the hierarchy)

$$\begin{aligned} &\left(i\frac{\partial}{\partial t_1} + \frac{\nabla_1^2}{2m} - U(1) \right) g^c(1, 1') = \delta(1 - 1') \\ &\pm \; i \int d2\, V(1-2) \left\{ g^c(1, 1')g^c(2, 2^+) \pm g^c(1, 2^+)g^c(2, 1') \right\} . \end{aligned} \tag{3.108}$$

This equation is a nonlinear integro-differential equation for the single-particle Green's function. One may see easily that (3.108) can be written in the shape

$$\left(i\frac{\partial}{\partial t_1}+\frac{\nabla_1^2}{2m}-U(1)\right)g^c(1,1^{'})-\int d\bar{1}\,\Sigma^{\rm HF}(1,\bar{1})g^c(\bar{1},1^{'})=\delta(1-1^{'})\,. \quad (3.109)$$

Here we introduced the Hartree–Fock self-energy

$$\Sigma^{\rm HF}(1,1')=\Sigma^{\rm H}(1,1')+\Sigma^{\rm ex}(1,1')\,. \quad (3.110)$$

The first (Hartree) contribution is

$$\Sigma^{\rm H}(1,1')=\pm i\,\delta(1-1')\int d2\,V(1-2)g^<(2,2^+)\,. \quad (3.111)$$

The second term accounts for exchange effects. It reads

$$\Sigma^{\rm ex}(1,1')=i\,V(1-1')g^<(1,1')=i\,V(\boldsymbol{r}_1-\boldsymbol{r}_1')\delta(t_1-t_1')g^<(1,1')\,. \quad (3.112)$$

A special property of $\Sigma^{\rm HF}(1,1')$ follows from the fact that it is proportional to $\delta(t_1-t_1')$, i.e., $\Sigma^{\rm HF}$ gives a contribution local in time. The Hartree–Fock approximation leads to a mean field theory of the many-particle system. Collisions are not included. But it seems to be obvious that the structure of (3.109) remains valid in the general theory. This will be shown in the next subsections. Further, it is important to remark that (3.109) is valid without any approximation with respect to the time.

A further simplification of the Hartree–Fock theory is possible if the exchange term is neglected in (3.107). Equation (3.109) can then be written in the form

$$\left(i\frac{\partial}{\partial t_1}+\frac{\nabla_1^2}{2m}-U^{\rm eff}(1)\right)g^c(1,1^{'})=\delta(1-1^{'})\,. \quad (3.113)$$

Here we introduced the effective potential by

$$U^{\rm eff}(1)=U(1)\pm i\int d2\,V(1-2)g^c(2,2^+)\,, \quad (3.114)$$

which is known as the Hartree or Vlasov field. This effective potential is the sum of the external field U and the average or self-consistent field produced by all the other particles of the system. The Hartree approximation thus describes the motion of independent particles in the mean (Vlasov) field of the other particles.

Subtracting the adjoint equation from the original equation (3.113), and introducing the variables defined by (3.32), it follows for $t_1'=t_1^+=t$

$$\left\{i\frac{\partial}{\partial t}+\frac{\nabla_{\boldsymbol{R}}\cdot\nabla_{\boldsymbol{r}}}{m}-\left[U^{\rm eff}\left(\boldsymbol{R}+\frac{\boldsymbol{r}}{2},t\right)-U^{\rm eff}\left(\boldsymbol{R}-\frac{\boldsymbol{r}}{2},t\right)\right]\right\}$$
$$\times\int\frac{d\omega}{2\pi}\,g^<(\boldsymbol{r}\omega,\boldsymbol{R}t)=0\,. \quad (3.115)$$

If $U^{\text{eff}}(\boldsymbol{R},t)$ is a slowly varying function of $\boldsymbol{R}$, we can insert the expansion

$$U^{\text{eff}}\left(\boldsymbol{R}\pm\frac{\boldsymbol{r}}{2},t\right)=U^{\text{eff}}(\boldsymbol{R},t)\pm\frac{\boldsymbol{r}}{2}\cdot\nabla_{\boldsymbol{R}}U^{\text{eff}}(\boldsymbol{R},t)\,.$$

After Fourier transformation and using the definition of the *Wigner function*

$$\int\frac{d\omega}{2\pi}g^{<}(\boldsymbol{p},\omega,\boldsymbol{R}t)=f(\boldsymbol{p}\boldsymbol{R}t)\,, \tag{3.116}$$

we find

$$\left(\frac{\partial}{\partial t}+\frac{\boldsymbol{p}}{m}\cdot\nabla_{\boldsymbol{R}}-\nabla_{\boldsymbol{R}}U^{\text{eff}}(\boldsymbol{R},t)\cdot\nabla_{\boldsymbol{p}}\right)f(\boldsymbol{p},\boldsymbol{R}t)=0$$

where the effective potential is given by

$$U^{\text{eff}}(\boldsymbol{R},t)=U(\boldsymbol{R},t)+\int\frac{d\boldsymbol{p}'}{(2\pi)^3}d\boldsymbol{R}'\,V(\boldsymbol{R}-\boldsymbol{R}')f(\boldsymbol{p}',\boldsymbol{R}'t)\,. \tag{3.117}$$

This is the well-known Vlasov equation which represents a collision-less kinetic equation for the Wigner distribution function $f(\boldsymbol{p},\boldsymbol{R}t)$. In Chap. 2, it was used to describe the dielectric properties of plasmas in the frame of the classical theory.

3.3.3 Functional Form of the Martin–Schwinger Hierarchy

In order to include correlations into the many-particle theory, the decoupling of the hierarchy has to be performed at a higher level. There are a number of different approximation schemes which may be employed. A good idea is to neglect the correlations between more than a certain number s of particles. In this way, we reduce the infinite set of coupled equations to a closed set of the first s of them. For $s=1$, the Hartree–Fock approximation follows, and for $s=2$, we get the binary collision approximation. We want to draw attention to the fact that, for $s>1$, initial conditions have to be adopted necessarily in order to fix the solutions of the first few equations. The result then is a kind of cluster expansion for the two-particle Green's function and thus for the r.h.s. of the first equation of the hierarchy.

However, before we consider such methods, a more general investigation of the structure of the Martin–Schwinger hierarchy and of its formal solutions is useful. We especially want to show that the infinite set of equations is equivalent to a closed functional differential equation. Let us start again with the Martin–Schwinger hierarchy for the Green's functions given by (3.99). The first equation can be written in the form

$$\begin{aligned}&\int d\bar{1}\,\left\{g_0^{-1}(1,\bar{1})-U(1,\bar{1})\right\}g^c(\bar{1},1';U)\\&=\;\delta(1-1')\pm i\int d2\,V(1-2)g^c(12,1'2^+;U)\,.\end{aligned} \tag{3.118}$$

Here we introduced, in a generalization of (3.97), a formal external potential $U(1,1')$ which is nonlocal in space, i.e., $U(1,1') = U(\boldsymbol{r}_1 t_1, \boldsymbol{r}'_1 t'_1)\delta(t_1 - t'_1)$. This potential will be given a physical meaning and specification after carrying out all formal manipulations. Further we used

$$g_0^{-1}(1,1') = \left(i\frac{\partial}{\partial t_1} + \frac{\nabla_1^2}{2m}\right)\delta(1-1').$$

As we can see from (3.118), any Green's function is a functional of the external potential. We will determine the functional $g(1,1';U)$ from the first equation of the hierarchy.

A solution of this equation may be found in the interaction picture with respect to the external potential $U(1,1')$. The field operator ψ_U in the Heisenberg picture and the operator ψ in the interaction picture are related by the unitary transformation

$$\psi_U(1) = S(t_0,t_1)\psi(1)S(t_1,t_0), \tag{3.119}$$

where the field operator ψ in the interaction picture obeys the equation of motion

$$\left(i\frac{\partial}{\partial t_1} + \frac{\nabla_1^2}{2m}\right)\psi(1) - \int d2\, V(1-2)\psi^\dagger(2)\psi(2)\psi(1) = 0. \tag{3.120}$$

The operators $\psi(1)$ and $\psi_U(1)$ coincide at $t_1 = t_0$. The evolution operator $S(t,t_0)$ for $t > t_0$ is given by

$$S(t,t_0) = \mathrm{T}\exp\left\{-i\int_{t_0}^{t} d2 d\bar{2}\, U(2,\bar{2})\psi^\dagger(2)\psi(\bar{2})\right\}. \tag{3.121}$$

Here, T is the chronological or time ordering operator, and S has the well-known properties

$$S(t,t') = S^{-1}(t',t) = S^\dagger(t',t)\,; \quad S(t,\bar{t})S(\bar{t},t') = S(t,t'). \tag{3.122}$$

By introduction of (3.119) in the definition of the causal Green's function, we get

$$g(1,1') = \frac{1}{i}\mathrm{Tr}\left\{\varrho \mathrm{T}(S(t_0,t_1)\psi(1)S(t_1,t'_1)\psi^\dagger(1')S(t'_1,t_0))\right\}. \tag{3.123}$$

The basic difficulty in this equation is obvious. Because $S(t,t_0) = S^\dagger(t_0,t)$ produces anti-chronological time order, we have a chronological (T) and an anti-chronological ($\tilde{\mathrm{T}}$) time order in this expression. With the properties (3.122), we rewrite (3.123) into

$$\mathrm{Tr}\left\{\varrho\left\{S(t_0,\infty)\mathrm{T}(S(\infty,t_0)\psi(1)\psi^\dagger(1'))\right\}\right\}. \tag{3.124}$$

Fig. 3.1. The Keldysh time contour

Only in the case of vacuum, ground state, or equilibrium averaging, we can eliminate the anti-chronological evolution in known manner using the adiabatic theorem. Then we have

$$g(1,1') = \frac{1}{i}\frac{\langle \mathrm{T}[S(\infty,t_0)\psi(1)\psi^\dagger(1')]\rangle}{\langle TS(\infty,-\infty)\rangle} . \tag{3.125}$$

The adiabatic theorem cannot be applied to non-equilibrium systems. For such systems, we may use an idea outlined in the papers by Schwinger (1961), Keldysh (1964), and by Craig (1968). From the time translation in the expression (3.123), it is obvious to introduce an oriented contour as shown in Fig. 3.1. In contrast, the (simple) time abscissa running from $-\infty$ to ∞ is referred to as *physical time axis.* Now we consider a variable 1 defined on this contour. Operators $\psi(1)$ which are located at the upper branch are developing in chronological sense (operator T), and operators which are located on the lower branch or anti-chronological branch, are developing in anti-chronological sense (operator $\tilde{\mathrm{T}}$). Accordingly, we introduce the operator T_C ordering along the oriented Keldysh contour. Then it is a simple matter to rewrite the expression (3.123) in the form

$$g(1,1') = \frac{1}{i}\mathrm{Tr}\left\{\varrho\mathrm{T}_C\left\{S(t_0,t_0)\psi(1)\psi^\dagger(1')\right\}\right\} , \tag{3.126}$$

where

$$S(t_0,t_0) = \mathrm{T}_C \exp\left\{-i\int_C d2d\bar{2}U(2,\bar{2})\psi^\dagger(2)\psi(\bar{2})\right\} . \tag{3.127}$$

Subsequently, the integral $\int_C$ stands for an integration along the Keldysh contour. Notice that the expression (3.126) is not only an expression for the single-particle causal Green's function, but contains the complete information about the many-particle system.

The function (3.126) is a very compact representation of all single-particle functions. By positioning the time variables t_1 and t_1' at the Keldysh contour, we get the different single-particle functions. When one time of this function is given on the chronological and the other on the anti-chronological branch, we obtain, for $U=0$, the usual two correlation functions $g^<(1,1')$ and $g^>(1,1')$. This is obvious due to the fact that, because of the time ordering on the Keldysh contour, times on the lower branch are always later than times on the upper branch. The causal $g^c(1,1')$ or anti-causal Green's functions $g^a(1,1')$ follow if the times are restricted to the chronological or anti-chronological branches, respectively.

To illustrate this, we denote times t located on the upper (+) branch by t_+ and those located on the lower branch (−) by t_-. Then we have

$$\begin{aligned} g(1_+,1'_+) &= g^c(1,1') \qquad & g(1_-,1'_+) = g^>(1,1') \\ g(1_-,1'_-) &= g^a(1,1') \qquad & g(1_+,1'_-) = g^<(1,1') \,. \end{aligned} \tag{3.128}$$

As mentioned, time integrations that occur in the equations for Green's functions have to be performed along the Keldysh time contour. For example, we consider the function $C(t,t')$ given by

$$C(t,t') = \int_C d\bar{t} A(t,\bar{t})B(\bar{t},t') \,. \tag{3.129}$$

The integration along the contour can then be written as

$$C(t,t') = \int_{t_0}^{\max(t,t')} d\bar{t} A(t,\bar{t})B(\bar{t},t')\Big|_{\text{upper}} - \int_{t_0}^{\max(t,t')} d\bar{t} A(t,\bar{t})B(\bar{t},t')\Big|_{\text{lower}} \,. \tag{3.130}$$

Here, $\max(t,t')$ takes the time later on the contour. For symmetry properties of the functions on the Keldysh contour, we refer to Danielewicz (1984).

Let us consider the identity

$$F(t) = \int_C d\bar{t}\, \delta_C(t-\bar{t})F(\bar{t}) \,. \tag{3.131}$$

According to this relation the δ_C-function is defined by

$$\delta_C(t-t') = \begin{cases} \delta(t-t') & \text{for} \quad t_+, t'_+ \,, \\ -\delta(t-t') & \text{for} \quad t_-, t'_- \,, \\ 0 & \text{for} \quad t_-, t'_+ \,, \\ 0 & \text{for} \quad t_+, t'_- \,. \end{cases}$$

The Heaviside function is $\Theta(t-t') = 1$ if the position of the time t on the contour is later than t', otherwise $\Theta(t-t') = 0$.

Though the interpretation of the Green's functions using the double-time Keldysh contour directly arises from physical considerations as presented above, for many purposes it is quite cumbersome.

A very elegant alternative interpretation has been given by Keldysh himself. Any quantity defined on the contour can be interpreted as a matrix the indices of which take the values + (−) if the respective time is located on the upper (lower) branch. Then we have for g

$$g(t,t') = \begin{pmatrix} g^{++}(t,t') & g^{+-}(t,t') \\ g^{-+}(t,t') & g^{--}(t,t') \end{pmatrix} = \begin{pmatrix} g^c(t,t') & g^<(t,t') \\ g^>(t,t') & g^a(t,t') \end{pmatrix} . \tag{3.132}$$

Analogously, any one-particle function is represented by a 2×2 matrix, any two-particle function by a 4×4 matrix and so on. The product of two functions has to be evaluated according to the rules of matrix multiplication keeping in mind that on the lower branch a minus sign occurs:

$$c^{\alpha\beta}(t,t') = \sum_{\gamma=+,-} \gamma \int d\bar{t}\, a^{\alpha\gamma}(t,\bar{t})\, b^{\gamma\beta}(\bar{t},t')\,. \qquad (3.133)$$

Now it is possible to interpret equations given on the contour as matrix equations, and the contour has become redundant.

Let us return to the single-particle Green's function $g(1,1';U)$ given by (3.126). An important property is that any higher Green's function can be obtained from (3.126) by taking the functional derivative with respect to the external potential $U(2,2')$. So it is easy to get the following relation, using the rules of functional derivation

$$\begin{aligned} i\frac{\delta g(1,1';U)}{\delta U(2',2)}\bigg|_{t_2=t_2'} &= \pm i\left\{g(12,1'2') - g(1,1')g(2,2')\right\}\bigg|_{t_2=t_2'} \\ &= \pm L(12,1'2')|_{t_2=t_2'}\,. \end{aligned} \qquad (3.134)$$

In the following, we will extend to the case $t_2 \neq t_2'$. This allows us to find more general equations for the two-particle Green's functions if the external potential U is set equal to zero at the end of the calculation. Especially, equations are obtained for four-point two-particle Green's functions.

Let us now come back to the first hierarchy equation (3.118). This equation is, of course, valid for the more general Green's function on the Keldysh contour. Introducing the relation (3.134) into this equation, one can obtain the following equation for the single particle Green's function

$$\begin{aligned} &\left\{i\frac{\partial}{\partial t_1} + \frac{\nabla_1^2}{2m})\right\} g(1,1';U) - \int d\bar{1}U(1,\bar{1})g(\bar{1}1';U) = \delta(1-1') \\ &\pm i\int d2\, V(1-2)\left\{g(1,1';U)g(2,2^+;U) \pm \frac{\delta g(1,1';U)}{\delta U(2^+,2)}\right\}\,. \end{aligned} \qquad (3.135)$$

This equation is formally closed. With (3.134) and (3.135), we have a very compact representation of the Green's function formalism. The functional $g(1,1';U)$ is equivalent to the full set of many-particle Green's functions, and (3.135) is equivalent to the Martin–Schwinger hierarchy on the Keldysh contour and is therefore a comprehension of equations for $g^>$, $g^<$, g^c, and g^a. Unfortunately, no techniques for solving functional differential equations exactly exist. But (3.135) may be used for formal considerations and manipulations.

3.3.4 Self-Energy and Kadanoff–Baym Equations

We recall the Hartree–Fock equation (3.109). It seems to be obvious that the structure of this equation is more general than it appears in the Hartree–Fock approximation. Thus, we try to obtain a closed equation for $g(1,1')$ from the first hierarchy equation (3.135) by introduction of the self-energy function

$$\int_C d\bar{1}\Sigma(1,\bar{1})g(\bar{1},1') = \pm i \int_C d2\, V(1,2)g(12,1'2^+)$$
$$= \pm i \int_C d2\, V(1,2)\left[g(1,1')g(2,2^+) \pm \frac{\delta g(1,1')}{\delta U(2^+,2)}\right]. \quad (3.136)$$

The self-energy $\Sigma(1,1')$ is here defined on the Keldysh contour and, therefore, we have the four functions $\Sigma^{\gtrless}(1,1')$ and $\Sigma^{c,a}(1,1')$ on the *physical time axis*. If we introduce this definition into the first equation of the Martin–Schwinger hierarchy, we immediately get a closed equation for the single-particle Green's function on the Keldysh contour

$$\int_C d\bar{1}\, [\, g_0^{-1}(1\bar{1}) - U(1\bar{1}) - \Sigma(1\bar{1})]\, g(\bar{1}1') = \delta(1-1')\,. \quad (3.137)$$

This famous equation was derived for the first time by Kadanoff and Baym and by Keldysh. It is a generalization of the Dyson equation of the field theory to quantum statistics. Equation (3.137) describes the time evolution of the real-time Green's function $g(1,1')$ under equilibrium and non-equilibrium conditions. Of course, all problems are now transferred to the self-energy. We therefore have to consider the properties of this function and to find a possibility to determine appropriate approximations for Σ.

Let us first consider the dependence of Σ on the initial value of the two-particle Green's function.

From (3.136) it is obvious that the self-energy has to fulfill the relation

$$\lim_{t_1=t_1'=t_0} \int_C d\bar{1}\Sigma(1,\bar{1})g(\bar{1},1') = \pm i \int d\boldsymbol{r}_2 V(\boldsymbol{r}_1-\boldsymbol{r}_2)g(\boldsymbol{r}_1\boldsymbol{r}_2,\boldsymbol{r}_1'\boldsymbol{r}_2';t_0) \quad (3.138)$$

with $g(\boldsymbol{r}_1\boldsymbol{r}_2,\boldsymbol{r}_1'\boldsymbol{r}_2';t_0)$ being the initial value of the two-particle Green's function. We write the initial binary density matrix in the following form

$$g(\boldsymbol{r}_1\boldsymbol{r}_2,\boldsymbol{r}_1'\boldsymbol{r}_2';t_0) = g(1,1')g(2,2')|_{t_0} \pm g(1,2')g(2,1')|_{t_0} + c(\boldsymbol{r}_1\boldsymbol{r}_2,\boldsymbol{r}_1'\boldsymbol{r}_2';t_0)\,, \quad (3.139)$$

where $c(\boldsymbol{r}_1\boldsymbol{r}_2,\boldsymbol{r}_1'\boldsymbol{r}_2';t_0)$ is the initial binary correlation. Consequently, it is necessary to introduce initial conditions, in order to make the self-energy unique.

Now we have to take into consideration that the integral over the Keldysh contour for a regular integrand vanishes in the limit $t,t' \to t_0$. Therefore, the equation (3.138) can be fulfilled only if the self-energy has a contribution proportional to a δ-function with respect to the times. Such a term is $\Sigma^{\mathrm{HF}}(1,\bar{1})$ given by (3.189). But this term only produces the uncorrelated contribution to the initial binary density matrix. This means that the self-energy must contain a second part with a δ-singularity. The general structure of the self-energy is thus given (Kremp et al. 2000; Semkat et al. 2000)

$$\Sigma(1,1') = \Sigma^{\mathrm{HF}}(1,1') + \Sigma^{\mathrm{corr}}(1,1') + \Sigma^{\mathrm{in}}(1,1'), \tag{3.140}$$

where

$$\Sigma^{\mathrm{in}}(1,1') = \Sigma^{\mathrm{in}}(1,\boldsymbol{r}_1', t_0)\delta(t_1' - t_0).$$

Using this structure of Σ in (3.137), we obtain for the Kadanoff–Baym equations (K–B equations)

$$\begin{aligned}&\int\limits_C d\bar{1}\,\{g_0^{-1}(1,\bar{1}) - U(1,\bar{1}) - \Sigma^{\mathrm{HF}}(1,\bar{1})\}\,g(\bar{1},1')\\ &= \delta(1-1') + \int\limits_C d\bar{1}\,\{\Sigma^{\mathrm{corr}}(1,\bar{1}) + \Sigma^{\mathrm{in}}(1,\bar{1})\}\,g(\bar{1},1').\end{aligned} \tag{3.141}$$

We analogously get for the adjoint equation

$$\begin{aligned}&\int\limits_C d\bar{1} g(1,\bar{1})\,\{g_0^{-1}(\bar{1},1') - U(\bar{1},1') - \Sigma^{\mathrm{HF}}(\bar{1},1')\}\\ &= \delta(1-1') + \int\limits_C d\bar{1} g(1,\bar{1})\,\{\Sigma^{\mathrm{corr}}(\bar{1},1') + \Sigma_{\mathrm{in}}(\bar{1},1')\},\end{aligned} \tag{3.142}$$

where

$$\Sigma_{\mathrm{in}}(\bar{1},1') = \delta(t_1 - t_0)\Sigma_{\mathrm{in}}(\boldsymbol{r}_1 t_0, 1'),$$

and the adjoint self energy $\hat{\Sigma}$ reads

$$\hat{\Sigma} = \Sigma^{\mathrm{HF}}(1,1') + \Sigma^{\mathrm{corr}}(1,1') + \Sigma_{\mathrm{in}}(1,1'). \tag{3.143}$$

In this form of K–B equations, the dependence of Σ on initial correlations is shown explicitly.

If we use the Bogolyubov condition (3.104) instead of the general initial condition (3.139), we immediately get $\Sigma^{\mathrm{in}} = 0$ and the lower limit of time integration becomes $t_0 = -\infty$. In this case, we recover the original K–B equations. This was shown in the paper by Kremp et al. (1985). Clearly, this is a strong restriction which is not justified for all physical situations. Here, we mention only short-time processes in laser produced plasmas, short-time fluctuations and long-living correlations such as bound states.

On the physical time axis, (3.141) is equivalent to the four equations to be written subsequently. By positioning both times t_1 and t_1' on the upper or lower branches of the contour, we get a first pair of equations for the causal $g^c(1,1')$ and the anti-causal Green's functions $g^a(1,1')$. The equation for the causal function reads

$$
\begin{aligned}
&\left(i\frac{\partial}{\partial t_1}+\frac{\nabla_1^2}{2m}\right)g^c(1,1')-\int d\bar{1}\Big[U(1,\bar{1})+\Sigma^{\rm HF}(1,\bar{1})\Big]g^c(\bar{1},1')\\
=\;&\int_{t_0}^{\infty} d\bar{1}\left[\Sigma^c(1,\bar{1})+\Sigma^{\rm in}(1,\bar{1})\right]g^c(\bar{1},1')\\
&-\int_{t_0}^{\infty} d\bar{1}\left[\Sigma^<(1,\bar{1})+\Sigma^{\rm in}(1,\bar{1})\right]g^>(\bar{1},1')\,.
\end{aligned}
\tag{3.144}
$$

The second pair of equations is that for the correlation functions $g^{\gtrless}$. The pair follows from fixing the time arguments of the Green's functions t_1 and t_1' on the opposite branches of the contour.

$$
\begin{aligned}
&\left(i\frac{\partial}{\partial t_1}+\frac{\nabla_1^2}{2m}\right)g^{\gtrless}(1,1')-\int d\bar{1}\Big[U(1,\bar{1})+\Sigma^{\rm HF}(1,\bar{1})\Big]g^{\gtrless}(\bar{1},1')\\
=\;&\int_{t_0}^{t_1} d\bar{1}\Big[\Sigma^>(1,\bar{1})-\Sigma^<(1,\bar{1})\Big]g^{\gtrless}(\bar{1},1')\\
&-\int_{t_0}^{t_1'} d\bar{1}\left[\Sigma^{\gtrless}(1\bar{1})+\Sigma^{\rm in}(1\bar{1})\right]\left[g^>(\bar{1},1')-g^<(\bar{1},1')\right].
\end{aligned}
\tag{3.145}
$$

For later purposes, we will also write the equations adjoint to (3.145)

$$
\begin{aligned}
&\left(-i\frac{\partial}{\partial t_1'}+\frac{\nabla_{1'}^2}{2m}\right)g^{\gtrless}(1,1')-\int d\bar{1}g^{\gtrless}(1,\bar{1})\Big[U(\bar{1},1')+\Sigma^{\rm HF}(\bar{1},1')\Big]\\
=\;&\int_{t_0}^{t_1} d\bar{1}\Big[g^>(1,\bar{1})-g^<(1,\bar{1})\Big]\left[\Sigma^{\gtrless}(\bar{1},1')+\Sigma_{\rm in}(\bar{1},1')\right]\\
&-\int_{t_0}^{t_1'} d\bar{1}g^{\gtrless}(1,\bar{1})\Big[\Sigma^>(\bar{1},1')-\Sigma^<(\bar{1},1')\Big].
\end{aligned}
\tag{3.146}
$$

Here, $\Sigma_{\rm in}$ is the adjoint of $\Sigma^{\rm in}$. Equations (3.145) and (3.146) are referred to as the Kadanoff–Baym equations in a more restricted sense. They determine the dynamics of the correlation functions $g^{\gtrless}$. The genuine coupling of the equations of motion for $g^>$ and $g^<$ is characteristic of non-equilibrium systems. Thus we point out again that these functions are independent under non-equilibrium conditions. In contrast to the original KBE, there are two important new properties which have to be underlined here: Equations (3.145) and (3.146) are valid for an arbitrary initial time point t_0, and they explicitly contain the influence of arbitrary initial correlations in the additional self-energy terms $\Sigma^{\rm in}$ and $\Sigma_{\rm in}$.

In Sects. 3.2.1 and 3.2.2, we introduced retarded and advanced quantities. In the formulation of equations of motion for the correlation functions, it is useful to take advantage of such functions.

We use the relations (3.49) and the definition (3.89) in the Kadanoff–Baym equations (3.145), (3.146), and in addition, we set up an equation of motion for the retarded and advanced Green's functions. Furthermore, we replace the *non*local external potential by a local one. We get the important system of equations

$$\left\{ i\frac{\partial}{\partial t_1} + \frac{\nabla_1^2}{2m} - U(1) \right\} g^{\gtrless}(1,1') - \int_{-\infty}^{+\infty} d\bar{1}\, \Sigma^R(1,\bar{1}) g^{\gtrless}(\bar{1},1')$$

$$= \int_{-\infty}^{+\infty} d\bar{1} \left[\Sigma^{\gtrless}(1\bar{1}) + \Sigma^{\text{in}}(1\bar{1}) \right] g^A(\bar{1},1') \,, \qquad (3.147)$$

$$\left\{ i\frac{\partial}{\partial t_1} + \frac{\nabla_1^2}{2m} - U(1) \right\} g^{R/A}(1,1') - \int_{-\infty}^{+\infty} d\bar{1}\, \Sigma^{R/A}(1,\bar{1}) g^{R/A}(\bar{1},1')$$

$$= \delta(1-1') \,. \qquad (3.148)$$

We see from the set (3.147) and (3.148) that $g^{R/A}$ are the Green's functions belonging to the correlation functions $g^{\gtrless}$ in the common mathematical sense. Taking the initial condition for the correlation function to be

$$g^{\gtrless}(1,1')\Big|_{t_0} = g^{\gtrless}\left(\boldsymbol{r}_1\boldsymbol{r}_1', t_0\right) \,,$$

the solution of the equation of motion (3.147) may be represented as (source representation)

$$g^{\gtrless}(1,1') = \int d\bar{\boldsymbol{r}}_1 d\bar{\bar{\boldsymbol{r}}}_1\, g^R(1,\bar{\boldsymbol{r}}_1 t_0) g^{\gtrless}(\bar{\boldsymbol{r}}_1 t_0, \bar{\bar{\boldsymbol{r}}}_1 t_0) g^A(\bar{\bar{\boldsymbol{r}}}_1 t_0, 1')$$

$$+ \int_{t_0}^{t_1} d\bar{1} \int_{t_0}^{t_1'} d\,\bar{\bar{1}}\, g^R(1,\bar{1}) \left[\Sigma^{\gtrless}(1\bar{1}) + \Sigma^{\text{in}}(1\bar{1}) \right] g^A(\bar{\bar{1}},1') \,. \qquad (3.149)$$

The first r.h.s. contribution describes the time evolution of the initial value. The second r.h.s. term represents the influence of the "source" $\Sigma^{\gtrless}(\bar{1}\;\bar{\bar{1}})$ on the correlation function $g^{\gtrless}(11')$ for the time intervals $\bar{\bar{t}}_1 < t_1'$; $\bar{t}_1 < t_1$ such that causality is fulfilled.

It is possible to assume (under conditions to be specified) that the *initial value contributions are damped out* what follow from the properties of $g^{R/A}$. Then we may write for a sufficiently long time after t_0

$$g^{\gtrless}(1,1') = \int_{t_0}^{\infty} d\bar{1} \int_{t_0}^{\infty} d\,\bar{\bar{1}}\; g^R(1,\bar{1})\Sigma^{\gtrless}(\bar{1},\bar{\bar{1}})g^A(\bar{\bar{1}},1')\,. \tag{3.150}$$

Here, the limits of integration refer to time integration.

Let us summarize the important result of this subsection. By introduction of the self-energy into the first equation for the Green's function $g(1,1')$ defined on the Keldysh contour, we achieved a closed system of equations for the dynamical behavior of an equilibrium or a non-equilibrium many-particle system. These equations are exact and fully equivalent to the Martin–Schwinger hierarchy. They have been derived without any approximation with respect to the time or other quantities like density, coupling parameter, etc.

3.3.5 Structure and Properties of the Self-Energy. Initial Correlation

In the previous section, the self-energy was introduced as the central quantity in the Kadanoff–Baym theory. Therefore, we will discuss some general properties of this function.

Introducing again the micro and macro variables

$$T = \frac{t_1 + t_1'}{2}; \; \boldsymbol{R} = \frac{\boldsymbol{r}_1 + \boldsymbol{r}_1'}{2}; \quad \tau = t_1 - t_1'; \; \boldsymbol{r} = \boldsymbol{r}_1 - \boldsymbol{r}_1'\,, \tag{3.151}$$

we have $\Sigma^{\gtrless}(1,1') = \Sigma^{\gtrless}(\boldsymbol{r}\tau, \boldsymbol{R}T)$, and after Fourier transformation with respect to the difference variables, we get

$$\Sigma^{\gtrless}(\boldsymbol{p}\omega, \boldsymbol{R}T) = \int d\boldsymbol{r} d\tau e^{-i\boldsymbol{p}\cdot\boldsymbol{r} + i\omega\tau} \Sigma^{\gtrless}(\boldsymbol{r}\tau, \boldsymbol{R}T)\,. \tag{3.152}$$

Like in the case of the single-particle Green's functions, the following spectral representation can be found for the retarded and advanced self-energies

$$\Sigma^{R/A}(\boldsymbol{p}\omega, \boldsymbol{R}T) = \Sigma^{\mathrm{HF}}(\boldsymbol{p}, \boldsymbol{R}T) + \int \frac{d\bar{\omega}}{2\pi} \frac{\Gamma(\boldsymbol{p}\bar{\omega}, \boldsymbol{R}T)}{\omega - \bar{\omega} \pm i\varepsilon}\,. \tag{3.153}$$

Here, we introduced the function

$$\begin{aligned} \Gamma(\boldsymbol{p}\omega, \boldsymbol{R}T) &= i\Sigma^{>}(\boldsymbol{p}\omega, \boldsymbol{R}T) - i\Sigma^{<}(\boldsymbol{p}\omega, \boldsymbol{R}T) \\ &= -2\mathrm{Im}\Sigma^R(\boldsymbol{p}\omega, \boldsymbol{R}T)\,. \end{aligned} \tag{3.154}$$

Lateron we will see that this function has the physical meaning of a single-particle damping which determines the life time of a single-particle state due to the influence of the surrounding medium. Like in the case of the single-particle Green's function, we can find the dispersion relation

$$\mathrm{Re}\Sigma^{\mathrm{R}}(\boldsymbol{p}\omega, \boldsymbol{R}T) = \Sigma^{\mathrm{HF}}(\boldsymbol{p}, \boldsymbol{R}T) - \mathrm{P} \int \frac{d\bar{\omega}}{\pi} \frac{\mathrm{Im}\Sigma^R(\boldsymbol{p}\bar{\omega}, \boldsymbol{R}T)}{\omega - \bar{\omega}}\,. \tag{3.155}$$

In Sect. 3.2.2, we found that, in thermodynamic equilibrium, the single-particle correlation functions $g^{\gtrless}$ are not independent from each other because of the KMS condition. The same result follows for the self-energies $\Sigma^{\gtrless}$. They are connected by the relation

$$\pm\Sigma^{<}(\boldsymbol{p},\omega) = e^{-\beta(\omega-\mu)}\Sigma^{>}(\boldsymbol{p},\omega)\,, \tag{3.156}$$

and we get the spectral representations

$$\pm i\Sigma^{<}(\boldsymbol{p},\omega) = \Gamma(\boldsymbol{p},\omega)f(\omega) \tag{3.157}$$

$$i\Sigma^{>}(\boldsymbol{p},\omega) = \Gamma(\boldsymbol{p},\omega)[1\pm f(\omega)]. \tag{3.158}$$

Here, $f(\omega)$ are Fermi or Bose functions given by (2.23).

In order to determine the self-energy function, we start from (3.136). Therefore, we need the functional derivative $\delta g/\delta U$. Fortunately, a simple procedure for the calculation of this quantity is available. In order to explain this procedure, we use the Kadanoff–Baym equations (3.141) and (3.142) for $t,t'>t_0$ and write (3.141) in the form

$$\int_C g_1^{-1}(1,\bar{1})g_1(\bar{1},1')d\bar{1} = \delta(1-1')\,. \tag{3.159}$$

Here, the inverse Green's function g^{-1} is given by

$$g_1^{-1}(1\bar{1}) = g_0^{-1}(1\bar{1}) - U(1\bar{1}) - \Sigma(1\bar{1})\,. \tag{3.160}$$

Functional differentiation of (3.159) for $t,t'>t_0$ easily yields

$$\delta(1-2')g_1(21') + \int\limits_C d\bar{1}\frac{\delta\Sigma(1\bar{1})}{\delta U(2'2)}g_1(\bar{1}1') = \int\limits_C d\bar{1}g_1^{-1}(1\bar{1})\frac{\delta g_1(\bar{1}1')}{\delta U(2'2)}\,. \tag{3.161}$$

The general solution of this equation and its adjoint can be written immediately

$$L(121'2') = \frac{\delta g_1(11')}{\delta U(2'2)} = g_1(12')g_1(21') \pm C(121'2')$$
$$+ \int\limits_C d\bar{1}d\,\bar{\bar{1}}\;g_1(1\bar{1})\frac{\delta[\Sigma^{\rm corr}(\bar{1}\,\bar{\bar{1}}) + \Sigma^{\rm in}(\bar{1}\,\bar{\bar{1}}) + \Sigma_{\rm in}(\bar{1}\,\bar{\bar{1}})]}{\delta U(2'2)}g_1(\bar{\bar{1}}\;1')\,, \tag{3.162}$$

where C is an arbitrary function which obeys the homogeneous equation, i.e.,

$$\int\limits_C d\bar{1}g_1^{-1}(1\bar{1})C(\bar{1}21'2') = 0\,, \tag{3.163}$$

and three similar equations for the other variables. The further procedure to find an equation for Σ is similar to the consideration given by Kadanoff and Baym (1962). But in generalization to this discussion, now the function $C(121'2')$ enables us to take into account the initial correlations. We will not explain here the details of the further calculation. One can find them in the papers by Semkat et al. (1999) and Semkat et al. (2000).

The result is a functional equation for the self-energy:

$$\begin{aligned}\Sigma(11') &= \pm i \int d2V(1-2)\Bigg\{\delta(1-1')g_1(22^+) \pm \delta(2-1')g_1(12^+) \\ &+ \int_C d\bar{1}d\bar{2}d\,\bar{\bar{2}}\; g_1(1\bar{1})g_1(2\bar{2})c(\bar{1}\bar{2}1'\,\bar{\bar{2}})g_1(\bar{\bar{2}}\;2^+) \\ &\pm \int_C d\bar{1}g_1(1\bar{1})\frac{\delta[\Sigma^{\mathrm{corr}}(\bar{1}\;\bar{\bar{1}}) + \Sigma^{\mathrm{in}}(\bar{1}\;\bar{\bar{1}}) + \Sigma_{\mathrm{in}}(\bar{1}\;\bar{\bar{1}})]}{\delta U(2^+2)}\Bigg\} \end{aligned} \tag{3.164}$$

with

$$\begin{aligned} c(\bar{1}\bar{2}\;\bar{\bar{1}}\bar{\bar{2}}) &= c(\bar{\boldsymbol{r}}_1 t_0, \bar{\boldsymbol{r}}_2 t_0, \bar{\bar{\boldsymbol{r}}}_1\; t_0, \bar{\bar{\boldsymbol{r}}}_2\; t_0) \\ &\times \delta(\bar{t}_1 - t_0)\delta(\bar{t}_2 - t_0)\delta(\bar{\bar{t}}_1 - t_0)\delta(\bar{\bar{t}}_2 - t_0)\,. \end{aligned} \tag{3.165}$$

An analogous equation follows readily for $\hat{\Sigma}$. For vanishing initial correlations, this equation is just the the functional integral equation derived by Kadanoff and Baym.

With (3.164), the self-energy is given as a functional of the interaction, the initial correlations and the single-particle Green's function, where the initial correlations are contained in the last term. From the definition of c, (3.163), it is obvious that this contribution is local in time with a δ-type singularity at $t = t'$. Additional terms of this structure arise from the functional derivative.

A further important property of the self-energy follows from comparing Σ, (3.164), with the corresponding expression for $\hat{\Sigma}$. One verifies that $\Sigma = \hat{\Sigma}$ for all times $t, t' > t_0$, what means, in particular, that for these times, a well defined inverse Green's function does exist.

Equation (3.164) is well suited to define approximations for the self-energy. By iteration, a perturbation series for Σ in terms of g, V and C can be derived which begins with

$$\begin{aligned} \Sigma^1(11') &= \pm i\delta(1-1')\int d2V(1-2)g_1(22^+) + iV(1-1')g_1(11') \\ &\pm i\int d2V(1-2)\int_C d\bar{1}d\bar{2}d\,\bar{\bar{2}}\; g_1(1\bar{1})g_1(2\bar{2})c(\bar{1}\bar{2}1'\,\bar{\bar{2}})g_1(\bar{\bar{2}}\;2^+)\,. \end{aligned} \tag{3.166}$$

It is now instructive to introduce Feynman diagrams which allow for the representation of formula (3.166). In general, the self-energy is now given by the graphs of Fig. 3.2, and (3.166) corresponds to the three first r.h.s. graphs. Here, we used the abbreviation $\tilde{\Sigma} = \Sigma^{\mathrm{corr}} + \Sigma^{\mathrm{in}} + \Sigma_{\mathrm{in}}$.

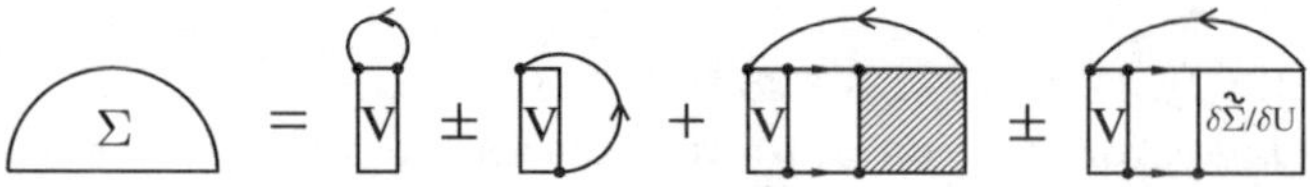

Fig. 3.2. Self-energy including initial correlations

In contrast to conventional diagram techniques, we have to introduce the initial correlations as a new basic element, drawn as a shaded rectangle. The second iteration step yields straightforwardly

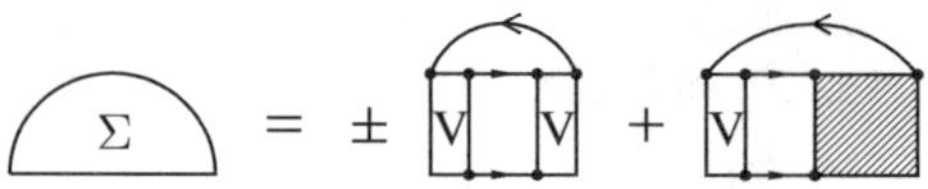

Fig. 3.3. Second order self-energy including initial correlations

The results of our analysis of the iteration scheme allow us to conclude that all contributions to the self-energy (all diagrams) fall into two classes: (I) the terms Σ^{HF} and Σ^{corr} which begin and end with a potential and (II), Σ^{in} – those which begin with a potential, but end with an initial correlation. This means, we have verified the structure (Kremp et al. 2000)

$$\begin{aligned} \Sigma(11') &= \Sigma^{\mathrm{HF}}(11') + \Sigma^{\mathrm{corr}}(11') + \Sigma^{\mathrm{in}}(11'), \\ \Sigma^{\mathrm{in}}(11') &= \Sigma^{\mathrm{in}}(1, \boldsymbol{r}_1' t_0)\delta(t_1' - t_0)\,, \end{aligned} \tag{3.167}$$

given by the relation (3.140). Interestingly, the same result was obtained by Danielewicz based on his perturbation theory for general initial states (Danielewicz 1984). The agreement of the two approaches becomes particularly obvious from the diagrammatic representation of Σ.

If one considers the first two iterations for the self-energy (3.166) more in detail, it becomes evident that, in the initial correlation contribution, in front of the function c, there appear just the ladder terms which lead to the buildup of the two-particle Green's function. Thus, obviously, the iteration "upgrades" the product of retarded single-particle propagators in the function C to a full two-particle propagator, in the respective order, i.e., Σ^{in} is of the form (Semkat et al. 2000)

$$\begin{aligned} &\Sigma^{\mathrm{in}}(11') = \pm i \int d2\, V(1-2) \int d\bar{\boldsymbol{r}}_1 d\bar{\boldsymbol{r}}_2 d\,\bar{\bar{\boldsymbol{r}}}_1\, d\,\bar{\bar{\boldsymbol{r}}}_2 \\ &\times g^R_{12}(12, \bar{\boldsymbol{r}}_1 t_0, \bar{\boldsymbol{r}}_2 t_0) c(\bar{\boldsymbol{r}}_1 t_0, \bar{\boldsymbol{r}}_2 t_0, 1', \bar{\bar{\boldsymbol{r}}}_2\, t_0) g^A_1(\bar{\bar{\boldsymbol{r}}}_2\, t_0, 2^+)\delta(t_1' - t_0)\,. \end{aligned} \tag{3.168}$$

The analytic properties of the retarded and advanced Green's functions give rise to a damping γ_{12} leading to a decay of the initial correlation term after a time of the order $t \sim 1/\gamma_{12} \sim \tau_{cor}$. Thus, there is no need at all to postulate Bogolyubov's weakening condition; for $t > \tau_{cor}$, the generalized Kadanoff–Baym equations switch from the initial regime into the kinetic, or Bogolyubov regime, "automatically".

The *generalized* Kadanoff–Baym equations presented here were for the first time derived by Danielewicz (1984) using a perturbation scheme generalized by inclusion of an initial condition. Initial correlations were already dealt with by Fujita (1966), Fujita (1971) and Hall (1975); however, no explicit results were given in those papers. Furthermore, initial correlations can be included by deformation of the Keldysh contour to the imaginary time axis (Danielewicz 1990; Wagner 1991; Zubarev et al. 1996; Morozov and Röpke 1999). Finally, initial correlations were considered in the paper by Kremp et al. (1997) on the basis of the Bogolyubov hierarchy for the density operators. An extensive presentation of the problem of initial conditions is found in papers by Resibois (1965) for classical statistical mechanics. In all papers mentioned, rules for perturbative schemes were developed for systems with initial correlations. In contrast, we considered the problem as an initial value problem for the Martin–Schwinger hierarchy on the basis of a non-perturbative technique using functional derivatives.

3.3.6 Gradient Expansion. Local Approximation

Let us now come back to the Kadanoff–Baym equations (3.145) and (3.146). One can easily see that the equations for the two-time correlation functions may also be written in the form

$$\int_{-\infty}^{\infty} d\bar{1}\, g^{R^{-1}}(1,\bar{1})\, g^{\gtrless}(\bar{1},1') = \int_{-\infty}^{\infty} d\bar{1}\, \Sigma^{\gtrless}(1,\bar{1})\, g^{A}(\bar{1},1')\,, \tag{3.169}$$

$$\int_{-\infty}^{\infty} d\bar{1}\, g^{\gtrless}(1,\bar{1})\, g^{A^{-1}}(\bar{1},1') = \int_{-\infty}^{\infty} d\bar{1}\, g^{R}(1,\bar{1})\, \Sigma^{\gtrless}(\bar{1},1')\,. \tag{3.170}$$

This set of equations has to be completed by the equation of motion for the retarded and advanced Green's functions (3.148).

In any of the equations of motion we have expressions of the structure

$$\int d\bar{x}_1 f\left(x_1, \bar{x}_1\right) u\left(\bar{x}_1, x_1'\right) \equiv I\left(x_1, x_1'\right)\,, \tag{3.171}$$

where $x_1 = (r_1, t_1)$. Under equilibrium conditions, the functions depend only on the difference variables, and the integrals are of the convolution type. In the general case, there is a dependence on both variables x_1 and x_1'. Here, it is useful to introduce the new variables

$$x = x_1 - x_1'\,, \quad X = \frac{1}{2}\left(x_1 + x_1'\right)\,, \quad \bar{x} = \bar{x}_1 - x_1'\,, \tag{3.172}$$

where $x = (\boldsymbol{r}, \tau)$ and $X = (\boldsymbol{R}, t)$. Then we get instead of (3.171)

$$I(x, X) = \int d\bar{x}\, f\left(x - \bar{x}, X + \frac{\bar{x}}{2}\right) u\left(\bar{x}, X + \frac{\bar{x} - x}{2}\right) . \tag{3.173}$$

Therefore, in non-equilibrium, there is a complicated coupling between the macro variables X and the micro variables x. If we may consider $\bar{x}$ and $\bar{x} - x$ to be negligible corrections to X, we get a local approximation for $I(x, X)$. In such an approximation, there is no coupling between X and x, i.e., between the dynamics on the microscopic and macroscopic scale. This local approximation is sufficient for the derivation of a kinetic equation of the Boltzmann type.

In a next step of approximation, it is possible to expand the functions with respect to $\bar{x}$ and $\bar{x} - x$, respectively, up to the first order. After Fourier transformation of $I(x, X)$ with respect to the difference variable x, we get

$$I(w, X) \approx f(w, X)u(w, X) + \frac{i}{2}\left(\frac{\partial f}{\partial w}\frac{\partial u}{\partial X} - \frac{\partial f}{\partial X}\frac{\partial u}{\partial w}\right) , \tag{3.174}$$

where w stands for momentum and frequency.

Applying this relation to (3.169) and (3.170), and subtracting the equations from each other, we get the first order gradient expansion of the Kadanoff–Baym equations (Kadanoff and Baym 1962)

$$\left[\mathrm{Re}\; g^{-1^R}, ig^{\gtrless}\right] - \left[i\Sigma^{\gtrless}, \mathrm{Re}\; g^R\right] = g^< \Sigma^> - g^> \Sigma^< . \tag{3.175}$$

Here we introduced the generalized Poisson brackets

$$[A, B] = \frac{\partial A}{\partial \omega}\frac{\partial B}{\partial t} - \frac{\partial A}{\partial t}\frac{\partial B}{\partial \omega} - \nabla_{\boldsymbol{p}} A \cdot \nabla_{\boldsymbol{R}} B + \nabla_{\boldsymbol{R}} A \cdot \nabla_{\boldsymbol{p}} B , \tag{3.176}$$

where $A = A(\boldsymbol{p}\omega, \boldsymbol{R}t)$, etc. It should be noticed that further first order expansion terms are contained on the r.h.s of (3.175) taking into account the internal structure of $\Sigma^{\gtrless}$ and $g^{\gtrless}$ (Bornath et al. 1996) (see Sect. 3.3).

To set up the kinetic equations for the correlation functions $g^{\gtrless}$, we also need the corresponding equation for the retarded/advanced Green's functions $g^{R/A}$. Here, it is sufficient to take (3.148) in the local approximation (Kremp et al. 1986)

$$\begin{aligned}\left(i\frac{\partial}{\partial \tau} + \frac{\nabla_{\boldsymbol{r}}^2}{2m} - U^{\mathrm{ext}}(\boldsymbol{R}, t)\right) g^{R/A}(\boldsymbol{r}\tau, \boldsymbol{R}t) = \delta(\boldsymbol{r})\delta(\tau) \\ + \int_{-\infty}^{\infty} d\bar{\boldsymbol{r}} d\bar{\tau}\; \Sigma^{R/A}(\boldsymbol{r} - \bar{\boldsymbol{r}}, \tau - \bar{\tau}, \boldsymbol{R}t)\, g^{R/A}(\bar{\boldsymbol{r}}\bar{\tau}, \boldsymbol{R}t) .\end{aligned} \tag{3.177}$$

This equation is valid up to the first order of the gradient expansion around $\boldsymbol{R}, t$. This can easily be seen by adding (3.148) for $g^{R/A}$ and its adjoint, both expanded up to linear terms of the gradient expansion. The result is the local shape given by (3.177).

Fourier transformation with respect to $\boldsymbol{r}$ and τ then yields (without external field U^{ext})

$$g^{R/A}(\boldsymbol{p}\omega, \boldsymbol{R}t) = \left(\omega - \frac{\boldsymbol{p}^2}{2m} - \Sigma^{R/A}(\boldsymbol{p}\omega, \boldsymbol{R}t) \pm i\varepsilon\right)^{-1} . \tag{3.178}$$

Furthermore, we have

$$\operatorname{Re} g^{R/A^{-1}}(\boldsymbol{p}\omega, \boldsymbol{R}t) = \omega - \frac{p^2}{2m} - \operatorname{Re}\Sigma^{R/A}(\boldsymbol{p}\omega, \boldsymbol{R}t) . \tag{3.179}$$

Let us return to the Kadanoff–Baym equations (3.175). In order to find a kinetic equation for the Wigner distribution function $f(\boldsymbol{p}, \boldsymbol{R}, t)$, we integrate (3.175) over the frequency ω and use the definition (3.35). We then get

$$\left(\frac{\partial}{\partial t} + \frac{\boldsymbol{p}}{m} \cdot \nabla_{\boldsymbol{R}} - \nabla_{\boldsymbol{R}} U^{\text{ext}} \cdot \nabla_{\boldsymbol{p}}\right) f - \int \frac{d\omega}{2\pi} \{[\operatorname{Re}\Sigma^R, \pm i g^<]$$
$$-[\operatorname{Re} g^R, \pm i\Sigma^<]\} = \pm \int \frac{d\omega}{2\pi} \left(g^<\Sigma^> - g^>\Sigma^<\right) , \tag{3.180}$$

which represents a very general kinetic equation. This equation will be discussed in detail in Chap. 7.

3.4 Green's Functions and Physical Properties

3.4.1 The Spectral Function. Quasi-Particle Picture

Very important information about the properties and the behavior of a many-particle system may be obtained from the spectral function. As we know from Sect. 3.2.3 this function is closely connected to the retarded (advanced) Green's functions. Let us consider now the retarded and advanced Green's functions in local approximation given by (3.78). From Sect. 3.2.3, we know their analytic properties. In particular, the analytical continuation to complex energies z is possible, i.e.,

$$g(\boldsymbol{p}z, \boldsymbol{R}t) = \left(z - \frac{\boldsymbol{p}^2}{2m} - \Sigma(\boldsymbol{p}z, \boldsymbol{R}t)\right)^{-1} , \tag{3.181}$$

where $g(z)$ is an analytical function of z in the upper half plane. The possible poles of $g^R(z)$ are determined by the analytical continuation of $g^R(z)$ into the lower half plane, and therefore, they are the poles of the spectral function $a(z)$ for $\text{Im}z < 0$. The latter is given by the discontinuity of $g(z)$ across the real axis

$$a(\boldsymbol{p}\,\omega, \boldsymbol{R}\,t) = i\left[g(\boldsymbol{p}\,\omega + i\varepsilon, \boldsymbol{R}\,t) - g(\boldsymbol{p}\,\omega - i\varepsilon, \boldsymbol{R}\,t)\right] . \tag{3.182}$$

Then follows

$$a(\boldsymbol{p}\,\omega, \boldsymbol{R}\,t) = \frac{\Gamma(\boldsymbol{p}\omega, \boldsymbol{R}t)}{\left(\omega - \frac{p^2}{2m} - \mathrm{Re}\Sigma^R(\boldsymbol{p}\omega, \boldsymbol{R}t)\right)^2 + \left(\frac{1}{2}\Gamma(\boldsymbol{p}\omega, \boldsymbol{R}t)\right)^2} , \tag{3.183}$$

where Γ is given by

$$\Gamma(\boldsymbol{p}\omega, \boldsymbol{R}t) = i\left[\Sigma^{>}(\boldsymbol{p}\omega, \boldsymbol{R}t) - \Sigma^{<}(\boldsymbol{p}\omega, \boldsymbol{R}t)\right] .$$

As will be seen, $a(\boldsymbol{p}\omega, \boldsymbol{R}t)$ contains all the information about the dynamic behavior of a particle in a strongly coupled plasma. Moreover, the spectral function determines the equilibrium correlation functions according to (3.63), and, therefore, all thermodynamic properties.

From our general discussion in Sect. 3.2, it is clear that $a(\boldsymbol{p}\omega, \boldsymbol{R}t)$ represents the weight of the spectrum of possible energies ω for a particle with a given momentum $\boldsymbol{p}$ in an interacting many-particle system. Here, the sum rule (3.65) ensures the normalization of the weight.

In order to demonstrate the meaning of the spectral function, we consider a simple approximation, i.e., we take $\mathrm{Re}\Sigma$ and Γ with simplified arguments, namely we set $\hbar\omega = p^2/2m$. This leads to

$$a(\boldsymbol{p}, \omega) = \frac{\Gamma(\boldsymbol{p})}{(\omega - E(\boldsymbol{p}))^2 + \frac{1}{4}\left(\Gamma(\boldsymbol{p})\right)^2} \tag{3.184}$$

where $E(p) = \frac{p^2}{2m} + \mathrm{Re}\Sigma^{\mathrm{R}}(p\omega)|_{\omega=p^2/2m}$. We see that the spectral function has a Lorentz-type shape with a width determined by $\Gamma(p) = \Gamma(p\omega)|_{\omega=p^2/2m}$. Γ is the damping of the single-particle state. This can be seen from the temporal behavior of the single-particle correlation function $g^{<}(t-t')$ which follows by Fourier transformation from (3.63) using (3.184). With the assumption $\Gamma \ll k_B T$, we may apply $f(\omega) = f(E)$. Then, $f(E)$ can be taken out of the integral, and the integration may be carried out to give

$$\pm i g^{<} = \exp(-iE(t-t'))\exp\left(-\frac{1}{2}\Gamma|t-t'|\right) f(E) . \tag{3.185}$$

In this approximation, the retarded and advanced Green's functions $g^{R/A}$ follow immediately from their definition to become

$$g^{R/A}(t-t') = \mp i\Theta(\pm(t-t'))\exp\left\{-i\left(E \mp \frac{i}{2}\Gamma\right)(t-t')\right\} . \tag{3.186}$$

In the case of noninteracting particles, we have $\Gamma = 0$ and $\mathrm{Re}\Sigma^R = 0$. Then, the spectral function reduces to

$$a(\boldsymbol{p}\,\omega, \boldsymbol{R}\,t) = 2\pi\delta\left(\omega - \frac{p^2}{2m}\right). \tag{3.187}$$

Using the Hartree–Fock approximation (3.110), we still have $\Gamma = 0$, however the single-particle energy is now

$$E^{\mathrm{HF}}(\boldsymbol{p}, \boldsymbol{R}t) = \frac{p^2}{2m} + \Sigma^{\mathrm{HF}}(\boldsymbol{p}, \boldsymbol{R}t). \tag{3.188}$$

The interaction produces a shift of the single-particle energy determined by

$$\Sigma^{\mathrm{HF}}(\boldsymbol{p}, \boldsymbol{R}\,t) = n(\boldsymbol{R}\,\boldsymbol{t})V(0) \pm \int \frac{d\boldsymbol{p}'}{(2\pi)^3} V(\boldsymbol{p}-\boldsymbol{p}') f(\boldsymbol{p}, \boldsymbol{R}\,t). \tag{3.189}$$

The spectral function reads in Hartree–Fock approximation

$$a(\boldsymbol{p}\omega, \boldsymbol{R}t) = 2\pi\delta\left(\omega - E^{\mathrm{HF}}(\boldsymbol{p}, \boldsymbol{R}t)\right). \tag{3.190}$$

It turns out that, in the approximation of free particles and in the Hartree–Fock scheme, a particle with momentum $\boldsymbol{p}$ has exactly one possible energy, and the corresponding quasiparticle state has an infinite life time. In any other case, we have $\Gamma \neq 0$, and the state is damped. This means that, if we consider an interacting many-particle system, the energy spectrum is always sufficiently complex so that $a(\boldsymbol{p}\omega, \boldsymbol{R}t)$ appears to have no delta function shape, but is a continuous function of ω given by (3.183). We have a spread of energies for a given momentum. However, more or less sharp peaks are expected at the zeros of the first bracket in the denominator of (3.183). Under the condition $\Gamma \ll \mathrm{Re}\Sigma^R$, the peaks are very sharp, i.e., they represent coherent and long-living excitations. Because these excitations behave in many aspects like free particles, we call them *quasiparticles*.

Let us consider the case $\Gamma \ll \mathrm{Re}\Sigma^R$ in more detail. Then the spectral function may be expanded into a Taylor series with respect to Γ, i.e.,

$$a(\omega) = a(\omega)\Big|_{\Gamma=0} + \Gamma\frac{\partial}{\partial\Gamma}a(\omega)\Big|_{\Gamma=0} + \cdots. \tag{3.191}$$

The first r.h.s. contribution is just the quasiparticle approximation. For $\Gamma \to 0$, we get from (3.183)

$$a(\boldsymbol{p}\omega, \boldsymbol{R}t)\Big|_{\Gamma=0} = 2\pi\,\delta\left(\omega - \frac{p^2}{2m} - \mathrm{Re}\Sigma^R(\boldsymbol{p}\omega, \boldsymbol{R}t)\right). \tag{3.192}$$

If $E(\boldsymbol{p}, \boldsymbol{R}t)$ is the solution of the dispersion relation

$$E(\boldsymbol{p}, \boldsymbol{R}t) = \frac{\boldsymbol{p}^2}{2m} + \mathrm{Re}\Sigma^R(\boldsymbol{p}\omega, \boldsymbol{R}t)\Big|_{\omega=E(\boldsymbol{p},\boldsymbol{R}t)}, \tag{3.193}$$

we can write

$$a(\boldsymbol{p}\omega, \boldsymbol{R}t)\Big|_{\Gamma=0} = Z \cdot 2\pi\,\delta\left(\omega - E(\boldsymbol{p}, \boldsymbol{R}t)\right) , \tag{3.194}$$

where we used the abbreviation

$$Z^{-1}(\boldsymbol{pR}, t) = 1 - \frac{\partial}{\partial\omega}\mathrm{Re}\Sigma^R(\boldsymbol{p}\omega, \boldsymbol{R}t)|_{\omega=E(\boldsymbol{p},\boldsymbol{R}t)} \,. \tag{3.195}$$

The quantity Z is usually called re-normalization factor.

It is obvious that in the case $\Gamma \to 0$, the spectral function is still described by a δ-function similar to the ideal and Hartree–Fock cases. In this way, $E(\boldsymbol{p}, \boldsymbol{R}t)$ is the energy of quasiparticles.

For the second r.h.s. term of (3.191), we apply

$$\Gamma\,\frac{\partial}{\partial\Gamma}a(\omega)\Big|_{\Gamma=0} = -\,\Gamma(\boldsymbol{p}\omega, \boldsymbol{R}t)\,\frac{\partial}{\partial\omega}\,\frac{\mathcal{P}}{\omega - E(\boldsymbol{p}, \boldsymbol{R}t)} \,. \tag{3.196}$$

In the expansion (3.191), it is not necessary to retain the re-normalization factor (denominator of (3.181)) completely; instead we linearize (3.181) and apply the dispersion relation (3.90). We then get from (3.191) for the spectral function

$$\begin{aligned} a(\boldsymbol{p}\omega, \boldsymbol{R}t) &= 2\pi\,\delta\left(\omega - E(\boldsymbol{p}, \boldsymbol{R}t)\right)\left\{1 + \frac{\partial}{\partial\omega}\mathcal{P}\int\frac{d\bar{\omega}}{2\pi}\,\frac{\Gamma(\boldsymbol{p}\bar{\omega}, \boldsymbol{R}t)}{\omega - \bar{\omega}}\Big|_{\omega=E}\right\} \\ &- \Gamma(\boldsymbol{p}\omega, \boldsymbol{R}t)\,\frac{\partial}{\partial\omega}\,\frac{P}{\omega - E(\boldsymbol{p}, \boldsymbol{R}t)} \,. \end{aligned} \tag{3.197}$$

We notice that (3.197) fulfills the sum rule (3.65); however, this statement only holds if the re-normalization factor of (3.194) was taken in the linearized form. The approximation (3.197) is called the *extended quasiparticle approximation* and was discussed in many papers (Stolz and Zimmermann 1979; Kremp (c) et al. 1984; Köhler 1995).

The sum rule cannot be fulfilled if the spectral function is used in the form (3.194). Therefore, the spectral function is often used in the approximation given by

$$a(\boldsymbol{p}\omega, \boldsymbol{R}t) = 2\pi\,\delta\left(\omega - E(\boldsymbol{p}, \boldsymbol{R}t)\right) . \tag{3.198}$$

Let us come back to the full spectral function (3.183). Its behavior is essentially characterized by the poles $z = E \mp i\Gamma/2$ in the complex energy plane. The poles of $a(z)$ in the lower half plane are poles of $g^R(z)$, too. Correspondingly, the poles in the upper half plane are poles of $g^A(z)$. For the determination of the poles, we have the dispersion relation

$$\omega - \left[\frac{p^2}{2m} + \mathrm{Re}\Sigma^R(\boldsymbol{p}\omega, \boldsymbol{R}t) \mp \frac{i}{2}\Gamma(\boldsymbol{p}\omega, \boldsymbol{R}t)\right] = 0 \,. \tag{3.199}$$

Here, the real part of the retarded self-energy is given by the Hartree–Fock term and the additional correlation part, i.e., $\mathrm{Re}\Sigma^R = \Sigma^{\mathrm{HF}} + \mathrm{Re}\Sigma^R_{cor}$. In the case $\Gamma \ll \mathrm{Re}\Sigma^R$, the location of the poles is approximately determined by

$$E(\boldsymbol{p}, \boldsymbol{R}t) = \frac{p^2}{2m} + \mathrm{Re}\Sigma^R(\boldsymbol{p}\omega, \boldsymbol{R}t)\Big|_{\omega=E(\boldsymbol{p},\boldsymbol{R}t)}, \tag{3.200}$$

$$\Gamma(\boldsymbol{p}, \boldsymbol{R}t) = -2\mathrm{Im}\Sigma^R(\boldsymbol{p}\omega, \boldsymbol{R}t)\Big|_{\omega=E(\boldsymbol{p},\boldsymbol{R}t)}. \tag{3.201}$$

These two equations determine the shape of the spectral function and, therefore, the properties of the quasiparticles. The energy of quasiparticles $E(\boldsymbol{p})$ determines the position of the maximum, and $\Gamma(\boldsymbol{p})$ describes the width of the peak, respectively. This quasiparticle energy is determined by the kinetic energy and by an additional energy shift which has its origin in the interaction of the single particle with the surrounding particles of the system. The latter is given by the real part of the retarded self-energy, i.e.,

$$\Delta(\boldsymbol{p}, \boldsymbol{R}t) = \mathrm{Re}\Sigma^R(\boldsymbol{p}\omega, \boldsymbol{R}t)\Big|_{\omega=E(\boldsymbol{p},\boldsymbol{R}t)}. \tag{3.202}$$

The quantity $\Gamma(\boldsymbol{p})$ determines the lifetime or the damping of the quasiparticles, respectively. We have to remember that the quasiparticle shift $\Delta(\boldsymbol{p})$ and quasiparticle damping $\Gamma(\boldsymbol{p})$ are not independent of each other but are related by the dispersion relation (3.199).

In many cases, it is more convenient to replace the momentum dependence of the quasiparticle shift $\Delta(\boldsymbol{p})$ by a momentum independent shift Δ. We call this procedure *rigid shift approximation* (Zimmermann 1988). For the determination of the rigid shift Δ we start from the spectral function

$$a(\boldsymbol{p}\omega, \boldsymbol{R}t) = 2\pi\,\delta\left(\omega - \frac{p^2}{2m} - \Delta(\boldsymbol{p}, \boldsymbol{R}t)\right).$$

Furthermore, we consider the system to be in thermodynamic equilibrium. According to (3.68) and (3.35), the expression for the number density then reads

$$\begin{aligned} n(\mu, T) &= \int \frac{d\boldsymbol{p}}{(2\pi)^3}\frac{d\omega}{(2\pi)} a(\boldsymbol{p}\,\omega) f(\omega - \mu) \\ &= \int \frac{d\boldsymbol{p}}{(2\pi)^3} \frac{1}{e^{\beta(\frac{p^2}{2m}+\Delta(\boldsymbol{p})-\mu)} \mp 1}. \end{aligned} \tag{3.203}$$

The rigid shift Δ is determined such that the expression for the density gives the same value if we use Δ instead of $\Delta(\boldsymbol{p})$ in (3.203), i.e.,

$$\begin{aligned} n(\mu, T) &= \int \frac{d\boldsymbol{p}}{(2\pi)^3} f\left(\frac{p^2}{2m} + \Delta(p) - \mu\right) \\ &= \int \frac{d\boldsymbol{p}}{(2\pi)^3} f\left(\frac{p^2}{2m} + \Delta - \mu\right). \end{aligned}$$

In first order with respect to the difference $(\Delta(\boldsymbol{p}) - \Delta)$, we find

$$\Delta = \frac{\int \frac{d\boldsymbol{p}}{(2\pi)^3} \, \mathrm{Re}\Sigma^R(\boldsymbol{p}, E(\boldsymbol{p})) \, \frac{\partial}{\partial \mu^{\mathrm{id}}} f(\frac{p^2}{2m} - \mu^{\mathrm{id}})}{\int \frac{d\boldsymbol{p}}{(2\pi)^3} \frac{\partial}{\partial \mu^{\mathrm{id}}} f(\frac{p^2}{2m} - \mu^{\mathrm{id}})} \,. \tag{3.204}$$

Here, is $E(\boldsymbol{p}) = \frac{p^2}{2m} + \Delta$, and we used for the chemical potential

$$\mu^{\mathrm{id}} = \mu - \Delta \,.$$

This means that, in rigid shift approximation, the quasiparticle shift Δ is equal to the interaction part μ^{int} of the chemical potential, i.e.,

$$\mu = \mu^{\mathrm{id}} + \mu^{\mathrm{int}} = \mu^{\mathrm{id}} + \Delta \,. \tag{3.205}$$

This result is in agreement with thermodynamic calculations of μ applying the *incomplete inversion* of $n(\mu) = n(\mu^{\mathrm{id}} + \mu^{\mathrm{int}})$ (Ebeling (a) et al. 1976). A simple expression for Δ based on a more elementary theory was given in Sect. 2.5. For the charged plasma particles of species a, the shift was found to be

$$\Delta_a = -\frac{\kappa_a e_a^2}{2} \,.$$

Let us now derive a kinetic equation for quasiparticles. For this purpose, we start from the Kadanoff–Baym equations (3.175) in first order gradient expansion. In the quasiparticle approximation with $\Gamma \to 0$, we can neglect the second Poisson brackets on the l.h.s. of (3.175). Then follows

$$\begin{aligned} &\left(1 - \frac{\partial}{\partial \omega} \mathrm{Re}\Sigma^R\right) \frac{\partial}{\partial t} \, i g^{\gtrless} + \nabla_{\boldsymbol{p}} \left(\frac{p^2}{2m} + \mathrm{Re}\Sigma^R\right) \nabla_{\boldsymbol{R}} \, i g^{\gtrless} \\ &+ \frac{\partial}{\partial t} \mathrm{Re}\Sigma^R \frac{\partial}{\partial \omega} \, i g^{\gtrless} - \nabla_{\boldsymbol{R}} \mathrm{Re}\Sigma^R \, \nabla_{\boldsymbol{p}} \, i g^{\gtrless} = g^{<} \Sigma^{>} - g^{>} \Sigma^{<} \,. \end{aligned} \tag{3.206}$$

This equation determines the two-time correlation functions $g^{\gtrless}$ and, therefore, all dynamical and statistical properties of a non-equilibrium system of quasiparticles. In many cases, however, it is more convenient to describe the quasiparticle system in terms of a one-particle distribution function. For this purpose we need the connection between the correlation function and the distribution function. In equilibrium, this connection is given by the spectral representation (3.63). For non-equilibrium situations, this leads to a complicated problem known as the reconstruction problem: One has to express $g^{\gtrless}$ in terms of the distribution function (Lipavský et al. 1986; Bornath et al. 1996). Here, we will solve this problem in the frame of an ansatz originally introduced by Kadanoff and Baym (1962). It represents a formal generalization of the spectral representation (3.63) to non-equilibrium systems. Generalizations of this ansatz will be discussed in Sect. 7.2. We write in quasiparticle approximation

$$\begin{aligned} \pm i g^{<}(\boldsymbol{p}\omega, \boldsymbol{R}t) &= a(\boldsymbol{p}\omega, \boldsymbol{R}t) \, f(\boldsymbol{p}, \boldsymbol{R}t) \,, \\ i g^{>}(\boldsymbol{p}\omega, \boldsymbol{R}t) &= a(\boldsymbol{p}\omega, \boldsymbol{R}t) \, [1 \pm f(\boldsymbol{p}, \boldsymbol{R}t)] \,, \end{aligned} \tag{3.207}$$

where the spectral function is given by the expression (3.192) and f denotes the distribution function of quasiparticles with the energy $E(\boldsymbol{p}, \boldsymbol{R}t)$ determined by the relation (3.193). Now we use the ansatz (3.207) in the equation (3.206) for $g^<$. The δ-function in (3.192) leads to the quasiparticle energy. Then $\mathrm{Re}\Sigma^R$ does not depend on the variables $\boldsymbol{R}$ and $\boldsymbol{p}$ explicitly only but also implicitly via $\omega = E(\boldsymbol{p}, \boldsymbol{R}t)$. Finally, one gets (Danielewicz 1984)

$$\left(\frac{\partial}{\partial t} + \nabla_{\boldsymbol{p}} E(\boldsymbol{p}, \boldsymbol{R}t)\, \nabla_{\boldsymbol{R}} - \nabla_{\boldsymbol{R}} E(\boldsymbol{p}, \boldsymbol{R}t)\, \nabla_{\boldsymbol{p}}\right) f(\boldsymbol{p}, \boldsymbol{R}t)$$
$$= Z(\boldsymbol{p}, \boldsymbol{R}t) \Big\{ - f(\boldsymbol{p}, \boldsymbol{R}t)\, i\Sigma^>(\boldsymbol{p}\omega, \boldsymbol{R}t)|_{\omega=E(\boldsymbol{p},\boldsymbol{R}t)}$$
$$\pm [1 \pm f(\boldsymbol{p}, \boldsymbol{R}t)]\, i\Sigma^<(\boldsymbol{p}\omega, \boldsymbol{R}t)|_{\omega=E(\boldsymbol{p},\boldsymbol{R}t)} \Big\} \tag{3.208}$$

with $Z(\boldsymbol{p}, \boldsymbol{R}t)$ given by (3.195). The role of this kinetic equation in connection with (3.175) and the reconstruction problem is discussed in detail by Bornath et al. (1996).

A simplification of the kinetic equation follows for $Z(\boldsymbol{p}, \boldsymbol{R}t) = 1$. This corresponds to use the spectral function (3.198) which fulfills the sum rule and neglecting the ω-dependence of $\mathrm{Re}\Sigma^R$. Then the relations (3.207) take the form

$$\begin{aligned} \pm i g^<(\boldsymbol{p}\omega, \boldsymbol{R}t) &= 2\pi\, \delta(\omega - E(\boldsymbol{p}, \boldsymbol{R}t))\, f(\boldsymbol{p}, \boldsymbol{R}t)\,, \\ i g^>(\boldsymbol{p}\omega, \boldsymbol{R}t) &= 2\pi\, \delta(\omega - E(\boldsymbol{p}, \boldsymbol{R}t))\, [1 \pm f(\boldsymbol{p}, \boldsymbol{R}t)]\,. \end{aligned} \tag{3.209}$$

In the following, these relations are called *Kadanoff–Baym ansatz* (KBA). Here, the single particle energy is $E(\boldsymbol{p}, \boldsymbol{R}t) = p^2/2m + \mathrm{Re}\Sigma(\boldsymbol{p}, \boldsymbol{R}t)$.

The kinetic equation (3.208) represents a generalized version of the Landau–Silin equation. It gives the quantum statistical foundation of the semi-phenomenological theory proposed by Landau to treat degenerate Fermi systems using the concept of quasiparticles.

In spite of the approximations involved, the kinetic equation (3.208) is still a rather general one. The l.h.s. describes the drift of quasiparticles accounting for medium effects in the energies $E(\boldsymbol{p}, \boldsymbol{R}t)$. The r.h.s. represents a general collision integral in terms of the self-energy functions $\Sigma^{\gtrless}$. The latter can be interpreted as scattering-out and scattering-in rates to be taken for the energy $\omega = E(\boldsymbol{p}, \boldsymbol{R}t)$. Appropriate approximations for the $\Sigma^{\gtrless}$ lead to the well-known kinetic equations of Landau, Boltzmann, and Lenard–Balescu. This will be shown later in Chap. 7.

3.4.2 Description of Macroscopic Quantities

The macroscopic properties of a plasma described by thermodynamic functions and transport quantities are determined by appropriate statistical averages. All the information needed for the explicit calculation of such averages are contained in correlation or Green's functions.

We will demonstrate this considering the number density, the mean value of energy and the equation of state. According to the definition of the correlation functions, the number density at $\boldsymbol{R}, t$ is given by

$$\begin{aligned} n(\boldsymbol{R},t) &= \int \frac{d\boldsymbol{p}}{(2\pi)^3} d\boldsymbol{r} \left\langle \psi^\dagger(\boldsymbol{R}-\frac{\boldsymbol{r}}{2},t)\,\psi(\boldsymbol{R}+\frac{\boldsymbol{r}}{2},t)\right\rangle \exp(-i\boldsymbol{p}\boldsymbol{r}) \\ &= \pm i \int \frac{d\boldsymbol{p}}{(2\pi)^3}\frac{d\omega}{2\pi} g^<(\boldsymbol{p}\omega,\boldsymbol{R}t)\,. \end{aligned} \quad (3.210)$$

If the Wigner distribution function is introduced by means of (3.35), we get

$$n(\boldsymbol{R},t) = \int \frac{d\boldsymbol{p}}{(2\pi)^3} f(\boldsymbol{p},\boldsymbol{R}t)\,. \quad (3.211)$$

Thus, the determination of the number density demands the knowledge of the single-particle correlation function $g^<$, or the knowledge of the Wigner distribution function, respectively. In the general case of non-equilibrium, these quantities have to be calculated from a kinetic equation of the form given by (3.175) or (3.180). In thermodynamic equilibrium, the situation is much simpler. The correlation function does not depend on the variables $\boldsymbol{R}$ and t. Moreover, the spectral representation is valid

$$\pm i g^<(\boldsymbol{p},\omega) = a(\boldsymbol{p},\omega)\, f(\omega)\,, \quad (3.212)$$

which is exact in thermodynamic equilibrium. With (3.212), for the number density, we get

$$n(\mu,T) = \int \frac{d\boldsymbol{p}}{(2\pi)^3}\frac{d\omega}{2\pi}\, a(\boldsymbol{p},\omega)\, f(\omega)\,, \quad (3.213)$$

where μ is the chemical potential and T is the temperature. $f(\omega)$ is a Bose- or Fermi-like distribution function, i.e.,

$$f(\omega) = \frac{1}{e^{\beta(\omega-\mu)} \mp 1}\,.$$

In thermodynamic equilibrium, the spectral function is given by

$$a(\boldsymbol{p},\omega) = \frac{\Gamma(\boldsymbol{p},\omega)}{\left(\omega - \frac{p^2}{2m} - \mathrm{Re}\Sigma^R(\boldsymbol{p},\omega)\right)^2 + \left(\frac{1}{2}\Gamma(\boldsymbol{p},\omega)\right)^2}\,. \quad (3.214)$$

We remember that $\mathrm{Re}\Sigma^R$ and $\Gamma = -2\mathrm{Im}\Sigma^R$ are not independent but connected by the dispersion relation(3.155). The shape of the statistical expression for the density given by (3.213) is similar to that of ideal particles. However, in interacting many-particle systems, we have a complicated distribution of the energy for a given momentum. This is accounted for by the spectral function $a(p,\omega)$ in (3.213).

In this way, the thermodynamic properties of an interacting many-particle system are determined by (3.213) and thus by the spectral function. By inversion of (3.213), we may determine the chemical potential $\mu = \mu(n, T)$. Further thermodynamic quantities may be deduced from the density as a function of the chemical potential. The pressure or the equation of state of the plasma follows from the relation

$$p(\mu, T) = \int_{-\infty}^{\mu} n(\mu', T)\, d\mu' \,. \tag{3.215}$$

The relations for the thermodynamic properties become especially simple if we use the quasiparticle approximation. According to the discussion given in Sect. 3.4.1, we have in lowest order from (3.197)

$$a(\boldsymbol{p}, \omega) = 2\pi\delta(\omega - E(\boldsymbol{p}))\,, \tag{3.216}$$

where the quasiparticle energy $E(\boldsymbol{p})$ follows from the solution of the dispersion relation (3.193). Insertion of (3.216) into (3.213) for the density yields

$$n(\mu, T) = \int \frac{d\boldsymbol{p}}{(2\pi)^3} \frac{1}{e^{\beta[E(\boldsymbol{p})-\mu]} \mp 1} \tag{3.217}$$

and for the equation of state

$$p(\mu, T) = k_B T \int \frac{d\boldsymbol{p}}{(2\pi)^3} \ln\left(1 \mp e^{-\beta[E(\boldsymbol{p})-\mu]}\right) . \tag{3.218}$$

In quasiparticle approximation, all thermodynamic properties are thus reduced to Bose/Fermi type integrals, if the quasiparticle energy is known. Equations (3.217) and (3.218) once again show the physical content of the quasiparticle concept. For $E(\boldsymbol{p}) = \frac{p^2}{2m}$, they coincide with the equations for ideal particles considered in Chap. 2. In the quasiparticle approximation, the interacting many-particle system is replaced by a system of independent quasiparticles. The interaction is, approximately, condensed in the self-energy correction to the free particle energy. For this reason, the quasiparticle approximation is useful and popular. The simplest approximation is the Hartree–Fock approximation, i.e., $E(\boldsymbol{p}) = E^{\mathrm{HF}}(\boldsymbol{p})$.

In the general case, the set (3.213) and (3.214) represent a complicated scheme. In particular, we have to determine the full self-energy, and the many-particle system may no longer be considered to be a system of noninteracting (quasi-) particles. This problem will be dealt with in Chap. 6. Here, we mention only the simplification which is possible if $\Gamma(\boldsymbol{p}, \omega)$ is small.

Using the expansion (3.197) in (3.213), the number density is given by

$$n(\mu, T) = \int \frac{d\boldsymbol{p}}{(2\pi)^3} f\left(E(\boldsymbol{p})\right) - \int \frac{d\boldsymbol{p}}{(2\pi)^3} \frac{d\omega}{2\pi}$$
$$\times \left\{ \Gamma(\boldsymbol{p}, \omega) \left[f(\omega) - f\left(E(\boldsymbol{p})\right) \right] \frac{\partial}{\partial \omega} \frac{\mathcal{P}}{\omega - E(\boldsymbol{p})} \right\} . \tag{3.219}$$

This equation is easily interpreted. The first r.h.s. term accounts for the contribution of independent quasiparticles while the second r.h.s. term describes the interaction between the quasiparticles, i.e., scattering processes, bound states, etc.

A further essential macro-physical quantity of strongly correlated plasmas is the average potential energy of the interaction. The operator of the interaction energy is a binary operator, the mean value of which $\langle V(t)\rangle$ has to be determined from

$$\begin{aligned}\langle V(t)\rangle &= \frac{1}{2}\int d\boldsymbol{r}_1 d\boldsymbol{r}_2 V(\boldsymbol{r}_1-\boldsymbol{r}_2)\\ &\quad\times\left\langle\psi^\dagger(\boldsymbol{r}_1,t)\,\psi^\dagger(\boldsymbol{r}_2,t)\,\psi(\boldsymbol{r}_2,t)\,\psi(\boldsymbol{r}_1,t)\right\rangle . \end{aligned}\tag{3.220}$$

This mean value may also be determined by the two-time single-particle correlation function $g^<(1,1')$. To show this we start from the equation of motion for $\psi(\boldsymbol{r},t)$ given by (3.96) and from the adjoint equation. Then one easily finds (Baym and Kadanoff 1961)

$$\begin{aligned}\langle V(t)\rangle &= \pm\frac{i}{4}\int d\boldsymbol{r}_1\left\{i\frac{\partial}{\partial t_1}-i\frac{\partial}{\partial t_1'}+\frac{\nabla_1^2}{2m}+\frac{\nabla_{1'}^2}{2m}\right\}\\ &\quad\times g^<(\boldsymbol{r}_1t_1,\boldsymbol{r'}_1t_1')\Big|_{\boldsymbol{r}_1=\boldsymbol{r}_1',t_1=t_1'} ,\end{aligned}\tag{3.221}$$

and after Fourier transformation, the corresponding energy density reads

$$\langle V(\boldsymbol{R},t)\rangle = \pm i\int\frac{d\boldsymbol{p}}{(2\pi)^3}\frac{d\omega}{2\pi}\frac{\omega-\frac{\boldsymbol{p}^2}{2m}}{2}\,g^<(\boldsymbol{p}\omega,\boldsymbol{R}t)\,. \tag{3.222}$$

In thermodynamic equilibrium, we may use the spectral representation (3.212), and we get

$$\frac{\langle \mathrm{V}\rangle}{\Omega} = \int\frac{d\boldsymbol{p}}{(2\pi)^3}\frac{d\omega}{2\pi}\frac{\omega-\frac{\boldsymbol{p}^2}{2m}}{2}\,a(\boldsymbol{p},\omega)f(\omega)\,. \tag{3.223}$$

Here, Ω is the volume of the system.

In quasiparticle approximation, the spectral function is given by (3.197). If the lowest order expression (3.216) is used, the result is in the case of thermodynamic equilibrium

$$\langle \mathrm{V}\rangle = \frac{\Omega}{2}\int\frac{d\boldsymbol{p}}{(2\pi)^3}\mathrm{Re}\Sigma^R(\boldsymbol{p},E(\boldsymbol{p}))\,f(E(\boldsymbol{p}))\,. \tag{3.224}$$

We remind that there is a spin sum in (3.224) which is dropped for simplicity. Of course, in (3.223), we may also apply the Γ-expansion of the spectral function given by (3.197).

It is known that the mean value of the potential energy is related to the equation of state via the charging formula

$$k_BT\left\{Z_G - Z_G^{\mathrm{id}}\right\} = -\int_0^1 \frac{d\lambda}{\lambda}\langle\lambda\mathrm{V}\rangle_\lambda$$
$$= p\Omega - p^{\mathrm{id}}\Omega\,, \qquad (3.225)$$

where Z_G denotes the grand canonical partition function. Using the expression (3.223) for $\langle\lambda\mathrm{V}\rangle_\lambda$, we get the charging formula already used by Debye to calculate the equation of state. With (3.223), we then have

$$\left(p - p^{\mathrm{id}}\right)\Omega = -\Omega\int_0^1 \frac{d\lambda}{\lambda}\int\frac{d\boldsymbol{p}}{(2\pi)^3}\frac{d\omega}{2\pi}\,\frac{\omega - \frac{\boldsymbol{p}^2}{2m}}{2}\,a_\lambda(\boldsymbol{p},\omega)\,f(\omega)\,. \qquad (3.226)$$

Here, the subscript λ reminds us of the fact that the coupling parameter is still an integration variable (for Coulomb systems this means $e^2 \to \lambda e^2$).

Starting from (3.226), the equation of state is obtained as a function of temperature and chemical potential. The equation of state in the variables n and T follows applying the relation

$$n\left(\mu, T\right) = \frac{\partial}{\partial\mu}p\left(\mu, T\right)\Big|_{T,\Omega}\,. \qquad (3.227)$$

We notice that in multi-component systems both the pressure and the density depend on the chemical potentials of all species. In this way, the relations (3.226) and (3.227) are the appropriate starting point to calculate the thermodynamic properties of plasmas.

Let us now consider the question of how to determine the mean value of the potential energy for systems in non-equilibrium. For this purpose, we start from the expression (3.221). It can be reformulated using the equations of motion (3.145) and (3.146) for the correlation function $g^<(1,1')$, and thus, (3.221) is expressed in terms of the self-energies, i.e.,

$$\langle\mathrm{V}\rangle - \langle\mathrm{V}\rangle^{\mathrm{HF}} = \pm i\frac{1}{4}\int_{t_0}^{t_1} d\bar{1}\Big\{g^>(1,\bar{1})\Sigma^<(\bar{1},1') - g^<(1,\bar{1})\Sigma^>(\bar{1},1')$$
$$+\Sigma^>(1,\bar{1})g^<(\bar{1},1') - \Sigma^<(1,\bar{1})g^>(\bar{1},1')\Big\}\Big|_{1=1'}\,. \qquad (3.228)$$

In this equation, we have again integrals of the type

$$I(x_1, x_1') = \int d\bar{x}_1 f\left(x_1, \bar{x}_1\right) u\left(\bar{x}_1, x_1'\right)\Big|_{x_1 = x_1'}\,. \qquad (3.229)$$

Following the line performed in Sect. 3.4.2, we get with the additional condition $x_1 = x_1'$,

$$I(X) = \int d\bar{x}\, f\left(-\bar{x}, X + \frac{\bar{x}}{2}\right) u\left(\bar{x}, X + \frac{\bar{x}}{2}\right) . \tag{3.230}$$

Consequently, the correlation part of the mean potential energy reads

$$\begin{aligned}
&\langle \mathrm{V}(t)\rangle^{\mathrm{corr}} = \langle \mathrm{V}(t)\rangle - \langle \mathrm{V}(t)\rangle^{\mathrm{HF}} \\
= \; &\pm i\frac{1}{4} \int\limits_{t_0-t_1}^{0} d\tau \Big\{ g^{>}(-\tau, t+\frac{\tau}{2})\, \Sigma^{<}(\tau, t+\frac{\tau}{2}) - g^{<}(-\tau, t+\frac{\tau}{2})\, \Sigma^{>}(\tau, t+\frac{\tau}{2}) \\
- \; &\Sigma^{<}(-\tau, t+\frac{\tau}{2})\, g^{>}(\tau, t+\frac{\tau}{2}) + \Sigma^{>}(-\tau, t+\frac{\tau}{2})\, g^{<}(\tau, t+\frac{\tau}{2}) \Big\} .
\end{aligned} \tag{3.231}$$

Here, only the time variables are written explicitly. With the expression (3.231), the temporal evolution of the correlation energy $\langle \mathrm{V}(t)\rangle^{\mathrm{corr}}$, in terms of the self-energies and the single-particle correlation functions, is given. It should be noticed that $\langle \mathrm{V}(t)\rangle^{\mathrm{corr}}$ at time t is determined by the complete history of the system since we have to integrate from $t_0 - t_1$ to 0. This means that we have a temporal evolution including memory effects.

Now we consider a τ-expansion of (3.231) which is of relevance for long periods of time. The result reads in the case $t_0 = -\infty$

$$\begin{aligned}
\langle \mathrm{V}(t)\rangle^{\mathrm{corr}} &= \pm i\frac{1}{4} \sum_{n=0}^{\infty} \frac{1}{n!} \frac{d^n}{dt^n} \int\limits_{-\infty}^{0} d\tau \left(\frac{\tau}{2}\right)^n \Big\{ g^{>}(-\tau, t)\, \Sigma^{<}(\tau, t) \\
&- g^{<}(-\tau, t)\, \Sigma^{>}(\tau, t) - \Sigma^{<}(-\tau, t)\, g^{>}(\tau, t) + \Sigma^{>}(-\tau, t)\, g^{<}(\tau, t) \Big\} .
\end{aligned} \tag{3.232}$$

After Fourier transformation with respect to the variable τ, we arrive at

$$\begin{aligned}
\langle \mathrm{V}(t)\rangle^{\mathrm{corr}} &= \pm i\frac{1}{4} \sum_{n=0}^{\infty} \frac{1}{n!} \frac{d^n}{dt^n} \int \frac{d\omega}{2\pi} \frac{d\bar{\omega}}{2\pi} \int\limits_{-\infty}^{0} d\tau \left(\frac{\tau}{2}\right)^n \\
&\times \Big\{ e^{i(\omega-\bar{\omega})\tau} - e^{-i(\omega-\bar{\omega})\tau} \Big\} \Big\{ \Sigma^{>}(\omega, t) g^{<}(\bar{\omega}, t) - \Sigma^{<}(\omega, t) g^{>}(\bar{\omega}, t) \Big\} .
\end{aligned} \tag{3.233}$$

In a further approximation, we will restrict ourselves to the lowest order contribution, and we use the principal value relation

$$P\left(\frac{1}{x}\right) = \int\limits_{0}^{\infty} \sin kx \, dk . \tag{3.234}$$

The expression for the mean value of the potential energy then takes the form

$$\langle \mathrm{V}(t)\rangle^{\mathrm{corr}} = \pm\frac{1}{2}\int \frac{d\omega}{2\pi}\frac{d\bar{\omega}}{2\pi}\frac{P}{\omega-\bar{\omega}}\Big\{\Sigma^{>}(\omega,t)g^{<}(\bar{\omega},t) - \Sigma^{<}(\omega,t)g^{>}(\bar{\omega},t)\Big\}\,. \tag{3.235}$$

This formula determines the long time behavior of the correlation energy in non-equilibrium systems. It turns out that, in the lowest order in τ, the equilibrium behavior is determined by the same expression.

It is possible to find other useful expressions for the mean value of the potential energy. In order to do this, we rewrite (3.235) into

$$\begin{aligned} \langle \mathrm{V}(t)\rangle^{\mathrm{corr}} &= \pm\frac{1}{2}\int \frac{d\omega}{2\pi}\frac{d\bar{\omega}}{2\pi}\frac{P}{\omega-\bar{\omega}}\Big\{\Big(\Sigma^{>}(\omega,t) - \Sigma^{<}(\omega,t)\Big)g^{<}(\bar{\omega},t) \\ &- \Sigma^{<}(\omega,t)\Big(g^{>}(\bar{\omega},t) - g^{<}(\bar{\omega},t)\Big)\Big\}\,. \end{aligned} \tag{3.236}$$

Now, we use the dispersion relations given by (3.90). Writing the variables $\boldsymbol{p}, \boldsymbol{R}$ explicitly, we get for the mean value of the total potential energy

$$\begin{aligned} \langle \mathrm{V}(\boldsymbol{R},t)\rangle &= \pm i\frac{1}{2}\int \frac{d\boldsymbol{p}}{(2\pi)^3}\frac{d\omega}{2\pi}\Big\{\mathrm{Re}\Sigma^{R}(\boldsymbol{p}\omega,\boldsymbol{R}t)g^{<}(\boldsymbol{p}\omega,\boldsymbol{R}t) \\ &+ \mathrm{Re}\; g^{R}(\boldsymbol{p}\omega,\boldsymbol{R}t)\Sigma^{<}(\boldsymbol{p}\omega,\boldsymbol{R}t)\Big\}\,. \end{aligned} \tag{3.237}$$

The question of conservation of the total energy was discussed by Kraeft et al. (1986), by Bornath et al. (1996) and by Kremp et al. (1997).

4. Systems with Coulomb Interaction

4.1 Screened Potential and Self-Energy

Up to now, we did not specify the binary interaction in the basic equations of quantum statistical theory. In this section, we want to come back to the main goal of this book, namely to the quantum statistical description of strongly coupled plasmas. The charged particles of a plasma interact via the Coulomb potential

$$V_{ab}(\boldsymbol{r}_1 - \boldsymbol{r}_2) = \frac{e_a e_b}{|\boldsymbol{r}_1 - \boldsymbol{r}_2|}\,, \tag{4.1}$$

where $e_a = Z_a e$ is the charge of species a with the charge number Z_a. The Hamiltonian of the plasma has the shape

$$\begin{aligned} \mathrm{H} &= \sum_a \int d\boldsymbol{r}\, \psi_a^\dagger(\boldsymbol{r},t) \left(-\frac{\hbar^2\nabla^2}{2m_a}\right) \psi_a(\boldsymbol{r},t) \\ &+ \frac{1}{2}\sum_{ab} \int d\boldsymbol{r} d\boldsymbol{r}' \psi_a^\dagger(\boldsymbol{r},t)\psi_b^\dagger(\boldsymbol{r}',t) V_{ab}\left(\boldsymbol{r}_1 - \boldsymbol{r}_2\right) \psi_b(\boldsymbol{r}',t)\psi_a(\boldsymbol{r}',t)\,. \end{aligned} \tag{4.2}$$

As already outlined in Chap. 2, many-particle systems with Coulomb interaction are characterized by a number of features resulting from the long range of the $1/r$ potential. Such long range potential causes that many particles always interact with each other simultaneously so that we have to expect collective effects. The most characteristic effect is the dynamical screening of the Coulomb potential which was discussed in its simplest version in Chap. 2. The screening leads to the fact that the interaction between two particles is modified by the surrounding particles. This is of great importance in statistical theory of charged many-particle systems.

Approximations for the determination of physical properties based on a cluster expansion in terms of two, three or few body clusters assuming bare Coulomb interaction lead to divergencies in the theory (Macke 1950; Mayer 1950).

A consequent many-particle theory of plasmas has to incorporate the screening of the Coulomb potential for physical reasons and, for formal reasons, in order to avoid divergencies. The basic equations of quantum statistics have to be reformulated such that the Coulomb potential is replaced by the screened potential (Bonch–Bruevich and Tyablikov 1961; DuBois 1968; Baym and Kadanoff 1961). In order to follow such a line we pick up the ideas of Chap. 3. According to such ideas an applied external field $U_a^{\rm ext}(11') = U_a(11')$ causes an inhomogeneous plasma, and thus a mean Coulomb field of all particles is produced. Consequently, we now have an effective potential given by (Kadanoff and Baym 1962)

$$\begin{aligned} U_a^{\rm eff}(11') &= U_a(11') + \Sigma_a^H(11') \\ &= U_a(11') \pm i \sum_b \int_C d2 V_{ab}(12) g_b(22^+) \delta(1-1') \,. \end{aligned} \tag{4.3}$$

This relation is defined on the Keldysh time contour to allow for a general description of equilibrium as well as non-equilibrium plasmas. Σ_a^H is the Hartree self-energy. The effective potential U_a^{eff} determines, in a decisive way, the properties and the behavior of a system with Coulomb interaction. It especially follows from the Dyson equation (3.137,3.141) that the single-particle Green's function functionally depends on U via $U^{\rm eff}$, i.e.,

$$g_a\left(11', U\right) = g_a\left(11', U^{\rm eff}[U]\right) \,. \tag{4.4}$$

Let us now consider the special features of the self-energy for plasmas. We start from the relation (3.136) and define the screened self-energy $\bar{\Sigma}_a(11')$ by

$$\int_C d\bar{1} \bar{\Sigma}_a(1\bar{1}) g_a(\bar{1}1') = \pm \sum_b \int_C d2 V_{ab}(12) L_{ab}(12, 1'2^+) \tag{4.5}$$

such that we have $\Sigma_a = \Sigma_a^H + \bar{\Sigma}_a$. The function L_{ab} is defined by (3.42).

The latter plays a central role for the determination of the self-energy.

Taking into account the U-dependence according to (4.4) and using the chain rule for functional differentiation we immediately get

$$L_{ab}(12, 1'2') = \pm i \frac{\delta g_a(11')}{\delta U_b(2'2)} = \sum_c \int_C d3 d3' \Pi_{ac}(13, 1'3') K_{bc}(23', 2'3) \,. \tag{4.6}$$

Here, two essential quantities were introduced which are of special relevance for systems with Coulomb interaction, namely the functional derivatives

$$\pm i \frac{\delta g_a(11')}{\delta U_b^{\rm eff}(2'2)} = \Pi_{ab}(12, 1'2') \,, \qquad \frac{\delta U_b^{\rm eff}(22')}{\delta U_a(1'1)} = K_{ab}(12, 1'2') \,, \tag{4.7}$$

which are referred to as polarization function and generalized dielectric response function, respectively. As will be discussed later, L_{ab} and K_{ab} give a measure of the response of the plasma due to the external fields, whereas Π_{ab} determines the response to the total (effective) one. Using the expression

(4.3) for the effective potential, we find an equation for the response function K_{ab}. Introducing this equation into (4.6), we get an interesting integral equation which relates Π_{ab} and L_{ab}

$$\begin{aligned} L_{ab}(12,1'2') &= \Pi_{ab}(12,1'2') \\ &+ \sum_{cd} \int_C d3d4 \Pi_{ac}(13,1'3^+) V_{cd}(34) L_{db}(42,4^+2') \,. \end{aligned} \quad (4.8)$$

Let us come back to the screened self-energy. If we insert (4.6) into the relation (4.5), we have

$$\int_C d\bar{1} \Sigma_a(1\bar{1}) g_a(\bar{1}1') = \pm \sum_{bc} \int_C d2d3d3' V_{ab}(12) K_{bc}(23', 2^+3) \Pi_{ac}(13,1'3') \,. \quad (4.9)$$

In this equation, we only need the special response function $K_{bc}(23', 2^+3)$. For this special case, we find from the integral equation for the more general function $K_{ab}(12,1'2')$

$$K_{ab}(12,1^+2') = K_{ab}(12)\delta(2-2') \quad ; \; K_{ab}(12) = \int d2' K_{ab}(12,1^+2') \,, \quad (4.10)$$

and we get

$$K_{ab}(12) = \delta(1-2)\delta_{ab} + \sum_c \int_C d3 L_{ac}(13,1^+3^+) V_{cb}(32) \,. \quad (4.11)$$

Then it is obvious to introduce a time dependent effective interaction potential between two particles located at $\boldsymbol{r}_1$ and $\boldsymbol{r}_2$:

$$V^s_{ab}(12) = \sum_c \int_C d3 V_{ac}(13) K_{cb}(32) \,. \quad (4.12)$$

With (4.12) and (4.11), we can write the relation for the screened self-energy in the form

$$\int_C d\bar{1}\, \Sigma_a(1\bar{1}) g_a(\bar{1}1') = \pm \sum_b \int_C d2\, V^s_{ab}(12) \Pi_{ab}(12,1'2^+) \,. \quad (4.13)$$

In this way, an essential result was derived. On behalf of (4.13), the self-energy Σ_a is determined by the screened potential $V^s_{ab}(12)$. Thus, the Dyson equation can be written in terms of V^s_{ab} and the Coulomb potential is eliminated from theory. In all equations, only the dynamically screened potential $V^s_{ab}(12)$ appears.

The screened potential is a fundamental quantity in the theory of plasmas. This was already discussed in Chap. 2 on an elementary level of description. With (4.12) and (4.11), the correct quantum statistical definition of this quantity is given, according to which, $V^s_{ab}(12)$ is the Coulomb potential screened by the dielectric response function $K_{ab}(12)$ being nonlocal in space and time. We may derive an integral equation for the direct calculation of V^s_{ab}. For this purpose, we start from the definition (4.12) in combination with the integral equation (4.11) for the two-point dielectric response function. Furthermore, we consider (4.8) and get the relation

$$\sum_d \int_C d3 L_{ad}(13, 1^+3^+) V_{db}(32) = \sum_d \int_C d3 \Pi_{ad}(13, 1^+3^+) V^s_{db}(32) . \quad (4.14)$$

Then the integral equation for V^s_{ab} is obtained having the shape

$$V^s_{ab}(12) = V_{ab}(12) + \sum_{cd} \int_C d3d4 V_{ac}(13) \Pi_{cd}(34, 3^+4^+) V^s_{db}(42) . \quad (4.15)$$

This important equation of plasma physics is usually referred to as *screening equation*. It allows for the following interpretation. The first r.h.s. term $V_{ab}(12)$ describes the bare Coulomb interaction between particles of species a and b. The second r.h.s. term accounts for the change of the Coulomb interaction as a consequence of the surrounding particles. The latter contribution leads to the fact that the potential V^s_{ab} becomes time dependent, and the range of interaction is reduced. For this reason, the potential V^s_{ab} is called the dynamically screened potential.

4.2 General Response Functions

From the basic equations of the previous subsection we found that the response behavior and the collective effects of screening are determined by the functions L_{ab}, K_{ab}, and Π_{ab} defined in a general way on the Keldysh time contour. For such reasons, it is worthwhile to deal with these functions in more detail. Let us first derive equations for the determination of these functions. Starting from (4.5), and using the chain rule of functional derivation, we get the following general Bethe–Salpeter equation for the function $L(12, 1'2')$:

$$\begin{aligned} L_{ab}(12, 1'2') &= \pm i\, g_a(12') g_a(21') \\ &+ \sum_c \int_C d3d3'd4d4'\, g_a(13') g_a(31') \Xi_{ac}(3'4', 34) L_{cb}(42, 4'2') , \end{aligned} \quad (4.16)$$

where the initial correlation term was neglected. The quantity Ξ_{ac} has the meaning of a general effective potential given by

$$\begin{aligned}\Xi_{ac}(3'4',34) &= \frac{\delta\Sigma_a(3',3)}{\delta g_c(4,4')}\\ &= \pm i\,\delta(3-3')\delta(4-4')V_{ac}(3'-4') + \frac{\delta\bar{\Sigma}_a(3',3)}{\delta g_c(4,4')}\end{aligned} \tag{4.17}$$

with $\bar{\Sigma}_a$ being the screened self-energy defined by (4.5).

To derive an equation for the polarization function Π_{ab}, we start from its defining equation (4.7). First, we rewrite

$$\Pi_{ab}(12,1'2') = \pm i\frac{\delta g_a(11')}{\delta U_b^{\text{eff}}(2'2)} = \mp i\int_C d3d3'\,g_a(13)\frac{\delta g_a^{-1}(33')}{\delta U_b^{\text{eff}}(2'2)}g_a(3'1')\,. \tag{4.18}$$

The functional derivative of g_a^{-1} is calculated by means of the Dyson equation

$$g_a^{-1}(11') = g_a^{0\,-1}(11') - U_a^{\text{eff}}(11') - \bar{\Sigma}_a(11')\,. \tag{4.19}$$

Then we get easily

$$\begin{aligned}\Pi_{ab}(12,1'2') &= \pm i\delta_{ab}g_a(12')g_a(21')\\ &\quad \pm i\int_C d3d3'\,g_a(13)\frac{\delta\bar{\Sigma}_a(33')}{\delta U_b^{\text{eff}}(2'2)}g_a(3'1')\,.\end{aligned} \tag{4.20}$$

The upper sign refers to Bose statistics and the lower one to Fermi statistics. We now consider that $\bar{\Sigma}_a$ does not depend explicitly on U^{eff}, but is a functional of the type $\bar{\Sigma}_a\left(g\left[U^{\text{eff}}\right]\right)$. Using the chain rule, we may eliminate the derivative with respect to U^{eff}. With the definition (4.7) for Π_{ab}, we get, in analogy to (4.16), a Bethe–Salpeter equation for the polarization function

$$\begin{aligned}\Pi_{ab}(12,1'2') &= \pm i\,g_a(12')g_a(21')\delta_{ab}\\ &+ \sum_c\int_C d3d3'd4d4'\,g_a(13')g_a(31')\,\frac{\delta\bar{\Sigma}_a(3'3)}{\delta g_c(44')}\,\Pi_{cb}(42,4'2')\,.\end{aligned} \tag{4.21}$$

These equations determine, in principle, the polarization function.

In (4.15), instead of the general function $\Pi_{ab}(12,1'2')$, only the two-time polarization function $\Pi_{ab}(12) = \Pi_{ab}(12,1^+2^+)$ occurs. For this special function, we immediately get from (4.18)

$$\Pi_{ab}(12) = \mp i\int_C d3d3'\,g_a(13)\Gamma_{ab}(32,3'2)g_a(3'1)\,. \tag{4.22}$$

Here, we introduced the vertex function Γ_{ab} by

$$\Gamma_{ab}(12,1'2) = \frac{\delta g_a^{-1}(11')}{\delta U_b^{\text{eff}}(22)}\,. \tag{4.23}$$

We now use the relations derived for Π_{ab} in order to find a further equation for the screened self-energy $\bar{\Sigma}_a$. It is useful to insert (4.20) into (4.13). We then get a functional equation for the determination of $\bar{\Sigma}_a$ having the shape

$$\begin{aligned} \bar{\Sigma}_a(11') &= i\,V^s_{aa}(11')g_a(11') \\ &\quad + i\sum_b \int_C d2d3\,V^s_{ab}(12)\,g_a(13)\frac{\delta\bar{\Sigma}_a(31')}{\delta U^{\text{eff}}_b(22)}\,. \end{aligned} \tag{4.24}$$

For V^s_{ab} given, the self-energy $\bar{\Sigma}_a$ may be determined in a similar way like from (4.5). However, the difference is that we now get a perturbative series in terms of the screened potential, i.e., collective effects are automatically accounted for, and the Coulomb divergencies are avoided as a consequence of the short range of the screened potential.

Finally let us consider once more the response function. As we will show later in more detail, $K_{ab}(12)$ determines the collective and response properties of the plasma. For later purpose, we still need the inverse response function $K^{-1}_{ab}(12)$. Obviously, the following holds true

$$K^{-1}_{ab}(12) = \frac{\delta U_b(22)}{\delta U^{\text{eff}}_a(11)} \tag{4.25}$$

where $\delta U(11') = \delta U(11)\delta(1-1')$ was used for the fields.

From (4.11) or using (4.3), the following relation between K^{-1} and Π can be derived,

$$K^{-1}_{ab}(12) = \delta(1-2)\delta_{ab} - \sum_c \int_C d3\,\Pi_{ac}(13,1^+3^+)V_{cb}(32)\,. \tag{4.26}$$

For $K_{ab}(12)$, we found (4.11) which provides a relation to the function L_{ab}. Thus, we have two basic equations for the characterization of the dielectric response function, namely the integral equation for K_{ab} given by (4.11), and the equation for K^{-1}_{ab} given by (4.26). From these equations, we can see that the response behavior and the collective effects of screening determined by K_{ab} and K^{-1}_{ab} are controlled by the two functions Π_{ab} and L_{ab}.

In order to summarize the result of the previous two subsections, let us list the fundamental system of equations for the description of systems with Coulomb interaction:

(i) The relations between the self-energy, the screened potential, and the polarization function given by (4.5) and (4.13).
(ii) The screening equation (4.15).
(iii) The Bethe–Salpeter equation (4.21) for the determination of the polarization function.

Together with the Kadanoff-Baym equations for the single-particle correlation and Green's functions, these equations completely determine the properties of plasmas. The coupling with the Kadanoff–Baym equation is considered in the next section.

4.3 The Kinetics of Particles and Screening. Field Fluctuations

Using the relations obtained in the previous section, the basic equations for the description of strongly coupled plasmas can be written as a system of two groups of equations defined on the Keldysh time contour. The first ones are the Kadanoff–Baym equations completed by the expressions for the self-energy (without initial correlations)

$$\left(i\frac{\partial}{\partial t_1} + \frac{\nabla_1^2}{2m_a}\right) g_a(11') = \delta(11') + \int_C d\bar{1}\left[U_a^{\text{eff}}(1\bar{1}) + \bar{\Sigma}_a(1\bar{1})\right] g_a(\bar{1}1'), \tag{4.27}$$

$$\bar{\Sigma}_a(11') = iV_{aa}^s(11')g_a(11') + i\sum_b \int_C d2d3V_{ab}^s(12)g_a(13)\frac{\delta\bar{\Sigma}_a(31')}{\delta U_b^{\text{eff}}(22)}. \tag{4.28}$$

These equations describe the kinetics of the plasma particles. The second group of equations determine the screened potential, $V_{ab}^s(1,2)$, i.e., the kinetics of screening and polarization is described by

$$V_{ab}^s(12) = V_{ab}(12) + \sum_{cd}\int_C d3d4V_{ac}(13)\Pi_{cd}(34)V_{db}^s(42) \tag{4.29}$$

and

$$\begin{aligned}\Pi_{ab}(12) = &\pm i\delta_{ab}g_a(12)g_a(21)\\ &\pm i\int_C d3d3'g_a(13)\frac{\delta\bar{\Sigma}_a(33')}{\delta U_b^{\text{eff}}(22)}g_a(3'1).\end{aligned} \tag{4.30}$$

Again, the upper sign refers to Bose statistics and the lower one to Fermi statistics. These two groups of equations form a coupled system of equations for the particles and the (longitudinal) fields which is formally closed. Unfortunately, it is not possible to solve these functional differential equations exactly. Equations (4.28) and (4.30) may be used, however, to generate approximate equations for g_a.

Obviously, a simple approximation including collective effects follows, if we neglect the integral terms. Then we arrive at

$$\bar{\Sigma}_a(12) = iV^s_{aa}(12)g_a(12)\,; \qquad \Pi_{ab}(12) = \pm i\delta_{ab}g_a(12)g_a(21). \tag{4.31}$$

This approximation for Π_{ab} is one of the standard approximations of many-particle theory, especially for plasma physics, and it is referred to as *random phase approximation* (RPA). The approximation given for $\bar{\Sigma}_a$ is called V^s-approximation. With (4.31), the set of (4.27)–(4.30) represents a self-consistent system of equations for the determination of any property of the plasma.

In order to get the physical contents of these equations, we return from the Keldysh contour to the physical time domain $-\infty \le t \le +\infty$. From the particle equation (4.27), we find the Kadanoff–Baym kinetic equations for the single-particle correlation functions $g_a^>$ and $g_a^<$. Positioning the times t_1 and t_1' on the two different branches of the contour, we get

$$\begin{aligned}\left(i\frac{\partial}{\partial t_1} + \frac{\nabla_1^2}{2m_a}\right) g_a^{\gtrless}(11') &= \int\limits_{-\infty}^{+\infty} d\bar{1}\, U_a^{\text{eff}}(1\bar{1})g_a^{\gtrless}(\bar{1}1') \\ &+ \int\limits_{-\infty}^{t_1} d\bar{1}\left[\bar{\Sigma}_a{}^>(1\bar{1}) - \bar{\Sigma}_a{}^<(1\bar{1})\right] g_a^{\gtrless}(\bar{1}1') \\ &- \int\limits_{-\infty}^{t_1'} d\bar{1}\bar{\Sigma}_a{}^{\gtrless}(1\bar{1})\left[g_a^>(\bar{1}1') - g_a^<(\bar{1}1')\right].\end{aligned} \tag{4.32}$$

Here, the limits of time integration are written explicitly.

From (4.28), we get the expressions for the self-energy functions $\Sigma_a^{\gtrless}$ in terms of the single-particle correlation functions $g_a^{\gtrless}(12)$ and the correlation functions of the screened potential $V_{ab}^{s\gtrless}(12)$. An equation for the $V_{ab}^{s\,\gtrless}$ follows from the general screening equation (4.29). The appropriate setting of the times t_1 and t_2 on the lower and upper branch of the contour leads to

$$\begin{aligned}V_{ab}^{s\gtrless}(12) &= \sum_{cd}\int\limits_{-\infty}^{+\infty} d3d4\left\{V_{ac}(13)\Pi^R_{cd}(34)V_{db}^{s\gtrless}(42)\right. \\ &+ \left. V_{ac}(13)\Pi^{\gtrless}_{cd}(34)V_{db}^{sA}(42)\right\}.\end{aligned} \tag{4.33}$$

For convenience, retarded and advanced quantities were introduced. Therefore, let us consider the latter quantities in more detail. Using the definition of the retarded and advanced screened potential

$$V_{ab}^{sR/A}(12) = V_{ab}(12) \pm \Theta[\pm(t_1 - t_2)](V_{ab}^{s>}(12) - V_{ab}^{s<}(12)), \tag{4.34}$$

we find the screening equation

$$V_{ab}^{sR/A}(12) = V_{ab}(1-2) + \sum_{cd} \int_{-\infty}^{+\infty} d3d4 V_{ac}(1-3) \Pi_{cd}^{R/A}(34) V_{db}^{sR/A}(42) \,. \quad (4.35)$$

In the following, the limits of time integration are dropped, i.e., the integral symbol "$\int$" without the index "C" means time integration over the physical time domain $-\infty \leq t \leq +\infty$.

The equations for $V_{ab}^{s\gtrless}$ and $V_{ab}^{sR/A}$ may also be written as differential equations. Applying the Laplacean, e.g., to (4.35), we easily get

$$\Delta_1 V_{ab}^{sR}(12) + 4\pi \sum_{cd} e_a e_c \int d3\, \Pi_{cd}^{R}(13) V_{db}^{sR}(32) = -4\pi e_a e_b\, \delta(1-2) \,. \quad (4.36)$$

In the same way, we get from (4.33)

$$\Delta_1 V_{ab}^{s\gtrless}(12) + 4\pi \sum_{cd} e_a e_c \int d3\, \Pi_{cd}^{R}(13) V_{db}^{s\gtrless}(32) \,. = -4\pi \sum_{cd} e_a e_c \int d3\, \Pi_{cd}^{\gtrless}(13) V_{db}^{sA}(32) \,. \quad (4.37)$$

The first of these equations is obviously a generalization of the Poisson Boltzmann equation of the elementary Debye theory of screening, introduced in Chap. 2. Therefore, the response function V_{ab}^{sR} has the physical meaning of the screened potential. In this sense, (4.37) represents the quantum statistical generalization of Debye screening. The retarded/advanced screened potential plays a central role in the theory, and it is useful to express all quantities, especially the correlation functions $V_{ab}^{s\gtrless}$ by the functions $V_{ab}^{sR/A}$. For this purpose, we take into account that the retarded screened potential is, in the mathematical sense, the Green's function to the inhomogeneous equation (4.37). Using this idea, it is easy to construct a formal solution for $V_{ab}^{s\gtrless}$ which can be written as

$$V_{ab}^{s\gtrless}(12) = V_{ab}^{0\gtrless}(12) + \sum_{cd} \int d3d4\, V_{ac}^{sR}(13) \Pi_{cd}^{\gtrless}(34) V_{db}^{sA}(42) \,. \quad (4.38)$$

The first term denotes the solution of the homogeneous equation, which is related to initial correlations. If the system is perturbed adiabatically or the times are large enough, this term will be damped out. For a physical interpretation of $V^{\gtrless}$ and of the equation (4.38), let us first consider the relation between the correlation functions $V_{ab}^{s\gtrless}$ and $L_{ab}^{\gtrless}$. By the appropriate setting of the times t_1 and t_2 on the upper and lower branches of the Keldysh time contour, we get from (4.14) and (4.15)

$$V_{ab}^{s\gtrless}(1,2) = \sum_{cd} \int d3d4\, V_{ac}(13) L_{cd}^{\gtrless}(34) V_{db}(42) \,. \tag{4.39}$$

Here, initial correlation terms were not taken into account. According to (3.44), $L_{cd}^{\gtrless}$ has, for $t_1' = t_1^+$ and $t_2' = t_2^+$, the physical meaning of the correlation function of density–density fluctuations; this means

$$V_{ab}^{s\gtrless}(12) = -i \sum_{cd} \int d3d4 V_{ac}(13) \left\langle \delta\rho_c(22)\delta\rho_d(11) \right\rangle V_{db}(42) \,. \tag{4.40}$$

Now we introduce the longitudinal field fluctuations due to the density–density fluctuations by

$$Ze\delta\boldsymbol{E}(1) = -\nabla_1 \int d2 V(1-2)\delta\varrho(2) \,. \tag{4.41}$$

Then immediately follows that $V_{ab}^{s\gtrless}$ determines the correlation function of the fluctuations of the electric field

$$Z^2e^2 \left\langle \delta\boldsymbol{E}(1)\delta\boldsymbol{E}(2) \right\rangle = i\nabla_1\nabla_2 V^{s>}(1,2) \,. \tag{4.42}$$

In many cases it is more useful to use a symmetrized correlation function of the longitudinal field fluctuations

$$Z^2e^2 \left\langle \overline{\delta\boldsymbol{E}\delta\boldsymbol{E}} \right\rangle = \frac{1}{2}\nabla_1\nabla_2(V^>(12) + V^<(12)) \,. \tag{4.43}$$

In this way, the physical meaning of the quantity is clear, and the relation (4.38) can be interpreted as a nonequilibrium fluctuation–dissipation theorem (DuBois 1968). We will consider the fluctuations in the plasma more in detail in Sect. 4.7.

Let us come back to the density–density correlation function. An equation for the functions $L_{ab}^{\gtrless}$ follows from (4.8). With the known rules valid on the Keldysh time contour, we get

$$\begin{aligned} L_{ab}^{\gtrless}(12) &= \Pi_{ab}^{\gtrless}(12) + \sum_{cd} \int d3d4 \left\{ \Pi_{ac}^{R}(13) V_{cd}(34) L_{db}^{\gtrless}(42) \right. \\ &+ \left. \Pi_{ac}^{\gtrless}(13) V_{cd}(34) L_{db}^{A}(42) \right\} . \end{aligned} \tag{4.44}$$

Now, retarded and advanced functions are introduced

$$L_{ab}^{R/A}(12) = \pm\Theta\left(\pm(t_1 - t_2)\right) \left[L_{ab}^{>}(12) - L_{ab}^{<}(12) \right] . \tag{4.45}$$

We easily find $L_{ab}^{R}(12) = L_{ba}^{A}(21)$, and correspondingly $V_{ab}^{sR}(12) = V_{ba}^{sA}(21)$. Using (4.44), we have the following

$$L_{ab}^{R/A}(12) = \Pi_{ab}^{R/A}(12) + \sum_{cd} \int d3d4\, \Pi_{ac}^{R/A}(13) V_{cd}(34) L_{db}^{R/A}(42)\,. \qquad (4.46)$$

We note again that the integral symbol "$\int$" means integration over the physical time domain $-\infty < t < \infty$ and over the space.

A more compact expression for the density–density correlation functions $L_{ab}^{\gtrless}$ may be obtained if the response function K^R is written as

$$K_{ab}^{R}(12) = \delta(1-2)\delta_{ab} + \sum_{c} \int d3\, L_{ac}^{R}(13) V_{cb}(32)\,. \qquad (4.47)$$

Furthermore, if we introduce the advanced function $K_{ab}^{A}(12) = K_{ba}^{R}(21)$, (4.44) can be rewritten as

$$L_{ab}^{\gtrless}(12) = \sum_{cd} \int d3d4\, K_{ac}^{R}(13) \Pi_{cd}^{\gtrless}(34) K_{db}^{A}(42)\,. \qquad (4.48)$$

Equations (4.48) and (4.38) represent general versions of the *fluctuation–dissipation theorems* valid for non-equilibrium systems, too. They describe the connection between the noise source of the fluctuations $\Pi_{ab}^{\gtrless}$ and the correlation functions of density fluctuations.

Inserting (4.48) into (4.39) and comparing with (4.38), we find for the retarded screened potential

$$V_{ab}^{sR}(12) \;=\; \sum_{c} \int d3 V_{ac}(13) K_{cb}^{R}(32)\,. \qquad (4.49)$$

By the relations given above, the screening properties of the plasma are described in terms of the response functions $K^R(12)$ and $K^A(12)$. Let us finally introduce the more familiar dielectric function $\varepsilon^{R/A}$. Instead of (4.49), the screened potential can be written in the form

$$V_{ab}^{sR}(12) \;=\; \int d3\, V_{ab}(13) \varepsilon^{R^{-1}}(32)\,. \qquad (4.50)$$

In this way, screening is described by the dielectric function of the plasma given by

$$\varepsilon^{R^{-1}}(12) \;=\; \delta(1-2) + \sum_{ab} \int d3\, L_{ab}^{R}(13)\, V_{ba}(32)\,. \qquad (4.51)$$

The response properties of the plasma determined by the general relations (4.6) and (4.7) can be expressed in terms of the dielectric function, too. Transforming these relations to the physical time domain, we find for the response to an external field U_a

$$\delta U_a^{\text{eff}}(1) = \int d2\, \varepsilon^{R^{-1}}(12)\, \delta U_a(2) \tag{4.52}$$

and

$$\delta n_a(1) = \sum_b \int d2\, L_{ab}^R(12)\, \delta U_b(2)\,. \tag{4.53}$$

From the second equation, it can be seen that the induced density $\delta n_a(1) = \pm i\delta g_a^<(1,1^+)$ is determined by the retarded density response function L_{ab}^R which is directly related to the inverse dielectric function according to (4.51). It should be noticed that the integration over t_2 in (4.53) is carried out from $-\infty$ to t_1. This follows from the Heaviside step function included in the definition (4.45) of $L_{ab}^R(12)$. Therefore, causality is ensured, i.e., the induced density at time t_1 is due to the external field for times $t_2 < t_1$.

If the density response to the effective potential is considered, we have

$$\delta n_a(1) = \sum_b \int d2\, \Pi_{ab}^R(12)\, \delta U_b^{\text{eff}}(2)\,. \tag{4.54}$$

The response function to the effective potential turns out to be the polarization function Π_{ab}^R. According to (4.35) and (4.50), it is connected with the dielectric function by

$$\varepsilon^R(12) = \delta(1-2) - \sum_{ab} \int d3\, \Pi_{ab}^R(13) V_{ba}(32)\,. \tag{4.55}$$

Like in Sect. 3.2.1 it is convenient to introduce variables describing processes on microscopic and macroscopic scales. These variables are defined as

$$\begin{aligned} t &= \frac{t_1+t_2}{2}, \quad \tau = t_1 - t_2 \\ \boldsymbol{R} &= \frac{\boldsymbol{r}_1+\boldsymbol{r}_2}{2}, \quad \boldsymbol{r} = \boldsymbol{r}_1 - \boldsymbol{r}_2\,. \end{aligned} \tag{4.56}$$

Then, it is useful to perform a Fourier transformation with respect to the difference variables. For the correlation functions of the screened potential, we have

$$V_{ab}^{s\gtrless}(\boldsymbol{q}\omega, \boldsymbol{R}t) = \int d\boldsymbol{r} d\tau e^{-i\boldsymbol{q}\boldsymbol{r}+i\omega\tau} V_{ab}^{s\gtrless}(\boldsymbol{r}\tau, \boldsymbol{R}t)\,. \tag{4.57}$$

In lowest order gradient expansion (local approximation), we get from (4.38) for the *fluctuation–dissipation theorem*

$$V_{ab}^{s\gtrless}(\boldsymbol{q}\omega, \boldsymbol{R}t) = \sum_{cd} V_{ac}^{sR}(\boldsymbol{q}\omega, \boldsymbol{R}t)\Pi_{cd}^{\gtrless}(\boldsymbol{q}\omega, \boldsymbol{R}t)V_{db}^{sA}(\boldsymbol{q}\omega, \boldsymbol{R}t)\,. \tag{4.58}$$

The equations for the retarded and advanced screened potentials follow from (4.35). For the retarded one, it reads

$$V_{ab}^{sR}(\boldsymbol{q}\omega,\boldsymbol{R}t) = V_{ab}(q)+\sum_{cd} V_{ac}^{sR}(\boldsymbol{q}\omega,\boldsymbol{R}t)\Pi_{cd}^{R}(\boldsymbol{q}\omega,\boldsymbol{R}t)V_{db}(q)$$
$$= \frac{V_{ab}(q)}{\varepsilon^{R}(\boldsymbol{q}\omega,\boldsymbol{R}t)}, \tag{4.59}$$

with the Coulomb potential $V_{ab} = 4\pi e_a e_b/q^2$. The non-equilibrium dielectric function is given by

$$\varepsilon^{R}(\boldsymbol{q}\omega,\boldsymbol{R}t) = 1-\sum_{ab}\Pi_{ab}^{R}(\boldsymbol{q}\omega,\boldsymbol{R}t)V_{ba}(q) \tag{4.60}$$

and

$$\varepsilon^{R^{-1}}(\boldsymbol{q}\omega,\boldsymbol{R}t) = 1+\sum_{ab}L_{ab}^{R}(\boldsymbol{q}\omega,\boldsymbol{R}t)V_{ba}(q)\,. \tag{4.61}$$

Of special importance for our further considerations is the imaginary part of the inverse dielectric function given by

$$\operatorname{Im}\varepsilon^{R^{-1}}(\boldsymbol{q}\omega,\boldsymbol{R}t) = -\frac{\operatorname{Im}\varepsilon^{R}(\boldsymbol{q}\omega,\boldsymbol{R}t)}{\left|\varepsilon^{R}(\boldsymbol{q}\omega,\boldsymbol{R}t)\right|^2}\,. \tag{4.62}$$

As will be shown in the next sections, $\mathrm{Im}\varepsilon^{R^{-1}}$ is closely related to the spectral function of density fluctuations describing the plasma excitation spectrum in the frame of many-particle theory.

Finally, let us consider how the relations given above are modified in thermodynamic equilibrium (Kadanoff and Baym 1962; Kraeft et al. 1986). In this case, all quantities depend on the difference variables only, and the relations get the same structure as in local approximation, but without the dependence on the macro variables R and t. Furthermore, the correlation functions are connected by the KMS relation. For $V_{ab}^{s\gtrless}$, the latter reads

$$V_{ab}^{s\,>}(\boldsymbol{q},\omega) = e^{\beta\omega}\,V_{ab}^{s\,<}(\boldsymbol{q},\omega)\,. \tag{4.63}$$

Using $2i\,\mathrm{Im}V_{ab}^{s\,R} = V_{ab}^{s\,>} - V_{ab}^{s\,<}$ and the expression (4.59), we find the spectral representation

$$V_{ab}^{s\,<}(\boldsymbol{q},\omega) = 2i\,V_{ab}(q)\,\mathrm{Im}\varepsilon^{R^{-1}}(\boldsymbol{q},\omega)\,n_B(\omega)$$
$$V_{ab}^{s\,>}(\boldsymbol{q},\omega) = 2i\,V_{ab}(q)\,\mathrm{Im}\varepsilon^{R^{-1}}(\boldsymbol{q},\omega)\left[1+n_B(\omega)\right], \tag{4.64}$$

with the Bose function $n_B(\omega) = [\exp(\beta\omega)-1]^{-1}$.

Let us come back to the connection between $V^{\gtrless}$ and the correlation function of the field fluctuations. Introducing (4.64) into (4.43), we easily get

$$\langle\delta\boldsymbol{E}\delta\boldsymbol{E}\rangle = 8\pi\,\mathrm{Im}\epsilon^{R-1}\left(\frac{1}{2}+n_B(\omega)\right). \tag{4.65}$$

This formula connects the field fluctuations with the dissipation function $\mathrm{Im}\epsilon^{R-1}$ and represents, therefore, a fluctuation–dissipation theorem (Klimontovich 1975).

4.4 The Dielectric Function of the Plasma. General Properties, Sum Rules

The general formalism of quantum statistical theory for systems with Coulomb interaction was developed in the previous sections of this chapter. Using the fact that the single-particle Green's function depends on an external potential via the effective potential, the applied functional derivation technique gave us the basic relations to describe equilibrium and non-equilibrium properties of strongly coupled plasmas. Finally, it was shown that the long range character of the Coulomb potential is accounted for by the dielectric function $\varepsilon^R(12)$. This makes the latter to be a central quantity in the many-particle theory of plasmas.

In this section, we will focus on some general properties of the dielectric function. For simplicity, the one component plasma model is used, such as the electron gas with a uniform background of positive charge.

Let us start from (4.50). For a one-component plasma, we have

$$V^{sR}(12) = \int d3\, V(13)\varepsilon^{R^{-1}}(32)$$

with the Coulomb potential $V(13) = \delta(t_1 - t_3)Z^2e^2/|\boldsymbol{r}_1 - \boldsymbol{r}_3|$.

It turns out that the dielectric function determines one of the most peculiar properties of plasmas, the screening of the long range Coulomb interaction between the charged particles in the system. Furthermore, we found the relations (4.52), (4.53), and (4.54) which relate the dielectric function to the response properties of the plasma. If only the time arguments are written explicitly, the relation (4.52) reads

$$\begin{aligned} \delta U^{\text{eff}}(t_1) &= \int_{-\infty}^{+\infty} dt_2\, \varepsilon^{R^{-1}}(t_1, t_2)\, \delta U(t_2) \\ &= \delta U(t_1) + \int_{-\infty}^{t_1} dt_2\, L^R(t_1, t_2) V\, \delta U(t_2)\,. \end{aligned} \tag{4.66}$$

Here, the induced part of the effective potential at time t_1 is due to the external field for times $t_2 < t_1$ which ensures causality.

In the following, the variables defined by (4.56) are introduced, and Fourier transformation with respect to the difference variables $\boldsymbol{r}$ and τ is performed. The local variables $\boldsymbol{R}$ and t are not written explicitly. Because $\varepsilon^{R^{-1}}(12)$ is a retarded quantity, the following spectral representation can be found

$$\varepsilon^{R^{-1}}(\boldsymbol{q}, \omega) = 1 + V(q) \int \frac{d\omega'}{2\pi} \frac{\hat{L}(\boldsymbol{q}, \omega')}{\omega - \omega' + i\varepsilon}\,, \tag{4.67}$$

where $V(q) = 4\pi Z^2 e^2/q^2$ ($\hbar = 1$) is the Fourier transform of the Coulomb potential, and $\hat{L}$ is the spectral function of density fluctuations given by

$$\hat{L}(\boldsymbol{q},\omega) = i\left(L^{>}(\boldsymbol{q},\omega) - L^{<}(\boldsymbol{q},\omega)\right). \tag{4.68}$$

If $\hat{L}(\boldsymbol{q},\omega)$ obeys a Hölder condition the retarded inverse dielectric function is determined by the well-known properties of the analytic continuation to complex energies z. These properties were already discussed in Sect. 3.2.3 for the single-particle Green's function. Let us write them again to derive some important consequences for the inverse dielectric function (Pines 1962; Pines and Nozieres 1958; Kadanoff and Baym 1962; Stolz 1974):

(i) ${\varepsilon^{R}}^{-1}(z,t)$ is an analytic function of the complex variable z in the upper half plane. Using the fact that $L^R(\boldsymbol{q},z) \sim 1/z^2$ for $z \to \infty$, we then have

$$\int\limits_{-\infty}^{+\infty} d\omega' \, \frac{{\varepsilon^{R}}^{-1}(\boldsymbol{q},\omega') - 1}{\omega' - \omega + i\varepsilon} = 0\,. \tag{4.69}$$

(ii) On the real axis, ${\varepsilon^{R}}^{-1}(\boldsymbol{q},z)$ is given by the Plemelj formula

$${\varepsilon^{R}}^{-1}(\boldsymbol{q},\omega) = 1 + V(q)\,\mathcal{P}\int\limits_{-\infty}^{+\infty} \frac{d\omega'}{2\pi}\,\frac{\hat{L}(\boldsymbol{q},\omega')}{\omega - \omega'} - \frac{i}{2}V(q)\hat{L}(\boldsymbol{q},\omega)\,, \tag{4.70}$$

where $\mathcal{P}$ denotes the principal value.

(iii) If $\hat{L}(\boldsymbol{q},\omega)$ has an analytic continuation it is possible to continue ${\varepsilon^{R}}^{-1}(\boldsymbol{q},\omega)$ into the lower half plane

$${\varepsilon^{R}}^{-1}(\boldsymbol{q},z) = {\varepsilon^{A}}^{-1}(\boldsymbol{q},z) - iV(q)\hat{L}(\boldsymbol{q},z) \quad (\mathrm{Im}\, z < 0) \tag{4.71}$$

with ${\varepsilon^{A}}^{-1}(\boldsymbol{q},z)$ being the corresponding advanced function.

From these properties, we find that ${\varepsilon^{R}}^{-1}$ is a complex quantity, and from (4.69), we have the following equations

$$\mathrm{Re}\,{\varepsilon^{R}}^{-1}(\boldsymbol{q},\omega) = 1 + \mathcal{P}\int\limits_{-\infty}^{+\infty} \frac{d\omega'}{\pi}\,\frac{\mathrm{Im}\,{\varepsilon^{R}}^{-1}(\boldsymbol{q},\omega')}{\omega' - \omega} \tag{4.72}$$

and

$$\mathrm{Im}\,{\varepsilon^{R}}^{-1}(\boldsymbol{q},\omega) = -\mathcal{P}\int\limits_{-\infty}^{+\infty} \frac{d\omega'}{\pi}\,\frac{\mathrm{Re}\,{\varepsilon^{R}}^{-1}(\boldsymbol{q},\omega') - 1}{\omega' - \omega}\,. \tag{4.73}$$

These important relations are known as *Kramers–Kronig dispersion relations* (Kronig 1926; Kramers 1927) . Furthermore, using the spectral representation (4.67) with (4.70), we get

$$\mathrm{Im}\varepsilon^{R^{-1}}(\omega,t) = -\frac{1}{2}V(q)\hat{L}(\boldsymbol{q},\omega)\,. \tag{4.74}$$

This equation relates the inverse dielectric function, being the central dielectric response function in the frame of a phenomenological theory, to the spectral function of density fluctuations which can be calculated explicitly by applying the methods of many-particle theory.

In the previous section, it was shown that there is a further (screened) response function, the polarization function $\Pi^R(\boldsymbol{q},\omega)$ which measures the density response to the (total) effective field, and not to the external one. According to (4.55), it is related directly to the dielectric function $\varepsilon^R(\boldsymbol{q},\omega)$. Because $\Pi^R(\boldsymbol{q},\omega)$ and $\varepsilon^R(\boldsymbol{q},\omega)$ describe the response properties due to the effective field, there is, in general, no way to apply the causality principle in order to find the analytic behavior of these quantities. On the other hand, causality arguments are usually applied to determine the analytic properties of the response functions $L^R(\boldsymbol{q},\omega)$ and $\varepsilon^{R^{-1}}(\boldsymbol{q},\omega)$ which measure the response of the system to the external field. For a more detailed discussion of this problem and its consequences we refer to several papers (Martin 1967; Izuyama 1973; Kirzhnitz 1976; Gorobchenko et al. 1989). It turns out that the analyticity of $\varepsilon^R(\boldsymbol{q},z)$ depends on the sign of $\varepsilon^R(\boldsymbol{q},0)$. The dielectric function is analytic in the upper half plane if the following condition is fulfilled

$$\varepsilon^R(\boldsymbol{q},0) > 0\,. \tag{4.75}$$

Otherwise, $\varepsilon^{R^{-1}}(\boldsymbol{q},z)$ can have zeros, and $\varepsilon(\boldsymbol{q},z)$ is nonanalytic. Thus, in the case $\varepsilon^R(\boldsymbol{q},0) > 0$, the dielectric function is analytic in the upper half plane with properties similar to that found for the inverse dielectric function. In local approximation, the spectral representation can be written as

$$\varepsilon^R(\boldsymbol{q},\omega) = 1 - V(q)\int\frac{d\omega'}{2\pi}\frac{\hat{\Pi}(\boldsymbol{q},\omega)}{\omega-\omega'+i\varepsilon} \tag{4.76}$$

with the screened spectral function $\hat{\Pi}(\boldsymbol{q},\omega)$

$$\hat{\Pi}(\boldsymbol{q},\omega) = i\left(\Pi^>(\boldsymbol{q},\omega) - \Pi^<(\boldsymbol{q},\omega)\right)\,. \tag{4.77}$$

As above, the macroscopic variables $\boldsymbol{R}$ and t are not written explicitly.

If $\varepsilon^R(\boldsymbol{q},z)$ is analytic for $\mathrm{Im}z > 0$, Kramers–Kronig relations can be found which read

$$\mathrm{Re}\,\varepsilon^R(\boldsymbol{q},\omega) = 1 + \mathcal{P}\int_{-\infty}^{+\infty}\frac{d\omega'}{\pi}\frac{\mathrm{Im}\,\varepsilon^R(\boldsymbol{q},\omega')}{\omega'-\omega} \tag{4.78}$$

and

$$\mathrm{Im}\,\varepsilon^R(\boldsymbol{q},\omega) = -\mathcal{P}\int_{-\infty}^{+\infty}\frac{d\omega'}{\pi}\frac{\mathrm{Re}\,\varepsilon^R(\boldsymbol{q},\omega') - 1}{\omega'-\omega}\,. \tag{4.79}$$

From the spectral representation (4.76), we find

$$\operatorname{Im}\varepsilon^R(\boldsymbol{q},\omega) = \frac{1}{2}V(q)\hat{\Pi}(\boldsymbol{q},\omega)\,. \tag{4.80}$$

There are further general properties of the dielectric function known as *sum rules* which represent a special class of exact results. They are given as frequency moments of the spectral functions of density fluctuations, and they allow us to check the quality of a given approximation to calculate the dielectric properties of the plasma.

The sum rules can be found using a variety of methods. To give an example, let us follow the main steps in deriving the well-known *f-sum rule.* As before in this section, the one-component plasma is considered. Furthermore, we assume the plasma to be in thermodynamic equilibrium, i.e., the response functions do not depend on the macroscopic variables $\boldsymbol{R}$ and t, and the averages are performed using the grand canonical density operator. We start from (4.68) which defines the spectral function of density fluctuations. Using (3.42) we find

$$\frac{1}{\mathcal{V}}\left\langle\left[\rho(\boldsymbol{q},t)\,\rho^\dagger(\boldsymbol{q},t')\right]\right\rangle = \int\limits_{-\infty}^{+\infty}\frac{d\omega}{2\pi}\hat{L}(\boldsymbol{q},\omega)\,e^{-i\omega(t-t')}\,, \tag{4.81}$$

where $[\ldots]$ denotes the commutator, and $\mathcal{V}$ is the volume. The Fourier transform of the particle density operator is

$$\rho(\boldsymbol{q},t) = \int d\boldsymbol{r}\,\psi^\dagger(\boldsymbol{r},t)\psi(\boldsymbol{r},t)\,e^{-i\boldsymbol{q}\cdot\boldsymbol{r}}\,. \tag{4.82}$$

Using the equation of motion for $\rho(\boldsymbol{q},t)$, we have (Stolz 1974; Mahan 1990)

$$\begin{aligned}\left\langle\left[\frac{\partial}{\partial t}\rho(\boldsymbol{q},t),\rho^\dagger(\boldsymbol{q},t')\right]\right\rangle &= \left\langle\left[\left[\rho(\boldsymbol{q},t)\,\mathrm{H}\right]\rho^\dagger(\boldsymbol{q},t')\right]\right\rangle \\ &= \mathcal{V}\int\limits_{-\infty}^{+\infty}\frac{d\omega}{2\pi i}\,\omega\,\hat{L}(\boldsymbol{q},\omega)\,e^{-i\omega(t-t')}\,. \end{aligned}\tag{4.83}$$

Here, H is the Hamiltonian of the one-component plasma. Now, we specialize the times to be $t=t'$. Then, the double commutator can be evaluated taking into account that $\rho(\boldsymbol{q},t)$ does not commute with the kinetic energy term of the total Hamiltonian. Subsequently we arrive at

$$\left[\left[\rho(\boldsymbol{q},t)\,\mathrm{H}\right]\rho^\dagger(\boldsymbol{q},t)\right] = \frac{q^2}{m}N\,, \tag{4.84}$$

where N is the number of particles in the system, and m is the particle mass. Using (4.84) in (4.83), we find the sum rule

$$\int \frac{d\omega}{2\pi}\,\omega\,\hat{L}(\boldsymbol{q},\omega) = \frac{q^2}{m}n \tag{4.85}$$

with $n = \langle N \rangle / \mathcal{V}$ being the average number density. Introducing the dielectric function according to

$$\hat{L}(\boldsymbol{q},\omega) = -\frac{2}{V(q)}\,\mathrm{Im}\varepsilon^{R^{-1}}(\boldsymbol{q},\omega)\,, \tag{4.86}$$

the *f-sum rule* takes the form

$$\int\limits_{-\infty}^{+\infty} \frac{d\omega}{\pi}\,\omega\,\mathrm{Im}\varepsilon^{R^{-1}}(\boldsymbol{q},\omega) = -\,\omega_{\mathrm{pl}}^2\,. \tag{4.87}$$

Here, ω_{pl}^2 is the square of the plasma frequency for a one component plasma

$$\omega_{\mathrm{pl}}^2 = \frac{4\pi n Z^2 e^2}{m}\,.$$

Further sum rules follow from the analytic behavior of $\varepsilon^{R^{-1}}(\boldsymbol{q},\omega)$ and of $\varepsilon^R(\boldsymbol{q},\omega)$, i.e., they may be derived using the Kramers–Kronig relations given above. We will present the results without a derivation. For a detailed analysis, we refer to the extensive literature existing in this field (Pines and Nozieres 1958; Singwi and Tosi 1981; Mahan 1990; Gorobchenko et al. 1989; Stolz 1974). There is a further frequency moment of the spectral function of density fluctuations, the so-called long wavelength *perfect screening sum rule*. In terms of the imaginary part of the inverse dielectric function it can be written as

$$\lim_{q\to 0}\int\limits_{-\infty}^{+\infty} \frac{d\omega}{\pi}\,\frac{\mathrm{Im}\varepsilon^{R^{-1}}(\boldsymbol{q},\omega)}{\omega} = -1\,. \tag{4.88}$$

Let us now consider the frequency moments of the screened spectral function (4.77) related to the polarization functions $\Pi^{\gtrless}(\boldsymbol{q},\omega)$. The sum rule analogous to (4.85) is

$$\int\limits_{-\infty}^{+\infty} \frac{d\omega}{2\pi}\,\omega\hat{\Pi}\,(\boldsymbol{q},\omega) = \frac{q^2}{m}\,n \tag{4.89}$$

and in terms of the dielectric function, we have

$$\int\limits_{-\infty}^{+\infty} \frac{d\omega}{\pi}\,\omega\,\mathrm{Im}\varepsilon^R(\boldsymbol{q},\omega) = \omega_{\mathrm{pl}}^2\,. \tag{4.90}$$

This equation is sometimes called the *conductivity sum rule*. A third exact relation is known as the *compressibility sum rule* given by

$$\lim_{q\to 0}\int_{-\infty}^{+\infty}\frac{d\omega}{\pi}\frac{\mathrm{Im}\varepsilon^R(\boldsymbol{q},\omega)}{\omega}=V(q)\,\varkappa\, n^2 \tag{4.91}$$

with $\varkappa$ being the isothermal compressibility defined by

$$\frac{1}{\varkappa}=n\frac{\partial}{\partial n}p=n^2\frac{\partial}{\partial n}\mu\,. \tag{4.92}$$

Here, p denotes the pressure, and μ is the chemical potential.

It is possible to derive further sum rules corresponding to higher frequency moments of the dielectric function (Gorobchenko et al. 1989). Of special importance is the third moment

$$\int_{-\infty}^{+\infty}\frac{d\omega}{\pi}\,\omega^3\,\mathrm{Im}\varepsilon^{R^{-1}}(\boldsymbol{q},\omega)=-M_3(\boldsymbol{q})\,. \tag{4.93}$$

While the first moment (*f-sum rule*) given by (4.87) does not depend on the interaction between the particles in the plasma, the third moment accounts for first order interaction effects. Explicit expressions for the quantity $M_3(\boldsymbol{q})$ related to the correlation functions of density fluctuations can be found in Puff (1965) and Mihara and Puff (1968). Finally, let us give some exact expressions for the dielectric function valid in the limiting cases of long wave lengths ($q\to 0$) and high frequencies ($\omega\to\infty$). From the Kramers–Kronig relation (4.78), we have

$$\mathrm{Re}\,\varepsilon^R(\boldsymbol{q},0)=1+\mathcal{P}\int_{-\infty}^{+\infty}\frac{d\omega'}{\pi}\frac{\mathrm{Im}\,\varepsilon^R(\boldsymbol{q},\omega')}{\omega'}\,, \tag{4.94}$$

and, using the compressibility sum rule (4.91) for the static dielectric function, we get

$$\lim_{q\to 0}\varepsilon^R(\boldsymbol{q},0)=1+V(q)n^2\varkappa\,. \tag{4.95}$$

In order to find the high frequency behavior, we start from the Kramers–Kronig relation (4.72), which can be written in the form

$$\mathrm{Re}\,\varepsilon^{R^{-1}}(\boldsymbol{q},\omega)=1+\mathcal{P}\int_{-\infty}^{+\infty}\frac{d\omega'}{\pi}\frac{\omega'}{\omega'^2-\omega^2}\,\mathrm{Im}\varepsilon^{R^{-1}}(\boldsymbol{q},\omega')\,. \tag{4.96}$$

For large ω, we have

$$\mathrm{Re}\,\varepsilon^{R^{-1}}(\boldsymbol{q},\omega)=1-\frac{1}{\omega^2}\int_{-\infty}^{+\infty}\frac{d\omega'}{\pi}\,\omega'\,\mathrm{Im}\,\varepsilon^{R^{-1}}(\boldsymbol{q},\omega')\;+\;\mathcal{O}(\omega^4)\,. \tag{4.97}$$

Now, the f-sum rule can be used in (4.97), and we easily find the following behavior for large ω

$$\lim_{\omega\to\infty} \mathrm{Re}\,\varepsilon^{R^{-1}}(\boldsymbol{q},\omega) = 1 + \frac{\omega_p^2}{\omega^2}\,, \tag{4.98}$$

and we get for the dielectric function $\varepsilon^R(\boldsymbol{q},\omega)$

$$\lim_{\omega\to\infty} \mathrm{Re}\,\varepsilon^{R}(\boldsymbol{q},\omega) = 1 - \frac{\omega_p^2}{\omega^2}\,. \tag{4.99}$$

The expressions (4.95) and (4.99) determine the general asymptotic behavior of the dielectric function for the one-component plasma considered. As will be shown in the following sections, the functions reflect the typical properties of plasmas, static screening of the long range Coulomb forces, and dynamic effects such as plasma oscillations.

4.5 The Random Phase Approximation (RPA)

4.5.1 The RPA Dielectric Function

Let us now consider an explicit expression for the dielectric function, namely the simple scheme of *random phase approximation* (RPA) (4.31) originally introduced by Bohm and Pines (1953). The expressions we derive for non-equilibrium plasmas will be considered in lowest order gradient expansion (local approximation). In contrast to the thermodynamic equilibrium, there is an additional dependence of the dielectric function on the macroscopic variables $\boldsymbol{R}$ and t. The latter dependence results from the single-particle correlation functions determined self-consistently by the Kadanoff–Baym kinetic equation.

We return to the general expression for the screened potential given by (4.50). In local approximation and after Fourier transformation with respect to the difference variables, the retarded/advanced screened potential between two particles of species a and b is given by

$$V_{ab}^{sR/A}(\boldsymbol{q}\omega,\boldsymbol{R}t) = \frac{V_{ab}(q)}{\varepsilon^{R/A}(\boldsymbol{q}\,\omega,\boldsymbol{R}t)}\,. \tag{4.100}$$

In this section, the Planck constant is written explicitly. Then, $V_{ab}(q) = 4\pi\hbar^2 e_a e_b/q^2$ is the Coulomb potential with $e_a = Z_a e$ being the charge of species a. As in the previous section, we will focus on the retarded dielectric function $\varepsilon^R(\boldsymbol{q}\omega,\boldsymbol{R}t)$. Therefore, we omit the superscript R used for retarded quantities, i.e., we write for the retarded dielectric function

$$\varepsilon^R(\boldsymbol{q}\omega,\boldsymbol{R}t) = \varepsilon(\boldsymbol{q}\omega,\boldsymbol{R}t)\,. \tag{4.101}$$

This notation will also be used for other retarded quantities.

Let us start with (4.60) which relates the dielectric function to the polarization function. Using the notation just introduced, we have

$$\varepsilon(\boldsymbol{q}\omega, \boldsymbol{R}t) = 1 - \sum_{ab} V_{ab}(q) \Pi_{ab}(\boldsymbol{q}\omega, \boldsymbol{R}t) \tag{4.102}$$

with the retarded polarization function given by

$$\Pi_{ab}(\boldsymbol{q}\omega, \boldsymbol{R}t) = i \int \frac{d\omega'}{2\pi} \frac{\Pi^{>}_{ab}(\boldsymbol{q}\omega', \boldsymbol{R}t) - \Pi^{<}_{ab}(\boldsymbol{q}\omega', \boldsymbol{R}t)}{\omega - \omega' + i\varepsilon} . \tag{4.103}$$

The basic equations for the determination of the polarization function defined on the Keldysh time contour were derived in Sect. 4.2. The simplest approximation for $\Pi_{ab}(12)$ was given by (4.31). It is usually referred to as the random phase approximation (RPA). In this approximation and in the case of Fermi particles, the polarization functions $\Pi^{\gtrless}_{ab}(12)$ are given by

$$\Pi^{\gtrless}_{ab}(12) \;=\; -i\delta_{ab}\, g^{\gtrless}_{a}(12) g^{\lessgtr}_{b}(21) . \tag{4.104}$$

Fourier transformation with respect to the relative variables leads to

$$\begin{aligned} &\Pi^{\gtrless}_{aa}(\boldsymbol{q}\omega, \boldsymbol{R}\,t) \\ &= -i \int \frac{d\boldsymbol{p}'}{(2\pi\hbar)^3} \frac{d\omega'}{2\pi}\, g^{\gtrless}_{a}(\boldsymbol{p}' + \boldsymbol{q}\,\omega' + \omega, \boldsymbol{R}\,t)\, g^{\lessgtr}_{a}(\boldsymbol{p}'\omega', \boldsymbol{R}\,t) . \end{aligned} \tag{4.105}$$

Now, we use the Kadanoff–Baym ansatz for the single-particle correlation functions given by (3.209). The spectral function is taken in the quasiparticle approximation (3.198). As in the previous section, we suppress the macroscopic variables $\boldsymbol{R}$ and t. Then, from (4.105), we have the following

$$\begin{aligned} i\Pi^{\gtrless}_{aa}(\boldsymbol{q}, \omega) \;&=\; \int \frac{d\boldsymbol{p}}{(2\pi\hbar)^3} 2\pi\delta\Big(\hbar\omega - E_a(\boldsymbol{p}+\boldsymbol{q}) + E_a(\boldsymbol{p})\Big) \\ &\quad \times f^{\gtrless}_{a}(\boldsymbol{p}+\boldsymbol{q})\, f^{\lessgtr}_{a}(\boldsymbol{p}) , \end{aligned} \tag{4.106}$$

where we introduced the abbreviations $f^{>}_{a} = 1 - f_a$ and $f^{<}_{a} = f_a$. The spin dependencies were taken into account explicitly.

Inserting (4.106) into (4.103), the following expression can be obtained for the retarded polarization function in RPA

$$\Pi_{aa}(\boldsymbol{q}, \omega) = \int \frac{d\boldsymbol{p}}{(2\pi\hbar)^3} \frac{f_a(\boldsymbol{p}) - f_a(\boldsymbol{p}+\boldsymbol{q})}{\hbar\omega + E_a(\boldsymbol{p}) - E_a(\boldsymbol{p}+\boldsymbol{q}) + i\varepsilon} . \tag{4.107}$$

It is easy to find other useful expressions for the RPA polarization function. For particles with $E_a(\boldsymbol{p}) = p^2/2m_a$, we get by a simple transformation of (4.107)

$$\Pi_{aa}(\boldsymbol{q},\omega) = \frac{q^2}{m_a} \int \frac{d\boldsymbol{p}}{(2\pi\hbar)^3} \frac{f_a(\boldsymbol{p})}{(\hbar\omega - \frac{\boldsymbol{p}\cdot\boldsymbol{q}}{m_a} + i\varepsilon)^2 - (q^2/2m_a)^2} \,. \tag{4.108}$$

Now, we are able to write the retarded RPA dielectric function in the form

$$\varepsilon(\boldsymbol{q},\omega) = 1 + \sum_{a,s_a^z} \frac{4\pi\hbar^2 e_a^2}{q^2} \int \frac{d\boldsymbol{p}}{(2\pi\hbar)^3} \frac{f_a(\boldsymbol{p}+\boldsymbol{q}) - f_a(\boldsymbol{p})}{\hbar\omega + E_a(\boldsymbol{p}) - E_a(\boldsymbol{p}+\boldsymbol{q}) + i\varepsilon} \,. \tag{4.109}$$

The sum runs over the plasma species. Furthermore, the spin is accounted for by the sum over the spins.

Equation (4.109) is a famous expression of plasma physics and was derived in the pioneering papers by Bohm and Pines (1953), Lindhard (1954), and by Klimontovich and Silin (1952). Applying the Dirac identity to the spectral representation (4.76), we find for the real part of the dielectric function

$$\mathrm{Re}\,\varepsilon(\boldsymbol{q},\omega) = 1 + \sum_{a,s_a^z} \frac{4\pi\hbar^2 e_a^2}{q^2} \mathcal{P} \int \frac{d\boldsymbol{p}}{(2\pi\hbar)^3} \frac{f_a(\boldsymbol{p}+\boldsymbol{q}) - f_a(\boldsymbol{p})}{\hbar\omega + E_a(\boldsymbol{p}) - E_a(\boldsymbol{p}+\boldsymbol{q})} \tag{4.110}$$

and for the imaginary part

$$\mathrm{Im}\,\varepsilon(\boldsymbol{q},\omega) = \sum_{a,s_a^z} \frac{4\pi\hbar^2 e_a^2}{q^2} \int \frac{d\boldsymbol{p}}{(2\pi\hbar)^3} \pi\hbar\,\delta\Big(\hbar\omega + E_a(\boldsymbol{p}) - E_a(\boldsymbol{p}+\boldsymbol{q})\Big) \times \Big\{ f_a(\boldsymbol{p}) - f_a(\boldsymbol{p}+\boldsymbol{q}) \Big\} \,. \tag{4.111}$$

With (4.110) and (4.111), the basic expressions are now given to study the dielectric properties of dense plasmas in RPA.

First, we consider a non-equilibrium plasma with distribution functions f_a assumed to be isotropic in momentum space. Let us start with the determination of the real part of the dielectric function. In (4.110), the integration over the angles can be carried out, and we get

$$\mathrm{Re}\,\varepsilon(\boldsymbol{q}\omega, \boldsymbol{R}t) = 1 - 2\pi \sum_{a,s_a^z} m_a \frac{4\pi\hbar^2 e_a^2}{q^2} \mathcal{P} \int_{-\infty}^{+\infty} \frac{dp}{(2\pi\hbar)^3} p f_a(p, \boldsymbol{R}t) \times \frac{1}{2q} \left\{ \ln\left(\frac{p a_B}{\hbar} - \frac{q a_B}{2\hbar} - \frac{m_a \omega a_B}{q} \right) + \ln\left(\frac{p a_B}{\hbar} - \frac{q a_B}{2\hbar} + \frac{m_a \omega a_B}{q} \right) \right\} . \tag{4.112}$$

Without further restrictions, this expression can be evaluated only numerically. In the same manner, we have to consider the imaginary part of the dielectric function. Again, it is possible to carry out the integration over the angles. Accounting for the properties of the delta function, we arrive at

$$\operatorname{Im}\varepsilon(\boldsymbol{q}\omega,\boldsymbol{R}t) = (2\pi)^2 \sum_{a,s_a^z} m_a \frac{4\pi\hbar^2 e_a^2}{q^3} \int\limits_{\frac{q}{2}-\frac{m_a\hbar\omega}{q}}^{\frac{q}{2}+\frac{m_a\hbar\omega}{q}} \frac{dp}{(2\pi\hbar)^3}\, p\, f_a(p,\boldsymbol{R}t)\,. \tag{4.113}$$

Let us now consider plasmas in thermodynamic equilibrium. For Fermi particles, the distribution functions are given by

$$f_a(p) = \frac{1}{e^{\beta(\frac{p^2}{2m_a}-\mu_a)}+1} \tag{4.114}$$

where μ_a is the chemical potential and $\beta = 1/k_BT$.

In the case of thermodynamic equilibrium, it is possible to find a simple analytic expression for the imaginary part of the dielectric function. The integral in (4.113) can be evaluated with the result (Gluck 1971)

$$\operatorname{Im}\varepsilon(\boldsymbol{q},\omega) = \sum_{a,s_a^z} \frac{m_a^2 e_a^2 k_BT}{\hbar q^3} \ln\left\{\frac{1+\exp\left\{\beta\left(-\frac{E_a^-}{2m_a}+\mu_a\right)\right\}}{1+\exp\left\{\beta\left(-\frac{E_a^+}{2m_a}+\mu_a\right)\right\}}\right\}. \tag{4.115}$$

Here, we introduced the abbreviation

$$E_a^{\pm} = \left(\pm\frac{q}{2} - \frac{m_a\hbar\omega}{q}\right)^2 . \tag{4.116}$$

Let us now discuss the RPA dielectric function for a plasma in thermodynamic equilibrium starting from the expressions (4.112) and (4.113). The real part was calculated from (4.112) performing the principal value integral numerically whereas the imaginary part could easily be obtained from the analytic formula (4.115). We want to mention that the real and imaginary parts are connected by the dispersion relation (4.72).

Essential features of the behavior of the RPA dielectric function may be already found if the electron gas model is considered. In Figs. 4.1 and 4.2, results are presented for a quantum electron gas at a density of $n = 4\times10^{21}\,\mathrm{cm}^{-3}$ and a temperature of $T = 22000\,\mathrm{K}$. These density–temperature values correspond to a coupling parameter $\Gamma_{ee} = 1.95$ and a Brueckner parameter $r_s = 7.4$.

First, we look at the results for the imaginary part of the dielectric function. As discussed later in more detail, this function can be interpreted as the damping of plasma excitations described by the spectral function (4.62). To get a picture of the global behavior, a 3d-plot is presented which shows $\operatorname{Im}\varepsilon(q,\omega)$ as a function of q and ω, respectively. There is a nondramatic behavior for high values of momentum and energy where the imaginary part

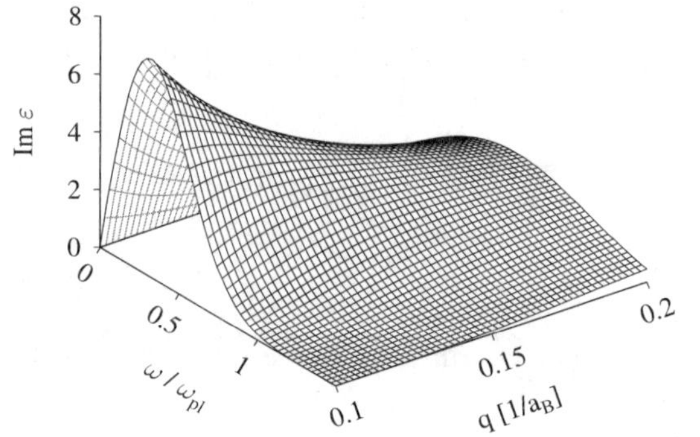

Fig. 4.1. The imaginary part of the RPA dielectric function for a quantum electron gas at a temperature of $T = 22000\,\mathrm{K}$ and density $n = 4 \times 10^{21}\,\mathrm{cm}^{-3}$ versus momentum q and frequency ω

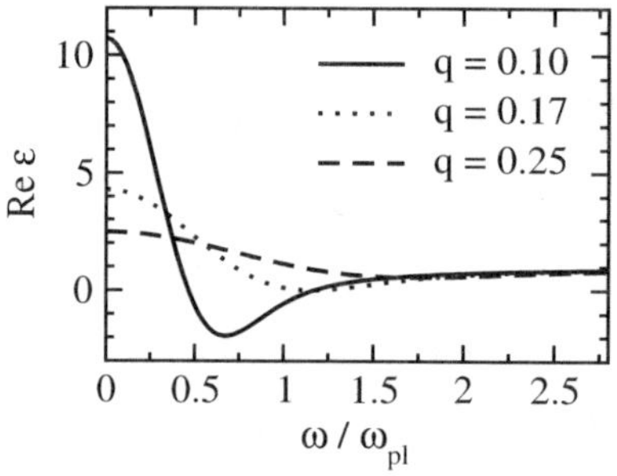

Fig. 4.2. The real part of the RPA dielectric function for a quantum electron gas at temperature $T = 22000\,\mathrm{K}$ and density $n = 4 \times 10^{21}\,\mathrm{cm}^{-3}$ versus frequency ω for three different values of momentum q given in units $\hbar/a_B$

simply becomes zero. Deviations from this behavior can be found in the region where q and ω are small. The consequences will be clear if the real part of the dielectric function is considered. Results for $\mathrm{Re}\,\varepsilon(\boldsymbol{q}, \omega)$ are shown in Fig. 4.2.

For larger values of q and ω, the real part of the RPA dielectric function becomes unity. In this region, plasma effects such as screening and collective excitations can be neglected. An interesting behavior is only found for small values of q and ω. Here, $\mathrm{Re}\,\varepsilon(\boldsymbol{q}, \omega)$ has a minimum. Furthermore, for q-values lower than a critical one, there are two zeros which are of special importance for the spectrum of collective excitations. Looking once more at Fig. 4.1, one observes relatively high values of $\mathrm{Im}\,\varepsilon(q, \omega)$ in the vicinity of the first zero of $\mathrm{Re}\,\varepsilon(q, \omega)$, whereas the imaginary part is very small at the second zero. This behavior suggests that the collective excitations corresponding to the second zero of $\mathrm{Re}\,\varepsilon(q, \omega)$ are only weakly damped. A detailed discussion of the dielectric function in connection with the collective excitations is given in Sect. 4.6.

From the second chapter we know that the properties of dense plasmas are essentially determined by quantum effects due to Fermi statistics. Therefore, it is interesting to see how the behavior of the dielectric function changes with the degeneracy parameter $n_a \Lambda_a^3$. In Fig. 4.3, the imaginary part of the

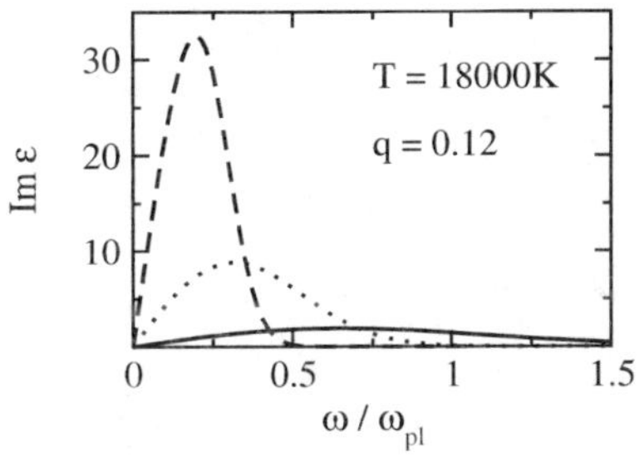

Fig. 4.3. Imaginary part of the dielectric function for an electron gas as a function of ω for a fixed momentum $q = 0.12\,\hbar/a_B$. Three values of the degeneracy parameter $n_e\Lambda_e^3$ were taken (*solid*: 0.2, *dotted*: 1, *dashed*: 5). The temperature is $T = 18000\,\mathrm{K}$

dielectric function is shown for an electron gas at different values of $n_e\Lambda_e^3$ representing a non-degenerate ($n_e\Lambda_e^3 < 1$) and a degenerate ($n_e\Lambda_e^3 > 1$) electron gas. The curves are given as a function of ω for a fixed value of q.

In order to have a better understanding of the behavior of the dielectric function, we look at the limiting cases with respect to the different variables n_a, T, ω and q.

4.5.2 Limiting Cases. Quantum and Classical Plasmas

Let us first consider quantum plasmas in thermodynamic equilibrium for $T \to 0$, i.e., in the highly degenerate case $n_a\Lambda_a^3 \gg 1$. In particular, we will discuss the results for an electron gas at T=0. In order to get the real part of the dielectric function, we remark that the Fermi distribution function becomes a step function $f_e(\boldsymbol{p}) = \Theta(p_F - p)$ with $p_F = \sqrt{2m_e\epsilon_F} = \hbar(3\pi^2 n_e)^{1/3}$ being the Fermi momentum. Then, after an elementary integration, we find

$$\mathrm{Re}\,\varepsilon(\boldsymbol{q},\omega) = 1 + \frac{4\pi e^2}{q^2}\frac{m_e\,p_F}{2\pi^2\,\hbar}\left\{1 + \frac{p_F}{2q}\left[1 - A_+^2\right]\ln\left|\frac{1+A_+}{1-A+}\right| - \frac{p_F}{2q}\left[1 - A_-^2\right]\ln\left|\frac{1+A_-}{1-A_-}\right|\right\}. \qquad (4.117)$$

Here, we introduced the abbreviation

$$A_\pm = \frac{m_e\hbar\omega}{qp_F} \pm \frac{q}{2p_F}\,.$$

To determine the imaginary part for T=0, we start from our previous result (4.115) valid for arbitrary degeneracy. It turns out that there are different regions where $\mathrm{Im}\,\varepsilon(\boldsymbol{q}\omega)$ is zero or nonzero, respectively. We have

$$\mathrm{Im}\,\varepsilon(\boldsymbol{q},\omega) = 0 \qquad (4.118)$$

for $\quad 2m_e\hbar\omega > q^2 + 2qp_F \quad$ and for $\quad q > 2p_F,\; 2m_e\hbar|\omega| < q^2 - 2qp_F.$

The nonzero values of $\mathrm{Im}\varepsilon(\boldsymbol{q}\omega)$ are given by

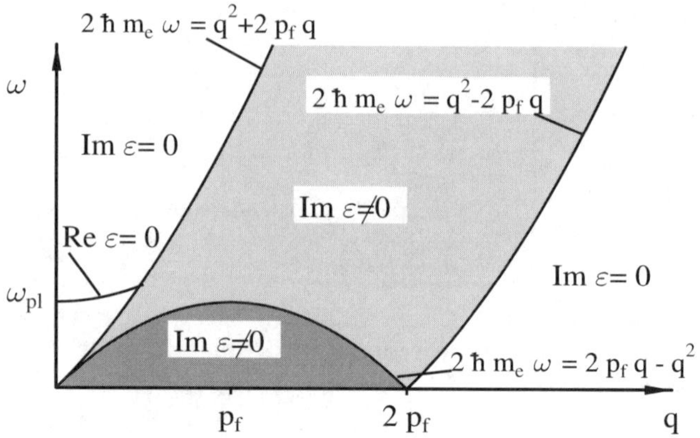

Fig. 4.4. Regions of the q-ω-plane with vanishing or nonvanishing imaginary parts of the RPA dielectric function at $T = 0$

$$\mathrm{Im}\,\varepsilon(\boldsymbol{q}, \omega) = \frac{4\pi e^2}{q^2} \frac{m_e^2\, \omega}{2\pi\, q} \tag{4.119}$$

for $\quad q < 2p_F,\ 2m_e\hbar|\omega| < |q^2 - 2qp_F|\,,$ and by

$$\mathrm{Im}\,\varepsilon(\boldsymbol{q}, \omega) = \frac{4\pi e^2}{q^2} \frac{m_e\, p_F^2}{4\pi\hbar q} \Big(1 \,-\, A_-^2 \Big) \tag{4.120}$$

for $\quad |q^2 - 2qp_F| < 2m_e\hbar|\omega| < |q^2 + 2qp_F|$.

In Fig. 4.4, the different regions are shown where the imaginary part of the dielectric function gives vanishing and non-vanishing contributions at T=0.

The other important limiting case is the non-degenerate plasma, i.e., $n_a \Lambda_a^3 \ll 1$. In that case Maxwell–Boltzmann statistics may be applied, and the Fermi distribution function of species a is replaced by

$$f_a(p) = \exp\left\{ -\beta \left(\frac{p^2}{2m_a} - \mu_a \right) \right\}, \tag{4.121}$$

with the chemical potential μ_a following from (2.18).

Now, the expressions (4.112) and (4.113) get a simpler shape. Especially, we see that the real part of the dielectric function can be written in terms of the confluent hypergeometric function ${}_1F_1(1, \frac{3}{2}, -z)$. The result is (Klimontovich and Kraeft 1974)

$$\mathrm{Re}\,\varepsilon(\boldsymbol{q}, \omega) \,=\, 1 \,-\, \sum_a \frac{\hbar^2 \kappa_a^2}{q^3}$$
$$\times \left\{ \sqrt{E_a^-}\; {}_1F_1\left(1, \frac{3}{2}, -\frac{E_a^-}{2m_a k_B T}\right) \,-\, \sqrt{E_a^+}\; {}_1F_1\left(1, \frac{3}{2}, -\frac{E_a^+}{2m_a k_B T}\right) \right\} \tag{4.122}$$

with $E_a^{\pm}$ given by (4.116), and $\kappa_a^2 = 4\pi n_a e_a^2/k_B T$. The function ${}_1F_1(1, \frac{3}{2}, -z)$ can be expressed by a Dawson integral according to

$$z^{1/2}\ {}_1F_1(1, \frac{3}{2}, -z) = e^{-z} \int\limits_0^{z^{1/2}} e^{t^2}\, dt\,. \tag{4.123}$$

For practical purposes, it is useful to have a fit formula for the above function which allows for a simple and rapid calculation. Such a formula can be constructed starting from the limiting behavior of (4.123) for small and large values of z (Zimmermann 1988). Results with relatively high accuracy can be obtained from

$${}_1F_1(1, \frac{3}{2}, -z) = \frac{1 + \frac{z}{3} + \frac{z^2}{10} + \frac{z^3}{42} + \frac{z^4}{218} + \frac{7z^5 + z^6}{9360}}{1 + z + \frac{z^2}{2} + \frac{z^3}{6} + \frac{z^4}{24} + \frac{z^5}{120} + \frac{z^6}{720} + \frac{z^7}{4680}}\,. \tag{4.124}$$

For non-degenerate plasmas, a simplification is possible for the imaginary part of the dielectric function, too. In the limiting case $n_a \Lambda_a^3 \ll 1$, the exponential functions in the logarithm of (4.115) are small, and we may perform an expansion. The leading terms then give

$$\mathrm{Im}\, \varepsilon(\boldsymbol{q}, \omega) = \sum\nolimits_a \frac{m_a^2 e_a^2 k_B T}{\hbar q^3} n_a \Lambda_a^3 \left\{ \exp\left(-\frac{E_a^-}{2 m_a k_B T} \right) - \exp\left(-\frac{E_a^+}{2 m_a k_B T} \right) \right\}. \tag{4.125}$$

The classical expression for the RPA dielectric function is found from (4.109) in the limit $\hbar \to 0$. We then have

$$\varepsilon(\boldsymbol{q}\omega, \boldsymbol{R}t) = 1 + \sum_a \frac{4\pi\hbar^2 e_a^2}{q^2\, m_a} \int d\boldsymbol{v}\ \frac{\boldsymbol{k} \cdot \frac{\partial}{\partial \boldsymbol{v}} f_a(\boldsymbol{v}, \boldsymbol{R}t)}{\omega - \boldsymbol{k} \cdot \boldsymbol{v} + i\varepsilon}\,, \tag{4.126}$$

where $\boldsymbol{v}$ is the particle velocity, $\boldsymbol{k} = \boldsymbol{q}/\hbar$ the wave number, and ω is the frequency. In thermodynamic equilibrium, the distribution function is a Maxwellian. In the velocity space, we have

$$f_a(\boldsymbol{v}) = n_a \left(\frac{m_a}{2\pi k_B T} \right)^{\frac{3}{2}} \exp\left(-\frac{m_a v^2}{2 k_B T} \right).$$

Then the expression (4.126) can be evaluated. For the real part, we get

$$\mathrm{Re}\, \varepsilon(\boldsymbol{k}, \omega) = 1 + \sum_a \frac{\kappa_a^2}{k^2} W\left(\frac{\sqrt{m_a}\omega}{\sqrt{(k_B T)}k} \right). \tag{4.127}$$

The function $W(z)$ is referred to as the dispersion function and is given by (see, e.g., Ichimaru (1992))

$$\begin{aligned} W(z) &= \frac{1}{\sqrt{2\pi}} \int_{-\infty}^{+\infty} dx \frac{x}{x - z - i\eta} e^{-\frac{x^2}{2}} \\ &= 1 - z^2 {}_1F_1\left(1, \frac{3}{2}; -\frac{z^2}{2}\right) + i\left(\frac{\pi}{2}\right)^{1/2} z \exp\left(-\frac{z^2}{2}\right) . \end{aligned} \tag{4.128}$$

Here, $\kappa_a = (4\pi n_a e_a^2/k_B T)^{1/2}$ is the special inverse screening length. For the imaginary part, we find in the classical limit

$$\mathrm{Im}\varepsilon(\boldsymbol{k}, \omega) = \sqrt{\pi} \sum_a \frac{\kappa_a^2 \omega}{k^3} \left(\frac{m_a}{2k_B T}\right)^{1/2} \exp\left(-\frac{\omega^2 m_a}{2k^2 k_B T}\right) . \tag{4.129}$$

In this form, the imaginary part of the dielectric function is known as Landau damping.

We still give the classical polarization function in RPA. The real part reads

$$\mathrm{Re}\Pi_{aa}(k, \omega) = \frac{n_a}{k_B T} \left\{1 - \left(\frac{\omega}{k}\right)^2 \frac{m_a}{k_B T} {}_1F_1\left(1, \frac{3}{2}; -\frac{\omega^2 m_a}{2k^2 k_B T}\right)\right\} , \tag{4.130}$$

and the imaginary part is given by

$$\mathrm{Im}\Pi_{aa}(k, \omega + i\epsilon) = \pi^{1/2} \frac{n_a}{k_B T} \frac{\omega}{k} \left(\frac{m_a}{2k_B T}\right)^{1/2} \exp\left(-\frac{\omega^2 m_a}{2k^2 k_B T}\right) . \tag{4.131}$$

Let us now consider the behavior of the dielectric function with respect to ω and q.

At small values of the momentum q or for large ω and finite q, it is convenient to start from the general expression (4.108) for the polarization function. In this case, we can expand $\mathrm{Re}\,\Pi_{aa}(q, \omega)$ with respect to q/ω. A simple calculation yields up to the order $(q/\omega)^2$

$$\begin{aligned} \mathrm{Re}\,\Pi_{aa}(\boldsymbol{q}, \omega) &= \frac{(2s_a + 1)}{\hbar^2 m_a} \left(\frac{q}{\omega}\right)^2 \\ &\times \int \frac{d\boldsymbol{p}}{(2\pi\hbar)^3} f_a(p) \left[1 + 2\frac{(\boldsymbol{p} \cdot \boldsymbol{q})}{\hbar m_a} \frac{1}{\omega} + \frac{p^2}{\hbar^2 m_a^2} \left(\frac{q}{\omega}\right)^2\right] . \end{aligned} \tag{4.132}$$

Here, ω denotes the frequency. The second term in the square brackets does not contribute to the integral. Taking into account that the distribution function is normalized with respect to the density, we get

$$\mathrm{Re}\,\Pi_{aa}(\boldsymbol{q}, \omega) = \frac{n_a}{\hbar^2 m_a} \left(\frac{q}{\omega}\right)^2 \left[1 + \frac{1}{\hbar^2 m_a^2} \left(\frac{q}{\omega}\right)^2 \langle p^2 \rangle_a\right] . \tag{4.133}$$

The real part of the dielectric function can then be written as

$$\mathrm{Re}\,\varepsilon(\boldsymbol{q}, \omega) = 1 - \frac{\Omega^2(\boldsymbol{q}, \omega)}{\omega^2} \tag{4.134}$$

with $\Omega(\boldsymbol{q},\omega)$ defined by

$$\Omega^2(\boldsymbol{q},\omega) = \sum_a \omega_a^2 \left[1 + \frac{1}{\hbar^2 m_a^2}\left(\frac{q}{\omega}\right)^2 \left\langle p^2 \right\rangle_a\right] . \tag{4.135}$$

Here, ω_a is the plasma frequency of species a

$$\omega_a = \left(\frac{4\pi e^2 n_a Z_a^2}{m_a}\right)^{1/2} . \tag{4.136}$$

Furthermore, we introduced the abbreviation

$$\left\langle p^2 \right\rangle_a = \frac{(2s_a+1)}{n_a} \int \frac{d\boldsymbol{p}}{(2\pi\hbar)^3} p^2 f_a(p) . \tag{4.137}$$

It is easy to express this mean value by the Fermi integrals defined by (2.26).

A simpler expression results in the limiting case $q \to 0$. Then we have the following well-known classical result

$$\mathrm{Re}\,\varepsilon(\boldsymbol{q},\omega) = 1 - \frac{\omega_{\mathrm{pl}}^2}{\omega^2} , \tag{4.138}$$

where ω_{pl} is the plasma frequency given by $\omega_{\mathrm{pl}}^2 = \sum_a \omega_a^2$. (4.138) is equal to the exact asymptotic formula for large ω found in Sect. 4.4 using the Kramers–Kronig relations in the case of the one-component plasma (OCP). This result is of great importance in plasma physics because it is the basic expression in the elementary theory of collective excitations.

Another important limiting case is statical screening in the long wave limit, this means that the RPA dielectric function is considered at zero frequency ω=0 and for small momenta $q \to 0$. We start from the expression (4.108) for the polarization function $\Pi_{aa}(\boldsymbol{q},\omega)$. For equilibrium distribution functions of the form $f_a(p) = f_a(p^2/2m_a - \mu_a)$, we get after simple algebra

$$\begin{aligned} \lim_{q\to 0} \Pi_{aa}^R(\boldsymbol{q},0) &= -(2s_a+1)\, 4\pi m_a \int_0^\infty \frac{dp}{(2\pi\hbar)^3} f_a(p) \\ &= -\frac{\partial n_a(\mu_a)}{\partial \mu_a} . \end{aligned} \tag{4.139}$$

In this limit, the RPA dielectric function is a real function and can be written as

$$\lim_{q\to 0} \varepsilon(\boldsymbol{q},0) = 1 + \frac{\hbar^2\kappa^2}{q^2} , \tag{4.140}$$

where $\kappa = 1/r_0$ is the inverse screening length given by

$$r_0^{-2} = 4\pi \sum_a e_a^2 \frac{\partial n_a(\mu_a)}{\partial \mu_a} . \tag{4.141}$$

This quantity was already discussed in Chap. 2. For instance, we get in the non-degenerate case $\partial n_a/\partial \mu_a = n_a/k_BT$, and (4.141) gives just the Debye screening length. If we compare this result for an OCP with the exact asymptotic expression (4.95), Debye screening corresponds to the application of the compressibility of a classical system of noninteracting particles.

With (4.140), we found the RPA dielectric function which describes static screening. For the screened potential, we then have

$$V_{ab}^s(\boldsymbol{q}, 0) = \frac{V_{ab}(q)}{\varepsilon(\boldsymbol{q}, 0)}, \tag{4.142}$$

and after Fourier transformation

$$V_{ab}^s(r) = \frac{e_a e_b}{r} e^{-r/r_0}. \tag{4.143}$$

This is the well-known expression for the Debye potential given as a function of the distance r between two interacting particles. This was already found in Chap. 2 in the frame of an elementary theory.

4.5.3 The Plasmon–Pole Approximation

Let us now consider the dielectric properties in the approximations given by (4.134) and (4.138). For this purpose we have to look at the quantity $\mathrm{Im}\, \varepsilon^{-1}(\boldsymbol{q}, \omega)$ which is given by

$$\mathrm{Im}\, \varepsilon^{-1}(\boldsymbol{q}, \omega) = - \frac{\mathrm{Im}\, \varepsilon(\boldsymbol{q}, \omega)}{[\mathrm{Re}\, \varepsilon(\boldsymbol{q}, \omega)]^2 + [\mathrm{Im}\, \varepsilon(\boldsymbol{q}, \omega)]^2}. \tag{4.144}$$

Again, we will restrict ourselves to the electron gas in thermodynamic equilibrium. From this model, main features of the dielectric properties of plasmas may be obtained.

The imaginary part of the dielectric function is small near the plasma frequency $\omega_{\mathrm{pl}} = \omega_e = (4\pi n_e e^2/m_e)^{1/2}$. Therefore, we find from (4.144)

$$\mathrm{Im}\varepsilon^{-1}(\boldsymbol{q}, \omega) = -\sum_K \frac{1}{\left|\frac{\partial}{\partial \omega} \mathrm{Re}\, \varepsilon(\boldsymbol{q}, \omega)|_{\omega=\omega_K}\right|} \pi\, \delta(\omega - \omega_K). \tag{4.145}$$

The frequency ω_K is determined by the dispersion relation

$$\mathrm{Re}\, \varepsilon(\boldsymbol{q}, \omega) = 1 - \frac{\Omega^2(\boldsymbol{q}, \omega)}{\omega^2} = 0. \tag{4.146}$$

The first step of iteration gives the solution (Kraeft et al. 1986)

$$\omega_{1/2} = \pm\omega(q) = \pm\sqrt{\Omega^2(\boldsymbol{q}, \omega_{\mathrm{pl}})}. \tag{4.147}$$

For an electron gas, we get after expansion of the square root

$$\omega(q) = \omega_{\rm pl} \left[1 + \frac{1}{2} \frac{q^2}{\hbar^2 m_e^2} \frac{\langle p^2 \rangle_e}{\omega_{\rm pl}^2} \right] . \tag{4.148}$$

Now, instead of (4.145), we may write

$$\mathrm{Im}\, \varepsilon^{-1}(\boldsymbol{q},\omega) = -\frac{\pi}{2}\omega(q) \Big[\delta\Big(\omega - \omega(q)\Big) - \delta\Big(\omega + \omega(q)\Big) \Big] . \tag{4.149}$$

This formula is valid near the plasmon pole, and thus it is referred to as plasmon pole approximation. For $q \to 0$, (4.149) reduces to $\omega_{1/2} = \pm\omega_{pl}$, and we get the classical result

$$\mathrm{Im}\, \varepsilon^{-1}(\boldsymbol{q},\omega) = -\frac{\pi}{2}\omega_{\rm pl} \Big[\delta\left(\omega - \omega_{\rm pl}\right) - \delta\left(\omega + \omega_{\rm pl}\right) \Big] . \tag{4.150}$$

The expression (4.150) is sufficiently simple and contains one essential physical effect: Oscillations of the plasma with frequency $\omega_{\rm pl}$. These collective excitations will be discussed in more detail in the next section.

The relation (4.150) together with the spectral representation (4.67) for $\varepsilon^{-1}(\boldsymbol{q},\omega)$ may be used in order to construct an expression for the response function. Then we have

$$\varepsilon^{-1}(\boldsymbol{q},\omega) = 1 + \frac{\omega_{\rm pl}^2}{(\omega + i\,\varepsilon)^2 - \omega_{\rm pl}^2} . \tag{4.151}$$

This expression just has poles at the plasma frequency $\pm\,\omega_{\rm pl}$, i.e., the spectrum is reduced to a single mode.

In Sect. 4.4, *sum rules* were derived for the dielectric function which represent exact results. Of course, approximations for the dielectric function do not obey the sum rules automatically. On the other hand, sum rules can be used to test the quality of a given approximation for $\varepsilon(\boldsymbol{q},\omega)$. For example, the plasmon pole approximation (4.150) which describes the important phenomenon of plasma oscillations fulfills the sum rule (4.87). Unfortunately, the validity of the plasmon pole approximation is restricted to the vicinity of the plasma frequency $\omega_{\rm pl}$. It does not describe static screening as can be seen from (4.138).

Now, we will construct an interpolation formula for $\mathrm{Im}\, \varepsilon^{-1}(\boldsymbol{q},\omega)$ which retains the simple structure in terms of δ-functions, but which fulfills the two conditions for $q \to 0$:

$$\varepsilon(0,\omega) = 1 - \frac{\omega_{\rm pl}^2}{\omega^2}; \qquad \varepsilon(\boldsymbol{q},0) = 1 + \frac{\hbar^2\kappa^2}{q^2} . \tag{4.152}$$

This means that the formula describes static screening and, because of the first condition of (4.152), it satisfies the sum rule (4.87). Such an approximation was proposed by Lundquist (1967) and is usually called single-pole approximation. It takes the form (Zimmermann 1976)

$$\mathrm{Im}\,\varepsilon^{-1}(\boldsymbol{q},\omega) = -\pi \frac{\omega_{\mathrm{pl}}^2}{2\,\omega(q)} \left[\delta\Big(\omega - \omega(q)\Big) - \delta\Big(\omega + \omega(q)\Big) \right] . \tag{4.153}$$

Here, the frequency $\omega(q)$ is a modification of (4.148) and is given by

$$\omega^2(q) = \omega_{\mathrm{pl}}^2 \left[1 + \frac{q^2}{\hbar^2 \kappa^2} \right] . \tag{4.154}$$

From the spectral representation (4.67), we then find for the response function $\varepsilon^{-1}(\boldsymbol{q},\omega)$ in single-pole approximation

$$\varepsilon^{-1}(\boldsymbol{q},\omega) = 1 + \frac{\omega_{\mathrm{pl}}^2}{(\omega + i\,\varepsilon)^2 - \omega^2(q)} . \tag{4.155}$$

The expression (4.151) follows immediately in the limiting case $q \to 0$, and for $\omega = 0$, statical screening is described according to (4.140). In this manner, (4.155) incorporates two important physical effects: (i) plasma oscillations and (ii) statical screening of the Coulomb potential.

4.6 Excitation Spectrum, Plasmons

Important many-body effects in strongly correlated plasmas are the short-range screened Coulomb interaction and the phenomenon of collective motion of the particles corresponding to plasma oscillations (plasmons). Let us now discuss the effect of plasma oscillations in more detail. In order to show how plasma oscillations are described in the frame of quantum statistical theory we start from relation (4.52) which determines the dielectric response properties of the plasma due to an applied external field $U_a(\boldsymbol{R}t)$. In local approximation and after Fourier transformation, we get for the induced effective field

$$\delta U_a^{\mathrm{eff}}(\boldsymbol{R}\,t) = \int \frac{d\boldsymbol{q}}{(2\pi\hbar)^3} \frac{d\omega}{2\pi} \frac{\delta U_a(\boldsymbol{q}\omega) \exp\left(\frac{i}{\hbar}\boldsymbol{q}\cdot\boldsymbol{R} - i\omega t\right)}{\varepsilon(\boldsymbol{q}\omega, \boldsymbol{R}\,t)} , \tag{4.156}$$

where $\varepsilon(\boldsymbol{q}\omega, \boldsymbol{R}t)$ is the retarded dielectric function.

From this relation, we find two important consequences:

(i) The response of the plasma is determined by the function $1/\varepsilon = \varepsilon^{-1}$.
(ii) For vanishing ε^{-1}, an arbitrarily small perturbation δU_a is sufficient to produce an oscillatory behavior of the response quantity $\delta U_a^{\mathrm{eff}}$.

This means that the zeros of the retarded dielectric function are the eigen frequencies of the plasma.

It turns out that the eigen frequencies $z(\boldsymbol{q})$ follow from the equation

$$\varepsilon(\boldsymbol{q}, z(\boldsymbol{q})) = 0 . \tag{4.157}$$

Again, we omit the macroscopic variables $\boldsymbol{R}$ and t for simplicity. Because $\varepsilon(\boldsymbol{q},\omega)$ is a complex function, the solutions $z(\boldsymbol{q})$ are complex frequencies, i.e., $z(\boldsymbol{q}) = \omega(\boldsymbol{q}) - i\gamma(\boldsymbol{q})$. Therefore, the collective modes are determined by the analytic properties of the retarded dielectric function as a function of complex z. Let us consider the response function $\varepsilon^{-1}(\boldsymbol{q},z)$. According to (4.157), this function has poles at $z(\boldsymbol{q})$. But, from Sect. 4.4, we know that, under certain conditions, the inverse retarded dielectric function is analytic in the upper half plane ($\mathrm{Im}\, z > 0$). Poles of $\varepsilon^{-1}(\boldsymbol{q},z)$ and therefore zeros of $\varepsilon(\boldsymbol{q},z)$ are possible only in the lower complex z-plane. This means that the zeros of the retarded dielectric function define the frequencies of stable collective excitations (Kraeft et al. 1986)

$$z(\boldsymbol{q}) = \omega(\boldsymbol{q}) - i\,\gamma(\boldsymbol{q}), \qquad \gamma(\boldsymbol{q}) > 0\,. \tag{4.158}$$

The collective excitation is a long living one under the additional condition that $\mathrm{Im}\,\varepsilon\,(\boldsymbol{q},\omega(\boldsymbol{q}))$ is sufficiently small, or

$$\frac{\gamma(\boldsymbol{q})}{\omega(\boldsymbol{q})} \ll 1 \tag{4.159}$$

In this case, one can find an approximate solution of the dispersion relation (4.157) in the following form:

The real part $\omega(\boldsymbol{q})$ of the plasmon frequency is calculated from

$$\mathrm{Re}\,\varepsilon\,(\boldsymbol{q},\omega(\boldsymbol{q})) = 0\,, \tag{4.160}$$

and the imaginary part $\gamma(\boldsymbol{q})$ follows by expansion of $\varepsilon(\boldsymbol{q},z)$ with respect to γ with the result

$$\gamma(\boldsymbol{q}) = \left.\frac{\mathrm{Im}\,\varepsilon(\boldsymbol{q},\omega)}{\frac{\partial}{\partial\omega}\,\mathrm{Re}\,\varepsilon(\boldsymbol{q},\omega)}\right|_{\omega=\omega(\boldsymbol{q})}\,. \tag{4.161}$$

It is known that a small external perturbation applied to a plasma can also produce instable collective modes. This means that the dispersion relation (4.157) has solutions

$$z(\boldsymbol{q}) = \omega(\boldsymbol{q}) - i\,\gamma(\boldsymbol{q}), \qquad \gamma(\boldsymbol{q}) < 0$$

which are located in the upper half plane. In the case that $\varepsilon(\boldsymbol{q},\omega)$ is an analytic function in the upper half plane, the number of zeros is determined by

$$N = \frac{1}{2\pi\, i}\int_C dz\,\frac{\frac{\partial}{\partial z}\,\varepsilon(\boldsymbol{q},z)}{\varepsilon(\boldsymbol{q},z)}$$

where the contour is given in Fig. 4.5. Of course, if $\varepsilon^{-1}(\boldsymbol{q},z)$ is also analytic in the upper half plane, we have $N = 0$. Instabilities are therefore connected with a nonanalytic behavior of the inverse dielectric function. A more detailed

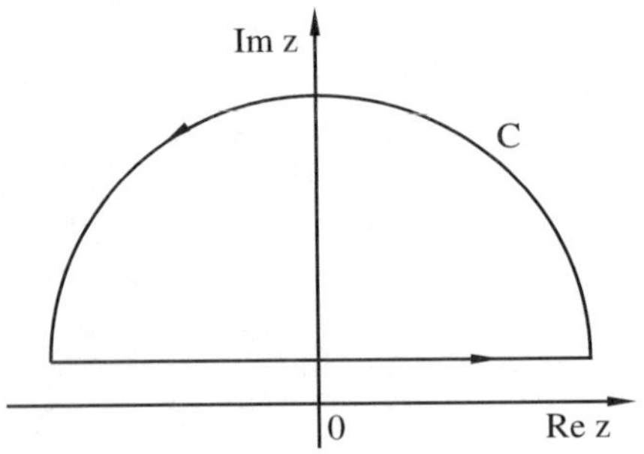

Fig. 4.5. Contour of integration

analysis shows that the collective modes are unstable if there exists a real $\omega = \omega_0$ for which the following relations are valid (Penrose 1960; Nyquist 1928)

$$\mathrm{Im}\,\varepsilon(\omega_0) = 0, \qquad \frac{\partial}{\partial\omega}\,\mathrm{Im}\,\varepsilon(\omega)|_{\omega=\omega_0} < 0, \qquad \mathrm{Re}\,\varepsilon(\omega_0) \leq 0\,.$$

To get further properties of the instable modes, the special form of the imaginary part of the dielectric function must be known (Balescu 1963; Mikhailovskii 1974).

Now, we come back to the approximate solution of the dispersion relation given by (4.160). Analytic results can be obtained only in limiting cases.

The discussion of plasma oscillations becomes rather complicated if one has to sum over species. For this reason we consider again the more simple situation of the *electron gas* with neutralizing background. We assume thermodynamic equilibrium and start our discussion with the expansion of the real part of the dielectric function given by (4.134). For the zeros, according to (4.147) and (4.148), we get

$$\omega(q) = \omega_{pl}\left[1 + \frac{1}{2}\frac{q^2}{\hbar^2 m_e^2}\frac{\langle p^2\rangle_e}{\omega_{pl}^2} + O(q^4)\right] \tag{4.162}$$

with $\omega_{pl} = \omega_e$ being the plasma frequency. The averaged square momentum $\langle p^2\rangle_e$ may be evaluated analytically in limiting situations. First, we consider a non-degenerate electron gas, i.e., $T \gg T_F$. In this case we have $\langle p^2\rangle_e = 3m_e k_B T$ and the following can be written

$$\begin{aligned}\omega(q) &= \omega_{pl}\left[1 + \frac{3k_BT}{2\hbar^2 m_e}\frac{q^2}{\omega_{pl}^2}\right] \\ &= \omega_{pl}\left[1 + \frac{3}{2\hbar^2}q^2 r_D^2\right]\end{aligned} \tag{4.163}$$

where $r_D = (4\pi n_e e^2/k_BT)^{-1/2}$ is the Debye screening length. This represents the classical result for the frequency of plasma waves under the condition $q^2 r_D^2/\hbar^2 \ll 1$ (Vlasov 1938).

In the case $q \to 0$, we get $\omega(q) = \omega_{pl}$ (Tonks and Langmuir 1929; Langmuir 1928).

The corresponding damping of the plasma waves may be obtained from (4.161). With (4.129) and (4.138), we have in the classical limit

$$\gamma(k) = \sqrt{\frac{\pi}{8}} \frac{\omega_{pl}}{(kr_D)^3} \exp\left[-\frac{1}{2(kr_D)^2} - \frac{3}{2}\right] . \tag{4.164}$$

Here, $k = q/\hbar$ denotes the wave number. Taking into account the condition $kr_D \ll 1$ we see that the damping is exponentially small. Therefore, the plasma waves correspond to stable and long living collective excitations. This result for weakly damped waves is known as Landau damping. Landau was the first who considered the solution of the dispersion relation to be complex.

We notice from (4.164) that the damping increases if the wave number goes to higher values. Especially, we have $\gamma \approx \omega_{pl}$ for $kr_D = 1$. But the expression (4.164) cannot be used here because it is restricted to the case $kr_D \ll 1$.

An analytical evaluation of $\langle p^2 \rangle_e$ is also possible for the highly degenerate electron gas, i.e $T \ll T_F$. In this limit we have $\langle p^2 \rangle_e = 3p_F^2/5$, and we find from (4.162)

$$\omega(q) = \omega_{pl} \left[1 + \frac{3}{10\hbar^2 m_e^2} \frac{(q\, p_F)^2}{\omega_{pl}^2}\right] = \omega_{pl} \left[1 + \frac{9}{10}(k\, r_{TF})^2\right] \tag{4.165}$$

with the Thomas–Fermi screening length $r_{TF} = (6\pi n_e e^2/\epsilon_F)^{-1/2}$.

Again, we determine the damping from (4.161). According to our approximation, the condition $2m_e\omega > q^2 + 2qp_F$ is valid, and the imaginary part of the dielectric function can be taken from (4.118). The result is that the damping vanishes for $T = 0$, and the plasmon has an infinite life time.

In the general case, analytical results for the dielectric function are not available, and we have to determine the characteristic features of the excitation spectrum by numerical evaluation of the expressions for the real and imaginary parts of the dielectric function. Then, the excitation spectrum of the plasma is well characterized by the spectral function

$$\operatorname{Im} \varepsilon^{R^{-1}}(\boldsymbol{q}\omega) = -\frac{\operatorname{Im} \varepsilon(\boldsymbol{q}\omega)}{[\operatorname{Re} \varepsilon(\boldsymbol{q}\omega)]^2 + [\operatorname{Im} \varepsilon(\boldsymbol{q}\omega)]^2} . \tag{4.166}$$

In the case of thermodynamic equilibrium, the real and imaginary parts of the RPA dielectric function can be calculated from (4.112) and (4.115). Numerical results are plotted in Fig. 4.6 for a quantum electron gas at a temperature of $T = 18000\,\mathrm{K}$ and density of $n = 1 \cdot 10^{22}\,\mathrm{cm}^{-3}$. For the momentum chosen, we observe two zeros of $\operatorname{Re} \varepsilon(\boldsymbol{q}, \omega)$. The mode corresponding to the smaller ω value is strongly damped because $\operatorname{Im} \varepsilon(\boldsymbol{q}, \omega)$ is large. For the second zero, the

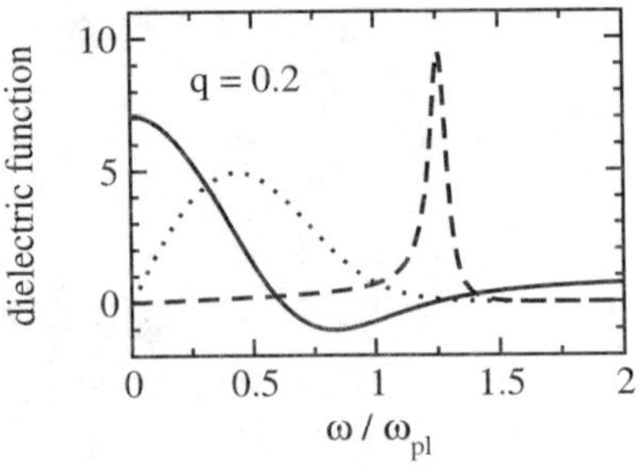

Fig. 4.6. Real part (*full line*), imaginary part (*dotted*) and the inverse dielectric function (spectral function) (*dashed*) in RPA for an electron gas versus frequency ω for a wave number $qa_B/\hbar = 0.2$. Temperature and density are $T = 18000\,\mathrm{K}$ and $n = 1 \times 10^{22}\,\mathrm{cm}^{-3}$

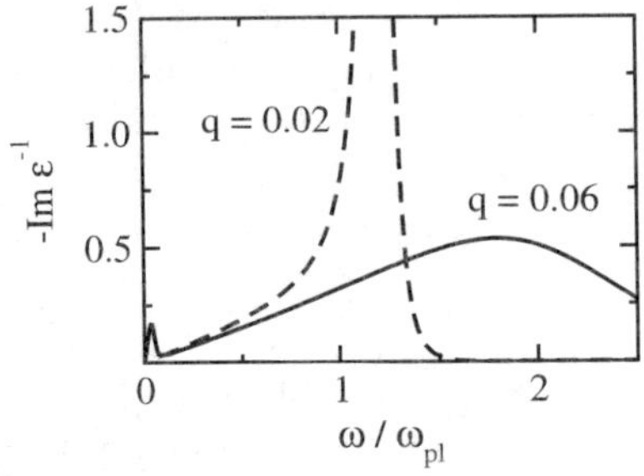

Fig. 4.7. Imaginary part of the inverse dielectric function for a hydrogen plasma as a function of the frequency for the wave numbers $qa_B/\hbar$=0.06 (*solid*) and 0.02 (*dashed*). The plasma is considered to be fully ionized at the temperature $T = 18000\,\mathrm{K}$ and the total electron density $n_e^{tot} = 10^{20}\,\mathrm{cm}^{-3}$

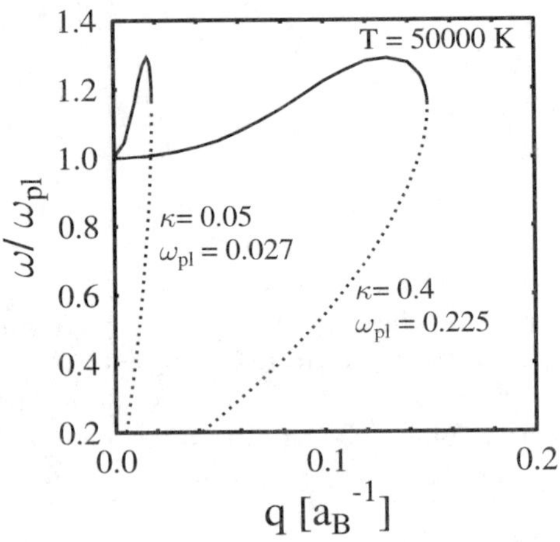

Fig. 4.8. The zeros $\omega(q)$ of the real part of the RPA dielectric function as a function of the wave number q. The resulting dispersion curves are shown for a hydrogen plasma with $T = 50000\,\mathrm{K}$ and two different values of the inverse screening length κ. The upper branch describes weakly damped excitations (plasmons) whereas the lower branch is located in the region of strong damping (acoustic mode)

imaginary part of the dielectric function is small. Here, the spectral function has a sharp peak which describes a long living collective mode, the plasmon.

A similar behavior of the spectral function $\mathrm{Im}\,\varepsilon^{-1}(\boldsymbol{q},\omega)$ is shown in Fig. 4.7 for a hydrogen plasma which is considered to be fully ionized. Again, we observe sharp peaks near the plasma frequency due to the collective modes. The small peaks in the low frequency range represent the contribution of the protons to the RPA dielectric function.

Zeros of $\mathrm{Re}\,\varepsilon(\boldsymbol{q},\omega)$ do not exist for any value of $\boldsymbol{q}$. We can demonstrate this by looking at the dispersion curve which presents the zeros $\omega(q)$ as a function of the momentum q. Dispersion curves for a fully ionized hydrogen plasma are shown in Fig. 4.8.

Three remarkable features can be found:

(i) Zeros $\omega(q)$ exist only below a critical momentum q_c. For a highly degenerate electron gas, the latter is given by

$$\left(\frac{q_c}{p_F}\right)^2 = \frac{4e^2 m_e}{\pi p_F}\left\{\left(1+\frac{q_c}{2p_F}\right)\ln\left(1+\frac{2p_F}{q}\right)-2\right\}.$$

In the general case, q_c has to be determined numerically.

(ii) For a given $q < q_c$, we have two values of $\omega(q)$.

(iii) The asymptotic behavior for the upper plasmon branch is given by (4.148), i.e.,

$$\omega(q) = \omega_{pl}\left[1+\frac{1}{2}\frac{q^2}{\hbar^2 m_e^2}\frac{\langle p^2\rangle_e}{\omega_{pl}^2}\right], \qquad (q\to 0).$$

Let us now discuss the physical meaning of the different features of the spectral function (4.166). To simplify the discussion, we restrict ourselves to the ground state ($T \ll T_F$) of an electron gas. We consider weak excitations, i.e., low frequencies. We then find single-particle excitations (electron–hole excitations at $T=0$) determined by

$$\hbar\omega = E(\boldsymbol{p}+\boldsymbol{q}) - E(\boldsymbol{p}) = \frac{\boldsymbol{q}\cdot\boldsymbol{p}}{m_e} + \frac{q^2}{2m_e}. \tag{4.167}$$

with $p < p_F$ and $|\boldsymbol{p}+\boldsymbol{q}| > p_F$.

The electron–hole excitations with momentum $\boldsymbol{q}$ form a pair continuum having sharp boundaries at $T=0$. These boundaries are found from (4.167) taking the maximum value $p = p_F$. Because $\boldsymbol{q}\cdot\boldsymbol{p} = q\,p\cos\left(\angle(\boldsymbol{q},\boldsymbol{p})\right)$, we have to consider the cases $q < 2p_F$ and $q > 2p_F$. For the boundaries, we then have the following

$$\begin{aligned} 0 \le\ & \hbar\omega \le -\frac{qp_F}{m_e}+\frac{q^2}{2m_e} && \text{if} \quad q < 2p_F\,,\\ -\frac{qp_F}{m_e}+\frac{q^2}{2m_e} \le\ & \hbar\omega \le \frac{qp_F}{m_e}+\frac{q^2}{2m_e} && \text{if} \quad q > 2p_F\,. \end{aligned} \tag{4.168}$$

Therefore, the values allowed for the single-particle excitation energies lie between two parabolas. They were already shown in Fig. 4.4 where we discussed the imaginary part of the dielectric function at $T=0$. These single-particle excitations determine the low-frequency part of the spectral function $\mathrm{Im}\,\varepsilon^{R^{-1}}(\boldsymbol{q},\omega)$. Notice that the single-particle excitations for $q\to 0$ are proportional to the momentum q, i.e.,

$$\hbar\,\omega(\boldsymbol{q}) = q\,\mathrm{v}_p \qquad \text{with} \qquad 0 \le \mathrm{v}_p \le \frac{p_F}{m_e}\,.$$

At higher frequencies, the behavior of the spectral function depends on the choice of q with respect to p_F. For $q < p_F$ (see Fig. 4.4), the frequency behavior is determined by the zero $\omega \approx \omega_{pl}$ of $\mathrm{Re}\,\varepsilon(\boldsymbol{q},\omega)$ and by the smallness of $\mathrm{Im}\,\varepsilon(\boldsymbol{q},\omega)$. The spectral function has a sharp peak near $\omega(q)$ which corresponds to collective excitations (plasmons). Another situation follows for $p_F < q < 2p_F$. Now, the plasmon spectrum overlaps the single-particle excitation spectrum. In this region, strong Landau damping occurs because the plasmons decay into single-particle excitations. The corresponding plasmon peak in the spectral function is widened. For $q > 2p_F$, the effects of the Coulomb interaction are no longer important.

At finite temperatures, there are no such sharp boundaries in the excitation spectrum. This can be seen from the behavior of the spectral function in Fig. 4.6. Qualitatively, the excitation spectrum remains similar to the $T = 0$ scenario.

At the end of this section, we want to make some remarks concerning a non-degenerate two-temperature plasma consisting of Z-fold charged ions with temperature T_i and electrons with temperature T_e. The ions give rise to an additional structure as shown in Fig. 4.7 for the imaginary part of the inverse dielectric function, here for the same temperature of electrons and ions.

For the ions, we apply (4.133) with (4.137). Under the assumption $T_e \gg T_i$, we approximately write for the electrons $\mathrm{Re}\Pi_{ee}(q,\omega) = -\frac{n_e}{k_BT}$. Then we get for the real part of the dielectric function (Bornath 2004)

$$\varepsilon(q,\omega) = 1 + \frac{\kappa_e^2}{\hbar^2 q^2} - \frac{\omega_{pi}^2}{\omega^2} - \frac{\omega_{pi}^4}{\omega^4}\frac{3\hbar^2 q^2}{\kappa_i^2}\,. \tag{4.169}$$

Here we used $\omega_{pa}^2 = 4\pi n_a e_a^2/m_a$ and $\kappa_a^2 = 4\pi n_a e_a^2/(k_BT_a)$. The dispersion relation corresponds to the zeros of (4.169). The ion-acoustic mode follows for $\kappa_e \ll \kappa_i$ and $q \ll \kappa_e$ to be $\omega(q) = \sqrt{(Zk_BT_e/m_i)}q$, where the square root expression is the speed of sound.

4.7 Fluctuations, Dynamic Structure Factor

The properties of plasmas are determined by quantum statistical averages of physical quantities. Of course, the value of such a quantity in a given microstate deviates from the statistical average, i.e., there are fluctuations around the average. If the physical quantity is represented by the operator A, the fluctuations can be characterized by

$$\delta\mathrm{A} = \mathrm{A} - \langle\mathrm{A}\rangle$$

where $\delta\mathrm{A}$ is the operator of fluctuations about the average $\langle\mathrm{A}\rangle$.

Fluctuations are important features of macroscopic many-particle systems, especially of plasmas. The reasons are the following ones:

(i) Fluctuations are connected with the correlations between the particles.
(ii) Fluctuations essentially determine the scattering processes of particles on a target plasma (Pines and Nozieres 1958). Therefore, particle scattering yields an important diagnostic tool.
(iii) The response properties of the system are closely related to fluctuations. This connection is expressed by fluctuation–dissipation theorems (Callen and Welton 1951; Kubo 1957)

Let us turn to the question of how fluctuations can be described in the framework of quantum statistical theory. As shown in Sect. 3.2, the fluctuation theory can be developed in terms of the two-time correlation functions $L^>_{ab}(1,2)$ and $L^<_{ab}(1,2)$ defined by (3.44). We have ($\hbar$=1)

$$iL^>_{ab}(12) = \langle \delta\varrho_a(1)\delta\varrho_b(2)\rangle \qquad iL^<_{ab}(12) = \langle \delta\varrho_a(2)\delta\varrho_b(1)\rangle \ , \tag{4.170}$$

with the density fluctuation operator $\delta\varrho_a(1) = \psi^\dagger_a(1)\psi_a(1) - \langle\psi^\dagger_a(1)\psi_a(1)\rangle$.

To begin, we consider a one-component plasma. The multi-component plasma will be dealt with at the end of this section.

For plasmas, the density fluctuations are closely connected to longitudinal field fluctuations. The relation is given by

$$Ze\delta\boldsymbol{E}(1) = -\nabla_1 \int d2 V(1-2)\delta\varrho(2) \tag{4.171}$$

and was already discussed in Sect. 4.3. There it was shown that the field–field correlation functions can be expressed by the screened potential

$$Z^2e^2\,\langle\delta\boldsymbol{E}(1)\delta\boldsymbol{E}(2)\rangle = i\nabla_1\nabla_2 V^{s>}(1,2)\,. \tag{4.172}$$

The functions $L^{\gtrless}(12)$ have all properties of two-time correlation functions as discussed in Sect. 3.2. In order to show this, we introduce the new variables $1,2 \to \boldsymbol{r}t, \boldsymbol{R}T$ and consider the Fourier transform of $L^{\gtrless}$ with respect to the difference variables $\boldsymbol{r}$ and τ denoted by $L^{\gtrless}(\boldsymbol{q}\omega, \boldsymbol{R}t)$.

In thermodynamic equilibrium, the Kubo–Martin–Schwinger condition can be applied to $L^{\gtrless}$, which then reads

$$L^>(\boldsymbol{q},\omega) = e^{\beta\omega}L^<(\boldsymbol{q},\omega)\,.$$

Here the relations $L^{\gtrless}(\boldsymbol{q},\omega) = L^{\lessgtr}(\boldsymbol{q},-\omega)$ are valid. Introducing the spectral function $\hat{L}(\boldsymbol{q},\omega)$ given by (4.68), we find the spectral representation for the density–density correlation functions

$$iL^<(\boldsymbol{q},\omega) = \hat{L}(\boldsymbol{q},\omega)n_B(\omega) \tag{4.173}$$

and

$$iL^>(\boldsymbol{q},\omega) = \hat{L}(\boldsymbol{q},\omega)\left[1+n_B(\omega)\right]\,. \tag{4.174}$$

As usual, the dynamic structure factor $S(\boldsymbol{q},\omega)$ is considered (Pines and Nozieres 1958; Ichimaru 1992). We introduce it by

$$\begin{aligned} S(\boldsymbol{q},\omega) &= \frac{i}{2\pi} L^{>}(\boldsymbol{q},\omega) \\ &= \frac{1}{2\pi}\frac{1}{\Omega}\int_{-\infty}^{\infty} d\tau\, \langle \delta\varrho(\boldsymbol{q},\tau)\delta\varrho(-\boldsymbol{q},0)\rangle\, e^{i\omega\tau}\,. \end{aligned} \tag{4.175}$$

In thermodynamic equilibrium, we get, from (4.174), the following important formula for the structure factor

$$S(\boldsymbol{q},\omega) = \frac{1}{\pi}\frac{1}{e^{-\beta\omega}-1}\mathrm{Im}L^{R}(\boldsymbol{q},\omega)\,. \tag{4.176}$$

This equation may be interpreted as a fluctuation–dissipation theorem.

For non-equilibrium systems, the dynamic structure factor is given by the more general expression

$$S(\boldsymbol{q}\,\omega,\boldsymbol{R}\,t) = \frac{i}{2\pi}L^{>}(\boldsymbol{q}\,\omega,\boldsymbol{R}\,t) \tag{4.177}$$

with an additional dependence on the macroscopic variables $\boldsymbol{R}$ and t.

In the literature, also the symmetrized structure factor is used. It can be defined as (Pines and Nozieres 1958)

$$\tilde{S}(\boldsymbol{q}\omega,\boldsymbol{R}t) = \frac{i}{4\pi}\left(L^{>}(\boldsymbol{q}\omega,\boldsymbol{R}t) + L^{<}(\boldsymbol{q}\omega,\boldsymbol{R}t)\right)\,. \tag{4.178}$$

In the same way, a symmetrized correlation function of the longitudinal field fluctuations can be introduced

$$\frac{Z^2e^2}{q^2}\left\langle\overline{\delta\boldsymbol{E}\delta\boldsymbol{E}}\right\rangle_{\boldsymbol{q}\omega,\boldsymbol{R}t} = \frac{i}{2}\left(V^{s>}(\boldsymbol{q}\omega,\boldsymbol{R}t) + V^{s<}(\boldsymbol{q}\omega,\boldsymbol{R}t)\right)\,, \tag{4.179}$$

which is related to the symmetrized structure factor according to (4.39). The dynamic structure factor describes the spectrum of density fluctuations and contains all relevant properties of the plasma. Now, we show that there is a close relation between the structure factor and the dielectric function.

We start from the rather general expressions of the density–density correlation functions given in Sect. 4.3. For the one-component plasma, we get from (4.177) and (4.48) in local approximation

$$\begin{aligned} S(\boldsymbol{q}\omega,\boldsymbol{R}t) &= \frac{i}{2\pi}K^{R}(\boldsymbol{q}\omega,\boldsymbol{R}t)\Pi^{>}(\boldsymbol{q}\omega,\boldsymbol{R}t)K^{A}(\boldsymbol{q}\omega,\boldsymbol{R}t) \\ &= \frac{i}{2\pi}\frac{\Pi^{>}(\boldsymbol{q}\omega,\boldsymbol{R}t)}{\left|\varepsilon^{R}(\boldsymbol{q}\omega,\boldsymbol{R}t)\right|^2}\,. \end{aligned} \tag{4.180}$$

Here, the relation $K^R = \varepsilon^{R^{-1}}$ was used with the (retarded) inverse dielectric function given by (4.61).

In thermodynamic equilibrium, we can use the relation (4.174) with (4.74). From (4.175) then follows

$$S(\boldsymbol{q},\omega) = \frac{1}{\pi V(q)} \frac{\mathrm{Im}\varepsilon^R(\boldsymbol{q},\omega)}{|\varepsilon^R(\boldsymbol{q},\omega)|^2} [1 + n_B(\omega)] . \tag{4.181}$$

This result represents one version of the fluctuation–dissipation theorem (Callen and Welton 1951; Kubo 1957; Klimontovich 1975) where the structure factor gives a measure of the density fluctuations, and the inverse dielectric function describes the dissipation in the response of the system. In a similar manner, we find for the symmetrized structure factor (Ichimaru 1992)

$$\tilde{S}(\boldsymbol{q},\omega) = -\frac{1}{\pi V(q)} \mathrm{Im}\varepsilon^{R^{-1}}(\boldsymbol{q},\omega) \left(\frac{1}{2} + n_B\left(\frac{\omega}{k_B T}\right)\right) , \tag{4.182}$$

and for the symmetrized correlation function of the field fluctuations

$$\langle\overline{\delta \boldsymbol{E} \delta \boldsymbol{E}}\rangle_{q\omega} = 8\pi^2 V(q) \tilde{S}(\boldsymbol{q},\omega) . \tag{4.183}$$

In the *classical limit*, the well-known formula holds

$$S(\boldsymbol{q},\omega) = \tilde{S}(\boldsymbol{q},\omega) = -\frac{k_B T}{\pi V(q)} \frac{1}{\omega} \mathrm{Im}\varepsilon^{R^{-1}}(\boldsymbol{q},\omega) . \tag{4.184}$$

From the expressions given above we see that the dynamic structure factor is essentially determined by the inverse dielectric function of the plasma. In RPA, the latter is given by (4.166) with (4.110) and (4.111). In this case, the polarization function is equal to the density–density correlation function of free Fermi particles, i.e.,

$$\Pi^{\gtrless}(12) = L^{0\gtrless}(12) = -i g^{\gtrless}(12) g^{\lessgtr}(21) . \tag{4.185}$$

Using this approximation, we find, from (4.180) and (4.179), the expressions for the structure factor and the correlation function of field fluctuations in RPA. For the latter, we get

$$\begin{aligned} \langle\overline{\delta \boldsymbol{E} \delta \boldsymbol{E}}\rangle_{q\omega}^{RPA} = 4\pi^2 \frac{V(q)}{|\varepsilon^{RPA}(\boldsymbol{q},\omega)|^2} \int \frac{d\boldsymbol{p}'}{(2\pi)^3} \delta\left(\omega - E(\boldsymbol{p}') + E(\boldsymbol{p}' - \boldsymbol{q})\right) \\ \times [1 - f(\boldsymbol{p}')] f(\boldsymbol{p}' - \boldsymbol{q}) + f(\boldsymbol{p}') [1 - f(\boldsymbol{p}' - \boldsymbol{q})] . \end{aligned} \tag{4.186}$$

Here, the variables $\boldsymbol{R}, t$ were dropped. In order to discuss the RPA expression for the dynamic structure factor, we consider first $S(\boldsymbol{q},\omega)$ for the noninteracting Fermi gas. Then we have

$$S^0(\boldsymbol{q},\omega) = \frac{i}{2\pi} L^{0>}(\boldsymbol{q},\omega), \tag{4.187}$$

and from (4.185) with (4.106), we get

$$\begin{aligned} S^0(\boldsymbol{q},\omega) &= \int \frac{d\boldsymbol{p}'}{(2\pi)^3} \delta\left(\omega - E(\boldsymbol{p}') + E(\boldsymbol{p}' - \boldsymbol{q})\right) \\ &\times \left[1 - f(\boldsymbol{p}')\right] f(\boldsymbol{p}' - \boldsymbol{q}). \end{aligned} \tag{4.188}$$

Let us proceed with systems in thermodynamic equilibrium. In this case, the expression (4.188) can be rewritten as

$$S^0(\boldsymbol{q},\omega) = \frac{1}{\pi V(q)} \mathrm{Im}\, \varepsilon(\boldsymbol{q},\omega) \left[1 + n_B(\omega)\right] \tag{4.189}$$

where $\mathrm{Im}\, \varepsilon(q,\omega)$ is the imaginary part of the retarded RPA dielectric function given by (4.115). With these relations, the dynamic structure factor in RPA can be represented in the form

$$S^{RPA}(\boldsymbol{q},\omega) = \frac{S^0(\boldsymbol{q},\omega)}{|\varepsilon^{RPA}(\boldsymbol{q},\omega)|^2} = \frac{S^0(\boldsymbol{q},\omega)}{|1 - V(q) L^0(\boldsymbol{q},\omega)|^2}. \tag{4.190}$$

This expression clearly demonstrates both the single particle excitation part and the collective part of the structure factor:

(i) The structure factor $S^0(q,\omega)$ determines the fluctuation spectrum arising from the excitations of single particle–hole pairs in a system without interactions.
(ii) The contribution $1/|\varepsilon^{RPA}(\boldsymbol{q},\omega|^2$ characterizes the change in the fluctuation spectrum due to dynamical screening in the interacting system. It essentially determines the properties of the total structure factor. Especially, collective excitations of the plasma are included.
(iii) In the case of static screening, the structure factor is

$$S(\boldsymbol{q},\omega) = \frac{S^0(\boldsymbol{q},\omega)}{|\varepsilon^{RPA}(\boldsymbol{q},0)|^2}. \tag{4.191}$$

In order to give a simple illustration of the single particle–hole excitation spectrum we consider the behavior of $S^0(\boldsymbol{q},\omega)$ for the highly degenerate case ($T \ll T_F$) (Pines and Nozieres 1958). We then see that the spectrum of pairs with momentum $\boldsymbol{q}$ forms a continuum as discussed already in Sect. 4.6.

In Fig. 4.9, the free part of the structure factor is shown for an electron gas at $T = 0$ for different values of momentum $\boldsymbol{q}$. The regions where $S^0(\boldsymbol{q},\omega)$ is nonzero are determined by the relations (4.168) (see Fig. 4.4), and they correspond to the allowed energies of pair excitations.

Of course, the pair excitation spectrum changes if the interaction is taken into account in the structure factor according to (4.190). Special features of

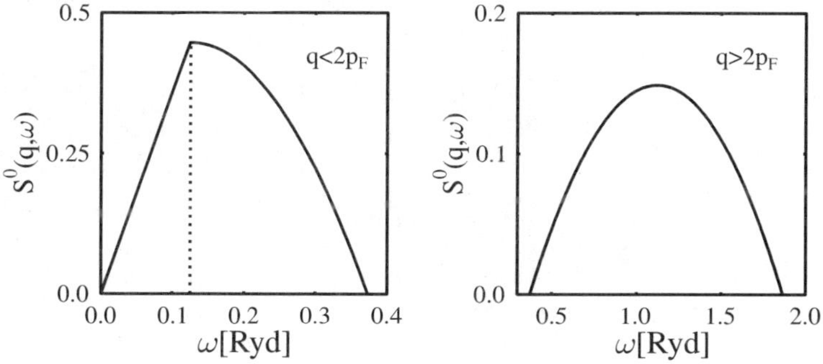

Fig. 4.9. Structure factor for a non-interacting electron gas at $T = 0$ for different momenta q in arbitrary units. The regions where $S^0(\boldsymbol{q}\omega)$ is nonzero correspond to the allowed energies of the pair excitations determined by the relations (4.168). $n_e = 10^{22}\,\mathrm{cm}^{-3}$, $q = p_F$ (**left**), $q = 3p_F$ (**right**)

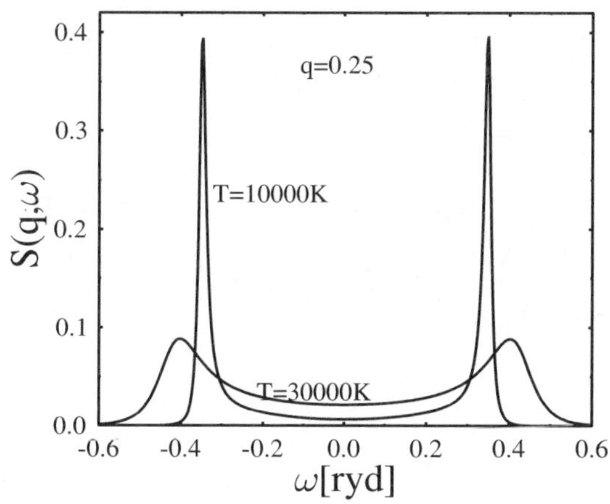

Fig. 4.10. Dynamic structure factor in RPA versus energy for a fixed wave number $qa_B/\hbar = 0.25$ in arbitrary units. An electron gas is considered at the density $n_e = 1 \times 10^{22}\,\mathrm{cm}^{-3}$ for two temperatures

dynamical screening are collective excitations of the plasma. This is demonstrated in Fig. 4.10 where the RPA dynamic structure factor is shown for an electron gas. According to (4.181) and (4.190), the structure factor shows a similar behavior as it was discussed already in Sect. 4.6 for the imaginary part of the inverse dielectric function in RPA. Here, the excitation spectrum of density fluctuations is essentially determined by the more or less broadened peaks due to the collective modes. At high densities these excitations dominate the density fluctuation spectrum. In this case we can use the results obtained in Sect. 4.6 for the case of small plasmon damping rates $\gamma(q)$. For $\gamma(q) \to 0$, we get

$$S(\boldsymbol{q},\omega) = \frac{1}{V(q)} \sum_K \frac{1}{\left|\frac{\partial}{\partial\omega}\mathrm{Re}\,\varepsilon(\boldsymbol{q},\omega)|_{\omega=\omega_K}\right|}\,\delta(\omega-\omega_K)\,[1+n_B(\omega)]\,. \quad (4.192)$$

This equation represents the plasmon pole approximation of the dynamic structure factor.

Let us finally consider a multi-component plasma. Then the structure factor is given by

$$S_{ab}(\boldsymbol{q},\omega) = \frac{1}{\pi}\frac{1}{e^{-\beta\omega}-1}\mathrm{Im}L^R_{ab}(\boldsymbol{q},\omega)\,. \tag{4.193}$$

The function L^R_{ab} has to be determined from (4.46). After Fourier transformation, we get an algebraic system of equations

$$L^{R/A}_{ab}(\boldsymbol{q},\omega) = \Pi^{R/A}_{ab}(\boldsymbol{q},\omega) + \sum_{cd}\Pi^{R/A}_{ac}(\boldsymbol{q},\omega)V_{cd}(q)L^{R/A}_{db}(\boldsymbol{q},\omega)\,. \tag{4.194}$$

For a two-component system consisting of electrons and ions, it is easy to find the solution of (4.194); we get (omitting the superscript R/A)(Kovalenko, Krasnii, and Trigger 1990)

$$\begin{aligned} L_{ee}(\boldsymbol{q},\omega) &= \frac{1}{\varepsilon(\boldsymbol{q},\omega)}\left\{\Pi_{ee} - V_{ii}\left[\Pi_{ee}\Pi_{ii} - \Pi_{ei}\Pi_{ie}\right]\right\}, \\ L_{ei}(\boldsymbol{q},\omega) &= \frac{1}{\varepsilon(\boldsymbol{q},\omega)}\left\{\Pi_{ei} + V_{ie}\left[\Pi_{ee}\Pi_{ii} - \Pi_{ei}\Pi_{ie}\right]\right\}, \end{aligned} \tag{4.195}$$

and analogous equations for L_{ii} and L_{ie}. The dielectric function $\varepsilon(\boldsymbol{q},\omega)$ reads for the two-component system

$$\varepsilon(\boldsymbol{q},\omega) = 1 - \sum_{a,b=e,i}V_{ab}\Pi_{ba} + \left[V_{ii}V_{ee} - V_{ei}^2\right]\left[\Pi_{ee}\Pi_{ii} - \Pi_{ei}\Pi_{ie}\right]\,. \tag{4.196}$$

For the Coulomb potential, the combination $\left[V_{ii}V_{ee} - V_{ei}^2\right]$ vanishes.

We now consider the approximation $\Pi_{ei} = 0$, which is fulfilled, e.g., for RPA. Obviously, the set of relations (4.195) then simplifies to the equation for a "one-component system". We have, e.g.,

$$L_{ii}(\boldsymbol{q},\omega) = \frac{\Pi_{ii}(\boldsymbol{q},\omega)}{1 - V^{\mathrm{eff}}_{ii}(\boldsymbol{q},\omega)\Pi_{ii}(\boldsymbol{q},\omega)}\,, \tag{4.197}$$

with an effective potential $V^{\mathrm{eff}}_{ii}(q\omega)$ defined by

$$V^{\mathrm{eff}}_{ii}(\boldsymbol{q},\omega) = \frac{V_{ii}(q)}{1 - V_{ee}(q)\Pi_{ee}(\boldsymbol{q},\omega)}\,. \tag{4.198}$$

This means, the ion–ion Coulomb potential is screened by the electron dielectric function only. In this approximation, the ion–ion structure factor is

$$S_{ii}(\boldsymbol{q},\omega) = \frac{1}{\pi}\frac{1}{\exp(\beta\omega)-1}\mathrm{Im}\frac{\Pi_{ii}(\boldsymbol{q},\omega)}{1 - V^{\mathrm{eff}}_{ii}(\boldsymbol{q},\omega)\Pi_{ii}(\boldsymbol{q},\omega)}\,. \tag{4.199}$$

Therefore, the electrons influence the structure factor S_{ii} only by the screening of the potential V_{ii}.

4.8 Static Structure Factor and Radial Distribution Function

So far we have described the spectrum of density fluctuations by the dynamic structure factor including the dependencies with respect to the space and the time variables. But, there is much information already contained in the *static structure factor*. Of special interest are the properties of the static structure factor for plasmas in thermodynamic equilibrium which will be considered in this section.

Then, the static structure factor is defined as

$$S_{ab}(\boldsymbol{q}) = \frac{1}{\sqrt{n_a n_b}} \int_{-\infty}^{\infty} d\omega\, S_{ab}(\boldsymbol{q}, \omega) \tag{4.200}$$

where $n_a = \langle N_a \rangle / \Omega$ is the number density. With (4.176), we have

$$S_{ab}(\boldsymbol{q}) = \frac{1}{\sqrt{n_a n_b}} \int_{-\infty}^{\infty} \frac{d\omega}{\pi} \frac{1}{e^{-\beta\omega} - 1} \mathrm{Im} L^R_{ab}(\boldsymbol{q}, \omega)\,. \tag{4.201}$$

In the *classical limit* we have $[e^{-\beta\omega} - 1]^{-1} = -\frac{k_B T}{\omega}$, and we may carry out the integration using the dispersion relation (3.90) for retarded/advanced functions. The result is

$$S_{ab}(\boldsymbol{q}) = -\frac{k_B T}{\sqrt{n_a n_b}} \mathrm{Re} L_{ab}(\boldsymbol{q}, 0)\,. \tag{4.202}$$

As an example, let us consider an electron–ion system. Furthermore, we consider $L^R_{ii}(\boldsymbol{q}, \omega)$ in the approximation (4.197). Then we easily find the ion–ion structure factor

$$S_{ii}(\boldsymbol{q}) = \frac{1}{1 - V^{\mathrm{eff}}_{ii}(\boldsymbol{q}) \Pi_{ii}(\boldsymbol{q}, 0)}\,. \tag{4.203}$$

This is an interesting result. Equation (4.203) is the static structure factor of the ion one-component system, where the ions interact with an effective potential. The effective potential describes the influence of the electrons and is just the ion–ion interaction screened by the electrons. In this way, a modified OCP for the ion component of the plasma is established.

In the long wave limit $\lim_{q \to 0} \Pi^{RPA}(\boldsymbol{q}, 0) = \Pi(0, 0)$, we can apply the relation (4.139). In this limit, (4.203) reduces to

$$S_{ii}(q) = \frac{q^2 + \kappa_e^2}{q^2 + \kappa_e^2 + \kappa_i^2}\,, \tag{4.204}$$

which simplifies, for the pure ion system, to

$$S_{ii}(q) = \frac{q^2}{q^2 + \kappa_i^2} \,. \tag{4.205}$$

Here, the inverse screening length κ_a, $a = e, i$ is determined by (4.141). This quantity changes from the Thomas–Fermi inverse screening length in the strongly degenerate plasma to the Debye one for the low density/high temperature (classical) plasma.

The static structure factor is closely related to the radial distribution function

$$n_a n_b g_{ab}(\boldsymbol{r}) = \left\langle \psi_a^\dagger(\boldsymbol{r}) \psi_b^\dagger(0) \psi_b(0) \psi_a(\boldsymbol{r}) \right\rangle \,. \tag{4.206}$$

From the definition (4.200) of the static structure factor, it follows

$$S_{ab}(\boldsymbol{k}) = \delta_{ab} + \sqrt{n_a n_b} \int d\boldsymbol{r} \, [g_{ab}(\boldsymbol{r}) - 1] \, e^{-i\boldsymbol{k}\cdot\boldsymbol{r}} \,. \tag{4.207}$$

In the Debye–Hückel approximation, the radial distribution function of a classical electron gas is given by the well-known expression (Falkenhagen 1971)

$$g(\boldsymbol{r}) = 1 - \frac{e^2}{k_B T} \frac{e^{-\kappa r}}{r} \,. \tag{4.208}$$

As outlined in the previous sections, the RPA represents a weak coupling approximation and, therefore, the Debye–Hückel expression (4.208) may be applied only to plasmas of weakly interacting particles. In order to include strong coupling effects in the density fluctuation spectrum described by the structure factor one has to go beyond the RPA scheme. Let us consider such a generalization first for a classical OCP. Starting point is (4.207) which relates the static structure factor to the radial distribution function. The latter is determined by the BBGKY hierarchy. If the direct correlation function $c(\boldsymbol{r})$ is introduced in the BBGKY hierarchy we find a closed equation for the total correlation function $h(\boldsymbol{r}) = g(\boldsymbol{r}) - 1$ (Hansen and McDonald 1986)

$$h(r) = c(r) + n \int d\boldsymbol{r}' c(|\boldsymbol{r} - \boldsymbol{r}'|) \, h(\boldsymbol{r}') \tag{4.209}$$

which is known as the *Ornstein–Zernicke relation.* A generalization to non-equilibrium situations can be found in Kremp (a) et al. (1983).

Now, the problem is to have an expression for the direct correlation function $c(\boldsymbol{r})$. In classical statistical mechanics, powerful schemes such as the Percus–Yevick (PY) and the hypernetted chain (HNC) approximations and modified versions thereof were developed to solve this problem (Hansen and McDonald 1986). For classical Coulomb systems, the HNC approximation was shown to be the most adequate scheme. It is determined by the closure relation

$$g(r) = \exp\left\{-\beta V(r) + h(r) - c(r)\right\} . \tag{4.210}$$

This expression, together with the Ornstein–Zernicke relation (4.209) form a closed set of equations to calculate the total and direct correlation functions for the classical OCP in thermodynamic equilibrium.

Further improvements of the closure relations may be obtained by including the bridge functions $B(\boldsymbol{r})$. This problem will be considered in Chap. 6 in connection with the thermodynamic properties of strongly coupled plasmas.

For quantum plasmas, the structure factor has to be calculated from the general expressions (4.175) or (4.177) using quantum many-particle theory. Quantum mechanics is included in the static structure factor on the RPA level if one starts from (4.201) and (4.112). Another way to incorporate quantum effects is shown by the method of quantum potentials introduced in Morita (1959). This allows for the application of methods of classical statistics, and quantum effects enter the theory by effective potentials determined by the quantum mechanical Slater sums (Kelbg 1963; Ebeling et al. 1967; Kremp and Kraeft 1968; Kraeft and Kremp 1968).

4.9 Dielectric Function Beyond RPA

The RPA is the simplest approximation to (4.30). In this approximation, correlations are not included. But it has to be expected that correlation effects influence essentially the collective properties of the plasma. We expect, e.g., a reduction of the plasmon energy and an increasing damping. In order to include correlations, vertex corrections have to be taken into account in (4.30). Such procedure is difficult since a number of constraints like conservation laws or sum rules have to be fulfilled. An important result in this direction is the inclusion of bound states in the dielectric function (Röpke and Der 1979). We will consider the Röpke–Der dielectric function in Sect. 7.7.3.

There are several other methods to construct a dielectric function *beyond RPA*. In accordance with the additional requirements, Hubbard (1958) and later Singwi et al. (1968), Singwi et al. (1970), Ichimaru (1973), Totsuji and Ichimaru (1974) and Ichimaru (1982) introduced a correction factor to the RPA which is known as local field correction. The dynamical behavior of the correction to the RPA in the long-wavelength limit was considered in time dependent mean field theory neglecting damping by Golden and Kalman (1979) and Kalman and Golden (1990). On the basis of the theorem of Nevanlinna, Adamyan and Tkachenko (1983) constructed the dielectric function using its frequency moments.

Let us consider here another idea. We remember that the classical dielectric function as was shown in Chap. 2 follows from the Vlasov equation. Of course it is possible to derive the RPA dielectric function from the quantum Vlasov equation (3.115), too (Klimontovich and Silin 1952). A simple generalization of this equation is to add a collision contribution in relaxation time approximation of the form $(f - f_0)/\tau = \nu(f - f_0)$ to the equation (3.115). Here,

f_0 is the equilibrium distribution function. Then the effect of collisions on the dielectric function is taken into account simply by the replacement $\imath\varepsilon \to i(\varepsilon + \nu)$. It was shown by Kliewer and Fuchs (1969) that such a simple generalization violates the local density conservation, and important sum rules.

To remove this defect, Mermin (1970) proposed to use a relaxation time approximation in which f_0 is a local equilibrium distribution

$$f_0 = \frac{1}{e^{\beta(E-\mu-\delta\mu)} + 1} \, . \tag{4.211}$$

The local chemical potential $\mu + \delta\mu$ is determined by the local density conservation. Without going into the details of the calculation, we give the final exppression of the Mermin dielectric function

$$\varepsilon^M(\boldsymbol{q},\omega) = 1 + \frac{(1+i\nu/\omega)\left(\varepsilon^{RPA}(\boldsymbol{q},\omega+i\nu) - 1\right)}{1 + (i\nu/\omega)\left(\varepsilon^{RPA}(\boldsymbol{q},\omega+i\nu) - 1\right) / \left(\varepsilon^{RPA}(\boldsymbol{q},0) - 1\right)} \, . \tag{4.212}$$

The Mermin dielectric function has the following limiting properties.

1. If $\nu \to 0\,, \varepsilon^M(\boldsymbol{q},\omega) = \varepsilon^{RPA}(\boldsymbol{q},\omega)$ becomes the RPA dielectric function ,

2. For $q \to 0$, the Drude formula is obtained
$$\varepsilon^M(0,\omega) = 1 - \frac{\omega_{pl}^2}{\omega(\omega+i\nu)} \, ,$$

3. For $\omega \to 0$, the well known statical result (4.140) follows
$$\varepsilon^M(\boldsymbol{q},0) = \varepsilon^{RPA}(\boldsymbol{q},0) \, ,$$
and for $q \to 0$, we get (4.140) with the Debye and Thomas–Fermi limits.

Of course we now need expressions for the collision frequency. In papers by Röpke (1998), Röpke and Wierling (1998), and Röpke and Wierling (1999) a generalized linear response theory of the dielectric function was given. In those papers, the Mermin dielectric function was reproduced. Moreover, with the linear response theory, the Mermin expression is generalized to dynamic collision frequencies instead of a static one (Selchow et al. 2001), and general expressions for the dynamic collision frequency were derived. Compare in this connection also the considerations in Sect. 9.1.4. In Born approximation for a statically screened Coulomb potential, the following expression is obtained

$$\nu^{Born}(\omega) = -i\,\frac{e^2(1/k_BT)^{\frac{3}{2}}\,n_e}{24\pi^{\frac{5}{2}}\,\varepsilon^2\,m_e^{\frac{1}{2}}} \int_0^\infty dy\,\frac{y^4}{(n^*+y^2)^2}$$
$$\times \int_{-\infty}^{\infty} dx\, e^{-(x-y)^2}\,\frac{1-e^{-4xy}}{xy(xy-\omega^*-i\varepsilon)} \, ,$$

(see Reinholz et al. (2000), Wierling et al. (2001), and Höll et al. (2004)). Here is $n^* = \hbar^2/(8m_e r_D^2 k_B T)$ and $\omega^* = \hbar\omega/4kT$.

We finally mention an idea of Kwong and Bonitz (2000) and Bonitz et al. (1999), which is close to the lines of this monograph. This approach starts from the Kadanoff–Baym equations (non-equilibrium Dyson equation) with an external inhomogeneous field U

$$g = g_0^0 + g_0^0(\Sigma + U)g \,. \qquad (4.213)$$

To separate correlation and field effects, we rewrite (4.213) into the two equations

$$g^0 = g_0^0 + g_0^0 \Sigma^0 g^0; \quad g = g^0 + g^0(\Sigma^1 + \Sigma^2 + \ldots + U)g \,, \qquad (4.214)$$

where g^0 and Σ are written in powers of the external field. In linear response, only Σ^1 is taken into account. Then it was shown that, already within a Hartree–Fock approximation for Σ^0, correlation effects are included in the response functions. Because the Kadanoff–Baym equations conserve the density and the full energy, the response functions, which follow from the solutions of these equations, automatically fulfill the sum rules.

4.10 Equations of Motion for Density–Density Correlation Functions. Schrödinger Equation for Electron–Hole Pairs

In the previous section, the structure factor was considered to describe the density fluctuation spectrum of plasmas. According to (4.177), the structure factor is directly related to the density–density correlation function $L^>(1,2)$. The latter is determined by the Bethe–Salpeter equation (4.16) for the more general four-point function $L(12,1'2')$ defined on the Keldysh time contour. If we restrict ourselves to a one component plasma, we get

$$L(12) = L^0(12) + \int_C d3d3'd4d4' \, g(13')g(31)\Xi(3'4',34)L(42,4'2) \qquad (4.215)$$

where the kernel Ξ is defined by (4.17). Furthermore, we introduced the free Green's function of density fluctuations which is (in the case of Fermi statistics)

$$L^0(12) = -i\, g(12)\, g(21) \,. \qquad (4.216)$$

Because of the dynamic character of Ξ, (4.215) is, in general, not a closed equation for $L(12)$. This difficulty does not appear if the simplest approximation is used, i.e.,

$$\Xi(3'4',34) = -i\,\delta(3-3')\delta(4-4')\,V(3'-4') \,. \qquad (4.217)$$

This gives the random phase approximation (RPA) for $L(12)$. The equations on the physical time axis then follow by an appropriate choice of the times on the upper and lower branches of the contour. This was already explained in Sect. 3.3.3, and was applied to the basic equations for $L^{\gtrless}$ in Sect. 4.3.

For the density–density correlation functions in RPA we get

$$\begin{aligned} L^{\gtrless}(12) &= L^{0\gtrless}(12) + \int\limits_{-\infty}^{+\infty} d3d4 \Big[L^{0R}(13)V(3-4)L^{\gtrless}(42) \\ &+ L^{0\gtrless}(13)V(3-4)L^{A}(42) \Big] , \end{aligned} \tag{4.218}$$

and the retarded and advanced functions read

$$L^{R/A}(12) = L^{0R}(12) + \int\limits_{-\infty}^{+\infty} d3d4 L^{R/A}(13)V(3-4)L^{0R/A}(42) . \tag{4.219}$$

Here, the limits of time integration were written explicitly.

It is easy to see that (4.218) and (4.219) represent the weak coupling approximation of the more general equations (4.44) and (4.46) derived in Sect. 4.3.

For non-equilibrium systems, we use the new variables defined by (4.56) and perform Fourier transformations with respect to the relative variables. Then, in local approximation, the solution of (4.218) is

$$\begin{aligned} L^{>}(\boldsymbol{q}\omega, \boldsymbol{R}t) &= \langle \delta\varrho\, \delta\varrho \rangle_{\boldsymbol{q}\omega, \boldsymbol{R}t} \\ &= \frac{L^{0>}(\boldsymbol{q}\,\omega, \boldsymbol{R}\,t)}{|\varepsilon^{R}(\boldsymbol{q}\,\omega, \boldsymbol{R}\,t)|^{2}} . \end{aligned} \tag{4.220}$$

According to (4.187,4.190), this expression is related to the dynamic structure factor. It should be noticed that the $L^{0\gtrless}$ are equal to the RPA polarization functions given by (4.106).

It is interesting to consider the connection between $L(12)$ and the more general function $L(12, 1'2')$. The latter is determined by the Bethe–Salpeter equation (4.16) in the particle–hole channel and has to be treated on the Keldysh contour. From this it is easy to find the Bethe–Salpeter equation for the retarded and advanced two-time Green's functions $L^{R/A}(12, 1'2')$ (with $t_1' = t_1^+$, $t_2' = t_2^+$). In RPA, we find

$$\begin{aligned} &\left(\omega + \epsilon\left(\boldsymbol{p}_1'\right) - \epsilon\left(\boldsymbol{p}_1\right)\right) L^{R/A}\left(\boldsymbol{p}_1\boldsymbol{p}_2\boldsymbol{p}_1'\boldsymbol{p}_2', \omega\right) \\ &= \left[f\left(\boldsymbol{p}_1'\right) - f\left(\boldsymbol{p}_1\right)\right](2\pi)^6 \delta\left(\boldsymbol{p}_1 - \boldsymbol{p}_2'\right)\delta\left(\boldsymbol{p}_2 - \boldsymbol{p}_1'\right) \\ &+ V\left(\boldsymbol{p}_1 - \boldsymbol{p}_1'\right)\left[f\left(\boldsymbol{p}_1'\right) - f\left(\boldsymbol{p}_1\right)\right] \int \frac{d\bar{\boldsymbol{p}}_1}{(2\pi)^3}\frac{d\bar{\boldsymbol{p}}_2}{(2\pi)^3} L^{R/A}\left(\bar{\boldsymbol{p}}_1\boldsymbol{p}_2\bar{\boldsymbol{p}}_2\boldsymbol{p}_2', \omega\right) . \end{aligned} \tag{4.221}$$

Here, the local approximation with respect to the times was considered, and Fourier transformations were performed. The macroscopic time t was dropped for simplicity.

Let us construct a solution of (4.221). For this purpose we consider the associated homogeneous equation

$$\Big(\omega_K + \epsilon\,(\boldsymbol{p}_1 - \boldsymbol{q}) - \epsilon\,(\boldsymbol{p}_1)\Big)\Phi_K\,(\boldsymbol{p}_1, \boldsymbol{p}_1 - \boldsymbol{q}) \\ - \quad [f\,(\boldsymbol{p}_1 - \boldsymbol{q}) - f\,(\boldsymbol{p}_1)]\,V(q)\int \frac{d\bar{\boldsymbol{p}}_1}{(2\pi)^3}\Phi_K\,(\bar{\boldsymbol{p}}_1\bar{\boldsymbol{p}}_1 - \boldsymbol{q}) = 0\,. \qquad (4.222)$$

For Fermi systems, this equation can be interpreted as the Schrödinger equation for the wave functions $\Phi_K\,(\boldsymbol{p}_1\boldsymbol{p}_1')$ of an electron–hole pair with the Hamiltonian

$$H = H^0_{11'} + (f_{1'} - f_1)\,V\,, \qquad (4.223)$$

where $H^0_{11'}$ determines the free electron–hole propagation, and the interaction term gives rise to pair excitations (Pines and Nozieres 1958; Stolz 1974; Danielewicz 1990; Haug and Koch 1993). The Hamiltonian (4.223) is not a hermitean operator. Therefore, the solutions of (4.222) do not form an orthogonal system. But, we have a Schrödinger equation for the dual wave functions $\tilde{\Phi}_K$:

$$(\omega_K + E\,(\boldsymbol{p}_1 - \boldsymbol{q}) - E\,(\boldsymbol{p}_1))\,\tilde{\Phi}_K\,(\boldsymbol{p}_1, \boldsymbol{p}_1 - \boldsymbol{q}) \\ - \quad V(q)\int \frac{d\bar{\boldsymbol{p}}_1}{(2\pi)^3}\,[f\,(\bar{\boldsymbol{p}}_1 - \boldsymbol{q}) - f\,(\bar{\boldsymbol{p}}_1)]\,\tilde{\Phi}_K\,(\bar{\boldsymbol{p}}_1, \bar{\boldsymbol{p}}_1 - \boldsymbol{q}) = 0\,. \qquad (4.224)$$

If we use the simple relation $H = (f_{1'} - f_1)H^\dagger(f_{1'} - f_1)$, for the wave functions we get

$$\tilde{\Phi}_K\,(\boldsymbol{p}_1\boldsymbol{p}_1') = \frac{F_K}{f\,(\boldsymbol{p}_1') - f\,(\boldsymbol{p}_1)}\Phi_K\,(\boldsymbol{p}_1\boldsymbol{p}_1') \qquad (4.225)$$

with the normalization constant F_K

$$F_K^{-1} = \int \frac{d\boldsymbol{p}_1}{(2\pi)^3}\frac{d\boldsymbol{p}_1'}{(2\pi)^3}\Phi_K^*\,(\boldsymbol{p}_1\boldsymbol{p}_1')\,[f\,(\boldsymbol{p}_1') - f\,(\boldsymbol{p}_1)]^{-1}\,\Phi_K\,(\boldsymbol{p}_1\boldsymbol{p}_1')\,. \qquad (4.226)$$

Thus the wave functions Φ_K and $\tilde{\Phi}_K$ form a bi-orthonormal system with the orthogonality and completeness relations

$$\Big\langle \Phi_K|\tilde{\Phi}_{K'}\Big\rangle = \delta_{KK'}\,, \qquad 1 = \sum_K |\tilde{\Phi}_K\rangle\langle\Phi_K|\,. \qquad (4.227)$$

Now, it is possible to construct a formal solution of the Bethe–Salpeter equation (4.221). With the relations (4.225), (4.227) and using the wave equation (4.222), we can write the following bilinear expansion of the density–density response function (see also Danielewicz (1990))

$$L^{R/A}\left(\boldsymbol{p}_1\boldsymbol{p}_2\boldsymbol{p}_1'\boldsymbol{p}_2'\omega\right) = \sum_K \frac{\Phi_K\left(\boldsymbol{p}_1\boldsymbol{p}_1'\right)\Phi_K^*\left(\boldsymbol{p}_2\boldsymbol{p}_2'\right)}{\omega-\omega_K\pm i\varepsilon}F_K\,. \qquad (4.228)$$

The K summation runs over the eigen-states of the effective Hamiltonian given by (4.223). These eigen-states describe the density fluctuation excitations and can be divided into two groups.

(i) There are states corresponding to the continuous spectrum of single pair excitations with the energies

$$\omega_K = E\left(\boldsymbol{p}_1\right) - E\left(\boldsymbol{p}_1-\boldsymbol{q}\right)\,.$$

Furthermore, multi-pair excitations are included in the continuous spectrum.

(ii) We have resonance states in the continuum, the plasmons, describing collective excitations.

The latter plasmon states appear as peaks in the spectral function (4.166) as a function of real ω. This was shown in detail in Sect. 4.6. In the limiting case $T=0$, these states form the discrete part of the spectrum.

At finite T and for $q<\kappa$ (κ – inverse screening length), we have sharp resonance peaks with small Landau damping. Therefore, we can treat these collective modes as particles, more precisely as damped quasiparticles (DuBois 1968). To describe the different types of excitations it is more useful to consider $L^{R/A}$ given by (4.228) as a function of complex frequencies z. Then the excitation spectrum corresponds to the singularities of $L^{R/A}$ in the complex z-plane. These singularities are

(i) a branch cut which corresponds to the continuous pair excitation spectrum, and

(ii) single poles $z_K=\omega(\boldsymbol{q})-i\gamma(\boldsymbol{q})$ corresponding to the discrete (complex) energies of plasmons.

To find the energies of the plasmons we can start from the wave equation (4.222). After integration over $\boldsymbol{p}_1$ we get

$$\int\frac{d\boldsymbol{p}_1}{(2\pi)^3}\Phi_K\left(\boldsymbol{p}_1\boldsymbol{p}_1-\boldsymbol{q}\right)\left\{1+V(q)\int\frac{d\bar{\boldsymbol{p}}_1}{(2\pi)^3}\frac{f\left(\boldsymbol{p}_1\right)-f\left(\boldsymbol{p}_1-\boldsymbol{q}\right)}{z-\epsilon\left(\boldsymbol{p}_1\right)+\epsilon\left(\boldsymbol{p}_1-\boldsymbol{q}\right)}\right\}=0\,. \qquad (4.229)$$

We see that this just gives the plasmon dispersion relation

$$\varepsilon^R\left(\boldsymbol{q},z_K(\boldsymbol{q})\right)=0\,. \qquad (4.230)$$

This means that the discrete complex energies are given by the zeros of the retarded dielectric function in the lower half plane (see Sect. 4.4).

Starting from the bilinear expansion (4.228), we find for the imaginary part of the retarded screened potential

$$\mathrm{Im}V^{sR}(\boldsymbol{q}\,\boldsymbol{q}',\omega) = -\sum_K V(q)\Phi_K(\boldsymbol{q})\,\Phi_K^*(\boldsymbol{q}') \times F_K\,V(q')\,\pi\,\delta(\omega-\omega_K) \tag{4.231}$$

where we introduced the wave functions

$$\int \frac{d\boldsymbol{p}}{(2\pi)^3}\Phi_K(\boldsymbol{p}\,\boldsymbol{p}-\boldsymbol{q}) = \Phi_K(\boldsymbol{q})\,. \tag{4.232}$$

From the wave equation (4.222), a formal solution can be found for the $\Phi_K(\boldsymbol{q})$

$$\Phi_K(\boldsymbol{q}) = K^R(\boldsymbol{q}\,\omega_K)\,\Phi_K^0(\boldsymbol{q})\,, \tag{4.233}$$

where the $\Phi_K^0(\boldsymbol{q})$ follow from the solutions of the wave equation (4.222) without the interaction term:

$$\Phi_K^0(\boldsymbol{q}) = \Phi_{\boldsymbol{p}\boldsymbol{p}'}^0(\boldsymbol{q}) = (2\pi)^3\delta(\boldsymbol{p}-\boldsymbol{p}'-\boldsymbol{q})\,. \tag{4.234}$$

The retarded response function K^R is defined by (4.47), and for the one-component plasma, we have $K^R = \varepsilon^{R^{-1}}$. It should be noted that (4.233) represents an integral equation for the wave functions $\Phi_K(\boldsymbol{q})$ similar to the Lippmann–Schwinger equation of scattering theory.

Now, we find from (4.231), in the spatially homogeneous case,

$$\mathrm{Im}V^{sR}(\boldsymbol{q}\omega) = \int \frac{d\boldsymbol{p}}{(2\pi)^3}\frac{d\boldsymbol{p}'}{(2\pi)^3}V^{sR}(\boldsymbol{q}\,\omega)\Phi_{\boldsymbol{p}\boldsymbol{p}'}^0(\boldsymbol{q})\Phi_{\boldsymbol{p}\boldsymbol{p}'}^{0\;*}(\boldsymbol{q})$$
$$\times\left[f(\boldsymbol{p})-f(\boldsymbol{p}')\right]V^{sA}(\boldsymbol{q}\omega)\pi\delta(\omega-\epsilon(\boldsymbol{p})+\epsilon(\boldsymbol{p}'))\,. \tag{4.235}$$

Here, the sum over the modes K had to be replaced by the momentum integrals over $\boldsymbol{p}$ and $\boldsymbol{p}'$. Let us now construct the bilinear expansion for the correlation functions $V^{s\gtrless}$ adopting the ansatz

$$iV^{s\gtrless}(\boldsymbol{q}\omega,t) = 2\,\mathrm{Im}V^{sR}(\boldsymbol{q}\omega,t)\,N_K^{\gtrless}(t) \tag{4.236}$$

with the imaginary part of the screened potential given by (4.235). The $N_K^{\gtrless}$ are given by

$$N_K^{>} = F_K(1+N_K)\,,\qquad N_K^{<} = F_K N_K\,, \tag{4.237}$$

where the N_K are bosonic occupation numbers. (For simplicity, the case $F_K > 0$ is considered only (Danielewicz 1990).)

In thermodynamic equilibrium, the N_K are given by the Bose function $N_K = n_B(\omega) = (e^{\beta\omega}-1)^{-1}$ with $\omega = \epsilon(\boldsymbol{p})-\epsilon(\boldsymbol{p}-\boldsymbol{q})$. Then we have the following

$$iV_s^{>}(\boldsymbol{q}\omega) = \mathrm{Im}\;V^{sR}(\boldsymbol{q}\omega)\,[1+n_B(\omega)] \tag{4.238}$$

with $V^{s>}(\boldsymbol{q},\omega) = V^{s<}(\boldsymbol{q},-\omega)$. This means that the occupation number of all parts of the excitation spectrum is determined by the same function $n_B(\omega)$.

In non-equilibrium, we have to have in mind that the weakly damped plasmons require a more rigorous kinetic description (DuBois 1968). An approximate scheme is to divide the q range in (4.235) into $|\boldsymbol{q}| < \kappa$ and $|\boldsymbol{q}| > \kappa$ with κ being the inverse screening length. There are sharp resonances for $|\boldsymbol{q}| < \kappa$, the weakly damped plasmons. Using the plasmon pole approximation given by (4.149), we get for $\omega > 0$

$$iV^{s>}(\boldsymbol{q}\,\omega, t) = -V(q)\,\omega(\boldsymbol{q}, t)\,\frac{\pi}{2}\,\delta\Big(\omega - \omega(\boldsymbol{q}, t)\Big)\,[1 + N_{\boldsymbol{q}}(t)]\,. \qquad (4.239)$$

Here, $N_{\boldsymbol{q}}(t)$ is interpreted as the distribution function of plasmons which has to be calculated from a plasmon kinetic equation (DuBois (1968); see also Chap. 7).

For $|\boldsymbol{q}| > \kappa$, the correlation functions $V^{S\gtrless}$ have to be considered in the form given by (4.236) and (4.235). This leads to the fluctuation–dissipation theorem (4.64) obtained directly from (4.218).

4.11 Self-Energy in RPA. Single-Particle Spectrum

In the previous sections, the most peculiar properties of plasmas, the screening and the collective effects were described by the dielectric function and the related quantities such as the screened potential, the polarization function, and the dynamic structure factor. Now, we will show how these quantities determine the self-energy which plays an essential role in many-particle theory. As discussed in Chap. 3, the self-energy decouples the Martin–Schwinger hierarchy and leads to formally closed equations in the shape of the Kadanoff–Baym equations for the single-particle correlation functions $g^{\gtrless}$ and for the Green's functions $g^{R/A}$. In this way, the self-energy determines the scattering-in and scattering-out rates for a particle in the medium as well as the spectrum of single-particle excitations, i.e., energy shifts and life times.

The investigation of the self-energy in the framework of real-time Green's function techniques has to be carried out along the lines discussed in Sect. 3.3.4. Having determined the real and the imaginary parts of the self-energy, we may write the expression (3.183) for the spectral function $a(\boldsymbol{p}\omega, \boldsymbol{R}t)$ in local approximation, i.e.,

$$a_a(\boldsymbol{p}, \omega) = \frac{\Gamma_a(\boldsymbol{p}, \omega)}{[\omega - p^2/2m_a - \Sigma_a^{\mathrm{HF}}(\boldsymbol{p}) - \mathrm{Re}\Sigma_a(\boldsymbol{p}, \omega)]^2 + [\frac{1}{2}\Gamma_a(\boldsymbol{p}, \omega)]^2}\,. \qquad (4.240)$$

Here, the macroscopic variables $\boldsymbol{R}, t$ were dropped for simplicity. From Sect. 3.4.1, we know that the poles of the spectral function for complex energies in the lower half plane, i.e., $z = E_a(\boldsymbol{p}) - i\gamma_a(\boldsymbol{p})$, determine the energy E_a and the damping γ_a of the single-particle excitations. If the damping is small, E_a and γ_a determine energy and life time of quasiparticles, respectively. In this case the poles have to be determined from

$$E_a(\boldsymbol{p}) = \frac{p^2}{2m_a} + \Sigma_a^{\mathrm{HF}}(\boldsymbol{p}) + \mathrm{Re}\bar{\Sigma}_a(\boldsymbol{p}, E(\boldsymbol{p}))\,, \tag{4.241}$$

and the damping is

$$\gamma_a(\boldsymbol{p}) = \left.\frac{2\mathrm{Im}\bar{\Sigma}_a(\boldsymbol{p}\omega)}{1 - \frac{\partial}{\partial\omega}\mathrm{Re}\bar{\Sigma}_a(\boldsymbol{p}\omega)}\right|_{\omega=E(\boldsymbol{p})}. \tag{4.242}$$

The imaginary part of the retarded self-energy is given by

$$-2\mathrm{Im}\Sigma(\boldsymbol{p},\omega) = \Gamma(\boldsymbol{p},\omega) = i\left(\Sigma_a^{>}(\boldsymbol{p},\omega) - \Sigma_a^{<}(\boldsymbol{p},\omega)\right)\,, \tag{4.243}$$

and the real part can be calculated from the dispersion relation (3.155). In the case that the spectral function is sharply peaked, i.e., if $\Gamma_a(\boldsymbol{p},\omega)$ is small, one could, instead, also discuss the spectral function for real frequencies. Then, the complex quasiparticle energies correspond to the peak position of the spectral function (real part), and the width of the peak (imaginary part) represents the quasiparticle damping.

In this section, we will consider the self-energy in the V^s-approximation with the polarization function used in RPA. Then we have, according to (4.31)

$$\Sigma_a^{\gtrless}(1,2) = i\,V_{aa}^{s\;\gtrless}(1,2) g_a^{\gtrless}(1,2) \qquad \Pi_{ab}^{\gtrless}(1,2) = \pm i\delta_{ab}\, g_a^{\gtrless}(1,2)\, g_a^{\lessgtr}(2,1) \tag{4.244}$$

and we get after Fourier transformation with respect to the difference variables

$$\begin{aligned}\Sigma_a^{\gtrless}(\boldsymbol{p}\omega,\boldsymbol{R}t) &= i\int \frac{d\boldsymbol{p}'}{(2\pi)^3}\frac{d\omega'}{2\pi}\frac{d\boldsymbol{p}''}{(2\pi)^3}\frac{d\omega'}{2\pi}\, V_{aa}^{s\gtrless}(\boldsymbol{p}'\omega',\boldsymbol{R}t)\, g_a^{\gtrless}(\boldsymbol{p}''\omega'',\boldsymbol{R}t)\\ &\times\; (2\pi)^3\delta(\boldsymbol{p}-\boldsymbol{p}'-\boldsymbol{p}'')\, 2\pi\delta(\omega-\omega'-\omega'')\,.\end{aligned} \tag{4.245}$$

These are the most general expressions for the self-energy in V^s-approximation. Together with the Kadanoff–Baym equations (4.32) and with the screening equation (4.33), we have a self-consistent system of equations for the description of the equilibrium and non-equilibrium properties of the plasma.

Let us now use the generalized fluctuation–dissipation theorem for $V_{aa}^{s\gtrless}$ given by (4.58) without the initial correlation term. From (4.244), in space-time representation, we then can write

$$\Sigma_a^{\gtrless}(1,2) = \mp\sum_c\int d3d4 V_{ac}^{s\;R}(1,3) g_c^{\gtrless}(3,4) g_c^{\lessgtr}(4,3) V_{ca}^{s\;A}(4,2) g_a^{\gtrless}(1,2)\,. \tag{4.246}$$

In the spatially homogeneous case, after Fourier transformation with respect to the difference variables in space, we get

$$
\begin{aligned}
\Sigma_a^{\gtrless}(\boldsymbol{p}, t_1 t_2) &= \mp \sum_c \int \frac{d\boldsymbol{p}'}{(2\pi)^3} \frac{d\boldsymbol{q}}{(2\pi)^3} dt_3 dt_4 \, V_{ac}^{s\,R}(\boldsymbol{q}, t_1 t_3) V_{ca}^{s\,A}(\boldsymbol{q}, t_4 t_2) \\
&\times \; g_a^{\gtrless}(\boldsymbol{p}-\boldsymbol{q}, t_1 t_2) \, g_c^{\gtrless}(\boldsymbol{p}'+\boldsymbol{q}, t_3 t_4) \, g_c^{\lessgtr}(\boldsymbol{p}', t_4 t_3) \,. \qquad (4.247)
\end{aligned}
$$

For an approximate treatment, we use the fluctuation–dissipation theorem for $V_{aa}^{s\gtrless}$ in local approximation given by (4.58). In RPA, we can write

$$
\begin{aligned}
V_{aa}^{s\gtrless}(\boldsymbol{q}\omega, \boldsymbol{R}t) &= \sum_b |V_{ab}^s(\boldsymbol{q}\omega, \boldsymbol{R}t)|^2 \, \Pi_{bb}^{\gtrless}(\boldsymbol{q}\omega, \boldsymbol{R}t) \\
&= \sum_b \left| \frac{V_{ab}(q)}{\varepsilon(\boldsymbol{q}\omega, \boldsymbol{R}t)} \right|^2 \Pi_{bb}^{\gtrless}(\boldsymbol{q}\omega, \boldsymbol{R}t) \qquad (4.248)
\end{aligned}
$$

with the retarded screened potential $V_{ab}^s(\boldsymbol{q}\omega, \boldsymbol{R}t) = V_{ab}(q)/\varepsilon(\boldsymbol{q}\omega, \boldsymbol{R}t)$. The dielectric function ε and the polarization functions $\Pi_{bb}^{\gtrless}$ are given by (4.102), (4.103), and (4.105). To eliminate the correlation functions $g^{\gtrless}(\boldsymbol{p}\omega, \boldsymbol{R}t)$, we use the Kadanoff–Baym ansatz (3.209) with the quasiparticle spectral function given by (3.198). Then, the self-energy function $\Sigma_a^{>}$ can be written as

$$
\begin{aligned}
i\Sigma_a^{>}(\boldsymbol{p}\omega, \boldsymbol{R}t) &= \sum_b \int \frac{d\boldsymbol{p}'}{(2\pi)^3} \frac{d\bar{\boldsymbol{p}}}{(2\pi)^3} \frac{d\bar{\boldsymbol{p}}'}{(2\pi)^3} \left| V_{ab}^s(\boldsymbol{p}-\bar{\boldsymbol{p}}, \omega - \bar{E}_a) \right|^2 \\
&\times \; (2\pi)^3 \, \delta(\boldsymbol{p}+\boldsymbol{p}'-\bar{\boldsymbol{p}}-\bar{\boldsymbol{p}}') 2\pi \, \delta(\omega + E_b' - \bar{E}_a - \bar{E}_b') \\
&\times \; f_b(\boldsymbol{p}', \boldsymbol{R}t) \, [1 - f_a(\bar{\boldsymbol{p}}, \boldsymbol{R}t)] \, [1 - f_b(\bar{\boldsymbol{p}}', \boldsymbol{R}t)] \,, \qquad (4.249)
\end{aligned}
$$

and for the self-energy function $\Sigma_a^{<}$ follows

$$
\begin{aligned}
i\Sigma_a^{<}(\boldsymbol{p}\omega, \boldsymbol{R}t) &= -\sum_b \int \frac{d\boldsymbol{p}'}{(2\pi)^3} \frac{d\bar{\boldsymbol{p}}}{(2\pi)^3} \frac{d\bar{\boldsymbol{p}}'}{(2\pi)^3} \left| V_{ab}^s(\boldsymbol{p}-\bar{\boldsymbol{p}}, \omega - \bar{E}_a) \right|^2 \\
&\times \; (2\pi)^3 \, \delta(\boldsymbol{p}+\boldsymbol{p}'-\bar{\boldsymbol{p}}-\bar{\boldsymbol{p}}') 2\pi \, \delta(\omega + E_b' - \bar{E}_a - \bar{E}_b') \\
&\times \; [1 - f_b(\boldsymbol{p}', \boldsymbol{R}t)] \, f_a(\bar{\boldsymbol{p}}, \boldsymbol{R}t) \, f_b(\bar{\boldsymbol{p}}', \boldsymbol{R}t) \,. \qquad (4.250)
\end{aligned}
$$

We want to remark that the Kadanoff–Baym ansatz (3.209) is a restricting approximation insofar as we neglect retardation which would still be included in the Lipavský ansatz (see (7.11) in Sect. 7.2).

In the case of Fermi particles, we get for the damping rate

$$
\begin{aligned}
&\Gamma_a(\boldsymbol{p}\omega, \boldsymbol{R}t) = i \int \frac{d\boldsymbol{q}}{(2\pi)^3} \frac{d\omega'}{2\pi} 2\pi\delta\left(\omega - \omega' - E_a(\boldsymbol{p}-\boldsymbol{q}, \boldsymbol{R}t)\right) \\
&\times \left\{ V_{aa}^{s>}(\boldsymbol{q}\omega', \boldsymbol{R}t) - \left(V_{aa}^{s>}(\boldsymbol{q}\omega', \boldsymbol{R}t) - V_{aa}^{s<}(\boldsymbol{q}\omega', \boldsymbol{R}t) \right) f_a(\boldsymbol{p}-\boldsymbol{q}, \boldsymbol{R}t) \right\} . \\
&\hfill (4.251)
\end{aligned}
$$

Here, the functions $f_a(\boldsymbol{p}, \boldsymbol{R}t)$ are, in general, non-equilibrium distribution functions and thus indicate the coupling to kinetic equations.

For further simplification, we consider now an equilibrium plasma. In this case, the f_a are the well-known Fermi distribution functions, and the relations (4.64) may be applied for determination of the correlation functions $V_{aa}^{s\gtrless}(\boldsymbol{q},\omega)$. We then find for $\Gamma_a(\boldsymbol{p},\omega)$

$$\begin{aligned}\Gamma_a(\boldsymbol{p},\omega) &= -\int \frac{d\boldsymbol{q}}{(2\pi)^3}\frac{d\omega'}{2\pi} 2\mathrm{Im}V_{aa}^s(\boldsymbol{q},\omega')\, 2\pi\delta\left(\omega-\omega'-E_a(\boldsymbol{p}-\boldsymbol{q})\right) \\ &\times \;\{1+n_B(\omega')-f_a(\boldsymbol{p}-\boldsymbol{q})\}\,. \end{aligned} \tag{4.252}$$

The corresponding real part of the retarded self-energy follows from the dispersion relation (3.155) and reads

$$\begin{aligned}\mathrm{Re}\Sigma_a(\boldsymbol{p},\omega) &= \Sigma_a^{\mathrm{HF}}(\boldsymbol{p}) + \mathrm{Re}\Sigma_a^{\mathrm{MW}}(\boldsymbol{p},\omega) \\ &= \Sigma_a^{\mathrm{HF}}(\boldsymbol{p}) - \mathcal{P}\int \frac{d\boldsymbol{q}}{(2\pi)^3}\frac{d\omega'}{2\pi}\, 2\mathrm{Im}V_{aa}^s(\boldsymbol{q},\omega') \\ &\times \;\frac{1+n_B(\omega')-f_a(\boldsymbol{p}-\boldsymbol{q})}{\omega-\omega'-E_a(\boldsymbol{p}-\boldsymbol{q})}\,, \end{aligned} \tag{4.253}$$

where the second r.h.s. contribution is called *Montroll–Ward* self-energy. Another useful representation for $\mathrm{Re}\Sigma_a$ may be obtained using the dispersion relation for the retarded screened potential which follows from (4.72). Then we take into account that one of the contributions contains a distribution function which does not depend on the frequency ω'. We get

$$\begin{aligned}\mathrm{Re}\Sigma_a(\boldsymbol{p},\omega) &= -\int \frac{d\boldsymbol{q}}{(2\pi)^3}\,\mathrm{Re}V_{aa}^s\left(\boldsymbol{q},\omega-E_a(\boldsymbol{p}-\boldsymbol{q})\right) f_a(\boldsymbol{p}-\boldsymbol{q}) \\ -\int \frac{d\boldsymbol{q}}{(2\pi)^3}\mathcal{P}\int &\frac{d\omega'}{2\pi}\,\frac{2\mathrm{Im}V_{aa}^s(\boldsymbol{q},\omega')}{\omega-\omega'-E_a(\boldsymbol{p}-\boldsymbol{q})}\left(1+n_B(\omega')\right)\,. \end{aligned} \tag{4.254}$$

The simplest approximation to (4.254) is the *Hartree–Fock self-energy* which does not depend on the frequency and reads

$$\Sigma_a^{\mathrm{HF}}(\boldsymbol{p}) = -\int \frac{d\boldsymbol{q}}{(2\pi)^3}\, V_{aa}(q) f_a(\boldsymbol{p}-\boldsymbol{q})\,. \tag{4.255}$$

The imaginary part may also be broken down in the same manner as (4.254).

An analytic evaluation of these expressions is possible only in limiting situations. In the low temperature region, i.e., at high degeneracy, one evaluates the formulae (4.254) and (4.252) and uses *Sommerfeld expansions*, (Kremp et al. 1972; Fennel et al. 1974). The self-energy is written as $\mathrm{Re}\Sigma_a = \mathrm{Re}\Sigma_a^1 + \mathrm{Re}\Sigma_a^2$, where $\mathrm{Re}\Sigma_a^1$ corresponds to the first r.h.s. term of (4.254) and includes the Hartree–Fock term. The remaining term $\mathrm{Re}\Sigma_a^2$ contains the *line-contribution* according to Quinn and Ferrel (1958) and Galitski and Migdal (1958), and a correction term. The T=0 – result was given in the latter two papers, and the Sommerfeld correction terms were derived by

Kremp et al. (1972) and by Fennel et al. (1974). We will not go into details of the calculation. The result for an electron gas near the Fermi surface is

$$\begin{aligned} \mathrm{Re}\Sigma_e^1(\boldsymbol{p}) &= \int \frac{d\boldsymbol{q}}{(2\pi)^3} f_e(\boldsymbol{p}-\boldsymbol{q}) \mathrm{Re} V_{ee}^s(\boldsymbol{q}, \omega - E_e(\boldsymbol{p}-\boldsymbol{q}))\Big|_{\omega=E(\boldsymbol{p})} \\ &= 0.166 r_s p_F^2 \left\{ -(x-1)(\ln r_s + 0.203) - 2 + \frac{\pi^2}{12}\left(\frac{k_B T}{\epsilon_F}\right)^2 \right. \\ &\times \left. \left[(x-1)\left(1.35 - \frac{1}{0.66 r_s} + \frac{1}{2}\ln r_s\right) + 0.4 - \frac{1}{2}\ln r_s \right] \right\} . \end{aligned} \tag{4.256}$$

Here, the abbreviation $x = p/p_F$ was used, where $p_F = \sqrt{2m_e\epsilon_F}$ is the Fermi momentum, ϵ_F is the Fermi energy defined by (2.32), and $r_s = (3/4\pi n_e a_B^3)^{1/3}$ is the Brueckner parameter. For the second r.h.s. contribution of (4.254), one gets

$$\begin{aligned} \mathrm{Re}\Sigma_e^2(\boldsymbol{p}) &= 0.062\ln r_s - 0.065 - (x-1)(0.101\ln r_s - 0.094) \\ &+ \left(\frac{k_B T}{\epsilon_F}\right)^2 \left[x(x-1)\left(\frac{0.152}{r_s} + 0.206\ln r_s - 0.031\right) \right. \\ &- \left. 0.125\ln r_s - \frac{0.076}{r_s} + 0.057 \right] . \end{aligned} \tag{4.257}$$

For the imaginary part, the result is given by

$$\begin{aligned} \Gamma_e(\boldsymbol{p}) &= 0.524\,(x-1)^2\, r_s^{3/2} + 1.522\left(\frac{k_B T}{\epsilon_F}\right)^2 \\ &\times \left[r_s^{-3/2} - 3.149\, r_s^{-2}\,(x-1) \right] . \end{aligned} \tag{4.258}$$

One can see that, for zero temperature, the single-particle damping vanishes at the Fermi surface so that we have quasiparticles of infinite life time. In this case, the single-particle spectral function $a(\boldsymbol{p},\omega)$ may exactly be a delta-distribution. At finite temperatures, there is always a nonzero damping so that the spectral function is always broadened. Therefore, the *quasiparticle concept* is applicable, strictly speaking, *only at zero temperature.*

For the non-degenerate case, one gets for an electron gas for small momenta (Kraeft et al. 1986)

$$\mathrm{Re}\Sigma_e(\boldsymbol{p}, \omega = p^2/2m_e) = -\frac{1}{2}\kappa_e e^2 \left(\frac{3}{4} - \frac{\pi}{32}\right) + \tilde{\gamma}_e \frac{p^2}{2m_e} \tag{4.259}$$

with $\kappa_e = (4\pi n_e e^2/k_B T)^{1/2}$ being the inverse screening length. The parameter

$$\tilde{\gamma}_e = 0.0248 \frac{\kappa_e e^2}{k_B T}$$

accounts for a correction to the kinetic energy leading to an effective mass of the electron.

Often it is convenient to use a momentum independent self-energy correction. Such expressions are referred to as *rigid shift approximations* and are determined according to (3.204) as a mean value of the self-energy. At high temperatures (non-degenerate plasma), the mean value is determined from (Kraeft et al. 1986; Zimmermann 1988)

$$\begin{aligned}\Delta_a &= \langle \mathrm{Re}\Sigma_a(\boldsymbol{p})\rangle \\ &= \frac{1}{n_a}\int \frac{d\boldsymbol{p}}{(2\pi)^3}\,\mathrm{Re}\Sigma_a(\boldsymbol{p}, E_a(\boldsymbol{p}))\, f_a(\boldsymbol{p})\,. \end{aligned} \tag{4.260}$$

The energy we take to be $E_a(\boldsymbol{p}) = p^2/2m_a$, and the distribution function is

$$f_a(\boldsymbol{p}) = n_a \Lambda_a^3\, e^{-\beta E_a(\boldsymbol{p})}$$

with n_a being the density and $\Lambda_a = (2\pi\hbar^2/m_a k_B T)^{1/2}$.

Let us find the averaged self-energy shift Δ_a for a particle of species a in a multi-component plasma in lowest order with respect to the density. For this purpose, the evaluation is carried out inserting the correlation part of the self-energy $\mathrm{Re}\Sigma^{\mathrm{MW}}$ into (4.260), which includes the quantum ring sum. Furthermore, we use the screening equation (4.59) where we take the screened potential on the r.h.s. to be a statically screened one, i.e.,

$$\mathrm{Im}V_{aa}^s(\boldsymbol{q},\omega) = \sum_b V_{ab}(q)\mathrm{Im}\Pi_{bb}(\boldsymbol{q},\omega)V_{ba}^s(\boldsymbol{q},0)\,.$$

The imaginary part of the polarization function in RPA follows from $2\mathrm{Im}\Pi_{bb} = -i(\Pi_{bb}^{>} - \Pi_{bb}^{<})$ with $\Pi_{bb}^{\gtrless}$ given by (4.106). We then get after a simple calculation

$$\begin{aligned}\Delta_a = \sum_b n_b \Lambda_b^3 \Lambda_a^3 \int \frac{d\boldsymbol{p}}{(2\pi)^3}\frac{d\boldsymbol{p}'}{(2\pi)^3}\frac{d\boldsymbol{q}}{(2\pi)^3} V_{ab}(q)\, V_{ab}^s(\boldsymbol{q}) \\ \times \frac{e^{-\beta E_b(\boldsymbol{p}'-\boldsymbol{q})}\, e^{-\beta E_a(\boldsymbol{p})}}{E_a(\boldsymbol{p}) - E_b(\boldsymbol{p}') + E_b(\boldsymbol{p}'-\boldsymbol{q}) - E_a(\boldsymbol{p}-\boldsymbol{q})}\,. \end{aligned} \tag{4.261}$$

The integrations may be done analytically if $V_{ab}^s(\boldsymbol{q})$ is taken to be the Debye potential (4.142) with (4.140) (Kraeft et al. 1986). The final result is

$$\Delta_a = -\sum_b \frac{2\pi n_b e_a^2 e_b^2}{k_B T \kappa}\,\mathrm{G}(\kappa\lambda_{ab}) \tag{4.262}$$

with the inverse screening length $\kappa = (4\pi \sum_c n_c e_c^2/k_B T)^{1/2}$ and the thermal wave length $\lambda_{ab} = (\hbar^2/2\mu_{ab}k_B T)^{1/2}$. The function $\mathrm{G}(y)$ is given by

$$
\begin{aligned}
\mathrm{G}(y) &= \frac{\sqrt{\pi}}{y}\left\{1-\exp\left(\frac{y^2}{4}\right)\left[1-\Phi\left(\frac{y}{2}\right)\right]\right\} \\
&= 1-\frac{\sqrt{\pi}}{4}\,y+O(y^2)\,,
\end{aligned}
\tag{4.263}
$$

with the error-function

$$
\Phi(x)=\frac{2}{\sqrt{\pi}}\int_0^x dt\, e^{-t^2}\,. \tag{4.264}
$$

The leading expression for the mean value of the self-energy is then

$$
\Delta_a = \langle \mathrm{Re}\Sigma_a(\boldsymbol{p})\rangle = -\frac{1}{2}\kappa e_a^2 + \mathcal{O}(n)\,. \tag{4.265}
$$

It should be noted that this expression coincides with the expression (2.102) for the self-energy derived in Sects. 2.5 and 2.6 on the basis of an elementary theory of strongly coupled plasmas.

At the end of this section, we will give some numerical results concerning the single-particle excitations described in RPA. The spectral function and related quantities are especially considered for a plasma in thermodynamic equilibrium, i.e., we start from the expressions (4.252) and (4.253). In a first step of the evaluation of these expressions, free single-particle Green's functions are used. Figure 4.11 shows the damping, the real part of the self-energy and the spectral function versus energy for a quantum electron gas

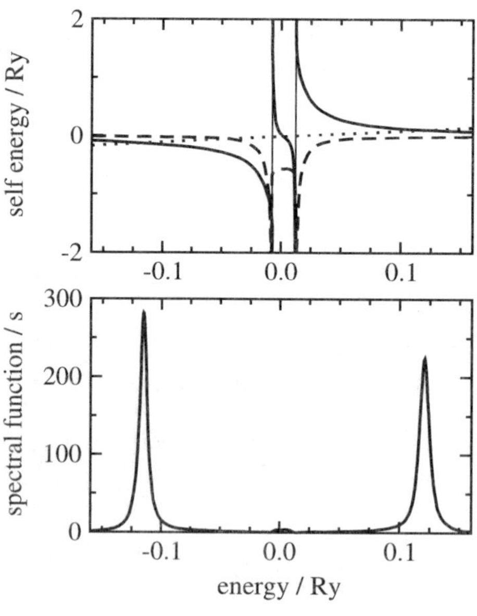

Fig. 4.11. Above: OCP self-energy in MW approximation for a fixed momentum $pa_B/\hbar = 0.05$. $T = 1\,\mathrm{Ry}$, $\kappa a_B = 0.01$. Single particle damping (*dashed line*); real part of the self energy (*solid line*), $\omega - p^2/2m - \Sigma_e^{\mathrm{HF}}(p)$ (*dotted line*). The *crossing points* with the *dotted line* correspond to the solutions of the dispersion relation (4.241). **Below**: The corresponding single-particle spectral function

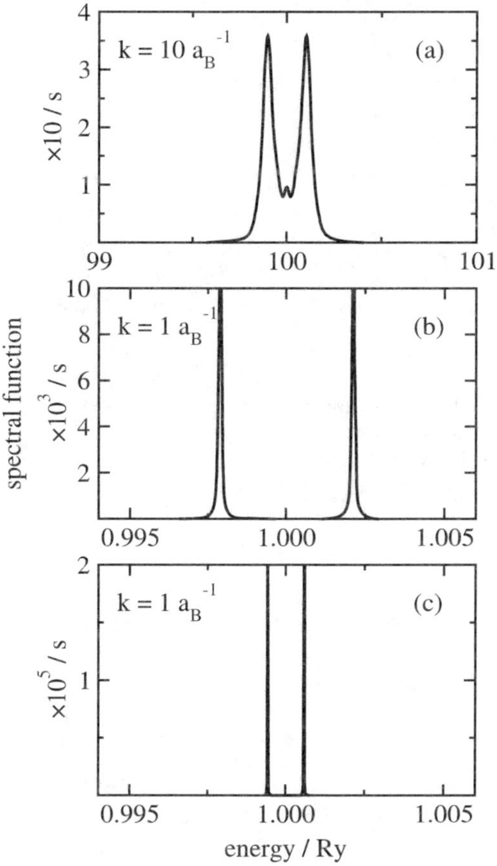

Fig. 4.12. Spectral function in RPA for an electron gas at three different values of the Debye quantity κ: $\kappa a_B = 10^{-2}$ (**a**), 10^{-5} (**b**), and 10^{-6} (**c**). Momentum $pa_B/\hbar = 10$ (**a**) and 1 (**b** and **c**). Temperature fixed to be $k_BT = 1\,\text{Ryd}$ or $T = 1.58 \times 10^5\,\text{K}$

(Fehr 1997; Fehr and Kraeft 1995). In order to determine the quasiparticle energies, the solutions of the dispersion relation (4.241) are presented, too. For the electron gas, there are 5 possible solutions which give un-physical characteristics of the single-particle excitation spectrum (Blomberg and Bergersen 1972). For an electron proton (hydrogen) plasma, we find even 7 solutions; however, only those of them can be considered to be physically relevant, for which $\text{Im}\Sigma(\boldsymbol{p}\omega)$ is small as can be seen in Fig. 4.11; see also Fehr and Kraeft (1995). Indeed, already for the electron gas, we get an unusual structure of the spectral function with three maxima given in Fig. 4.12. This demonstrates the restricted applicability of the simple RPA scheme used in the evaluation presented which corresponds to a first iteration step in determining the spectral function. We may expect that this approximation could be more appropriate at very high densities where we have a strongly degenerate and weakly nonideal plasma.

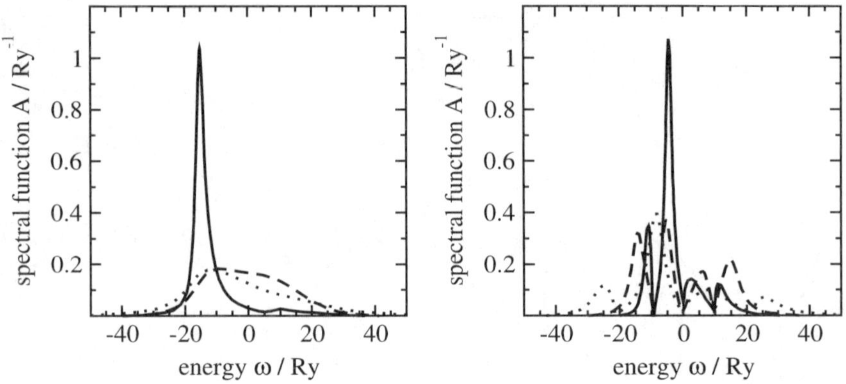

Fig. 4.13. Spectral function determined self-consistently by including self-energy and vertex corrections under solar conditions (H–He) (**left**), see Wierling and Röpke (1996). RPA quasiparticle picture for hydrogen (**right**), see Fehr (1997). Fixed momentum $pa_B/\hbar = 0.21$, $T_{\rm sun} = 15.8 \times 10^6$ K, $n_{e,\rm sun} = 6.2 \times 10^{25}\,\rm cm^{-3}$. *Full line*: $n_{e,\rm sun}$, $T_{sun}/5$; *dashed*: $n_{e,\rm sun}/5$, T_{sun}; *dotted*: $n_{e,\rm sun}$, T_{sun} (Wierling and Röpke (1996); Fehr 1997)

As just mentioned, the RPA scheme used represents a simple approximation. Especially, the application of free single-particle Green's functions in the V^s-approximation of the self-energy and the neglect of vertex corrections lead to an inconsistent approximation to treat the single-particle excitation spectrum for nonideal plasmas. Therefore, one has to apply a consistent approach including additional correlation terms such as T-matrix ones in the self-energy. Attempts of this type were made in the work by Fehr (1997), by Fehr and Kraeft (1995), by Röpke and Wierling (1998), Schepe et al. (1998), and by Schepe (2001). Results obtained in this way are presented in Fig. 4.13. To make the changes more visible, the spectral function calculated from the self-consistent scheme is compared with that obtained in the approximation discussed above (first step RPA). The self-consistent approximation beyond the RPA leads to one well developed broadened peak in the spectral function.

The results obtained in this section for the single-particle properties were based on the RPA and extended versions of it. The special features of these approximations are the inclusion of collective excitations connected with dynamical screening. Strong correlation effects such as multiple scattering and bound states were not taken into account and have to be incorporated in improved approaches. The appropriate scheme to include such effects is the binary collision approximation discussed in the next chapter.

5. Bound and Scattering States in Plasmas. Binary Collision Approximation

In the preceding chapter, we considered the Coulomb interaction and found the RPA to be a physically relevant approximation. The RPA describes the simultaneous interaction of many plasma particles, i.e., the collective behavior. However, this approximation is not sufficient for the discussion of further essential plasma properties, e.g., such as two-particle bound and scattering states beyond the Born approximation. The simplest approximation to take into account bound states, is the ladder approximation for the two-particle Green's function. Let us therefore consider in this chapter the properties of the two-particle Green's function, its connection to the two-particle states, and the equation of motion for these quantities. Equations of motion for the two-particle Green's function are usually called Bethe–Salpeter equations. In order to consider two-particle states in plasmas, many-particle effects like screening, self-energy and Pauli-blocking have to be taken into account. Therefore, we especially have to consider the screened ladder approximation.

5.1 Two-Time Two-Particle Green's Function

Information about two-particle properties of the plasma are given by the two-particle Green's function. The two-particle Green's function determines the self-energy in the Dyson equation of the single-particle Green's function, and it determines the dynamical behavior of a pair of particles in an interacting many-particle system. Let us therefore start with the consideration of this function. Corresponding to (3.41), the two-particle Green's function is defined by

$$g_{ab}(121'2') = \frac{1}{i^2}\left\langle T\{\Psi_a(1)\Psi_b(2)\Psi_b^\dagger(2')\Psi_a^\dagger(1')\}\right\rangle . \tag{5.1}$$

In order to describe non-equilibrium systems, too, the time ordering in (5.1) has to be defined at the Keldysh contour Fig. 3.1; see also Fig. 5.1. The discussion of the two-particle Green's function (5.1) is similar to that of the single-particle Green's function as outlined in Sect. 3.3. But now we have 16 different possibilities to place the four times t_1, t_2, t_1', t_2' at the upper and lower branches of the contour. The expression (5.1) is, therefore, a compact

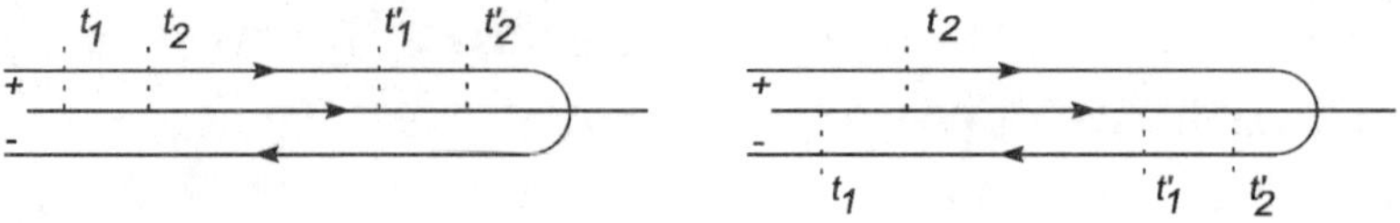

Fig. 5.1. Keldysh contour: Location of times for $g_{ab}^{++++}(t_1, t_2, t'_1, t'_2)$ (**left**) and for $g_{ab}^{-+--}(t_1, t_2, t'_1, t'_2)$ (**right**)

representation of sixteen different functions $g_{ab}^{\alpha\beta\gamma\delta}(t_1, t_2, t'_1, t'_2)$ with the Greek superscripts + or −. Here we used the notation of Sect. 3.3.

Let us consider some examples. If we locate the four times at the upper branch, we get the two-particle causal Green's function $g_{ab}^{++++}(t_1, t_2, t'_1, t'_2)$ (see Fig. 5.1, left). Other examples are the anti-causal Green's function g_{ab}^{----} and mixed arrangements, e.g., g_{ab}^{-+--}, see Fig. 5.1, right.

In the function $g_{ab}^{++++}(t_1, t_2, t'_1, t'_2)$, the time ordering according to (5.1) has to be considered. There are 24 different possibilities to arrange the four times t_1, t_2, t'_1, t'_2 at the upper branch of the contour. These possibilities correspond to the different arrangements of the field operators in (5.1). Corresponding time ordering operations are necessary for two or three out of the four times at the upper/lower branches of the contour, e.g., for g_{ab}^{-+--}. The two-particle Green's function is, therefore, a rather complicated quantity.

Fortunately, in many applications, it is sufficient to consider simpler special cases, namely

$$g_{ab}(12, 1'2')|_{\substack{t_1 = t_2 = t \\ t'_1 = t'_2 = t'}} , \tag{5.2}$$

and

$$g_{ab}(12, 1'2')|_{\substack{t_1 = t'_1 = t \\ t_2 = t'_2 = t'}} . \tag{5.3}$$

The function (5.2) describes the properties of a pair of particles. This case is usually referred to as *particle–particle channel.* The function (5.3) describes the density correlations, and, especially for Fermi particles, this case is called the *particle–hole channel.*

Let us first consider (5.2). In spite of the specialization of the times in the two-particle Green's function given now, there are still 16 different possibilities to arrange the times t_1, t_2, t'_1, t'_2 on the branches of the Keldysh contour.

Fixing the times $t_1 = t_2 = t$ on the upper and $t'_1 = t'_2 = t'$ on the lower branch of the Keldysh contour, respectively, one gets the correlation function

$$g_{ab}^{++--}(x_1x_2t, x'_2x'_1t') = \frac{1}{i^2}\left\langle \Psi_a^\dagger(t')\Psi_b^\dagger(t')\Psi_b(t)\Psi_a(t)\right\rangle = g_{ab}^{<}(t, t') . \tag{5.4}$$

The correlation function $g_{ab}^{<}$ is closely connected to the binary density matrix

$$g_{ab}^{++--}(x_1x_2t, x'_2x'_1t) = n^2 \langle x_1x_2|F_{ab}(t)|x'_2x'_1\rangle$$

and is, therefore, of special importance for the statistical properties of the many-particle system. In the following, we do not always indicate the variables x_i.

Another case follows for $t_1 = t$ and $t_1' = t'$ at the upper and $t_2 = t$ and $t_2' = t'$ at the lower branch

$$\begin{aligned} g_{ab}^{+--+}(x_1x_2t, x_2'x_1't') &= \theta(t-t')(\pm)\frac{1}{i^2}\left\langle \Psi_a^\dagger(t')\Psi_a(t)\Psi_b(t)\Psi_b^\dagger(t')\right\rangle \\ &+\theta(t'-t)(\pm)\frac{1}{i^2}\left\langle \Psi_b(t)\Psi_b^\dagger(t')\Psi_a^\dagger(t')\Psi_a(t)\right\rangle . \end{aligned} \tag{5.5}$$

Further important functions are

$$\begin{aligned} g_{ab}^{-++-}(x_1x_2t, x_2'x_1't') &= \theta(t-t')(\pm)\frac{1}{i^2}\left\langle \Psi_b^\dagger(t')\Psi_b(t)\Psi_a(t)\Psi_a^\dagger(t')\right\rangle \\ &+\theta(t'-t)(\pm)\frac{1}{i^2}\left\langle \Psi_a(t)\Psi_a^\dagger(t')\Psi_b^\dagger(t')\Psi_b(t)\right\rangle , \end{aligned} \tag{5.6}$$

and

$$g_{ab}^{--++}(x_1x_2t, x_2'x_1't') = \frac{1}{i^2}\left\langle \Psi_a(t')\Psi_b(t')\Psi_b^\dagger(t)\Psi_a^\dagger(t)\right\rangle = g_{ab}^{>}(t,t') . \tag{5.7}$$

These four functions (5.4)–(5.7) are determined by six correlation functions. It is possible to show that all other functions $g_{ab}^{\alpha\beta\gamma\delta}$ with the Greek superscripts equal to + or − can be expressed by these six quantities. Similar to the single-particle case, we can introduce retarded and advanced Green's functions by linear combinations of correlation functions. A simple possibility to define such functions is, in analogy to the one-particle case, given by

$$g_{ab}^{R/A}(x_1x_2t, x_2'x_1't') = \pm\theta[\pm(t-t')]\{g_{ab}^{>}(t,t') - g_{ab}^{<}(t,t')\} . \tag{5.8}$$

As usual, the variables $\tau = t - t'$ and $T = (t+t')/2$ are introduced, and the Fourier transform with respect to the difference time τ is considered. Then we get, quite similar to the one-particle functions, a spectral representation of the retarded/advanced two-particle Green's functions

$$g_{ab}^{R/A}(x_1x_2, x_2'x_1'\omega, T) = -i\int \frac{d\bar\omega}{2\pi}\frac{a_{ab}(\bar\omega, T)}{\omega - \bar\omega \pm i\varepsilon} . \tag{5.9}$$

Here, the two-particle spectral function $a_{ab}(\omega, T)$ is given by

$$\begin{aligned} a_{ab}(x_1x_2, x_2'x_1'\omega, T) &= i\int d\tau e^{i\omega\tau}\left\langle [\Psi_a^\dagger(t')\Psi_b^\dagger(t'), \Psi_b(t)\Psi_a(t)]_\mp \right\rangle_{\tau,T} \\ &= \int d\tau e^{i\omega\tau}\{g_{ab}^{>}(\tau,T) - g_{ab}^{<}(\tau,T)\} . \end{aligned} \tag{5.10}$$

We find immediately that the spectral function satisfies the sum rule

$$\int \frac{d\omega}{2\pi} a_{ab}(x_1 x_2, x_2' x_1' \omega, T) = \delta(r_1 - r_1')\delta(r_2 - r_2')$$
$$\pm \left\langle \Psi_a(r_1)\Psi_a^\dagger(r_1')\right\rangle_T \delta(r_2 - r_2') \pm \left\langle \Psi_b(r_2)\Psi_b^\dagger(r_2')\right\rangle_T \delta(r_1 - r_1') . \quad (5.11)$$

Like in the case of the single-particle spectral function, the two-particle one describes the dynamical properties of a pair of particles in an interacting many-particle system. In the general case, it is a continuous function of ω. In certain situations, one may observe more or less pronounced peaks. The physical meaning of these peaks may be seen, e.g., for many-particle systems at very low densities. In such cases, we have delta-function like peaks, i.e., $a_{ab}(\omega) \sim \delta(\omega - E^0_{\alpha P})$, where $E^0_{\alpha P}$ represents the energy spectrum of an isolated pair of particles. In strongly correlated systems, the peaks are widened. The location of the peaks determines the two-particle excitation energies, while the width corresponds to the damping of the excitations. If the spectral function is considered in the complex energy plane, there are singularities which correspond to the peaks for real ω.

Further discussion is quite similar to the discussion of the single-particle Green's function. The function $g^R(z, T)$ is an analytic function in the upper half $z-$plane, and the analytical continuation into the lower half plane is

$$g^R_{ab}(z, T) = g^A_{ab}(z, T) + a_{ab}(z, T) , \quad \mathrm{Im} z < 0 . \quad (5.12)$$

The singularities of $g^R_{ab}(z, T)$ for $\mathrm{Im} z < 0$ are therefore determined by the singularities of $a_{ab}(z, T)$. As mentioned above, the singularities are of physical importance because they determine the position and the width of the peaks of the spectral function, and therefore they determine the energy and the life time of the long living pair excitations of the plasma, especially of the atoms.

The correlation functions $g^{\gtrless}_{ab}(\omega)$ obey, in the thermodynamic equilibrium, the KMS relation

$$g^>_{ab}(\omega) = e^{\beta(\omega - \mu_a - \mu_b)} g^<_{ab}(\omega) . \quad (5.13)$$

Using the definition of a_{ab}, we get the spectral representation

$$\begin{aligned} g^>_{ab}(\omega) &= a_{ab}(\omega, T)[1 + n_{ab}(\omega)] , \\ g^<_{ab}(\omega) &= a_{ab}(\omega, T) n_{ab}(\omega) . \end{aligned} \quad (5.14)$$

Here, n_{ab} is the Bose function given by

$$n_{ab}(\omega) = \frac{1}{e^{\beta(\omega - \mu_a - \mu_b)} - 1} . \quad (5.15)$$

Like in the case of the single-particle correlation function, we have, roughly speaking, a subdivision into statistical and dynamical properties. The Bose function $n_{ab}(\omega)$ contains the statistical information, while the spectral function $a_{ab}(\omega)$ describes the dynamical one.

Regardless of the nice properties of the retarded Green's functions and their spectral functions, we are faced with several shortcomings. Because of the sum rule (5.11), it is not possible, like in the single-particle case, to interpret the spectral function as a weight function with the total weight unity. Further one can show that the equation of motion for the retarded Green's function has a complicated inhomogeneity depending on the one-particle occupation numbers. Therefore, an interpretation as a time propagator is not possible. Finally, for vanishing interaction, the retarded function is given by

$$g_{ab}^{0\,R/A}(t,t') = \pm\theta[\pm(t-t')]\{g_a^>(t,t')g_b^>(t,t') - g_a^<(t,t')g_b^<(t,t')\} \qquad (5.16)$$

instead of $g_{ab}^{0\,R/A} = g_a^{R/A} g_b^{R/A}$.

For this reason, it is interesting to look for other possibilities to construct retarded and advanced Green's functions. An idea how to construct retarded and advanced Green's functions follows from the necessary requirement for the free Green's function

$$g_{ab}^{0R} = g_a^{0R} g_b^{0R} = \Theta(t-t')[g_a^{+-} g_b^{+-} - g_a^{+-} g_b^{-+} - g_a^{-+} g_b^{+-} + g_a^{-+} g_b^{-+}] \,. \qquad (5.17)$$

Therefore, it is obvious to define the following retarded Green's function (Bornath et al. 1999)

$$G_{ab}^R(t,t') = i\Theta(t-t')[g_{ab}^{++--} - g_{ab}^{+--+} - g_{ab}^{-++-} + g_{ab}^{--++}] \,, \qquad (5.18)$$

where the correlation functions are given by (5.4)–(5.7). We remark that an equivalent definition was given by Schmielau (2003)

$$G_{ab}^R(t,t') = \sum_{\bar\alpha\bar\beta} \alpha\beta g_{ab}^{\alpha\beta\bar\alpha\bar\beta} \,, \qquad (5.19)$$

which is sometimes more useful. Here, $\alpha, \beta, \cdots$ are Keldysh indices.

Another interesting representation may be achieved in form of nested commutators. We get

$$G_{ab}^R(r_1 r_2 t, r_1' r_2' t')$$
$$= \Theta(t-t')\frac{1}{i}\left\langle \left[\Psi_a^\dagger(r_1',t'), \left[\Psi_b^\dagger(r_2',t'), \Psi_b(r_2,t)\Psi_a(r_1,t)\right]_- \right]_\mp \right\rangle \,. \qquad (5.20)$$

Such a nested commutator structure was found for the first time by Rajagopal and Majumdar (1970) in their analysis of the double dispersion relation of three-time Matsubara Green's functions.

The corresponding advanced function is defined by

$$G_{ab}^{A}(r_1r_2t, r_1'r_2't')$$
$$= \Theta(t'-t)(-i)[g_{ab}^{++--} - g_{ab}^{+--+} - g_{ab}^{-++-} + g_{ab}^{--++}] . \quad (5.21)$$

Now it is easy to show that the functions $G_{ab}^{R/A}$ have the following useful properties:

1. The relation between the Hermitean conjugated quantities reads
$$G_{ab}^{R}(r_1r_2t, r_1'r_2't') = [G_{ab}^{A}(r_1'r_2't', r_1r_2t)]^* ; \quad (5.22)$$
2. both functions have the property of crossing symmetry
$$G_{ab}^{R/A}(r_1r_2t, r_1'r_2't') = G_{ba}^{R/A}(r_2r_1t, r_2'r_1't') ; \quad (5.23)$$
3. from the retarded and advanced Green's functions, we may define the spectral function a_{ab} by the relation
$$iG_{ab}^{R} - iG_{ab}^{A} = \hat{a}_{ab}(r_1r_2t, r_1'r_2't') . \quad (5.24)$$
4. Because of the equal-time commutation relations, we have for $t = t'$
$$\hat{a}_{ab}(r_1r_2t, r_1'r_2't')_{t=t'} =$$
$$\delta(r_1 - r_1')\delta(r_2 - r_2') \pm \delta_{ab}\delta(r_1 - r_2')\delta(r_2 - r_1') . \quad (5.25)$$
5. In the limit of free particles, G_{ab}^{R} reduces to
$$G_{ab}^{R} = g_a^R g_b^R .$$

Now we introduce the variables τ and T and take the Fourier transform $G_{ab}^{R/A}(\omega, T)$ with respect to the difference variable τ. Then, in the usual way, we get the spectral representation

$$G_{ab}^{R/A}(r_1r_2, r_1'r_2', \omega, T) = \int \frac{d\bar{\omega}}{2\pi} \frac{\hat{a}_{ab}(r_1r_2, r_1'r_2', \bar{\omega}, T)}{\omega - \overline{\omega} \pm i\varepsilon} . \quad (5.26)$$

The spectral representation enables us to discuss the two-particle excitations, similar to the considerations given above for $g_{ab}^{R/A}$, as peaks in the spectral function $a_{ab}(\omega, T)$ or as singularities of $G_{ab}^{R}(z, T)$ for Im$z < 0$.

From the property 4, we get a sum rule for the spectral function

$$\int \hat{a}_{ab}(\omega, T)\frac{d\omega}{2\pi} = \delta(r_1 - r_1')\delta(r_2 - r_2') \pm \delta_{ab}\delta(r_1 - r_2')\delta(r_2 - r_1') , \quad (5.27)$$

which is simpler than relation (5.11). Applying the Dirac identity to the relation (5.26), we easily get the following interesting relations:

$$G^R_{ab}(\omega,T) - G^A_{ab}(\omega,T) = -i\hat{a}_{ab}(r_1 r_2, r_1' r_2' \omega, T)\,, \tag{5.28}$$

$$G^R_{ab}(\omega,T) + G^A_{ab}(\omega,T) = 2i\mathcal{P}\int \frac{d\bar{\omega}}{2\pi} \frac{G^R_{ab}(\bar{\omega},T) - G^A_{ab}(\bar{\omega},T)}{\omega - \bar{\omega}}\,. \tag{5.29}$$

From the property 1, it is clear that the difference of the retarded and advanced functions is the anti-Hermitean, while the sum is the Hermitean part of these functions. Therefore the relation (5.29) may be interpreted as general dispersion relation.

5.2 Bethe–Salpeter Equation in Dynamically Screened Ladder Approximation

Let us now consider the question how to determine the two-particle Green's functions. In principle, the two-particle Green's function is determined by the second equation of the Martin–Schwinger hierarchy. One achieves an equation more conveniently by performing a functional derivation with respect to the external potential $U(1)$ in Dyson's equation. We start from the definition (3.134) and use (3.137). After functional derivation with respect to U we obtain the equation

$$\begin{aligned} L(121'2') &= \pm i g(12')g(21') \\ &+ \int g(1\bar{1}) \frac{\delta\Sigma(\bar{1}\bar{2})}{\delta U(2'2)} g(\bar{2}1'))d\bar{1}d\bar{2}\,. \end{aligned} \tag{5.30}$$

The self-energy depends on U only through its dependence on $g[U]$, i.e., $\Sigma = \Sigma(g[U])$. Therefore, we can apply the chain rule for the functional differentiation. Then we get an integral equation for the function $L(121'2')$ (see Baym and Kadanoff (1961))

$$L(121'2') = \pm i g(12')g(21') + \int g(1\bar{1})g(\bar{2}1')\Xi(\bar{1}4\bar{2}3)L(3242')d\bar{1}d\bar{2}d3d4\,. \tag{5.31}$$

Here we introduced an effective potential Ξ by the definition

$$\Xi(\bar{1}4\bar{2}3) = \frac{\delta\Sigma(\bar{1}\bar{2})}{\delta g(34)}\,. \tag{5.32}$$

In Feynman diagrams, this equation is shown in Fig. 5.2. From the point of view of diagrams, Ξ is the sum of all amputated irreducible two-particle diagrams in the "vertical channel" (electron–hole channel for fermions). Equation (5.31) is useful for the time specialization $t_1 = t_1'$, $t_2 = t_2'$, i.e., for considerations of density–density correlations and is referred to as the Bethe–Salpeter equation in the vertical channel.

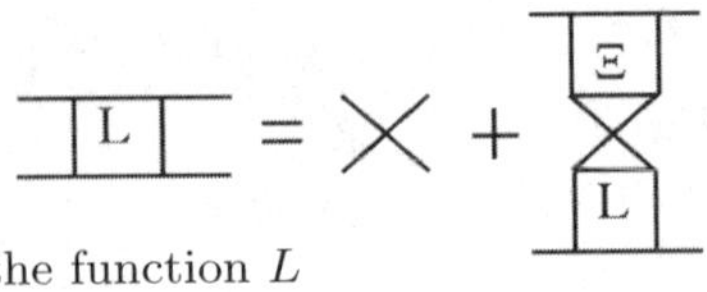

Fig. 5.2. Equation for the function L

For the dynamics of pairs of particles, it is more convenient to introduce a two-particle interaction W in the "horizontal channel" (particle–particle channel). This interaction is defined by the Bethe–Salpeter equation

$$\begin{aligned} g_{12}(121'2') &= g(11')g(22') \pm g(12')g(21') \\ &+ \int g(1\bar{1})g(2\bar{2})W(\bar{1}\bar{2}\bar{3}\bar{4})g_{12}(\bar{3}\bar{4}1'2')d\bar{1}d\bar{2}d\bar{3}d\bar{4}\,. \end{aligned} \quad (5.33)$$

In terms of Feynman diagrams, (5.33) is given in Fig. 5.3. From the point of view of diagrams, W is the sum of all amputated irreducible two-particle diagrams in the "horizontal channel".

Fig. 5.3. The two-particle Green's function in particle–particle channel

A comparison of the two Bethe–Salpeter equations (5.33) and (5.31) leads to an equation for the effective two-particle interaction

$$\begin{aligned} &\pm \int g(1\bar{1})\frac{\delta\Sigma(\bar{1}\bar{2})}{\delta U(2'2)}g(\bar{2}1')d\bar{1}d\bar{2} = \pm \int g(1\bar{1})g(\bar{2}1')\Xi(\bar{1}4\bar{2}3)L(3242')d\bar{1}d\bar{2}d3d4 \\ &= \int g(13)g(24)W(34\bar{3}\bar{4})g_{12}(\bar{3}\bar{4}1'2')d3d4d\bar{3}d\bar{4}\,. \end{aligned} \quad (5.34)$$

For a given approximation of the self-energy, we may determine the block $\Xi(\bar{1}4\bar{2}3)$. Then, (5.34) is an implicit equation for the block $W(34\bar{3}\bar{4})$.

In principle, the exact relations (5.30) and (5.33) define the same two-particle function $g_{12}(121'2') = -iL_{12}(121'2') + g(11')g(22')$. For practical purposes, however, they are of different advantage for the approximate determination of the 4-point function g_{12} in the 24 possible channels.

The Bethe–Salpeter equation (5.33) may also be written in the shape

$$\int \left(g_{12}^{0\,-1}(12\bar{1}\bar{2}) + W(12\bar{1}\bar{2})\right) g_{12}(\bar{1}\bar{2}1'2')d\bar{1}d\bar{2} = \delta(11')\delta(22')\,, \quad (5.35)$$

which means that we introduced the inverse two-particle Green's function by

$$g_{12}^{-1}(12\bar{1}\bar{2}) = g_{12}^{0\,-1}(12\bar{1}\bar{2}) - W(12\bar{1}\bar{2})\,. \quad (5.36)$$

We see that this is a form analogous to the Dyson equation (3.137).

A useful often applied property of the Bethe–Salpeter equation follows if we subdivide the effective two-particle interaction W in an *arbitrary* manner into

$$W = W^{(1)} + W^{(2)} . \tag{5.37}$$

Then it is easy to show that we get, instead of (5.33), the pair of equations (we omit the variables)

$$g_{12} = g_{12}^{(1)} + g_{12}^{(1)} W^{(2)} g_{12} , \tag{5.38}$$

$$g_{12}^{(1)} = g_{12}^{(0)} + g_{12}^{(0)} W^{(1)} g_{12}^{(1)} . \tag{5.39}$$

Let us now determine the effective interaction W for plasmas. The self-energy for Coulomb systems is represented as

$$\Sigma(12) = \Sigma^H(12) + \bar{\Sigma}(12) , \tag{5.40}$$

where the Hartree contribution Σ^H is given by (3.111). For the determination of $\bar{\Sigma}$, we restrict ourselves to the V^s approximation

$$\bar{\Sigma}(12) = V^s(12) g(12) . \tag{5.41}$$

For the determination of W from (5.34) we have to consider the functional derivative

$$\frac{\delta \Sigma(\bar{1}\bar{2})}{\delta U(2'2)} = \int \frac{\delta \Sigma(\bar{1}\bar{2})}{\delta g(56)} \frac{\delta g(56)}{\delta U(2'2)} d5d6 . \tag{5.42}$$

Considering the Hartree contribution of (5.42), we get immediately

$$\frac{\delta \Sigma^H(\bar{1}\bar{2})}{\delta U(2'2)} = \pm i \int \delta(\bar{1}-\bar{2}) \delta(5-6) V(\bar{1}6) L(5262') d5d6 . \tag{5.43}$$

Regarding the further derivatives we restrict ourselves to the first order terms with respect to the screened potential V^s. For this purpose, in (5.43), we take into account the relation (4.14), and we apply the approximation $L = gg$. In this way we arrive at

$$\frac{\delta \Sigma^H(\bar{1}\bar{2})}{\delta U(2'2)} = \pm \delta(\bar{1}\bar{2}) \int d3 V^s(\bar{1}3) g(23) g(32') . \tag{5.44}$$

For the $\bar{\Sigma}$ contribution to (5.42), we get easily

$$\frac{\delta \bar{\Sigma}(\bar{1}\bar{2})}{\delta U(2'2)} = i \int \delta(\bar{1}-5) \delta(\bar{2}-6) V^s(\bar{1}6) L(5262') d5d6 + O(V^{s^2}) . \tag{5.45}$$

Again we approximate $L = gg$. Introducing (5.43) and (5.45) into (5.42), we finally obtain

$$\frac{\delta\Sigma(\bar{1}\bar{2})}{\delta U(2'2)} = iV^s(\bar{1}\bar{2})g(2\bar{2})g(\bar{1}2') \pm \delta(\bar{1}\bar{2})\int d3V^s(\bar{1}3)g(23)g(32')\,. \tag{5.46}$$

Now we come back to (5.34). We insert (5.46) into the l.h.s., and, at the r.h.s., we use the approximation

$$g_{12}(121'2') = g(11')g(22') \pm g(12')g(21') \tag{5.47}$$

for the two-particle Green's function. Then, by comparison of both sides, we obtain an expression for the effective potential

$$W(121'2') = iV^s(12)\delta(1-1')\delta(2-2')\,. \tag{5.48}$$

Applying this effective potential in (5.33), we get the BSE in dynamically screened ladder approximation

$$\begin{aligned} g_{12}(121'2') &= g(11')g(22') \pm g(12')g(21') \\ &+ i\int g(1\bar{1})g(2\bar{2})V^s(\bar{1}\bar{2})g_{12}(\bar{1}\bar{2}1'2')d\bar{1}d\bar{2}\,. \end{aligned} \tag{5.49}$$

Initial correlations were neglected here. Equation (5.49) is conveniently illustrated in terms of Feynman diagrams, see Fig. 5.4.

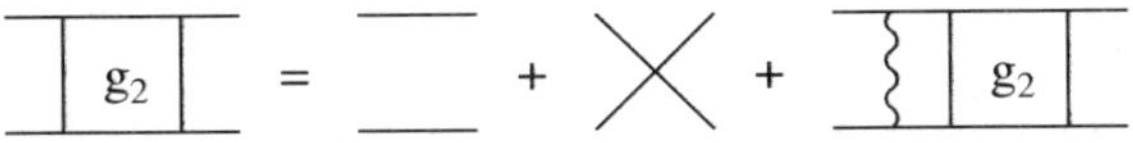

Fig. 5.4. The Bethe–Salpeter equation

The single-particle Green's function occurring in (5.49) is given by the Dyson–Schwinger equation

$$g_a^{-1}(1,1') = g_a^{0-1}(1,1') - \bar{\Sigma}_a(1,1')\,, \tag{5.50}$$

where $g_a^{0-1}(1,1')$ reads

$$g_a^{0-1}(1,1') = S_a(1)\,\delta(1-1')\,, \qquad S_a(1) = i\frac{\partial}{\partial t_1} + \frac{\nabla_1^2}{2m_a} - \Sigma^H(1)\,. \tag{5.51}$$

For further considerations, it is convenient to transform (5.49) into the corresponding differential equation. For this purpose we apply the inverse single-particle Green's function g_a^{-1} to the relation (5.49) and add the corresponding equation resulting from the application of g_b^{-1}. Furthermore, we use the Dyson–Schwinger equation (5.50) and arrive at (Kraeft et al. 1973; Kraeft et al. 1975; Ebeling (a) et al. 1976; Kraeft et al. 1986; Zimmermann et al. 1978; Bornath et al. 1999)

$$\begin{aligned} &\Big[S_a(1) + S_b(2)\Big]g_{ab}(12,1'2') - \int_C d\bar{1}d\bar{2}\Big[\bar{\Sigma}_a(1,\bar{1})\delta(2-\bar{2}) \\ &+\bar{\Sigma}_b(2,\bar{2})\delta(1-\bar{1})\Big]g_{ab}(\bar{1}\bar{2},1'2') = g_a(1,1')\delta(2-2') + g_b(2,2')\delta(1-1') \\ &+i\int_C d\bar{1}d\bar{2}\Big[g_a(1,\bar{1})\delta(2-\bar{2}) + g_b(2,\bar{2})\delta(1-\bar{1})\Big]V_{ab}^s(\bar{1},\bar{2})g_{ab}(\bar{1}\bar{2},1'2')\,. \end{aligned} \tag{5.52}$$

For simplicity, the exchange term was dropped. We now consider the special case $t_1 = t_2 = t$, $t_1' = t_2' = t'$, i.e., we deal with the two-time Green's function $g_{ab}(t,t')$. Then (5.52) turns into

$$
\begin{aligned}
&(i\frac{\partial}{\partial t} - \mathrm{H}^0_{ab})g_{ab}(t,t') \\
&-\int_C d\bar{t}_1 d\bar{t}_2 [\bar{\Sigma}_a(t,\bar{t}_1)\delta(t-\bar{t}_2) + \bar{\Sigma}_b(t,\bar{t}_2)\delta(t-\bar{t}_1)]\; g_{ab}(\bar{t}_1\bar{t}_2, t_1't_2')|_{t_1'=t_2'} \\
&-i\int_C d\bar{t}_1 d\bar{t}_2 [g_a(t,\bar{t}_1)\;\delta(t-\bar{t}_2) + g_b(t,\bar{t}_2)\;\delta(t-\bar{t}_1)] \\
&\times V^s_{ab}(\bar{t}_1\bar{t}_2) g_{ab}(\bar{t}_1\bar{t}_2, t_1't_2')|_{t_1'=t_2'} \;=\; -i\mathrm{N}_{ab}(t)\delta(t-t')\,.
\end{aligned}
\qquad (5.53)
$$

Here, N_{ab} is the phase space occupation factor

$$
\mathrm{N}_{ab}(t) = ig_a^>(t,t) + ig_b^<(t,t) = 1 \pm f_a(t) \pm f_b(t)\,, \qquad (5.54)
$$

and $\mathrm{H}^0_{\mathrm{ab}}$ is the Hamiltonian of two (free) particles including single-particle potentials.

In fact, with (5.53), an equation of motion for the two-particle two-time Green's function is found. But, there is still a problem. Because of the dynamical self-energy and of the dynamically screened potential, this equation also contains the two-particle Green's function with three time arguments. This means that equation (5.53) is not a closed equation for the determination of the two-time Green's function $g_{ab}(t,t')$. This problem will be considered in Sect. 5.8.

5.3 Bethe–Salpeter Equation for a Statically Screened Potential

First we will consider the simpler case of a static self-energy and a statically screened potential, i.e.,

$$
\Sigma_a(t,t') = \Sigma_a\delta(t-t'); \quad V_{ab}(t,t') = V_{ab}\delta(t-t')\,. \qquad (5.55)
$$

Then the theory is essentially simplified. In the statical case we find, instead of (5.53), the equation

$$
\left[i\frac{\partial}{\partial t} - \mathrm{H}^{\mathrm{eff}}_{ab}(t) - \mathrm{N}_{ab}(t)\,V^s_{ab}\right] g_{ab}(t,t') = -i\mathrm{N}_{ab}(t)\,\delta(t-t')\,, \qquad (5.56)
$$

where V^s is the statical limit of the dynamically screened potential which reads in lowest order

$$
V^s_{ab}(q) = \frac{4\pi e_a e_b}{q^2+\kappa^2}\,. \qquad (5.57)
$$

We remark that (5.56) is not only valid for plasmas and (5.57) may be replaced by any static potential $V(q)$.

The Hamiltonian $\mathrm{H}_{ab}^{\mathrm{eff}}$ can be used in the approximation

$$\mathrm{H}_{ab}^{\mathrm{eff}} = \mathrm{H}_{ab}^{0} + \Sigma_a + \Sigma_b \,. \tag{5.58}$$

Here, Σ_a is the statical self-energy given by

$$\Sigma_a = \Sigma_a^{\mathrm{HF}} + \Delta_a(n,T) \,. \tag{5.59}$$

Δ_a is the rigid shift approximation of the Montroll–Ward part of $\bar{\Sigma}$ (4.253). In lowest order we have $\Delta_a = -(\kappa e_a^2)/2$. If we neglect the phase space factors and the exchange self-energy, the static self-energy reduces to the rigid shift. The relation (5.56) then reduces to the Ecker–Weizel model used in Sect. 2.7. Further, H_{ab}^{0} is the free Hamiltonian, i.e.,

$$\left\langle \boldsymbol{r}_1\boldsymbol{r}_2 \left|\mathrm{H}_{ab}^{0}\right| \boldsymbol{r}_1'\boldsymbol{r}_2' \right\rangle = \left(-\frac{\nabla_1^2}{2m_a} - \frac{\nabla_2^2}{2m_b} \right) \delta(\boldsymbol{r}_1 - \boldsymbol{r}_1')\,\delta(\boldsymbol{r}_2 - \boldsymbol{r}_2') \,. \tag{5.60}$$

Equation (5.56) is defined on the Keldysh contour. Therefore, the equation represents a compact set of equations for the correlation functions $g_{ab}^{\gtrless}(t,t')$, and the causal and anti-causal Green's functions. Using the rules of time ordering on the contour, we find the equation for the correlation functions

$$\left[i\frac{\partial}{\partial t} - \mathrm{H}_{ab}^{\mathrm{eff}}(t) - \mathrm{N}_{ab}(t)\,\mathrm{V}_{ab}^{s} \right] g_{ab}^{\gtrless}(t,t') = 0 \,. \tag{5.61}$$

From the definition of the retarded and advanced Green's functions (5.8), and using (5.61), we get

$$\left[i\frac{\partial}{\partial t} - \mathrm{H}_{ab}^{\mathrm{eff}} - \mathrm{N}_{ab}(t)\,\mathrm{V}_{ab}^{s} \right] g_{ab}^{R/A}(t,t') = -i\mathrm{N}_{ab}(t)\,\delta(t-t') \,. \tag{5.62}$$

This equation is of central importance. With the substitution $\tilde{g}_{ab}^{R/A}(tt') = ig_{ab}^{R/A}(tt')/N_{ab}(t)$, we get from (5.62)

$$\left[i\frac{\partial}{\partial t} - \mathrm{H}_{ab}^{\mathrm{eff}} - \mathrm{N}_{ab}(t)\,\mathrm{V}_{ab}^{s} \right] \tilde{g}_{ab}^{R/A}(t,t') = \delta(t-t') \,, \tag{5.63}$$

that means, $\tilde{g}_{ab}^{R/A}$ is the usual propagator Green's function. We can use $\tilde{g}_{ab}^{R/A}$ in order to find the general solution of (5.61) for the construction of $g_{ab}^{R/A}$. Accounting for the initial condition

$$\lim_{t,t'\to t_0} g_{ab}^{\gtrless}(t,t') = g_{ab}^{\gtrless}(t_0,t_0) \,, \tag{5.64}$$

the general solution is written in the form

$$g_{ab}^{\gtrless}(t,t') = \tilde{g}_{ab}^{R}(t,t_0)\, g_{ab}^{\gtrless}(t_0,t_0)\, \tilde{g}_{ab}^{A}(t_0,t') \,. \tag{5.65}$$

A further important point is that (5.62) determines the retarded and advanced two-particle Green's functions, and therefore the dynamical properties of the two-particle problem. In order to consider this aspect it is convenient to introduce again the sum and difference variables T and τ. Then, we take the Fourier transform of (5.62) with respect to τ and transform the equation into the momentum representation. In local approximation including the exchange term, we get

$$\begin{aligned}
&\Big(E_a(\boldsymbol{p}_1,T) + E_b(\boldsymbol{p}_2,T) - \omega\Big) g_{ab}^{R/A}\left(\boldsymbol{p}_1\boldsymbol{p}_2\boldsymbol{p}_1'\boldsymbol{p}_2',\omega T\right) \\
&+N_{ab}(\boldsymbol{p}_1\boldsymbol{p}_2,T) \int V_{ab}^{s}\left(\boldsymbol{p}_1 - \bar{\boldsymbol{p}}_1\right)(2\pi)^3\delta\left(\boldsymbol{p}_1+\boldsymbol{p}_2-\bar{\boldsymbol{p}}_1-\bar{\boldsymbol{p}}_2\right) \\
&\times g_{ab}^{R/A}\left(\bar{\boldsymbol{p}}_1\bar{\boldsymbol{p}}_2\boldsymbol{p}_1'\boldsymbol{p}_2',\omega T\right)\frac{d\bar{\boldsymbol{p}}_1 d\bar{\boldsymbol{p}}_2}{(2\pi)^6} = -i\,(2\pi)^3\Big[\delta\left(\boldsymbol{p}_1-\boldsymbol{p}_1'\right)\delta\left(\boldsymbol{p}_2-\boldsymbol{p}_2'\right) \\
&\pm\delta_{ab}\,\delta\left(\boldsymbol{p}_1-\boldsymbol{p}_2'\right)\delta\left(\boldsymbol{p}_2-\boldsymbol{p}_1'\right)\Big] N_{ab}(\boldsymbol{p}_1\boldsymbol{p}_2,T)\,.
\end{aligned} \tag{5.66}$$

Here, the spatially homogeneous case was considered. The quasiparticle energies are

$$\begin{aligned}
E_a(\boldsymbol{p}_1,T) &= \frac{p_1^2}{2m_a} + \Sigma_a^{\mathrm{HF}}(\boldsymbol{p}_1,T) + \Delta(n,T)\,, \\
\Sigma_a^{\mathrm{HF}}(\boldsymbol{p},T) &= \sum_b n_b V_{ab}(\boldsymbol{q}=0) \pm \int \frac{d\boldsymbol{p}'}{(2\pi)^3} V_{aa}(\boldsymbol{p}-\boldsymbol{p}') f_a(\boldsymbol{p}',T)
\end{aligned} \tag{5.67}$$

with the number density n_b of species b and $V_{ab}(\boldsymbol{q}=0) = \int d\boldsymbol{r}\, V_{ab}(\boldsymbol{r})$. In the form (5.66), the equation is often called *Bethe–Goldstone equation.* The Bethe–Goldstone equation describes the propagation of a pair of particles a and b in a strongly correlated many-particle system.

To summarize, in a dense plasmas, the influence of the medium on the two-particle properties is taken into account by the following effects:

(i) The so-called Pauli blocking or phase space occupation effect. Here, the two-particle scattering produced by V_{ab}^{s} is restricted by the factor N_{ab}.
(ii) There is a self-energy correction to the kinetic energy given by the Hartree–Fock contribution and the static lowest order Montroll–Ward term.
(iii) Static screening is taken into account by the Debye potential.

A more rigorous discussion of the BSE for dynamical screening for strongly coupled plasmas will be given in Sect. 5.8. We remark that the BSE is widely used, too, for systems with short range interactions, like nuclear matter.

5.4 Effective Schrödinger Equation. Bilinear Expansion

It is interesting from the physical and mathematical points of view to consider the associated homogeneous Bethe–Salpeter equation. The homogeneous equation corresponding to (5.66) reads

$$\Big(E_a(\boldsymbol{p}_1) + E_b(\boldsymbol{p}_2) - E\Big)\psi_E\,(\boldsymbol{p}_1\boldsymbol{p}_2) + N_{ab}(\boldsymbol{p}_1\boldsymbol{p}_2)\int \frac{d\bar{\boldsymbol{p}}_1 d\bar{\boldsymbol{p}}_2}{(2\pi)^3}$$
$$\times V^s_{ab}\,(\boldsymbol{p}_1 - \bar{\boldsymbol{p}}_1)\,\delta\,(\boldsymbol{p}_1 + \boldsymbol{p}_2 - \bar{\boldsymbol{p}}_1 - \bar{\boldsymbol{p}}_2)\,\psi_E\,(\bar{\boldsymbol{p}}_1\bar{\boldsymbol{p}}_2) = 0\,. \tag{5.68}$$

The dependence on the macroscopic time T was suppressed for simplicity. In operator notation, the equation can be written as

$$(\mathrm{H}^{\mathrm{eff}}_{ab} + \mathrm{N}_{ab}\mathrm{V}_{ab})\,|\psi_E\rangle = E\,|\psi_E\rangle\,.$$

We use the effective Hamiltonian $\mathrm{H}^{\mathrm{eff}}_{ab}$ in Hartree–Fock approximation with the single-particle energies given by (5.67).

After some rearrangements we can write (5.68) in the form

$$\left(\frac{p_1^2}{2m_a} + \frac{p_2^2}{2m_b} - E\right)\psi_E\,(\boldsymbol{p}_1\boldsymbol{p}_2) + \int \frac{d\boldsymbol{q}}{(2\pi)^3}V_{ab}(q)\psi_E\,(\boldsymbol{p}_1 - \boldsymbol{q}\,\boldsymbol{p}_2 + \boldsymbol{q})$$
$$= -\sum_c n_c\Big(V_{ac}(0) + V_{bc}(0)\Big)$$
$$\mp \int \frac{d\boldsymbol{q}}{(2\pi)^3}\Big[V_{aa}(q)f_a(\boldsymbol{p}_1 - \boldsymbol{q}) + V_{bb}(q)f_a(\boldsymbol{p}_2 + \boldsymbol{q})\Big]\psi_E\,(\boldsymbol{p}_1\boldsymbol{p}_2)$$
$$-\Big(N_{ab}(\boldsymbol{p}_1\boldsymbol{p}_2) - 1\Big)\int \frac{d\boldsymbol{q}}{(2\pi)^3}V_{ab}(q)\psi_E\,(\boldsymbol{p}_1 - \boldsymbol{q}\,\boldsymbol{p}_2 + \boldsymbol{q})\,. \tag{5.69}$$

As can easily be seen, the left hand side of this equation is just the Schrödinger equation of an isolated pair of particles. The many-particle effects are condensed on the r.h.s. of (5.69). The first two terms account for the Hartree self-energy and the Hartree–Fock exchange term, and the third one corresponds to phase space occupation effects (Pauli blocking). Obviously, the homogeneous Bethe–Salpeter equation may be interpreted therefore as the many-particle version of the two-particle Schrödinger equation.

Let us review the solution of the Schrödinger equation for the isolated two-particle problem with a spherically symmetric interaction potential. As discussed already in Sect. 2.7, we have two kinds of solutions.

(i) There are scattering states with energies

$$E^0_{\boldsymbol{pP}} = \frac{p^2}{2\mu_{ab}} + \frac{P^2}{2M}$$

where $\boldsymbol{p}, \boldsymbol{P}$ are the relative and total momenta, and the masses are $\mu_{ab} = m_a m_b/(m_a + m_b),\;\; M = m_a + m_b\,.$

The normalization condition of the scattering states reads

$$\langle \psi^0_{\boldsymbol{pP}} | \psi^0_{\boldsymbol{p}'\boldsymbol{P}'} \rangle = (2\pi)^6 \delta(\boldsymbol{p}-\boldsymbol{p}')\delta(\boldsymbol{P}-\boldsymbol{P}') .$$

(ii) We have bound state solutions with energies

$$E^0_{j,\boldsymbol{P}} = E^0_j + \frac{P^2}{2M} ,$$

where $j = n, \ell, m$ denotes the set of internal quantum numbers.

The bound states are normalized as

$$\langle \psi^0_{j,\boldsymbol{P}} | \psi^0_{j',\boldsymbol{P}'} \rangle = (2\pi)^3 \delta(\boldsymbol{P}-\boldsymbol{P}')\delta_{j,j'} .$$

The solutions of the isolated two-particle problem satisfy the following completeness relation

$$1 = \sum_{\alpha \boldsymbol{P}} |\psi^0_{\alpha \boldsymbol{P}}\rangle \langle \psi^0_{\alpha \boldsymbol{P}}| .$$

The sum over α includes bound ($\alpha = j$) and scattering states ($\alpha = \boldsymbol{p}$).

In order to discuss the physical meaning of the solutions determined by the in-medium Schrödinger equation (5.69), we write

$$(\mathrm{H}^0_{ab} + \mathrm{V}_{ab} - E_{\alpha \boldsymbol{P}}) |\psi_{\alpha \boldsymbol{P}}\rangle = -\mathrm{H}^{\mathrm{medium}}_{ab} |\psi_{\alpha \boldsymbol{P}}\rangle . \tag{5.70}$$

We will assume that the r.h.s. of (5.70) is small. First order perturbation theory gives the corrected binding energy ($\alpha = n\ell m$)

$$\begin{aligned} E_{\alpha \boldsymbol{P}} &= E^0_{\alpha \boldsymbol{P}} + \Delta E_{\alpha \boldsymbol{P}} \\ \Delta E_{\alpha \boldsymbol{P}} &= \langle \psi^0_{\alpha \boldsymbol{P}} | \mathrm{H}^{\mathrm{medium}}_{ab} | \psi^0_{\alpha \boldsymbol{P}} \rangle = \Delta E^{Self}_{\alpha \boldsymbol{P}} + \Delta E^{Pauli}_{\alpha \boldsymbol{P}} . \end{aligned} \tag{5.71}$$

Let us consider the contributions to the energy shift $\Delta E_{\alpha \boldsymbol{P}}$. The first one comes from the Hartree–Fock self-energies:

$$\begin{aligned} \Delta E^{\mathrm{self}}_{\alpha \boldsymbol{P}} = \mp \int \frac{d\boldsymbol{p}_1}{(2\pi)^3} \frac{d\boldsymbol{p}_2}{(2\pi)^3} \frac{d\boldsymbol{q}}{(2\pi)^3} \psi^{0*}_{\alpha \boldsymbol{P}}(\boldsymbol{p}_1 \boldsymbol{p}_2) \Big[V_{aa}(q) f_a(\boldsymbol{p}_1 - \boldsymbol{q}) \\ + V_{bb}(q) f_b(\boldsymbol{p}_2 + \boldsymbol{q}) \Big] \psi^0_{\alpha \boldsymbol{P}}(\boldsymbol{p}_1 \boldsymbol{p}_2) + \sum_c n_c \Big(V_{ac}(0) + V_{bc}(0) \Big) . \end{aligned} \tag{5.72}$$

The second one accounts for phase space occupation effects (Pauli blocking)

$$\begin{aligned} \Delta E^{\mathrm{Pauli}}_{\alpha \boldsymbol{P}} = \mp \int \frac{d\boldsymbol{p}_1}{(2\pi)^3} \frac{d\boldsymbol{p}_2}{(2\pi)^3} \frac{d\boldsymbol{q}}{(2\pi)^3} \psi^{0*}_{\alpha \boldsymbol{P}}(\boldsymbol{p}_1 \boldsymbol{p}_2) \Big[f_a(\boldsymbol{p}_1) + f_b(\boldsymbol{p}_2) \Big] \\ \times V_{ab}(q) \, \psi^0_{\alpha \boldsymbol{P}}(\boldsymbol{p}_1 - \boldsymbol{q} \boldsymbol{p}_2 + \boldsymbol{q}) . \end{aligned} \tag{5.73}$$

Then, there is an interesting compensation between the phase space occupation term and the Hartree–Fock self-energy contribution

$$\Delta E^{\rm self}_{\alpha \boldsymbol{P}} + \Delta E^{\rm Pauli}_{\alpha \boldsymbol{P}} = \int \frac{d\boldsymbol{p}_1}{(2\pi)^3}\frac{d\boldsymbol{p}_2}{(2\pi)^3}\frac{d\boldsymbol{q}}{(2\pi)^3} V(q)\Big[f_a(\boldsymbol{p}_1) + f_b(\boldsymbol{p}_2)\Big]$$
$$\times \psi^{0*}_{\alpha \boldsymbol{P}}(\boldsymbol{p}_1 - \boldsymbol{q}\,\boldsymbol{p}_2 + \boldsymbol{q})\Big\{\psi^{0}_{\alpha \boldsymbol{P}}(\boldsymbol{p}_1\boldsymbol{p}_2) - \psi^{0}_{\alpha \boldsymbol{P}}(\boldsymbol{p}_1 - \boldsymbol{q}\,\boldsymbol{p}_2 + \boldsymbol{q})\Big\} \qquad (5.74)$$

For bound states, the low lying states are localized and thus extended in the momentum space. Therefore, the wave functions vary only weakly with momentum, and thus the terms in the second line in (5.74) compensate each other to a large extent. The result is that the bound state energy shifts down only slightly. However, for scattering states, the wave functions are sharply peaked in the momentum space and there is no compensation.

Let us discuss the energy shift given by (5.71) with (5.72) and (5.73). To determine the position of the continuum edge we take $\boldsymbol{p} = 0$ and $\boldsymbol{P} = 0$. Due to the fact that the energy spectrum of scattering states is equal to that of free quasiparticles, we can start from (5.69) neglecting the potential terms with V_{ab} (the last terms on both sides of the equation). For the shift of the continuum edge, we get

$$\Delta E_{cont} = \Sigma^{\rm HF}_a(\boldsymbol{p}_1 = 0) + \Sigma^{\rm HF}_b(\boldsymbol{p}_2 = 0)\,. \qquad (5.75)$$

In general, a $\boldsymbol{P}$-dependence of the continuum edge has to be considered.

From this we find that only the self-energy term contributes to ΔE_{cont}. Therefore, the influence of many-particle effects is different for the continuum edge and for the bound states. We can expect that the continuum edge shifts more rapidly than the bound state energy.

Now we come back to the case of plasmas. As pointed out in the previous section, (5.68) can be applied to plasmas using the single-particle energies (5.67) and the statically screened Coulomb potential (5.57). Because of the compensation between Pauli blocking and Hartree–Fock self energy, (2.127) can be applied approximately. The resulting behavior of the energy spectrum was shown already in Fig. 2.6.

The physical interpretation of this result is very interesting. At the so-called Mott density, the bound state of the pair of particles a and b vanishes, i.e., it breaks up. More rigorously, the bound state merges into the continuum. Such effect can be observed in strongly correlated many-particle systems, such as nuclear matter, dense gaseous plasmas, and electron–hole plasmas in highly excited semiconductors. The results are drastic changes in thermodynamic and transport properties. The Mott effect in dense plasmas will be discussed in more detail in the following chapters.

Let us come back to the complete effective Schrödinger equation (5.68). In operator form we have

$$\mathrm{H}_{ab}\,|\psi_E\rangle = E\,|\psi_E\rangle\;. \qquad (5.76)$$

The total two-particle Hamiltonian reads

$$\mathrm{H}_{ab} = \mathrm{H}_{ab}^{\mathrm{eff}} + \mathrm{N}_{ab}\mathrm{V}_{ab} \tag{5.77}$$

with the free quasiparticle Hamiltonian $\mathrm{H}_{ab}^{\mathrm{eff}}$ given by (5.58). The Hamiltonian (5.77) is a non-Hermitean one, because the adjoint operator $\mathrm{H}_{ab}^{\dagger}$ reads

$$\mathrm{H}_{ab}^{\dagger} = \mathrm{H}_{ab}^{\mathrm{eff}} + \mathrm{V}_{ab}\mathrm{N}_{ab}\,. \tag{5.78}$$

Consequently, we have a second Schrödinger equation (Danielewicz 1990)

$$\mathrm{H}_{ab}^{\dagger}\left|\tilde{\psi}_E\right\rangle = \tilde{E}\left|\tilde{\psi}_E\right\rangle . \tag{5.79}$$

There is a simple connection between H_{ab} and $\mathrm{H}_{ab}^{\dagger}$, and between the eigenvectors, namely

$$\mathrm{H}_{ab}^{\dagger} = \mathrm{N}_{ab}^{-1}\mathrm{H}_{ab}\mathrm{N}_{ab}\,; \qquad \left|\tilde{\psi}_E\right\rangle = \mathcal{F}_E \mathrm{N}_{ab}^{-1}\left|\psi_E\right\rangle . \tag{5.80}$$

The quantity $\mathcal{F}_E$ follows from the normalization condition $\left\langle\tilde{\psi}_E|\psi_E\right\rangle = 1$, i.e.,

$$\mathcal{F}_E^{-1} = \left\langle\psi_E\right|\mathrm{N}_{ab}^{-1}\left|\psi_E\right\rangle . \tag{5.81}$$

As a consequence, the eigenvalues of H_{ab} and $\mathrm{H}_{ab}^{\dagger}$ are the same. One may show that there exists the following orthogonality and completeness relations

$$\left\langle\psi_E|\tilde{\psi}_{\bar{E}}\right\rangle = \delta_{E\bar{E}}\,; \quad 1 = \int dE\left|\psi_E\right\rangle\left\langle\tilde{\psi}_E\right| . \tag{5.82}$$

The solutions of the different Schrödinger equations form thus a *bi-orthogonal system.*

Now we are able to construct a formal solution of the Bethe–Salpeter equation (5.62) or (5.66). In operator form it follows that

$$g_{ab}^{R/A}(\omega) = -i\,\frac{\mathrm{N}_{ab}}{\omega - \mathrm{H}_{ab} \pm i\varepsilon}\,. \tag{5.83}$$

Using the completeness relation (5.82), we find the bilinear expansion of the retarded and advanced two-particle Green's function. We may replace $\left|\tilde{\psi}_{\alpha\boldsymbol{P}}\right\rangle$ with the help of (5.80). The bilinear expansion in terms of the eigenvectors $\left|\psi_{\alpha\boldsymbol{P}}\right\rangle$ can then be written as

$$g_{ab}^{R/A}(\omega) = -i\sum_{\alpha\boldsymbol{P}}\frac{\left|\psi_{\alpha\boldsymbol{P}}\right\rangle\left\langle\psi_{\alpha\boldsymbol{P}}\right|}{\omega - E \pm i\varepsilon}\,\mathcal{F}_E\,. \tag{5.84}$$

The analytical properties of these functions are well known. Especially, the analytical continuation of the retarded two-particle Green's function $g_{ab}^{R}(E)$

has a branch cut at the positive real axis corresponding to scattering states. The bound states manifest themselves, in our simple approximation, as poles at the negative real axis. The two-particle spectral function is related to the retarded and advanced two-particle Green's functions by $a_{ab} = i^2 \left(g_{ab}^R - g_{ab}^A\right)$. With (5.84), we get

$$a_{ab}(\omega, T) = \sum_{\alpha \boldsymbol{P}} |\psi_{\alpha \boldsymbol{P}}\rangle \langle \psi_{\alpha \boldsymbol{P}}| \, \mathcal{F}_{\alpha \boldsymbol{P}}(T) \, 2\pi\delta\Big(\omega - E_{\alpha \boldsymbol{P}}(T)\Big) . \tag{5.85}$$

This spectral function determines all dynamical two-particle properties of the plasma in binary collision approximation. To include the statistical information, the correlation functions $g_{ab}^{\gtrless}$ have to be considered. So we have the ansatz

$$i^2 g_{ab}^{\gtrless}(\omega T) = \sum_{\alpha P} |\psi_{\alpha P}\rangle \langle \psi_{\alpha \boldsymbol{P}}| \, N_{\alpha \boldsymbol{P}}^{\gtrless}(T) \, 2\pi\delta\Big(\omega - E_{\alpha \boldsymbol{P}}(T)\Big) . \tag{5.86}$$

As a consequence of (5.85), the functions $N_{\alpha P}^{\gtrless}$ have to fulfill the relation

$$N_{\alpha \boldsymbol{P}}^{>} - N_{\alpha \boldsymbol{P}}^{<} = \mathcal{F}_{\alpha \boldsymbol{P}} . \tag{5.87}$$

We may introduce the Bosonic occupation number $N_{\alpha P}$ of two-particle states:

$$N_{\alpha \boldsymbol{P}}^{<} = N_{\alpha \boldsymbol{P}} \mathcal{F}_{\alpha \boldsymbol{P}} ; \qquad N_{\alpha \boldsymbol{P}}^{>} = \left(1 + N_{\alpha \boldsymbol{P}}\right) \mathcal{F}_{\alpha \boldsymbol{P}} . \tag{5.88}$$

Using the KMS-relation, the occupation number is, in thermodynamic equilibrium, $N_{\alpha P} = n_{ab}(\omega)$, with $n_{ab}(\omega)$ being the Bose function given by (5.15). In non-equilibrium, $N_{\alpha P}$ retains its dependence on the states and, thus, cannot be taken out of the sum in (5.86).

5.5 The *T*-Matrix

In many cases it is more convenient to use the T-matrix instead of the two-particle Green's function. In fact, in binary ladder approximation, important quantities such as cross sections, scattering rates and single-particle damping can be expressed directly in terms of the T-matrix. The T-matrix is defined on the Keldysh contour by

$$\begin{aligned} g_{ab}(121'2') &= g_a(11')g_b(22') \pm g_a(12')g_b(21') \\ &+ \int d\bar{1}d\bar{2}d\bar{3}d\bar{4} g_a(1\bar{1})g_b(2\bar{2})T_{ab}(\bar{1}\bar{2}\bar{3}\bar{4}) \left(g_a(\bar{3}1')g_b(\bar{4}2') \pm g_a(\bar{3}2')g_b(\bar{4}1')\right) . \end{aligned} \tag{5.89}$$

This equation shows that T determines the correlation part of the two-particle function g_{12}. We get the function $T(\bar{1}\bar{2}\bar{3}\bar{4})$ from the correlated part of g_{12} by amputation of four external single-particle lines.

Therefore, it is possible to express the correlation properties of the system in terms of the T-matrix just introduced. For example, it is readily verified from (3.136) that the self-energy is given by

$$\begin{aligned}\Sigma(11') &= \Sigma^{HF} \pm i \int d\bar{1}d\bar{2}d\bar{3}d\bar{4}V(1\bar{1}) \\ &\times \; g_1(1\bar{2})g_1(\bar{1}\bar{3})(T(\bar{2}\bar{3}1'\bar{4}) \pm T(\bar{2}\bar{3}\bar{4}1'))g_1(\bar{4}\bar{1}) \,. \qquad (5.90)\end{aligned}$$

General equations for the determination of the T-matrix may be found immediately from the corresponding equation (5.33) for the two-particle Green's function. We will consider here the ladder approximation (5.49). In this approximation, we have

$$\begin{aligned}T_{ab}(12,1'2') &= V_{ab}(1-2)\delta(1-1')\delta(2-2') \\ &+ \; i\int_C d\bar{1}d\bar{2}\, V_{ab}(1-2)g_a(1,\bar{1})g_b(2,\bar{2})T_{ab}(\bar{1}\bar{2},1'2') \,. \qquad (5.91)\end{aligned}$$

It is useful to introduce the two-particle Green's function $\bar{g}_{ab}$ without exchange. Then we get

$$\begin{aligned}T_{ab}(12,1'2') &= V_{ab}(1-2)\delta(1-1')\delta(2-2') \\ &\quad +iV_{ab}(1-2)\bar{g}_{ab}(12,1'2')V_{ab}(1'-2') \,. \qquad (5.92)\end{aligned}$$

From these equations, we find the structure

$$\begin{aligned}T_{ab}(12,1'2') &= \langle 12|\mathrm{T}_{ab}|1'2'\rangle \\ &= \langle \boldsymbol{r}_1\boldsymbol{r}_2\,|\mathrm{T}_{ab}(t_1,t_1')|\,\boldsymbol{r}_1'\boldsymbol{r}_2'\rangle\,\delta(t_1-t_2)\delta(t_1'-t_2') \,. \qquad (5.93)\end{aligned}$$

Equation (5.91) can be represented in a diagrammatic form,

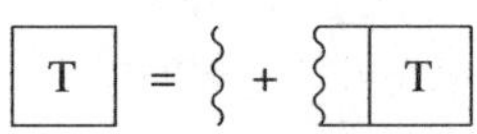

Fig. 5.5. T-matrix ladder equation

and the iterative solution leads to the series

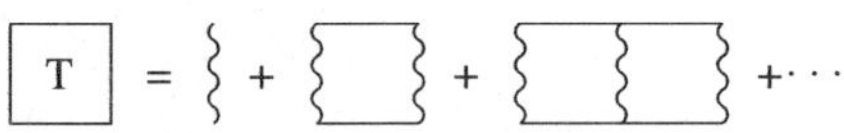

Fig. 5.6. T-matrix series

The T-matrix turns out to be a power series with respect to the interaction potential. Any of the terms is topologically characterized as an amputated binary ladder diagram.

The equations for the T-matrix given so far are determined on the Keldysh time contour. Now we follow the scheme to pass over from the Keldysh contour to the physical time axis (see Sect. 5.1). The equation for the T-matrices

$T^{\gtrless}$ follows from (5.91) and (5.92) by positioning of the times on opposite branches of the contour. We get

$$T_{ab}^{\gtrless}(t,t') = i \int_{-\infty}^{+\infty} d\bar{t} \Big(V_{ab}\, \mathcal{G}_{ab}^{R}(t,\bar{t})\, T_{ab}^{\gtrless}(\bar{t},t') + V_{ab}\, \mathcal{G}_{ab}^{\gtrless}(t,\bar{t})\, T_{ab}^{A}(\bar{t},t') \Big).$$

$$= iV_{ab}\, \bar{g}_{ab}^{\gtrless}(t,t')\, V_{ab}\,. \tag{5.94}$$

Here, the operator notation was used. Furthermore, we introduced the correlation functions

$$\mathcal{G}_{ab}^{\gtrless}(t,t') = g_a^{\gtrless}(t,t') g_b^{\gtrless}(t,t')\,. \tag{5.95}$$

Then, the simplest approximation for $T_{ab}^{\gtrless}(t,t')$ is the Born approximation

$$T_{ab,\mathrm{Born}}^{\gtrless}(t,t') = iV_{ab}\mathcal{G}_{ab}^{\gtrless}(t,t')V_{ab}\,. \tag{5.96}$$

Using (5.94), from the definition of the retarded and advanced T-matrices $T_{ab}^{R/A}(t,t')$ we find

$$T_{ab}^{R/A}(t,t') = V_{ab}\delta(t-t') + i \int_{-\infty}^{+\infty} d\bar{t}\, V_{ab}\, \mathcal{G}_{ab}^{R/A}(t,\bar{t})\, T_{ab}^{R/A}(\bar{t},t')$$

$$= V_{ab}\delta(t-t') + iV_{ab}\, \bar{g}_{ab}^{R/A}(t,t')\, V_{ab}\,. \tag{5.97}$$

This is a central equation for the ladder approximation. For very low densities, it reduces to the well-known Lippmann–Schwinger equation of scattering theory. In the general case, (5.97) describes in-medium scattering, and $T_{ab}^{R/A}$ represents a generalization of the T-matrix used in scattering theory.

The BSE for the retarded and advanced two-particle Green's functions without exchange can be written as

$$\bar{g}_{ab}^{R/A}(t,t') = \mathcal{G}_{ab}^{R/A}(t,t') + i \int_{-\infty}^{+\infty} d\bar{t}\, \mathcal{G}_{ab}^{R/A}(t,\bar{t}) V_{ab}\, \bar{g}_{ab}^{R/A}(\bar{t},t')$$

$$= \mathcal{G}_{ab}^{R/A}(t,t') + i \int_{-\infty}^{+\infty} d\bar{t}\, \bar{g}_{ab}^{R/A}(t,\bar{t}) V_{ab}\, \mathcal{G}_{ab}^{R/A}(\bar{t},t')\,. \tag{5.98}$$

From (5.98) together with (5.97), further useful relations of scattering theory may be derived. For this purpose, we introduce generalized Møller operators. The retarded one is defined as

$$\Omega_{ab}^{R}(t,t') = \delta(t-t') + i\, \bar{g}_{ab}^{R}(t,t')\, V_{ab}\,. \tag{5.99}$$

Furthermore, we use the relation $\Omega^A_{ab}(t,t') = \Omega^{R\,\dagger}_{ab}(t',t)$. Using (5.97) and (5.98), we get

$$\Omega^R_{ab}(t,t') = \delta(t-t') + i \int\limits_{-\infty}^{+\infty} d\bar{t}\, \mathcal{G}^R_{ab}(t,\bar{t})\, T^R_{ab}(\bar{t},t')$$

$$V_{ab}\Omega^R_{ab}(t,t') = T^R_{ab}(t,t') \qquad \Omega^A_{ab}(t,t')V_{ab} = T^A_{ab}(t,t')\,. \qquad (5.100)$$

With the help of these relations, compact expressions can be derived for the two-particle correlation functions $\bar{g}^{\gtrless}_{ab}(t,t')$ and for the T-matrices $T^{\gtrless}_{ab}(t,t')$. From (5.61), instead of (5.65), neglecting the initial correlation term, we find

$$\bar{g}^{\gtrless}_{ab}(t,t') = \int\limits_{-\infty}^{+\infty} d\bar{t}d\tilde{t}\, \Omega^R_{ab}(t,\bar{t})\, \mathcal{G}^{\gtrless}_{ab}(\bar{t},\tilde{t})\, \Omega^A_{ab}(\tilde{t},t')\,. \qquad (5.101)$$

The quantity $\bar{g}^{<}_{ab}(t,t)$ gives the known expression for the two-particle density operator in binary collision approximation derived under the assumption of the weakening of initial correlations. Inserting the solution (5.101) into (5.94) and using (5.100), an important relation between the T-matrices $T^{\gtrless}_{ab}$ and $T^{R/A}_{ab}$ can be obtained;

$$T^{\gtrless}_{ab}(t,t') = i \int\limits_{-\infty}^{+\infty} d\bar{t}d\tilde{t}\, T^R_{ab}(t,\bar{t})\, \mathcal{G}^{\gtrless}_{ab}(\bar{t},\tilde{t})\, T^A_{ab}(\tilde{t},t')\,. \qquad (5.102)$$

This relation gives a generalization of the optical theorem of usual scattering theory. Of course, it can also be derived directly from the first line in (5.94).

Due to the central role of the T-matrices for the ladder approximation, we will consider further properties of this quantity. In *this* section, we use (in contrast to other sections) the macroscopic time $T = (t+t')/2$ and the microscopic time $\tau = t - t'$, and we perform the Fourier transformation with respect to τ. Then the following spectral representation is valid

$$T^{R/A}_{ab}(\omega,T) - \mathrm{V}_{ab} = i \int \frac{d\bar{\omega}}{2\pi}\, \frac{T^{>}_{ab}(\bar{\omega},T) - T^{<}_{ab}(\bar{\omega},T)}{\omega - \bar{\omega} \pm i\varepsilon}\,. \qquad (5.103)$$

From this representation, we get analytic properties very similar to those of $g^{R/A}_{ab}$ discussed in Sect. 3.2.3. Especially, we get a *subtracted dispersion relation*, namely

$$\mathrm{Re}\, T^R_{ab}(\omega,T) - V_{ab} = \mathcal{P} \int \frac{d\bar{\omega}}{\pi}\, \frac{\mathrm{Im}\, T^R_{ab}(\bar{\omega},T)}{\omega - \bar{\omega}}\,, \qquad (5.104)$$

where the imaginary part is given by

$$2i\,\mathrm{Im}\,T^R_{ab}(\omega,T) = T^>_{ab}(\omega,T) - T^<_{ab}(\omega,T)\,. \tag{5.105}$$

From (5.102), we find in lowest order gradient expansion (local approximation) and after Fourier transformation with respect to the difference variables

$$T^{\gtrless}_{ab}(\omega,T) = iT^R_{ab}(\omega,T)\mathcal{G}^{\gtrless}_{ab}(\omega,T)T^A_{ab}(\omega,T)\,. \tag{5.106}$$

In the case of thermodynamic equilibrium, we find the spectral representation

$$\begin{aligned} i\,T^>_{ab}(\omega) &= -2\,\mathrm{Im}T^R_{ab}(\omega)\,[1+n_{ab}(\omega)] \\ i\,T^<_{ab}(\omega) &= -2\,\mathrm{Im}T^R_{ab}(\omega)\,n_{ab}(\omega) \end{aligned} \tag{5.107}$$

with the Bose function $n_{ab}(\omega)$ given by (5.15).

Next, we want to show that there follows a bilinear expansion of the T-matrix starting from the bilinear expansion of $g^{R/A}_{ab}$ and $g^{\gtrless}_{ab}$. Inserting (5.84) into (5.97), we get

$$T^R_{ab}(\omega,T) - V_{ab} = \sum_{\alpha \boldsymbol{P}} \frac{V_{ab}\,|\psi_{\alpha \boldsymbol{P}}\rangle\langle\psi_{\alpha \boldsymbol{P}}|\,V_{ab}}{\omega - E_{\alpha \boldsymbol{P}} + i\varepsilon}\,\mathcal{F}_{\alpha \boldsymbol{P}}(T)\,. \tag{5.108}$$

In some cases, it is more convenient to take another expression for $T^R_{ab}(\omega,T)$ which may be achieved by elimination of V_{ab} from (5.108) using the eigenvalue equation for the state vectors $|\psi_{\alpha \boldsymbol{P}}\rangle$.

The imaginary part of the T-matrix can be obtained using the Dirac identity. From (5.108), can write

$$\mathrm{Im}T^R_{ab}(\omega,T) = -\pi\sum_{\alpha \boldsymbol{P}} V_{ab}\,|\psi_{\alpha \boldsymbol{P}}\rangle\langle\psi_{\alpha \boldsymbol{P}}|\,V_{ab}\,\delta(\omega - E_{\alpha \boldsymbol{P}})\,\mathcal{F}_{\alpha \boldsymbol{P}}(T)\,. \tag{5.109}$$

For complex ω, the retarded T-matrix has poles at the bound state energies in the lower half plane.

In order to consider the statistical properties, let us come back to $T^{\gtrless}_{ab}$. Similar to the representation of the two-particle correlation functions given by (5.86), we use the Kadanoff–Baym ansatz

$$i\,T^{\gtrless}_{ab}(\omega,T) = \sum_{\alpha \boldsymbol{P}} V_{ab}\,|\psi_{\alpha \boldsymbol{P}}\rangle\langle\psi_{\alpha \boldsymbol{P}}|\,V_{ab}\,2\pi\delta(\omega - E_{\alpha \boldsymbol{P}})\,N^{\gtrless}_{\alpha \boldsymbol{P}}(T)\,. \tag{5.110}$$

We can split up this expression into a bound state part for $\alpha = j$ and a scattering part for $\alpha = \boldsymbol{p}$. For the scattering states $|\psi_{\boldsymbol{pP}}\rangle = |\psi^{(+)}_{\boldsymbol{pP}}\rangle$, the following identity holds

$$|\psi^{(+)}_{\boldsymbol{pP}}\rangle = \Omega^R_{ab}(E_{\boldsymbol{pP}})\,|\psi^{\mathrm{free}}_{\boldsymbol{pP}}\rangle\,, \tag{5.111}$$

where $|\Psi^{\mathrm{free}}_{\boldsymbol{pP}}\rangle$ are the states determined by the effective Schrödinger equation of two noninteracting quasiparticles. With (5.100), the following expression can be obtained

$$
\begin{aligned}
i\,T^{\gtrless}_{ab}(\omega,T) &= i^2\,T^R_{ab}(\omega,T)\,\mathcal{G}^{\gtrless}_{ab}(\omega,T)\,T^A_{ab}(\omega,T) \\
&+ \sum_{j\boldsymbol{P}} \mathrm{V}_{ab}\,|\psi_{j\boldsymbol{P}}\rangle\langle\psi_{j\boldsymbol{P}}|\,V_{ab}\;2\pi\,\delta(\omega - E_{j\boldsymbol{P}})\,N^{\gtrless}_{j\boldsymbol{P}}(T)\,. \qquad (5.112)
\end{aligned}
$$

The first term comes from the scattering states, and the second one from the bound states. Therefore, a separation of the T-matrix into the scattering and bound state parts is given. To get (5.112), we used

$$
i^2\mathcal{G}^{\gtrless}_{ab}(\omega,T) = \sum_{\boldsymbol{pP}} |\psi^{free}_{\boldsymbol{pP}}\rangle\langle\psi^{free}_{\boldsymbol{pP}}|\;2\pi\,\delta(\omega - E_{\boldsymbol{pP}})N^{\gtrless}_{\boldsymbol{pP}}(T)\,. \qquad (5.113)
$$

Neglecting the quantity $\mathcal{F}_{\alpha\boldsymbol{P}}(T)$ given by (5.81), the occupation numbers become

$$
\begin{aligned}
N^{<}_{\boldsymbol{pP}}(T) &= f_a(\boldsymbol{p}_1,T)f_b(\boldsymbol{p}_2,T) \\
N^{>}_{\boldsymbol{pP}}(T) &= \Big(1 \pm f_a(\boldsymbol{p}_1,T)\Big)\Big(1 \pm f_b(\boldsymbol{p}_2,T)\Big)\,. \qquad (5.114)
\end{aligned}
$$

Finally, let us discuss the T-matrix for a simple model of two interacting charged particles in a hydrogen plasma. In this model, which we have already discussed in Sect. 5.3, we account for the influence of the surrounding plasma medium on the two-particle dynamics by a statically screened Coulomb potential (5.57) and by the Debye self-energy shift $\Delta_a = -\kappa e_a^2/2$. The homogeneous Bethe–Salpeter equation then takes the form

$$
\Big(\mathrm{H}^0_{ab} + \Delta_{ab} + \mathrm{V}^s_{ab} - E_{\alpha\boldsymbol{P}}\Big)\,|\psi_{\alpha\boldsymbol{P}}\rangle = 0\,, \qquad (5.115)
$$

where V^s_{ab} is the Debye potential, and $\Delta_{ab} = \Delta_a + \Delta_b$. Using relative and center-of-mass variables $\boldsymbol{P} = \boldsymbol{p}_1 + \boldsymbol{p}_2$, $\boldsymbol{p} = (m_2\boldsymbol{p}_1 - m_1\boldsymbol{p}_2)/M$, the two-particle T-matrix can be written in the non-degenerate case

$$
\langle\boldsymbol{p}_1\boldsymbol{p}_2|T_{ab}(\omega)|\bar{\boldsymbol{p}}_2\bar{\boldsymbol{p}}_1\rangle = (2\pi\hbar)^3\,\langle\boldsymbol{p}|T_{ab}(\omega)|\bar{\boldsymbol{p}}\rangle\,\delta(\boldsymbol{P} - \bar{\boldsymbol{P}})\,. \qquad (5.116)
$$

Then the imaginary part of the T-matrix of relative motion can be written as

$$
\mathrm{Im}\langle\boldsymbol{p}\,|\mathrm{T}^R_{ab}(\omega)|\,\boldsymbol{p}'\rangle = -\sum_{\alpha}\langle\boldsymbol{p}\,|\mathrm{V}^s_{ab}|\psi_\alpha\rangle\langle\psi_\alpha|\mathrm{V}^s_{ab}|\,\boldsymbol{p}'\rangle\pi\delta(\omega - E_\alpha)\,. \qquad (5.117)
$$

After partial wave expansion, for the scattering part of (5.117) we find

$$
\mathrm{Im}\langle\boldsymbol{p}\,|\mathrm{T}^R_{ab}(\omega)|\,\boldsymbol{p}'\rangle^{\mathrm{sc}} = \frac{4\pi^2\mu_{ab}}{\hbar}\,\frac{1}{p_\omega}\sum_{l=0}^{\infty}(2l+1)\,T_l(p)\,T_l^*(p')\,P_l(\hat{\boldsymbol{p}}\cdot\hat{\boldsymbol{p}}')\,. \qquad (5.118)
$$

Here, $\hat{\boldsymbol{p}}$ and $\hat{\boldsymbol{p}}'$ are the unit vectors of $\boldsymbol{p}$ and $\boldsymbol{p}'$. The transition elements $T_l(p)$ are given by

$$T_l(p) = \int_0^\infty dr\; r\; j_l\left(\frac{pr}{\hbar}\right) V^s_{ab}(r)\; u_{p_\omega,l}(r)\,, \tag{5.119}$$

where μ_{ab} is the reduced mass, j_l are the spherical Bessel functions and $p_\omega = \sqrt{2\mu_{ab}\omega}$. To get (5.118), we used a partial wave expansion for the scattering wave function. For the bound state part of (5.117) we get

$$\begin{aligned}\mathrm{Im}\langle \boldsymbol{p}\,|\mathrm{T}^R_{ab}(\omega)|\,\boldsymbol{p}'\rangle^{\mathrm{b}} &= -\sum_{nlm} \pi\delta(\omega - E_{nlm}) \\ &\quad\times \left| \int d\boldsymbol{r}\, e^{-\frac{i}{\hbar}\boldsymbol{p}\cdot\boldsymbol{r}}\, V^s_{ab}(r) \frac{u_{n,l}(r)}{r} Y^m_l(\theta,\varphi) \right|^2 .\end{aligned} \tag{5.120}$$

The wave functions $u_\alpha(r)$ are solutions of the radial Schrödinger equation

$$\left[\frac{\hbar^2}{2\mu_{ab}}\frac{d^2}{dr^2} - \Delta_{ab} - \frac{\hbar^2 l(l+1)}{2\mu_{ab}r^2} - V^s_{ab}(r) + E_\alpha\right] u_\alpha(r) = 0\,. \tag{5.121}$$

We want to point out that the expression (5.117) determines the imaginary part of the *off-shell* T-matrix, i.e., we take $\omega \neq E_p, E_{p'}$. Especially, bound states are included in the *off-shell* T-matrix. In the special case $\omega = E_p = E_{p'}$, the T-matrix is on-shell, and it is directly related to the scattering cross section (see Sect. 5.6).

The main task of determining the imaginary part of the T-matrix from (5.118) and (5.120) is the calculation of the wave functions $u_\alpha(r)$. This was carried out by numerical solution of (5.121) using the Numerov algorithm.

Results for the imaginary part of the *off-shell* T-matrix for the electron proton scattering are given in Fig. 5.7 for different values of the inverse screening length (Kremp et al. 1998; Schmielau 2001). To present the bound state part, an artificial damping was introduced which provides a broadening of the bound state peak. For certain values of κ, only one bound state exists. The bound state manifests itself as a peak separated from the continuum. With increasing κ-values (increasing density), the bound state energy is lowered. But this lowering is weaker than the continuum lowering. At the critical value $\kappa = 1.19a_B^{-1}$, the ground state vanishes merging into the scattering continuum (Rogers 1971). The momentum and energy dependencies of the imaginary part of the T-matrix for $\kappa = 0.5a_B^{-1}$ are demonstrated in Fig. 5.8.

With the imaginary part of the T-matrix, we are able to determine the real part from the dispersion relation (5.104). The result is shown in Fig. 5.9.

By further approximations, we can simplify the expressions for the imaginary part of the T-matrix. Let us first consider the scattering part. In Born approximation, we replace the scattering states by free ones, and we get

$$\mathrm{Im}\langle \boldsymbol{p}\,|\mathrm{T}^R_{ab}(\omega)|\,\boldsymbol{p}'\rangle^B = -\int \frac{d\bar{\boldsymbol{p}}}{(2\pi)^3}\, V^s_{ab}(\boldsymbol{p}-\bar{\boldsymbol{p}}) V^s_{ab}(\bar{\boldsymbol{p}}-\boldsymbol{p}')\, \pi\, \delta(\omega - E_{\bar{\boldsymbol{p}}})\,, \tag{5.122}$$

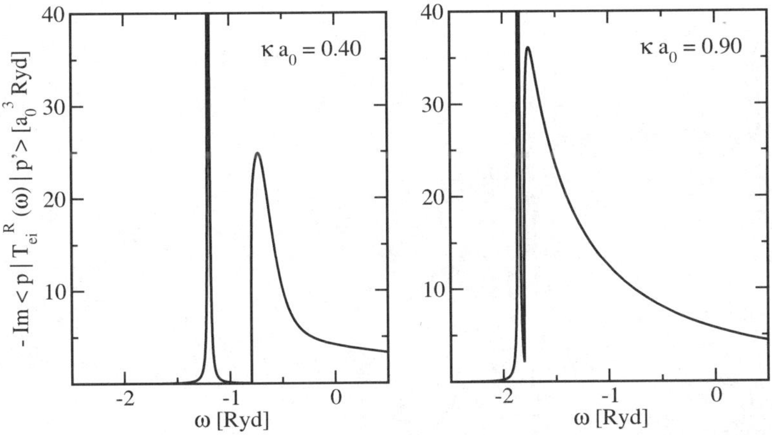

Fig. 5.7. Imaginary part of the off-shell T-matrix as a function of the energy variable ω for different values of the inverse screening length: $\kappa = 0.40\, a_B^{-1}$ (**left**), $\kappa = 0.9\, a_B^{-1}$ (**right**). The momenta in the initial and final state are $p = p' = \hbar/a_B$ with $\cos(\boldsymbol{p}, \boldsymbol{p}') = 0.7$. The artificial damping in the bound state part is $\Gamma = 0.01\,\mathrm{Ryd}$

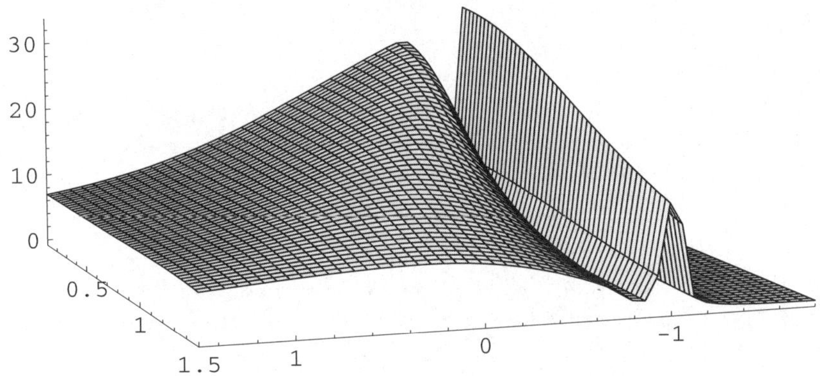

Fig. 5.8. Momentum-energy surface for the imaginary part of the off-shell T-matrix for the inverse screening length $\kappa = 0.5\, a_B^{-1}$. The momenta were taken to be $p = p'$ with $\cos(\boldsymbol{p}, \boldsymbol{p}') = 0.7$. The energy axis runs from right to left (ryd), the momentum from behind (a_B^{-1})

where $E_{\boldsymbol{p}} = \bar{p}^2/2\mu_{ab} + \Delta_{ab}$. With the Debye potential (5.57), we may write

$$\mathrm{Im}\langle \boldsymbol{p}\,|\mathrm{T}_{ab}^{R}(\omega)|\,\boldsymbol{p}'\rangle^{B} = -\pi \int_0^\infty \frac{d\bar{p}}{(2\pi)^3}\bar{p}^2 \int_{-1}^{+1} du \int_0^{2\pi} d\varphi\, \delta(\omega - E_{\bar{p}})$$
$$\times \frac{(4\pi e_a e_b)^2}{(p^2 - 2p\bar{p}u + \bar{p}^2 + \kappa^2)(\bar{p}^2 + p'^2 - 2p'\bar{p}v + \kappa^2)}\,, \qquad (5.123)$$

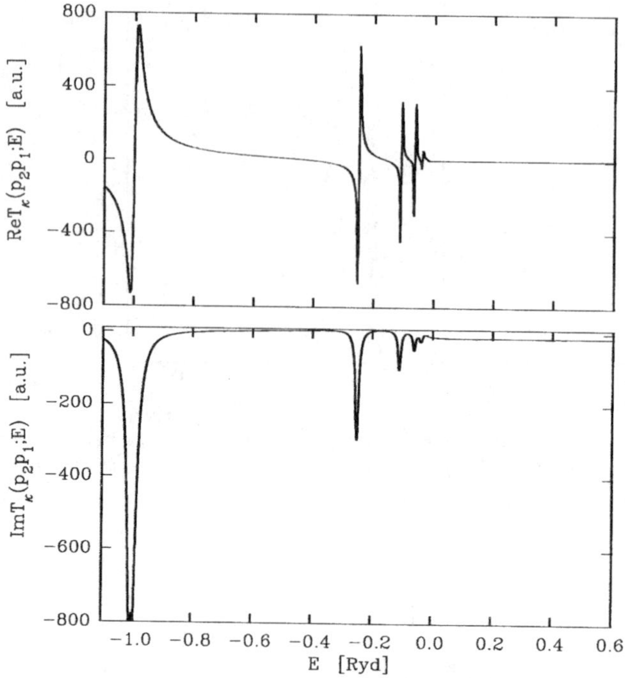

Fig. 5.9. Real and imaginary parts of the off-shell T-matrix for the inverse screening length $\kappa = 0.243 a_B^{-1}$. The momenta in the initial and final states are $p_1 = \hbar/a_B$, $p_2 = 0.6\hbar/a_B$ with $\cos(\boldsymbol{p}_1, \boldsymbol{p}_2) = 0.5$. There are five bound states of the s-wave

where $v = \cos(\bar{\boldsymbol{p}}, \boldsymbol{p}')$, $w = \cos(\boldsymbol{p}, \boldsymbol{p}')$ and $u = \cos(\boldsymbol{p}, \bar{\boldsymbol{p}})$. Here one has to consider the relation $v = uw + \sqrt{(1-u^2)}\sqrt{(1-w^2)} cos\varphi$.

Two of the integrations in (5.123) may be carried out. After some calculation, we have

$$\mathrm{Im}\langle \boldsymbol{p} | \mathrm{T}^R_{ab}(\omega) | \boldsymbol{p}' \rangle^B = 8\pi e^4 \mu_{ab} \sqrt{2\mu_{ab}\omega} \int_0^1 dt \frac{1}{A + Bt + Ct^2} \Big|_{q=\sqrt{2\mu_{ab}\omega}} . \tag{5.124}$$

Here we used Heaviside units with $e^2/2 = 2\mu_{ab} = 1$. The abbreviations are

$$A = (q^2 + \kappa^2 + p'^2)^2 - 4q^2 p'^2 \qquad C = (p'^2 - p^2)^2 - 4q^2(\boldsymbol{p} - \boldsymbol{p}')^2 .$$

$$B = 2\Big[(q^2 + \kappa^2 + p'^2)(p^2 - p'^2) - 4q^2(pp'w - p'^2)\Big] .$$

The integration of (5.124) may be carried out analytically. However, the numerical integration turns out to be more convenient.

An approximation for the bound state part of the T-matrix can be obtained if unscreened wave functions are used for the atomic states. For the hydrogen-like ground state contribution to the T-matrix, we easily find from (5.120)

$$\mathrm{Im}\langle \boldsymbol{p}|\mathrm{T}^R_{ab}(\omega)|\boldsymbol{p}'\rangle^{\mathrm{b}} = -\frac{(4\pi)^2 e^4 a_B}{(1+k^2)(1+k'^2)}\,\delta(\omega - E_{10})\,, \tag{5.125}$$

where $k = pa_B/\hbar$, $k' = p'a_B/\hbar$, and E_{10} is the atomic ground state energy.

Calculations for the determination of T-matrices in electron–hole plasmas were carried out in Schmielau et al. (2000) and Schmielau et al. (2001).

5.6 Two-Particle Scattering in Plasmas. Cross Sections

In the following we consider a non-degenerate plasma described by (5.115). With the knowledge of the T-matrix discussed in the previous section, it is possible to determine cross sections which are essential quantities to describe the scattering of two-particles in the system. For this case, the on-shell T-matrix is needed which is directly related to the differential cross section. The latter gives a measure of the scattering probability into a solid angle element $d\Omega$. It is defined by (Taylor 1972)

$$\begin{aligned}\frac{d\sigma_{ab}}{d\Omega} &= \frac{(2\pi)^4\hbar^2 m_{ab}^2}{(2\pi\hbar)^6}\left|\langle \mathbf{p}|T^R_{ab}(E_p)|\mathbf{p}'\rangle\right|^2_{p=p'}\,,\\ &= \left|f(p,\theta,\phi)\right|^2\,,\end{aligned} \tag{5.126}$$

where $\mathbf{p}$ and $\mathbf{p}'$ are the relative momenta and m_{ab} is the reduced mass. Additionally, we introduced the scattering amplitude $f(p,\theta,\phi)$ with θ being the scattering angle between $\mathbf{p}$ and $\mathbf{p}'$. Integrating over the angles, we get the total cross section

$$\sigma^{tot}_{ab}(p) = \int \frac{d\sigma_{ab}}{d\Omega}\,d\Omega\,. \tag{5.127}$$

Furthermore, we may define transport cross sections which play an essential role in transport theory:

$$Q^{(m)}_{ab}(p) = \int_0^{2\pi} d\phi \int_{-1}^{1} d\cos\theta(1-\cos^m\theta)\frac{d\sigma_{ab}}{d\Omega}\,. \tag{5.128}$$

Of special importance for our further considerations is the transport cross section with $m = 1$. In this case, we use the notation $Q^{(1)}_{ab} = Q^T_{ab}$. Furthermore, in connection with the contributions of electron–electron scattering to the electrical conductivity (see Sect. 9.2), we have to consider the cross section with $m = 2$, too.

The determination of the cross section from the T-matrix can be done by solving the Lippmann–Schwinger equation (5.145). This approach is of particular advantage if a perturbation theory may be applied which is possible for weak interaction or large scattering energies. In such situations, the first Born approximation is already a satisfying approximation. For a Debye potential (2.72), the differential cross section in first Born approximation is given by

$$\frac{d\sigma_{ab}(p,\theta)}{d\Omega} = \left(\frac{2m_{ab}e_a e_b}{4p^2\sin^2\frac{\theta}{2} + \hbar^2\kappa^2}\right)^2, \tag{5.129}$$

with κ being the inverse Debye screening length. With (5.129), we get for the transport cross section defined by formula (5.128) with $m = 1$

$$Q^T_{ab} = Q^{(1)}_{ab} = \frac{2\pi\hbar^4}{p^4 a_B^2}\left[\ln(1+4y) - \frac{4y}{1+4y}\right], \qquad y = \frac{p^2}{\hbar^2\kappa^2}. \tag{5.130}$$

If one has to go beyond the first Born approximation and if the interaction potential is spherically symmetric, the method of partial wave expansion is an appropriate method to calculate the cross sections. The scattering amplitude then takes the form (Taylor 1972)

$$f(p,\theta) = \frac{\hbar}{2ip}\sum_{l=0}^{\infty}(2l+1)[S_l(p) - 1]P_l(\cos\theta), \tag{5.131}$$

where $S_l(p)$ is the partial wave S-matrix. Because the S-matrix is unitary, the quantity S_l can be written as

$$S_l(p) = e^{2i\delta_l(p)}, \tag{5.132}$$

which defines the scattering phase shift $\delta_l(p)$. The latter is considered below. According to the relation (5.126) the partial wave expansion of the differential cross section reads

$$\begin{aligned} \frac{d\sigma_{ab}}{d\Omega} &= \frac{\hbar^2}{p^2}\sum_{ll'}(2l+1)(2l'+1)e^{i(\delta_l - \delta_{l'})}\sin\delta_l\sin\delta_{l'} \\ &\times\ P_l(\cos\theta)P_{l'}(\cos\theta). \end{aligned} \tag{5.133}$$

Using the orthogonality relations for the Legendre Polynomials one gets for the total cross section

$$\sigma^{tot}_{ab}(p) = \frac{4\pi\hbar^2}{p^2}\sum_{l=0}^{\infty}(2l+1)\sin^2\delta_l(p). \tag{5.134}$$

In similar manner, one can find the phase shift representations for the transport cross sections defined by (5.135). For the case $m = 1$, it follows

$$Q_{ab}^{T} = Q_{ab}^{(1)} = \frac{4\pi\hbar^2}{p^2} \sum_{l=0}^{\infty} (l+1) \sin^2\left(\delta_l(p) - \delta_{l+1}(p)\right). \tag{5.135}$$

The cross section with $m = 2$, for the case of electron–electron scattering, can be written as

$$Q_{ee}^{(2)} = \frac{4\pi\hbar^2}{p^2} \sum_{l=0}^{\infty} \frac{(l+1)(l+2)}{2l+3} \left[1 - \frac{(-1)^l}{2}\right] \sin^2\left(\delta_l(p) - \delta_{l+2}(p)\right). \tag{5.136}$$

In the latter expression, exchange processes were taken into account.

Thus, the scattering amplitude and the cross sections are expressed in terms of the scattering phase shifts. As shown in quantum scattering theory (Taylor 1972; Joachain 1975; Newton 1982), the phase shifts are determined by the scattering solutions $u_l = rR_l$ of the radial Schrödinger equation. We will assume that the two-body interaction potential is of finite range d. More precisely, the influence of the potential is negligible for $r > d$. From the demand for the continuity of the logarithmic derivative of u_l at $r = d$, it follows for the phase shifts

$$\tan \delta_l(k) = \frac{\hat{j}_l{}'(kd)\, u_l(k,d) \; - \; \hat{j}_l(kd)\, u_l{}'(k,d)}{\hat{n}_l{}'(kd)\, u_l(k,d) \; - \; \hat{n}_l(kd)\, u_l{}'(k,d)}, \tag{5.137}$$

with $k = p/\hbar$ being the wavenumber, and $\hat{j}_l(z)$ and $\hat{n}_l(z)$ are the Riccati-Bessel and Riccati–Neumann functions (Abramowitz and Stegun 1984). The prime means the derivative with respect to r at $r = d$, i.e., $\hat{j}_l{}'(k,d) = d\hat{j}_l(kr)/dr|_{r=d}$, etc.

Once the scattering phase shifts are known, the cross sections may be determined from their partial wave expansions. In order to get the $\delta_l(k)$, one has, in general, to solve numerically the radial Schrödinger equation. An effective method to solve this problem is given by the Numerov algorithm (Numerov 1923; Chow 1972).

For the description of two-particle scattering processes in plasmas, let us consider the transport cross section $Q_{ab}^{T} = Q_{ab}^{(1)}$ which will be of importance for many subsequent calculations. We use again the simple model considered in the previous section. The scattering wave function and thus the phase shifts are determined then by the effective Schrödinger equation (5.193) where the influence of the plasma medium on the two-particle states is accounted for by the Debye self-energy shifts and by the statically screened Coulomb (Debye) potential. We write the corresponding radial Schrödinger equation as

$$\left[\frac{d^2}{dr^2} - \frac{l(l+1)}{r^2} - \frac{2m_{ab}}{\hbar^2} V_{ab}^{S}(r) + k^2\right] u_l(k,r) = 0\,. \tag{5.138}$$

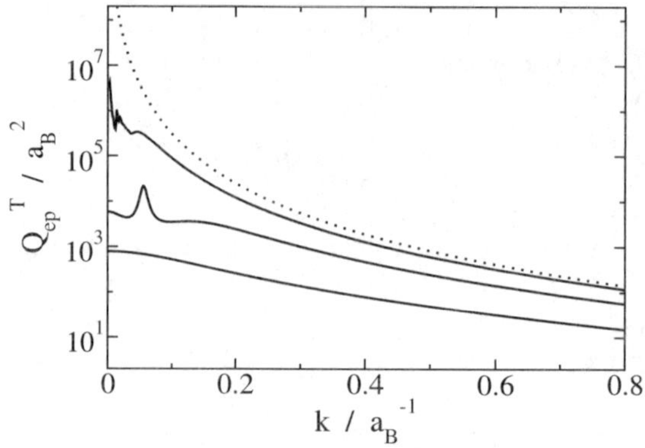

Fig. 5.10. Transport cross section of electron–proton scattering in a hydrogen plasma as a function of the wave number for different values of the inverse screening length $\kappa = \left(8\pi n_e e^2/k_B T\right)^{1/2}$. *Solid curves*: T-matrix approximation for $\kappa = 0.01 a_B^{-1}$, $\kappa = 0.1 a_B^{-1}$, $\kappa = 1.0 a_B^{-1}$ (from above to below). *Dotted curve*: First Born approximation for $\kappa = 0.01 a_B^{-1}$

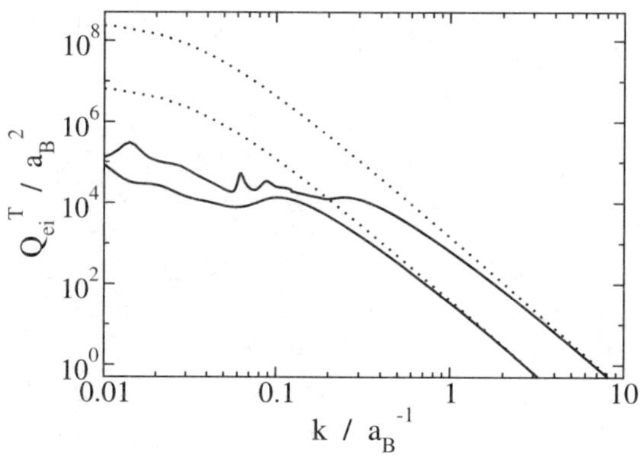

Fig. 5.11. Transport cross section versus wave number for electron–ion scattering in Born (*dotted*) and T-matrix approximation (*solid*). The ion charge number is $Z = 7$ for the upper pair of lines and $Z = 1$ for the lower one. The inverse Debye screening length is $\kappa = 0.1 a_B^{-1}$

Here, the wave number is $k = \sqrt{\frac{2m_{ab}}{\hbar^2}(E - \Delta_{ab})}$. In order to get the phase shifts from (5.138) we use the Numerov algorithm accounting for the behaviour $u_l(k,r) \sim r^{l+1}/(2l+1)$ for $r \to 0$ of the regular solution. The wave function is determined up to a point $r = d$, at which the ratio of the potential and kinetic energy is smaller than 10^{-6}. In the calculation of the cross section, the sum over l is truncated at some $l = l_0$ for which the contribution is less than 10^{-5} related to the total value (Gericke 2000). In Fig. 5.10, results for the transport cross section Q_{ep}^T of electron–proton scattering in a nonideal hydrogen plasma are shown for different values of the inverse screening length. We observe a lowering with increasing screening. The peaks in the T-matrix curves at low energies follow from resonance states. These resonances are typical quantum effects describing the contribution of former bound states which merged into the continuum at their respective Mott points. For the purpose of comparison, the Born approximation is presented for the case $\kappa = 0.01a_B^{-1}$. Results for the transport cross section Q_{ei}^T of electron–ion scattering considering different ion charge numbers Z are shown in Fig. 5.11. Here, only the contribution of the screening by the free electrons is taken into account in the Debye length. The cross sections are displayed versus wave number in T-matrix and Born approximations, respectively. While for large wave numbers both approximations merge into each other, the contributions of the higher-order ladder terms in the T-matrix reduce the cross section for small k values. These deviations increase with increasing ion charge number due to strong correlations. Again, the peaks in the low energy range of the T-matrix cross sections are due to resonance states in the electron–ion scattering.

5.7 Self-Energy and Kadanoff–Baym Equations in Ladder Approximation

In Sect. 4.1, the self-energy was introduced to get a formally closed equation for the single-particle Green's function. The self-energy is a central quantity of many-particle theory. It gives the scattering rates in the Kadanoff–Baym kinetic equations, and it determines the single-particle excitation spectrum. On the Keldysh time contour, we have

$$\pm i \sum_b \int_C d2\, V_{ab}(1-2)\, g_{ab}(12, 1'2^+) = \int_C d\bar{1}\, \Sigma_a(1,\bar{1})\, g_a(\bar{1},1') \,. \qquad (5.139)$$

Introducing the T-matrix according to (5.91) and taking into account (5.89), we find for the self-energy in ladder approximation

$$\Sigma_a(1,1') = \pm i \sum_b \int_C d\bar{2} \Big[\langle 12|\mathrm{T}_{ab}|1'\bar{2}\rangle \pm \delta_{ab} \langle 12|\mathrm{T}_{ab}|\bar{2}1'\rangle \Big]\, g_b(\bar{2}, 2^+) \,. \qquad (5.140)$$

Here, the second term in the brackets on the r.h.s. accounts for the exchange effects in the case of two identical particles $a = b$. It will be dropped in the following for simplicity.

The contour expression (5.140) represents a compact form of the different components of the self-energy on the physical time axis. For one time on the upper and one on the lower branch of the contour, respectively, we can write (cf. (5.93))

$$\Sigma_a^{\gtrless}(1,1') = \pm i \sum_b \int d\bar{\boldsymbol{r}}_2 \left\langle \boldsymbol{r}_1\boldsymbol{r}_2 \,|\mathrm{T}_{ab}^{\gtrless}(t_1,t_1')|\, \boldsymbol{r'}_1\bar{\boldsymbol{r}}_2 \right\rangle g_b^{\lessgtr}(\bar{\boldsymbol{r}}_2 t_1', \boldsymbol{r}_2 t_1) \,. \tag{5.141}$$

The simplest approximation is the Born approximation. We have to replace $\mathrm{T}_{ab}^{\gtrless}(t_1, t_1')$ by (5.96). In momentum representation we get

$$\begin{aligned}\Sigma_a^{\gtrless}(\boldsymbol{p}, tt') \;&=\; \pm i \sum_b \int \frac{d\boldsymbol{p}'}{(2\pi)^3}\frac{d\boldsymbol{q}}{(2\pi)^3} |V_{ab}(\boldsymbol{q})|^2 \\ &\times\; g_a^{\gtrless}(\boldsymbol{p}+\boldsymbol{q}, tt') g_b^{\gtrless}(\boldsymbol{p}'-\boldsymbol{q}, tt') g_b^{\lessgtr}(\boldsymbol{p}', t't) \,. \end{aligned} \tag{5.142}$$

Equation (5.142) is a frequently used formula.

In order to express the self-energy functions in terms of the retarded and advanced T-matrices, we use the optical theorem given by (5.102)

$$\Sigma_a^{\gtrless}(t,t') = \mp \sum_b \mathrm{Tr}_2 \int_{-\infty}^{+\infty} d\bar{t}\, d\bar{\bar{t}}\, \mathrm{T}_{ab}^{R}(t,\bar{t})\, \mathrm{T}_{ab}^{A}(\bar{\bar{t}},t')\, g_a^{\gtrless}(\bar{t},\bar{\bar{t}})\, g_b^{\gtrless}(\bar{t},\bar{\bar{t}})\, g_b^{\lessgtr}(t',t) \,. \tag{5.143}$$

Here, short-hand operator notation was used with "Tr" being the trace.

Now we can write the Kadanoff–Baym kinetic equations in ladder approximation. The general scheme is the following; we have two groups of equations. In the first group are the kinetic equations for the single-particle correlation functions

$$\begin{aligned}\left(i\frac{\partial}{\partial t_1} + \frac{\nabla_1^2}{2m_a}\right) g_a^{\gtrless}(1,1') \;&-\; \int d\bar{r}_1 \Sigma_a^{\mathrm{HF}}(\boldsymbol{r}_1 t_1, \bar{\boldsymbol{r}}_1 t_1)\, g_a^{\gtrless}(\bar{\boldsymbol{r}}_1 t_1, \boldsymbol{r}' t_1') \\ &= \int_{-\infty}^{t_1} d\bar{1} \left[\Sigma_a^{>}(1,\bar{1}) - \Sigma_a^{<}(1,\bar{1})\right] g_a^{\gtrless}(\bar{1},1') \\ &- \int_{-\infty}^{t_1'} d\bar{1}\, \Sigma_a{}^{\gtrless}(1,\bar{1}) \left[g_a^{>}(\bar{1},1') - g_a^{<}(\bar{1},1')\right] ,\end{aligned} \tag{5.144}$$

with the self-energies $\Sigma_a^{\gtrless}$ given in terms of the T-matrices according to (5.141) and (5.143). The second group of equations determines the T-matrices from the optical theorem (5.102) and from the Lippmann–Schwinger equation

$$\mathrm{T}_{ab}^{R/A}(t,t') = \mathrm{V}_{ab}\, \delta(t-t') + i \int_{-\infty}^{+\infty} d\bar{t}\, \mathrm{V}_{ab}\, \mathcal{G}_{ab}^{R/A}(t,\bar{t})\, \mathrm{T}_{ab}^{R/A}(\bar{t},t') \,. \tag{5.145}$$

The self-consistent system of equations allows the determination of all the properties of equilibrium and non-equilibrium many-particle systems in ladder approximation.

Let us consider the expression for the self-energy more in detail. For this purpose, a spatially homogeneous system is assumed. From (5.141), after Fourier transformation with respect to the relative time and using momentum representation, we get

$$\begin{aligned}&\Sigma_a^{\gtrless}(\boldsymbol{p}_1\omega,T)\\&=\pm\frac{i}{V}\sum_b\int\frac{d\boldsymbol{p}_2}{(2\pi)^3}\frac{d\bar{\omega}}{2\pi}\,\langle\boldsymbol{p}_1\boldsymbol{p}_2\,|\,\mathrm{T}_{ab}^{\gtrless}(\omega+\bar{\omega},T)\,|\,\boldsymbol{p}_1\boldsymbol{p}_2\rangle\,g_b^{\lessgtr}(\boldsymbol{p}_2\bar{\omega},T)\,.\end{aligned}\tag{5.146}$$

The upper sign refers to Bose particles, the lower one to Fermi particles. Using the lowest order gradient expansion (local approximation) of the optical theorem, i.e., (5.106), we get

$$\begin{aligned}&i\Sigma_a^{\gtrless}(\boldsymbol{p}_1\omega_1,T)\\&=\pm\frac{2}{V}\sum_b\int\frac{d\boldsymbol{p}_2d\omega_2}{(2\pi)^4}\frac{d\bar{\boldsymbol{p}}_1d\bar{\omega}_1}{(2\pi)^4}\frac{d\bar{\boldsymbol{p}}_2d\bar{\omega}_2}{(2\pi)^4}\,\pi\delta(\omega_1+\omega_2-\bar{\omega}_1-\bar{\omega}_2)\\&\times|\,\langle\boldsymbol{p}_1\boldsymbol{p}_2|T_{ab}^{R}(\omega_1+\omega_2,T)|\bar{\boldsymbol{p}}_1\bar{\boldsymbol{p}}_2\rangle\,|^2 ig_b^{\lessgtr}(\boldsymbol{p}_2\omega_2,T)ig_a^{\gtrless}(\bar{\boldsymbol{p}}_1\bar{\omega}_1,T)ig_b^{\gtrless}(\bar{\boldsymbol{p}}_2\bar{\omega}_2,T)\,.\end{aligned}\tag{5.147}$$

Now we will consider the expression (5.146) in quasiparticle approximation using (3.209). Furthermore, inserting the bilinear expansion of the T-matrices given by (5.112), we can write for $\Sigma_a^{>}$

$$\begin{aligned}i\,\Sigma_a^{>}(\boldsymbol{p}_1\omega)&=\frac{2}{V}\sum_b\int\frac{d\boldsymbol{p}_2}{(2\pi)^3}\frac{d\bar{\boldsymbol{p}}_1}{(2\pi)^3}\frac{d\bar{\boldsymbol{p}}_2}{(2\pi)^3}\,\pi\,\delta(\omega+E_2-\bar{E}_1-\bar{E}_2)\\&\times\left|\langle\boldsymbol{p}_1\boldsymbol{p}_2\,|\mathrm{T}_{ab}^{R}(\omega+E_2)|\,\bar{\boldsymbol{p}}_1\bar{\boldsymbol{p}}_2\rangle\right|^2\,f_a^{>}(\bar{\boldsymbol{p}}_1)\,f_b^{>}(\bar{\boldsymbol{p}}_2)f_b^{<}(\boldsymbol{p}_2)\\&+\frac{2}{V}\sum_b\sum_{jP}\int\frac{d\boldsymbol{p}_2}{(2\pi)^3}\,\pi\,\delta(\omega+E_2-E_{jP})\\&\times|\langle j\boldsymbol{P}|\mathrm{V}_{ab}|\,\boldsymbol{p}_1\boldsymbol{p}_2\rangle|^2\,N_{j\boldsymbol{P}}^{>}\,f_b^{<}(\boldsymbol{p}_2)\,.\end{aligned}\tag{5.148}$$

Here, the time variables are dropped. Furthermore, the abbreviations $f_a^{<}(\boldsymbol{p})=f_a(\boldsymbol{p})$ and $f_a^{>}(\boldsymbol{p})=1\pm f_a(\boldsymbol{p})$ are used. For the single-particle energies, we write $E_1=E_a(\boldsymbol{p}_1)$, $E_2=E_b(\boldsymbol{p}_2)$. A similar expression follows from (5.146) for the self-energy function $\Sigma_a^{<}$.

The first term on the r.h.s. of (5.148) is the contribution of two-particle scattering states, and the second one is that of two-particle bound states. If the energy variable is taken on-shell to be $\omega=E_1$, the bound state part vanishes, and the single-particle self-energies $\Sigma_a^{\gtrless}$ reduce to the scattering-in and scattering-out rates in ladder approximation.

In the general case of non-equilibrium systems, both quantities $\Sigma_a^{>}$ and $\Sigma_a^{<}$ are independent, whereas in thermodynamic equilibrium, they are connected by the KMS relation. As already shown in Sects. 3.2.2 and 3.3.5, the KMS condition leads to the spectral representation

$$\begin{aligned} \pm i\,\Sigma_a^<(\boldsymbol{p},\omega) &= \Gamma_a(\boldsymbol{p},\omega) f_a(\omega) \\ i\,\Sigma_a^>(\boldsymbol{p},\omega) &= \Gamma_a(\boldsymbol{p},\omega)\big[1 \pm f_a(\omega)\big] \end{aligned} \tag{5.149}$$

with $f_a(\omega) = [e^{\beta(\omega-\mu_a)} \mp 1]^{-1}$ being the Bose- or Fermi-like functions, and $\Gamma_a = i(\Sigma_a^> - \Sigma_a^<)$ the single-particle damping. In thermodynamic equilibrium, the following expression for the damping can be found from (5.146)

$$\begin{aligned} \Gamma_a(\boldsymbol{p}_1,\omega) &= -\frac{2}{V}\sum_b \int \frac{d\boldsymbol{p}_2}{(2\pi)^3} \mathrm{Im}\,\langle \boldsymbol{p}_1\boldsymbol{p}_2|T^R_{ab}(\omega+E_2)|\boldsymbol{p}_1\boldsymbol{p}_2\rangle \\ &\times \Big[f_b(E_2) \mp n_{ab}(\omega+E_2)\Big]\,, \end{aligned} \tag{5.150}$$

where n_{ab} is the two-particle Bose function given by (5.15). Again the quasi-particle approximation was applied to get (5.150), and for the T-matrices $T_{ab}^{\gtrless}$ the KMS relation (5.107) was used. Inserting the bilinear expansion of the T-matrix given by (5.109) into (5.150), we arrive at

$$\begin{aligned} &\Gamma_a(\boldsymbol{p}_1,\omega) = -\frac{2}{V}\sum_b\sum_\alpha \int \frac{d\bar{\boldsymbol{P}}}{(2\pi)^3}\frac{d\boldsymbol{p}_2}{(2\pi)^3}\,\big|\langle \boldsymbol{p}_1\boldsymbol{p}_2|V_{ab}(\omega+E_2)|\alpha\,\bar{\boldsymbol{P}}\rangle\big|^2 \\ &\times\pi\delta(\omega+E_2-E_{\alpha\bar{\boldsymbol{P}}})\,\mathcal{F}_{\alpha\bar{\boldsymbol{P}}}\Big[f_b(E_2) \mp n_{ab}(\omega+E_2)\Big]\,. \end{aligned} \tag{5.151}$$

The sum over α runs over all two-particle states, i.e., $|\alpha\rangle = |j\rangle$ for bound states, and $|\alpha\rangle = |\bar{\boldsymbol{p}}+\rangle$ for scattering states.

5.8 Dynamically Screened Ladder Approximation

In the previous sections, we considered the influence of the plasma on the two-particle properties, describing the interaction by the statically screened Coulomb potential and accounting for the static lowest order Montroll–Ward term in the self-energy shifts.

Of course, a more rigorous consideration of the two-particle properties in strongly coupled plasmas requires the inclusion of dynamical screening. Then, one has to start from the Bethe–Salpeter equation in dynamically screened ladder approximation derived in Sect. 5.2 in form of an integral equation (5.49) or in the corresponding differential equation on the Keldysh contour given by (taking into account the definition (3.131) for the special δ-distribution)

$$\begin{aligned} &(i\frac{\partial}{\partial t} - H^0_{ab})g_{ab}(t,t') - \int_C d\bar{t}_1 d\bar{t}_2\Big[\,\bar{\Sigma}_a(t,\bar{t}_1)\,\delta(t-\bar{t}_2) \\ &+\bar{\Sigma}_b(t,\bar{t}_2)\,\delta(t-\bar{t}_1)\Big]\,g_{ab}(\bar{t}_1\bar{t}_2,t') = -i\,N_{ab}(\,t\,)\,\delta(t-t') \\ &+i\int_C d\bar{t}_1 d\bar{t}_2\Big[\,g_a(t,\bar{t}_1)\,\delta(t-\bar{t}_2) + g_b(t,\bar{t}_2)\,\delta(t-\bar{t}_1)\Big] \\ &\times V^s_{ab}(\bar{t}_1,\bar{t}_2)\,g_{ab}(\bar{t}_1\bar{t}_2,t')\,. \end{aligned} \tag{5.152}$$

Exchange terms were neglected for simplicity. $N_{ab}(t)$ is the phase space occupation factor, and H^0_{ab} is the free two-particle Hamiltonian. In fact, with (5.152), an equation of motion for the two-time Green's function is found. But, this is not a closed equation for the determination of $g_{ab}(t,t')$.

As already discussed earlier, because of the dynamical self-energies and the dynamically screened potential, (5.152) contains the two-particle Green's function with three time arguments, too. The question arises whether it is possible to find approximately a BSE which involves only two-time functions.

There have been different attempts in the past to achieve such an equation; we mention Kilimann et al. (1977), Zimmermann et al. (1978), Haug and Thoai (1978), Schäfer and Treusch (1986), Manzke (b) et al. (1998), and Manzke (a) et al. (1998). All these papers made use of the Shindo approximation in frequency or in time variables (Shindo 1970) to obtain a closed equation.

On the basis of this closed BSE, an effective Schrödinger equation was derived which gives important many-particle corrections to the Schrödinger equation of the isolated two-particle problem. We mention the weak density dependence of the bound state energies as a result of a compensation between the different many-particle effects and the lowering of the continuum edge due to the dynamical self-energy.

In spite of these interesting results, these approaches have some serious shortcomings especially for degenerate plasmas.

(i) There occur contributions of the kind $N_{ab}^{-1} = (1-f_a-f_b)^{-1}$ which lead to artificial singularities.
(ii) There are static contributions to the plasma Hamiltonian which have no clear physical meaning.
(iii) The retarded/advanced Green's functions $g_{ab}^{R/A}$ used in such approaches have all the deficits considered in Sect. 5.1.

The solution of this problem was given by Bornath et al. (1999). In that paper, the BSE (5.152) for the real-time Green's function was considered avoiding the Shindo approximation. Here, we follow the scope of that paper.

Let us start with a discussion of the integral terms. The self-energy is assumed to be given in V^s-approximation. Then we get for the matrix elements of the integral terms

$$\begin{aligned} &I^{\alpha\beta\alpha'\beta'}(tt') \\ &= i\sum_{\bar\beta=\pm}\bar\beta\int_c d\bar t g_b^{\beta\bar\beta}(t,\bar t)\Big[\bar V_{bb}^{\beta\bar\beta}(t,\bar t) + \bar V_{ab}^{\bar\beta\alpha}(t,\bar t)\Big]g_{ab}^{\bar\beta\alpha\beta'\alpha'}(t,\bar t,t't') \\ &+ i\sum_{\bar\alpha=\pm}\bar\alpha\int_c d\bar t g_a^{\alpha\bar\alpha}(t,\bar t)\Big[\bar V_{aa}^{\alpha\bar\alpha}(t,\bar t) + \bar V_{ab}^{\bar\alpha\beta}(\bar t,t)\Big]g_{ab}^{\bar\alpha\beta\alpha'\beta'}(\bar t,t,t't')\,. \end{aligned} \quad (5.153)$$

The contributions of (5.153) are given in Feynman diagrams in Figs. 5.12 and 5.13. We consider the perturbation expansion of $I^{\alpha\beta\alpha'\beta'}(tt')$ with respect

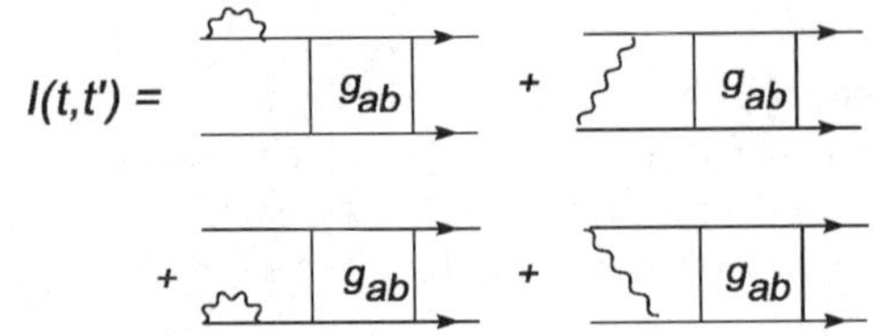

Fig. 5.12. Diagram representation of $I(tt')$ given in (5.153)

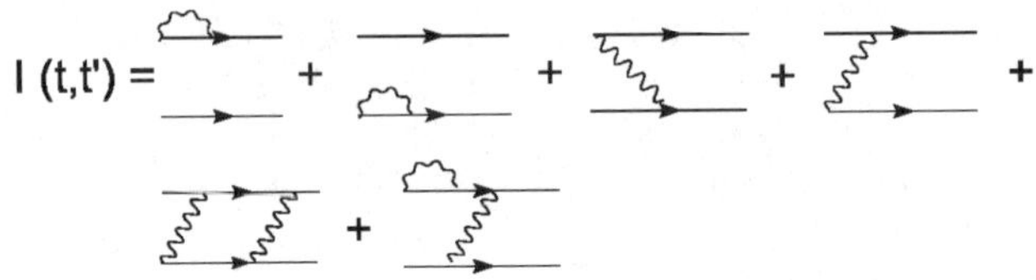

Fig. 5.13. Diagram expansion of $I(tt')$

to the dynamically screened potential and the free single-particle Green's function. The diagram representation of this series is shown in Fig. 5.13. It is possible to classify these diagrams into reducible and irreducible ones. Such a classification has to be done carefully due to the time dependence of the dynamically screened potential. Let us introduce a diagram as reducible if it decays into disconnected parts by cutting a pair of single-particle Green's functions $g^0_{ab}(t_1, t_2, t'_1, t'_2) = g_a(t_1, t'_1) g_b(t_2, t'_2)$ only which overlap in time, i.e., if there exists a time t_m for which we have, e.g., $t_1 < t_m < t'_1$ and $t_2 < t_m < t'_2$. Otherwise it is irreducible. Then we know that the sum of all diagrams is created by the sum of irreducible diagrams only. In the following, we restrict ourselves to irreducible diagrams of first order only. Let us consider the fifth diagram of Fig. 5.13 more in detail. We can assume that there is an "overlap" of the potentials, because such contributions are, in the sense discussed above, irreducible second-order diagrams (Bornath et al. 1999). Therefore, we have to analyze the diagram given in Fig. 5.14. Because of the dynamically screened potential, both single-particle functions overlap only partially. Therefore, the free two-particle Green's function depends, in general, on four times. In order to construct a two-time free two-particle Green's function we make use of the semi-group properties of the free single-particle propagators. This means, we may write for any time $\bar{t}$ with $t > \bar{t} > t'$ (Bornath et al. 1999)

$$g^R_a(x_1, t, x'_1, t') = i \int d\bar{x}_1 \, g^R_a(x_1, t, \bar{x}_1, \bar{t}) \, g^R_a(\bar{x}_1, \bar{t}, x'_1, t') \,. \tag{5.154}$$

In the following we will use the shorter operator notation. We write instead of (5.154)

$$g^R_a(t, t') = i g^R_a(t, \bar{t}) g^R_a(\bar{t}, t') \,. \tag{5.155}$$

Further we obtain the relations for the correlation functions

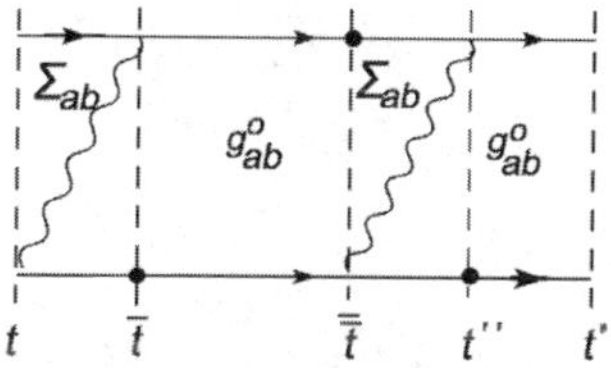

Fig. 5.14. Reducible second-order diagram. *Dashed lines* mean temporal segments

$$\begin{aligned} g_a^{\gtrless}(t,t') &= i g_a^R(t,\bar{t}) g_a^{\gtrless}(\bar{t},t') \quad \text{for } t > \bar{t} > t' \,, \\ g_a^{\gtrless}(t,t') &= (-i) g_a^{\gtrless}(t,\bar{t}) g_a^A(\bar{t},t') \quad \text{for } t < \bar{t} < t' \,. \end{aligned} \tag{5.156}$$

Then we can write (Schmielau 2003)

$$g_a^{\alpha\beta}(t,t') = i \sum_{\gamma=\pm} \gamma g_a^{\alpha\gamma}(t,\bar{t}) g_a^{\gamma\beta}(\bar{t},t') \quad \text{for } t < \bar{t} < t' \text{ or } t' < \bar{t} < t \,. \tag{5.157}$$

With this relation, it is possible to insert vertices at the begin and at the end of the overlapping, see Fig. 5.14. In this way, the free two-particle Green's function $g^0_{ab}(\bar{t},\bar{t}_1) = g_a(\bar{t},\bar{t}_1) g_b(\bar{t},\bar{t}_1)$ can be cut out. What remains are irreducible interaction contributions. In a similar way, the other second-order diagrams of Fig. 5.13 may be analyzed.

We introduce the sum of the irreducible diagrams of first order in analogy to the single-particle case, now as the *two-particle self-energy* Σ_{ab}. If we know the irreducible self-energy $\Sigma^{\alpha\beta\gamma\delta}$, the sum of all dynamical ladder diagrams is given by

$$\begin{aligned} I^{\alpha\beta\alpha'\beta'}(t,t') &= \sum_{\gamma\delta=\pm} \gamma\delta \int d\bar{t} \Sigma_{ab}^{\alpha\beta\gamma\delta}(t,\bar{t}) g_{ab}^{\gamma\delta\alpha'\beta'}(\bar{t},t') \,, \\ \Sigma_{ab}^{\alpha\beta\gamma\delta}(t,\bar{t}) &= i\, g_a^{\alpha\gamma}(t,\bar{t}) \Big[V_{aa}^{\alpha\gamma}(t,\bar{t}) + V_{bb}^{\beta\delta}(t,\bar{t}) \\ &\quad + V_{ab}^{\alpha\delta}(t,\bar{t}) + V_{ba}^{\beta\gamma}(t,\bar{t}) \Big] g_b^{\beta\delta}(t,\bar{t}) \,. \end{aligned} \tag{5.158}$$

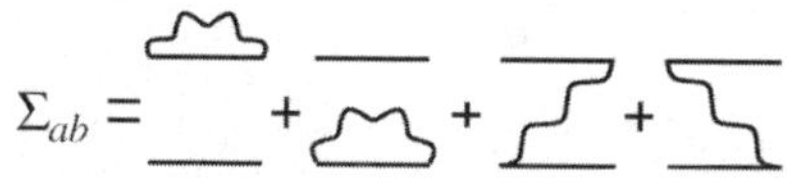

Fig. 5.15. Two-particle self-energy in first order with respect to the screened potential

The diagrammatic structure of the self-energy in first order with respect to V^s is shown in Fig. 5.15. We now introduce (5.158) into (5.152) and get the important equation

$$\begin{aligned} &\left(i\frac{\partial}{\partial t} - H^0_{ab}\right) g_{ab}^{\alpha\beta\alpha'\beta'}(t,t') + \sum_{\gamma\delta=\pm} \gamma\delta \int_{-\infty}^{+\infty} d\bar{t}\; \Sigma_{ab}^{\alpha\beta\gamma\delta}(t,\bar{t}) g_{ab}^{\gamma\delta\alpha'\beta'}(\bar{t},t') \\ &= -i\, N_{ab}(t)\, \alpha\, \delta_{\alpha\alpha'}\, \beta \delta_{\beta\beta'}\, \delta(t-t') \,. \end{aligned} \tag{5.159}$$

The structure of this equation is much more general than one can see from the derivation just presented. All considerations, and especially the insertion of vertices according to (5.157), are possible for arbitrary reducible diagrams. Furthermore, is is possible to use more general expressions for W in (5.35). Now it is interesting to introduce, in analogy to the single-particle case, retarded and advanced quantities according to the definitions (5.18) and (5.19). After longer but straightforward rearrangements, using (4.34) for the retarded screened potential and repeated applications of (3.48) (or (3.89)), and of (3.49), we find the following set of equations

$$\begin{aligned}(i\frac{\partial}{\partial t} - H_{ab})g_{ab}^{\alpha\beta\gamma\delta}(t,t') - \int_{-\infty}^{+\infty} d\bar{t}\ \Sigma_{ab}^{R}(t,\bar{t})\ g_{ab}^{\alpha\beta\gamma\delta}(\bar{t},t') \\ = -i\,N_{ab}(\,t\,)\alpha\,\delta_{\alpha\gamma}\beta\delta_{\beta\delta}\ \delta(t-t') + \int_{-\infty}^{+\infty} d\bar{t}\ \Sigma_{ab}^{\alpha\beta\gamma\delta}(t,\bar{t})\ G_{ab}^{A}(\bar{t},t')\,;\end{aligned} \tag{5.160}$$

$$(i\frac{\partial}{\partial t} - H_{ab})G_{ab}^{R}(t,t') - \int_{-\infty}^{+\infty} d\bar{t}\ \Sigma_{ab}^{R}(t,\bar{t})G_{ab}^{R}(\bar{t},t') = \delta(t-t')\,, \tag{5.161}$$

where $H_{ab} = H_{ab}^{0} + V_{ab}$ is the two-particle Hamiltonian, and the retarded two-particle self-energy is defined in analogy to g_{ab}^{R} by

$$\Sigma_{ab}^{R}(t,t') = \Sigma_{ab}^{0}(t)\delta(t-t') + \theta(t-t') \sum_{\gamma,\delta=\pm} \gamma\delta\ \Sigma_{ab}^{\alpha\beta\gamma\delta}(t,t') \tag{5.162}$$

with a contribution local in time, $\Sigma_{ab}^{0}(t) = \Sigma_{a}^{\mathrm{HF}}(t)+\Sigma_{b}^{\mathrm{HF}}(t)+[N_{ab}-1]V_{ab}$. The equations (5.160) and (5.161) have a structure similar to the non-equilibrium Dyson equations (3.147) and (3.148) in the single-particle case.

We can proceed in the known manner to find a general solution for the two-particle two-time correlation functions

$$\begin{aligned}g_{ab}^{\gtrless}(t,t') &= G_{ab}^{R}(t,t_0)\,g_{ab}^{\gtrless}(t_0,t_0)\,G_{ab}^{A}(t_0,t') \\ &\quad + \int_{-\infty}^{+\infty} d\bar{t}d\,\bar{\bar{t}}\ G_{ab}^{R}(t,\bar{t})\ \Sigma_{ab}^{\gtrless}(\bar{t},\bar{\bar{t}})\,G_{ab}^{A}(\bar{\bar{t}},t')\,.\end{aligned} \tag{5.163}$$

The first term is the general solution of the homogeneous equation including the initial condition for the temporal evolution of $g_{ab}^{\gtrless}(t,t')$. In the case of a static effective potential and a static self-energy correction, this term gives the expression (5.65) obtained in Sect. 5.3. The second term in (5.163) is a source contribution with the sources given by the dynamical quantities $\Sigma_{ab}^{\gtrless}$. The correlation function $g_{ab}^{<}(t,t')|_{t=t'}$ is of special importance because it is the two-particle density matrix in dynamically screened ladder approximation.

Let us now consider the two-particle self-energy Σ_{ab} given by (5.158). For $\alpha=\beta=+$ and $\gamma=\delta=-$, we get, for example, the important quantities $\Sigma_{ab}^{\gtrless}$. The retarded two-particle self-energy in first order with respect to the dynamically screened potential follows from (5.158) and (5.162). After some algebra we get

$$\Sigma_{ab}^{R}(t,t') - \Sigma_{ab}^{0}\delta(t-t') = ig_b^R(t,t')\; i[V_{aa}^{>}(t,t') + V_{ab}^{>}(t,t')]\;\; g_a^R(t,t')$$
$$+ig_b^{<}(t,\bar{t})\, i\,[V_{aa}^{R}(t,t') + V_{ab}^{R}(t,t')]\;\, g_a^R(t,t') + (a \leftrightarrow b)\,. \tag{5.164}$$

The retarded two-particle self-energy consists of two parts of different physical meaning. The first part does not contain an interaction between the particles a and b. These terms are due to the single-particle self-energy. The second part contains the interaction between a and b and describes the modification of the Coulomb interaction by the many-particle system. Therefore, let us divide the two-particle self-energy in the following form

$$\Sigma_{ab}^{R}(t,t') = \Delta_{ab}^{R}(t,t') + \Delta V_{ab}^{\mathrm{effR}}(t,t')\,. \tag{5.165}$$

Both quantities have to be used for the description of the behavior of a pair of particles in dense strongly coupled plasmas. The effective potential describes the interaction between two charged particles accounting for the surrounding medium. The bare Coulomb interaction is modified by dynamic screening, by retardation effects and by phase space occupation and is given by

$$\Delta V_{ab}^{\mathrm{eff\ R}} = V_{ab}^{\mathrm{eff\ R}}(t,t') - V_{ab}\delta(t-t') = ig_b^R(t,t')\; iV_{ab}^{>}(t,t')\, g_a^R(t,t')$$
$$+ig_b^{<}(t,t')\, iV_{ab}^{R}(t,t')\, g_a^R(t,t') + (a \leftrightarrow b)\,. \tag{5.166}$$

The self-energy Δ_{ab} changes the kinetic energies of the particles. It represents an effective two-particle energy shift, given in terms of the dynamical single-particle self-energies Σ_a and Σ_b in V^s-approximation

$$\Delta_{ab}^{R}(t,t') = \left[i\Sigma_a^R(t,t')g_b^R(t,t') + i\Sigma_b^R(t,t')g_a^R(t,t')\right]. \tag{5.167}$$

Then we can write the equation for the retarded two-particle Green's function in the following form

$$\begin{aligned}(i\frac{\partial}{\partial t} - H_{ab})\, G_{ab}^R(t,t') \quad &- \quad \int_{-\infty}^{+\infty} d\bar{t}\left[\Delta_{ab}^R(t,\bar{t}) + \Delta V_{ab}^{\mathrm{effR}}(t,\bar{t})\right] G_{ab}^R(\bar{t},t') \\ &= \quad \delta(t-t').\end{aligned} \tag{5.168}$$

Together with the Dyson equation for the single-particle Green's function, (5.160) and (5.161) are a self-consistent system of equations for the determination of the single-particle Green's function $g_a(t,t')$ and the two-time two-particle Green's function $g_{ab}(t,t')$. These equations completely determine the one- and two-particle properties of an equilibrium and of a non-equilibrium plasma in dynamically screened ladder approximation.

For the further considerations, we still have to find explicit expressions for the two-particle self-energy, for the effective potential and for the effective two-particle energy shifts. Let us consider these quantities more in detail in the next section.

5.9 The Bethe–Salpeter Equation in Local Approximation. Thermodynamic Equilibrium

In order to find the dynamical properties of a pair of particles in a dense strongly coupled plasma, we have to determine the retarded (advanced) Green's function from the BSE (5.161). Of course, this equation is highly complicated. A special feature is the nonlocal behavior if the system is considered to be in non-equilibrium. We simplify the equation by a gradient expansion. First, we add (5.161) and the equation of motion with the Schrödinger operator acting on the primed variables. Then the times $\tau = t-t'$ and $T = (t+t')/2$ are introduced. We proceed in the usual manner. In lowest order of the gradient expansion, there is a complete decoupling with respect to τ and T. In momentum representation, and if the spatially homogeneous case is considered, we get

$$\begin{aligned} &\Big[E_a(\boldsymbol{p}_1) + E_b(\boldsymbol{p}_2) + \Delta_{ab}^{R/A}(\boldsymbol{p}_1\boldsymbol{p}_2,\omega) - \omega\Big] G_{ab}^{R/A}(\boldsymbol{p}_1\boldsymbol{p}_2\boldsymbol{p}'_1\boldsymbol{p}'_2,\omega) \\ &-N(p_1,p_2)\int \frac{d\boldsymbol{q}}{(2\pi)^3} V_{ab}(\boldsymbol{q}) G_{ab}^{R/A}(\boldsymbol{p}_1-\boldsymbol{q},\boldsymbol{p}_2+\boldsymbol{q},\boldsymbol{p}'_1,\boldsymbol{p}'_2,\omega) \\ &-\int \frac{d\boldsymbol{q}}{(2\pi)^3} \Delta V_{ab}^{\mathrm{eff}\,R/A}(\boldsymbol{p}_1\boldsymbol{p}_2\boldsymbol{q},\omega) G_{ab}^{R/A}(\boldsymbol{p}_1-\boldsymbol{q},\boldsymbol{p}_2+\boldsymbol{q},\boldsymbol{p}'_1,\boldsymbol{p}'_2,\omega) \\ = \; &(2\pi)^6\, \delta(\boldsymbol{p}_1-\boldsymbol{p}'_1)\,\delta(\boldsymbol{p}_2-\boldsymbol{p}'_2) \end{aligned} \tag{5.169}$$

with $E_a = \frac{p^2}{2m_a} + \Sigma_a^{\mathrm{HF}}$.

All quantities in (5.169) depend on the time T which was suppressed for simplicity. Equation (5.169) describes the two-particle properties in local approximation accounting for the influence of the surrounding non-equilibrium medium. The first term on the l.h.s. corresponds to a free but effective two-particle Hamiltonian with the single-particle energies $E_a = p_1^2/2m_a$ and with the energy shifts $\Delta_{ab}^{R/A}$. The further terms describe the interaction between the particles a and b. They are given by the Coulomb potential modified by the phase space occupation factors N_{ab} relevant for dense degenerate systems, and by an additional contribution $\Delta V_{ab}^{\mathrm{eff}\,R/A}$ which describes the influence of the dynamical screening and of the retardation on the interaction.

For further treatment, it is necessary to have explicit expressions for the energy shift and for the effective potential in the considered local approximation. The retarded and advanced effective potentials occurring in (5.169) are given by (5.166). Now we have to consider this expression in local approximation and to take the Fourier transform with respect to the difference time. Further, we take into account the dispersion rules for the Fourier transform of the retarded Green's function (5.26) and the retarded screened potential. After some algebra, in momentum representation, we arrive at

$$\Delta V_{ab}^{\mathrm{eff}\,R/A}(\boldsymbol{p}_1\boldsymbol{p}_2\,\boldsymbol{q},\omega\,T) = V_{ab}^{\mathrm{eff}\,R/A}(\boldsymbol{p}_1\boldsymbol{p}_2\,\boldsymbol{q},\omega\,T) - V_{ab}(q)N_{ab}(p_1,p_2)$$

$$= \int \frac{d\omega_1 d\omega_2 d\bar{\omega}}{(2\pi)^3} \frac{iA_a(\boldsymbol{p}_1 - q, \bar{\omega})}{\omega - \omega_1 - \omega_2 - \bar{\omega} \pm i\varepsilon} \{ iA_a(\boldsymbol{p}_2, \omega_1)\, iV_{ab}^{>}(\boldsymbol{q}, \omega_2)$$
$$+ig_b^{<}(\boldsymbol{p}_2, \omega_1)\, i[V_{ab}^{>}(\boldsymbol{q}, \omega_2) - V_{ab}^{<}(\boldsymbol{q}, \omega_2)]$$
$$+\{\boldsymbol{p}_1 \leftrightarrow \boldsymbol{p}_2,\, a \leftrightarrow b, \boldsymbol{q} \leftrightarrow -\boldsymbol{q}\}\} . \tag{5.170}$$

In a similar way, we can determine the corresponding expression for the two-particle self-energy shift. Starting from (5.167), we find

$$\Delta_{ab}^{R/A}(\boldsymbol{p}_1\boldsymbol{p}_2, \omega T)$$
$$= - \int \frac{d\omega_1 d\omega_2 d\bar{\omega}}{(2\pi)^3} \int \frac{d\boldsymbol{q}}{(2\pi)^3} \frac{A_a(\boldsymbol{p}_1\bar{\omega})}{\omega - \omega_1 - \omega_2 - \bar{\omega} \pm i\varepsilon}$$
$$\times\{iV_{bb}^{>}(\boldsymbol{q}, \omega_1)\, ig_b^{>}(\boldsymbol{p}_2 - \boldsymbol{q}, \omega_2) \; - V_{bb}^{<}(\boldsymbol{q}\, \omega_1)\, ig_b^{<}(\boldsymbol{p}_2 - \boldsymbol{q}, \omega_2)\}$$
$$+ [a \leftrightarrow b, \boldsymbol{q} \leftrightarrow -\boldsymbol{q}, \boldsymbol{p}_1 \leftrightarrow \boldsymbol{p}_2] \; . \tag{5.171}$$

The spectral function A_b is given by

$$A_b(\boldsymbol{p}, \omega) = i[g_b^{>}(\boldsymbol{p}, \omega) - g_b^{<}(\boldsymbol{p}, \omega)] \, .$$

It is easy to show that, in the case of a symmetrical plasma with $Z_a = -Z_b$, the self-energy and the effective potential are connected by the relation

$$\Delta_{ab}^{R/A}(\boldsymbol{p}_1\boldsymbol{p}_2, \omega T) = \int \frac{d\boldsymbol{q}}{(2\pi)^3} V_{ab}^{\mathrm{eff}\, R/A}(\boldsymbol{p}_1 + \boldsymbol{q}, \boldsymbol{p}_2 - \boldsymbol{q}, \boldsymbol{q}, \omega T) \, . \tag{5.172}$$

The self-energy $\Delta_{ab}^{R/A}$ and the effective potential are determined by the single-particle correlation functions $g_a^{\gtrless}(\boldsymbol{p}, \omega T)$ which are given by the Kadanoff–Baym equations. With the help of the Kadanoff–Baym ansatz

$$\begin{aligned} \pm ig_a^{<}(\boldsymbol{p}, \omega T) &= 2\pi\delta\Big(\omega - E_a(\boldsymbol{p}, T)\Big) f_a(\boldsymbol{p}, T), \\ ig_a^{>}(\boldsymbol{p}, \omega T) &= 2\pi\delta\Big(\omega - E_a(\boldsymbol{p}, T)\Big) \Big[1 \pm f_a(\boldsymbol{p}, T)\Big] , \end{aligned} \tag{5.173}$$

we may express the effective potential (5.170) and the two-particle shift (5.171) approximately in terms of the Wigner function. We will not write the resulting expressions. With the relations (5.170) and (5.171), an important result was derived. These two equations describe the modifications of the two-particle properties due to the influence of the surrounding plasma. These modifications are:

(i) The bare Coulomb potential is replaced by a dynamically screened potential.
(ii) Dynamic screening produces a retardation in the interaction. The result is the propagator in the effective potential.
(iii) Scattering processes are restricted by the phase space occupation factors (Pauli blocking).

(iv) Non-equilibrium phenomena are included by the single-particle distribution functions.

Let us consider the case of thermodynamic equilibrium. In equilibrium, only the dynamical properties have to be determined, i.e., only the equation for $g_{ab}^{R/A}$ has to be considered. All other quantities follow from spectral representations. Further it is convenient to take into account that we have in the equilibrium case

$$\begin{aligned} iV_{ab}^{s<}(q\omega) &= -2V_{ab}(q)\mathrm{Im}\varepsilon^{R^{-1}}(q\omega)n_B(\omega) \\ iV_{ab}^{s>}(q\omega) &= -2V_{ab}(q)\mathrm{Im}\varepsilon^{R^{-1}}(q\omega)\Big[1+n_B(\omega)\Big], \end{aligned} \tag{5.174}$$

with $n_B(\omega) = (\exp(\beta\omega) - 1)^{-1}$ being the Bose function without chemical potential. For the effective potential we find (Bornath et al. 1999)

$$\begin{aligned} \Delta V_{ab}^{\mathrm{eff}\,R/A}(\boldsymbol{p}_1\boldsymbol{p}_2\boldsymbol{q},\omega) &= -V_{ab}(\boldsymbol{q})\int_{-\infty}^{\infty}\frac{d\omega_1}{\pi}\mathrm{Im}\varepsilon^{-1}(\boldsymbol{q},\omega_1) \\ &\times \left\{\frac{1\pm f_b(\boldsymbol{p}_2)+n_B(\omega_1)}{\omega-\omega_1-E_b(\boldsymbol{p}_2)-E_a(\boldsymbol{p}_1-\boldsymbol{q})\pm i\varepsilon}\right. \\ &\left.+\frac{1\pm f_a(\boldsymbol{p}_1)+n_B(\omega_1)}{\omega+\omega_1-E_a(\boldsymbol{p}_1)-E_b(\boldsymbol{p}_2+\boldsymbol{q})\pm i\varepsilon}\right\}. \end{aligned} \tag{5.175}$$

The expression (5.175) has to be compared to results given in earlier work (Kilimann et al. 1977; Zimmermann et al. 1978; Schäfer et al. 1986). We showed that there are no additional static contributions beyond the Hartree–Fock level. Furthermore, no Pauli blocking expression occurs in the denominator of (5.175). In the limit of a non-degenerate system, however, our results are in agreement with the former ones.

It is interesting to study some special cases of the effective potential $\Delta V_{ab}^{\mathrm{eff,R}}$ and of the shift Δ_{ab}^{R}.

An important case is the static limit. We assume that

$$\omega \gg \omega_2 - \epsilon(p_1) - \epsilon(p_2)\,. \tag{5.176}$$

Roughly speaking, ω is the plasmon energy, and ω_2 is the two-particle scattering or bound state energy. Therefore, (5.176) requires that the plasmon energy is much larger than the energy difference between a two-particle energy and the energy of a free pair. With (5.176), we have approximately

$$\begin{aligned} \Delta V_{ab}^{\mathrm{eff}} &= -V_{ab}(q)\int_{-\infty}^{\infty}\frac{d\omega_1}{\pi}\frac{\mathrm{Im}\epsilon^{R-1}(q\omega_1)}{\omega_1}\{1\pm f_b(p_2)+n_B(\omega_1)\} \\ &+ [a\leftrightarrow b, 1\leftrightarrow 2]\,. \end{aligned} \tag{5.177}$$

Now we use the dispersion relation for V_{ab}^{R} which follows from (4.72), and the fact that $\mathrm{Im}\epsilon^{R-1}/\omega_1$ is an even function of ω_1, while $n_B(\omega_1)+1/2$ is odd. Then we get

$$\Delta V_{ab}^{\mathrm{eff}}(\boldsymbol{p}_1\boldsymbol{p}_2 q) = \{V_{ab}^s(q,\omega=0) - V_{ab}(q)\}\left(1 \pm f_a(p_1) \pm f_b(p_2)\right) . \qquad (5.178)$$

Here, $V_{ab}^s(\omega=0)$ is the statically screened potential.

In a similar way we obtain Δ_{ab}^R

$$\Delta_{ab}^R(\boldsymbol{p}_1\boldsymbol{p}_2) = \int \frac{d\boldsymbol{q}}{(2\pi)^3}\Big[\{V_{aa}^s(q,\omega=0) - V_{aa}(q)\}\left[\pm f_a(\boldsymbol{p}_1+\boldsymbol{q}) + \frac{1}{2}\right]$$

$$+ \{V_{bb}^s(q,\omega=0) - V_{bb}(q)\}\left[\pm f_b(\boldsymbol{p}_2+\boldsymbol{q}) + \frac{1}{2}\right]\Big] . \qquad (5.179)$$

If we introduce (5.178) and (5.179) into (5.169), we get the statical Bethe–Salpeter equation (5.62), and, in the non-degenerate case, the Ecker–Weizel model, see (2.134).

A second interesting approximation for the expressions (5.170) and (5.171) may be obtained if we restrict ourselves to the plasmon contribution in $V_{ab}^{s\gtrless}$, i.e., we apply the plasmon pole approximation. In the non-equilibrium case, one has to start from (4.236) instead of (5.174). Using (4.149), we have

$$iV_{ab}^{s<}(\boldsymbol{q},\omega T) = \pi\omega_{pl}\left[\delta(\omega-\omega_{pl}) - \delta(\omega+\omega_{pl})\right]V_{ab}(\boldsymbol{q})n(\boldsymbol{q},T) , \qquad (5.180)$$

where $\omega_{pl}^2 = \sum_a (4\pi Z_a^2 e^2 n_a(T)/m_a)$ is the square of the plasma frequency, and $n(\boldsymbol{q},T)$ is the non-equilibrium plasmon distribution function. The special advantage of the plasmon pole approximation is that the ω-integration in (5.175) can be performed, and we get

$$V_{ab}^{\mathrm{eff}\,R/A}(\boldsymbol{p}_1\boldsymbol{p}_2\boldsymbol{q},\omega T)$$

$$= V_{ab}(q)\Big\{1 + \frac{\omega_{pl}}{2}\Big[\frac{1+n(\boldsymbol{q},T) \pm f_b(\boldsymbol{p}_2)}{\omega-\omega_{pl}-E_b(\boldsymbol{p}_2,T)-E_a(\boldsymbol{p}_1-\boldsymbol{q},T) \pm i\varepsilon}$$

$$- \frac{1+n(\boldsymbol{q},T) \pm f_b(\boldsymbol{p}_2)}{\omega+\omega_{pl}-E_b(\boldsymbol{p}_2,T)-E_a(\boldsymbol{p}_1-\boldsymbol{q},T) \pm i\varepsilon}$$

$$+(a \leftrightarrow b, p_1 \leftrightarrow p_2, q \leftrightarrow -q)\Big]\Big\} . \qquad (5.181)$$

This expression allows for the following physical interpretation. The first term corresponds to the decay of a correlated two-particle state into free particles with absorption of a plasmon. The second contribution describes the inverse process. In thermodynamic equilibrium, the plasmon occupation is described by the Bose function $n_B(\omega)$.

For further considerations, it is convenient to introduce an operator notation. The Bethe–Salpeter equation (5.169) can then be written as

$$(\mathrm{H}^0_{ab} + V_{ab} - \omega)\, G^R_{ab}(\omega, T) + \mathrm{H}^{pl}_{ab}(\omega, T)\, G^R_{ab}(\omega, T) \;=\; i\,. \tag{5.182}$$

For simplicity, we consider only the retarded Green's function. The influence of the surrounding plasma on the two-particle properties is described by the in-medium part $\mathrm{H}^{pl}_{ab}(\omega T)$. According to (5.169), the operator is given by

$$H^{\mathrm{pl}}_{ab} = \Sigma^{\mathrm{HF}}_a + \Sigma^{\mathrm{HF}}_b + (N-1)V_{ab} + \Delta^R_{ab}(\omega) + \Delta V^{\mathrm{effR}}_{ab}(\omega)\,. \tag{5.183}$$

The plasma Hamiltonian is a non-hermitean one and can be written as

$$\mathrm{H}^{pl}_{ab}(\omega T) = \mathcal{H}\mathrm{H}^{pl}_{ab}(\omega T) + i\mathcal{A}\mathrm{H}^{pl}_{ab}(\omega T)\,. \tag{5.184}$$

For the hermitean part $\mathcal{H}\mathrm{H}$, we have

$$\mathcal{H}\mathrm{H}^{pl}_{ab}(\,\omega\, T\,) = \overline{\mathrm{H}}^{pl}_{ab}(\,T\,) - P\int \frac{d\bar\omega}{\pi}\,\frac{\mathcal{A}\mathrm{H}^{pl}_{ab}(\,\bar\omega\, T\,)}{\omega\,-\,\bar\omega}\,. \tag{5.185}$$

Here, $\overline{\mathrm{H}}^{pl}_{ab}(T)$ is the static part which follows from the static contributions of the energy shift and of the effective potential. The anti-hermitean part $\mathcal{A}\mathrm{H}$ of the plasma Hamiltonian is

$$\mathcal{A}\mathrm{H}^{pl}_{ab}(\omega T) = \mathcal{A}\,\Delta^R_{ab}(\omega T) + \,\mathcal{A}\mathrm{V}^{\mathrm{eff}}_{ab} R(\omega T)\,. \tag{5.186}$$

In order to simplify the considerations, the plasma is assumed to be in thermodynamic equilibrium. We restrict ourselves to the non-degenerate case. We then get

$$\begin{aligned}
\mathcal{A}V^{\mathrm{eff},R}_{ab}(\boldsymbol{p}_1\boldsymbol{p}_2 q,\omega) &= V_{ab}(q)\Big\{i\mathrm{Im}\,\varepsilon^{R^{-1}}\Big(q,\omega - E_b(\boldsymbol{p}_2) - E_a(\boldsymbol{p}_1-\boldsymbol{q})\Big)\\
&\times\Big[1 + n_B\Big(\omega - E_b(\boldsymbol{p}_2) - E_a(\boldsymbol{p}_1-\boldsymbol{q})\Big) \pm \frac{1}{2}[f_b(\boldsymbol{p}_2) + f_a(\boldsymbol{p}_1-\boldsymbol{q})]\Big]\\
&-\mathcal{P}\int\frac{d\omega_2}{\pi}\mathrm{Im}\varepsilon^{-1}(q,\omega_2)\frac{\pm\frac{1}{2}[f_b(\boldsymbol{p}_2) - f_a(\boldsymbol{p}_1-\boldsymbol{q})]}{\omega-\omega_2 - E_b(\boldsymbol{p}_2) - E_a(\boldsymbol{p}_1-\boldsymbol{q})}\\
&+(\,a\leftrightarrow b,\ \boldsymbol{p}_1\leftrightarrow\boldsymbol{p}_2,\ \boldsymbol{q}\leftrightarrow -\boldsymbol{q}\,)\Big\}\,.
\end{aligned} \tag{5.187}$$

For the imaginary part of the energy shift Δ^R_{ab}, we find

$$\mathcal{A}\,\Delta^R_{ab}(\boldsymbol{p}_1\boldsymbol{p}_2,\omega) = \Gamma_a\Big(\boldsymbol{p}_1,\omega - E_b(\boldsymbol{p}_2)\Big) + \Gamma_b\Big(\boldsymbol{p}_2,\omega - E_a(\boldsymbol{p}_1)\Big)\,. \tag{5.188}$$

Here, $\Gamma_a = i(\Sigma^>_a - \Sigma^<_b)$ is the single-particle damping. In the approximation considered it is given by

$$\begin{aligned}
\Gamma_a(\boldsymbol{p}_1,\omega) \;&=\; -i\int\frac{d\boldsymbol{q}}{(2\pi)^3}V_{aa}(q)\mathrm{Im}\,\varepsilon^{R^{-1}}\Big(\boldsymbol{q},\omega - E_a(\boldsymbol{p}_1-\boldsymbol{q})\Big)\\
&\times\Big[1 + n_B\big(\omega - E_a(\boldsymbol{p}_1-\boldsymbol{q})\big)\Big]\,.
\end{aligned} \tag{5.189}$$

5.10 Perturbative Solutions. Effective Schrödinger Equation

From the two-particle Green's function determined by the BSE (5.169), we can find the properties of a pair of particles accounting for the influence of the surrounding plasma medium. Now, it is necessary to develop techniques for the solution of this equation. For this purpose, we consider the associated homogeneous BSE. The corresponding homogeneous equation to (5.169) can be written as

$$\Big(\mathrm{H}^0_{ab} + \mathrm{V}_{ab} - E_{\alpha \boldsymbol{P}}(\omega, T)\Big)\, |\psi_{\alpha \boldsymbol{P}}(\omega T)\rangle + \mathrm{H}^{pl}_{ab}(\omega T)\, |\psi_{\alpha \boldsymbol{P}}(\omega T)\rangle = 0\,. \quad (5.190)$$

It represents an eigenvalue equation of the total Hamiltonian $\mathrm{H}_{ab} = \mathrm{H}^0_{ab} + \mathrm{V}_{ab} + \mathrm{H}^{pl}_{ab}$ with the possible energies $E_{\alpha \boldsymbol{P}}$ and the eigen-states $|\psi_{\alpha \boldsymbol{P}}\rangle$. Here $\boldsymbol{P} = \boldsymbol{p}_a + \boldsymbol{p}_b$ is the total momentum, and α denotes the set of internal quantum numbers.

In the case $\mathrm{H}^{pl}_{ab} = 0$, this equation reduces to the Schrödinger equation for two isolated charged particles, i.e.,

$$\Big(\mathrm{H}^0_{ab} + \mathrm{V}_{ab} - E_{\alpha \boldsymbol{P}}\Big)\, |\psi^0_{\alpha \boldsymbol{P}}\rangle = 0\,. \quad (5.191)$$

As discussed earlier, this well-known equation has the following types of solutions:

(i) The scattering states $\left|\psi^0_{\boldsymbol{p}\boldsymbol{P}}\right\rangle$ with energies

$$E_{\boldsymbol{p}\boldsymbol{P}} = \frac{p^2}{2\mu_{ab}} + \frac{P^2}{2M}\,.$$

(ii) If an electron ion pair is considered, we have bound states $\left|\psi^0_{j\boldsymbol{P}}\right\rangle$ ($j = n, l, m$) with energies

$$E^0_{nlm} = E^0_n = -\frac{\mu_{ab} Z^2 e^4}{2\hbar^2}\frac{1}{n^2}\,.$$

Here, Z denotes the charge number of the atomic nucleus.

In the general case, the effective Schrödinger equation contains the non-equilibrium in-medium part $\mathrm{H}^{pl}_{ab}(\omega, T)$ of the Hamiltonian as given by (5.183). Inserting the expression (5.183) in (5.190), one finds

$$\begin{aligned}&\Big(\mathrm{H}^0_{ab} + \mathrm{V}_{ab} - E_{\alpha \boldsymbol{P}}(\omega, T)\Big)\, |\psi_{\alpha \boldsymbol{P}}(\omega, T)\rangle \\ &= -\Big(\Delta^R_{ab}(\omega, T) + \mathrm{V}^{\mathrm{eff}\,R}_{ab}(\omega, T) - \mathrm{V}_{ab}\Big)\, |\psi_{\alpha \boldsymbol{P}}(\omega, T)\rangle\,. \quad (5.192)\end{aligned}$$

Equation (5.192) may be interpreted as the many-particle version of the ordinary Schrödinger equation describing the two-particle problem in plasmas.

The isolated two-particle problem is given by the l.h.s. of (5.192). Many-particle effects are condensed on the r.h.s. given by the following contributions:

1. There is a two-particle self-energy correction Δ_{ab}^R given by (5.171). It consists of the exchange self-energy (Hartree–Fock contribution) and of a correlation part.
2. The last term on the r.h.s. contains the deviations from the bare Coulomb interaction determined by the effective potential $V_{ab}^{\mathrm{eff}\,R}$. The latter is given by (5.175). This contribution takes into account (i) the dynamical screening of the Coulomb potential, (ii) the retardation of the dynamical screened interaction, and (iii) the phase space occupation effects described by the Pauli blocking term N_{ab}.

Let us consider the effective Schrödinger equation for a pair of particles with opposite charges ($e_a = -e_b$). This case is of importance for hydrogen and electron–hole plasmas in semiconductors. Using the relation (5.172), the effective Schrödinger equation then reads

$$
\begin{aligned}
&\Big(E_a(\boldsymbol{p}_1) + E_b(\boldsymbol{p}_2) - E_{\alpha \boldsymbol{P}}(\omega)\Big)\psi_{\alpha \boldsymbol{P}}(\boldsymbol{p}_1\boldsymbol{p}_2,\omega) \\
&+\int \frac{d\boldsymbol{q}}{(2\pi)^3} V_{ab}(q)\psi_{\alpha \boldsymbol{P}}(\boldsymbol{p}_1-\boldsymbol{q}\,\boldsymbol{p}_2+\boldsymbol{q},\omega) \\
= \; &-\int \frac{d\boldsymbol{q}}{(2\pi)^3} V_{ab}(q)\Big\{\Big[N_{ab}(p_1p_2)-1\Big]\psi_{\alpha P}(\boldsymbol{p}_1-\boldsymbol{q}\,\boldsymbol{p}_2+\boldsymbol{q},\omega) \\
&-\Big[N_{ab}(p_1p_2)-1\Big]\psi_{\alpha P}(\boldsymbol{p}_1\boldsymbol{p}_2,\omega)\Big\} \\
&-\int \frac{d\boldsymbol{q}}{(2\pi)^3}\Delta V_{ab}^{\mathrm{eff}\,R}(\boldsymbol{p}_1\boldsymbol{p}_2\boldsymbol{q},\omega)\Big\{\psi_{\alpha P}(\boldsymbol{p}_1\boldsymbol{p}_2,\omega) \\
&-\psi_{\alpha P}(\boldsymbol{p}_1-\boldsymbol{q}\,\boldsymbol{p}_2+\boldsymbol{q},\omega)\Big\}\,, \qquad (5.193)
\end{aligned}
$$

where $\Delta V_{ab}^{\mathrm{eff}\,R} = V_{ab}^{\mathrm{eff}\,R} - N_{ab}V_{ab}$. The first term in the first curly brackets on the r.h.s. accounts for Pauli blocking, and the second one for the exchange self-energy correction (Hartree–Fock). The first term of the second curly brackets (third line) stems from the correlation part of Δ_{ab}^R. The contribution of the effective potential is given by the last line.

It is interesting to look at the combined effect of the different contributions in the effective Schrödinger equation (5.193). Especially, if the shift $\boldsymbol{q}$, occurring in two of the wave functions at the r.h.s., is $\boldsymbol{q} \approx 0$, the curly brackets vanish on the r.h.s. of (5.193). Therefore, there is a compensation between Pauli blocking and the exchange part of the Hartree–Fock self-energy. Furthermore, compensation acts between the correlation part of the two-particle self-energy shift, and the effective potential (the last two terms in (5.193)). In fact, this requires a consistent treatment of all these contributions in order to calculate the two-particle properties from the effective Schrödinger equation.

The compensation just mentioned acts strongly for a low-lying bound state because the wave function of such a state is sharply localized in coordinate space whereas the Fourier transform is a slowly varying function of $\boldsymbol{q}$. We mention here that the compensation does not act according to the scheme just described for particles with different charges $e_a \neq -e_b$, see Röpke et al. (1978).

Due to the many-particle effects included, the effective Schrödinger equation (5.192) is very complicated. Especially, the plasma Hamiltonian H_{ab}^{pl} is a non-hermitean operator. Therefore, attention has to be paid to the physical interpretation of the eigenvalues and the eigenfunctions. To give a first overview (Kilimann et al. 1983; Bornath et al. 1999), we consider the case

$$\mathcal{A}\,\mathrm{H}_{ab}^{pl}(\omega + i\varepsilon, T) < \mathcal{H}\,\mathrm{H}_{ab}^{pl}(\omega + i\varepsilon, T)\,. \tag{5.194}$$

In first approximation, there follows the hermitean eigenvalue problem

$$\left(\mathrm{H}_{ab}^{0} + \mathrm{V}_{ab} - E_{n\boldsymbol{P}}(\omega) + \mathcal{H}\mathrm{H}_{ab}^{pl}(\omega, T)\right) |\psi_{n\boldsymbol{P}}(\omega, T)\rangle = 0\,. \tag{5.195}$$

Here, $E_{nP}(\omega, T)$ are the real eigenvalues of the effective Hamiltonian $\mathrm{H} + \mathrm{H}_{ab}^{pl}(\omega, T)$. However, they do not directly give the two-particle energy spectrum.

The two-particle spectrum follows from the two-particle spectral function a_{ab}. In the representation with respect to the eigenstates $|\Psi_{nP}(\omega T)\rangle$, the Bethe–Salpeter equation (5.182) reads

$$[\omega - E_{nP}(\omega)]\, g_{nn'}^{R}(\boldsymbol{P}\omega) - \sum_{m} \mathcal{A}H_{nm}(\boldsymbol{P}\omega) g_{mn'}(\boldsymbol{P}\omega) = \delta_{nn'}\,. \tag{5.196}$$

Under the condition (5.194), the solution to (5.196) was found to be (Kilimann et al. 1983; Kraeft et al. 1986; Bornath et al. 1999)

$$\begin{aligned} g_{nn'}^{R}(\boldsymbol{P}\omega) = {} & \frac{\delta_{nn'}}{\omega + i\epsilon - E_{nP}(\omega) + i\Gamma_{nn}(\boldsymbol{P}\omega)} \\ & + \frac{-i\Gamma_{nn'}(\boldsymbol{P}\omega)(1 - \delta_{nn'})}{[\omega + i\epsilon - E_{nP}(\omega) + i\Gamma(\boldsymbol{P}\omega)]\,[\omega + i\epsilon - E_{n'P}(\omega) + i\Gamma_{nn'}(\boldsymbol{P}\omega)]}\,, \end{aligned} \tag{5.197}$$

with $\Gamma_{nn'}(\omega) = i\mathcal{A}H_{nn'}(\omega)$.

Now we consider the case $\Gamma_{nn'} = 0$ (coherent part). We then get for the spectral function

$$a_{nn}(\boldsymbol{P}\omega) = \frac{2\Gamma_{nn}(\boldsymbol{P}\omega)}{[\omega - E_{n\boldsymbol{P}}(\omega)]^2 + \Gamma_{nn}^2(\boldsymbol{P}\omega)}\,. \tag{5.198}$$

According to (5.198), the spectrum of the two-particle excitation follows from the equation

$$\omega = E_{n\boldsymbol{P}}(\omega)\,.$$

In a first step, we solve (5.195) applying first order perturbation theory with respect to ReH^{pl}_{ab}. Then, for the spectrum of the two-particle bound states ($\alpha = j = n, l, m$) we have

$$\begin{aligned} E_{\alpha\boldsymbol{P}}(\omega, T) &= E^0_{\alpha\boldsymbol{P}} + \left\langle \psi^0_{\alpha\boldsymbol{P}} | \mathcal{H}\mathrm{H}^{pl}_{ab}(\omega, T) | \psi^0_{\alpha\boldsymbol{P}} \right\rangle \\ &= E^0_{\alpha\boldsymbol{P}} + \left\langle \psi^0_{\alpha\boldsymbol{P}} | \overline{H}^{pl}_{ab}(T) - P \int \frac{d\bar{\omega}}{\pi} \frac{\mathcal{A}\mathrm{H}^{pl}_{ab}(\bar{\omega}, T)}{\omega - \bar{\omega}} | \psi^0_{\alpha\boldsymbol{P}} \right\rangle . \end{aligned} \tag{5.199}$$

In momentum representation, in terms of the unperturbed wave functions $\psi^0_{\alpha\boldsymbol{P}}(\boldsymbol{p}_1\boldsymbol{p}_2)$ we arrive at

$$\begin{aligned} &\left\langle \psi^0_{\alpha\boldsymbol{P}} | \mathcal{H}\mathrm{H}^{pl}_{ab}(\omega) | \psi^0_{\alpha\boldsymbol{P}} \right\rangle = \int \frac{d\boldsymbol{p}_1}{(2\pi)^3} \frac{d\boldsymbol{p}_2}{(2\pi)^3} \psi^{0*}_{\alpha\boldsymbol{P}}(\boldsymbol{p}_1\boldsymbol{p}_2) \mathcal{H}\, \Delta^R_{ab}(\boldsymbol{p}_1\boldsymbol{p}_2, \omega) \\ &\times \psi^0_{\alpha\boldsymbol{P}}(\boldsymbol{p}_1\boldsymbol{p}_2) + \int \frac{d\boldsymbol{p}_1}{(2\pi)^3} \frac{d\boldsymbol{p}_2}{(2\pi)^3} \frac{d\boldsymbol{q}}{(2\pi)^3} \psi^{0*}_{\alpha\boldsymbol{P}}(\boldsymbol{p}_1\boldsymbol{p}_2) \Big[\mathcal{H}\, V^{\mathrm{eff}\,R}_{ab}(\boldsymbol{p}_1\boldsymbol{p}_2\boldsymbol{q}, \omega) \\ &- V_{ab}(q) \Big] \psi^0_{\alpha\boldsymbol{P}}(\boldsymbol{p}_1 - \boldsymbol{q}, \boldsymbol{p}_2 + \boldsymbol{q})\,. \end{aligned} \tag{5.200}$$

If we consider the special case of opposite charges ($e_a = -e_b$), from (5.193) we have

$$\begin{aligned} &\left\langle \psi^0_{\alpha\boldsymbol{P}} | \mathcal{H}\mathrm{H}^{pl}_{ab}(\omega) | \psi^0_{\alpha\boldsymbol{P}} \right\rangle = \int \frac{d\boldsymbol{p}_1}{(2\pi)^3} \frac{d\boldsymbol{p}_2}{(2\pi)^3} \frac{d\boldsymbol{q}}{(2\pi)^3} V_{ab}(q) \Big[f_a(\boldsymbol{p}_1) + f_b(\boldsymbol{p}_2) \Big] \\ &\times \psi^0_{\alpha\boldsymbol{P}}(\boldsymbol{p}_1 - \boldsymbol{q}, \boldsymbol{p}_2 + \boldsymbol{q}) \Big[\psi^0_{\alpha\boldsymbol{P}}(\boldsymbol{p}_1\boldsymbol{p}_2) - \psi_{\alpha\boldsymbol{P}}(\boldsymbol{p}_1 + \boldsymbol{q}, \boldsymbol{p}_2 - \boldsymbol{q}) \Big] \\ &+ \int \frac{d\boldsymbol{p}_1}{(2\pi)^3} \frac{d\boldsymbol{p}_2}{(2\pi)^3} \frac{d\boldsymbol{q}}{(2\pi)^3} \mathcal{H}\Delta V^{\mathrm{eff}\,R}_{ab}(\boldsymbol{p}_1\boldsymbol{p}_2\boldsymbol{q}, \omega) \Big[f_a(\boldsymbol{p}_1) + f_b(\boldsymbol{p}_2) \Big] \\ &\times \psi^0_{\alpha\boldsymbol{P}}(\boldsymbol{p}_1 - \boldsymbol{q}, \boldsymbol{p}_2 + \boldsymbol{q}) \Big[\psi^0_{\alpha\boldsymbol{P}}(\boldsymbol{p}_1\boldsymbol{p}_2) - \psi_{\alpha\boldsymbol{P}}(\boldsymbol{p}_1 + \boldsymbol{q}, \boldsymbol{p}_2 - \boldsymbol{q}) \Big]\,. \end{aligned} \tag{5.201}$$

This expression gives the first-order correction to the energy eigenvalue of the isolated bound state due to many-particle effects. We will discuss this in more detail. The first term on the r.h.s. comes from Pauli blocking and exchange self-energy, whereas the second contribution consists of the correction following from the correlation part of the self-energy and from the effective screened potential. In every case, the quadratic terms $|\psi^0_{\alpha\boldsymbol{P}}(\boldsymbol{p}_1\boldsymbol{p}_2)|^2$ are due to self-energy contributions, and the mixed terms $\psi^0_{\alpha\boldsymbol{P}}(\boldsymbol{p}_1\boldsymbol{p}_2)\psi^0_{\alpha\boldsymbol{P}}(\boldsymbol{p}'_1\boldsymbol{p}'_2)$ are due to the effective interaction potential.

From the effective Schrödinger equation (5.193), we see that there is an essential difference between the bound states and the scattering states. For bound states, the low lying states are localized, and thus spread in the momentum space. The wave function is a slowly varying function with respect to the momenta, and compensation can act considerably. Therefore, a weak dependence of the bound state energies on density and temperature is expected. For these low lying states, the influence of many-particle effects can be ignored in a wide density range.

On the other hand, the scattering states are sharply peaked in momentum space. Of special importance is the determination of the continuum edge. As discussed in Sect. 5.4, we can use the fact that the energy spectrum of the scattering states is equal to that of the free quasiparticles determined by the effective Schrödinger equation (5.195) without the effective potential in $\mathcal{H}\mathrm{H}^{pl}_{ab}$. Then, only the self-energy correction (quadratic terms) contribute to the shift of the continuum edge. For $\boldsymbol{p} = 0$ and $\boldsymbol{P} = 0$, the shift of the continuum edge follows from

$$E_{cont}(\omega) = \mathcal{H}\,\Delta^R_{ab}(\boldsymbol{p}_1 = 0\,\boldsymbol{p}_2 = 0, \omega)\,,$$

where ω has to be taken as a solution of the dispersion relation; an approximation is given by (5.223). Here, the Hartree–Fock exchange part, the additional static contributions and the correlation part of the two-particle self-energy in V^s-approximation are taken into account.

5.11 Numerical Results

In Sect. 5.10, we discussed the properties of the solution of the homogeneous Bethe–Salpeter equation (5.190). There, we assumed the deviations of the energy levels from those of an isolated pair to be small; the justification for this assumption will be given (*a posteriori*) in this section. Details of the numerical efforts are found in Kraeft (a) et al. (1990), Fehr and Kraeft (1994), Kraeft et al. (1995), Kraeft and Fehr (1997), and Kraeft et al. (1999). In Sect. 5.10, a perturbation procedure was proposed which takes a Hamiltonian with a simple two-particle potential to be that for the unperturbed problem. Such (simple) potentials are, e.g., the Coulomb potential or the Ecker–Weizel potential (2.134) (Ecker and Weizel 1956), respectively. For earlier work we refer to Ecker and Kröll (1966).

In this section, we start from the eigenvalue problem (5.190) where the plasma Hamiltonian $H^{p\ell}_{ab}(\omega T)$ is given by (5.183) and the effective potential $V^{\mathrm{eff}}_{ab}(\omega T)$ by (5.175), the latter being connected to the effective self-energy Δ_{ab} by (5.172).
In order to simplify the numerical effort we apply the non-degenerate version for the effective potential V^{eff}_{ab} (5.175), i.e., we replace $1 \pm f_a$ by 1.
Let us consider a further-going approximation to this potential. Using relative and center-of-mass coordinates with the center-of-mass momentum $\boldsymbol{P} = 0$, the potential reads

$$V_{ab}^{\text{eff}}(\boldsymbol{p}\boldsymbol{q}\boldsymbol{\omega}) = V_{ab}(\boldsymbol{q})\Big\{1 - \int_{-\infty}^{+\infty} \frac{d\omega'}{\pi} \text{Im}\varepsilon^{-1}(\boldsymbol{q}\omega')$$
$$\times \left[\frac{1+n_B(\omega')}{\omega - \omega' - E_a(\boldsymbol{p}) - E_b(-\boldsymbol{p}-\boldsymbol{q})} + \frac{1+n_B(\omega')}{\omega - \omega' - E_a(\boldsymbol{p}+\boldsymbol{q}) - E_b(-\boldsymbol{p})}\right]\Big\}. \tag{5.202}$$

From (5.202), in the long wave limit $p \to 0$, $q \to 0$ and in the static case $\omega \to 0$ we get

$$V_{ab}^{\text{eff}}(q0) = V_{ab}(q) \left\{ 1 + 2 \int_{-\infty}^{+\infty} \frac{d\omega}{\pi} \frac{\text{Im}\varepsilon^{-1}(q\omega)}{\omega} (1 + n_B(\omega)) \right\}. \tag{5.203}$$

The imaginary part of the inverse dielectric function is an odd function with respect to the frequency. Consequently we have, in combination with the odd part of the Bose function,

$$\int_{-\infty}^{+\infty} \frac{\text{Im}\varepsilon^{-1}(\omega)}{\omega} \left(\frac{1}{2} + n_B(\omega)\right) d\omega = 0\,. \tag{5.204}$$

The remaining contribution of (5.203) is the spectral representation of the real part of the inverse dielectric function in static approximation. Thus, (5.203) may by written as

$$V_{ab}^{\text{eff}}(q0) = V_{ab}(q) \frac{q^2}{q^2 + \kappa_D^2} = V_{ab}^D(q)\,, \tag{5.205}$$

which is nothing else than the Debye potential discussed in Chap. 2.
This means that (5.202) includes the statically screened Debye potential in the long wave static limit in the low density (classic) case. Let us now deal with the numerical solution of the eigenvalue problem (5.193). We follow the lines of the papers by Fehr and Kraeft (1994) and Seidel et al. (1995). We summarize our approximations:

(i) We drop the Pauli blocking factors in the effective potential and thus we restrict ourselves to not too high densities.
(ii) We consider only the case $\boldsymbol{P} = 0$, i.e., we neglect the motion of the center of mass of the pair considered. Thus, we neglect the considerations which are of importance for the discussion of spectral lines (Seidel 1977); however, the numerical effort to include the center-of-mass motion would be too big.
(iii) As already discussed in section 5.10, we assume the imaginary part of the plasma Hamiltonian to be small as compared to the real part, cf. (5.194).

Then the following equation has to be considered for the relative motion

$$
\begin{aligned}
&\left(\frac{\boldsymbol{p}^2}{2m_{ab}} - z\right) \psi^{ab}_{n\ell m}(\boldsymbol{p}, z) + \int \frac{d\boldsymbol{q}}{(2\pi\hbar)^3} V_{ab}(\boldsymbol{q}) \psi^{ab}_{n\ell m}(\boldsymbol{p}+\boldsymbol{q}, z) \\
&- \int \frac{d\boldsymbol{q}}{(3\pi\hbar)^3} V_{ab}(\boldsymbol{q}) \left[\left(N_{ab}(\boldsymbol{p}) - 1\right) \psi^{ab}_{n\ell m}(\boldsymbol{p}+\boldsymbol{q}, z) \right. \\
&\left. - \left(N_{ab}(\boldsymbol{p}+\boldsymbol{q}) - 1\right) \psi^{ab}_{n\ell m}(\boldsymbol{p}, z) \right] \\
&+ \int \frac{d\boldsymbol{q}}{(2\pi\hbar)^3} \Delta V^{\text{eff}}_{ab}(\boldsymbol{p}\boldsymbol{q}, z) \left[\psi^{ab}_{n\ell m}(\boldsymbol{p}+\boldsymbol{q}, z) - \psi^{ab}_{n\ell m}(\boldsymbol{p} z) \right] = 0 \,. \qquad (5.206)
\end{aligned}
$$

Here, N_{ab} are the Pauli blocking factors which account for the Hartree Fock self-energy and its "compensating partner".

The effective potential entering (5.206) is given by (5.202) with the definition $V^{\text{eff}} = V + \Delta V^{\text{eff}}$.

Equation (5.206) is an hermitean eigenvalue problem as the imaginary part of the plasma Hamiltonian was dropped. We will come back to the imaginary part later.

The l.h.s. of (5.206) represents the Schrödinger equation for the relative motion of an isolated pair. According to the discussion of Sect. 5.8, the first integral term on the r.h.s. of (5.206) represents the Pauli blocking and the Hartree–Fock self-energy correction. The second integral contribution on the r.h.s. stands for the dynamical effective potential correction and the dynamical self-energy correction.

Let us now give some details concerning the perturbation procedure. We assume the solution of (5.206) to be close to that of the (static) Debye problem and that the perturbation is small. The perturbation becomes larger in the region of the Mott transition, as we will see from the results. The equation to be considered reads

$$
\begin{aligned}
&\left(\frac{\boldsymbol{p}^2}{2m_{ab}} - \hbar z\right) \psi_{ab}(\boldsymbol{p}, z) + \int \frac{d\boldsymbol{q}}{(2\pi\hbar)^3} V_{ab}(\boldsymbol{q}) \psi_{ab}(\boldsymbol{p}+\boldsymbol{q}, z) \\
&= - \int \frac{d\boldsymbol{q}}{(2\pi\hbar)^3} \left\{ V_{ab}(\boldsymbol{q})[-f_a(\boldsymbol{p}) - f_b(\boldsymbol{p})] \psi_{ab}(\boldsymbol{p}+\boldsymbol{q}, z) \right. \\
&\left. - V_{ab}(\boldsymbol{q})[f_a(\boldsymbol{p}+\boldsymbol{q}) + f_b(\boldsymbol{p}-\boldsymbol{q}) \psi_{ab}(\boldsymbol{p}, z) \right\} \\
&+ \int \frac{d\boldsymbol{q}}{(2\pi\hbar)^3} \Delta^{\text{eff}}_{ab}(\boldsymbol{q}\boldsymbol{p}z) [\psi_{ab}(\boldsymbol{p}+\boldsymbol{q}, z) - \psi_{ab}(\boldsymbol{p}, z)] \,. \qquad (5.207)
\end{aligned}
$$

We carry out first order perturbation theory for the real part of the two-particle energies and take the Hamiltonian, e.g., to be

$$
H_{ab} = H^{\text{Debye}}_{ab} + H^{\text{plasma}}_{ab} \,, \qquad (5.208)
$$

leading to the energy shift

$$E' = \frac{{}^D\left\langle \psi(\boldsymbol{p}) | H_{ab}^{\text{plasma}} | \psi(\boldsymbol{p}) \right\rangle^D}{{}^D\left\langle \psi(\boldsymbol{p}) | \psi(\boldsymbol{p}) \right\rangle^D} \tag{5.209}$$

and

$$E = E^{\text{Debye}} + E' \,. \tag{5.210}$$

In this example, the l.h.s. of (5.207) with the Debye potential instead of the Coulomb one has to be considered in the unperturbed case, i.e.,

$$\left(\frac{\boldsymbol{p}^2}{2m_{ab}} - \hbar z\right)\psi_{ab}(\boldsymbol{p}, z) + \int \frac{d\boldsymbol{q}}{(2\pi\hbar)^3} V_{ab}^D(\boldsymbol{q})\psi_{ab}(\boldsymbol{p}+\boldsymbol{q}, z) = 0 \,, \tag{5.211}$$

and the r.h.s. of the following (5.212) is the perturbation

$$\begin{aligned}
&\left(\frac{\boldsymbol{p}^2}{2m_{ab}} - \hbar z\right)\psi_{ab}(\boldsymbol{p}, z) + \int \frac{d\boldsymbol{q}}{(2\pi\hbar)^3} V_{ab}^D(\boldsymbol{q})\psi_{ab}(\boldsymbol{p}+\boldsymbol{q}, z) \\
&= \int \frac{d\boldsymbol{q}}{(2\pi\hbar)^3} [V_{ab}^D(\boldsymbol{q}) - V_{ab}(\boldsymbol{q})]\psi_{ab}(\boldsymbol{p}+\boldsymbol{q}, z) \\
&- \int \frac{d\boldsymbol{q}}{(2\pi\hbar)^3} \{V_{ab}(\boldsymbol{q})[-f_a(\boldsymbol{p}) - f_b(\boldsymbol{p})]\psi_{ab}(\boldsymbol{p}+\boldsymbol{q}, z) \\
&- V_{ab}(\boldsymbol{q})[f_a(\boldsymbol{p}+\boldsymbol{q}) + f_b(\boldsymbol{p}-\boldsymbol{q})\psi_{ab}(\boldsymbol{p}, z)\} \\
&+ \int \frac{d\boldsymbol{q}}{(2\pi\hbar)^3} \Delta_{ab}^{\text{eff}}(\boldsymbol{q}\boldsymbol{p}z)[\psi_{ab}(\boldsymbol{p}+\boldsymbol{q}, z) - \psi_{ab}(\boldsymbol{p}, z)] \,.
\end{aligned} \tag{5.212}$$

As compared to (5.207), (5.212) has, on its r.h.s., the additional term including $V^D - V$. The Debye eigenfunctions are determined by a simple Numerov algorithm (in position space) and are then transformed into momentum space. The eigenvalue shifts are determined according to (5.209). The perturbative part of the Hamiltonian is represented by the r.h.s. of (5.212). We want to mention that, instead of the Debye potential, the Coulomb potential could have also been chosen for the unperturbed case. Then, of course, the perturbative Hamiltonian corresponds to the r.h.s. of (5.207). In Table 5.1, we present numerical results for the $1s$ state of hydrogen, see Fehr and Kraeft (1994).

Now we consider the numerical solution of the full equation (5.206). The wave function $\psi_{n\ell m}^{ab}$ in (5.206) is therefore broken down in terms of hydrogen wave functions according to Seidel et al. (1995)

$$\psi_{n\ell m}(\boldsymbol{p}) = \sum_{n'}\sum_{\ell'}\sum_{m'} a_{n'\ell' m'}^{n\ell m}\psi_{n'\ell' m'}^{c}(\boldsymbol{p}) \,, \tag{5.213}$$

where ψ^c are hydrogen (Coulomb) wave functions.

The breaking down into spherical harmonics leads to a simplification, namely with

$$\psi_{n\ell m}(\boldsymbol{p}) = \psi_{n\ell}(p) Y_\ell^m(\vartheta\varphi) \,, \tag{5.214}$$

Table 5.1. Hydrogen 1s bound state energies. Comparison of the results of Seidel et al. (1995) with those of this section (5.209) in different approximations

κ	Debye	$n_e[cm^{-3}]$	Arndt (1993)	(1)	(2)	E_{cont}
$1E-06$	-1.0000	1.343E11	-1.0000	-1.0000	-1.0000	-2E-06
.001	-1.0000	1.343E17	-1.0000	-1.0000	-1.0000	-0.0632
.010	-1.0001	1.343E19	-1.0001	-1.0000	-1.0000	-0.1995
.020	-1.0006	5.370E19	-1.0002	-1.0004	-1.0004	-0.2816
.025	-1.0009	8.391E19	-1.0003	-1.0006	-1.0006	-0.3168
.050	-1.0036	3.356E20	-1.0026	-1.0022	-1.0021	-0.4426
.100	-1.0175	1.343E21	-1.0100	-1.0093	-1.0091	-0.6211
.200	-1.0538	5.370E21	-1.0382	-1.0351	-1.0399	-0.8690
.250	-1.0821	8.391E21	-1.0588	-1.0541	-1.0565	-0.9716
.300	-1.1135	1.208E22	-1.0847	-1.0801	-1.0784	-1.0577

we arrive at

$$\psi_{n\ell m}(\boldsymbol{p}) = \sum_{n'} a^{n\ell m}_{n'\ell m} \psi^c_{n'\ell m}(\boldsymbol{p}) \,. \tag{5.215}$$

With this simplification and with the known properties of the Coulomb wave functions $\Psi^c_{n\ell m}$, we get, from (5.206), a system of equations for the determination of the coefficients $a^{n\ell m}_{n'\ell m}$

$$\sum_{n''} a^{n\ell m}_{n''\ell m} \mathcal{C}(n'n''\ell) = z_{n\ell} a^{n\ell m}_{n'\ell m} \,. \tag{5.216}$$

The system (5.216) represents an eigenvalue problem from which the energy eigenvalues $z_{n\ell}$ and the coefficients $a^{n\ell m}_{n'\ell m}$ for the decomposition (5.215) have to be determined (Seidel et al. 1995). The quantity $\mathcal{C}$ is a functional containing the *Gegenbauer polynomials* and the quantities n and ΔV^{eff} occurring in (5.206).

The explicit determination of $\mathcal{C}$ is straightforward but cumbersome. It was presented in detail by Seidel et al. (1995). The problems are, e.g., the summation over n'' in (5.216) and the integration over the plasmon peaks occurring in ΔV^{eff} in (5.206) via the imaginary part of the inverse dielectric function $\mathrm{Im}\varepsilon^{-1}$, see (5.202).

Concerning the first question, n''_{max} depends on density and temperature and is, in typical examples, of the order of 10, see Seidel et al. (1995).

For the second question, it is useful to substitute the integration over plasmon poles occurring in (5.202) by a modified Kramers–Kronig relation. While the latter reads (for ω_0 given)

$$P \int_{\infty}^{+\infty} \frac{\mathrm{Im}\varepsilon^{-1}(\boldsymbol{p}\omega)}{\omega - \omega_0} d\omega = \pi(\mathrm{Re}\varepsilon^{-1}(\boldsymbol{p}\omega_0) - 1) \,, \tag{5.217}$$

one may use the modification (Kraeft (b) et al. 1990)

$$\mathcal{P}\int_{-\infty}^{+\infty} \frac{n_B(\omega)\mathrm{Im}\varepsilon^{-1}(\boldsymbol{p}\omega)}{\omega-\omega_0}d\omega = \frac{1}{A\omega_0}\left(\mathrm{Re}\varepsilon^{-1}(\boldsymbol{p}\omega_0) - \mathrm{Re}\varepsilon^{-1}(\boldsymbol{p}0)\right)$$
$$+\left(n_B(\omega_0) - \frac{1}{A\omega_0}\right)\left(\mathrm{Re}\varepsilon^{-1}(\boldsymbol{p}\omega_0) - 1\right)$$
$$-2\pi A\omega_0 \sum_{k=1}^{\infty} \frac{\varepsilon^{-1}\left(i\frac{2\pi k}{A}\right) - 1}{(A\omega_0)^2 + (2\pi k)^2}\,. \tag{5.218}$$

The sum in (5.218) converges rapidly and is usually small. The quantity A is arbitrary within some limits. With (5.218), the ω integration may be performed; however, the subsequent momentum integration is faced with the peaks of $\mathrm{Re}\varepsilon^{-1}$ which must be handled carefully.

Another possibility is the performance of the integration of the l.h.s. integral in (5.218) using the sum rule for $\mathrm{Im}\varepsilon^{-1}$, namely according to (4.87)

$$\int_{-\infty}^{+\infty} \omega \mathrm{Im}\varepsilon^{-1}(\omega)d\omega = -\pi\omega_p^2\,. \tag{5.219}$$

Then we may write (Kraeft (b) et al. 1990; Fehr 1997)

$$\int_{-\infty}^{\infty} \frac{1}{Y - Y_1^0} B(X,Y)dY =$$
$$\int_{0}^{\infty} \left\{\frac{1}{Y - Y_1^0} - \frac{Y}{Y_{pl}}\frac{n_B(Y_{pl})}{n_B(Y)(Y_{pl} - Y_1^0)}\right\} B(X,Y)dY$$
$$+\int_{-\infty}^{0} \left\{\frac{1}{Y - Y_1^0} - \frac{Y}{Y_{pl}}\frac{n_B(-Y_{pl})}{n_B(Y)(Y_{pl} + Y_1^0)}\right\} B(X,Y)dY$$
$$-\frac{\pi}{2}\left\{\frac{n_B(Y_{pl})}{Y_{pl}(Y_{pl} - Y_1^0)} + \frac{n_B(-Y_{pl})}{Y_{pl}(Y_{pl} + Y_1^0)}\right\}\,. \tag{5.220}$$

Here we used the abbreviation $B(X,Y) = \mathrm{Im}\varepsilon^{-1}(X,Y)n_B(Y)$, and $\pm Y_{pl}$ are the extrema of $\mathrm{Im}\varepsilon$ located near the solution of the dispersion relation $\mathrm{Re}\varepsilon(X, \pm Y_{pl}) = 0$. The frequency Y_1^0 is an arbitrary value. The frequency Y and the momentum X are normalized to the plasma frequency and to the thermal wave-number, respectively. For the application of (5.220), only the position Y_{pl} of the peak enters the calculation while the integration itself is deferred to the sum rule, which, in the variables just mentioned, reads

$$\int_{-\infty}^{+\infty} dY Y \mathrm{Im}\varepsilon^{-1}(X,Y) = -\pi \,.$$

We now give some examples for level shifts for different states, densities, and temperatures both for hydrogen (Seidel et al. 1995) and for electron–hole systems in optically excited semiconductors (Arndt 1993; Arndt et al. 1996).

As already discussed earlier, we only observe a weak density dependence of the bound state levels justifying *a posteriori* the application of perturbation techniques discussed above; see also section 5.10.

We are now going to consider the damping of the energy levels corresponding to the imaginary part of the plasma Hamiltonian according to (5.222). In agreement with Sect. 5.10, we assume the imaginary part $\mathrm{Im}H_{ab}^{p\ell}$ to be small as compared to the real one. This assumption is justified practically always for systems in which bound states do exist; only for those densities at which the bound state levels merge into the continuum, the role of the imaginary part is essential and leads to a level broadening and to short life times of the bound states; see the discussion below in connection with the continuum edge and the Mott effect.

The imaginary parts of the complex two-particle energies are determined by perturbation theory of first order, namely

$$\begin{aligned}
\mathrm{Im}z_{n\ell}^{ab} &= \left\langle \psi_{n\ell m} \left| \mathrm{Im}H_{ab}^{\mathrm{pl}} \right| \psi_{n\ell m} \right\rangle \\
&= \int \frac{d\boldsymbol{p}d\boldsymbol{q}}{(2\pi\hbar)^6} \mathrm{Im}H_{ab}^{\mathrm{pl}} \left(\boldsymbol{p}-\boldsymbol{q}, \boldsymbol{p}+\boldsymbol{q}, -\boldsymbol{p}-\boldsymbol{q}, z_{n\ell}^{ab}\right) \psi_{n\ell m}^{*}(\boldsymbol{p}) \psi_{n\ell m}(\boldsymbol{p}+\boldsymbol{q}) \,.
\end{aligned} \tag{5.221}$$

In the right hand expression, the functions $\psi_{n\ell m}$ and the energies $z_{n\ell}^{ab}$ are the eigenfunctions and (real) eigenvalues of the (5.206) determined and discussed above. The imaginary part of the plasma hamiltonian is given according to (5.186)–(5.188) and reads (with Pauli blocking replaced by unity)

$$\begin{aligned}
&\mathrm{Im}H_{ab}^{\mathrm{plas}}(\boldsymbol{p}, -\boldsymbol{p}, \boldsymbol{p}+\boldsymbol{q}, -\boldsymbol{p}-\boldsymbol{q}, \omega) \\
&= \delta_{\boldsymbol{q},0} \int \frac{d\bar{\boldsymbol{q}}}{(2\pi\hbar)^3} \left[\mathrm{Im}V_{aa}^{s}\left(\boldsymbol{q}, \hbar\omega - E_a(\boldsymbol{p}) + E_b(\boldsymbol{p}+\bar{\boldsymbol{q}})\right)\right. \\
&\times \left(1 + n_B\left(\hbar\omega - E_a(\boldsymbol{p}) + E_b(\boldsymbol{p}+\bar{\boldsymbol{q}})\right)\right) + \left.\{a \longleftrightarrow b, p \longleftrightarrow -\boldsymbol{p}\}\right] \\
&+ \left[\mathrm{Im}V_{ab}^{s}\left(\boldsymbol{q}, \hbar\omega - E_a(\boldsymbol{p}+\boldsymbol{q}) - E_b(-\boldsymbol{p})\right)\right. \\
&\times \left(1 + n_B\left(\hbar\omega - E_a(\boldsymbol{p}+\boldsymbol{q}) - E_b(-\boldsymbol{p})\right)\right) \\
&+ \left.\{a \longleftrightarrow b, \boldsymbol{p} \longleftrightarrow -\boldsymbol{p}, \boldsymbol{q} \longleftrightarrow -\boldsymbol{q}\}\right] \,.
\end{aligned} \tag{5.222}$$

Here, the plasmon peak discussed above does not have to be integrated; on behalf of the identity $1/(\omega - \omega_0 - i\varepsilon) = \mathcal{P}/(\omega - \omega_0) + i\pi\delta(\omega - \omega_0)$, the ω integration was carried out already.

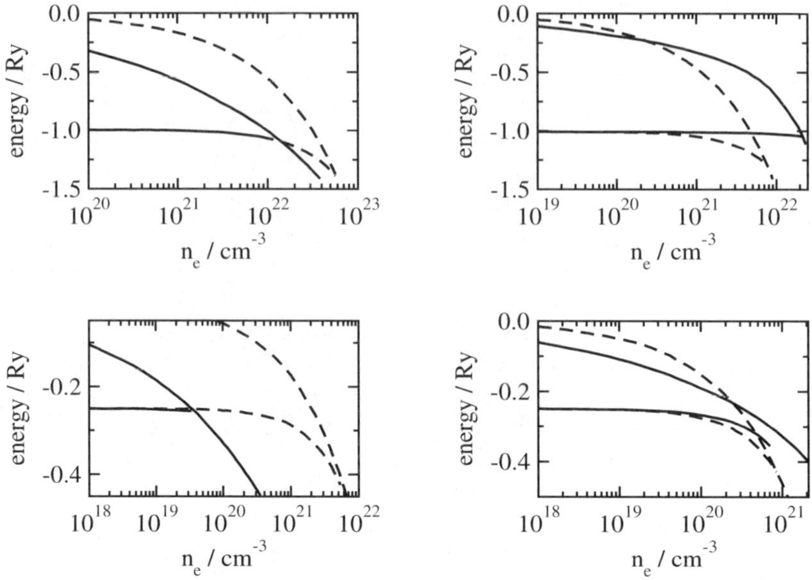

Fig. 5.16. Continua and 1s- (**upper figures**) and 2s- (**lower figures**) bound states energies of hydrogen for $k_BT = 1$ Ryd (**left figures**) and $k_BT = 0.125$ Ryd (**right figures**). *Full lines*: Full dynamical RPA screening (Arndt 1993); *dashed lines*: full static (RPA) screening inclusive quantum corrections (see static version of (4.122)) (Arndt 1993; Seidel, Arndt, and Kraeft 1995)

The imaginary part of the eigenvalues was determined for the examples represented in Fig. 5.17. As outlined by Seidel et al. (1995), both the level shifts and the damping of the levels are not in agreement with the observation on spectral lines. There, higher-order approximations have to be taken into account for the determination of two-particle properties; see, e.g., Griem (1974), Günter et al. (1991) and Griem (1997). We observe that the damping of the energy levels is essential for higher densities only.

Let us now discuss the question of the continuum edge. The compensation between Pauli blocking and Hartree–Fock self-energy, and between the dynamical effective screening and the dynamic self-energy observed in the case of bound state levels, does not work for the determination of the continuum edge. While the bound state wave functions are sharply peaked in position space and thus extended in momentum space, the scattering state wave functions are extended in position space and sharply peaked in momentum space. Thus, the compensation in (5.206) corresponding to the small relative weight for small q does not occur in the continuum case.

As already discussed in Sect. 5.9, it is sufficient to consider (5.206) under the condition that the (effective) interaction potential is neglected, and only the corresponding self-energies are taken into account. Furthermore, we as-

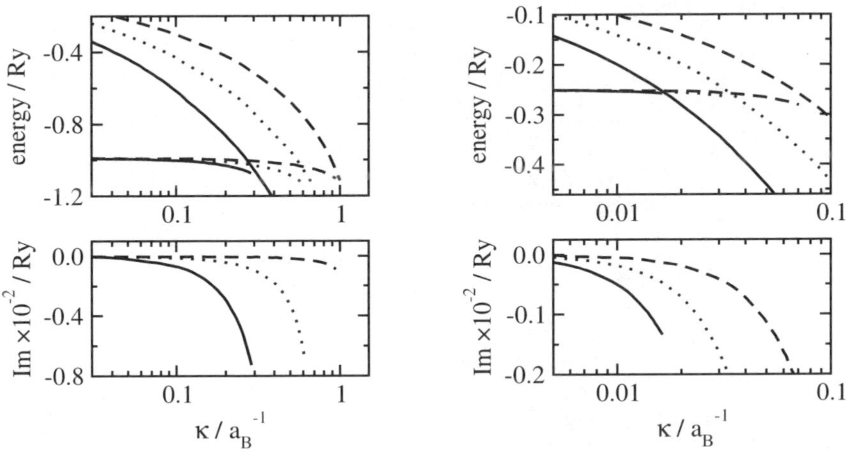

Fig. 5.17. Continua, bound state energies and damping (Im parts) for hydrogen as a function of the inverse screening length. 1s (**left figures**), 2s (**right figures**). Temperatures: *Full line*: 1 ryd, *dotted*: 0.5 ryd, *dashed*: 0.25 ryd (full dynamical RPA screening Arndt (1993), Seidel et al. (1995))

sume the continuum *edge* to be given for zero momenta of the two particles in question. We arrive at (Arndt 1993)

$$\begin{aligned} z_{ab}^{\text{cont}} &= \Sigma_a^{HF}(p=0) + \Sigma_b^{HF}(p=0) \\ &+ \Sigma_a^{MW}\left(0, z_{ab}^{\text{cont}} - E_b(p=0)\right) \\ &+ \Sigma_b^{MW}\left(0, z_{ab}^{\text{cont}} - E_a(p=0)\right) . \end{aligned} \tag{5.223}$$

Here, $\Sigma^{HF}(p)$ and $\Sigma^{MW}(p,\omega)$ are the Hartree–Fock and Montroll–Ward self-energy expressions given in Sects. 3.3 and 4.11.

An extended discussion of the question of the continuum edge was given by Fehr and Kraeft (1994) and by Fehr (1997). Numerical examples are given in Figs. 5.16 and 5.17. Here the main numerical problem is connected with the fact that, in solving (5.223) self-consistently, one is faced with the fact that Σ^{MW} has a pole structure which is difficult to handle in iterative procedures; see, e.g., Fehr (1997) and Wierling (1997).

In order to have an estimate of what a very simple theory would give, we indicated the continuum edge according to the Ecker–Weizel Hamiltonian after (2.134). In all cases considered, the density dependence of energy levels is relatively weak while there is a more pronounced density dependence of the continuum states. Thus, at certain density, there is a crossover of continuum and discrete states thus leading to a merging of the bound state levels into the continuum. This effect was discussed above and is referred to as the Mott effect.

6. Thermodynamics of Nonideal Plasmas

6.1 Basic Equations

For the quantum statistical approach to the thermodynamic properties of strongly correlated plasmas, we have different possibilities, outlined in Sect. 3.4.2. A very convenient relation is (3.215). This relation determines the density as a function of the chemical potentials μ_c of the species c. The inversion of this expression gives $\mu_a(\{n_c\}, T)$ and, therefore, thermodynamic information such as the pressure, the free energy, etc., in grand canonical description.

The direct determination of the pressure (equation of state, EOS) is possible from the charging formula (3.225)

$$p - p_0 = -\frac{1}{\Omega} \int_0^1 \frac{d\lambda}{\lambda} \langle \lambda V \rangle_\lambda \, . \tag{6.1}$$

Here, p_0 is the pressure of the noninteracting system discussed in Chap. 2. Together with

$$n_a = \frac{\partial p}{\partial \mu_a} |_\beta \, , \tag{6.2}$$

equation 6.1 determines the pressure and, thus, the density as a function of the chemical potential μ and of the temperature T. From these two equations, the chemical potential may be eliminated, and we get the pressure as a function of the density. According to (6.1), one has to determine the mean value of the potential energy, discussed in detail in Sect. 3.4.2, formulae (3.220)–(3.223). Recall the definition of the self-energy on the Keldysh contour (3.136). Taking into account (3.220), we get the useful representation

$$\langle \lambda V \rangle_\lambda = \pm \frac{1}{2} \sum_a (2s_a + 1) \int_C d\bar{1} \Sigma_a(1 - \bar{1}) g_a(\bar{1} - 1^+) \tag{6.3}$$

for the mean value of the potential energy; this means, the determination of thermodynamic properties is closely connected to that of the self-energy. Another possibility to write the mean value of the potential energy is the application of the two-particle Green's function, namely

$$\langle \lambda V \rangle_\lambda = \frac{1}{2\Omega} \sum_{ab} \int d1 d2 \lambda V_{ab}(12) g_{ab}(12, 1^{++}2^{+})|_{t_2=t_1^+} \,. \tag{6.4}$$

Let us remember here that, according to the considerations in Sect. 3.4.2, especially (3.222), $\langle \lambda V \rangle_\lambda$ is not identical with the correlation energy; see, e.g., Kremp and Kraeft (1968), Kraeft and Kremp (1968), and Kraeft et al. (2002).

In order to demonstrate the peculiarities of the EOS for plasmas, let us consider the binary collision approximation for the self-energy. Then Σ is given by (5.140). If the T-matrix is determined from the Lippmann–Schwinger equation (5.91) with the bare Coulomb potential, (6.3) may be represented in terms of the following Coulomb ladder diagrams

$$\langle \lambda V \rangle_\lambda = \pm\frac{1}{2} \tag{6.5}$$

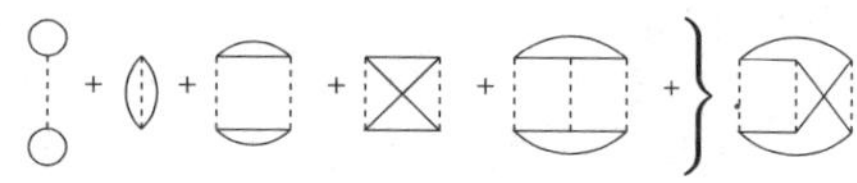

Introducing this result into (6.1), the EOS follows in the approximation of the second virial coefficient (for more details see Sect. 6.4.1). Due to the long range behavior of the Coulomb potential, we then have the well-known Coulomb divergencies. Indeed, it is easy to show that the direct contributions of the orders e^4 and e^6 are divergent. This means that the binary collision approximation with the bare Coulomb potential is not appropriate to describe the thermodynamic properties of plasmas. It should be noticed that, in the classical limit, additional divergencies appear (for zero distances) beginning with the order e^6 due to the point-like character of the plasma particles. In the quantum statistical treatment such divergencies do not exist because of the Heisenberg uncertainty relation. This relation produces a natural cut-off determined by the thermal wavelength.

In order to solve the problem of the Coulomb divergencies, we have to take into account that the long range behavior of the Coulomb potential leads to collective effects such as screening and plasma oscillations. This was discussed in Chap. 4. Following the scheme developed there, we are able to eliminate the Coulomb potential from (3.136) and to express the mean value of the potential energy in terms of the screened one. With the relation (4.13), we arrive at

$$\langle \lambda V \rangle = \langle \lambda V \rangle^H \pm \frac{1}{2} \sum_{ab} \int_C d\bar{1} V^s_{ab}(1-\bar{1}) \Pi_{ab}(\bar{1}-1^+) \,. \tag{6.6}$$

Because the system is assumed to be in equilibrium state and to be homogeneous, all quantities depend only on the difference variables $t_1 - \bar{t}_1$ and $\boldsymbol{r}_1 - \bar{\boldsymbol{r}}_1$, respectively. For convenience, we may choose $t_1 = 0$, $\boldsymbol{r}_1 = 0$.

All the equations are given on the Keldysh contour indicated in Sect. 3.3.3 by Fig. 3.1. Fourier transformation of V^s and Π with respect to $\bar{\boldsymbol{r}}_1$ and $\bar{t}_1$, and passing to the physical time axis, we get

$$\langle \lambda V \rangle = \langle \lambda V \rangle^{\mathrm{H,HF}} + \frac{1}{2} \sum_{ab} \int \frac{d\omega d\bar{\omega}}{(2\pi)^2} \frac{d\boldsymbol{p}}{(2\pi)^3} \int_0^\infty \left(e^{i(\omega-\bar{\omega})\bar{t}} - e^{-i(\omega-\bar{\omega})\bar{t}} \right) d\bar{t}$$
$$\times \left\{ V_{ab}^{>}(\boldsymbol{p}\omega) \Pi_{ab}^{<}(\boldsymbol{p}\bar{\omega}) - V_{ab}^{<}(\boldsymbol{p}\omega) \Pi_{ab}^{>}(\boldsymbol{p}\bar{\omega}) \right\} . \quad (6.7)$$

The superscript H, HF means the direct Hartree (H), and the exchange Hartree–Fock (HF) contributions. As we will discuss in Sect. 6.3, the Hartree term vanishes in electro-neutral systems. The Hartree–Fock contribution will be discussed there, too; it reads

$$\frac{1}{\Omega} \langle \lambda V \rangle_\lambda^{\mathrm{HF}} = \frac{1}{2} \sum_a (2s_a + 1) \int \frac{d\boldsymbol{p}_1 d\boldsymbol{p}_2}{(2\pi)^6} \frac{d\omega d\bar{\omega}}{2\pi^2}$$
$$\times V_{aa}(\boldsymbol{p}_1 - \boldsymbol{p}_2) g_a^{<}(\boldsymbol{p}_1\omega) g_a^{<}(\boldsymbol{p}_2\bar{\omega}) . \quad (6.8)$$

Using the property of the principal value, we find the result

$$\frac{1}{\Omega} \langle \lambda V \rangle = \frac{1}{\Omega} \langle \lambda V \rangle^{\mathrm{H,HF}} \pm \frac{1}{2} \sum_{ab} (2s_a + 1)(2s_b + 1) \int \frac{d\boldsymbol{p}}{(2\pi)^3} \frac{d\omega d\bar{\omega}}{(2\pi)^2}$$
$$\times \frac{\mathcal{P}}{\omega - \omega'} \left\{ V_{ab}^{>}(\boldsymbol{p}\omega) \Pi_{ab}^{<}(\boldsymbol{p}\bar{\omega}) - V_{ab}^{<}(\boldsymbol{p}\omega) \Pi_{ab}^{>}(\boldsymbol{p}\bar{\omega}) \right\} . \quad (6.9)$$

A formula of this type was given, e.g., also by Kadanoff and Baym (1962). Starting from this expression, we may derive a scheme for the determination of the EOS which is known as the dielectric formalism. We apply the connections (4.64) and use the dispersion relations (4.72), (4.78). It is easy to transform the expression (6.9) into

$$\frac{1}{\Omega} \langle \lambda V \rangle = \frac{1}{\Omega} \langle \lambda V \rangle^{\mathrm{H,HF}}$$
$$\pm \int \frac{d\boldsymbol{p}}{(2\pi)^3} \int \frac{d\omega}{2\pi} n_B(\omega) \left\{ \mathrm{Im} \varepsilon^{-1R}(\boldsymbol{p}\omega) + \mathrm{Im} \varepsilon^R(\boldsymbol{p}\omega) \right\} , \quad (6.10)$$

which is well-known as the dielectric formula given first by Pines and Nozieres (1958), and by Hubbard (1958). This exact relation determines an interesting connection between the thermodynamic and the dielectric properties of the plasma. That is, of course, not surprising because of the sum-rules for the dielectric function.

Finally we want to pay attention to the relation

$$\frac{1}{\Omega} \langle \lambda V \rangle = \frac{1}{2} \sum_a \int \frac{d\boldsymbol{p}}{(2\pi)^3} \mathrm{Re} \Sigma_a(\boldsymbol{p}, E_a(\boldsymbol{p})) f_a(E_a(\boldsymbol{p})) , \quad (6.11)$$

where we applied the simple quasiparticle approximation (3.192). Unfortunately, the expression (6.11) does not even give the correct classical value $\frac{1}{\Omega} \langle \lambda V \rangle = n\kappa e^2$ derived in Sect. 2.6. But it is possible to show that, by application of the quasiparticle spectral function (3.197) in formula (3.223), the correct classical result is achieved.

6.2 Screened Ladder Approximation

As the basic equations may not be evaluated exactly in general, we have to develop appropriate approximation schemes for the determination of thermodynamic functions for the plasma. As discussed earlier, the corresponding approximations have, first of all, to account for screening, both for physical and mathematical reasons. In general, we have, further, to deal with partially ionized plasmas. Therefore, we have to consider bound states between the plasma particles. Bound states essentially influence the properties of a plasma, especially also the EOS. In the preceding subsection we learned that the simplest approximation taking into account bound states is the binary collision approximation. In systems with Coulomb interaction, we have binary collisions between particles which interact via a screened potential. Consequently, we have to use, in such situation, the screened binary collision approximation.

In order to derive the EOS in screened ladder approximation, we start from (6.6), which reads in terms of diagrams

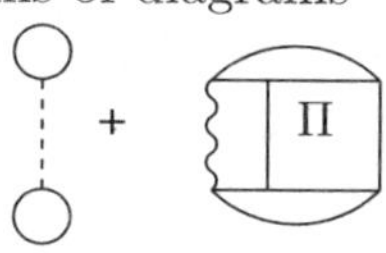

Therefore, one has to determine the polarization function in screened ladder approximation. In order to solve this problem, it is convenient to apply (4.8) and (4.14). Then, the former may be transformed into the more appropriate form

$$\begin{aligned}\Pi_{ab}(12,1'2') &= g_{ab}(12,1'2') - g_a(11')g_a(22') \\ &- \int \Pi_{ab}(1,\bar{1},2,\bar{2})V^s_{ac}(\bar{1},\bar{2})\Pi_{ab}(\bar{1},1'\bar{2},2')\end{aligned} \qquad (6.12)$$

or diagrammatically

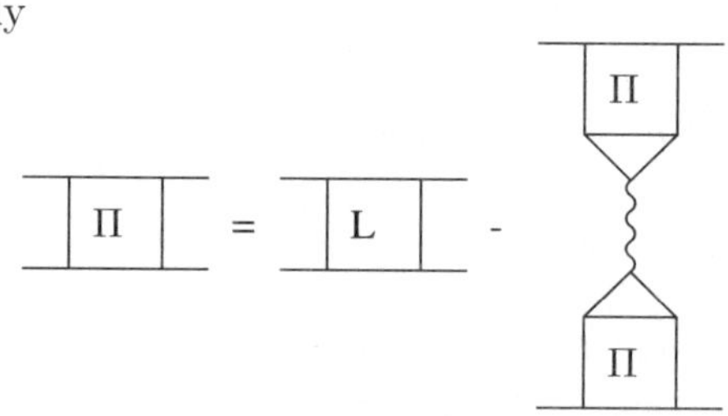

From the diagrams we see that a ladder approximation follows for Π if in the integral term of (6.12) the polarization functions are replaced by their lowest approximation

$$\Pi_{ab}(12,1'2') = \pm i\, g_a(12')g_a(21')\,,$$

and $L_{ab}(12,1'2') = g_{ab}(12,1'2') - g_a(11')g_b(11')$ is determined by the Bethe–Salpeter equation in dynamically screened ladder approximation. It follows a simple connection between the two-particle Green's function g_{ab} and the polarization function Π_{ab}

$$\Pi_{ac}(12,1'2') = g_{ac}(12,1'2') - g_a(1,1')g_c(2,2')$$
$$- \int d\bar{1}d\bar{2}\, g_a(1,\bar{1})g_c(2,\bar{2})V^s_{ac}(\bar{1},\bar{2})g_a(\bar{1},1')g_c(\bar{2},2')\,. \tag{6.13}$$

Now we are able to determine the EOS in dynamically screened ladder approximation. Introducing (6.13) for Π_{ac} into (6.6) and using the result in (6.1), we may write

$$p - p_0 = -\frac{1}{2\Omega}\int_0^1 \frac{d\lambda}{\lambda}\left\{ \cdots \right\}. \tag{6.14}$$

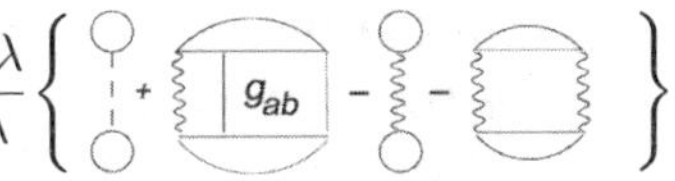

The diagrams determine the screened second virial coefficient. It can be seen that, in contrast to the usual ladder sum considered before, in the screened ladder sum, two terms are missing. Therefore, the ladder-sum representation of the EOS has the form (Kraeft et al. 1973; Ebeling (a) et al. 1976)

$$p - p_0 = -\frac{1}{2\Omega}\int_0^1 \frac{d\lambda}{\lambda}\left\{ \cdots \right\}. \tag{6.15}$$

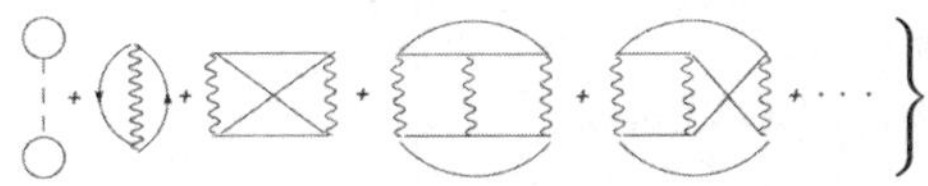

Here, the diagrams with less then four wavy lines are divergent for a pure Coulomb potential, while the higher order ladder terms remain convergent. The general screening procedure developed in Sects. 4.1, 4.2 renormalizes these divergent terms. In a more elementary consideration, this procedure is known as the ring summation. Let us demonstrate this with the second diagram, the screened Hartree–Fock contribution. Using the iterative solution of the screening (4.15), this diagram is given by the following infinite sum of divergent ring type Coulomb contributions leading to a convergent sum:

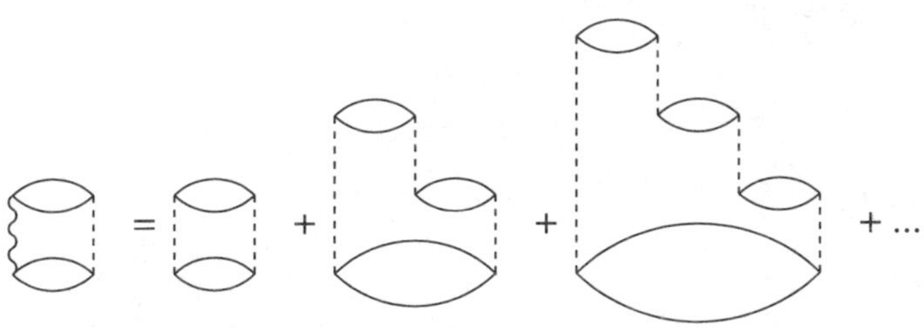

The ring summation idea was for the first time performed by Mayer (1950) in classical statistical mechanics in order to determine the sum of states for ions in an electrolyte solution, and it represents a rigorous foundation of the more elementary Debye–Hückel theory. The quantum mechanical ring summation was first given by Macke (1950), Gell-Mann and Brueckner (1957) and by Montroll and Ward (1958).

The screened Hartree-Fock contribution determines, together with the Hartree contribution, the first correction to the ideal gas result. Using again the screening equation, we can decompose this contribution in the following way

The first diagram is the Hartree term. The second term is the convergent Coulomb–Hartree-Fock result, and the third term is usually called the Montroll–Ward contribution. The Montroll–Ward term contains the most important peculiarities of the EOS for systems with Coulomb interaction, i.e., it represents the quantum ring sum. We will consider this interesting term in more detail in the next section.

6.3 Ring Approximation for the EOS. Montroll–Ward Formula

6.3.1 General Relations

We start with the simplest corrections to the ideal EOS, i.e., with the Hartree–Fock and Montroll–Ward contributions. Then, the EOS reads in diagrams

$$p - p_0 = \int_0^1 \frac{d\lambda}{\lambda} \Big\{$$ 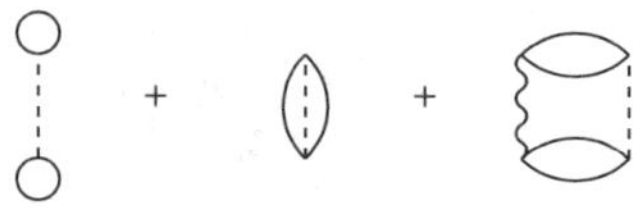 $$\Big\} = p^{\mathrm{H}} + p^{\mathrm{HF}} + p^{\mathrm{MW}} \,. \quad (6.16)$$

In this equation, any node of two Green's functions and one potential means integration over time and space.

The ideal pressure

$$p_0 = \sum_a \frac{2s_a + 1}{\beta \Lambda_a^3} I_{3/2}(\beta \mu_a) \tag{6.17}$$

was discussed in Chap. 2.

Equation (6.16) already reflects the typical peculiarities of the EOS of plasmas. It is divergent for the pure Coulomb potential. This divergence is avoided by screening, which leads to the fact that the resulting EOS is not simply a power series in terms of the density, as will be shown below.

Moreover, (6.16) is an appropriate approximation for applications where it is possible to omit higher orders in the coupling parameter. Such situations are met at sufficiently high densities, i.e., in the case of strong degeneracy, and at high temperatures and low densities, which is the case for weak degeneracy. In such situations, bound states do not play a role.

Let us first consider the Hartree– and Hartree–Fock contributions. On behalf of their special property to be local in time, we may immediately write for the Hartree term using the spectral representation (3.63) with the spectral function (3.70)

$$\frac{1}{\Omega}\langle\lambda V\rangle_\lambda^{\mathrm{H}} = \frac{1}{2}\sum_{ab}(2s_a+1)(2s_b+1)$$
$$\times\int\frac{d\boldsymbol{p}_1 d\boldsymbol{p}_2}{(2\pi)^6}V_{ab}(0)(\pm i)f_a(\boldsymbol{p}_1)(\pm i)f_b(\boldsymbol{p}_2)\,. \tag{6.18}$$

Then we get

$$\frac{1}{\Omega}\langle\lambda V\rangle_\lambda^{\mathrm{H}} = \frac{1}{2}\sum_{ab}n_a n_b V_{ab}(0)\,. \tag{6.19}$$

We remark that $V_{ab}(0)$ is divergent for Coulomb potentials. However, in electro-neutral systems, we have

$$\sum_c n_c e_c = 0\,,\quad \sum_{ab}n_a n_b e_a e_b = 0\,,$$

and thus the expression (6.19) vanishes. In one component plasmas (OCP), such as the electron gas, the existence of a neutralizing background is assumed to compensate the divergent Hartree term. In the following text, the Hartree term will not be considered.

The next diagram in (6.16) is the Hartree–Fock exchange contribution which is of typical quantum character.

According to the scheme applied above, we arrive at

$$\frac{1}{\Omega}\langle\lambda V\rangle_\lambda^{\mathrm{HF}} = \frac{1}{2}\sum_a(2s_a+1)\int\frac{d\boldsymbol{p}_1 d\boldsymbol{p}_2}{(2\pi)^6}V_{aa}(\boldsymbol{p}_1-\boldsymbol{p}_2)f_a(\boldsymbol{p}_1)f_a(\boldsymbol{p}_2)\,. \tag{6.20}$$

As was shown by DeWitt (1961), this term can be transformed into the expression

$$\frac{1}{\Omega}\langle\lambda V\rangle_\lambda^{\mathrm{HF}} = \sum_a\frac{2s_a+1}{\Lambda_a^4}e_a^2\int\limits_{-\infty}^{\alpha_a}d\alpha' I^2_{-1/2}(\alpha')\,, \tag{6.21}$$

where $\alpha_a = \mu_a/(k_B T)$. The Fermi-integral I_ν is defined by (see Chap. 2)

$$I_\nu(\alpha) = \frac{1}{\Gamma(\nu+1)}\int\limits_0^\infty dx\frac{x^\nu}{\exp(x-\alpha)+1}\,. \tag{6.22}$$

The Hartree–Fock contribution to the EOS in (6.16) then reads

$$p^{\mathrm{HF}}(\{\mu_c\}) = \sum_a\frac{2s_a+1}{\Lambda_a^4}e_a^2\int_{-\infty}^{\beta\mu_a}dx I^2_{-1/2}(x)\,. \tag{6.23}$$

The argument μ_c means that the pressure is given as a function of the chemical potentials.

The last diagram of (6.16) is the physically most interesting contribution. It accounts for correlations and is called the *Montroll–Ward* term. Using relation (6.10) with the *RPA* dielectric function, we find for the Montroll–Ward contribution to the pressure

$$p^{\mathrm{MW}} = \pm \int \frac{d\lambda}{\lambda} \int \frac{d\boldsymbol{p}}{(2\pi)^3} \frac{d\omega}{2\pi} n_B(\omega) \left\{ \mathrm{Im}\varepsilon^{-1R}(\boldsymbol{p}\omega) + \mathrm{Im}\varepsilon^{R}(\boldsymbol{p}\omega) \right\} . \tag{6.24}$$

Let us first consider the part involving $\mathrm{Im}\varepsilon^{-1}(p\omega)$. We start from the relation (4.62). In RPA, we have

$$\mathrm{Im}\varepsilon^{-1}(\boldsymbol{p}\omega) = \frac{\sum_a \lambda V_{aa}(\boldsymbol{p})\mathrm{Im}\Pi_{aa}(\boldsymbol{p}\omega)}{(1-\sum_c \lambda V_{cc}(\boldsymbol{p})\mathrm{Re}\Pi_{cc}(\boldsymbol{p}\omega))^2 + (\sum_c \lambda V_{cc}(\boldsymbol{p})\mathrm{Im}\Pi_{cc}(\boldsymbol{p}\omega))^2} . \tag{6.25}$$

The charging procedure in (6.24) may be carried out to give

$$\int_0^1 \frac{d\lambda}{\lambda} \mathrm{Im}\varepsilon^{-1}(\boldsymbol{p}\omega) = \begin{cases} \arctan \frac{\sum_c V_{cc}(\boldsymbol{p})\mathrm{Im}\Pi_{cc}(\boldsymbol{p}\omega)}{(1-\sum_c V_{cc}\mathrm{Re}\Pi_{cc}(\boldsymbol{p}\omega))} , & \mathrm{Re}\varepsilon > 0 , \\ \arctan \frac{\sum_c V_{cc}(\boldsymbol{p})\mathrm{Im}\Pi_{cc}(\boldsymbol{p}\omega)}{(1-\sum_c V_{cc}\mathrm{Re}\Pi_{cc}(\boldsymbol{p}\omega))} - \pi , & \mathrm{Re}\varepsilon < 0 . \end{cases} \tag{6.26}$$

Here we applied the relation (4.102). The charging of the Imε-term is trivial. We make use of the property

$$n_B(\omega) + \frac{1}{2} = -(n_B(-\omega) + \frac{1}{2}) = \frac{1}{2} \coth \frac{\beta\hbar\omega}{2} , \tag{6.27}$$

and of the fact that the function *arctan* is an odd function, and Imε is an odd function of ω, while Reε is even. Then we arrive at

$$p_{\mathrm{MW}} = -\int \frac{d\boldsymbol{p}}{(2\pi)^3} \int_0^\infty \frac{d\omega}{2\pi} \coth \frac{\beta\hbar\omega}{2} \left[\arctan \frac{\mathrm{Im}\varepsilon(\boldsymbol{p}\omega)}{\mathrm{Re}\varepsilon(\boldsymbol{p}\omega)} - \mathrm{Im}\varepsilon(\boldsymbol{p}\omega) \right] . \tag{6.28}$$

This relation is useful as a starting point for numerical evaluations at any degeneracy. In (6.28), the dielectric function has to be taken in random phase approximation. The angle integration may be carried out trivially as the (equilibrium) dielectric function depends only on the modulus of $\boldsymbol{p}$. The remaining integration has to be done numerically. In Fig. 6.1, the result is given for the evaluation of the Montroll–Ward contribution of the electron gas as a function of the logarithm of the fugacity $\alpha_e = \ln \tilde{z}_e = \beta\mu_e$. Especially, one can see how the numerical results coincide with the non-degenerate results (negative α-values) and that for highly degenerate systems (positive α) to be dealt with in the subsequent subsections.

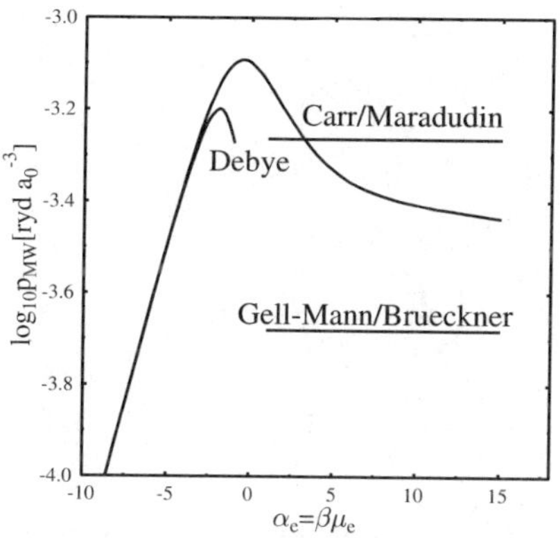

Fig. 6.1. Montroll–Ward pressure for an electron gas as a function of the fugacity $\alpha_e = \beta\mu_e$ for a fixed density $r_s = 4$. The weakly degenerate result (Debye) is given by (6.48), the result for the highly degenerate case is given by (6.65) (Gell-Mann–Brueckner, *horizontal line*), and by (6.66) (Carr and Maradudin, *horizontal line*)

Now we collect the contributions (6.17), (6.23), (6.28, and 6.16). Then we get the EOS in Montroll–Ward approximation

$$p = \sum_a \frac{2s_a+1}{\beta\Lambda_a^3} I_{3/2}(\beta\mu_a) + \sum_a \frac{2s_a+1}{\Lambda_a^4} e_a^2 \int_{-\infty}^{\beta\mu_a} dx I_{-1/2}^2(x) + p^{\mathrm{MW}}(\beta\mu_a) \,. \tag{6.29}$$

Expression 6.29 determines the EOS in terms of the chemical potential as the independent variable. In addition to (6.23), we get, from (6.2), the density as a function of the chemical potential

$$n_a = \frac{\partial p}{\partial\mu_a} = \frac{2s_a+1}{\Lambda_a^3} I_{1/2}(\beta\mu_a) + \frac{2s_a+1}{\Lambda_a^4} \beta e_a^2 I_{-1/2}^2(\beta\mu_a) + \frac{\partial}{\partial\mu_a} p^{\mathrm{MW}} \,. \tag{6.30}$$

The relations (6.29), (6.30) determine pressure and density in the framework of the grand canonical ensemble.

The chemical potential as a function of the density follows from the inversion of (6.30), while we get the pressure from (6.29) eliminating the chemical potential via (6.30).

Let us consider an approximate inversion of (6.30) with respect to μ_a. We write for this reason

$$\mu_a = \mu_a^{\mathrm{id}} + \mu_a^{\mathrm{int}} \,.$$

Here, the ideal part is

$$\beta\mu_a^{\mathrm{id}} = \alpha_a(n_a, T) \,.$$

The function $\alpha_a(n_a, T)$ is well-known and follows by inversion from (2.24) or from interpolation formulae like (2.33).

We consider μ_a^{int} to be a small perturbation. Consequently, we expand the expressions $n_a(\mu_a^{\mathrm{id}} + \mu_a^{\mathrm{int}})$ and $I_{3/2}(\beta(\mu_a^{\mathrm{id}} + \mu_a^{\mathrm{int}}))$ into Taylor series and apply the relation (2.27). Then, for the chemical potential, we get

$$\mu_a(n_a, T) = k_B T \alpha_a - \frac{e_a^2}{\Lambda_a} I_{-1/2}(\alpha_a) - \frac{\Lambda_a^3}{(2s_a+1) I_{-1/2}(\alpha_a)} \frac{\partial}{\partial \alpha_a} p^{\mathrm{MW}} . \quad (6.31)$$

Introducing (6.31) into the Taylor expansion of $I_{3/2}(\mu^{\mathrm{id}} + \mu^{\mathrm{int}})$, we get, from (6.29), the pressure as a function of the density

$$p = \sum_a \frac{2s_a+1}{\beta \Lambda_a^3} I_{3/2}(\alpha_a) + \sum_a \frac{2s_a+1}{\Lambda_a^4} e_a^2 \Big[\int_{-\infty}^{\alpha_a} I_{-1/2}^2(x) dx$$
$$- I_{1/2}(\alpha_a) I_{-1/2}(\alpha_a) \Big] + p^{\mathrm{MW}} - \sum_a \frac{I_{1/2}(\alpha_a)}{I_{-1/2}(\alpha_a)} \times \frac{\partial}{\partial \alpha_a} p^{\mathrm{MW}} . \quad (6.32)$$

The relations (6.29), (6.30) provide an *exact* representation of the pressure up to the Montroll–Ward contribution in the grand canonical ensemble. In contrast, with (6.32), we have an approximation which is referred to as *incomplete inversion.*

We still give the expressions for the free energy. For this purpose, we apply the relation

$$F = \sum_a N_a \mu_a - pV , \quad (6.33)$$

or the corresponding formula for the density of the free energy

$$f = \sum_a n_a \mu_a - p . \quad (6.34)$$

For the free energy density we get

$$f = \frac{F}{V} = \sum_a \frac{2s_a+1}{\beta \Lambda_a^3} \left[\alpha_a I_{1/2}(\alpha_a) - I_{3/2}(\alpha_a) \right]$$
$$- \sum_a \frac{(2s_a+1) e_a^2}{2\Lambda_a^4} \int_{-\infty}^{\alpha_a} I_{-1/2}^2(a) da - p^{\mathrm{MW}}(\{\alpha_c\}) . \quad (6.35)$$

The analytic evaluation of thermodynamic functions in Montroll–Ward approximation is possible only in limiting situations $n\Lambda^3 \ll 1$ and $n\Lambda^3 \gg 1$, respectively. These limiting situations are dealt with in the following subsection. For arbitrary degeneracy, a numerical evaluation is necessary. The Hartree–Fock contribution is expressed in terms of Fermi integrals which may be evaluated numerically without difficulty. The evaluation of the Montroll–Ward contribution (6.28) is more complicated and more interesting as well. This term is essentially determined by the dielectric properties of the plasma. The numerical evaluation was carried out by Jakubowski, Kraeft, and Stolzmann and later by Fromhold–Treu and Riemann (Kraeft and Jakubowski 1978; Stolzmann et al. 1989; Riemann 1997).

We discuss the EOS showing isotherms of the pressure (6.32) for the electron gas model and for the more realistic electron–proton plasma. On

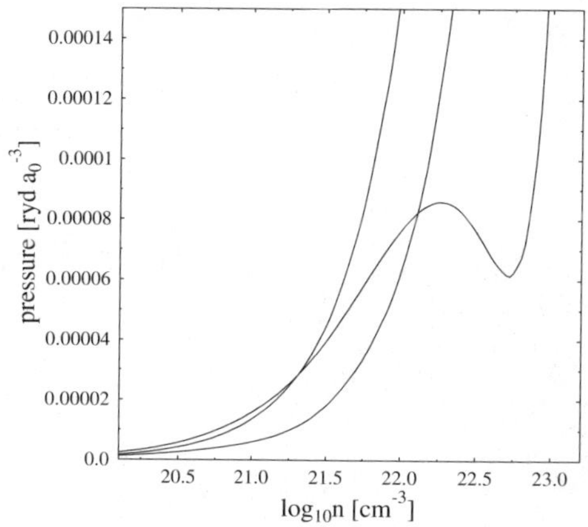

Fig. 6.2. Isotherms of the pressure at $T = 14000\,\mathrm{K}$. Free particles (**upper**), interacting electrons, and e–p plasma (**lower curve** with loop)

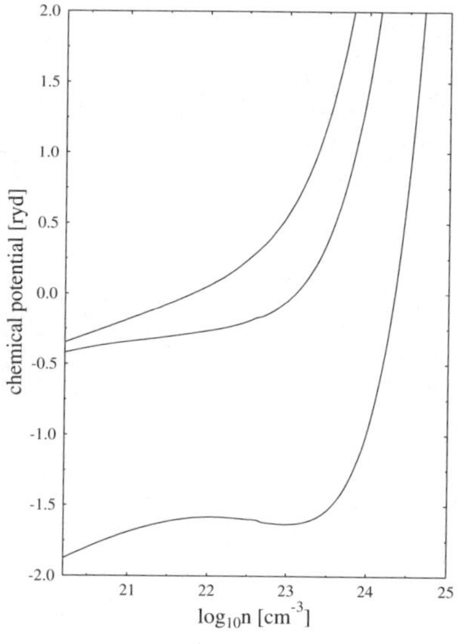

Fig. 6.3. Isotherms of μ_e^{id}, μ_e^{MW} and of $\mu = \mu_e + \mu_p$ (**from above**) for $T = 14000\,\mathrm{K}$

behalf of the large difference between electron and proton masses, m_e and m_p, the properties of electrons and protons are quite different. The protons are practically non-degenerate even in regions where the electrons are highly degenerate (cf. the density–temperature plane Fig. 2.8).

In Fig. 6.2, the pressure of the electron gas is shown in comparison to that of the ideal electron gas, in order to estimate the influence of the Montroll–Ward term. In addition, the full pressure of the e–p plasma is given, in order to demonstrate the role of the protons.

The chemical potential is shown in Fig. 6.3, again for free and interacting electrons, and for the H-plasma, $\mu = \mu_e + \mu_p$. Isotherms for the chemical potential of the e–p plasma are plotted in Fig. 6.4. At low temperatures, the curves exhibit a Van der Waals-loop.

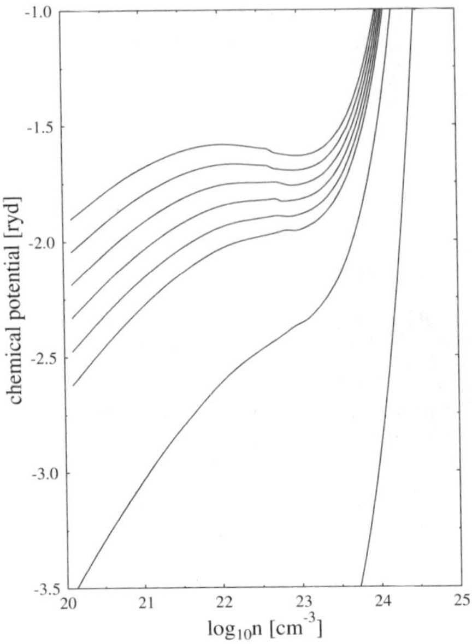

Fig. 6.4. Isotherms of the chemical potential of an H-plasma for various temperatures; **from above**: 14, 15, 16, 17, 18, 19, 25, and 50 × 1000 K

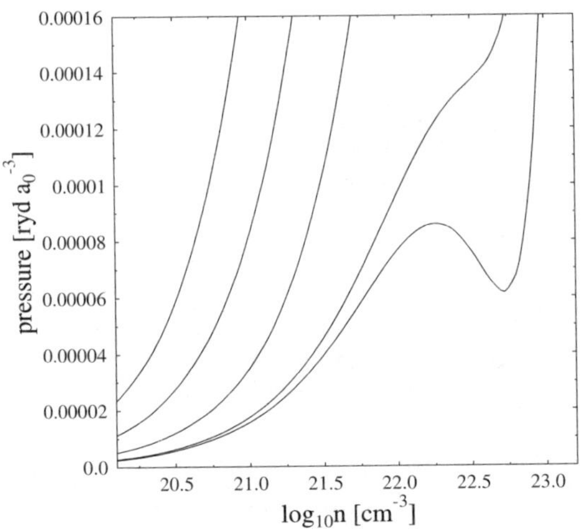

Fig. 6.5. Isotherms of the pressure of an H-plasma for various temperatures; **from below**: 14, 15, 25, 50, and 50 × 1000 K

The temperature dependence of the pressure is demonstrated in Fig. 6.5 by a sequence of isotherms.

As already discussed in Chap. 2 in the frame of the simple Debye–Hueckel model, we observe the appearance of a critical temperature T_c, i.e., the pressure shows Van der Waals-loops, which represent instabilities and may be connected, in principle, with phase transitions, e.g., with the *plasma phase transition*, representing a coexistence between two differently ionized phases. Such phase transition was not verified experimentally so far. However, in

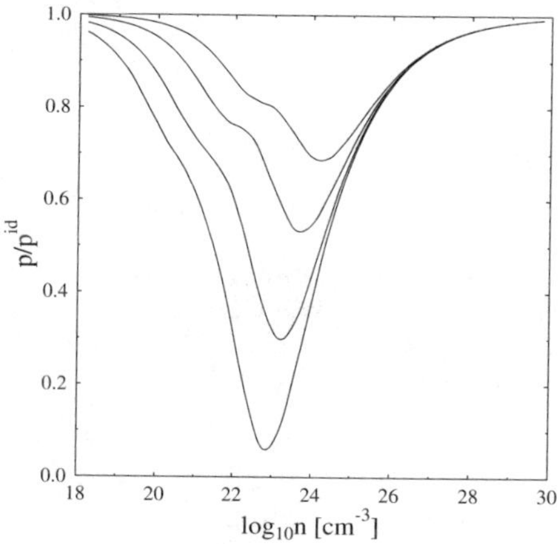

Fig. 6.6. Isotherms of the relative pressure of an H-plasma for various temperatures; **from below**: 15,25,50, and $50 \times 1000\,\mathrm{K}$

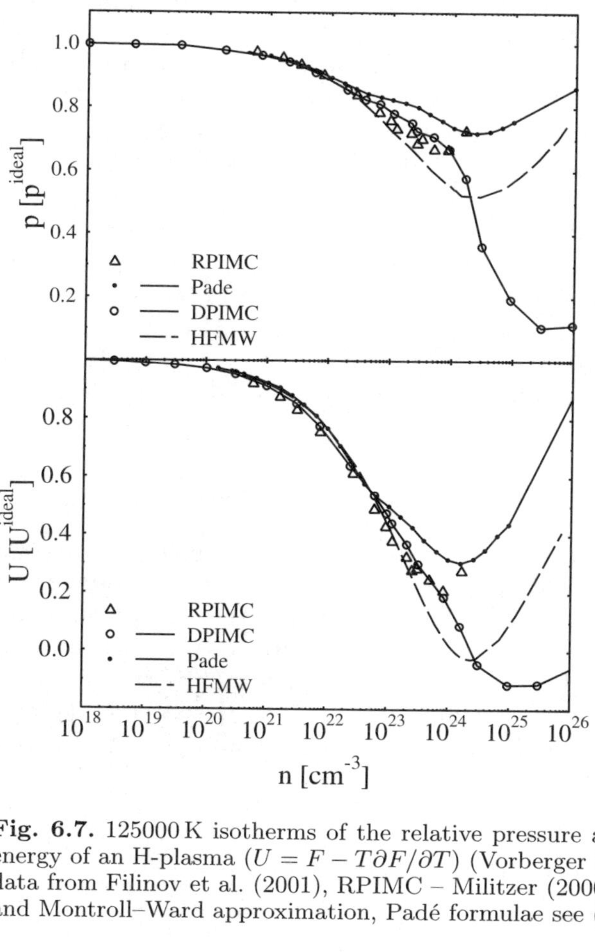

Fig. 6.7. 125000 K isotherms of the relative pressure and of the relative internal energy of an H-plasma ($U = F - T\partial F/\partial T$) (Vorberger 2005). DPIMC – numerical data from Filinov et al. (2001), RPIMC – Militzer (2000), HFMW – Hartree–Fock and Montroll–Ward approximation, Padé formulae see (6.70), Sect. 6.3.4

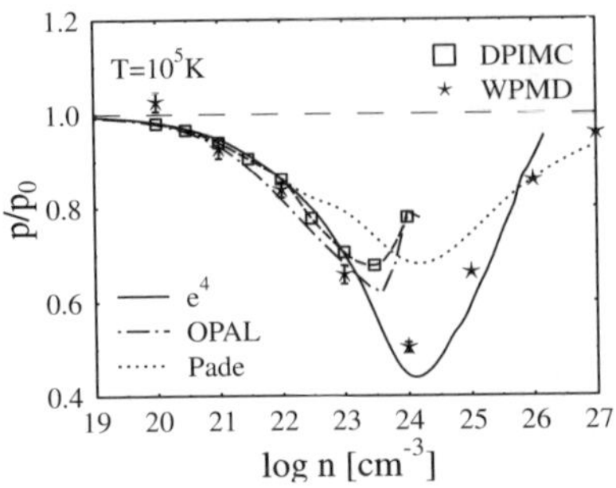

Fig. 6.8. Isotherms of the pressure. e^4 – Vorberger (2005), WPMD – Knaup et al. (2001), OPAL – ACTEX-OPAL (2000). For explanations see also Fig. 6.7

view of our simple approximation, the interpretation just given is questionable. The formalism applied accounts only for weak coupling in the plasma. For temperatures of $14000K$, at which Van der Waals-loops are observed, there are strong correlations, and especially the formation of bound states (H-atoms and H_2-molecules) has to be taken into account. Consequently, the incomplete inversion becomes doubtful; for a further discussion of the problem, we have to consider improved approximations.

The next Fig. 6.6 shows the relative pressure

$$\frac{p}{p_0} = 1 + \frac{p^{\mathrm{HF}} + p^{\mathrm{MW}}}{p_0}$$

with p according to (6.32) for different temperatures. In the high density limit, the pressure becomes temperature independent, i.e., all isotherms approach the ground state behavior. In Fig. 6.7, a comparison is given to data from *Monte Carlo path integral calculations*, cf. Filinov et al. (2001). At higher densities, larger deviations are observed. In Fig. 6.8, another example is given.

6.3.2 The Low Density Limit (Non-degenerate Plasmas)

In the following two subsections we want to consider limiting cases. First we will deal with the situation $n\lambda^3 \ll 1$, this means the case of a non-degenerate plasma. This limiting case was intensely dealt with in papers by Vedenov and Larkin (1959), Trubnikov and Elesin (1964), DeWitt (1966), Ebeling et al. (1967), Kraeft and Kremp (1968), Hoffmann and Ebeling (1968a), (Hoffmann and Ebeling 1968b), Ebeling (a) et al. (1976), and in Kraeft et al. (1988). In the weakly degenerate case, the Fermi integrals are given by the fugacity expansion (2.28)

$$I_j(\tilde{z}) = \tilde{z} - \frac{1}{2^{j+1}}\tilde{z}^2 + \cdots, \tag{6.36}$$

where the fugacity $\tilde{z}$ is defined by

$$\tilde{z}_a = e^{\beta\mu_a}.$$

It is often more convenient to use z instead of the fugacity $\tilde{z}$. The advantage of the modified fugacity z is that it coincides with the density if we take the non-degenerate limit, i.e.,

$$z_a = \tilde{z}_a(2s_a + 1)\frac{1}{\Lambda_a^3} \,. \tag{6.37}$$

We give the relation between different thermal wavelengths

$$\Lambda_a = \frac{h}{(2\pi m_a k_B T)^{1/2}} = (2\pi)^{1/2}\lambda_{aa} \,, \qquad \lambda_{ab} = \frac{\hbar}{(2m_{ab}k_B T)^{1/2}} \,. \tag{6.38}$$

With the expansion (6.36), Fermi statistics is replaced by Boltzmann statistics including, in addition, first quantum corrections. Using (6.36) in the expression (6.17), we get the pressure for the ideal non-degenerate plasma

$$p_0 = \sum_a \frac{2s_a + 1}{\beta\Lambda_a^3}\left\{\tilde{z}_a - \frac{\tilde{z}_a^2}{2^{5/2}}\right\} \,. \tag{6.39}$$

The first correction to the ideal pressure is the *Hartree–Fock* contribution. With (6.36), we get from (6.32)

$$p^{\rm HF} = -\frac{1}{2}\sum_a \frac{2s_a + 1}{\Lambda_a^4} e_a^2 \tilde{z}_a^2 \,. \tag{6.40}$$

With the modified fugacities introduced above, we get for the Hartree–Fock pressure

$$p^{\rm HF} = -\pi\sum_a \frac{z_a^2\lambda_{aa}^2 e_a^2}{2s_a + 1} \,. \tag{6.41}$$

The next contribution in (6.16) is the more complicated Montroll–Ward term $p^{\rm MW}$. For its evaluation, we start from (4.31), this means, we have to determine the self-energy $\Sigma^{\gtrless}$ given by the diagram (Ebeling (a) et al. (1976) and references quoted therein).

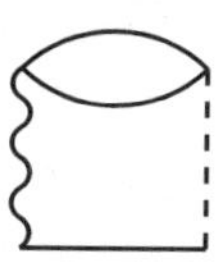

The Montroll–Ward diagram has to be considered on the Keldysh contour.

We go to the physical time axis. This can be done easily because we are interested in the low density limit (high temperature case). In this situation, it is sufficient to use a statically screened potential. For the self-energy, we then have

$$\Sigma_a^{\gtrless}(\boldsymbol{p}\omega) = \sum_b (2s_b+1) \int \frac{d\bar{\boldsymbol{p}}d\bar{\boldsymbol{p}}'d\boldsymbol{p}'}{(2\pi)^9} \frac{d\bar{\omega}d\bar{\omega}'d\omega'}{(2\pi)^3}$$
$$\times (2\pi)^3\delta(\boldsymbol{p}+\boldsymbol{p}'-\bar{\boldsymbol{p}}-\bar{\boldsymbol{p}}')2\pi\delta(\omega+\omega'-\bar{\omega}-\bar{\omega}')$$
$$\times V_{ab}^s(\boldsymbol{p}-\bar{\boldsymbol{p}})V_{ab}(\boldsymbol{p}-\bar{\boldsymbol{p}})g_b^{\gtrless}(\boldsymbol{p}'\omega')g_b^{\gtrless}(\bar{\boldsymbol{p}}\bar{\omega})g_a^{\gtrless}(\bar{\boldsymbol{p}}'\bar{\omega}') . \tag{6.42}$$

Next we use the quasiparticle approximation and introduce the self-energy (6.42) into (6.11). Then we get the Montroll–Ward contribution to the mean potential energy

$$\frac{1}{\Omega}\langle\lambda V\rangle_\lambda^{\mathrm{MW}} \pm \sum_{ab}(2s_a+1)(2s_b+1)\int \frac{d\boldsymbol{p}d\boldsymbol{q}d\boldsymbol{p}'}{(2\pi)^9}$$
$$\times V_{ab}^s(q)V_{ab}(q)\frac{f_a(\boldsymbol{p})f_b(\boldsymbol{p}')-f_a(\boldsymbol{p}'+\boldsymbol{q})f_b(\boldsymbol{p}-\boldsymbol{q})}{\varepsilon_b(\boldsymbol{p}-\boldsymbol{q})+\varepsilon_a(\boldsymbol{p}'+\boldsymbol{q})-\varepsilon_b(\boldsymbol{p}')-\varepsilon_a(\boldsymbol{p})} . \tag{6.43}$$

The statically screened potential V^s is determined by the static limit of the *RPA* dielectric function (4.122) and reads

$$V_{ab}^s(q) = \frac{4\pi e_a e_b}{q^2+\sum_c \kappa_c^2\, {}_1F_1\left(1,\frac{3}{2};-\lambda_{cc}^2 q^2/8\right)} . \tag{6.44}$$

Here, ${}_1F_1(..)$ is the confluent hypergeometric function, and κ_c is the special Debye quantity

$$\kappa_c^2 = \frac{4\pi n_c e_c^2}{k_B T} . \tag{6.45}$$

For further evaluation we can use the Boltzmann distribution $f(\boldsymbol{p})$. Then, in (6.43), all integrations up to the one over the modulus of q may be carried out analytically. The result for spin-$\frac{1}{2}$ particles is

$$\frac{1}{\Omega}\langle\lambda V\rangle_\lambda^{\mathrm{MW}} = -\frac{4}{k_B T}\sum_{ab} z_a z_b e_a^2 e_b^2 I_{ab}^{\mathrm{MW}} \tag{6.46}$$

where

$$I_{ab}^{\mathrm{MW}} = \int_0^\infty dq \frac{{}_1F_1(1,\frac{3}{2};-\lambda_{ab}^2 q^2/4)}{q^2+\sum_c \kappa_c^2\, {}_1F_1(1,\frac{3}{2};-\lambda_{cc}^2 q^2/8)} . \tag{6.47}$$

This expression is the general result for the low density limit. The evaluation can only be done numerically. An approximate analytical evaluation is possible if the denominator of (6.47) is linearized according to ${}_1F_1 = 1+({}_1F_1-1)$. Equation (6.47) is correct up to the order $z^{5/2}$, however, the order z^3 included in (6.47) is not correct on behalf of the neglect of the dynamics in V^S in (6.42), (6.43) and due to the approximation of the correlation functions. The analytical evaluation of (6.47) leads to

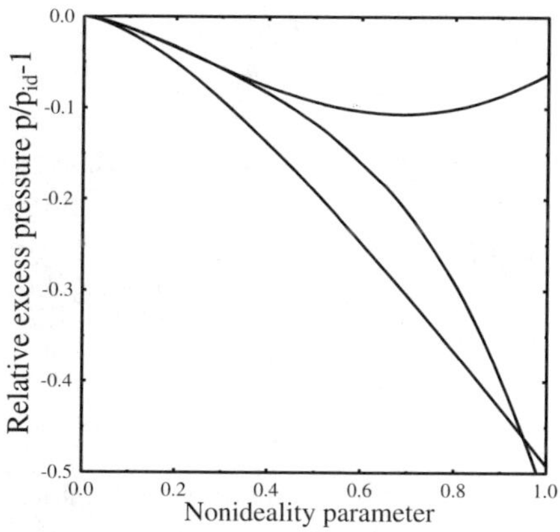

Fig. 6.9. Relative excess pressure $p/p_0 - 1$ of an H-plasma at low density in different approximations as a function of the nonideality parameter Γ (see Chap. 2 or (6.76) for fixed density, $r_s = 1$. **From above**: Up to order n^2, up to $n^{5/2}$ (see Sect. 6.4.2), and up to $n^{3/2}$ (limiting law)

$$\frac{1}{\Omega} \langle \lambda V \rangle |_{\lambda}^{\mathrm{MW}} = -\frac{k_B T}{8\pi} \sum_{ab} \frac{\kappa_a^2 \kappa_b^2}{\kappa_D}$$

$$\times \left\{ 1 - \frac{\sqrt{\pi}}{4} \kappa_D \lambda_{ab} + \frac{1}{6} \left[\kappa_D^2 \lambda_{ab}^2 + \frac{1}{4} \sum_c \kappa_c^2 \lambda_{cc}^2 \right] \right\} . \tag{6.48}$$

The first term in the curly brackets is the Debye limiting law. The next one is a quantum correction of order z^2, while the last two terms are $z^{5/2}$ terms. There are still more terms of the order $z^{5/2}$ corresponding to higher screening contributions (*beyond RPA*) which will be discussed later (Sect. 6.4.2) (Riemann 1997). The contribution to the pressure reads in this approximation

$$\beta p^{\mathrm{MW}} = \frac{\kappa^3}{12\pi} - \frac{1}{4} \pi^{3/2} \sum_{ab} z_a z_b \lambda_{ab}^3 \xi_{ab}^2 . \tag{6.49}$$

Here, the usual Born parameter was introduced

$$\xi_{ab} = -\beta e_a e_b / \lambda_{ab} = -\frac{e_a e_b}{\hbar} \sqrt{\frac{2 m_{ab}}{kT}} . \tag{6.50}$$

Now we are ready to summarize our results up to the Montroll–Ward approximation in the non-degenerate limit. The fugacity expansion of the pressure is given by (Bartsch and Ebeling 1975; Ebeling (a) et al. 1976)

$$\beta p(\{z_c\}) = \sum_a z_a + \frac{\kappa^3}{12\pi} + 2\pi \sum_{ab} z_a z_b \lambda_{ab}^3 \left\{ -\frac{\sqrt{\pi}}{8} \xi_{ab}^2 \right.$$

$$\left. - \frac{\delta_{ab}}{2 s_a + 1} \left(\frac{\sqrt{\pi}}{4} + \frac{\xi_{ab}}{2} \right) \right\} , \tag{6.51}$$

while the density reads

$$n_a(z) = z_a + \frac{\kappa^3}{8\pi} + 2\pi \sum_b z_a z_b \lambda_{ab}^3 \left\{ -\frac{\sqrt{\pi}}{8}\xi_{ab}^2 - \frac{\delta_{ab}}{2s_a+1}\left(\frac{\sqrt{\pi}}{4} + \frac{\xi_{ab}}{2}\right)\right\}. \tag{6.52}$$

Like in the general case, we perform an incomplete inversion and arrive at the EOS in Montroll–Ward approximation in the non-degenerate limit, i.e., we have the pressure as a function of the density

$$p(n) = \sum_a n_a - \frac{\kappa^3}{24\pi} - 2\pi \sum_{ab} n_a n_b \lambda_{ab}^3 \left\{ -\frac{\sqrt{\pi}}{8}\xi_{ab}^2 - \frac{\delta_{ab}}{2s_a+1}\left(\frac{\sqrt{\pi}}{4} + \frac{\xi_{ab}}{2}\right)\right\}. \tag{6.53}$$

In former work, the pressure was determined from the canonical distribution. Especially, the method of Slater sums and quantum potentials was developed (Kelbg 1963; Ebeling et al. 1967; Kremp and Kraeft 1968; Kraeft and Kremp 1968).

For the chemical potential, we easily get

$$\mu_a = \mu_a^{id} - \frac{\kappa e^2}{2} - 2\pi \sum_b n_b \lambda_{ab}^3 \Big\{ -\sqrt{\pi}\frac{\xi_{ab}^2}{8} - \frac{\delta_{ab}}{2s_a+1}\left(\frac{\sqrt{\pi}}{4} + \frac{\xi_{ab}}{2}\right)\Big\}. \tag{6.54}$$

6.3.3 High Density Limit. Gell-Mann–Brueckner Result

For the highly degenerate situation, it is useful to start from the representation (6.28). According to several papers, it is advantageous to carry out the frequency integration along the imaginary frequency axis rather than along the real one. This is possible on behalf of the analytical properties of the dielectric function which lead to the relation

$$\int_0^\infty \mathrm{Im}\chi(k\omega)d\omega = \frac{1}{2}\int_{-\infty}^{+\infty} d\omega\chi(ki\omega)\,, \tag{6.55}$$

where

$$\chi = \varepsilon^{-1} - 1\,. \tag{6.56}$$

A similar relation is valid for $\varepsilon(k\omega)$. We may write for the Bose function

$$\lim_{T\to 0} n_B(\omega) = \begin{cases} -1\,,\ \omega < 0 \\ \ \ 0\,,\ \omega \ge 0 \end{cases}. \tag{6.57}$$

Then we get

$$p_{\mathrm{MW}} = -\int_0^1 \frac{d\lambda}{\lambda}\int \frac{d\boldsymbol{p}}{(2\pi)^3}\int_0^\infty \frac{d\omega}{2\pi}\left\{\mathrm{Im}\varepsilon^{-1}(\boldsymbol{p}\omega) + \mathrm{Im}\varepsilon(\boldsymbol{p}\omega)\right\} \tag{6.58}$$

and

$$p_{\mathrm{MW}} = -\frac{1}{2}\int_0^1 \frac{d\lambda}{\lambda}\int \frac{d\boldsymbol{p}}{(2\pi)^3}\int_{-\infty}^{+\infty}\frac{d\omega}{2\pi} \times \left\{\frac{\sum_c \lambda V_{cc}(\boldsymbol{p})\Pi_{cc}(\boldsymbol{p}i\omega)}{1-\sum_c \lambda V_{cc}(\boldsymbol{p})\Pi_{cc}(\boldsymbol{p}i\omega)} + \sum_c \lambda V_{cc}(\boldsymbol{p})\Pi_{cc}(\boldsymbol{p}i\omega)\right\}. \quad (6.59)$$

Now the charging procedure may be carried out. The result is

$$p_{\mathrm{MW}} = \int_{-\infty}^{+\infty}\frac{d\omega}{2\pi}\int \frac{d\boldsymbol{p}}{(2\pi)^3} \times \left\{\ln\left[1-\sum_c V_{cc}(\boldsymbol{p})\Pi_{cc}(\boldsymbol{p}i\omega)\right] + \sum_c V_{cc}(\boldsymbol{p})\Pi_{cc}(\boldsymbol{p}i\omega)\right\}. \quad (6.60)$$

For the further evaluation, one may use an approximation for the polarization function which is due to Gell-Mann and Brueckner; see Kraeft and Stolzmann (1979) and Kraeft and Rother (1988). The general RPA expression for Π, for $T=0$, reads in the case of an electron–ion system

$$\Pi_a(k,i\omega) = \frac{m_a\sqrt{k_F^e k_F^i}}{4\pi^2 z}K_a(z,\zeta)\,, \quad (6.61)$$

$$z = \frac{k}{2\sqrt{k_F^e k_F^i}}\,; \qquad \zeta = \frac{\omega}{k}\sqrt{\frac{m_e m_i}{k_F^e k_F^i}}\,.$$

Here we introduced the function

$$K_a(z,\zeta) = \frac{1}{2}\left(\frac{k_F^{a2}}{k_F^e k_F^i} + \frac{m_a^2}{m_e m_i}\zeta^2 - z^2\right)\ln\left|\frac{(c_1-z)^2+c_2\zeta^2}{(c_1+z)^2+c_2\zeta^2}\right| -2c_1 z + 2\sqrt{c_2}z\zeta\left[\arctan\left(\frac{c_1-z}{\sqrt{c_2}\zeta}\right) + \arctan\left(\frac{c_1+z}{\sqrt{c_2}\zeta}\right)\right], \quad (6.62)$$

with

$$c_1 = \frac{k_F^a}{\sqrt{k_F^e k_F^i}}\,;\; c_2 = \frac{m_a^2}{m_e m_i}\,;\; k_F \text{ – Fermi momenta.}$$

For the derivation of the Gell-Mann–Brueckner result, it is sufficient to apply the Gell-Mann–Brueckner approximation for Π which reads

$$\Pi_a^{\mathrm{GB}}(i\zeta) = -\frac{m_a k_F^a(2s_a+1)}{2\pi^2}R_a(\zeta) \quad (6.63)$$

with

$$R_a(\zeta) = 1 - A_a\zeta\arctan\frac{1}{A_a\zeta}\,, \quad (6.64)$$

$$A_a = \frac{m_a}{k_F^a}\sqrt{\frac{k_F^e k_F^i}{m_e m_i}}\,.$$

Using (6.63), the pressure according to (6.60) may be evaluated. Only the divergent term is achieved analytically (ln-term), while the remaining part has to be done numerically. The result for the electron gas in Gell-Mann–Brueckner approximation reads

$$p_{\mathrm{MW}} = -n(0.0622\ln r_s - 0.142)\,, \tag{6.65}$$

where r_s is the Brueckner parameter. Carr and Maradudin (1964) added two next order terms, and the pressure then reads

$$p_{\mathrm{MW}} = -n(0.0622\ln r_s - 0.142 - 0.0054 r_s \ln r_s - 0.015 r_s)\,. \tag{6.66}$$

Temperature corrections were given by Sommerfeld expansions in Kremp et al. (1972), Fennel et al. (1974).

6.3.4 Padé Formulae for Thermodynamic Functions

For practical purposes, it is useful to present thermodynamic functions in terms of Padé formulae. Such formulae were given in Ebeling et al. (1981) and Ebeling et al. (1991). Following the latter references, we write for the excess free energy of the electron gas

$$f^P = \frac{f_{\mathrm{D}} - \frac{1}{4}(\pi\beta)^{-1/2}\bar{n} + 8\bar{n}^2 f_{\mathrm{GB}}}{1 + 8\ln\left[1 + \frac{3}{64\sqrt{2}}(\pi\beta)^{1/4}\bar{n}^{1/2}\right] + 8\bar{n}^2}\,, \tag{6.67}$$

and for the interaction part of the chemical potential

$$\mu^P = \frac{\mu_{\mathrm{D}} - \frac{1}{2}(\pi\beta)^{-1/2}\bar{n} + 8\bar{n}^2 \mu_{\mathrm{GB}}}{1 + 8\ln\left[1 + \frac{1}{16\sqrt{(2)}}(\pi\beta)^{1/4}\bar{n}^{1/2}\right] + 8\bar{n}^2}\,. \tag{6.68}$$

In (6.67), (6.68), we used Heaviside units and the density $\bar{n}_e = n\Lambda_e^3$. One can easily verify that (6.67), (6.68) meet the limiting situations of the non-degenerate system and that of the highly degenerate case as well. This goal is achieved if, in the formulae (6.67), (6.68), the low density behavior is fixed by

$$f_{\mathrm{D}} = -\frac{2}{3}(\pi\beta)^{-1/4}\bar{n}^{1/2}\,, \tag{6.69}$$

and

$$\mu_{\mathrm{D}} = -(\pi\beta)^{-1/4}\bar{n}^{1/2}\,. \tag{6.70}$$

The two formulae (6.69), (6.70) represent the Debye limiting laws for the free energy and for the chemical potential, respectively; cf. Chap. 2. In order

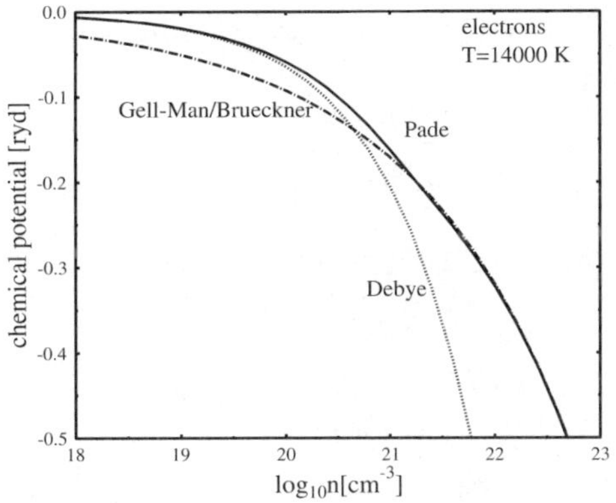

Fig. 6.10. Correlation part of the electron chemical potential as a function of the density in comparison with the Debye (*dotted line*) (6.70) and Gell-Mann–Brueckner (*dash-dot*) (6.72) approximations. $T = 14000\,\mathrm{K}$. Only in a small density region, the exact curve (in agreement with the Padé approximation) serves as an interpolation between the limiting situations

to meet the highly degenerate region, we introduced the (slightly modified) Gell-Mann–Brueckner approximations for the free energy

$$f_{\mathrm{GB}} = -\frac{0.9163}{r_s} - 0.08883\ln\left[1 + \frac{4.9262}{r_s^{0.7}}\right] \approx -\frac{0.9163}{r_s} + 0.0622\ln r_s\,, \quad (6.71)$$

and for the chemical potential, respectively

$$\mu_{\mathrm{GB}} = -\frac{1.2217}{r_s} - 0.08883\ln\left[1 + \frac{6.2208}{r_s^{0.7}}\right] \approx -\frac{1.2217}{r_s} + 0.0622\ln r_s\,. \quad (6.72)$$

In (6.71), (6.72), the expansion of the logarithm for large r_s leads to the Gell-Mann–Brueckner expression of the free energy and of the chemical potential. These approximations were given, too.

The Brueckner parameter r_s is given by $r_s = (3/(4\pi n))^{1/3}/a_B$.

While the Hartree–Fock term, i.e., the $1/r_s$ term in (6.71), (6.72), was retained unaffected, the additional terms in these equations and in (6.67), (6.68) were modified, or fitted, respectively, such that (6.67), (6.68) meet the numerical data *in between*, where the analytical limiting formulae are not applicable. In Fig. 6.10, an example is given for the application of Padé formulae. The exact result was determined according to Sect. 6.3.1 and is in agreement with the latter.

We will still give the corresponding formulae for a gas of protons. Due to the fact that the proton mass is considerably higher, the protons are still Boltzmann particles where the electrons are fully degenerate. This leads to the fact that, in the low density regime, the proton formulae are practically the same as the electron ones, however, in the high density limit, we adjust the formulae to (classical) Monte Carlo (MC) data.

For the free energy density, we have

$$-\frac{f_p}{k_BTn_p} = \frac{(-f_p^{\mathrm{int}}/k_BTn_p)_{\mathrm{D}}[1 - a\tilde{n}_p^{2/3}(f_p^{\mathrm{int}}/k_BTn_p)_{\mathrm{MC}}]}{1 - a\tilde{n}_p^{1/2}\left[\tilde{n}_p^{1/2}/(\frac{f_p^{\mathrm{int}}}{k_BTn_p})_{\mathrm{D}} + \tilde{n}_p^{1/6}(\frac{f_p^{\mathrm{int}}}{k_BTn_p})_{\mathrm{D}}\right]}\,. \quad (6.73)$$

For applications, we need the chemical potential, which correspondingly reads

$$-\frac{\mu_p}{k_BT} = \frac{(-\mu_p^{\rm int}/k_BT)_{\rm D}[1-2a\tilde{n}_p^{2/3}(\mu_p^{\rm int}/k_BTn_p)_{\rm MC}]}{1-2a\tilde{n}_p^{1/2}\left[\tilde{n}_p^{1/2}/(\frac{\mu_p^{\rm int}}{k_BT})_{\rm D}+\tilde{n}_p^{1/6}(\frac{\mu_p^{\rm int}}{k_BT})_{\rm D}\right]}\,. \tag{6.74}$$

We now have to explain the abbreviations used. The function a depends only on the temperature

$$a = \sqrt{\pi^3 k_BT}\Big\{\frac{1}{2}\Big[1+\sqrt{\frac{k_BT}{4\pi}}\exp\Big(\frac{\sqrt{\pi}/2}{\ln(4/k_BT)^{1/6}-2/\sqrt{k_BT}}\Big)\Big]-0.29931\Big\}\,. \tag{6.75}$$

A dimensionless density n was introduced applying the Landau length which is used in the plasma parameter Γ

$$\tilde{n}_p = \frac{8}{(k_BT)^3}n_p \qquad \Gamma = \left(\frac{4}{3}\pi\tilde{n}_p\right)^{1/3}\,. \tag{6.76}$$

The Debye approximations (low density) for the free energy density and for the chemical potential are given by

$$(-f_p^{\rm int}/k_BTn_p)_D = 2.1605\tilde{n}_p^{1/2}\,, \tag{6.77}$$

and

$$(-\mu_p^{\rm int}/k_BT)_D = \frac{3}{2}2.1605\tilde{n}_p^{1/2}\,. \tag{6.78}$$

For the high density region, the formulae are fitted to (numerical) Monte Carlo data. We use

$$\begin{aligned}(f_p^{\rm int}/k_BTn_p)_{MC} &= -0.8946\Gamma+3.266\Gamma^{1/4}-0.5012\ln\Gamma-2.809\\ &\quad -\frac{r_s\tilde{n}_p^{1/3}}{1+r_s^2}\left[0.0933+1.0941\tilde{n}_p^{-1/4}-0.343\tilde{n}_p^{-1/3}\right]\,,\end{aligned} \tag{6.79}$$

and for the chemical potential

$$\begin{aligned}(\mu_p^{\rm int}/k_BT)_{MC} &= -1.1928\Gamma+3.5382\Gamma^{1/4}-0.5012\ln\Gamma-2.9761\\ &\quad -\frac{r_s\tilde{n}_p^{1/3}}{1+r_s^2}\left[0.0933+0.8206\tilde{n}_p^{-1/4}-0.2287\tilde{n}_p^{-1/3}\right]\,.\end{aligned} \tag{6.80}$$

The pressure for an H-plasma is then given by

$$p = p_e + p_p = n_e\mu_e + n_p\mu_p - f_e - f_p\,. \tag{6.81}$$

In Figs. 6.2 to 6.6, the expressions (6.67), (6.68) are represented. In addition, some numerical results are indicated in Fig. 6.7.

So far, the Padé formulae were constructed on the level of the Montroll–Ward approximation. It is possible to include ladder type diagrams and multiply charged ions, too, for the application to systems with stronger correlations and partially ionized systems, e.g., to be used in Sect. 6.5.6 and in Sect. 9.2.

Such formulae were given in Ebeling et al. (1991) and may be written as follows.

We use the subdivision for the free energy of an electron ion system

$$f_{Coul} = f_e + f_p + f_{ei} \,, \tag{6.82}$$

representing the electron gas, the ion gas and the electron–ion interaction, respectively.

In these Padé formulae, the parameters are defined as follows with n_e-electron density and n_i- ion densities

$$\tau = k_B T/Ryd, \qquad n_+ = \sum_i n_i, \qquad \xi_i = \frac{n_i}{n_+}, \qquad \gamma_i = \frac{m_e}{m_i},$$
$$\tilde{n} = n_e \Lambda_e^3, \qquad \hat{n} = \frac{8n_+}{\tau^3}, \qquad \bar{n} = \hat{n}\langle z\rangle \left\langle z^{5/3}\right\rangle^3, \qquad \xi = \frac{n_e}{n_+}, \tag{6.83}$$

and the charge average of arbitrary functions is defined as

$$\langle f(z)\rangle = \sum_i f(z_i)\xi_i \,. \tag{6.84}$$

We make the following choice for the electron part

$$f_e = -n_e \frac{f_0 \tilde{n}^{1/2} + f_3 \tilde{n} - f_2 \tilde{n}^2 \epsilon(r_s)}{1 + f_1 \tilde{n}^{1/2} + f_2 \tilde{n}^2} \,. \tag{6.85}$$

We used the abbreviations

$$\begin{aligned} f_0 &= \frac{2}{3}\left(\frac{\tau}{\pi}\right)^{1/4}, \qquad f_1 = \frac{\sqrt{2}/8 + 0.25\sqrt{\tau/\pi} - \tau/(8\sqrt{2})}{f_0}, \\ f_2 &= 3.0, \qquad\qquad f_3 = \frac{1}{4}\left(\frac{\tau}{\pi}\right)^{1/2}, \\ \epsilon(r_s) &= -\frac{0.9163}{r_s} - 0.1244 \ln\left(\frac{1 + 0.3008\sqrt{r_s} + 2.117/\sqrt{r_s}}{1 + 0.3008\sqrt{r_s}}\right). \end{aligned} \tag{6.86}$$

For the ion contribution, we have

$$f_i/k_B T = -n_+ \frac{q_0 \hat{n}^{1/2} + q_2 \bar{n}^{3/2} E_1(\bar{n})}{1 + q_1 (\hat{n}\tau)^{1/2} + q_2 \bar{n}^{3/2}} \tag{6.87}$$

with the coefficients

$$q_0 = \frac{2\sqrt{\pi}}{3}\left\langle z^2\right\rangle^{3/2}, \quad q_1 q_0 = \frac{\pi^{3/2}}{8}\sum_i\sum_j \xi_i\xi_j z_i^2 z_j^2(\gamma_i+\gamma_j)^{1/2}, \quad q_2 = 1000,$$
$$E_1(\bar{n}) = 1.4474\bar{n}^{1/3} - 4.2944\bar{n}^{1/12} + 0.6712\bar{n}^{-1/12} + 0.2726\ln(\bar{n}) + 2.983. \tag{6.88}$$

For the electron–ion term, the following formula is used

$$f_{ei}/k_BT = -n_+ \frac{Q_0\hat{n}^{1/2} + Q_2\tilde{n}^{3/2}E_2(\bar{n}, r_s, \tau)}{1+Q_1(\hat{n}\tau)^{1/2}+Q_2\tilde{n}^{3/2}+Q_4\hat{n}^{1/2}\ln(1+Q_5/\hat{n}^{1/2})}, \tag{6.89}$$

and the quantities are

$$Q_0 = \frac{2\pi^{1/2}}{3}\left[(\xi+\left\langle z^2\right\rangle)^{3/2} - \xi^{3/2} - \left\langle z^2\right\rangle^{3/2}\right], \quad Q_1 Q_0 = \frac{\pi^{3/2}}{2}\xi\left\langle z^2\right\rangle - q_1 q_0,$$
$$Q_2 = 1.0, \quad Q_4 Q_0 = 1.0472\left[\left\langle z^3\right\rangle - \xi\right]^2, \quad Q_5 = \exp\left[-\frac{Q_3}{Q_4}\right],$$
$$Q_3 Q_0 = 0.5236\left[\left\langle z^3\right\rangle - \xi\right]\left\{4\left\langle z^3\ln(z)\right\rangle + \left[\left\langle z^3\right\rangle - \xi\right]\ln(29.09\left[\left\langle z^2\right\rangle + \xi\right])\right\},$$
$$E_2(\bar{n}, r_s, \tau) = \frac{0.8511 r_s\bar{n}^{1/12}}{(1+0.3135\bar{n}^{-1/12})(1+1.137 r_s\tau^{1/2})}$$
$$+\frac{r_s\bar{n}^{1/3}(0.0726+0.0161 r_s)}{1+0.0887 r_s^2}. \tag{6.90}$$

Other thermodynamic quantities may be deduced from the free energy using thermodynamic relations. The pressure, e.g., follows from $p = -\partial F/\partial V|_{T=const}$.

6.4 Next Order Terms

6.4.1 e^4-Exchange and e^6-Terms

We go back to (6.15). So far we discussed the first two diagrams. In the preceding section we discussed the Montroll–Ward term (MW) the diagram of which has (at least) two potential lines. In the low density limit, the MW term is of the order $n^{3/2}$ corresponding to the Debye limiting law. The next order contribution contained in the MW term is n^2e^4. There is still the exchange term of the same order n^2e^4, and moreover the anomalous exchange term of order e^4, however the latter is of order n^3.

The terms mentioned are determined from formula (6.15) and read in diagrammatic form

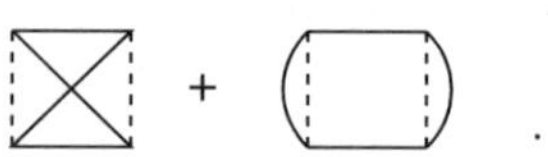

(6.91)

For the evaluation, we replaced the screened potentials by Coulomb ones. This is possible because the expressions remain convergent by such a replacement. The normal e^4-exchange term may be evaluated analytically in the weakly degenerate and in the highly degenerate cases. For low degeneracy, we have for the contribution to the mean value of the potential energy (Ebeling (a) et al. 1976)

$$\frac{1}{\mathcal{V}} \langle V \rangle_{E4N} = \beta\pi\sqrt{2} \ln 2 \sum_a \frac{z_a^2 e_a^4 \Lambda_a}{2s_a + 1} \,. \tag{6.92}$$

Here, the volume is denoted by $\mathcal{V}$. In the highly degenerate case, the corresponding expression was derived by Onsager, Mittag, and Stephen (1982) and reads

$$\frac{1}{\mathcal{V}} \langle V \rangle_{E4N} = \beta \frac{\pi}{2} \left[\frac{\ln 2}{3} - \frac{3}{2\pi^2} \zeta(3) \right] \sum_a \frac{(2s_a + 1) n_a e_a^4}{\Lambda_a^2} \,, \tag{6.93}$$

where $\zeta(x)$ is the Riemann zeta function. The normal e^4-exchange term was numerically evaluated for arbitrary degeneracy only recently (Vorberger et al. 2004).

The analytical evaluation of the anomalous exchange term is possible in the limiting situations. In the low degenerate case, we get an expression of order n^3 which reads

$$\frac{1}{\mathcal{V}} \langle V \rangle_{E4A} = -\beta \frac{2\pi}{3\sqrt{3}} \sum_a \frac{z_a^3 e_a^4 \Lambda_a^4}{(2s_a + 1)^2} \,. \tag{6.94}$$

This contribution does not have to be considered in the low density limit. In the high degeneracy case, the corresponding result reads

$$\frac{1}{\mathcal{V}} \langle V \rangle_{E4A} = -\beta \frac{3}{\pi} \sum_a \frac{(2s_a + 1) n_a e_a^4}{\Lambda_a^2} \,. \tag{6.95}$$

We want to mention that this contribution is of the same order as (6.93). The anomalous e^4 exchange term provides the necessary compensation at $T = 0$ (see Fetter and Walecka (1971), Stolzmann and Kraeft (1979)).

The numerical evaluation is possible for any degeneracy, too. The result is given in Riemann (1997).

Besides the Montroll–Ward term, there is still the e^6-term, the divergency of which has to be overcome by classical screening and then leads to a non-analytic contribution, namely $\ln n$. We write the diagrams with (at least) three potential lines. These terms are the three rung ladder terms including exchange terms. The diagrams have to be derived from (6.15) and have the shape

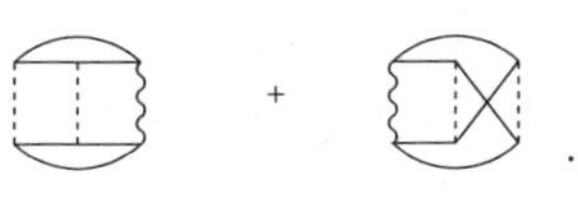

(6.96)

Again, we replaced two of the screened potentials by bare Coulomb ones. The remaining screened potential ensures the convergency. We mention here that there are still more terms of the low density limit order e^6, which are exchange terms of the types

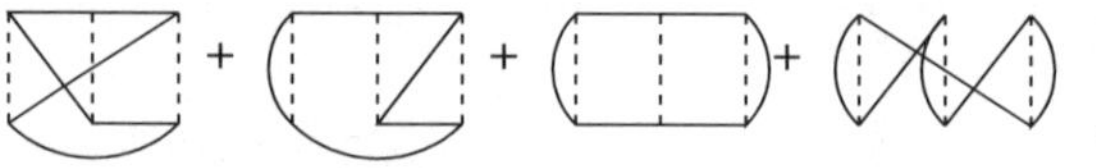

(6.97)

We do not consider these terms (6.97) as they are at least of the order n^3. However, there are terms with (at least) four potential lines. Due to the special character of the Coulomb potential, there is a lowering effect of the orders which reduces the diagrams from e^8 down to n^2e^6 and, thus, they have to be taken into account in the low density limit. In diagrammatic form, we have from (6.14)

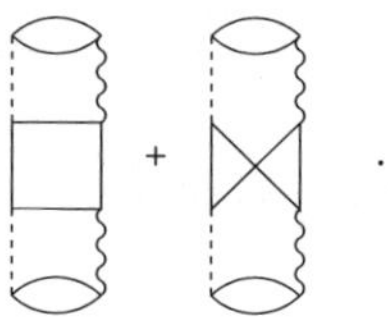

(6.98)

These terms have to be considered if one is interested in the correct expression for thermodynamic functions up to the order n^2. We do not want to go into the details of the (analytical) evaluation and refer to the literature (Ebeling (a) et al. 1976; Kraeft and Jakubowski 1978). The result is in the non-degenerate situation

$$\begin{aligned} \frac{1}{\mathcal{V}}\langle V\rangle_{D,E6} &= k_BT\pi\sum_{ab} z_a z_b \lambda_{ab}^3 \xi_{ab}^3 \left(\ln\kappa_D\lambda_{ab} + \frac{C}{2} + \ln 3 - \frac{1}{2} + \frac{1}{6}\right) \\ &- \beta^2\frac{\pi^3}{12}\sum_a \frac{z_a^2(Z_a e_a)^6}{2s_a+1}\,. \end{aligned} \quad (6.99)$$

The last contribution corresponds to the exchange e^6-term in lowest density order. The remaining e^6 exchange terms are anomalous ones and start with z^3. We have to mention that, at arbitrary degeneracy, none of the e^6 terms have been evaluated numerically so far.

6.4.2 Beyond Montroll–Ward Terms

In applications, e.g., in helio-seismology, it is of interest to have precise EOS data especially in the low density region. This is the case because the speed of sound, which determines the modes of solar oscillations, depends very sensitively on the EOS.

For this reason, it is desirable to have the precise coefficient of the density order $n^{5/2}$. For this purpose, we generalize the expression leading to the MW result. This means we have to take more general expressions for the dynamically screened potential V_s and for the polarization function Π, respectively.

In terms of Feynman graphs, it turns out to be necessary to consider the following diagrams for the mean value of the potential energy

$$\langle V \rangle_{\mathrm{BMW}} = \qquad (6.100)$$

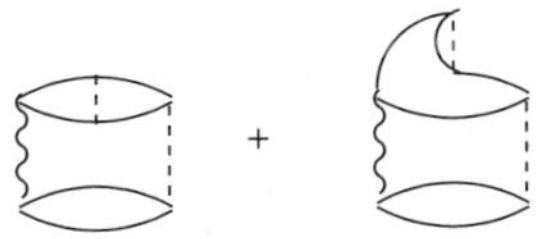

The diagrams correspond to self-energy contributions (6 pieces) and to vertex ones (3 pieces). The diagrams evolve from improvements of the simple RPA bubble in the simple RPA diagram (4+2 diagrams) and from the improvement of the polarization function in the screened potential in the simple RPA diagram (2+1 diagrams). In the low density limit, we get the following result (Riemann 1997)

$$\langle V \rangle_{\mathrm{BMW}} = -k_B T \frac{\kappa_D^3}{8\pi} \sum_a \frac{5}{2} \frac{(\kappa_a \lambda_{aa})^2}{2s_a + 1} \frac{\kappa_a^2}{\kappa_D^2} \,. \qquad (6.101)$$

Here we used

$$\kappa_a^2 = \frac{4\pi z_a e_a^2}{k_B T} \,.$$

Now we are able to give an expression for the pressure up to the order $(e^2 z)^{5/2}$ which reads

$$\begin{aligned}
\beta p(z,T) &= \sum_a z_a + \frac{\kappa^3}{12\pi} \\
&+ 2\pi \sum_{ab} z_a z_b \lambda_{ab}^3 \left[-\frac{\sqrt{\pi}}{8} \xi_{ab}^2 + \delta_{ab} \frac{(-1)^{2s_a}}{2s_a + 1} \left(\frac{\sqrt{\pi}}{4} + \frac{\xi_{ab}}{2} + \frac{\sqrt{\pi} \ln 2}{4} \xi_{ab}^2 \right) \right] \\
&+ \frac{\kappa^3}{12\pi} \sum_a \left[\frac{1}{8} \kappa_a^2 \lambda_{aa}^2 + \frac{3}{2(2s_a + 1)} \left(\kappa_a^2 \lambda_{aa}^2 - \frac{z_a \Lambda_a^3}{\sqrt{2}} \right) \frac{\kappa_a^2}{\kappa^2} \right] . \qquad (6.102)
\end{aligned}$$

The contributions up to z^2 were given in Ebeling (a) et al. (1976). For more details concerning the contributions of the order $z^{5/2}$ see Kraeft et al. (1998) and Kraeft et al. (2000). After inversion from fugacities to densities (incomplete inversion, see transition from (6.51) to (6.53) and later in Sect. 6.5.5) we get for the EOS (Riemann 1997; Kraeft et al. 2000)

$$\begin{aligned}
\beta p(n,T) &= \sum_a n_a - \frac{\kappa^3}{24\pi} \\
&- 2\pi \sum_{ab} n_a n_b \lambda_{ab}^3 \left[-\frac{\sqrt{\pi}}{8} \xi_{ab}^2 + \delta_{ab} \frac{(-1)^{2s_a}}{2s_a + 1} \left(\frac{\sqrt{\pi}}{4} + \frac{\xi_{ab}}{2} + \frac{\sqrt{\pi} \ln 2}{4} \xi_{ab}^2 \right) \right] \\
&- \frac{\kappa^3}{12\pi} \sum_a \left[\frac{3}{8} \kappa_a^2 \lambda_{aa}^2 + \frac{9}{4(2s_a + 1)} \left(\kappa_a^2 \lambda_{aa}^2 - \frac{n_a \Lambda_a^3}{\sqrt{2}} \right) \frac{\kappa_a^2}{\kappa^2} \right] . \qquad (6.103)
\end{aligned}$$

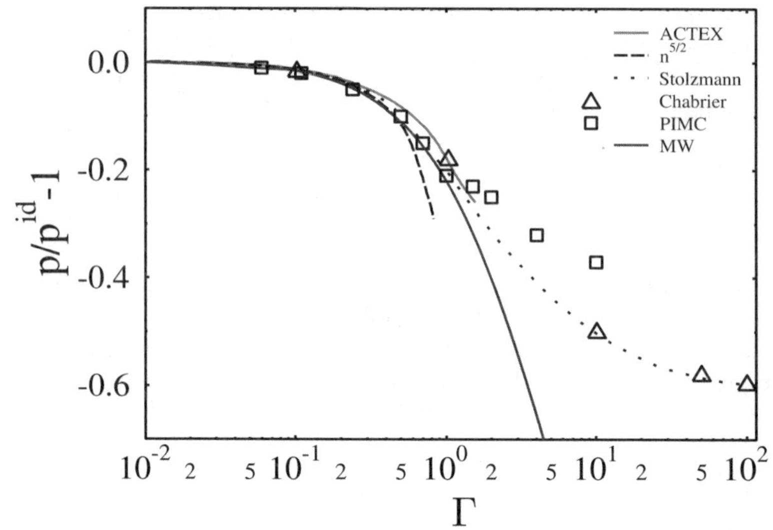

Fig. 6.11. Low density pressure as a function of Γ for a fixed density. *Upper full line* – ACTEX-OPAL (2000); *lower full line* – MW (Vorberger 2005; Vorberger et al. 2004); *dashes* – full expression (6.102) (Riemann et al. 1995); *dots* – Padé approximation (Stolzmann and Blöcker 1996); *triangles* – Chabrier (1990); *squares* – PIMC (Militzer and Ceperley 2001)

The direct terms of the orders $\hbar$ and $\hbar^2$ were given already by DeWitt (1962). Results of the order $n^{5/2}$ were also derived elsewhere; see Ebeling (1993), Alastuey et al. (1994a), Alastuey et al. (1994b), Alastuey and Perez (1996), Brown and Yaffe (2001), and references quoted therein. Exact low density expansions are of interest in helioseismology, see Christensen-Dalsgaard and Däppen (1992).

In Fig. 6.11 numerical results are shown for the low density expansion of the pressure, i.e., for the EOS, in different approximations.

6.5 Equation of State in Ladder Approximation. Bound States

6.5.1 Ladder Approximations of the EOS. Cluster Coefficients

In order to include stronger correlations and especially bound states into the EOS, we have to extend our approximation scheme beyond the weak coupling terms considered in the previous sections. It was shown in Chap. 5 that strong correlations such as bound states and multiple scattering can be described in the frame of ladder approximation.

We know from the previous section that the first Born terms of the ladder sum $n \leq 3$ are divergent for the long-range Coulomb interaction. Therefore, we introduced the dynamically screened ladder approximation.

As shown in Sect. 5.8, this is, in the case of dynamical screening, a very complicated approximation. However, in order to overcome the Coulomb divergencies, we need this approximation only for the first Born contributions of the order $n \leq 3$. The higher order terms are not divergent, and therefore it is meaningful to consider the ladder approximation as a *statically* screened approximation, or, for the Coulomb potential, as a subtracted ladder sum.

Moreover, the thermodynamics in ladder approximation is of importance for systems with short range interactions such as dense gases, liquids and nuclear matter. Therefore we will consider this type of approximation more in detail. Especially, two- and three-particle correlations are included leading to a cluster expansion of the EOS up to the third cluster coefficient.

To obtain the pressure we use the charging procedure. This means, we start from the relations (6.1) and (6.3). Here, the central quantity is the time specialized two-particle Green's function $g_{ab}(12, 1^{++}2^{+})|_{t_2=t_1^+} = g^{<}_{ab}(t_1, t_1)$. For simplicity, we will drop the space variables. The charging formula then reads

$$p - p_0 = \frac{1}{2\Omega} \sum_{a,b} \int_0^1 \frac{d\lambda}{\lambda} \mathrm{Tr}_{ab}\, \lambda V_{ab}\, g^{<}_{ab}(t_1, t_1; \lambda) \tag{6.104}$$

with p_0 being the ideal pressure given by (6.17). Therefore, we need approximations for the equilibrium two-particle density matrix $g_{ab}(t_1 - t_1')|_{t_1=t_1'}$. This quantity is connected to the Fourier transform of the two-particle correlation function $g^{<}_{ab}(t_1 - t_1')$ by

$$g^{<}_{ab}(t_1 - t_1')|_{t_1=t_1'} = \int \frac{d\omega}{2\pi}\, g^{<}_{ab}(\omega)\,. \tag{6.105}$$

Here we are interested in cluster expansions for $g^{<}_{ab}$ and for the pressure. The first contribution for such cluster expansion is the binary collision approximation $g^{<L}_{ab}$ which was considered intensely in Chap. 5.

An extension to three-particle collisions to be included into $g^{<}_{ab}$ was derived, e.g., by Klimontovich et al. (1987). In thermodynamic equilibrium, we have

$$g^{<}_{ab}(t, t) = g^{\mathrm{L}<}_{ab}(t, t) \mp i \sum_{c} \mathrm{Tr}_c \Big\{ g^{\mathrm{L}<}_{abc}(t, t) - g^{\mathrm{L}<}_{(ab)c}(t, t) \Big\} \tag{6.106}$$

for $t = t'$. To specify the ladder approximation we introduced the superscript L. Here, $g^{\mathrm{L}<}_{abc}$ accounts for the three-particle collisions, and $g^{\mathrm{L}<}_{(ab)c} = g^{\mathrm{L}<}_{ab} g^{<}_{c}$ are disconnected terms. Below, we will use the functions $\mathcal{G}^{\kappa\gtrless}(t, t')$ given in Table 6.1. There the superscript L is dropped. With this notation, we have $g^{\mathrm{L}<}_{(ab)c}(t, t') = \mathcal{G}^{3<}_{abc}(t, t')$.

Table 6.1. Quantities in the three-particle cluster coefficient

κ	$\mathcal{G}^{\kappa\gtrless}_{abc}$	V^{κ}_{abc}	V_{κ}
0	$g^{\gtrless}_a g^{\gtrless}_b g^{\gtrless}_c$	$V_{ab} + V_{ac} + V_{bc}$	0
1	$g^{\gtrless}_a g^{\gtrless}_{bc}$	$V_{ab} + V_{ac}$	V_{bc}
2	$g^{\gtrless}_b g^{\gtrless}_{ac}$	$V_{ab} + V_{bc}$	V_{ac}
3	$g^{\gtrless}_c g^{\gtrless}_{ab}$	$V_{ac} + V_{bc}$	V_{ab}

Using (6.105) and (6.106) in (6.104), we arrive at

$$\begin{aligned} p - p_0 &= \frac{1}{2\Omega} \sum_{a,b} \int_0^1 \frac{d\lambda}{\lambda} \mathrm{Tr}_{ab} \int \frac{d\omega}{2\pi} \lambda V_{ab} \Big\{ g^{\mathrm{L}<}_{ab}(\omega;\lambda) \\ &\quad \mp i \sum_c \mathrm{Tr}_c \Big[g^{\mathrm{L}<}_{abc}(\omega;\lambda) - g^{\mathrm{L}<}_{(ab)c}(\omega;\lambda) \Big] \Big\} . \end{aligned} \tag{6.107}$$

In this way, a cluster expansion for the pressure is obtained taking into account two- and three-particle ladder contributions. In the following, we introduce cluster coefficients according to

$$\beta(p - p_0) = \sum_{ab} z_a z_b \hat{b}_{ab} + \sum_{abc} z_a z_b z_c \hat{b}_{abc} , \tag{6.108}$$

where $z_a = \tilde{z}_a (2s_a + 1)/\Lambda_a^3$ are the modified fugacities with $\tilde{z}_a = \exp(\beta\mu_a)$ and $\beta = 1/k_B T$. Furthermore, we have $\hat{b}_{ab} = b_{ab} - b^0_{ab}$.

The second cluster coefficient

$$\hat{b}_{ab} = \frac{1}{z_a z_b} \frac{\beta}{2\Omega} \int_0^1 \frac{d\lambda}{\lambda} \mathrm{Tr}_{ab} \int_{-\infty}^{\infty} \frac{d\omega}{2\pi} \lambda V_{ab}\, g^{\mathrm{L}<}_{ab}(\omega;\lambda) \tag{6.109}$$

describes the influence of binary correlations on the pressure. The three-particle ladder contribution to the EOS is included in the third cluster coefficient $\hat{b}_{abc}$. Introducing the three-particle interaction potential $V^0_{abc} = V_{ab} + V_{ac} + V_{bc}$, we get

$$\begin{aligned} \hat{b}_{abc} &= \frac{1}{z_a z_b z_c} \frac{\mp i\beta}{3!\,\Omega} \int_0^1 \frac{d\lambda}{\lambda} \mathrm{Tr}_{abc} \\ &\quad \times \int_{-\infty}^{\infty} \frac{d\omega}{2\pi} \Big\{ \lambda V^0_{abc}\, g^{\mathrm{L}<}_{abc}(\omega;\lambda) - \sum_{\kappa=1}^{3} \lambda V_\kappa\, \mathcal{G}^{\kappa<}_{abc}(\omega;\lambda) \Big\} . \end{aligned} \tag{6.110}$$

The two-particle potentials V_κ and the correlation functions $\mathcal{G}^{\kappa<}_{abc}(t,t')$ are collected in Table 6.1. For further considerations, we use the spectral representation of the correlation functions for systems in thermodynamic equilibrium (KMS relations). For the two-particle correlation function in ladder approximation, we get

$$i^2 g_{ab}^{\mathrm{L}<}(\omega) = -n_{ab}(\omega)\, 2\mathrm{Im} i g_{ab}^R(\omega)\,, \tag{6.111}$$

where n_{ab} is the Bose-like function

$$n_{ab}(\omega) = \frac{1}{e^{\beta(\omega-\mu_a-\mu_b)}-1}\,. \tag{6.112}$$

According to our considerations in Sect. 5.3, the retarded Green's function g_{ab}^R is given by

$$i g_{ab}^R(\omega) = \frac{N_{ab}}{\omega - H_{ab}^{0eff} - N_{ab}V_{ab} + i\varepsilon}\,, \tag{6.113}$$

with $N_{ab} = 1 \pm f_a \pm f_b$ being the phase space occupation factor. Inserting the spectral representation (6.111) into (6.109), after a partial integration we get

$$\begin{aligned} \hat{b}_{ab} &= -\frac{1}{z_a z_b}\frac{1}{2\Omega}\int_0^1 \frac{d\lambda}{\lambda}\int_{-\infty}^{+\infty}\frac{d\omega}{\pi}\ln\left|1-e^{-\beta(\omega-\mu_a-\mu_b)}\right| \\ &\quad \times \mathrm{Im}\,\mathrm{Tr}_{ab}\left\{\lambda V_{ab}\frac{\partial}{\partial\omega} i g_{ab}^R(\omega)\right\}. \end{aligned} \tag{6.114}$$

Our next task is to carry out the charging procedure. For this purpose we neglect the self-energy shift in the effective Hamiltonian. The integration with respect to λ may be performed if we apply the identity

$$N_{ab}V_{ab}\frac{\partial}{\partial\omega} g_{ab}^R(\omega;\lambda) = -\frac{\partial}{\partial\lambda} g_{ab}^R(\omega;\lambda)\,. \tag{6.115}$$

Finally, the charging procedure leads to the following compact expression for the second cluster coefficient

$$\hat{b}_{ab} = \frac{1}{z_a z_b}\frac{1}{2\Omega}\int_{-\infty}^{+\infty}\frac{d\omega}{\pi}\ln\left|1-e^{-(\beta\omega-\mu_a-\mu_b)}\right|\mathrm{Im}F_{ab}(\omega+i\varepsilon)\,. \tag{6.116}$$

Here we introduced the "trace of the resolvent" $F_{ab}(z)$ by

$$F_{ab}(z) = i\mathrm{Tr}_{ab}\left\{\left[g_{ab}(z) - g_{ab}^0(z)\right]\frac{1}{N_{ab}}\right\}, \tag{6.117}$$

with the function $g_{ab}(z)$ being the analytical continuation defined by (6.113).

With (6.116), an interesting representation of the second cluster coefficient for degenerate quantum gases is found. The statistical and dynamical information contained in this expression are separated in the following manner. The statistical one is mainly given by the logarithm-term which demonstrates the Bose character of a pair of particles. The dynamical information about the two-particle system is contained in the trace of the resolvent where phase space occupation terms are included in the Hamiltonian.

Using $\mathrm{Im}\, ig^R_{ab}(\omega) = -\pi\delta(\omega - H^0_{ab} - N_{ab}V_{ab})$, we get an expression for $\hat{b}_{ab}$ identical to that obtained by Boercker and Dufty (1979) using reduced density operators. In the non-degenerate case, we get the well-known expression

$$\hat{b}_{ab} = \frac{1}{2\Omega}\frac{\Lambda_a^3}{(2s_a+1)}\frac{\Lambda_b^3}{(2s_b+1)}\mathrm{Tr}_{ab}\left\{e^{-\beta H_{ab}} - e^{-\beta H^0_{ab}}\right\}. \tag{6.118}$$

Let us now consider the third cluster coefficient $\hat{b}_{abc}$ given by (6.110) using the following spectral representation of the three-particle correlation function

$$\pm i^3 g^{L<}_{abc}(\omega) = -n_{abc}(\omega) 2\mathrm{Im} i^2 g^R_{abc}(\omega)\,. \tag{6.119}$$

Here, the retarded three-particle Green's function reads

$$i^2 g^R_{abc}(\omega) = \frac{1}{\omega - H^0_{abc} - V^0_{abc} + i\varepsilon}\,. \tag{6.120}$$

where phase space occupation effects are not included. Applying similar manipulations as for the second cluster coefficient, we find

$$\hat{b}_{abc} = -\frac{1}{3!\, z_a z_b z_c \Omega}\int\limits_{-\infty}^{+\infty}\frac{d\omega}{\pi}\ln\left|1 + e^{-\beta(\omega-\mu_a-\mu_b-\mu_c)}\right| \mathrm{Im}F_{abc}(\omega+i\varepsilon)\,. \tag{6.121}$$

The trace of the three-particle resolvent takes the form

$$F_{abc}(z) = i^2\mathrm{Tr}_{abc}\left\{g_{abc}(z) - g^0_{abc}(z) - \sum_{\kappa=1}^{3}\left[g^\kappa_{abc}(z) - g^0_{abc}(z)\right]\right\}. \tag{6.122}$$

We have $g^\kappa_{abc}(\omega + i\varepsilon) = -(\omega - H^0_{abc} - V_\kappa + i\epsilon)^{-1}$ with V_κ given in Table 6.1. Again, we observe a separation of dynamical and statistical information. Here, the logarithm-term demonstrates the Fermi character of the three-particle complex. The expansion for small fugacities leads to the known expression of the third cluster coefficient for non-degenerate quantum systems, i.e.,

$$\begin{aligned}\hat{b}_{abc} &= \frac{1}{3!\,\Omega}\frac{\Lambda_a^3}{(2s_a+1)}\frac{\Lambda_b^3}{(2s_b+1)}\frac{\Lambda_c^3}{(2s_c+1)} \\ &\quad\times\mathrm{Tr}_{abc}\left\{\left(e^{-\beta H_{abc}} - e^{-\beta H^0_{abc}}\right) - \sum_{\kappa=1}^{3}\left(e^{-\beta H^\kappa_{abc}} - e^{-\beta H^0_{abc}}\right)\right\}\end{aligned} \tag{6.123}$$

with the Hamiltonian $H^\kappa_{abc} = H^0_{abc} + V_\kappa$ in the channel κ.

An important and frequently used approximation of the EOS is the approximation of the second cluster coefficient. Let us, therefore, consider this

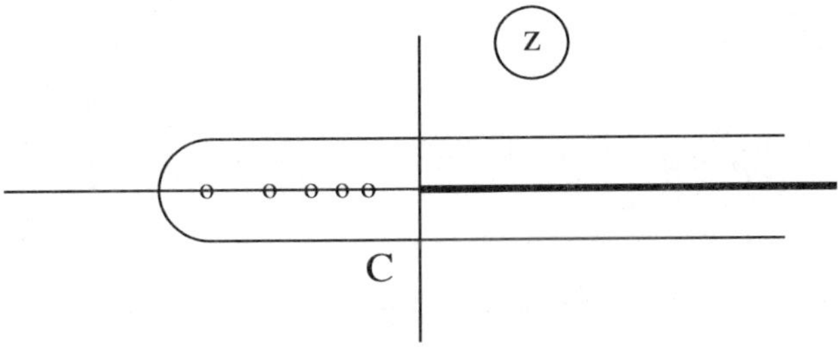

Fig. 6.12. Contour of integration in the complex energy plane

quantity more in detail. The second cluster coefficient given by (6.116) is essentially determined by the trace of the resolvent. Using the Bethe–Salpeter equation ($g_{ab} = g^0_{ab} + ig^0_{ab}T_{ab}g^0_{ab}$), we immediately get a representation in terms of the T-matrix

$$F_{ab}(z) = -i\mathrm{Tr}\Big\{ \Big(\frac{\partial}{\partial z} g^0_{ab}(z) \Big) T_{ab}(z) \Big\} . \tag{6.124}$$

From the bilinear expansion of T_{ab} or g_{ab} given in Sect. 5.4 and 5.5, we find the following interesting properties of the trace of the resolvent:

(i) $F_{ab}(z)$ has poles at the negative ω-axis at E^{ab}_n, i.e., at the positions of the bound state energies of the two particles a and b.
(ii) $F_{ab}(z)$ has a branch cut along the positive ω-axis corresponding to the energies of scattering states.

Consequently, $\hat{b}_{ab}$ may be given by the complex representation (see also Kraeft et al. (1969))

$$\hat{b}_{ab} = \frac{1}{z_a z_b} \frac{1}{2\Omega} \int_C \frac{dz}{2\pi i} \ln \left| 1 - e^{-\beta(z-\mu_a-\mu_b)} \right| F_{ab}(z) \tag{6.125}$$

with $F_{ab}(z)$ given by (6.117) and C being the contour of integration. There are various interesting and useful representations of the second cluster coefficient. We will derive some of them here. For this reason we start from (6.116) and consider $\mathrm{Im}F_{ab}(\omega + i\varepsilon)$. One can easily verify that it may be written in the shape

$$\begin{aligned} \mathrm{Im}F_{ab}(\omega+i\varepsilon) \;&=\; -\mathrm{Tr}_{ab}\Big\{ \frac{d}{d\omega}\Big[\mathrm{Im}\, ig^0_{ab}(\omega+i\varepsilon)\,\mathrm{Re}T_{ab}(\omega+i\varepsilon)\Big] \\ &-\; \mathrm{Im}\, ig^0_{ab}(\omega+i\varepsilon)\,\frac{d}{d\omega}\mathrm{Re}T_{ab}(\omega+i\varepsilon) \\ &+\; \frac{d}{d\omega}\mathrm{Re}\, ig^0_{ab}(\omega+i\varepsilon)\,\mathrm{Im}T_{ab}(\omega+i\varepsilon)\Big\} . \end{aligned} \tag{6.126}$$

Replacing the derivative $d\mathrm{Re}T/d\omega$ by means of the *differentiated optical theorem* following from (5.102), one readily verifies that $\mathrm{Im}F_{ab}(\omega+i\varepsilon)$ has the form

$$\mathrm{Im} F_{ab}(\omega + i\epsilon) = -\mathrm{Tr}_{ab}\left\{ \frac{d}{d\omega}\Big[\mathrm{Im}\, i g^0_{ab}(\omega+i\varepsilon)\, \mathrm{Re} T_{ab}(\omega+i\varepsilon)\Big] \right.$$
$$\left. -2\mathrm{Im}\Big[\mathrm{Im}\, i g^0_{ab}(\omega+i\varepsilon)\, \frac{d}{d\omega} T^*_{ab}(\omega+i\varepsilon)\, \mathrm{Im}\, i g^0_{ab}(\omega+i\varepsilon) T_{ab}(\omega+i\varepsilon)\Big]\right\}. \tag{6.127}$$

We get a very compact representation of $\hat{b}_{ab}$ if we introduce the S-matrix by

$$S_{ab}(\omega) - 1 \;=\; 2i\, \mathrm{Im}\, i g^0_{ab}(\omega+i\varepsilon)\, T_{ab}(\omega+i\varepsilon)\,. \tag{6.128}$$

After some algebra, an S-matrix representation follows for the second cluster coefficient

$$\hat{b}_{ab} = \frac{1}{z_a z_b}\frac{1}{2\Omega}\int\limits_{-\infty}^{+\infty} d\omega \ln\left|1 - e^{-\beta(\omega-\mu_a-\mu_b)}\right| \, Tr_{ab}\, \frac{1}{2}\,\mathrm{Im}\left[\frac{dS^{-1}(\omega)}{d\omega}\, S(\omega)\right]. \tag{6.129}$$

This expression was first given by Dashen et al. (1969) for the special case of non-degenerate many-particle systems. One may find the more general expression in the paper by Kremp (a) et al. (1984). Let us discuss expression (6.129). On behalf of the relation

$$\mathrm{Im}\, i g^0_{ab}(\omega+i\varepsilon)|_{\varepsilon\to 0} = -\frac{\varepsilon N_{ab}}{(\omega - H^0_{ab})^2+\varepsilon^2} = -\pi\delta(\omega - H^0_{ab}) N_{ab}\,, \tag{6.130}$$

one could argue that $\hat{b}_{ab}$ is determined by S- or T-matrix elements given *on-shell.* But, this is true only if the T-matrix itself is not singular. As shown by its bilinear expansion (see Sect. 5.5), the T-matrix remains finite for the scattering spectrum. However, it has singularities if bound states exist. In the latter case, there are isolated poles outside the scattering spectrum at the bound state energies E_n. Therefore, it is convenient to separate the bound state part. Using the bilinear expansion of $T_{ab}(\omega{+}i\varepsilon)$, or that of $g_{ab}(\omega{+}i\varepsilon)$, respectively, for the region with bound states, i.e., for frequencies from $\omega = -\infty$ up to the continuum, we have

$$\mathrm{Im} F^{\mathrm{bound}}_{ab}(\omega+i\varepsilon) = \sum_{\boldsymbol{P},n} \pi\delta(\omega - E^{ab}_{n\boldsymbol{P}})\sigma_{ab,n}\,. \tag{6.131}$$

The energies are $E^{ab}_{n\boldsymbol{P}} = E_n + P^2/2M$ with P being the center-of-mass momentum and M the total mass, respectively. The quantity $\sigma_{ab,n}$ is given by Kremp (a) et al. (1984)

$$\sigma_{ab,n} = (2s_a+1)(2s_b+1)\left\{1 \pm \frac{\delta_{ab}}{2s_b+1} e^{i\phi n}\right\}. \tag{6.132}$$

The second term in the curly brackets accounts for the exchange part where the upper sign refers to Bose particles and the lower one to Fermi particles.

If an energy gap exists between discrete and continuous eigenvalues of the energy spectrum we may separate bound and scattering contributions according to

$$\hat{b}_{ab} = b_{ab}^{\text{bound}} + b_{ab}^{\text{scatt}} \tag{6.133}$$

with the bound state contribution

$$b_{ab}^{\text{bound}} = -\frac{1}{2\Omega}\frac{1}{z_a z_b}\sum_{\boldsymbol{P},n} \ln\left|1 - e^{-\beta(E_n + \frac{P^2}{2M} - \mu_a - \mu_b)}\right| \sigma_{ab,n} \tag{6.134}$$

and with the scattering state part

$$b_{ab}^{\text{scatt}} = \frac{1}{z_a z_b}\frac{1}{2\Omega}\int\limits_0^\infty \frac{d\omega}{\pi} \ln\left|1 - e^{-\beta(\omega - \mu_a - \mu_b}\right| Tr_{ab}\frac{1}{2}\,\text{Im}\left[\frac{dS^{-1}}{d\omega} S\right] . \tag{6.135}$$

For the scattering state spectrum, $T_{ab}(\omega + i\epsilon)$ remains finite and, therefore, we may apply the relation (6.130) in (6.127) and (6.128), respectively. Then one can see clearly that in (6.135) only the *on-shell* S-matrix appears. We will demonstrate this fact explicitly. For this purpose, it is useful to apply again the T-matrix instead of the S-matrix. With (6.128) we get, by a straightforward calculation, the following *on-shell representation* of the EOS (Kremp (a) et al. 1984)

$$\begin{aligned} p - p_0 = &-\frac{k_B T}{2\Omega}\sum_{ab}\sum_{\boldsymbol{P},n} \ln\left|1 - e^{-\beta(E_{n\boldsymbol{P}}^{ab} - \mu_a - \mu_b)}\right| \sigma_{ab,n} \\ &+\frac{k_B T}{2\Omega}\sum_{ab}(2s_a+1)(2s_b+1)\sum_{\boldsymbol{p}_a\boldsymbol{p}_b} N_{ab}\Big\{\beta n_{ab}(E_{ab}) \\ &\times \text{Re}\,\langle \boldsymbol{p}_a\boldsymbol{p}_b|T_{ab}(E_{ab})|\boldsymbol{p}_b\boldsymbol{p}_a\rangle^{\pm} + \pi \ln\left|1 - e^{-\beta(E_{ab} - \mu_a - \mu_b)}\right| \sum_{\boldsymbol{p}'_a\boldsymbol{p}'_b} \delta(E_{ab} - E'_{ab}) \\ &\times \text{Im}\Big[\langle \boldsymbol{p}_a\boldsymbol{p}_b|\,\frac{d}{dE_{ab}}T^*_{ab}(E_{ab} + i\varepsilon)\,|\boldsymbol{p}'_b\boldsymbol{p}'_a\rangle \\ &\times N'_{ab}\,\langle \boldsymbol{p}'_a\boldsymbol{p}'_b|\,T_{ab}(E_{ab} - i\varepsilon)\,|\boldsymbol{p}_b\boldsymbol{p}_a\rangle^{\pm}\Big]\Big\} , \end{aligned} \tag{6.136}$$

where $n_{ab}(E_{ab})$ is the Bose function given by (6.112) with $E_{ab} = p_a^2/2m_a + p_b^2/2m_b$. The expression represents a very general shape of the EOS in binary collision approximation. Many-particle effects such as degeneracy occur at different places:

(i) We have a Bose-type shape of the two-particle sum of states both in the bound and in the scattering state contributions of the pressure. Especially, the bound state contribution is different from zero only for interaction potentials with an attractive part. One can see that the bound states, in general, behave approximately like Bose particles. The singularity of this term for the case

$$E^{ab}_{n\boldsymbol{P}} = E_n + \frac{\boldsymbol{P}^2}{2M} = \mu_a + \mu_b \tag{6.137}$$

is well-known to be connected with the Bose–Einstein condensation, here as that of *bound states of pairs* of particles.

(ii) The factors N_{ab} describe phase space occupation effects of two interacting particles in a dense system.
For two Fermi particles it describes the so-called Pauli blocking.

(iii) There is a density dependence of the energy levels E_n and of the T-matrix T_{ab} following from the phase space occupation terms N_{ab} included in the effective Schrödinger equation and in the generalized Lippmann–Schwinger equation.

The EOS (6.136) is a generalization of the EOS given in the textbook by Landau and Lifshits (1977). If we neglect degeneracy effects and replace the T-matrix in well-known manner by the scattering amplitude, we get the second virial (cluster) coefficient (Landau and Lifshits 1977). A T-matrix representation for non-degenerate systems was given in different papers (Baumgartl 1967; Kremp et al. 1971; Boercker and Dufty 1979). In the case of zero temperature, the theory includes that of the ground state of Fermi systems developed by Galitski and Migdal (1958).

Let us now consider the EOS under the further approximation that the phase space occupation effects concerning the particles a and b are neglected while the Bose character of bound and scattering states is retained. Then the factors N_{ab} are equal to unity, and it is convenient to separate the motion of the center-of-mass with

$$E_{ab} = \frac{p^2}{2m_{ab}} + \frac{P^2}{2M}\,, \tag{6.138}$$

where $\boldsymbol{p}$ is the relative momentum and m_{ab} is the reduced mass, respectively. Then, in momentum representation, we get the following equation

$$\begin{aligned} b^{\text{scatt}}_{ab} &= \frac{(2s_a+1)}{z_a}\frac{(2s_b+1)}{z_b}\int\limits_0^\infty \frac{d\omega}{2\pi}\int \frac{d\boldsymbol{P}}{(2\pi\hbar)^3}\,d\boldsymbol{p}\,d\bar{\boldsymbol{p}} \\ &\times\ \ln\left|1-e^{-\beta(\omega-\mu_a-\mu_b)}\right|\,\frac{1}{2}\,\mathrm{Im}\left[\langle\boldsymbol{p}|\frac{dS^{-1}(\omega)}{d\omega}|\bar{\boldsymbol{p}}\rangle\langle\bar{\boldsymbol{p}}|S(\omega)|\boldsymbol{p}\rangle\right]. \end{aligned} \tag{6.139}$$

It is often useful to determine b^{scatt}_{ab} in angular momentum decomposition. For this purpose, one uses the partial wave series for the S-matrix given by

$$\langle\boldsymbol{p}|S(\omega)|\bar{\boldsymbol{p}}\rangle = \frac{1}{p\,m_{ab}}\,\delta(\omega - E_p - \frac{P^2}{2M})\sum_l \frac{2l+1}{4\pi}P_l(\cos\theta)S_l(\omega) \tag{6.140}$$

with $\cos\theta = \boldsymbol{p}\cdot\bar{\boldsymbol{p}}/(p\bar{p})$ and $S_l(\omega) = \exp(2i\delta_l(\omega))$. In the *on-shell* case considered, S^{-1} is the same operator as $S^\dagger$. Then the series (6.140) gives rise to a double series for the second virial coefficient as a sum of products of Legendre polynomials. If this is integrated over all angles, the orthogonality of the Legendre polynomials

$$\int_{-1}^{1} d\cos\theta P_l(\cos\theta)P_{l'}(\cos\theta) = \frac{2}{2l+1}\delta_{ll'} \tag{6.141}$$

leads to the cancellation of cross terms. Furthermore, we may carry out certain integrations due to the occurrence of delta functions. After some algebra, the scattering contribution is found to be

$$\begin{aligned} b_{ab}^{\text{scatt}} &= -\frac{(2s_a+1)}{z_a}\frac{(2s_b+1)}{z_b}\int\frac{d\boldsymbol{P}}{(2\pi\hbar)^3}\sum_{l=0}^{\infty}(2l+1)\left[1\pm\delta_{ab}\frac{(-1)^l}{2s_a+1}\right] \\ &\times \frac{1}{2\pi}\int dE \ln\left|1-e^{-\beta(E+\frac{P^2}{2M}-\mu_a-\mu_b)}\right|\frac{d}{dE}\delta_l(E)\,. \end{aligned} \tag{6.142}$$

In the Boltzmann case, the integration with respect to the total momentum can be performed, and together with the bound state contribution (6.134), we get the famous and well-known Beth–Uhlenbeck representation of the second cluster coefficient, see Beth and Uhlenbeck (1936), Landau and Lifshits (1977), Huang (1963),

$$\begin{aligned} \hat{b}_{ab} &= 4\pi^{3/2}\lambda_{ab}^3\sum_0^{\infty}(2l+1)\left[1\pm\delta_{ab}\frac{(-1)^l}{2s_a+1}\right] \\ &\times\left\{\sum_{n\geq l+1}e^{-\beta E_{nl}}+\frac{1}{\pi}\int_0^{\infty}e^{-\beta E}\frac{d}{dE}\delta_l(E)\right\} \end{aligned} \tag{6.143}$$

with the thermal wave length $\lambda_{ab} = (\hbar^2/2m_{ab}k_BT)^{1/2}$.

Let us come back to the complex representation (6.125) of $\hat{b}_{ab}$. After separation of the center-of-mass motion, we may write

$$\hat{b}_{ab} = \frac{1}{z_a z_b}\frac{1}{2\Omega}\int_C\frac{dz}{2\pi i}\int\frac{d\boldsymbol{P}}{(2\pi\hbar)^3}\ln\left|1-e^{-\beta(z+\frac{P^2}{2M}-\mu_a-\mu_b)}\right|F_{ab}^{\text{rel}}(z)\,. \tag{6.144}$$

Here, C is the contour of integration around the spectrum of H_{ab}^{rel}. For the trace of the resolvent $F_{ab}^{\text{rel}}(z)$, we have

$$F_{ab}^{\text{rel}}(z) = (2s_a+1)(2s_b+1)\sum_{l=0}^{\infty}(2l+1)\left[1\pm\delta_{ab}\frac{(-1)^l}{2s_a+1}\right]F_l(z)\,. \tag{6.145}$$

The partial wave contribution $F_l(z)$ has the following properties:

(i) $F_l(z)$ has simple poles located at the bound state energies E_{nl} of $H_{ab}^{\rm rel}$;
(ii) $F_l(z)$ has a branch cut along the scattering energy spectrum of $H_{ab}^{\rm rel}$ with the jump

$$\rho_l(E) = F_l(E+i\varepsilon) - F_l(E-i\varepsilon) = -2i\,\frac{d}{dE}\delta_l(E)\,. \tag{6.146}$$

From these properties, we can find

$$F_l(z) = \frac{d}{dz}\ln D_l(z)\,, \tag{6.147}$$

where $D_l(z)$ is the Jost function for complex energies. The latter is related to the S-matrix elements according to

$$S_l(E) = (-1)^l\,\frac{D_l(E-i\varepsilon)}{D_l(E+i\varepsilon)}\,.$$

The complex representation (6.144) has several advantages. Bound states do not appear explicitly, i.e., bound and scattering states are treated together. Furthermore, general properties of $\hat{b}_{ab}$ may be investigated on the basis of its complex representation (Kraeft et al. 1969; Kremp et al. 1971).

6.5.2 Bound States. Levinson Theorem

In the previous subsection, we expressed the EOS in ladder approximation in terms of scattering quantities. Such representations are very useful for the numerical evaluation and for the investigation of general properties of the second virial coefficient b_{ab}. Of special interest is the influence of the bound states on the EOS. The possibility of the formation of bound states and the number of bound states are determined by the sign and by the strength of the coupling between the particles, or, more simple, by the coupling parameter λ. This fact may be discussed, for example, for the simple square-well potential and for the Debye potential. We know from textbooks of quantum mechanics that bound states occur (or vanish) at critical values of $V_0 d^2$ for the square well potential and critical values of r_D/a_B for the Debye potential.

It may be expected that the bound state part of the second viral coefficient has discontinuities at the critical values of the coupling parameter. There it may be further expected that one might observe drastic changes in the EOS.

This behavior was discussed for the square well potential by Kremp et al. (1977) and by Gau et al. (1981). We will consider here the behavior of the second virial coefficient for the Debye potential.

The scattering part, the bound state part, and the total second virial coefficient were calculated as functions of r_D/a_B. They are presented in Figs. 6.13 and 6.14. The results are given as $B_{ep}/(4^{3/2}\lambda_{ep}^3)$.

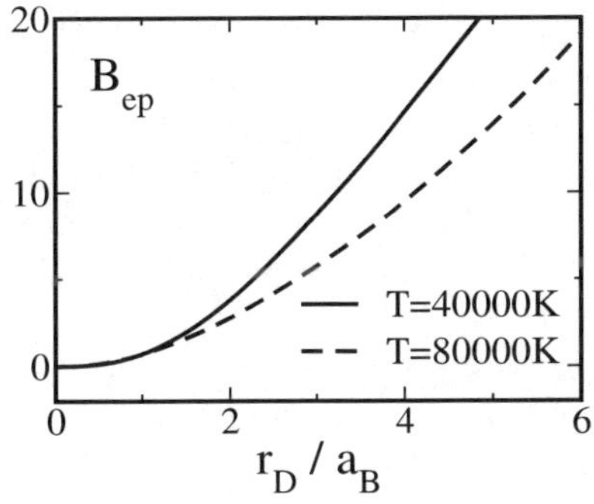

Fig. 6.13. The electron-proton second virial coefficient as a function of the coupling parameter r_D/a_B for different temperatures

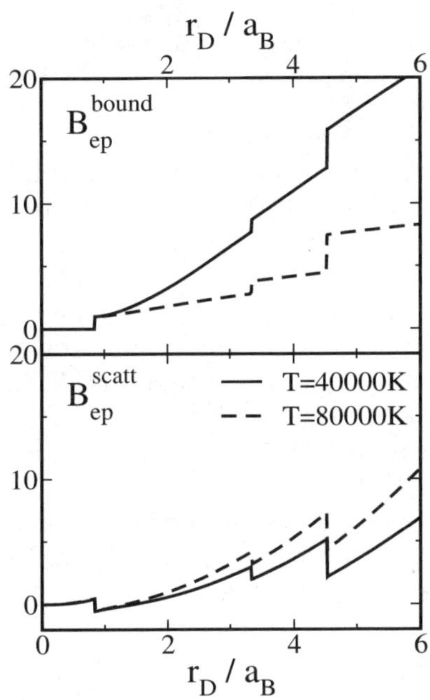

Fig. 6.14. The bound and scattering parts of the second virial coefficient as a function of the coupling parameter r_D/a_B for the electron–proton interaction

As expected, b_{ab}^{bound} shows discontinuities, if new bound states appear. It is interesting and surprising that b_{ab}^{sc} shows identical jumps, too, however with opposite sign. As a result, we have a compensation of the discontinuities, and the total virial coefficient is a smooth function of the coupling parameter r_D/a_B. This result is far more general. Indeed, one can show the following theorem to be valid (Kremp et al. 1971):

Theorem:

If the potential $V(r)$ satisfies the conditions

$$\int_0^\infty drr|V(r)| < \infty; \qquad \int_0^\infty drr^2|V(r)| < \infty \,,$$

the second virial coefficient is an analytic function of the coupling parameter.

From its analyticity with respect to λ, there follows the possibility of an expansion with respect to λ, i.e., the 2nd virial coefficient may be determined by perturbation methods with respect to powers of the interaction parameter λ both in the cases $\lambda > 0$ and $\lambda < 0$. This means in particular that the perturbation method is applicable for situations in which bound states appear. This is in contrast to the fact that it is impossible to determine bound state eigenfunctions by perturbation theory.

As a consequence of the analyticity of $b_l(T, \lambda)$, we may determine $b_l(-|\lambda|)$ from $b_l(|\lambda|) = \bar{b}_l^{\mathrm{sc}}(|\lambda|)$ by replacing $|\lambda|$ by $-|\lambda|$, i.e.,

$$b_l(-|\lambda|) = \bar{b}_l^{\mathrm{sc}}(-|\lambda|) = b_l^{\mathrm{bound}}(-|\lambda|) + b_l^{\mathrm{sc}}(-|\lambda|)\,.$$

This means, $b_l^{\mathrm{sc}}(-|\lambda|)$ may be subdivided in the following manner

$$b_l^{\mathrm{sc}}(-|\lambda|) = \bar{b}_l^{\mathrm{sc}}(-|\lambda|) - b_l^{\mathrm{bound}}(-|\lambda|)\,.$$

Consequently, the scattering contribution of $b_l(-|\lambda|)$ consists of the analytic part $\bar{b}_l^{\mathrm{sc}}(-|\lambda|)$ which arises from the continuation of $b_l(|\lambda|)$, and of a part b_l^{bound} which compensates the bound state contribution. The compensation between the bound states and a part of the scattering states follows straightforwardly from the analyticity of $b_l(\lambda)$.

In the past, this problem of compensation between the bound states and a part of the scattering states was discussed by several authors. This problem was dealt with by Gibson (1971), Ebeling (1969a), Ebeling (1969b), Rogers (1971), Petschek (1971), Kremp and Kraeft (1972), and by Nussenzveig (1973).

For special potentials, the compensation between bound and scattering states may be discussed in more detail. We assume that the potential fulfills the condition (which is the case, e.g., for the Coulomb potential)

$$\frac{d}{dk}\delta_l(k,\lambda) = (\mathrm{sign}\lambda)\left|\frac{d}{dk}\delta_l(k,\lambda)\right|\,. \tag{6.148}$$

Under this condition, the function b_l is given by

$$b_l(\lambda) = \Theta(-\lambda)f_1(|\lambda|) + (\mathrm{sign}\lambda)f_2(|\lambda|)\,. \tag{6.149}$$

Here, the first part is the bound state contribution, and the second one is the scattering state contribution. The function $b_l(\lambda)$ is analytic, and thus we have

$$b_l(\lambda) = a_0 + a_1\lambda + a_2\lambda^2 + \cdots\,.$$

Let us first determine the function f_2. For this purpose, we consider the repulsive case ($\lambda > 0$), and obtain

$$b_l(|\lambda|) = f_2(|\lambda|) = a_0 + a_1|\lambda| + a_2|\lambda|^2 + \cdots\,.$$

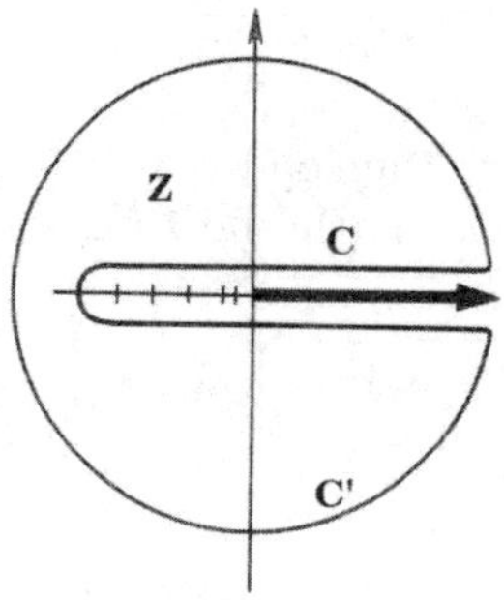

Fig. 6.15. Contour of integration in the complex z-plane

The bound state part is assumed to have the series

$$f_1(|\lambda|) = b_0 + b_1|\lambda| + b_2|\lambda|^2 + \cdots .$$

Writing all expressions for the attractive case ($\lambda < 0$), we get

$$a_0 - a_1|\lambda| + a_2|\lambda|^2 - a_3|\lambda|^3 \pm \cdots = b_0 + b_1|\lambda| + b_2|\lambda|^2 + b_3|\lambda|^3 + \cdots$$
$$-(a_0 + a_1|\lambda| + a_2|\lambda|^2 + a_3|\lambda|^3 + \cdots) .$$

Thus, we arrive at

$$b_0 = 2a_2, \quad b_1 = 0, \quad b_2 = 2a_2, \quad b_3 = 0 \cdots . \tag{6.150}$$

These latter equations may be regarded as a prescription for the calculation of the bound state part of $b_l(\lambda)$ from scattering quantities only.

The effect of compensation can be demonstrated explicitly by the Nussenzveig-pole-representation of the S-matrix (Nussenzveig 1973; Kremp et al. 1977; Gau et al. 1981), or by the higher order Levinson theorems.

Higher order Levinson theorems follow from the discussion of the expression

$$\int_C \frac{dz}{2\pi} z^N F(z) = I , \tag{6.151}$$

where $F(z)$ is the trace of the resolvent (6.117). For $N = 0$, this equation corresponds to the 2nd virial coefficient in the limit $T \to \infty$. The contour of integration encircling the real axis of the complex z-plane is given in Fig. 6.15. For the further discussion of I, it would be useful to replace the contour C by the contour C' which represents a circle with $|z| \to \infty$. From the theory of analytic functions, it is well-known that this is possible under the condition (Jordan lemma)

$$\lim_{z\to\infty} F(z) z^{N+1} = 0 . \tag{6.152}$$

In order to study the behavior of $F(z)z^{N+1}$ for large $|z|$, let us consider the perturbation expansion

$$F(z) = \sum_{m=1}^{\infty} F^m(z) = \sum_{m=1}^{\infty} Tr\left[\frac{1}{H_0 - z}\left(-V\frac{1}{H_0 - z}\right)^m\right]. \tag{6.153}$$

Evidently, because of the first terms of this expansion, the condition (6.152) is not fulfilled for $F(z)$. Therefore, it is obvious to define the new function $F^{>p}(z)$ by subtraction of the first p terms

$$F^{>p}(z) = F(z) - \sum_{m=1}^{p} F^m(z). \tag{6.154}$$

Then, depending on the potential and on N, a number $p(N) = p$ may be determined such that the Jordan lemma is fulfilled for $F^{>p}$, and we immediately have the interesting relation

$$\int_{C'} dz z^N F^{>p}(z) = 0. \tag{6.155}$$

On the other hand, we can again use the contour C and the properties of the trace of the resolvent. Then it follows

$$\sum_n E_n^N + \int dE E^N \rho^{<p}(E) = 0, \tag{6.156}$$

where ρ is given by (6.146). These equations are a system of sum rules known as the higher order Levinson theorems (Bollé 1979).

Levinson theorems are very important in the scattering theory. Moreover, they are extremely useful to study the properties of the second virial coefficient (6.116); see Ebeling (a) et al. (1976), Kremp and Kraeft (1972), Bollé (1979).

In order to apply the Levinson theorems, we have to find the relation between p and N. For this purpose, we have to study the high energy behavior of the terms of the expansion (6.153). From the definition of $F^{>p}(z)$, it may be expected that the asymptotic behavior for large z is determined by z^{-p-2}, i.e., we have $p = N$. But this is not correct because $\mathrm{Tr}V^m$ is divergent. Therefore, we have to investigate this problem more carefully.

Let us consider

$$\begin{aligned} F^m(z) &= Tr\left(\frac{1}{H_0 - z}\left[V\frac{1}{H_0 - z}\right]^m\right) \\ &= \int \frac{d\boldsymbol{p}}{(2\pi)^3}\left\langle \boldsymbol{p}\left|\frac{1}{H_0 - z}\left(V\frac{1}{H_0 - z}\right)^m\right|\boldsymbol{p}\right\rangle. \end{aligned} \tag{6.157}$$

Using the completeness of the states $|r\rangle$ and taking into account the simplification

$$V(r)V(\bar{r})V(\bar{\bar{r}})\cdots = V^m(r),$$

we immediately get

$$F^m(z) = 4\pi \int_0^\infty \frac{d\boldsymbol{p}}{(2\pi)^3} \frac{p^2}{(\mathbf{p}^2/2\mu - z)^{m+1}} \int d\boldsymbol{r} V^m(r)\,. \tag{6.158}$$

The integral over momenta can be carried out, and we have

$$F^m(z) \to -(z)^{-m+1/2} \frac{(2m-3)!!}{2^{m-1} m!} \int \frac{d\boldsymbol{r}}{8\pi} |V(r)|^m\,. \tag{6.159}$$

For $m = 1$, this relation already gives the exact result

$$F^1 = -\left(-z^{-1+1/2}\right) \frac{1}{8\pi} \int d\boldsymbol{r} V(r)\,. \tag{6.160}$$

Of course these relations are valid only for a class of potentials for which the integrals over $|V(r)|^m$ are well defined. For such potentials, the relation between p and N is determined by $p = N + 1$.

In a plasma, the relevant potential is the screened Coulomb potential. For this potential, the integrals are well defined only for $m = 1, 2$. According to (6.159), the correct result for $m = 1$ is

$$F^1(z) = \frac{1}{\sqrt{z}} \frac{e^2 r_0^2}{2}\,. \tag{6.161}$$

The asymptotic behavior for $m = 2$ is given by

$$F^2(z) \to \frac{1}{z\sqrt{z}} \int \frac{dr}{8\pi} V^2(r)\,. \tag{6.162}$$

Because of the singularity of the screened Coulomb potential at $r = 0$, the higher integrals are divergent. A more careful discussion for the screened Coulomb potential gives, for $m = 3$, the asymptotic behavior

$$F^3(z) \to \frac{1}{-z^{5/2}} \ln z\,. \tag{6.163}$$

Furthermore, we can find from considerations in Sect. 6.5.3, formulae (6.183) and (6.193) for $m > 3$

$$F^m(z) = const \frac{i\zeta(m-2)}{2} (-z)^{-m/2-1}\,. \tag{6.164}$$

Using these results, the $p = p(N)$ relation for systems with screened Coulomb interaction is given by the following table

N	0	1	2	3	4
p	1	2	3	6	8

(6.165)

Now we are able to consider the sum rules (6.156) in more detail. Let us first discuss the case $N = 0$. Then we obtain from (6.156)

$$\sum_n 1 = -\int_0^\infty dE\rho^{>1}(E)\,. \tag{6.166}$$

For the further calculations, it is convenient to introduce the function

$$\sigma(E) = \int_0^E dE'\rho(E') \tag{6.167}$$

and the corresponding quantities $\sigma^{>p}(E)$. It is clear that $\sigma^{>p}$ has the following properties:

$$\lim_{E\to\infty} \sigma^{>p}(E) = 0\,, \qquad \lim_{E\to\infty} E^N\sigma^{>p}(E) = 0\,. \tag{6.168}$$

For spherically symmetric potentials, $\sigma(E)$ is just the sum of the scattering phases

$$\sigma(E) = \sum_{\ell=0}(2\ell+1)\frac{1}{\pi}\delta_\ell(E) = \sum(2\ell+1)\sigma_\ell(E)\,. \tag{6.169}$$

Combining the last relations with (6.166), we get the result

$$\sum_n 1 = -\sigma^{>1}(E)_{E=0} = \sigma(0)\,. \tag{6.170}$$

Here, we took into account the relation $\sigma^1(0) = 0$. We are now ready to establish the first order Levinson theorem. For a spherically symmetric potential, the phase shifts satisfy the conditions

$$\begin{aligned} N &= \sum_\ell (2\ell+1)\frac{1}{\pi}\delta_\ell(0)\,, \\ n_\ell &= \frac{1}{\pi}\delta_\ell(0)\,. \end{aligned} \tag{6.171}$$

Here, N denotes the number of bound states, and n_ℓ is the number of bound states with the angular momentum number ℓ. These relations are a very important result of scattering theory. Moreover, Levinson's theorem is equivalent to the following properties of the second virial coefficient:

$$\lim_{T\to\infty} b_\ell = \lim_{T\to\infty} (b_\ell^{\text{bound}} + b_\ell^{\text{sc}}) = 0\,; \tag{6.172}$$

on the other hand we have

$$\lim_{T\to\infty} b_\ell^{\text{bound}} = n_\ell \qquad \lim_{T\to\infty} b_\ell^{\text{sc}} = -\frac{1}{\pi}\delta_\ell(0)\,. \tag{6.173}$$

Again it can be seen that both the bound state part and the scattering part of the second virial coefficient are no continuous functions of the interaction parameter at $\lambda = 0$. They show a discontinuity with a step n_ℓ. But the full virial coefficient is a continuous function. Bound and scattering parts are, therefore, of equal importance.

For plasmas, it is interesting to consider the next order of Levinson theorems, too. In the case $N = 1$, from the general set of Levinson theorems (6.156) we arrive at

$$\sum_n E_n = -\int_0^\infty dE E\rho^{>2}(E)\,. \qquad (6.174)$$

Taking into account the properties of $\sigma^{>p}(E)$, we find after a partial integration

$$\begin{aligned}\sum_n E_n &\equiv \int_0^\infty dE\ \sigma^{>2}(E) \\ &= \sum_{\ell=0}(2\ell+1)\frac{1}{\pi}\int_0^\infty dE\delta_\ell(E) - \int_0^\infty dE(\sigma^1(E)+\sigma^2(E))\,. \qquad (6.175)\end{aligned}$$

This equation is the second order Levinson theorem and can be interpreted as a sum rule for the bound state energies.

As already outlined, the Levinson theorems are appropriate to analyze the second virial coefficient. In particular, the relation between bound and scattering states may be investigated.

We start with the primary subdivision of b_2 for non-degenerate systems, i.e., with the Beth–Uhlenbeck formula (6.143)

$$\begin{aligned}b_2 &= 4\pi^{3/2}\lambda_{ab}^3\sum_\ell(2\ell+1)\left[1\pm\frac{(-1)^\ell}{2s_a+1}\delta_{ab}\right] \\ &\times \left\{\sum_{n\geq\ell+1}^{\infty} e^{-\beta E_{n\ell}} + \int_0^\infty e^{-\beta E}\varrho_\ell(E)dE\right\}\,. \qquad (6.176)\end{aligned}$$

In order to apply Levinson's theorems, we introduce $\varrho^{>1}(E)$ and get

$$b_\ell = \sum_{n\geq\ell+1} e^{-\beta E_{n\ell}} + \int_0^\infty e^{-\beta E}\varrho_\ell^{>1}(E)dE + \int_0^\infty e^{-\beta E}\varrho_\ell^1(E)dE\,, \qquad (6.177)$$

where b_ℓ is defined by the expression in curly bracket of (6.176). After a partial integration and application of (6.169),(6.171), we immediately get

$$b_\ell = \sum_{n \geq \ell+1} \left(e^{-\beta E_{n\ell}} - 1\right) + \beta \int_0^\infty e^{-\beta E} \sigma_\ell^{>1}(E) dE + \int_0^\infty e^{-\beta E} \varrho_\ell^1(E) dE \,. \quad (6.178)$$

After a further partial integration in the last integral and taking into account $\sigma^1(0) = 0$, we get a new subdivision of the second virial coefficient. We find

$$b_\ell = \sum_{n \geq \ell+1} \left(e^{-\beta E_{n\ell}} - 1\right) + \frac{\beta}{\pi} \int_0^\infty dE e^{-\beta E} \delta_\ell(E) \,. \quad (6.179)$$

Thus, we saw that scattering state contributions lead to a re-normalization of the sum of bound states.

This procedure may be continued using higher order Levinson theorems. For this purpose we write (6.179) in the shape

$$b_\ell = \sum_{n \geq \ell+1} \left(e^{-\beta E_{n\ell}} - 1\right) + \beta \int_0^\infty dE e^{-\beta E} \sigma_\ell^{>2}(E) + b_\ell^{(1)} + b_\ell^{(2)} \,. \quad (6.180)$$

Then another partial integration is performed, and the second order Levinson theorem is applied; we arrive at a further essential decomposition of the second virial coefficient

$$\begin{aligned} b_\ell &= \sum_{n \geq \ell+1} \left(e^{-\beta E_{n\ell}} - 1 + \beta E_{n\ell}\right) \\ &\quad -\beta^2 \int_0^\infty dE \int_E^\infty d\bar{E} \exp(-\beta E) \sigma_\ell^{>2}(\bar{E}) + b_\ell^{(1)} + b_\ell^{(2)} \,. \quad (6.181) \end{aligned}$$

This procedure may be further continued. We find that the different stages of Levinson theorems correspond to different arbitrary subdivisions of the second virial coefficient. The complete quantity of the total virial coefficient is not affected by such manipulations. However, depending on the problem, the different subdivisions may be of different use. For systems with Coulomb interaction, formula (6.181) will be of special importance. The expression

$$Z^{PL} = \sum_\ell \sum_{n \geq \ell+1} (2\ell + 1) \left\{e^{-\beta E_{n\ell}} - 1 + \beta E_{n\ell}\right\} \quad (6.182)$$

is known to be the Planck–Larkin sum of bound states Z^{PL} and has the following useful properties:

(i) Z^{PL} is continuous if a bound state disappears.
(ii) Z^{PL} is convergent even for the Coulomb potential.
(iii) At low temperatures, the sum of bound states Z^{PL} only is a rough approximation for the complete second virial coefficient.

6.5.3 The Second Virial Coefficient for Systems of Charged Particles

In order to illustrate the results of the previous section, let us consider the second virial coefficient for a pure Coulomb system. We start form (6.125). In the non-degenerate case, after performance of a partial wave expansion and integration over center-of-mass momenta, this equation takes the shape

$$\begin{aligned} b_{ab}(T) &= 4\pi^{3/2}\lambda_{ab}^3 \sum_{\ell=0}^{\infty}(2\ell+1)\left[1 \pm \frac{(-1)^\ell}{2s_a+1}\right] \\ &\times \frac{1}{2\pi i}\int_C e^{-\beta z} F_\ell(z)dz\,. \end{aligned} \tag{6.183}$$

From the scattering theory on Coulomb potentials (see Ebeling (a) et al. (1976), Bertero and Viano (1967)), we have for the function $F_\ell(z)$

$$F_\ell(z) = -\frac{d}{dz}\ln D_\ell^c(z) - \frac{\alpha\hbar}{2\sqrt{2m_{ab}}}\frac{1}{4\pi}\int_0^\infty dE\, \frac{1-\ln\left(\frac{2R}{\hbar}\sqrt{2m_{ab}E}\right)}{(E-z)E^{3/2}}\Bigg|_{R\to\infty}\,. \tag{6.184}$$

Here, $D_\ell^c(z)$ is the Jost function in the complex energy plane $D_\ell^c(z) = f_\ell^c(k)$; $z = \hbar^2k^2/(2m_{ab})$. The function $D_\ell^c(z)$ is given by

$$D_\ell^c(z) = \frac{C_\ell}{\Gamma\left(\ell+1+\frac{ie_ae_b\sqrt{2m_{ab}}}{2\hbar\sqrt{z}}\right)}\,, \tag{6.185}$$

and the logarithmic derivative reads

$$-\frac{d}{dz}\ln D_\ell^c(z) = \frac{d}{dz}\ln\Gamma\left(\ell+1+\frac{ie_ae_b\sqrt{2m_{ab}}}{2\hbar\sqrt{z}}\right) \tag{6.186}$$

with Γ being Euler's Gamma-function. As compared to the result achieved earlier (6.147), we now have an additional term which arises from the characteristic behavior of Coulomb scattering states.

From the well known properties of the function Γ, we get the following behavior of $F_\ell(z)$

(i) $F_\ell^c(z)$ is an analytic function.
(ii) $F_\ell^c(z)$ has a branch cut along the positive real axis, and has the jump there, namely

$$\begin{aligned} \varrho_\ell^c(E) &= \frac{1}{\pi}\left\{\frac{d}{dE}\eta_\ell(E)\right. \\ &\left.-\frac{\alpha\hbar}{2\sqrt{2m_{ab}}}\frac{d}{dE}\frac{1}{\sqrt{E}}\ln(2R\sqrt{2m_{ab}E}/\hbar)\Big|_{R\to\infty}\right\}. \end{aligned} \tag{6.187}$$

Here, $\alpha = 2m_{ab}e_ae_b/\hbar^2$.

(iii) $F_\ell^c(z)$ has poles at

$$z = E_{\ell n} = -\frac{e_a^2 e_b^2 m_{ab}}{2\hbar^2 n^2}; \quad n = 1, 2, 3, \cdots,$$

with the principal quantum number n.

One main purpose of this subsection is the investigation of the nature of the divergencies which occur. The logarithmic derivative of Γ is equal to Euler's ψ function $\psi(x) = \frac{d}{dx} \ln \Gamma(x)$ the properties of which we are going to apply. The second virial coefficient may then be written as a contour integral (with $t = \beta z$, $u = \beta E$, see Fig. 6.15)

$$\begin{aligned} b_{ab} &= 2\pi\lambda_{ab}^3 \sum_{\ell=0}^{\infty} (2\ell+1) \left[1 \pm \delta_{ab} \frac{(-1)^\ell}{2s_a+1}\right] \\ &\times \left\{ \frac{1}{2\sqrt{\pi}} \int_C e^{-t} \frac{\xi_{ab}}{2\pi i} \int_0^\infty \frac{1 - \ln\left(\frac{2R\sqrt{u}}{\lambda_{ab}}\right)}{(u-t)u^{3/2}} |_{R\to\infty} du dt \right. \\ &+ \left. \frac{1}{4\sqrt{\pi}} \int_C dt e^{-t} t^{-3/2} \xi_{ab}\, \psi\left(\ell + 1 - \frac{i}{2\sqrt{t}} \xi_{ab}\right) \right\}. \end{aligned} \tag{6.188}$$

The Born parameter ξ is given by (6.50). For the evaluation of (6.188), it is useful to expand the ψ-function in terms of $x = -i\xi_{ab}/(2\sqrt{t})$. We apply the following relations

$$\begin{aligned} \psi(\ell+1+x) &= \psi(\ell+1) + \sum_{p=1}^{\infty} \frac{x^p}{p!} \psi^{(p)}(\ell+1) \\ \psi(\ell+1) &= -C - \sum_{k=1}^{\infty} \left(\frac{1}{\ell+1+k} - \frac{1}{k+1} \right) \\ \psi^{(p)}(\ell+1) &= (-1)^{p+1} p! \sum_{k=0}^{\infty} \frac{1}{(\ell+1+k)^{p+1}}, \quad p \neq 0. \end{aligned} \tag{6.189}$$

The exponents in brackets are derivatives.

It is known that the (direct) contributions of the virial coefficient up to the order e^6, or ξ_{ab}^3, respectively, are divergent if screening is not taken into account. On the other hand, the higher orders beginning with e^8 or ξ_{ab}^4, do not have (classical) long range divergencies, and the (classical) short range divergencies are avoided automatically by quantum effects. For practical purposes, we denote the terms up to ξ^3 by $P'b_{ab}$, and all higher orders by $P''b_{ab}$. We are going to determine them separately and discuss the general features below. The higher order terms are written as

$$P''b_{ab} = 2\pi\lambda_{ab}^3 \frac{1}{4\sqrt{\pi}} \int_C dt\ t^{-3/2} e^{-t} f(\sqrt{t},\, \xi_{ab},\, s_a)\,, \tag{6.190}$$

where we introduced the auxiliary function

$$f(\sqrt{t},\, \xi_{ab},\, s_a) = \sum_{\ell=0}^{\infty} (2\ell+1) \left[1 \pm \delta_{ab} \frac{(-1)^\ell}{(2s_a+1)}\right] \xi_{ab} \sum_{m=3}^{\infty} \frac{x^m}{m!} \psi^{(m)}(\ell+1)\,. \tag{6.191}$$

As $x \sim \xi_{ab}$, this expression obviously begins with ξ_{ab}^4 and contains all higher orders in the coupling parameter $e_a e_b$. Using (6.189), this auxiliary function (6.191) may be rewritten in several steps

$$\begin{aligned} &f(\sqrt{t},\, \xi_{ab},\, s_a) \\ &= \sum_{\ell=0}^{\infty} (2\ell+1) \left[1 \pm \delta_{ab} \frac{(-1)^\ell}{2s_a+1}\right] \frac{\xi_{ab}}{x} \sum_{p=4}^{\infty} (-x)^p \sum_{k=0}^{\infty} \frac{1}{(\ell+1+k)^p} \\ &= \frac{\xi_{ab}}{x} \sum_{p=4}^{\infty} (-x)^p \sum_{\tau=1}^{\infty} \frac{1}{\tau^p} \sum_{\ell=0}^{\tau-1} \left[1 \pm \delta_{ab} \frac{(-1)^\ell}{2s_a+1}\right] (2\ell+1) \\ &= \frac{\xi_{ab}}{x} \sum_{p=4}^{\infty} (-x)^p \left\{ \sum_{\tau=1}^{\infty} \frac{1}{\tau^{p-2}} \pm \delta_{ab} \frac{1}{2s_a+1} \sum_{\tau=1}^{\infty} \frac{(-1)^{\tau+1}}{\tau^{p-1}} \right\}. \end{aligned} \tag{6.192}$$

The τ-summations are replaced by Riemann ζ-functions which lead to

$$\sum_{\tau=1}^{\infty} \frac{1}{\tau^p} = \zeta(p)\,, \quad p > 1\,; \quad \sum_{\tau=1}^{\infty} \frac{(-1)^{\tau+1}}{\tau^{p-1}} = (1-2^{2-p})\zeta(p-1)\,.$$

Then, only the summation over the powers of the Born parameter remains, and, for the convergent part of the second virial coefficient for Coulomb systems, we get

$$\begin{aligned} P''b_{ab} &= 2\pi^{3/2}\lambda_{ab}^3 \sum_{p=4}^{\infty} \left(\frac{\xi_{ab}}{2}\right)^p \Big[\zeta(p-2) \\ &\pm \frac{\delta_{ab}}{2s_a+1} \left(1-2^{2-p}\right) \zeta(p-1)\Big] \frac{1}{2\pi i} \int_C dt\ e^{-t} (-t)^{-\frac{p}{2}-1}\,. \end{aligned} \tag{6.193}$$

The integral over the contour may now be carried out using the representation of the Gamma function

$$\frac{1}{\Gamma(y)} = \frac{1}{2\pi i} \int_{0+} dt\ e^{-t} (\alpha t)^{-y}\,; \quad \alpha = -1\,, \quad \sqrt{\alpha} = -i\,.$$

In this representation, the positive real axis has to be encircled in the mathematical negative sense. As the integrals in (6.193) are free of poles, the path of integration may be identified to that occurring in the definition of $1/\Gamma$, and we get

$$\begin{aligned} P''b_{ab} &= 2\pi^{3/2}\lambda_{ab}^3 \sum_{p=4}^{\infty} \frac{1}{\Gamma\left(\frac{p}{2}+1\right)} \left(\frac{\xi_{ab}}{2}\right)^p \\ &\times \left[\zeta(p-2) \pm \frac{\delta_{ab}}{2s_a+1}\left(1-2^{2-p}\right)\zeta(p-1)\right]. \end{aligned} \tag{6.194}$$

This expression is convergent and, in particular, does not contain Coulomb divergencies; see, e.g., Kraeft et al. (1973).

Let us now consider the lower order terms which are known to exhibit divergencies at long distances. These divergencies are avoided by the use of a screened potential. Let us consider these contributions in detail. It is a consequence of (6.188) that the first three powers of ξ_{ab} which were not included in (6.190), (6.191), read

$$\begin{aligned} P'b_{ab} &= 2\pi\lambda_{ab}^3 \left\{ \frac{\xi_{ab}}{2} \sum_{\ell=0}^{\infty} (2\ell+1)\left[1 \pm \delta_{ab}\frac{(-1)^\ell}{2s_a+1}\right] \frac{1}{2\sqrt{\pi}} \right. \\ &\times \int_c dt e^{-t} \left[\frac{1}{\pi i} \int_0^\infty \frac{1-\ln(2R\sqrt{u}/\lambda_{ab})}{(u-t)u^{3/2}} |_{R\to\infty} du \right. \\ &+ \left. \frac{1}{t^{3/2}} \left(\sum_{k=0}^{\infty} \frac{1}{k+1} - \sum_{k=0}^{\infty} \frac{1}{\ell+k+1} - C \right) \right] \\ &+ \frac{\sqrt{\pi}}{\Gamma(2)} \left(\frac{\xi_{ab}}{2}\right)^2 \left[\sum_{k=1}^{\infty} 1 + \frac{\delta_{ab}}{2s_a+1} \ln 2 \right] \\ &+ \left. \frac{\sqrt{\pi}}{\Gamma\left(\frac{5}{2}\right)} \left(\frac{\xi_{ab}}{2}\right)^3 \left[\sum_{k=1}^{\infty} \frac{1}{k} \pm \frac{\delta_{ab}}{2s_a+1} \frac{1}{2} \xi(2) \right] \right\}. \end{aligned} \tag{6.195}$$

This expression includes the orders ξ, ξ^2 and ξ^3, among them "direct" and exchange terms. The latter exist only for $a = b$ and can easily be recognized by the factor $\delta_{ab}/(2s_a+1)$. The exchange terms are convergent, however, the "direct" terms are divergent. The term linear in ξ which is referred to as the Hartree term, is cancelled in electro-neutral systems, the divergence of the ξ^2 and ξ^3 "direct" terms can be seen from

$$\sum_{k=1}^{\infty} 1 \to \infty\,, \qquad \sum_{k=1}^{\infty} \frac{1}{k} \to \infty\,.$$

While the exchange contributions of the orders ξ, ξ^2, ξ^3 may be determined from (6.195) along the lines given leading to (6.194), the diverging "direct" terms of the orders ξ^2 and ξ^3 are characteristic of Coulomb systems, i.e., the second virial coefficient does not exist for Coulomb systems. The problem is dealt with in the usual manner by taking into account additional contributions leading to screening and thus to convergent results. The lower order "direct" contributions including screening were dealt with already in Sect. 6.3. We want to repeat that the terms of higher orders $\xi^n, n \geq 4$, are convergent.

We are now going to discuss the results we achieved so far. We can compose the second virial coefficient from the contributions (6.195) and (6.194) and write

$$b_{ab}(\lambda) = 2\pi^{3/2}\lambda_{ab}^3 \sum_{p=1}^{\infty} a_p \left(\frac{\xi_{ab}}{2}\right)^p . \tag{6.196}$$

According to the peculiarity of the Coulomb potential, the coefficient a_1 is rather complicated and corresponds to the first three lines of the r.h.s. of (6.195). The coefficient a_1 accounts for the Hartree and for the Hartree–Fock contributions. The Hartree term is divergent, however, on behalf of electroneutrality, it is compensated by the other species or by a neutralizing background. The Hartree–Fock exchange term is convergent. It was discussed in Sect. 6.3.1 and is given by (6.40). It leads to $a_1^{\mathrm{exch}} = 1$. The next term a_2 is given by

$$a_2 = \frac{1}{\Gamma(2)} \sum_{k=1}^{\infty} \pm \frac{\delta_{ab}}{2s_s + 1} \ln 2 .$$

Of course, the direct term is divergent, as a reflection of the long range character of the Coulomb interaction.

The general form of the coefficients reads for $p \geqslant 3$

$$a_p = \frac{\zeta(p-2)}{\Gamma(\frac{p}{2}+1)} \pm \frac{\delta_{ab}}{2s_a+1}(1-2^{2-p})\frac{\zeta(p-1)}{\Gamma(\frac{p}{2}+1)} , p \geq 3 . \tag{6.197}$$

Here, as already discussed after (6.195), the direct term a_3 is divergent, too, as $\zeta(1) \to \infty$.

For the further consideration, we give still the connection between a_2 and the bound state energy

$$a_2 \left(\frac{\xi_{ep}}{2}\right)^2 = -\frac{1}{2} \sum_n n^2 \frac{E_n}{kT} . \tag{6.198}$$

Here, we took into account the formula for the Coulomb levels $E_n = -e^4\mu_{ab}/(2\hbar^2 n^2)$.

It is interesting to consider the structure of the second virial coefficient especially in the case of attractive potentials $\lambda < 0$, $\xi_{ab} > 0$.In this case, it is useful to subdivide the second virial coefficient into bound and scattering contributions. Because the Coulomb potential has the property (6.148), it is possible to write the second virial coefficient in the following form

$$b(\xi_{ab}) = 2\pi^{3/2}\lambda_{ab}^3 \sum_{p=1}^{\infty} a_p(\frac{\xi_{ab}}{2})^p = \Theta(\xi_{ab}) f_1(|\xi_{ab}|) - (sign\xi_{ab}) f_2(-|\xi_{ab}|) . \tag{6.199}$$

According to the consideration carried out in Sect. 6.5.2, the function $f_2(-|\xi_{ab}|)$ is determined from the condition

$$b(\xi_{ab}) = f_2(-|\xi_{ab}|) \quad \text{for} \quad \xi_{ab} < 0$$

to be

$$f_2(-|\xi_{ab}|) = 2\pi^{3/2}\lambda_{ab}^3 \sum_{p=1}^{\infty} a_p \left(\frac{-|\xi_{ab}|}{2}\right)^p .$$

The function f_1 corresponds to the bound states and is assumed to have the series expansions

$$f_1(\xi_{ab}) = \sum b_n {\xi_{ab}}^n . \tag{6.200}$$

Then we have the connection, in agreement with the previous consideration (6.150),

$$b_n = \begin{cases} 2a_n, & n \ - \ \text{even;} \\ 0, & n \ - \ \text{odd}, \end{cases}$$

and the bound state contribution is given by

$$f_1(\xi_{ab}) = 4\pi^{3/2}\lambda_{ab}^3 \sum_{k=1}^{\infty} \frac{\zeta(2k-2)}{\Gamma(k+1)} \left(\frac{\xi_{ab}}{2}\right)^{2k} . \tag{6.201}$$

Using $\Gamma(k+1) = k!$, $\zeta(2k-2) = \sum_{s=1}^{\infty} 1/s^{2k-2}$, and the Coulomb levels $-\beta E_n = \xi_{ep}^2/(4n^2)$, for the bound state part we get the expected expression

$$f_1(\xi_{ab}) = 4\pi^{3/2}\lambda_{ab}^3 \sum_{s=1}^{\infty} s^2(e^{-\beta E_s} - 1) . \tag{6.202}$$

We have to mention that (6.202) is divergent.

For further discussions of the ladder approximation, we consider the expression (6.194)

$$P''b = 4\pi^{3/2}\lambda_{ab}^3 \Bigg[\theta(\xi_{ab}) \sum_{s=1}^{\infty} s^2(e^{-\beta E_s} - 1)$$

$$-\text{sign}(\xi_{ab}) \sum_{p=1}^{\infty} a_p \left(\frac{-|\xi_{ab}|}{2}\right)^p - \sum_{p=1}^{3} a_p \left(\frac{\xi_{ab}}{2}\right)^p \Bigg] .$$

This equation may easily be rearranged using formula (6.198) to give

$$P''b = 4\pi^{3/2}\lambda_{ab}^3 \left[\theta(\xi_{ab}) \sum_{s=1}^{\infty} s^2 (e^{-\beta E_s} - 1 + \beta E_s) \right.$$

$$\left. -sign(\xi_{ab}) \sum_{p=4}^{\infty} a_p \left(\frac{-|\xi_{ab}|}{2} \right)^p \right] . \qquad (6.203)$$

Our relation corresponds to expression (6.181) derived applying the Levinson theorem. Again we observe the appearance of the famous Planck-Larkin sum of states.

6.5.4 Equation of State in Dynamically Screened Ladder Approximation

Let us come back to the equation of states for plasmas. We start our further consideration with the screened ladder approximation (6.15) derived in Sect. 6.2.

So far we considered this expression essentially in Montroll–Ward approximation. In this approximation, the main physical effect of the dynamical screening is taken into account. However, there are important physical effects which are not included in the Montroll–Ward approximation. In many cases, we have to deal with partially ionized plasmas. Therefore, bound states occur between the plasma particles. Bound states are of special interest. They essentially influence the properties of a plasma. In the preceding subsection we learned that the simplest approximation to take into account bound states is the ladder approximation. Because of the Coulomb divergencies in plasmas, we need a *screened* ladder approximation.

From the representation (6.15), it can be seen that, in contrast to the consideration in the previous section, in the screened version, the full ladder sum does not exist; two terms are missing. Taking into account that the contributions of the orders e^4 and e^6 are divergent for a Coulomb potential and including the considerations in Sect. 6.4, the following subdivision of the sum of diagrams is convenient

$$p - p_0 = \frac{1}{2\Omega} \sum_{ab} \int \frac{d\lambda}{\lambda} \Big\{$$ 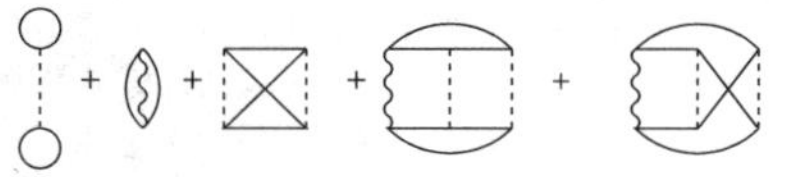 $$\Big\}$$

$$+ \frac{1}{2\Omega} \sum_{ab} \int \frac{d\lambda}{\lambda} \Big\{$$ 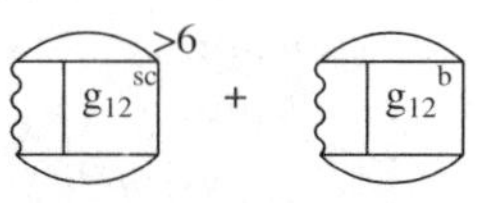 $$\Big\} . \qquad (6.204)$$

While the contribution of the second line remains convergent in the limit $V^s \to V$, the second and the fourth first line terms are divergent.

All these contributions were discussed in detail in the preceding subsections. Let us write the final result for the equation of state in the following form:

$$p - p_0 = p^{\mathrm{H}} + p^{\mathrm{RPA}} + p_4 + p_6 + \sum_{ab}{}' \tilde{z}_a \tilde{z}_b (b_{ab}^{\mathrm{sc}>3} + b_{ab}^{\mathrm{bound}}) \,. \tag{6.205}$$

The contributions $p^{\mathrm{HF}}, p^{\mathrm{RPA}}, p_4, p_6$ were discussed in detail in Sect. 6.3. An analytical determination of these terms is possible only in the low density limit. In the general case, we have to use numerical methods for the evaluation of the Padé formulae. The remaining terms are screened ladder contributions. Ladder contributions were intensely studied in the preceding subsection. We especially know that these terms describe the bound state contributions to the pressure. Using the result of the previous subsection, we obtain in the general (degenerate) case

$$b_{ab}^{\mathrm{bound}} = -\frac{1}{2\tilde{z}_a \tilde{z}_b \Omega} \sum_n \int \frac{d\boldsymbol{P}}{(2\pi)^3} \ln |1 - \exp(-\beta[E_{n\boldsymbol{P}} - \mu_a - \mu_b])| \,, \tag{6.206}$$

and

$$\begin{aligned} b_{ab}^{\mathrm{sc}} &= \frac{1}{2\tilde{z}_a \tilde{z}_b \Omega} \int \frac{d\boldsymbol{p}_1 d\boldsymbol{p}_2}{(2\pi)^6} \left(\frac{1}{kT} n_{ab}(E_{ab}) \mathrm{Re} \, \langle \boldsymbol{p}_1 \boldsymbol{p}_2 | T(E_{ab}) | \boldsymbol{p}_2 \boldsymbol{p}_1 \rangle \right. \\ &\quad + i\pi \ln |1 - \exp(-\beta[E_{ab} - \mu_a - \mu_b])| \int \frac{d\bar{\boldsymbol{p}}_1 d\bar{\boldsymbol{p}}_2}{(2\pi)^6} \delta(E_{ab} - \bar{E}_{ab}) N_{ab} \\ &\quad \left. \times \mathrm{Im} \frac{d}{dE} \langle \boldsymbol{p}_1 \boldsymbol{p}_2 | T(E_{ab}) | \bar{\boldsymbol{p}}_2 \bar{\boldsymbol{p}}_1 \rangle \, N_{ab} \, \langle \bar{\boldsymbol{p}}_1 \bar{\boldsymbol{p}}_2 | T(E_{ab}) | \boldsymbol{p}_2 \boldsymbol{p}_1 \rangle \right) \,. \end{aligned} \tag{6.207}$$

As an input to this equation, we need the two-particle energies E_{ab} and the on-shell T-matrix. For the determination of these two-particle properties, we have to use the Bethe–Salpeter equation in dynamically screened ladder approximation.

The result (6.205), (6.206), (6.207) is a very general EOS because the plasma correlation effects occur at several places. We have quasiparticle energies E_a in $E_{ab} = E_a + E_b$, we have dynamical screening, and there are degeneracy effects in the Bose type distribution function of the two-particle bound states, in the scattering state contributions, and in the Pauli blocking as well. As already discussed, the bound state contribution to (6.207) has singularities at $E_n + \frac{P^2}{2M} = \mu_a + \mu_b$ which are connected in well-known manner with the Bose–Einstein condensation of the bound states. In a hydrogen plasma, this contribution is responsible for the possibility of the Bose condensation of hydrogen atoms in spin polarized hydrogen.

In order to get the EOS $p = p(n)$, the fugacities z have to be eliminated from (6.205) using

$$n_a = z_a \frac{\partial}{\partial z_a} \frac{p}{k_B T} \,. \tag{6.208}$$

We mention here again that there is another way to determine thermodynamic functions. This way will be sketched only in principle. One may start from the formula (3.213) and apply the extended quasiparticle approximation (3.197) for the spectral function $a(\boldsymbol{p}\omega)$, and the dynamically screened ladder approximation for the self-energy Σ. Then we get for the density (see Kremp et al. (1993))

$$
\begin{aligned}
n_a(\mu_a T) = & \int \frac{d\boldsymbol{p}}{(2\pi)^3} f(\varepsilon_a) + \sum_b \sum_{nl} (2l+1) \int \frac{d\boldsymbol{P}}{(2\pi)^3} n_{ab}^B \left(\frac{P^2}{2M} + E_n \right) \\
& + \sum_b \int \frac{d\boldsymbol{p}}{(2\pi)^3} \frac{d\boldsymbol{p}'}{(2\pi)^3} \frac{\partial}{\partial \varepsilon_a} n_{ab}^B(\varepsilon_a + \varepsilon_b)(1 - f_a - f_b) \\
& \times \mathrm{Re} \, \langle \boldsymbol{p}\boldsymbol{p}' | T(\varepsilon_a + \varepsilon_b) | \boldsymbol{p}'\boldsymbol{p} \rangle \\
& - \sum_b \int_0^\infty \frac{d\omega}{\pi} n_{ab}^B(\omega) \int \frac{d\boldsymbol{p} d\boldsymbol{p}'}{(2\pi)^6} \frac{d\bar{\boldsymbol{p}} d\bar{\boldsymbol{p}}'}{(2\pi)^6} N_{ab} \delta(\omega - \varepsilon_a - \varepsilon_b) \\
& \times \mathrm{Im} \left\{ \langle \boldsymbol{p}\boldsymbol{p}' | T(\omega + i\epsilon) | \bar{\boldsymbol{p}}'\bar{\boldsymbol{p}} \rangle \, \bar{N}_{ab} \delta(\omega - \bar{\varepsilon}_a - \bar{\varepsilon}_b) \, \langle \bar{\boldsymbol{p}}\bar{\boldsymbol{p}}' | T'(\omega - i\epsilon) | \boldsymbol{p}'\boldsymbol{p} \rangle \right\} .
\end{aligned}
\tag{6.209}
$$

This formula determines any thermodynamic quantity, too. But this result is more general than formula (6.205) on behalf of the fact that the quasiparticle energy was taken into account in *RPA*. The equivalence to our result (6.205) is achieved in the expansion of the distribution functions in (6.209) with respect to the quasiparticle energy.

For many applications in plasma physics, it is sufficient to consider the low density case $n\lambda^3 < 1$ only. In this case, we may write an explicit expression for the plasma EOS. We collect the low density results for the different contributions to (6.205) from the preceding subsections. The result is a modified fugacity expansion of the EOS up to the order z^2

$$
\begin{aligned}
\frac{p(z)}{kT} = & \sum_a z_a + \frac{K^3}{12\pi} + \sum_{ab} z_a z_b \Big\{ -\frac{\pi}{3} (\beta e_a e_b)^3 \ln(K\lambda_{ab}) + \frac{\pi}{2} \beta^3 e_a^2 e_b^2 \\
& + 2\pi \lambda_{ab}^3 \Big[-\frac{\xi^2}{8} \sqrt{\pi} - \frac{\xi^3}{6} \left(\frac{C}{2} + \ln 3 - \frac{1}{2} \right) \\
& + \delta_{ab} \frac{(-1)^{2s_a}}{2s_a + 1} \left(\frac{\sqrt{\pi}}{4} + \frac{\xi}{2} + \frac{\sqrt{\pi}}{4} \xi^2 \ln 2 + \frac{\pi^2}{72} \xi^3 \right) \Big] \Big\} \\
& + \sum_{ab} z_a z_b (b_{ab}^{\mathrm{bound}} + b_{ab}^{\mathrm{sc}>3}) \, .
\end{aligned}
\tag{6.210}
$$

Here we used

$$
\xi_{ab} = -\frac{e_a e_b}{kT \lambda_{ab}} , \; z_a = \frac{2s_a + 1}{\Lambda_a^3} \tilde{z}_a \, , \; K^2 = \frac{4\pi}{kT} \sum_a e_a^2 z_a \, . \tag{6.211}
$$

The first term of (6.210) is the non-degenerate ideal gas result, and the next contribution is the famous Debye–Hückel limiting law already discussed in

Chap. 2. Further, there occurs a $z^2 \ln z$ contribution which follows from the e^6 direct term. With the Debye and the $\ln z$ terms we observe a deviation from the usual fugacity expansion. This modification is due to the screening of the Coulomb divergencies.

The bound state and the scattering state parts of the second virial coefficient have, for central symmetric potentials, the known Beth–Uhlenbeck form. Corresponding to the considerations in Sect. 6.5.1, we find for the bound state part

$$b_{ab}^{\text{bound}} = -\frac{4\pi^{3/2}}{(2s_a+1)(2s_b+1)}\lambda_{ab}^3 \sum_{l=0}^{\infty}(2l+1) \sum_{n\geq l+1} \exp(-\beta E_{nl}) \,. \tag{6.212}$$

The subtracted scattering part is given by

$$b_{ab}^{\text{sc}>3} = \frac{4\pi^{3/2}}{(2s_a+1)(2s_b+1)}\lambda_{ab}^3 \int_{\Delta}^{\infty} dE \exp(-\beta E)\rho^{>3}(E) \,. \tag{6.213}$$

Here, we used the abbreviation

$$\rho^{>3}(E) = \sum (2l+1) \left[1 \pm \delta_{ab}\frac{(-1)^l}{2s_a+1}\right] \frac{1}{\pi}\frac{d}{dE}\delta_l(E)^{>3} \,. \tag{6.214}$$

According to the results of Sect. 6.5.2, there also exist other possibilities for the subdivision of the second virial coefficient. We mention (6.179) and (6.181). The latter possibility leads to the introduction of the Planck–Larkin sum of bound states.

In the grand canonical description, the pressure and other thermodynamic quantities are functions of the chemical potentials, or of the fugacities, respectively, and of the temperature and the volume. Then, from the pressure in the grand canonical ensemble, we may derive any thermodynamic function, e.g., we get the density as a function of the chemical potential by differentiation of (6.210) with respect to the fugacity according to (6.208). Using (6.210), we arrive at

$$\begin{aligned} n_a = z_a - \frac{\kappa e^2}{2} - 2\sum_b z_b \Big[&-\frac{\pi}{3}(\beta e_a e_b)^3 \ln(\kappa\lambda_{ab}) + \frac{\pi}{2}\beta^3 e_a^2 e_b^2 \\ &+2\pi\lambda_{ab}^3 \Big\{ -\frac{\xi_{ab}^2}{8}\sqrt{\pi} - \frac{\xi_{ab}^3}{6}\big(\frac{C}{2} + \ln 3 - \frac{1}{2} + 1\big) \\ +\delta_{ab}\frac{(-1)^{2s_a}}{2s_a+1} &\left(\frac{\sqrt{\pi}}{4} + \frac{\xi_{ab}}{2} + \frac{\sqrt{\pi}}{4}\xi_{ab}^2 \ln 2 + \frac{\pi^2}{72}\xi_{ab}^3\right) \Big\} \Big] \\ &-2\sum_b z_b (b_{ab}^{\text{bound}} + b_{ab}^{\text{sc}>3}) \,. \end{aligned} \tag{6.215}$$

Because the contribution to the second virial coefficient, $b^{>3}$, is convergent for the pure Coulomb potential, we can neglect screening. In such case,

$b^{>3} = P''b$ is given by the results of the previous section. The EOS may be written as

$$\frac{p}{k_B T} = \sum_a z_a + \frac{\kappa^3}{12\pi} + \sum_{ab} z_a z_b \left[\frac{\pi}{2}(\beta e_a e_b)^3 \ln(\kappa\lambda_{ab}) + \frac{\pi}{2}\beta^3 e_a^2 e_b^4 + 2\pi\lambda_{ab}^3 K_0(\xi_{ab}, s_a)\right] + \mathcal{O}(z^{5/2}\ln z)\,. \quad (6.216)$$

Here, the virial function K_0 is given by

$$K_0(\xi) = Q(\xi) \pm \delta_{ab}\frac{(-1)^{s_a}}{s_a+1}E(\xi)\,. \quad (6.217)$$

The direct quantum mechanical contribution reads

$$Q(\xi) = -\frac{\sqrt{\pi}}{8}\xi^2 - \frac{\xi^3}{6}\left(\frac{C}{2} + \ln 3 - \frac{1}{2}\right) + \sum_{p=4} \frac{\sqrt{\pi}\zeta(p-2)}{\Gamma(\frac{p}{2}+1)}\left(\frac{\xi}{2}\right)^p . \quad (6.218)$$

The exchange contribution follows to

$$E(\xi) = \frac{\sqrt{\pi}}{4} + \frac{\xi}{2} + \sqrt{\pi}\ln 2\left(\frac{\xi}{2}\right)^2 + \frac{\pi^2}{9}\left(\frac{\xi}{2}\right)^3 + \sum_{p=4} \frac{\sqrt{\pi}(1-2^{2-p})\zeta(p-1)}{\Gamma(\frac{p}{2}+1)}\left(\frac{\xi}{2}\right)^p . \quad (6.219)$$

In Figs. 6.16 and 6.17, the functions $Q(\xi)$, $E(\xi)$ and $K_0(\xi)$ are represented graphically.

It may easily be verified that K_0 has the following representation, too,

$$K_0(\xi_{ab}, s_a) = 2\sqrt{\pi}\sigma(\xi_{ab}) + K_0^*(\xi_{ab}, s_a)\,. \quad (6.220)$$

Here, $\sigma(\xi_{ab})$ is the Planck-Larkin partition function given by (6.203), and $K_0^*(\xi_{ab}, s_a)$ is the remaining scattering contribution

$$K_0^*(\xi_{ab}, s_a) = -(\text{sign}\xi_{ab})Q(-\xi_{ab}) + \frac{1}{4}\sqrt{\pi}\Theta(\xi_{ab})\xi_{ab}^2 + \frac{(-1)^{2s_a}}{2s_a+1}\delta_{ab}E(\xi_{ab})\,. \quad (6.221)$$

For large positive Born parameters $\xi_{ab} > 0$, bound states give the dominant contribution to the pressure. In this case, the following asymptotic formula is valid

$$K_0^\infty(\xi_{ab}) = \sqrt{\pi}\Theta(\xi_{ab})\left(2\sigma(\xi_{ab}) - \frac{\xi_{ab}^2}{4}\right) - \frac{\xi_{ab}^3}{6}\left(\ln|\xi_{ab}| + 2C + \ln 3 - \frac{11}{6}\right) . \quad (6.222)$$

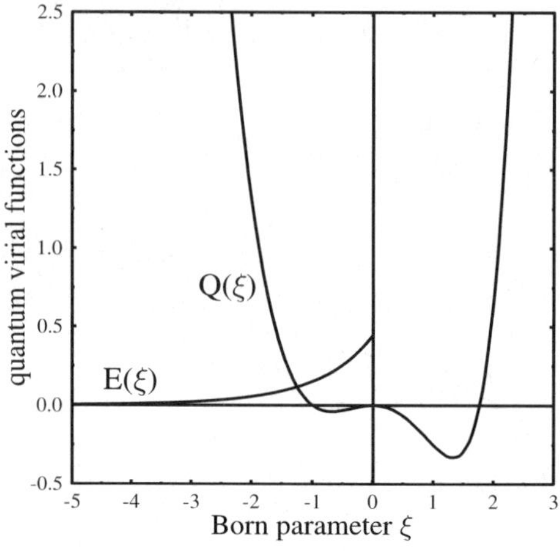

Fig. 6.16. Quantum virial functions Q and E

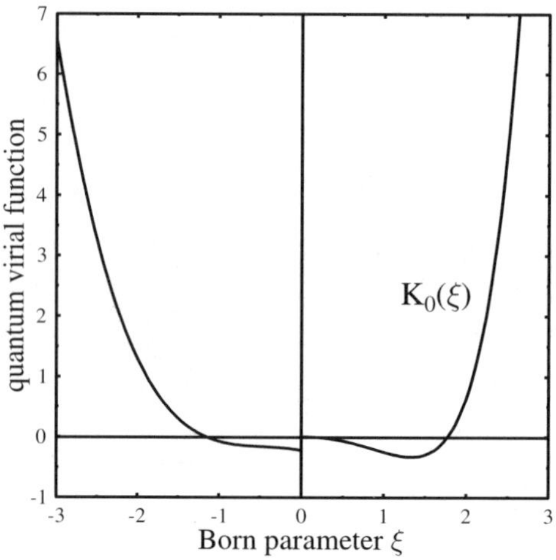

Fig. 6.17. Quantum virial function K_0

We use this relation in (6.216). For hydrogen, we have $e_p = -e_e$, and, consequently, the contributions of order ξ_{ab}^3 cancel each other after summation over species. The remaining terms contain the step function $\Theta(\xi_{ab})$, i.e., they contribute only for charges with different sign. Then we arrive at

$$\frac{p}{k_BT} = z_e + z_p + \frac{1}{12\pi}\kappa^3\left(1 - \frac{3}{16}\sqrt{\pi}\kappa\lambda_{ep}\right) + \frac{1}{32\pi}\kappa^2\kappa_4^2 + z_p z_e 8\pi^{3/2}\sigma(T)\,. \tag{6.223}$$

where

$$\kappa_4^2 = 4\pi\sum_b z_b\beta^2 e_b^4\,.$$

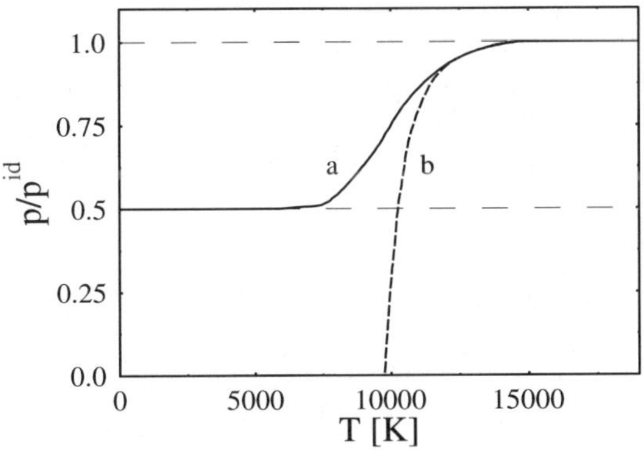

Fig. 6.18. Comparison of fugacity (**a**) and density (**b**) expansion. $n = 10^{15}\,\mathrm{cm}^{-3}$

The set of equations (6.210) and (6.215) is a parameter representation of the equation of state as a function of the density. This parameter representation implicitly determines the thermodynamic properties of the plasma as a function of the temperature and the density.

Let us calculate the pressure on the basis of (6.210) and (6.215), see Bartsch and Ebeling (1975). In Fig. 6.18 we presented the quantity p/p_0 for a non-degenerate H-plasma in density– and fugacity–expansion. Both curves in Fig. 6.18 decrease from unity, but only the fugacity curve tends to one half at higher densities, i.e., the pressure decreases from that of a fully ionized electron–proton plasma to the pressure of a gas of H-atoms with half the density. This means that the equation of state given by the equations (6.210), (6.215) is rather well suited for a partially ionized H-plasma. However, it is important to remark that the Mott effect and the degeneracy at high densities were not taken into account so far.

6.5.5 Density Expansion of Thermodynamic Functions of Non-degenerate Plasmas

Let us now consider the fugacity expansion of the EOS (6.210) in more detail. We especially will derive explicit expressions for the density expansion. The basic equations of the quantum mechanical fugacity expansion may be written as

$$\frac{p}{kT} = \sum_j b_j z^j ; \;\; n = \sum_j j b_j z^j \, . \tag{6.224}$$

Here, the first cluster coefficients are given by (6.109). The properties of the fugacity expansion are well-known from textbooks on statistical physics (Hill 1956), (Huang 1963). We know that the inversion of the fugacity expansion for the density n is given by

$$z = n \exp \left(- \sum_j \beta_j n^j \right) \, . \tag{6.225}$$

The coefficients β_j are the irreducible cluster coefficients which are connected to b_j via the relations

$$\begin{aligned} b_2 &= \frac{\beta_1}{2}\,, \\ b_3 &= \frac{\beta_1^2}{2} + \frac{\beta_2}{3}\,, \\ b_4 &= \frac{2\beta_1^3}{3} + \beta_1\beta_2 + \frac{\beta_3}{4}\,, \\ &\ldots\,. \end{aligned} \tag{6.226}$$

With the help of the relations (6.225), the fugacity may be eliminated from the expansion of the pressure, and we get a density expansion of the equation of state, which is usually called virial expansion

$$\frac{p}{kT} = n\left(1 - \sum_k \frac{k}{k+1}\beta_k n^k\right)\,. \tag{6.227}$$

The structure of (6.225) implies a rearrangement of the fugacity series for the pressure (6.224). For this purpose, the function $S(z)$ is defined by representing the sum of the irreducible contributions to the cluster coefficients b, namely

$$S(z) = \sum_j \frac{\beta_{j-1}}{j} z^j\,. \tag{6.228}$$

In papers by Rogers and DeWitt (1973) it was shown that the equation of state is determined entirely by the function $S(z)$,

$$\frac{p}{kT} = z + S(z) + \sum_m \frac{z}{m!}\left(\frac{\partial}{\partial z}z\right)^{m-2}\left(\frac{\partial}{\partial z}S(z)\right)^m\,. \tag{6.229}$$

According to (6.229), the contributions to the equation of state are subdivided into irreducible and reducible ones.

The function S is of great importance. It determines any thermodynamic property of the system. If, in the defining equation (6.228), the fugacities z are replaced by densities n, one gets the Mayer S-function $S(n)$ (Mayer 1950). Applying this function, one immediately arrives at the relations

$$\begin{aligned} z &= n\exp\left(\frac{\partial}{\partial n}S(n)\right)\,, \\ \frac{p}{kT} &= n + S(n) - n\frac{\partial}{\partial n}S(n)\,, \\ \frac{F - F_0}{VkT} &= -S(n)\,. \end{aligned} \tag{6.230}$$

Here, F is the free energy, and F_0 that of non-interacting particles.

Consequently, the central task turns out to be the approximate determination of the Mayer function S.

We still write the generalization of the preceding relations for a two-component system. The Mayer function then reads (Mayer 1950)

$$S(z_a z_b) = \sum_J \sum_{j_a j_b} \frac{z^{j_a}}{j_a} \frac{z^{j_b}}{j_b} \beta_{j_a j_b} = \sum_J \frac{z^J}{J} \beta_J \,, \tag{6.231}$$

with $J = j_a + j_b$; J signifies a multinomial expansion. For the fugacity expansion of the equation of state, the following generalization is valid

$$\begin{aligned} \frac{p}{kT} &= z_a + z_b + S(z_a z_b) + \sum_{m_a=2} \frac{z_a}{m_a!} \frac{\partial}{\partial z_a} z_a^{m_a-1} \left(\frac{\partial S}{\partial z_a} \right)^{m_a} \\ &+ \sum_{m_b=2} \frac{z_b}{m_b!} \frac{\partial}{\partial z_b} z_b^{m_b-1} \left(\frac{\partial S}{\partial z_b} \right)^{m_b} + z_a z_b \frac{\partial S}{\partial z_a} \frac{\partial^2 S}{\partial z_a \partial z_b} + \cdots \,. \end{aligned} \tag{6.232}$$

Further relations have to be generalized accordingly. We still write the density expansion of the equation of state which reads

$$\frac{p}{kT} = n_a + n_b + S(\{n_c\}) - n_a \frac{\partial S(\{n_c\})}{\partial n_a} - n_b \frac{\partial S(\{n_c\})}{\partial n_b} \,. \tag{6.233}$$

We mention again that, with equations of this type, a formal scheme is given for the determination of equilibrium properties of non-degenerate many-particle systems. Again, the task is the approximate determination of the Mayer function S. According to our scheme, an approximate expression for $S(z)$ may be deduced from (6.210). If we consider this approximation now to be the function S as a function of the density, we get (the subscript at ξ will be omitted, $\xi_{ab} = \xi$)

$$\begin{aligned} S(n) = \frac{\kappa^3}{12\pi} &+ \sum_{ab} n_a n_b \Bigg(-\frac{\pi}{3} (\beta e_a e_b)^3 \ln(\kappa \lambda_{ab}) + \frac{\pi}{2} \beta^3 e_a^2 e_b^2 \\ &+ 2\pi \lambda_{ab}^3 \Bigg\{ -\frac{\xi^2}{8} \sqrt{\pi} - \frac{\xi^3}{6} \left(\frac{C}{2} + \ln 3 - \frac{1}{2} \right) \\ &\pm \delta_{ab} \frac{(-1)^{2s_a}}{2s_a + 1} \left(\frac{\sqrt{\pi}}{4} + \frac{\xi}{2} + \frac{\sqrt{\pi}}{4} \xi^2 \ln 2 + \frac{\pi^2}{72} \xi^3 \right) \Bigg\} \Bigg) \\ &+ \sum_{ab} n_a n_b \left(b_{ab}^{\mathrm{bound}} + b_{ab}^{\mathrm{sc}>3} \right) . \end{aligned} \tag{6.234}$$

One might argue that the function S is divergent for Coulomb systems. However, as we saw in the preceding section, the divergencies were overcome by partial summation. From (6.233) and (6.234), we may determine the equation of state and further thermodynamic quantities as a function of density and temperature. The EOS is given by

$$\begin{aligned}\frac{p(n)}{kT} &= \sum_a n_a - \frac{\kappa^3}{24\pi} - \sum_{ab} n_a n_b \Bigg(-\frac{\pi}{3}(\beta e_a e_b)^3 \ln(\kappa\lambda_{ab}) + \frac{\pi}{2}\beta^3 e_a^2 e_b^2 \\ &\quad + 2\pi\lambda_{ab}^3 \Bigg\{ -\frac{\xi^2}{8}\sqrt{\pi} - \frac{\xi^3}{6}\left(\frac{C}{2} + \ln 3 - \frac{1}{2} + 1\right) \\ &\quad \pm\delta_{ab}\frac{(-1)^{2s_a}}{2s_a+1}\left(\frac{\sqrt{\pi}}{4} + \frac{\xi}{2} + \frac{\sqrt{\pi}}{4}\xi^2 \ln 2 + \frac{\pi^2}{72}\xi^3\right)\Bigg\}\Bigg) \\ &\quad - \sum_{ab} n_a n_b (b_{ab}^{\mathrm{bound}} + b_{ab}^{\mathrm{sc}>3}),\end{aligned} \tag{6.235}$$

while the free energy and the chemical potential read

$$F^{\mathrm{corr}}(n,T) = -VkTS(n), \tag{6.236}$$

$$\begin{aligned}\mu_a^{\mathrm{corr}} &= -\frac{\kappa e^2}{2} - 2\sum_b n_b \Bigg(-\frac{\pi}{3}(\beta e_a e_b)^3 \ln(\kappa\lambda_{ab}) + \frac{\pi}{2}\beta^3 e_a^2 e_b^2 \\ &\quad + 2\pi\lambda_{ab}^3 \Bigg\{ -\frac{\xi^2}{8}\sqrt{\pi} - \frac{\xi^3}{6}\left(\frac{C}{2} + \ln 3 - \frac{1}{2} + 1\right) \\ &\quad \pm\delta_{ab}\frac{(-1)^{2s_a}}{2s_a+1}\left(\frac{\sqrt{\pi}}{4} + \frac{\xi}{2} + \frac{\sqrt{\pi}}{4}\xi^2 \ln 2 + \frac{\pi^2}{72}\xi^3\right)\Bigg\}\Bigg) \\ &\quad - 2\sum_b n_b (b_{ab}^{\mathrm{bound}} + b_{ab}^{\mathrm{sc}>3}).\end{aligned} \tag{6.237}$$

We remark that the fact that the correlation part of the free energy as a function of the density coincides with the negative correlation part of the pressure as a function of the fugacity is referred to as *the golden rule of statistical physics.*

It is interesting to compare the fugacity expansion of the pressure with the corresponding density expansion (6.235). Obviously, the fugacity representation gives a more realistic description of the behavior of plasmas with bound states. The density representation (6.235) gives negative values for the pressure as a consequence of the exponential divergence of the atomic partition function σ at low temperatures, see Fig. 6.18. In the expansion at low fugacities, this term does not lead to difficulties as the pre-factor of σ, $z_p z_e$, vanishes in the same manner as σ increases. All terms of the fugacity expansion remain finite at low temperatures.

6.5.6 Bound States and Chemical Picture. Mott Transition

Up to now we considered the plasma as a system of charged particles in scattering and bound states. We will call this description the physical picture.

In this picture, pressure and density are given by the fugacity expansions (6.210) and (6.215). We showed that these equations provide an appropriate description for plasmas with the possibility of the formation of bound states. In contrast, the density expansions following by incomplete inversion lead to negative pressures as a consequence of exponential divergencies at low temperatures. To get a density expansion under such conditions, we switch to the *chemical picture*. In the limit of low densities, two-particle bound states are stable entities. For this reason, it is convenient to use the following interpretation

$$\textit{Bound States} = \text{New Particles } (\textit{Atoms})$$

Therefore, instead of a system consisting of two elementary particles, we consider now a system of three components, which are the free particles a and b and the atoms (ab), what means we now have a partially ionized plasma. For the transformation into this description, it is necessary to introduce fugacities for the new particles. For this purpose let us write the EOS in the following form

$$\begin{aligned} \beta\, p \;&=\; \sum_{a=e,i} (2s+1) \int \frac{d\boldsymbol{p}}{(2\pi\hbar)^3} \ln\left|1 + \tilde{z}_a e^{-\beta E_a}\right| \\ &+ \sum_{n} (2s+1)^2 \int \frac{d\boldsymbol{P}}{(2\pi\hbar)^3} \ln\left|1 - \tilde{z}_e \tilde{z}_i e^{-\beta E_{P\,n}}\right| \\ &+ \beta p^{\rm HF} + \beta p^{\rm MW} + \beta p^4 + \beta p^6 + \sum_{a,b=e,i} z_a z_b b_{ab}^{sc^{>3}} \,. \end{aligned} \tag{6.238}$$

We see from (6.238) that the contributions corresponding to free particles and to bound states are not very different from each other in their mathematical shape. Therefore, it is obvious to define the fugacity of the new particles by

$$\tilde{z}_H^n = \tilde{z}_e \tilde{z}_i e^{-\beta E_n} \,. \tag{6.239}$$

Then the Bose contributions of the bound state part of (6.238) are transformed into the Bose pressure of ideal atoms; that means the bound states contribute to the EOS like ideal Bose particles. In this chemical picture, the EOS has the following form

$$\begin{aligned} \beta\, p \;&=\; \sum_{a=e,i} (2s+1) \int \frac{d\boldsymbol{p}}{(2\pi\hbar)^3} \ln\left|1 + \tilde{z}_a e^{-\beta E_a}\right| \\ &+ \sum_{n} (2s+1)^2 \int \frac{d\boldsymbol{P}}{(2\pi\hbar)^3} \ln\left|1 - \tilde{z}_H^n e^{-\frac{P^2}{2M}}\right| \\ &+ \beta p^{\rm HF} + \beta p^{\rm MW} + \beta p^4 + \beta p^6 + \sum_{a,b=e,i} z_a z_b b_{ab}^{sc^{>3}} \,. \end{aligned} \tag{6.240}$$

These formulae contain the correct physics for small densities. They give the thermodynamic properties of a partially ionized plasmas in the ionization equilibrium

$$e + i \rightleftharpoons (ei) = H. \tag{6.241}$$

We have to take into account that, in this picture, the fugacities z_a, z_b, and z_{ab} are not independent of each other. They are connected by the relation (6.239). The latter connection determines the composition of the system and plays the role of a mass action law. It is equivalent to the following relation between the chemical potentials

$$\mu_H^n = \mu_e + \mu_i - E_n^{ei}\,, \tag{6.242}$$

this means the well-known thermodynamic condition for the ionization equilibrium considered already in Chap. 2. In order to complete the chemical picture, we need the connection between the fugacities and the densities. The number densities of the new species as a function of the fugacities are now determined by the relations

$$n_e = z_e \frac{\partial}{\partial z_e}(\beta p), \quad n_i = z_i \frac{\partial}{\partial z_i}(\beta p), \quad n_H^n = z_H^n \frac{\partial}{\partial z_H^n}(\beta p). \tag{6.243}$$

In this manner, we transformed the description from the physical picture into a chemical one merely by definition of the fugacity of new particles. In order to demonstrate this scheme in a more explicit way let us determine the density of the Bose like H-atoms. From the last of the equations (6.243), we get

$$n_H = \sum_n (2s+1)^2 \int \frac{d\boldsymbol{P}}{(2\pi\hbar)^3} \frac{z_H^n e^{-\beta \frac{P^2}{2M}}}{1 + z_H^n e^{-\beta \frac{P^2}{2M}}}\,.$$

Further, from the transformation rule (6.239), we get

$$z_H^n = e^{-\beta(E_n - \mu_e - \mu_p)}\,.$$

The chemical potentials μ_a have to be determined by inversion of the first two relations (6.243). The application of the Padé formulae (see Sect. 6.3.4) is very useful. Of course, now these quantities are functions of the densities n^* of the free particles, i.e., unbound electrons and protons. Next we introduce, in the same manner as in Chap. 2, the degree of ionization and get

$$n_e = \alpha n_e^{\text{tot}}; \quad n_H = (1-\alpha) n_e^{\text{tot}}; \quad \frac{1-\alpha}{\alpha} = \frac{n_H}{n_e}\,,$$

with the total electron density

$$n_e^{\text{tot}} = n_e + n_H\,.$$

With these relations we are able to determine α as a function of the free or of the total densities of the electrons. This procedure is simple for a non-degenerate H-plasma. Then we have

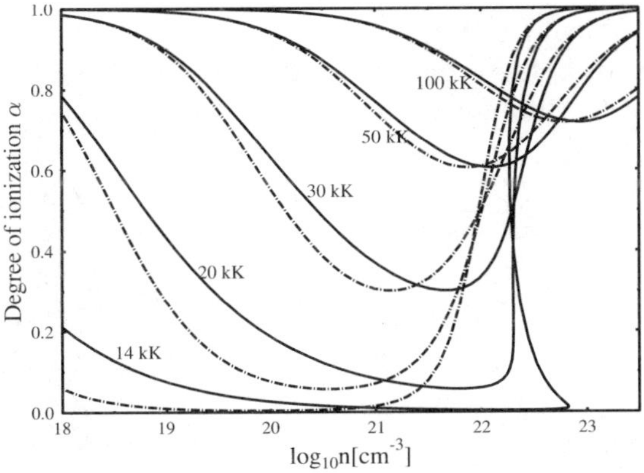

Fig. 6.19. Isotherms of the degree of ionization using (6.244). *Full lines*: abscissa – total electron density; *dash-dot*: abscissa – free electron density

$$\tilde{z}_H = \frac{n_H \Lambda_H^3}{(2s+1)^2}; \quad \tilde{z}_a = \frac{n_a \Lambda_a^3}{2s+1} \exp(\beta \Delta\mu_a)\,.$$

We immediately arrive at the usual form of the Saha equation for non-ideal plasmas, where we consider only the ground state

$$\frac{1-\alpha}{\alpha^2} = n_e^{\text{tot}} \Lambda_e^3 \exp(-\beta\{E_1 - \Delta\mu_e - \Delta\mu_p\})\,, \tag{6.244}$$

where $\Delta\mu_e$ is the interaction part of the chemical potential. Let us now determine the degree of ionization numerically. We use the Padé formulae (6.68) for the interaction part of the chemical potential. The energy levels have to be determined from the Bethe–Salpeter equation and are approximately independent of the density as known from Chap. 5. The results are given in Fig. 6.19. Again we observe, as a consequence of the nonideality, a very strong increase of the degree of ionization up to $\alpha = 1$, i.e., the Mott transition. It is interesting to use the same procedure with for the density. Then we have

$$n_a(\mu, T) = \int \frac{d\boldsymbol{p}}{(2\pi\hbar)^3} f_a(p) + \sum_{b,n} \int \frac{d\boldsymbol{p}}{(2\pi\hbar)} n_{ab,n}^B(P) + \sum_b z_a z_b b_{ab}^{sc}\,. \tag{6.245}$$

As we already discussed, these relations are more general than the corresponding relations which follow from the first (6.243). With (6.245), we are able to give a complete description of the equilibrium properties of a partially ionized plasma that means that we can determine the composition and the EOS of the system. In order to derive the results of Sect. 6.5.4 from our general theory let us consider the equation of state in the simple approximation (6.223). After transformation into the chemical picture, we get

$$\beta p = z_e + z_p + z_H + \frac{K^3}{12\pi}\left(1 - \frac{3}{16}\sqrt{\pi}K^3\lambda_{ei}\right) + \frac{K^2}{32\pi}K_4^2\,. \tag{6.246}$$

Here we used $K^2 = \frac{4\pi}{k_BT}\sum_a z_a e_a^2$ and $K_4^2 = \frac{4\pi}{(k_BT)^2}\sum_b z_b e_b^4$. The fugacities (and chemical potentials) have to be determined using (6.208). Solving the resulting equations by iteration, we find for the fugacities

$$\ln z_a = \ln n_a - \frac{1}{2}\beta\kappa e^2\left[1 - \frac{\sqrt{\pi}}{4}\kappa\lambda_{ei}\right] \tag{6.247}$$

and for the chemical potentials

$$\mu_a = -kT\ln\left(\frac{n_a\Lambda_a}{2s+1}\right) - \frac{1}{2}\beta\kappa e^2\left[1 - \frac{\sqrt{\pi}}{4}\kappa\lambda_{ei}\right]. \tag{6.248}$$

It is useful to consider the interaction terms as the beginning of a geometrical series. Then we get an extension of the form

$$\ln z_e = \ln n_e - \frac{\frac{1}{2}\beta\kappa e^2}{1 + \frac{1}{8}\kappa\Lambda}. \tag{6.249}$$

This equation is the Debye Hückel form discussed in Chap. 2. Introducing the relation between densities and fugacities we get the Saha equation

$$\frac{n_H}{n_e n_i} = \Lambda_e^3\sigma(T)\exp\left(\frac{-\beta\kappa e^2}{1 + \frac{1}{8}\kappa\Lambda}\right). \tag{6.250}$$

Here, $\sigma(T)$ is a simplified version of (6.182) including only the ground state. In this way, we found quantum-statistical arguments for the more intuitive foundation of the Saha equation given in Chap. 2.

The experimental determination of electron density and temperature in solid-density plasmas was performed, e.g., in papers by Gregori et al. (2003), Glenzer (a) et al. (2003), and Gregori et al. (2004). In Fig. 6.20, we show the experiments given by Glenzer (b) et al. (2003) together with theoretical interpretations. The curve labeled by LASNEX applies techniques of Desjarlais

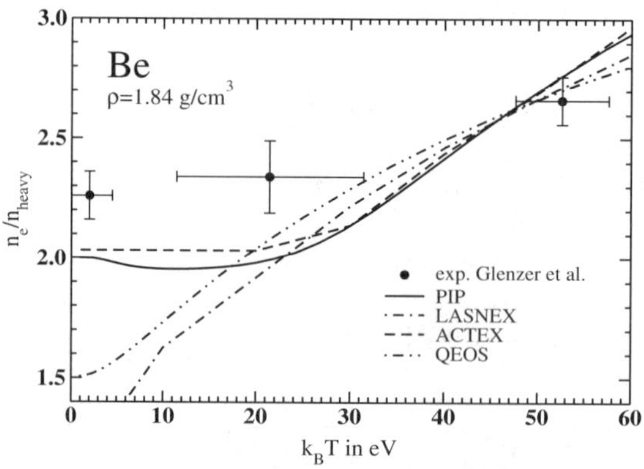

Fig. 6.20. Ionization state of Be. Experimental values from Glenzer (b) et al. (2003) and theoretical models (see text)

et al. (2002), QEOS means quantum EOS (More 1988). The LASNEX and QEOS curves only include free electrons. For ACTEX, see Rogers and Young (1997), Rogers (2000). PIP means partially ionized plasma. Both PIP and ACTEX include bound states. The PIP calculations performed by Kuhlbrodt (2003) are based on mass action laws and Padé formulae given in Sect. 6.3.4, formulae (6.82)–(6.90).

6.6 Thermodynamic Properties of the H-Plasma

6.6.1 The Hydrogen Plasma

In this section we will consider the thermodynamic properties of the H-plasma on the basis of the general theory of equilibrium properties developed in the preceding sections. The H-plasma is a simple but very important and interesting many-particle system. Hydrogen is the simplest and, at the same time, the most abundant element in the cosmos. It has been a subject of great interest due to the importance for such problems as astrophysics, inertial confinement fusion, and our fundamental understanding of condensed matter.

Because of the formation of bound states and of phase transitions, the hydrogen plasma exists in many different states and phases.

A first overview of the behavior can be seen in Fig. 6.21. The behavior of hydrogen is essentially dependent on temperature and density or pressure, respectively. At high temperatures, hydrogen is a simple system consisting of electrons and protons in scattering states. At lower temperatures, we observe the formation of H-atoms, this means we have a partially ionized H-plasma. The EOS in these regions is well investigated in the classical paper by Montroll and Ward (1958) and in further papers by DeWitt (1961), Rogers (1971), Kelbg (1963), and in Ebeling (a) et al. (1976), and is given in the previous sections of this chapter.

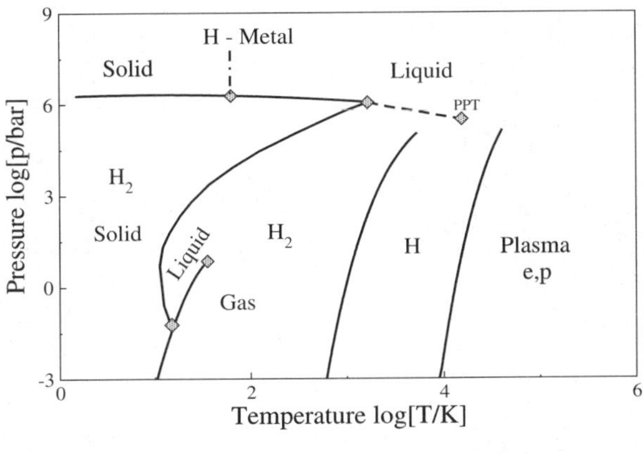

Fig. 6.21. Phase diagram of hydrogen. PPT: critical point of the hypothetical plasma phase transition

The region where atoms dominate the behavior of the system is very narrow. Additionally there are molecular ions such as H_2^+ and H^-. For temperatures lower than 1.5×10^3 K one has to account for the formation of molecules. First extensions in this region later were given by Rogers and DeWitt (1973); see also (Aviram et al. 1975; Aviram et al. 1976; Schlanges and Kremp 1982; Robnik and Kundt 1983; Haronska et al. 1987).

At pressures below 13 bar, a line of coexistence between a molecular liquid and a gas phase with the critical data $p_c = 13\,\text{bar}$, $T_c = 33\,\text{K}$ is observed.

Finally, with further decreasing temperature, hydrogen crystallizes and forms a dielectric H_2-solid. This low density behavior has been well investigated, too. More interesting, however, is the behavior at higher pressures. As pointed out already in Sect. 2.6, there are theoretical arguments for a phase transition between two differently ionized H-plasmas (Norman and Starostin 1968; Ebeling (a) et al. 1976; Ebeling and Richert 1985; Haronska et al. 1987; Saumon and Chabrier 1989). In the vicinity of the critical point, this phase transition is essentially determined by the Coulomb interaction and the ionization equilibrium.

At lower temperatures, the neutral particles are important (Schlanges and Kremp 1982; Robnik and Kundt 1983; Haronska et al. 1987; Saumon and Chabrier 1989; Saumon and Chabrier 1991; Saumon and Chabrier 1992; Juranek and Redmer 2000; Beule et al. 2001). The transition to the metallic state should be possible for the H_2-solid, too. This transition was first discussed by Wigner and Huntington (1935) and later by many authors.

Due to the rapid development of the experimental technique and also due to the results of theoretical simulations in the last years, the thermodynamic properties of dense plasmas such as the EOS have been a subject of considerable interest.

Very successful methods to produce and to investigate dense plasmas are shock wave experiments; see, e.g., Fortov (2003). In such an experiment, a driving force is utilized to press a pusher with the velocity $\boldsymbol{u}_1$ into the plasma with the initial state given by ρ_1, p_1, T_1. The impact generates a wave. As a result of the nonlinearity of the hydrodynamic equations, the waves steepen as they propagate and form shocks, i.e., thin layers, propagating through the system, separating two steady states of different density and temperature. The steepening is balanced by dispersion and/or dissipation. The ultra high pressure of the plasma is generated by compression and heating of the matter in the front of the intense shock wave.

From hydrodynamics it is well known that we have conservation laws of the mass density ρ, for the pressure p and for the energy E at the front. The conservation laws allow us to compute relations (*Rankine–Hugoniot-relations*) between the variables of the two steady states

$$\begin{aligned} \rho_1 \boldsymbol{u}_1 &= \rho_2 \boldsymbol{u}_2 \,, \\ p_1 + \rho_1 u_1^2 &= p_2 + \rho_2 u_2^2 \,, \end{aligned}$$

$$E_1 + \frac{p_1}{\rho_1} + \frac{u_1^2}{2} = E_2 + \frac{p_2}{\rho_2} + \frac{u_2^2}{2} . \tag{6.251}$$

Equations (6.251) represent a connection between the measurable kinematic parameters of the shock wave, i.e., between the velocity of the pusher $\boldsymbol{u}_1$ and that of the shock front $\boldsymbol{u}_2$, and the thermodynamic functions of the dense plasma. So the relations (6.251) are the basis for the experimental determination of the EOS.

The results of shock wave experiments are usually represented by an Hugoniot

$$(E_1 - E_2) + \frac{1}{2}\left(\frac{1}{\rho_1} - \frac{1}{\rho_2}\right)(p_1 + p_2) = 0 , \tag{6.252}$$

which follows from (6.251) after elimination of the shock-parameters $\boldsymbol{u}_i$. As the Hugoniot may also be determined by the EOS theoretically, there exists the possibility of an immediate comparison between theory and experiment.

Prior to 1997, shock wave experiments on hydrogen and deuterium plasmas had been limited to pressures below 20 GPa, which are accessible by conventional gas guns (Nellis et al. 1983). First results in a regime of extreme pressures and temperatures were provided in 1997 by the Nova laser driven shock wave experiments which reached pressures of up to 340 GPa (DaSilva et al. 1997). Four years later, Knudson et al. (2001) measured significant deviations from the Nova laser results, using a Z-pinch as a driver.

The current situation is shown in the Fig. 6.22. Here, the solid line is the Hugoniot which follows from the SESAME equation of state (Lyon and Johnson 1987), and the dashed line was derived by Ross (1998) from a linear mixing model. Further we present the data from the gas gun experiments

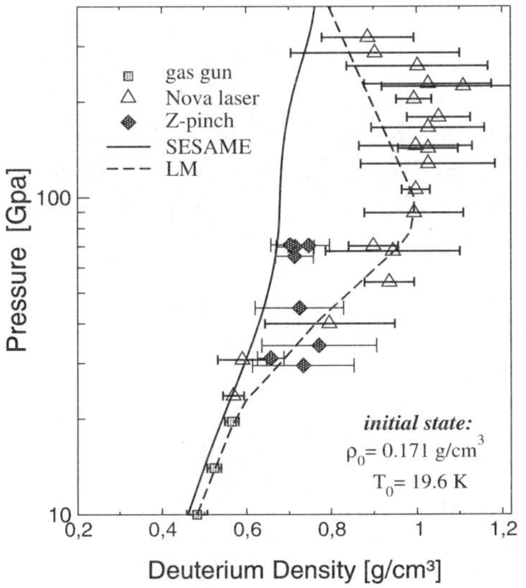

Fig. 6.22. Several experimental Hugoniot data and the SESAME Hugoniot. *Solid line*: Kerley (1983), Lyon and Johnson (1987), *dashes*: Ross (1996), *triangles*: Nellis et al. (1983), *squares*: DaSilva et al. (1997), *diamonds*: Knudson et al. (2001), Knudson et al. (2003)

(Nellis et al. 1983; Holmes et al. 1995), the laser driven (DaSilva et al. 1997; Collins et al. 1998) and the magnetically driven (Z-pinch) (Knudson et al. 2001) shock wave experiments. It is found that

- Laser Hugoniot experiments show higher compressibilities than predicted by SESAME data.
- Magnetically driven Hugoniot experiments give compressibilities closer to SESAME results.
- Laser Experiments are in agreement with the linear mixing model.

Therefore experimental, and semi-empirical results do not give a unique picture.

Let us now consider the problem from the point of view of the more rigorous quantum statistical theory. Of course, the results of the preceding sections of this chapter can not describe the Hugoniot over the full range of density and pressure. The relation gives only the asymptotic behavior at higher temperatures. As we mentioned already, the reason is the following. For lower temperatures, the neutral particles are more and more important, and finally, we have a neutral gas of H-atoms and H_2-molecules. That means, in order to understand the EOS or the Hugoniot over the full range of pressure, density and temperature, we have to include the formation of molecules. This will be considered in the next subsection.

To conclude, we want to mention that electron–hole plasmas including excitons as bound states represent a system with many analogies to hydrogen plasmas. The thermodynamics of electron–hole plasmas is dealt with, e.g., in Zimmermann (1988) and in Kraeft et al. (1986). However, in this monograph, such plasmas are touched only briefly.

6.6.2 Fugacity Expansion of the EOS. From Physical to Chemical Picture

If we want to account for the formation of molecules, the EOS has at least to be extended up to the order z^4. Then, four-particle processes may be described such as the scattering between H-atoms and the formation of molecules; the corresponding EOS should have the form

$$\begin{aligned}\beta p \;=\; & \sum_{a=e,i} (2s_a+1) \int \frac{d\boldsymbol{p}}{(2\pi\hbar)^3} \ln\left|1+\tilde{z}_a e^{-\beta E_a}\right| \\ & +(\beta p^{HF} + \beta p^{MW} + \beta p^4 + \beta p^6) \\ & + \sum_{a,b=e,i} z_a z_b (b_{ab}^{bound} + b_{ab}^{sc^{>3}}) + \sum_{a,b,c=e,i} z_a z_b z_c \,\hat{b}_{abc} \\ & + \sum_{a,b.c,d=e,i} z_a z_b z_c z_d \,\hat{b}_{abcd} + \cdots . \qquad (6.253)\end{aligned}$$

Here, $z_a = \tilde{z}_a(2s_a+1)/\Lambda_a^3$ are the normalized fugacities with $\tilde{z}_a = \exp(\beta\mu_a)$ and $\hat{b}_{ab,}; \hat{b}_{abc}$ and $\hat{b}_{abcd}$ are the cluster coefficients for the case of a two-component system. In our notation, the cluster coefficients $\hat{b}$ do not include ideal parts in the case of identical particles, i.e., $\hat{b}_{aa} = b_{aa} - b^0_{aa}$, etc. The ideal parts are summed up by the first term on the r.h.s of (6.253). The second cluster coefficient was considered in detail in Sect. 6.5. For non-degenerate systems, the third and the fourth cluster coefficients have the shape

$$\hat{b}_{abc} = \frac{1}{3!\Omega} \int \frac{dE}{\pi} \frac{e^{-\beta(E-\mu_a-\mu_b-\mu_c)}}{z_a z_b z_c} \operatorname{ImTr}\left[g^R_{abc}(E)\right]_{\text{conn}} , \tag{6.254}$$

$$\hat{b}_{abcd} = -\frac{1}{4!\Omega} \int \frac{dE}{\pi} \frac{e^{-\beta(E-\mu_a-\mu_b-\mu_c-\mu_d)}}{z_a z_b z_c z_d} \operatorname{ImTr}\left[g^R_{abcd}(E)\right]_{\text{conn}} . \tag{6.255}$$

In these expressions, the dynamical and the statistical information appear more or less separately. The dynamical information of the three- and four-particle problems, respectively, is given by the retarded Green's functions, where $[g^R]_{\text{conn}}$ means the connected parts as given, e.g., for the three-particle case in Sect. 6.5.1 in the trace of the resolvents. The statistical information, however, is expressed in terms of the Boltzmann factor. In general, there we have Bose- or Fermi-type logarithms.

Let us first treat the sum of the third cluster coefficients. Some considerations concerning the third cluster coefficient were already given. Out of the sum of the cluster coefficients, we will select only those contributions representing bound states of the particles *eep* or *pep* and scattering between bound e–p pairs and free electrons or protons, respectively. Consequently, we consider terms of the type $\hat{b}_{eep}$ ($\hat{b}_{pep}$) which appear three times in the sum over the species. In particular, $\hat{b}_{eep}$ is determined by (see also Sect. 6.5.1)

$$\begin{aligned} \operatorname{ImTr}_{eep}\, [g^R_{eep}]_{\text{conn}} &= \operatorname{ImTr}\Big[g^R_{eep} - g^{0R}_{eep} - (g^R_{e(ep)} - g^{0R}_{eep}) \\ &\quad -(g^R_{e(ep)} - g^{0R}_{eep}) - (g^R_{(ee)p} - g^{0R}_{eep})\Big] . \end{aligned} \tag{6.256}$$

Here is, e.g., $i^2 g^R_{e(ep)}(E) = (E - H^0_{eep} - V_{ep} + i\varepsilon)^{-1}$ one of the channel Green's functions of the corresponding three-particle problem. Let us consider the full three-particle Green's function using the completeness and normalization properties of the eigen states of the three-particle Hamiltonian (Taylor 1972). Then we can write

$$g^R_{aep}(E) = \sum_n \int d\boldsymbol{P}\, \frac{|n\boldsymbol{P}\rangle\,\langle \boldsymbol{P}n|}{E - E^{aep}_{n\boldsymbol{P}} + i\varepsilon} + \sum_\kappa \int d(\kappa\alpha)\, \frac{|\kappa,\alpha+\rangle\,\langle +\alpha,\kappa|}{E - E^\kappa_{aep} + i\varepsilon} , \tag{6.257}$$

where $|n\boldsymbol{P}\rangle$ are the three-particle bound states with the energies $E^{aep}_{n\boldsymbol{P}}$, and $|\kappa,\alpha+\rangle$ are the scattering states classified with respect to the asymptotic channels $|\kappa,\alpha\rangle$. Here, κ is the channel number and α denotes the corresponding set of dynamical variables. The $|\kappa,\alpha+\rangle$ represent states accounting for scattering of three unbound particles ($\kappa=0$), and for scattering between one free particle and a two-particle sub-cluster ($\kappa=1,2$). A classification of the scattering channels for the three-particle problem can be found in Table 7.1. For simplicity, we will assume the bound e–p pairs (H-atoms) to be in the ground state only with the binding energy $E_1=-13.6\,\mathrm{eV}$.

With (6.256) and (6.257), we are able to separate the three-particle bound state and scattering contributions between free electrons or protons and e–p sub-clusters. For this goal, as a consequence from the expression given above, we have to consider the bound states and the contributions from the channels 1 and 2 which are physically not distinct. If we furthermore apply the substitution $E\to E+E_1$, the three-particle contribution to the pressure takes the form

$$3\,\hat{b}_{aep}=\frac{1}{z_a z_e z_p}\Bigg\{\sum_n g_n\int\frac{d\boldsymbol{P}}{(2\pi\hbar)^3}\,e^{-\beta(E^{aep}_{nP}-\mu_a-\mu_e-\mu_p)}$$
$$-\frac{1}{\Omega}\int\frac{dE}{\pi}e^{-\beta(E+E_1-\mu_a-\mu_e-\mu_p)}\,\mathrm{Im}\mathrm{Tr}\int d\alpha\,\frac{|1,\alpha+\rangle\,\langle+\alpha,1|-|1,\alpha\rangle\,\langle\alpha,1|}{(E+E_1)-E^1_{aep}+i\varepsilon}\Bigg\}. \tag{6.258}$$

Here, n denotes the set of quantum numbers of the three-particle bound state, g_n is the statistical weight and $\boldsymbol{P}$ the total momentum. Furthermore, we have

$$|1,\alpha\rangle=|\boldsymbol{p}_a\rangle|1s\,\boldsymbol{P}_{ep}\rangle\,;\qquad E^1_{aep}=\frac{p_a^2}{2m_a}+\frac{P_{ep}^2}{2M_{ep}}+E_1 \tag{6.259}$$

with $a=e,p$. The atomic ground state is denoted by $|1s\boldsymbol{P}_{ep}\rangle$. The physical interpretation is obvious. The first term in (6.258) represents the contribution of three-particle bound states to the pressure, i.e., with $a=e$, it accounts for H^--particles and, for $a=p$, it describes molecular ions H_2^+. The second term gives the contribution of scattering states. Therefore, we write

$$3\,\hat{b}_{aep}=b^{bound}_{aep}+b^{sc}_{aep}\,. \tag{6.260}$$

Expressions for the bound state parts of the third and fourth cluster coefficients were given already by (Rogers and DeWitt 1973). In the next step, we analyze the contribution of the fourth cluster coefficient to the pressure. The formation of H_2-molecules and the scattering between two H-atoms is described by b_{epep}, which occur 6 times in the sum over the species:

$$6\,\hat{b}_{epep}=-\frac{1}{2!2!\Omega}\int\frac{dE}{\pi}\frac{e^{-\beta(E-2\mu_e-2\mu_p)}}{z_e^2 z_p^2}\,\mathrm{Im}\mathrm{Tr}\left[g^R_{epep}(E)\right]_{\mathrm{conn}}\,. \tag{6.261}$$

From this expression, we consider the contributions only, which correspond to four-particle bound states and to scattering states between two bound e–p pairs in the ground state. This channel is denoted by $\kappa = \kappa_{HH}$. Considerations similar to the three-particle case lead to

$$
\begin{aligned}
6\,\hat{b}_{epep} &= b^{bound}_{epep} + b^{sc}_{epep} \\
&= \frac{1}{z_e^2 z_p^2}\frac{1}{2}\Bigg\{\sum_n g_n \int \frac{d\boldsymbol{P}}{(2\pi\hbar)^3} e^{-\beta(E^{epep}_{n\boldsymbol{P}} - 2\mu_e - 2\mu_p)} \\
&\quad - \frac{1}{\Omega}\int \frac{dE}{\pi} e^{-\beta(E+2E_1-2\mu_e-2\mu_p)} \\
&\quad \times \mathrm{Im}\mathrm{Tr}\int d\alpha\, \frac{|\kappa_{HH},\alpha+\rangle\langle +\alpha,\kappa_{HH}| - |\kappa_{HH},\alpha\rangle\langle\alpha,\kappa_{HH}|}{(E+2E_1) - E^{HH}_{epep} + i\varepsilon}\Bigg\},
\end{aligned}
\tag{6.262}
$$

where

$$
|\kappa_{HH},\alpha\rangle = |1s\,\boldsymbol{P}_{ep}\rangle|1s\,\boldsymbol{P}'_{ep}\rangle; \quad E^{HH}_{epep} = \frac{P^2_{ep}}{2M} + \frac{P'^2_{ep}}{2M} + 2E_1\,. \tag{6.263}
$$

Like in Sect. 6.5.6, again, the bound state contributions lead to the idea of introducing the chemical picture, i.e., to consider the bound states to be new species (*composite particles*): The two-particle bound states are H-atoms, the three-particle ones are H^-- and H_2^+-ions, respectively, and the four-particle bound states are, of course, H_2-molecules. In order to realize the transition to the chemical picture, the fugacities of the new species have to be defined such that the bound state contributions appear like ideal particles in the EOS. Assuming the non-degenerate case, this goal is achieved by means of the transformations

$$
\begin{aligned}
z_H &= z_e z_p \Lambda_e^3 e^{-\beta E_1} & z_{H_2} &= z_e^2 z_p^2 b^{bound}_{epep} = z_H z_H b^{bound}_{HH} \\
z_{H^-} &= z_e z_H b^{bound}_{eH} & z_{H_2^+} &= z_p z_H b^{bound}_{pH}\,.
\end{aligned}
\tag{6.264}
$$

In the definition of z_H, we used $m_e \ll m_p$. From the transformations follow that the binding energies in b^{bound}_{aH} are given relatively to that of the hydrogen atom, and in b^{bound}_{HH} relatively to that of two atoms. With such definitions, (6.253) takes the form of an EOS of a multi-component system, the components of which being in chemical equilibrium

$$
\begin{aligned}
\beta p &= \sum_{a=e,p}(2s_a+1)\int\frac{dp}{(2\pi\hbar)^3}\ln\left|1+\tilde{z}_a e^{-\beta E_a}\right| + \sum_c z_c \\
&\quad + \beta p^{HF} + \beta p^{MW} + \beta p^4 + \beta p^6 + \sum_{a,b=e,p} z_a z_b b^{sc>3}_{ab} \\
&\quad + z_e z_H b^{sc}_{eH} + z_p z_H b^{sc}_{pH} + z_H\, z_H\, b^{sc}_{HH}\,.
\end{aligned}
\tag{6.265}
$$

The first sum in the first line runs over all free charged particles, and the second one over the composite particle ($c = \mathrm{H}, \mathrm{H}^-, \mathrm{H}_2^+, \mathrm{H}_2$). Interaction corrections occur as scattering parts of screened second cluster coefficients of elementary charged particles and of cluster coefficients of composite particles given by

$$b^{sc}_{aH} = -\frac{1}{\Omega}\frac{\Lambda_a^3\Lambda_p^3}{(2s+1)^3}\int\frac{dE}{\pi}e^{-\beta E}\,\mathrm{Im}\mathrm{Tr}\int d\alpha\,\frac{|1,\alpha+\rangle\langle+\alpha,1|-|1,\alpha\rangle\langle\alpha,1|}{(E+E_1)-E^1_{aep}+i\varepsilon}$$

$$b^{sc}_{HH} = -\frac{1}{2\Omega}\frac{\Lambda_p^6}{(2s_e+1)^2(2s_p+1)^2}\int\frac{dE}{\pi}e^{-\beta E} \times \mathrm{Im}\mathrm{Tr}\int d\alpha\,\frac{|\kappa_{HH},\alpha+\rangle\langle+\alpha,\kappa_{HH}|-|\kappa_{HH},\alpha\rangle\langle\alpha,\kappa_{HH}|}{(E+2E_1)-E^{HH}_{epep}+i\varepsilon}\,. \qquad (6.266)$$

Here, E^1_{aep} and E^{HH}_{epep} are given by (6.259) and (6.263) with $E_j = E_{j'} = E_1$. The calculation of the cluster coefficients b^{sc}_{aH} and b^{sc}_{HH} requires the solution of the corresponding three- and four-particle scattering problems assuming bound electron–proton pairs in the ground state. However, within an approximate treatment they can be reduced to second cluster coefficients. This will be shown in the following section considering the fourth cluster coefficient.

The EOS derived thus describes the observed behavior of a low density hydrogen plasma with stable atoms, molecules, etc. In the chemical picture, the bound state parts of the cluster coefficients contribute to the ideal pressure. In this sense, the bound states contribute both in this simple way and in additional scattering contributions.

In the chemical picture, the (new) fugacities are restricted by the relations (6.264). These conditions control the chemical equilibrium and, thus, the composition of the plasma. In the grand canonical ensemble, the condition for the chemical equilibrium follows immediately from the interpretation of bound states as new particles. No additional extremum principle is needed. It is obvious that such (chemical) picture is meaningful only so far as the bound states have a sufficiently long life time. This important question is dealt with, e.g., for the two-particle bound states, considering an effective Schrödinger equation (see Sect. 5.9).

Finally, we may conclude that the EOS given above provides an approximate description of the partially ionized plasma. For degrees of ionization near or equal to unity, we get the EOS of the fully ionized plasma. This situation was dealt with in Sects. 6.3 and 6.4. In the limiting situation with a degree of ionization equal to zero, we get the EOS of the neutral H-atom and H_2-molecular gas.

6.6.3 The Low-Density H-Atom Gas

In order to study the role of the neutral particles, we now consider the limiting case of a gas of interacting H-atoms only. Then the fugacities of free electrons and protons are equal to zero, and the EOS simplifies to one for a mixture of H-atoms and H_2-molecules being in chemical equilibrium. The composition of such a gas is determined by (6.264). The thermodynamic properties follow from equation (6.265) for $z_e = z_p = 0$, i.e., from

$$\beta p = z_H + z_{H_2} + z_H z_H b_{HH}^{sc}\,, \tag{6.267}$$

where the fugacity of molecules is defined by

$$z_{H_2} = z_H z_H b_{HH}^{bound}\,. \tag{6.268}$$

Here, the interaction contributions are determined by the fourth cluster coefficient. For bound electron–proton pairs in the ground state, we have according to (6.262)

$$b_{epep}^{bound} + b_{epep}^{sc} = \Lambda_e^6 e^{-2\beta E_1^{ep}} \left(b_{HH}^{bound} + b_{HH}^{sc}\right). \tag{6.269}$$

The bound state contribution describes the formation of H_2-molecules according to (6.268). The contribution of scattering processes is given by the second line of (6.266). Let us consider the fourth cluster coefficient including bound and scattering states more in detail. For this purpose, we introduce the trace of the resolvent F_{HH}, and we get

$$\hat{b}_{HH} = \frac{\Lambda_p^6}{(2s_e+1)^2(2s_p+1)^2}\frac{1}{2\Omega}\int \frac{dE}{\pi}\, e^{-\beta E}\,\mathrm{Im}F_{HH}(E+i\varepsilon) \tag{6.270}$$

with $b_{HH} = b_{HH}^{bound} + b_{HH}^{sc}$. To evaluate this expression, it is useful to determine the trace of the resolvent in the coordinate representation (spins are dropped for simplicity)

$$F_{HH}(E+i\varepsilon) = -\mathrm{Im}\int dX\,dx \sum_{\nu,\Gamma} \frac{|\Psi_{\nu,\Gamma}(X,x)|^2 - \left|\Psi_{\nu,\Gamma}^0(X,x)\right|^2}{E + 2E_1 - E_{\nu,\Gamma} + i\varepsilon}\,. \tag{6.271}$$

Here, X and x denote a set of coordinates of the protons and electrons, respectively. The functions $\Psi_{\nu,\Gamma}(X,x)$ are the eigenfunctions of the four-particle Hamiltonian with ν, Γ being the label of the corresponding set of quantum numbers. Because of the great difference of the masses $m_e/m_p \ll 1$, it is possible to separate the motion of the electrons and protons, this means we use the adiabatic approximation (Born–Oppenheimer separation). In adiabatic approximation, the scattering and bound state eigenfunctions can be written in the form

$$\Psi_{\nu,\Gamma}(X,x) = \Phi_\nu^\Gamma(X)\,\varphi_\Gamma(X,x)\,. \tag{6.272}$$

The functions $\varphi_\Gamma(X,x)$ are the eigenfunctions of the electron system depending on the positions of the heavier particles. The state of the heavier particles for a given electronic state is described by $\Phi_\nu^\Gamma(X)$.

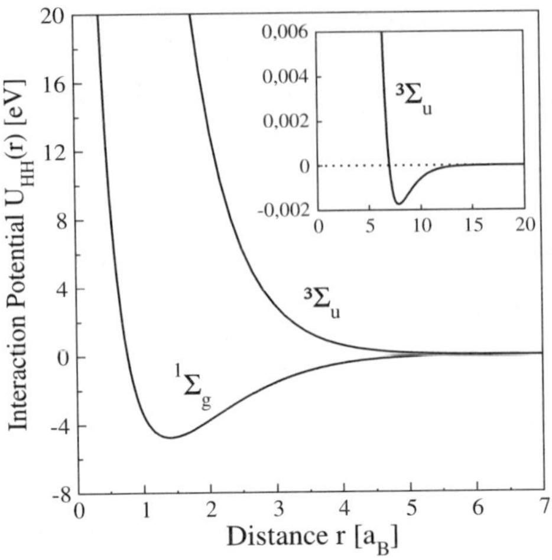

Fig. 6.23. Interaction energy $U^{s(t)}_{HH}(r)$ of two hydrogen atoms. *Singlet state*: $^1\Sigma_g$; *triplet state*: $^3\Sigma_u$. The insert shows details of the triplet potential

Within a perturbation calculation, first the equation for the electronic states for fixed protons is solved. This problem is the hydrogen molecule problem. The solution is well known from textbooks. We get singlet states φ^s for antiparallel spins and triplet states φ^t for parallel spins. The electronic energy eigenvalues are

$$\varepsilon^{s(t)}(R) = U^{s(t)}_{HH}(R) + 2E_1 \,, \tag{6.273}$$

where R denotes the distance between the protons. Very accurate calculations of $U^{s(t)}(R)$ were given by Kolos and Wolniewicz (1965) and are presented in Fig. 6.23. Following the scheme of the Born–Oppenheimer separation, we now obtain a two-particle Schrödinger equation for the protons:

$$\left(\mathrm{H}^0_{pp} + \varepsilon^{s(t)}(R) - E^{s(t)}_\nu\right) \Phi^{s(t)}_\nu = 0 \,. \tag{6.274}$$

The result is that the protons move in an effective potential $\varepsilon^{s(t)}(R)$ including the p–p Coulomb repulsion and the electronic energy. After separation of the center of mass motion, and with the ansatz $\psi^{s(t)}_\nu(\boldsymbol{R}) = R^{s(t)}_{El}(R)Y^m_l(\vartheta,\varphi)$ for the wave function of relative motion, one gets a radial Schrödinger equation for $R^{s(t)}_{El}$. The equation for the singlet state has bound and scattering solutions. Because an analytical solution is not available, we solve the Schrödinger equation for the bound states numerically (Bezkrovniy (a) et al. 2004). The result is the spectrum of vibrational and rotational energy levels $E^s_\nu = 2E_1 + E^s_{nl}$ of the H_2 molecule in the electronic ground state $^1\Sigma^+_g$. In Fig. 6.24, results are shown for E^s_{nl} of hydrogen and deuterium molecules for $l = 0$. The results are in excellent agreement with the experimental values given by Herzberg and Howe (1959). Numerical calculations were also carried out by Wolniewicz (1966) and by Waech and Bernstein (1967).

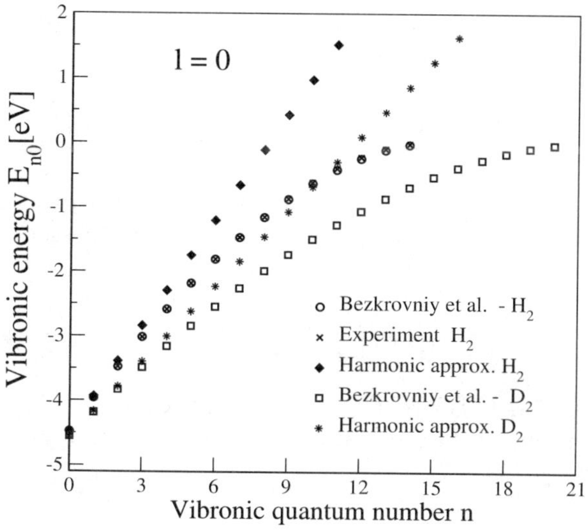

Fig. 6.24. Vibrational energy levels of hydrogen and deuterium for $l = 0$. Results from numerical solution of (6.274) – *open circles and squares*; harmonic approximation (6.275) – *diamonds and asterisks*; experimental data – *crosses*

In many cases, the energy levels are treated in the harmonic oscillator approximation. Then the bound state energies are given by

$$E^s_{nl} = U^s_{HH}(R_0) + \left(n + \frac{1}{2}\right)\hbar\omega + \frac{\hbar^2 l(l+1)}{2mR_0} . \tag{6.275}$$

Here, $U^s_{HH}(R_0) = -4.74\,\text{eV}$, $\hbar\omega = 0.54\,\text{eV}$, and $R_0 = 0.74 \times 10^{-8}\,\text{cm}$ is the position of the minimum of U^s_{HH}. However, this simple approximation of the interaction potential leads to the neglect of terms important especially for larger displacements from R_0. Moreover, the harmonic oscillator approximation leads to an infinite number of energy levels while the molecule *dissociates* already at a finite energy.

The equation for the triplet case has only scattering solutions.

Let us return to the trace of the resolvent. By summation over the electronic quantum numbers and integration over the electron coordinates, we have

$$F_{HH}(E + i\varepsilon) = -\text{Im}\Bigg\{ \int dX \sum_\nu \frac{|\Phi^s_\nu(X)|^2 - |\Phi^0_\nu(X)|^2}{E + 2E_1 - E^s_\nu + i\varepsilon}$$
$$+3 \int dX \sum_\nu \frac{|\Phi^t_\nu(X)|^2 - |\Phi^0_\nu(X)|^2}{E + 2E_1 - E^t_\nu + i\varepsilon} \Bigg\} . \tag{6.276}$$

Here, $F_{HH}(E + i\varepsilon)$ is just the trace of the resolvent with the well-known shape of that of a second cluster coefficient. Therefore, together with the effective Schrödinger equation (6.274), and using anti-symmetrized states for the protons, the fourth cluster coefficient is transformed into a second cluster coefficient, and we can use the relations of Sect. 6.5.1. We get

$$\hat{b}_{HH} = \frac{1}{4}\hat{b}^s_{HH} + \frac{3}{4}\hat{b}^t_{HH} = b^{bound}_{HH} + b^{sc}_{HH}\,, \tag{6.277}$$

with $b^{s(t)}_{HH}$ being the contribution of the the electrons with antiparallel (parallel) spins. The bound state contribution to b_{HH} comes from the singlet state. For the triplet state, there is only a scattering part. Therefore,

$$b^{sc}_{HH} = \frac{1}{4}b^{sc,s}_{HH} + \frac{3}{4}\hat{b}^t_{HH}\,. \tag{6.278}$$

Especially, we can find a Beth–Uhlenbeck representation-like formula (6.143). After performing the center of mass integrations, the cluster coefficients take the form (Schlanges and Kremp 1982)

$$\begin{aligned}
\hat{b}^s_{HH} &= 4\pi^{3/2}\,\lambda^3_{HH}\sum_{l=0}^{\infty}(2l+1)\left[1-\frac{(-1)^l}{2s_p+1}\right] \\
&\quad\times\left\{\sum_{n\geq l+1} e^{-\beta E^s_{nl}} + \frac{1}{\pi}\int_0^\infty e^{-\beta E}\frac{d}{dE}\,\delta^s_l(E)\right\} \\
\hat{b}^t_{HH} &= 4\pi^{3/2}\,\lambda^3_{HH}\sum_{l=0}^{\infty}(2l+1)\left[1-\frac{(-1)^l}{2s_p+1}\right]\frac{1}{\pi}\int_0^\infty e^{-\beta E}\frac{d}{dE}\,\delta^t_l(E)\,.
\end{aligned} \tag{6.279}$$

The two-particle quantities such as bound state energies and phase shifts have to be determined from the proton equation (6.274). They follow by numerical solution from the corresponding radial Schrödinger equation. For the discussion of b^{sc}_{HH}, we calculated the scattering parts of b^s_{HH} and b^t_{HH} from the corresponding Beth–Uhlenbeck formulas. The results are represented in Fig. 6.25.

Now, we return to the equation of state given by (6.267) and (6.268). For many applications it is more useful to consider the thermodynamic functions as a function of the density. For this purpose, we need the Mayer function $S(n)$ considered in Sect. 6.5.5. In our case, this function is very simple and given by

$$S(n) = \frac{\beta_1}{2}\,n^2_H \;\;;\;\; \beta_1 = 2\,b^{sc}_{HH}\,. \tag{6.280}$$

Next we use the relations of Sect. 6.5.5 and obtain

$$\beta p = n_H + n_{H_2} - n^2_H b^{sc}_{HH}\,. \tag{6.281}$$

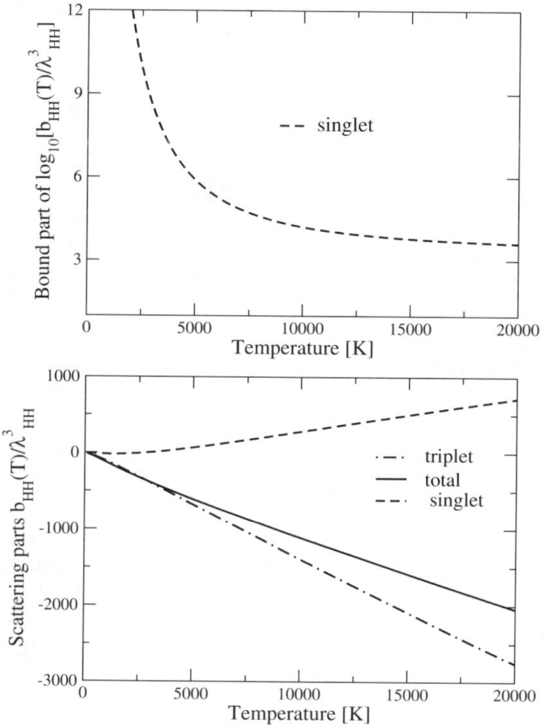

Fig. 6.25. Second cluster coefficient b_{HH} of a H-gas. Bound part (**above**); scattering part (**below**): The total, singlet and triplet scattering contributions are shown

Here, b^{sc}_{HH} represents a second virial coefficient given by the scattering contribution (6.278) with (6.279). For the fugacities and the interaction part of the atomic chemical potential we have

$$\begin{aligned} z_H &= n_H \exp(-2n_H b^{sc}_{HH}) \qquad z_{H_2} = n_{H_2} \\ \beta\mu^{int}_H &= -2n_H b^{sc}_{HH}\,. \end{aligned} \tag{6.282}$$

For the determination of the composition, we get a mass action law (MAL) given by

$$\frac{n_{H_2}}{n_H n_H} = b^{bound}_{HH} \exp(-4n_H b^{sc}_{HH})\,, \tag{6.283}$$

with the density balance

$$n = n_H + 2n_{H_2}\,. \tag{6.284}$$

These relations describe the behavior of the H-atom gas at low densities.

Let us consider the EOS and the composition of the H-gas in this low density approximation. Using the results for b_{HH}, the degree of dissociation

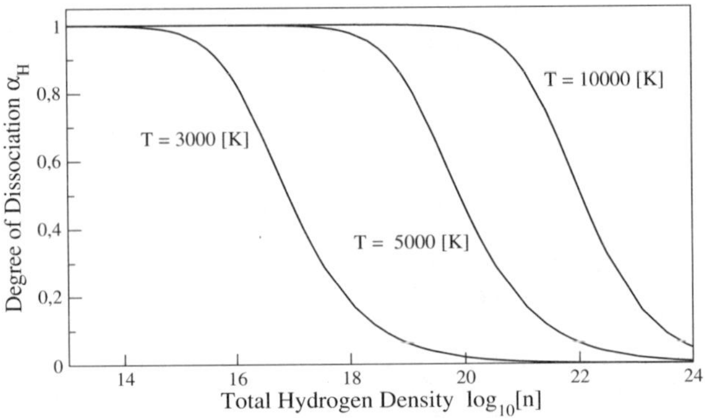

Fig. 6.26. Isotherms of the degree of dissociation as a function of the total density

$\alpha_H = n_H/(n_H + 2n_{H_2})$ is calculated from (6.283) and is represented for different temperatures in Fig. 6.26. With increasing density, the transition from an ideal atomic gas ($\alpha_H = 1$) to an ideal molecular gas ($\alpha_H = 0$) is described. Correspondingly, for isotherms of the pressure one gets at low densities $p/nk_BT = 1$ and, at high densities, this ratio tends to the value 1/2. Of course, pressure dissociation at high densities can not be described on this level of approximation.

6.6.4 Dense Fluid Hydrogen

For higher densities, we have to take into account the interaction between the H-atom and H_2-molecules and the interaction between the H_2-molecules. We will not derive these terms from first principles like for the atom-atom interaction. The further procedure is clear. We have to add the second cluster coefficients for these contributions to (6.281) with interaction potentials which have to be determined from the quantum theory of intermolecular interactions.

An extension to higher orders with respect to the density is difficult in the quantum statistical frame. Let us, therefore, consider the following idea: In classical statistical mechanics, the EOS is given by the virial formula (Hill 1956; Hansen and McDonald 1986)

$$\beta p = \sum_a n_a - \frac{2\pi}{3k_BT} \sum_{ab} n_a n_b \int_0^\infty g_{ab}(r) \frac{dU_{ab}(r)}{dr} r^3 dr \,, \tag{6.285}$$

where $g_{ab}(r)$ is the radial distribution function. For the determination of this function in high quality, we have effective methods like the hyper-netted chain (HNC) approximation, the Percus-Yevick (PY) approximation and others.

Since the atoms and the molecules are heavy particles, the EOS (which is basically a quantum one) may be extended by the classical scattering contributions of these particles. Then (6.281), given in the chemical picture, is generalized in the following way

$$\beta p = \sum_{a=H,H_2} n_a - n_H^2 (b_{HH}^{\mathrm{sc}} - b_{HH}^{cl}) - \frac{2\pi}{3k_B T} n_H^2 \int_0^\infty g_{HH}(r) \frac{dU_{HH}(r)}{dr} r^3 dr \,. \tag{6.286}$$

Here, the classical cluster coefficient

$$b_{HH}^{cl} = 2\pi \int_0^\infty \left(e^{-\beta U_{HH}(r)} - 1 \right) r^2 dr$$

has to be subtracted because this contribution is already contained in the virial formula. At this point, it should be noticed that the potential U_{HH} in the third term of (6.286) has to be chosen such that only scattering contributions are taken into account (see below). As a final generalization, we include the H–H_2 and H_2–H_2 interactions, and the equation for the pressure takes the form

$$\beta p = \sum_{a=H,H_2} n_a - \sum_{a,b=H,H_2} n_a n_b \Big\{ (b_{ab}^{\mathrm{sc}} - b_{ab}^{cl}) - \frac{2\pi}{3k_B T} \int_0^\infty g_{ab}(r) \frac{dU_{ab}(r)}{dr} r^3 dr \Big\} . \tag{6.287}$$

Similarly we may express other thermodynamic functions by the radial distribution function. So we get, e.g., for the interaction part of the chemical potential (Hill 1956)

$$\beta \mu_a^{int} = -\sum_b n_b \Big\{ 2(b_{ab}^{\mathrm{sc}} - b_{ab}^{cl}) - \frac{4\pi}{k_B T} \int_0^1 \int_0^\infty g_{ab}(r,\lambda) U_{ab}(r) \, r^2 dr d\lambda \Big\} , \tag{6.288}$$

where λ is the coupling parameter. Together with the mass action law

$$n_{H_2} = n_H^2 b_{HH}^{bound} \exp\left\{ \beta(2\mu_H^{int} - \mu_{H_2}^{int} \right\} , \tag{6.289}$$

we get a complete description of the H-atom gas.

In order to obtain classical density corrections to the quantum-mechanical EOS, we have, therefore, to determine the radial distribution function. An equation for this function may be obtained from the second equation of the Martin–Schwinger hierarchy. We take the time-diagonal equation in the classical limit and consider the equilibrium case. The result of these straightforward operations is the well-known second equation of the BBGKY-hierarchy

$$k_B T \frac{\partial}{\partial \boldsymbol{r}_a} g_{ab} + g_{ab} \frac{\partial}{\partial \boldsymbol{r}_a} U_{ab} + \sum_c n_c \int d\boldsymbol{r}_c \frac{\partial}{\partial \boldsymbol{r}_a} U_{ac}\, g_{abc} = 0 \,. \tag{6.290}$$

This equation is a complicated integro-differential equation. The main problem is to find an approximation for a closure relation. There exist many ideas like density expansions, superposition approximations, etc., for the solution to this problem. An overview is given in textbooks, e.g., by Hill (1956), by Münster (1969), and by Hansen and McDonald (1986).

A more general and powerful formalism for the calculation of g_{ab} is the direct correlation function method. With the direct correlation function, we may derive a formally closed equation for the correlation function of very simple structure. A derivation of such equation starting from (6.290) was proposed by Kremp (a) et al. (1983). Following this paper, we introduce the direct correlation function c_{ab} by

$$\begin{aligned} & g_{ab} \frac{\partial}{\partial \boldsymbol{r}_a} U_{ab} + \sum_c n_c \int d\boldsymbol{r}_c \frac{\partial}{\partial \boldsymbol{r}_a} U_{ac}\, g_{abc} \\ & = -k_B T \frac{\partial}{\partial \boldsymbol{r}_a} \left(c_{ab} + \sum_c \int c_{ac}\, h_{cb}\, d\boldsymbol{r}_c \right) . \end{aligned} \tag{6.291}$$

Here, $h_{ab} = g_{ab} - 1$ is the total correlation function. Introducing this definition into (6.290), we get the well-known Ornstein–Zernicke equation

$$h_{ab}(r) = c_{ab}(r) + \sum_c n_c \int c_{ac}(|\boldsymbol{r} - \boldsymbol{r}'|)\, h_{cb}(\boldsymbol{r}')\, d\boldsymbol{r}' \,. \tag{6.292}$$

The Ornstein–Zernicke equation is, knowing the direct correlation function, a closed integral equation with a convolution structure for the binary distribution function. Therefore, we can immediately find a formal solution in Fourier space. The most important problem now is to find approximations for the direct correlation function.

A good approximation for the solution to this problem is the HNC-approximation given by

$$g_{ab}(r) = \exp\left\{ -\frac{U_{ab}(r)}{k_B T} + h_{ab}(r) - c_{ab}(r) \right\} . \tag{6.293}$$

If the potentials are given for the interactions U_{HH}, U_{HH_2}, and $U_{H_2H_2}$, we have now a closed system of equations for the determination of both the correlation function and the direct correlation function.

The interaction of two hydrogen atoms in the ground state is, in principle, given by (6.273). It depends on the spin of the electrons, i.e., we have to consider two different states, the singlet state $^1\Sigma$ and the triplet one $^3\Sigma$. As it was shown in the previous section, this gives no problems in the approximation of the quantum mechanical 2nd cluster coefficient (6.267) and (6.268) with (6.277) and (6.279). Especially, in these formulae, the bound states are separated, and a double counting is avoided. These bound states may be separated in the classical theory, too, by a reduction of the phase space to classical scattering states only. This was shown for the 2nd virial coefficient in a paper by Kremp and Bezkrowniy (1996).

In the classical theory of integral equations, the spin dependence of the H–H interaction represents some complication. For a discussion, we make use of the results for the cluster coefficient presented in Fig. 6.25. As can be seen from the behavior of the curves, the scattering part of the singlet contribution represents only a small correction at low temperatures. This means, the singlet cluster coefficient is dominated by its bound state part. Furthermore, the scattering contribution of the total atom–atom cluster coefficient is essentially determined by the triplet contribution (which has a scattering part only). Moreover, we can assume the interaction between the H-atoms to be mediated by a *triplet-like potential* which avoids double counting of H_2-contributions in the EOS. An atom–atom potential determined by averaging the singlet and triplet potential was used by Saumon and Chabrier (1991). But in many cases, a modified *Buckingham* EXP6 potential (Hirschfelder et al. 1954) is used

$$U_{ab}(r) = \frac{\epsilon_{ab}}{\alpha_{ab} - 6} \left[\exp\left(\alpha_{ab}(1 - \frac{r}{r^*_{ab}}) \right) - \alpha_{ab} \left(\frac{r^*_{ab}}{r} \right)^6 \right] , \tag{6.294}$$

with the parameters

$$\alpha_{HH} = 13.0, \quad \epsilon_{HH}/k_B = 20.0\,K, \quad r^*_{HH} = 1.40\,\text{Å}\,,$$

proposed by Ree (1988). In order to find the interaction between two hydrogen molecules in the ground state, we have to determine the ground state energy of this system. This is a complicated problem of quantum chemistry. The main result is that the energy of interaction is a sum of three terms of different physical meaning:

$$U_{H_2H_2} = U^{\text{stat}}_{H_2H_2} + U^{\text{val}}_{H_2H_2} + U^{\text{dis}}_{H_2H_2} . \tag{6.295}$$

Here, the first part is the electrostatic interaction of the multi-poles of the molecules, the second term is the short range valence energy arising from the overlapping of the wave functions, and the last term is the long range dispersion energy. All these quantities are complicated functions of the configuration of the eight particles, and we especially obtain an angular dependence of the

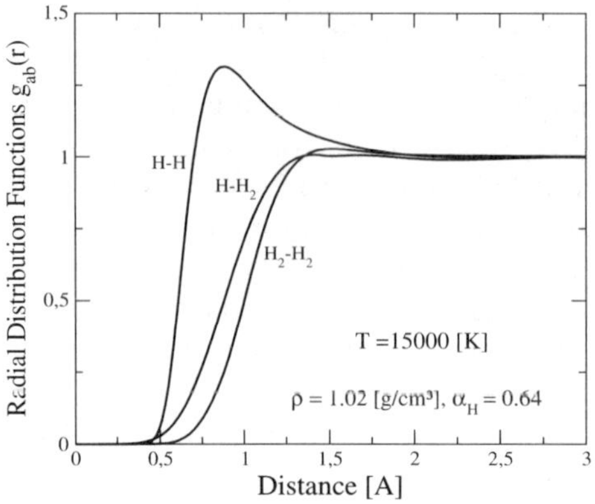

Fig. 6.27. Radial distribution functions for fluid hydrogen

interaction. The problem of fitting the angular behavior may be avoided by performing an average over the angular variables. After a spherical averaging, the potential (6.295) reduces to

$$\bar{U}_{H_2H_2} = \bar{U}^{\text{val}}_{H_2H_2} + \bar{U}^{\text{dis}}_{H_2H_2} , \tag{6.296}$$

where the short range strong repulsive valence part is well approximated by an exponential behavior, and the long range attractive part is sufficiently well described by an inverse sixth-power function. Thus an ansatz of the form of a *modified Buckingham potential* given by (6.294) is theoretically well established, too. The constants may be determined with high accuracy by *ab initio* quantum mechanical calculations. Starting from these values, an improved agreement with experimental results may be obtained by fitting the parameters to experimental data. In the paper by Ross et al. (1983), data from shock experiments were used for such a fit with the result $\epsilon_{H_2H_2}/k_B = 36.4\,\text{TK}$, $\alpha_{H_2H_2} = 11.1$, $r^*_{H_2H_2} = 3.43\,\text{Å}$.

Let us now consider the interaction between an H-atom and an H_2-molecule. Again we have a valence and a dispersion part, and, for the analytical representation, the modified Buckingham potential may be used. Determining the parameters using the Lorentz-Berthelot mixing rules, one gets $\epsilon_{HH_2}/k_B = 26.98\,\text{K}$, $\alpha_{HH_2} = 12.01$, $r^*_{HH_2} = 2.415\,\text{Å}$. For distances $\boldsymbol{r} \to 0$, where the EXP6 potential shows non-physical behavior, we used the extrapolation procedure proposed by Juranek and Redmer (2000). The coefficients for this extrapolation formula can be found in the original paper just mentioned.

Now we are able to solve the Ornstein-Zernike equation (6.292) in HNC-approximation. Results for the distribution function are shown in Fig. 6.27. Pressure and composition of the H–H_2-mixture follow from (6.287), (6.288), and (6.289). In the following, we will call this scheme HNC&MAL method. For the practical calculation of the interaction parts of the chemical poten-

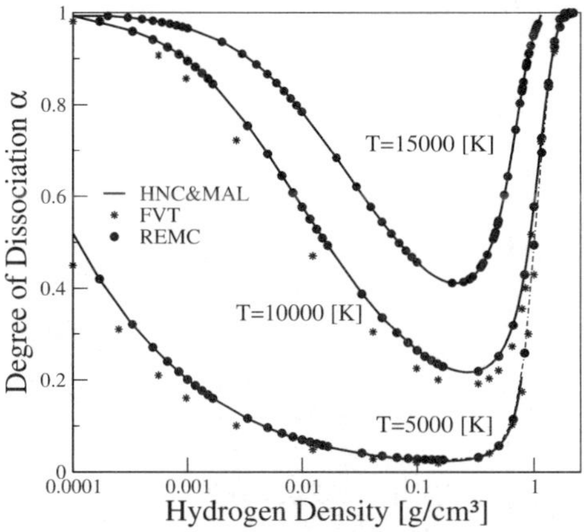

Fig. 6.28. Degree of dissociation α for fluid hydrogen calculated by HNC& MAL and REMC methods. Comparison to FVT Juranek et al. (2002) is also given. At $T = 5000\,\mathrm{K}$, the HNC&MAL result is terminated at the density $0.78\,\mathrm{g/cm}^3$. This indicates the region where we did not find solutions to the Ornstein-Zernicke equation

tials, we used an expression given in terms of the total and of the direct correlation functions (Hansen et al. 1977)

$$\beta\mu_a^{\mathrm{int}} = \sum_b \int d\boldsymbol{r}\, \{h_{ab}(r)\,[h_{ab}(r) - c_{ab}(r)] - c_{ab}(r)\}\ . \tag{6.297}$$

Results for the degree of dissociation $\alpha = n_H/(n_H + 2n_{H_2})$ as a function of the total hydrogen density for different temperatures are shown in Fig. 6.28.

Further we show in this figure results obtained by the Reaction Ensemble Monte Carlo (REMC) method (Bezkrovniy (a) et al. 2004) and the fluid variational theory (FVT) (Juranek and Redmer 2000; Juranek et al. 2002). There is an excellent agreement between HNC&MAL results and REMC calculations for the most temperatures and densities. For lower temperatures, we have found that there are regions in the n-T-plane, like in the case of $T = 5000\,\mathrm{K}$, where no solutions of the Ornstein–Zernicke equation exist while the REMC calculations give well defined results. Therefore, REMC is an appropriate method to complete the investigation of the hydrogen gas mixture at lower temperatures. Comparison to the degree of dissociation using FVT shows a good agreement at $5 \times 10^4\,\mathrm{K}$, but for the temperature $T = 10^5\,\mathrm{K}$, results of FVT are below ours. This means that the molecular partition function in harmonic approximation used in the FVT approach produces more molecules than our procedure does at given temperature and density.

An important feature of HNC&MAL, FVT and REMC methods is the correct quantum mechanical treatment of the H_2-molecule problem and, therefore, of the H_2-bound state contribution to the EOS, while the scattering contributions are considered classically or semi-classically because of the large masses of the particles.

In all cases, we observe a strong increase of the degree of dissociation at higher densities due to the strong interactions of the particles in the dense fluid hydrogen. This important effect is known as pressure dissociation. First

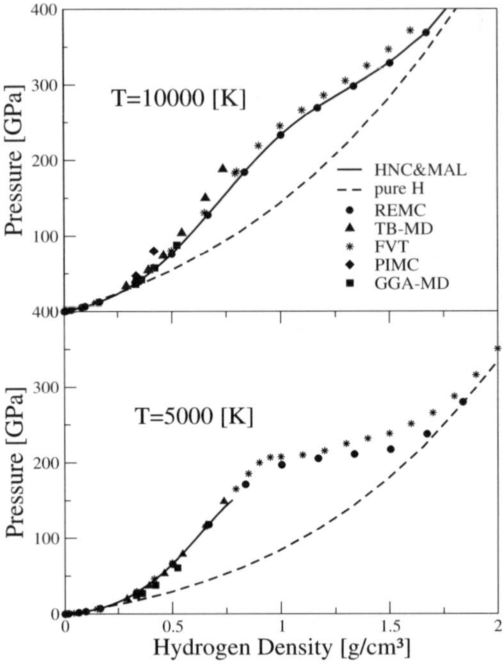

Fig. 6.29. Hydrogen pressure vs. density at $T = 10000\,\mathrm{K}$ and $T = 5000\,\mathrm{K}$. The results using HNC&MAL and REMC are compared to other theories. The curve obtained by using HNC&MAL terminates at the same density as explained in Fig. 6.28. TB-MD – Lenosky et al. (2000), GGA-MD –Collins et al. (2001)

theoretical investigations of this effect were performed in the papers by Aviram et al. (1975) and Aviram et al. (1976).

Next we consider the pressure (Bezkrovniy (a) et al. 2004; Bezkrovniy (b) et al. 2004). The isotherms for $T = 5 \times 10^4\,\mathrm{K}$ and $T = 10^5\,\mathrm{K}$ are shown in Fig. 6.29. Comparing the results for the pressure calculated by REMC and HNC&MAL, we found that they almost coincide for all temperatures and densities, except for regions where HNC does not give solutions of the Ornstein-Zernicke equation. At the temperature $T = 5 \times 10^4\,\mathrm{K}$, both REMC and HNC&MAL show very good agreement with the results obtained within FVT (Juranek and Redmer 2000; Juranek et al. 2002). At $T = 10^5\,\mathrm{K}$ and densities up to $0.5\,\mathrm{g/cm}^3$, our results and those of FVT are close to each other. In the papers by Militzer and Ceperley (2000), Militzer et al. (2001), and Militzer and Ceperley (2001), path integral Monte Carlo simulations (PIMC) were used, and higher pressures for this temperature were obtained as compared to results of REMC and HNC&MAL schemes.

The dissociation from molecular to atomic hydrogen occurs continuously, and this, in turn, leads to changing the pressure of mixtures of two components to the pressure of pure atomic hydrogen as shown in Fig. 6.29. Because of the restrictions of the pure hydrogen–gas mixture considered here, the atomic hydrogen does not undergo phase transitions to a metallic one with increasing pressure and remains as a dense atomic fluid.

For the further consideration we need still the internal energy. The internal energy per atom (in eV) is given by

$$E = \left(\frac{3}{2}k_BT + x_{H_2}E_{mol} + U^{int}\right)/(2 - x_H) - 13.6, \tag{6.298}$$

where U^{int} was calculated within REMC or HNC&MAL methods from

$$U^{int}/Nk_BT = 2\pi\frac{\hat{n}}{k_BT}\sum_a\sum_b x_a x_b \int_0^\infty U_{ab}(r)g_{ab}(r)r^2dr, \tag{6.299}$$

where $x_c = n_c/\hat{n}$ with $\hat{n} = \sum_b n_b$ is the fraction of the component c. The contribution of the internal degrees of freedom of a molecule to the internal energy is

$$E_{mol} = \frac{\sum_{n,l}(2l+1)E_{nl}\exp(-\beta E_{nl})}{\sum_{n,l}(2l+1)(\exp(-\beta E_{nl}) - 1)}. \tag{6.300}$$

Like in the case of density and pressure, the results for the internal energy, calculated by REMC and HNC&MAL, practically can not be distinguished (Bezkrovniy (a) et al. 2004). The results of the FVT (Juranek and Redmer 2000; Juranek et al. 2002) (re-scaled to the ground state energy of the isolated hydrogen atom) are also very close to our data.

Now we are able to derive the Hugoniot curve from the Hugoniot equation (6.252). For the calculation of the Hugoniot curve, the initial state was taken to be liquid deuterium with the initial density $\rho_0 = 0.171\,\mathrm{g/cm^3}$. This value is typical for experimental conditions (Nellis et al. 1983; Holmes et al. 1995; DaSilva et al. 1997; Collins et al. 1998; Mostovych et al. 2000; Ross and Yang 2001; Knudson et al. 2001; Knudson et al. 2003). We ignore the very small initial pressure p_0, using $p_0 = 0$. The initial energy was taken to be $E_0 = -15.875\,\mathrm{eV/atom}$ which is the sum of the ionization energy of the deuterium atom $I_0 = -13.6\,\mathrm{eV/atom}$ and half the dissociation energy of the deuterium molecule $D_0/2 = -2.275\,\mathrm{eV/atom}$.

In this way, we get the Hugoniot presented in Fig. 6.30 by the solid line (Bezkrovniy (a) et al. 2004; Bezkrovniy (b) et al. 2004). The behavior of the Hugoniot can be easily understood if we take into account the asymptotical behavior of a shock of an ideal classical gas. For $p \gg p_0$, it is given by (Landau and Lifshits 1977)

$$\frac{n}{n_0} = \frac{\gamma+1}{\gamma-1}, \qquad \frac{T}{T_0} = \frac{\gamma+1}{\gamma-1}\frac{p}{p_0} \tag{6.301}$$

with $\gamma = c_p/c_v$. Here, n_0 is the number density corresponding to the mass density ρ_0. According to (6.301), the fraction T/T_0 increases along with p/p_0 unrestrictedly, while the compressibility of an ideal gas is restricted by the asymptotes. From thermodynamics we know that for an atomic gas $\gamma = 5/3$ and for a gas consisting of diatomic molecules $\gamma = 7/5$. Therefore,

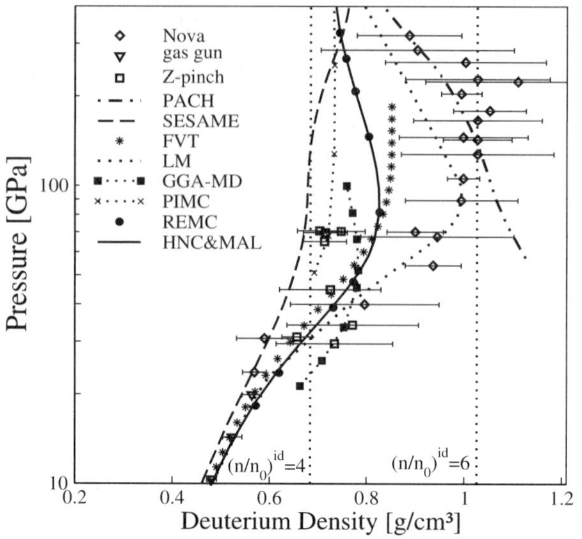

Fig. 6.30. Comparison of experimental results with Hugoniots derived from different EOS models. Nova – Nellis et al. (1983), gas gun – DaSilva et al. (1997), Z-pinch – Knudson et al. (2001), Knudson et al. (2003), PACH – Juranek et al. (2001), LM – Ross (1998), GGA-MD – Collins et al. (2001), PIMC – Militzer and Ceperley (2000), Militzer et al. (2001)

the maximum compression of an atomic gas is $n/n_0 = 4$, and $n/n_0 = 6$ for a gas of diatomic molecules because of the additional degrees of freedom.

Consequently, starting at lower pressures and temperatures, with increasing pressure, the molecular gas reaches a maximum compression which is, according to our calculations, $n/n_0 = 4.82$. Then the pressure dissociation starts as we can see in Fig. 6.29. An atomic gas evolves, and the Hugoniot approaches, along the dashed branch of the curve, the asymptotic limiting value of the compression, $n/n_0 = 4$, corresponding to an ideal classical atomic gas. At this point we note that this formally correct asymptotic behavior can only be met with the mass action constant based on the exact solution of the Schrödinger equation for the H_2-molecule problem (Bezkrovniy (a) et al. (2004)). The harmonic oscillator approximation shifts the Hugoniot to values $n/n_0 > 4$.

Of course, the asymptotic behavior obtained within our model corresponding to an ideal atomic gas with $n/n_0 = 4$ is not really met, as the atoms are ionized and the atomic gas turns into the plasma state. The Hugoniot curve following for a fully ionized plasma using Padé formulae for the thermodynamic quantities such as given in Sect. 6.3.4 is denoted by PACH (Juranek et al. 2001).

As we expected, the solutions to the Hugoniot equation (6.252) for both REMC and HNC&MAL differ only by a small amount. In the low pressure region, our results are close to the gas gun experiment data given by Nellis et al. (1983) and to the Z-pinch data by Knudson et al. (2001), Knudson et al. (2003). With increasing pressure, there are deviations. The maximum compression predicted by our calculation is $n/n_0 = 4.82$ which differs from the Z-pinch results, and at the same time from the laser driven experimental value $n/n_0 = 6$. It should be noted that for pressures higher than 30 GPa the differences between data obtained by the different experimental techniques increase. On one hand, the data obtained by Belov et al. (2002), who used the high explosive sphere experimental technique, and those of Knudson et al. (2001), Knudson et al. (2003), lie close to each other. On the other hand, higher compressibility is achieved by laser driven experiments (DaSilva et al. 1997; Collins et al. 1998).

Comparison of our results to other theoretical methods and computer simulations shows good agreement with FVT (Juranek et al. 2002) and with the linear mixing model (LM) (Ross 1998) in the low pressure regime. Differences at higher temperature are due to the different treatment of the vibration and rotation states of the H_2-molecule. Discrepancies to the results according to (Lyon and Johnson 1987), which are widely used as standard EOS, increase with increasing pressure. The values for the Hugoniot calculated within PIMC (Militzer and Ceperley 2000; Militzer et al. 2001) lie close to the line $n/n_0 = 4$. Unfortunately, there are no simulation data at low pressure, and therefore, the overlap in the region of validity of the HNC&MAL approach is small. Here, the influence of molecules becomes important which seems to be difficult to describe within the PIMC calculations. But, the PIMC calculations become more and more accurate for higher temperatures where the system is partially ionized. The results given by Collins et al. (2001) within the GGA-MD method, show a maximum compressibility $n/n_0 = 4.58$ which is close to our value. However, the general position of the GGA-MD Hugoniot curve in the $p - \rho$ plane in Fig. 6.30 differs substantially form our REMC and HNC&MAL results.

6.7 The Dense Partially Ionized H-Plasma

Up to now, we considered limiting cases of the complete EOS of H-plasmas. On one hand, the partially ionized plasma was dealt with in screened ladder approximation (6.15) and the thermodynamic functions such as the electron chemical potential μ_e, the chemical potential of protons μ_p, and the plasma pressure p_{plasma} were determined in this approximation. On the other hand, we investigated the thermodynamic properties of the neutral H–H_2 gas. After separation of the bound state part in second virial coefficients, the pressure p_{gas} and the chemical potentials of the atoms μ_H and of the H_2-molecules μ_{H_2} were determined using the HNC scheme in (6.287) or, alternatively, the Reaction Ensemble Monte Carlo method.

Let us now discuss the complete EOS. For this aim, it is useful to move to the chemical picture like in Sects. 6.5.6 and 6.6.2. Then, the pressure is composed as

$$p = p_{\mathrm{pl}} + p_{\mathrm{pl,gas}} + p_{\mathrm{gas}} \,. \tag{6.302}$$

To get the plasma composition, we start from the transformations (6.239) and (6.264). We get the coupled MAL's

$$\begin{aligned} \frac{n_H}{n_p} &= \Lambda_e^3 \exp\left(-\beta E_1\right) \exp\left[\beta\{\mu_e^{\mathrm{id}} + \mu_e^{\mathrm{int}} + \mu_p^{\mathrm{int}} - \mu_H^{\mathrm{int}}\}\right] , \\ \frac{n_{H_2}}{n_H^2} &= b_{HH}^{\mathrm{bound}} \exp[\beta\{\mu_H^{int} + \mu_H^{int} - \mu_{H_2}^{int}\}] , \end{aligned} \tag{6.303}$$

where only the electrons are considered to be degenerate.

In the chemical picture, the pressure p_{pl} is determined merely by the scattering contributions. The scattering contributions of the screened ladder approximation may be described in sufficient quality by Padé formulae, and we may apply the scheme of Sect. 6.3.4 for the determination of the plasma contributions in (6.302) and (6.303).

For the further discussion, especially of the contribution p_{pl}, we look at the $n-T$-plane shown in Sect. 2.9. The region of strong correlation is limited by $r_s = 1$, and by $\Gamma = 1$. We see from Fig. 6.31 that, e.g., for the temperatures higher than about 3×10^4 K, the pressure is essentially determined by p_{pl} in screened ladder approximation. For lower temperatures of about $T = 2 \times 10^4$ K, there are strong correlations connected with the formation of atoms and molecules. Therefore, now all terms in the EOS (6.302) are of importance, and the screened ladder approximation for p_{pl} looses more and more its validity with decreasing temperature. It only determines the asymptotic behavior with respect to high and low densities. In this area, p_{pl} has to be determined via numerical simulations, e.g., by path integral Monte Carlo techniques, or by quantum molecular dynamics (Militzer and Ceperley 2000; Filinov et al. 2001; Knaup et al. 2001; Filinov et al. 2003). With further lowering of the temperature, we finally reach a neutral gas of H-atoms and H_2-molecules.

Let us first consider a simplified determination of the thermodynamic functions for the complete system. In this scheme, we neglect, in (6.302), the $p_{\mathrm{pl,gas}}$-term, the molecule–ion interaction, and the formation of complexes beyond the hydrogen molecule. As was done in earlier work for the H–H, H–H_2 and H_2–H_2 interactions, we use effective temperature dependent hard core diameters (Aviram et al. 1975; Haronska et al. 1987; Schlanges et al. 1995). The procedure to find the effective diameter is the following: We determine the second virial coefficient with the real potential and put it equal to the coefficient of a hard core gas;

$$B_{aa}(T) = -2\pi \int_0^\infty r^2 \left(\exp(-\frac{U_{aa}(r)}{kT}) - 1\right) dr = \frac{2\pi}{3} d_{aa}^3(T) \,. \tag{6.304}$$

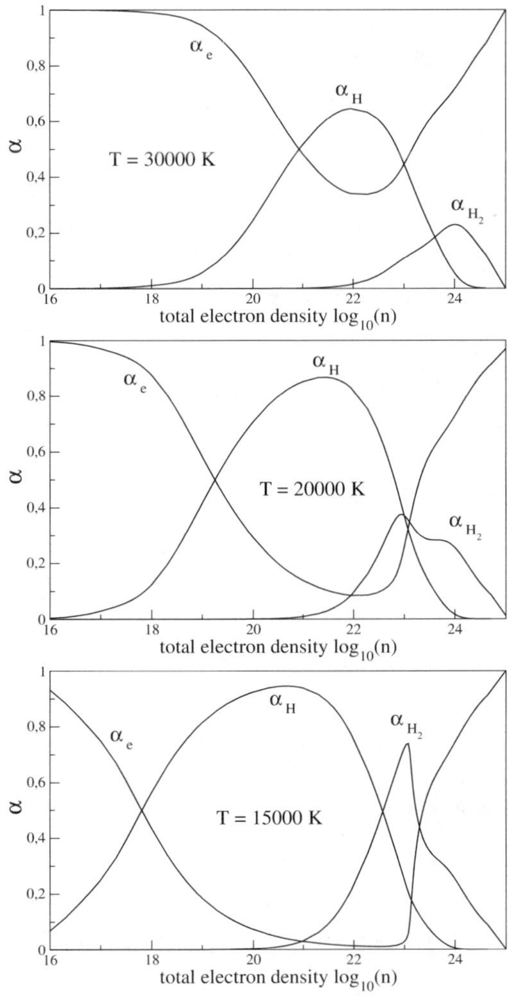

Fig. 6.31. The composition of hydrogen plasmas for different temperatures. Hard core model

In Table 6.2, we find effective diameters for some temperatures. Then we apply the formula given by Mansoori et al. (1971) for fluid mixtures of hard spheres for the calculation of the chemical potentials and the free energy.

With the chemical potentials we are able to determine the plasma composition in the ionization–dissociation model considered. In Fig. 6.31, the results of the numerical solution of the coupled mass action laws (6.303) are shown for three temperatures. Let us consider the isotherm for 15000 K. At low densities, the plasma is fully ionized. With increasing density, we observe the formation of H-atoms. For densities above $10^{21}\,\mathrm{cm}^{-3}$, the atom fraction decreases due to the formation of hydrogen molecules. Finally, at the Mott density around $n = 10^{23}\,\mathrm{cm}^{-3}$, we observe a strong decrease of the number

Table 6.2. Hard sphere diameters

T(K)	$d_H(10^{-8}$cm)	$d_{H_2}(10^{-8}$cm)
15000	1.59	1.89
16200	1.57	1.87
20000	1.51	1.80
25000	1.44	1.73
30000	1.38	1.66

Table 6.3. Plasma phase transition in an H-plasma and critical data

	T_c [10^3 K]	p_c [GPa]	ϱ_c [g/cm^3]
Ebeling and Sändig (1973)	12,6	95	0.95
Ebeling (a) et al. (1976)	12,6	3.6	
Robnik and Kundt (1983)	19,0	24	0.14
Ebeling and Richert (1985)	16,5	22,8	0.13
Haronska et al. (1987)	16.5	95	0.43
Saumon and Chabrier (1992)	15,3	61,4	0.35
Schlanges et al. (1995)	14,9	72,3	0.29
Reinholz et al. (1995)	15	26	0.2

densities of molecules and atoms, and a sharp increase of the degree of ionization due to the lowering of the ionization energy. This corresponds to the Mott effect in the model of partially ionized plasmas; see Sect. 2.8, Chap. 5, and Sect. 6.5.6. This behavior is known as pressure dissociation and pressure ionization. Physically, the Mott transition is an insulator–metal transition. This transition is not necessarily a phase transition. This is the case only if the Mott transition is connected with an instability of the thermodynamic functions.

Furthermore, the three pictures show the development of the plasma composition with decreasing temperatures. For the temperature of 15000 K, up to densities of 10^{24} cm^{-3}, the H-plasma mainly consists of neutrals.

Consider now the chemical potential and the pressure of the H-plasma in the simple hard core model. We clearly observe van-der-Waals loops below a critical temperature, i.e., we find a region of instability. As mentioned in Sect. 2.7, the van-der-Waals loop indicates the possibility of the existence of a phase transition. In addition to the calculations in Sect. 2.7 and in Sect. 6.3.1, here we took into account the neutrals and their interaction.

In the hard sphere model presented here the critical data would be

$$T_c = \ 16200\,\mathrm{K}, \qquad n_c = 1.39 \times 10^{23}\,\mathrm{cm}^{-3}, \qquad p_c = 0.33\,\mathrm{Mbar}.$$

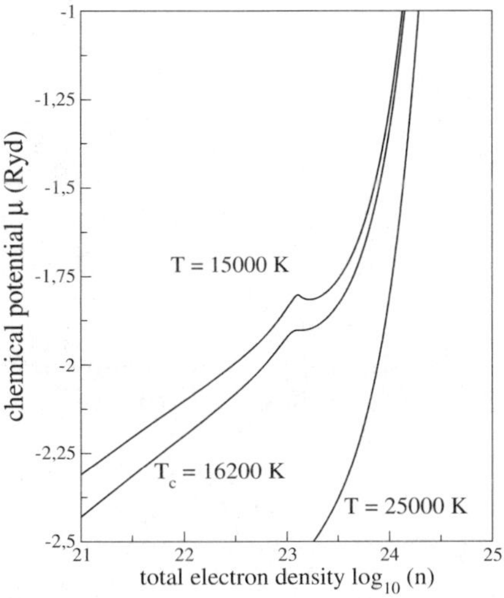

Fig. 6.32. Isotherms of the chemical potential of a hydrogen plasma (hard core model)

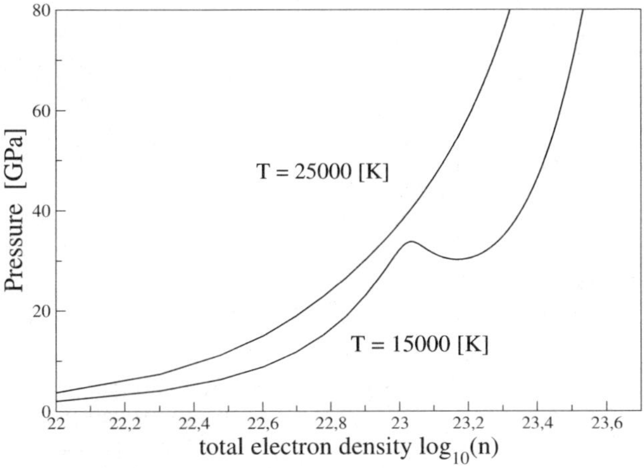

Fig. 6.33. Pressure isotherms of a hydrogen plasma (hard core model)

As mentioned already, first calculations and discussions of this plasma phase transition were performed by Norman and Starostin (1968), by Ebeling and Sändig (1973), and Ebeling (a) et al. (1976). Since then, numerous works were published. Especially in these papers, the H-atoms and H_2-molecules and their interactions were included. In Table 6.3, a survey is given mentioning papers concerning the plasma phase transition and the critical data obtained in these works.

The plasma phase transition is a first order phase transition and, therefore, connected to the coexistence of two phases. The region of coexistence may be obtained in well-known manner applying a Maxwell construction to

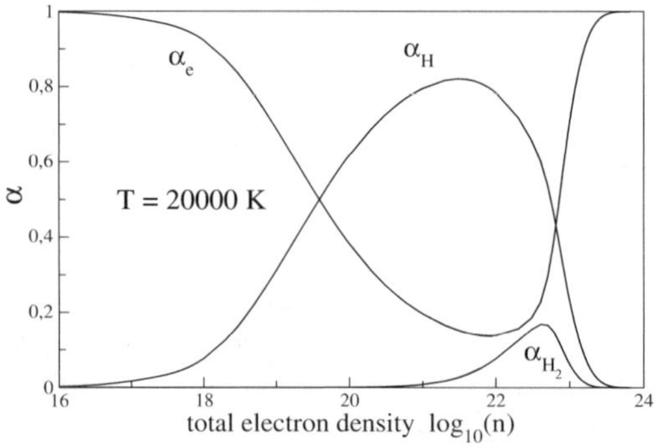

Fig. 6.34. Plasma composition for 20000 K (HNC+MAL)

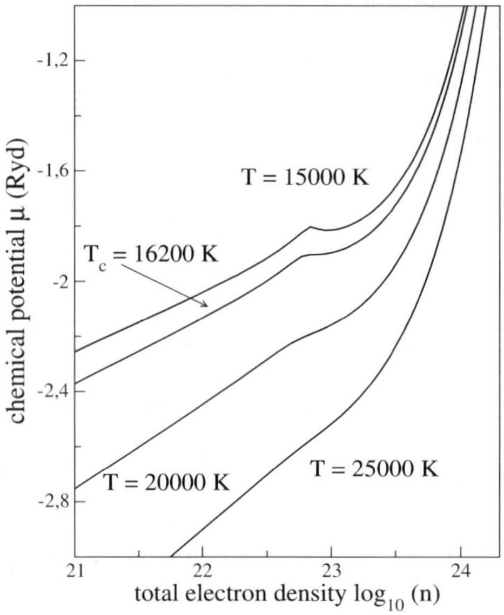

Fig. 6.35. Isotherms of the chemical potential for a hydrogen plasma

the van-der-Waals loops of the isotherms of the pressure or of the chemical potential.

Then the densities of the two phases are determined by the lower and upper borders of the coexistence region. Since the upper density, in the neighborhood of the critical point, is higher than the Mott density, one of the two phases is partially ionized, while the other is completely ionized (liquid metal like). In this case, the Mott transition is, therefore, a Mott phase transition.

We now come back to the more realistic interaction potentials between the neutrals used in the preceding section. Again, the Padé formulae are applied for the plasma, but for the neutral gas parts, like in the preceding section, the HNC approximation is applied. Numerical results for the plasma composition are given in Fig. 6.34. The physical contents of this picture is

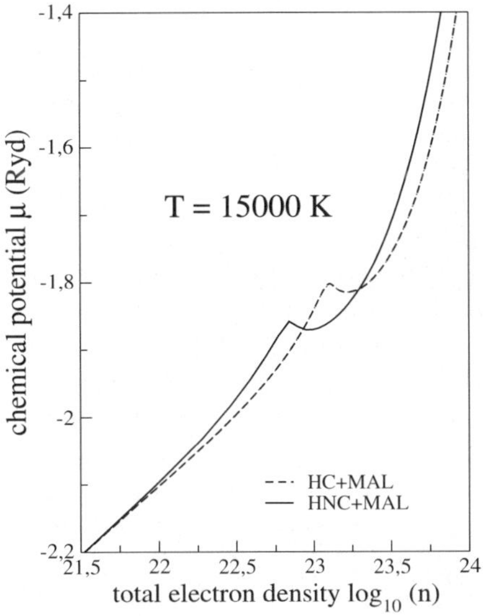

Fig. 6.36. Isotherms of the chemical potential for a hydrogen plasma: Comparison of results for hard core (HC) and soft core potentials (HNC)

the same as before. We find only slight quantitative corrections to the hard sphere model. Furthermore, in Fig. 6.35, we find isotherms of the chemical potential. Again a van-der-Waals loop appears. The critical data are (the critical pressure p_c was not determined)

$$T_c = 16200\,\mathrm{K}, \qquad n_c = 0.79 \times 10^{23}\,\mathrm{cm}^{-3}.$$

It is interesting to remark that the instability occurs in all calculations beginning from the simple Debye model and the screened ladder approximation to the more complex models with neutrals and their interactions. The instability is, therefore, produced by the plasma contributions and seems to be a universal property of systems with Coulomb interaction. This observation is emphasized by the fact that the plasma phase transition and the Mott transition occur in other systems with Coulomb interaction.

We mention the electron–hole plasma in highly excited semiconductors. Many of the properties of this interesting system follow simply by a *re-scaling* of the hydrogen results. Especially, there are theoretical reasons for the existence of the Mott transition and the plasma phase transition. These problems were intensively investigated in papers by Rice (1977), Hensel et al. (1977), Keldysh (1964), Zimmermann (1988), Ebeling (b) et al. (1976), by Smith and Wolfe (1986), and by Simon et al. (1992).

Other interesting systems are the alkali-atom plasmas. Results of extensive experimental investigations are found in the monograph by Hensel and Warren (1999). The theoretical description including phase- and Mott-transitions in alkali plasmas has been carried out in many papers, e.g., by Redmer and Röpke (1989) and by Redmer and Warren Jr. (1993). An overview and further references can be found in the report by Redmer (1997).

Instabilities and a kind of Mott transition occur also in the classical theory of Coulomb systems such as ions in an electrolyte solution. In papers by Ebeling and Grigo (1980) and Ebeling and Grigo (1982), a phase transition in electrolytes was described for the first time. A more extensive discussion of phase transitions in classical Coulomb systems may be found in papers by Fisher and Levin (1993) and Fisher (1996). A classical theory of the density dissociation was developed by Kremp and Bezkrowniy (1996).

In all calculations presented in this book, the thermodynamic functions of the partially ionized plasma are used in the screened ladder approximation. As already pointed out, this approximation is valid only for weakly nonideal plasmas. Therefore, it can not be excluded that the instability is a defect of this approximation. This idea is supported by the fact that, up to now, there is not any experimental evidence for the phase transition, and numerical simulations like RPIMC do not show instabilities.

With regard to the above discussions, it is interesting to compare the screened ladder approximation with ab initio computer simulations. There are many different methods. Quantum molecular dynamical simulations based on a density functional theory are usually applied to investigate the atomic and molecular region (Collins et al. 2001; Bonev et al. 2004). The wave packet molecular dynamics also includes the region of the fully ionized plasma (Knaup et al. 2001). Here we will not discuss these methods and refer to the work cited.

The *Path Integral Monte Carlo* (PIMC) method is another first principle method which is widely used for the investigation of the EOS of hydrogen. It is an exact solution of the many-body quantum problem for a finite system in thermodynamic equilibrium.

The idea of PIMC is the following (Feynman and Hibbs (1965)): Any thermodynamic property of a two-component plasma with N_e electrons and N_p protons at a temperature T and volume V is defined by the partition function $Z(N_e, N_p, V, T)$:

$$Z(N_e, N_p, V, T) = \frac{1}{N_e! N_p!} \sum_\sigma \int_V dR \left\langle R, \sigma | e^{-\beta \widehat{H}_N} | \sigma, R \right\rangle , \tag{6.305}$$

where R is a set of coordinates of the protons and the electrons, and σ is the spin of protons and electrons. Using the group properties of the density operator in imaginary time, the partition function (6.305) can be expressed as a product of high density operators $e^{-\beta \widehat{H}} = (e^{-\tau \widehat{H}})^n$ where the imaginary time step is $\tau = \beta/n$. In the position representation, we get in this way

$$\rho(q, r, \sigma; T) = \int dq_1 dr_1 \dots dq_{n-1} dr_{n-1}$$
$$\times \left\langle R, \sigma | e^{-\tau \widehat{H}_N} | \sigma, R_1 \right\rangle \cdots \left\langle R_{n-1}, \sigma | e^{-\tau \widehat{H}_N} | \sigma, R \right\rangle .$$

From the Hausdorff formulae follows, in the limit $\tau \to 0$, the possibility of the factorization of $e^{-\tau\hat{H}} = e^{-\tau T}e^{-\tau V}$. Then we get the Feynman path integral formula for the density matrix

$$\left\langle R,\sigma|e^{-\beta\hat{H}_N}|\sigma,R\right\rangle = \left(\frac{h}{\sqrt{2\pi m/\tau}}\right)^{-3Nn} \int dR_1 \dots dR_{n-1}$$
$$\times \exp\left[-\sum_{i=1}^{n}\tau\left\{\frac{m_i}{2}\frac{(R_{i-1}-R_i)^2}{\hbar^2\tau^2}+\frac{1}{2}(V(R_{i-1})+V(R_i))\right\}\right].$$

For identical particles, we have to take into account the spin-statistic theorem, i.e., the state vectors of Bose particles have to be symmetric, and those of Fermi particles antisymmetric. For Fermi particles, this is connected with the so-called *sign problem.* The treatment of the sign problem makes the main difference between the restricted path integral method (RPIMC) used by Militzer and Ceperley (2000) and the direct path integral method (DPIMC) by Filinov et al. (2000).

Let us consider the calculation reported in the latter paper. In this simulation, for the high temperature density matrix, an effective quantum pair potential was used, which is finite at zero distance (Filinov et al. 2003). It was obtained by Kelbg (1963) as a result of a first-order perturbation calculation of the diagonal elements two particle density matrix. These results for the Hugoniot determined by Filinov et al. (2003) together with selected other theoretical results are shown in Fig. 6.37. The lowest temperature shown in this figure for the DPIMC is 15625 K.

The behavior of the Hugoniot determined in screened ladder approximation (condensed in the Padé formulas) is shown in Fig. 6.37, too. It coincides only asymptotically with the *ab initio* RPIMC and DPIMC calculations and, with decreasing temperature, deviates considerably from those results. The Hugoniot calculated within the ACTEX theory which is not shown here exhibits a similar behavior (Rogers and Young 1997).

The main reason for the failure of the analytical theories is obvious. First it is a perturbation theory which is valid only for weakly nonideal plasmas. Further, for lower temperatures, the neutral particles, i.e., H-atoms and H_2−molecules, become more and more important, and we have a strongly coupled dense gas or liquid.

In order to correctly describe the quantum mechanics of the formation of molecules at lower temperatures with PIMC it is necessary to take many beads, i.e., to take a large number of parts in the decomposition of (6.305). In such region, PIMC calculations become very time consuming.

The natural proposal which appears for this region is to use the asymptotic property of the path integral which, for particles having large masses, goes over into the classical partition function (Feynman and Hibbs 1965). For such systems, the classical Monte Carlo scheme can be applied. A good

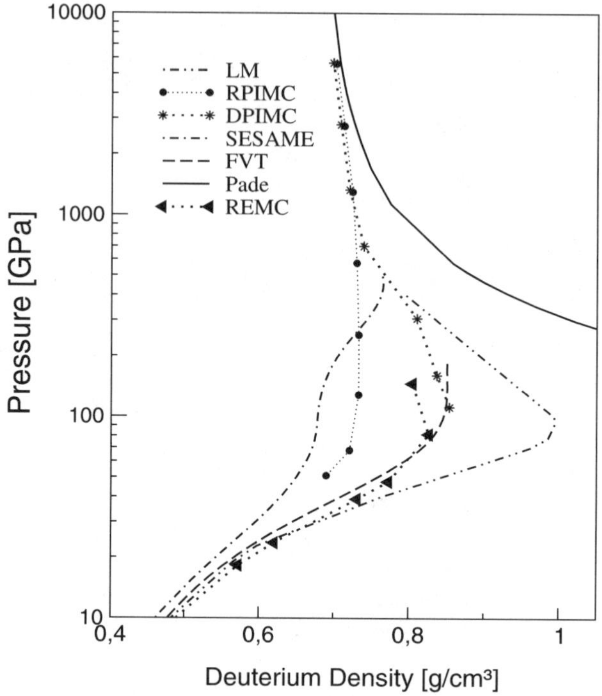

Fig. 6.37. Comparison of the Padé Hugoniot with several theoretical results. For explanations, see Fig. 6.30. DPIMC — direct PIMC (Filinov et al. (2000)), RPIM — restricted PIMC (Militzer and Ceperley (2000), Militzer et al. (2001))

version of the classical Monte Carlo scheme is the Reaction Ensemble Monte Carlo technique (REMC) (Smith and Triska 1994; Johnson et al. 1994). This method incorporates the quantum mechanical description of bound states, while the scattering states are treated classically.

As was shown by Bezkrovniy (a) et al. (2004), REMC describes rather well the low temperature region, and we have a good agreement with the gas gun experiments by Nellis et al. (1983). On the basis of the REMC, results are obtained much easier as compared to those gained by calculations using molecular dynamics based on a density functional theory; see Bonev et al. (2004).

In order to get a unified picture of the efforts of DPIMC and REMC, we use the fact that REMC turnes out to be the limiting case of DPIMC at low temperatures. Therefore, it is obvious to use the results of both methods in order to construct a Hugoniot which is valid in the entire range of compression. For the construction of the combined Hugoniot within DPIMC and REMC approaches, we carefully have to analyze the region where the Hugoniots produced by the two methods can be connected to each other.

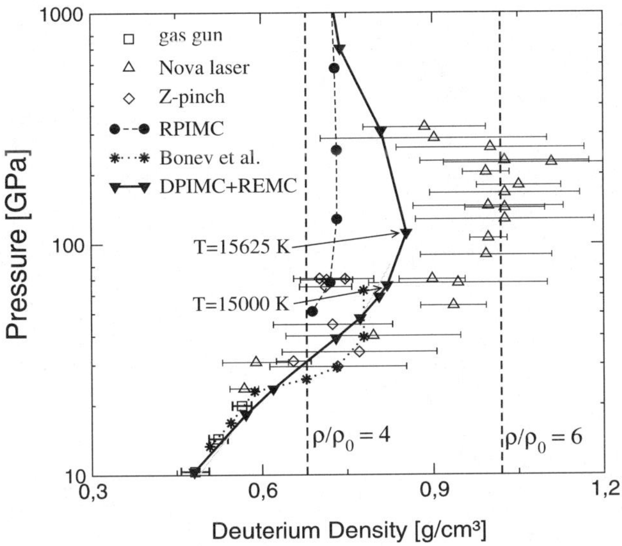

Fig. 6.38. Hugoniot-combination of DPIMC and REMC. For explanations see Figs. 6.30 and 6.37; Bonev et al. (2004)

As we can see from Fig. 6.38, the Hugoniot calculated within DPIMC ends at the point $15625K$. At this temperature, the largest contributions to the EOS are given by molecular states. As natural continuation of the DPIMC Hugoniot, we take the point corresponding to a temperature of 15000 K produced by REMC. We want to stress here that no interpolation procedure is used. Just two points at 15625 K of DPIMC and 15000 K of REMC are connected to each other. The final Hugoniot is plotted in Fig. 6.38 and shows a maximum compressibility of approximately 4.75 as compared to the initial deuterium density and is located close to the theoretical results obtained by other authors. Further numerical experiments on hydrogen and deuterium were performed by Bagnier et al. (2000) and by Clérouin and Dufreche (2001), and recent experimental results on D_2 are given by Knudson et al. (2004).

Extended work on the EOS of partially ionized plasmas in stellar envelopes was published by Hummer and Mihalas (1988), Mihalas et al. (1988), and Däppen et al. (1988).

7. Nonequilibrium Nonideal Plasmas

7.1 Kadanoff–Baym Equations. Ultra-fast Relaxation in Dense Plasmas

The non-equilibrium properties of dense strongly coupled plasmas are, like the thermodynamic ones, essentially determined by correlation and quantum effects such as dynamical screening, self-energy, Pauli-blocking, bound states and lowering of the ionization energy. Furthermore, in dense plasmas created by high intensity femto-second laser pulses, ultra-fast processes play an important role. We will show which generalizations of conventional kinetic theories have to be performed in order to give an adequate description of the non-equilibrium phenomena of strongly coupled plasmas.

We know from Chap. 2 that the non-equilibrium properties of many-particle systems can successfully be described, in a variety of cases, by kinetic equations of the Boltzmann-type

$$\left\{\frac{\partial}{\partial t}+\frac{\boldsymbol{p}_1}{m_a}\frac{\partial}{\partial \boldsymbol{R}}-\frac{\partial}{\partial \boldsymbol{R}}U_a(\boldsymbol{R}t)\frac{\partial}{\partial \boldsymbol{p}_1}\right\}f_a(\boldsymbol{p}_1,\boldsymbol{R}t)=I_a(\boldsymbol{p}_1,\boldsymbol{R}t)\,. \qquad (7.1)$$

Here, $I_a(\boldsymbol{p}_{1,}\boldsymbol{R}t)$ is the collision integral which can be used in different approximations depending on the physical situation. In Chap. 2 we presented the Born- and T-matrix expressions given by (2.172) and (2.174).

Kinetic equations of the Boltzmann-type (7.1) are very fundamental. They describe the irreversible relaxation to the equilibrium state starting from arbitrary initial conditions. Furthermore, they are the basic equations of transport theory. In spite of the fundamental character of Boltzmann-like kinetic equations, there exist many problems and substantial shortcomings:

(i) Ultra-fast processes, i.e., the behavior of the system for times t smaller than the correlation time τ_{corr} cannot be described correctly.

(ii) Because of the energy conserving δ-function in the collision integrals I_a, explicitly given by (2.172) or (2.174), the kinetic equations conserve the kinetic energy $\langle T\rangle$ only instead of the total energy $\langle H\rangle=\langle T\rangle+\langle V\rangle$. This is unphysical, especially for strongly correlated many-particle systems.

(iii) Furthermore, as a consequence of the on-shell character of the T-matrix, bound states (atoms) cannot be accounted for in Boltzmann-type collision integrals.

However, it is well known (Prigogine 1963; Kadanoff and Baym 1962; Zwanzig 1960; Bärwinkel 1969a; Klimontovich 1982), that these defects of Boltzmann-type kinetic equations are essentially connected with restricting assumptions with respect to the time. These assumptions are the condition of weakening of initial correlations and the neglect of retardation (memory effects).

Very general kinetic equations without the shortcomings mentioned above are the Kadanoff–Baym kinetic equations discussed in Chap. 3. Equations of similar generality were derived by Prigogine and Resibois, Zwanzig, and others. In this chapter, we will apply the general approach of real-time Green's function techniques to study ultra-fast processes, quantum and correlation effects, and the formation and decay of bound states in dense nonideal plasmas.

In particular, to overcome the shortcomings of Boltzmann-type kinetic equations, we will start here from the Kadanoff–Baym equations given by (3.145). For a multi-component system, they read

$$\left(i\hbar\frac{\partial}{\partial t_1}+\frac{\hbar^2\nabla_1^2}{2m_a}-U_a(1)\right)g_a^{\gtrless}(11')-\int d\boldsymbol{r}_1\Sigma_a^{\mathrm{HF}}\left(\boldsymbol{r}_1\bar{\boldsymbol{r}}_1t_1\right)g_a^{\gtrless}\left(\bar{\boldsymbol{r}}_1t_1\boldsymbol{r}_1't_1'\right)$$
$$=\int_{t_0}^{t_1}d\bar{1}\left[\Sigma_a^{>}(1\bar{1})-\Sigma_a^{<}(1\bar{1})\right]g_a^{\gtrless}(\bar{1}1')$$
$$-\int_{t_0}^{t_1'}d\bar{1}[\Sigma_a^{\gtrless}(1\bar{1})+\Sigma_a^{\mathrm{in}}(1\bar{1})]\left[g_a^{>}(\bar{1}1')-g_a^{<}(\bar{1}1')\right]. \tag{7.2}$$

These equations determine the time evolution of the two-time single-particle correlation functions $g_a^{\gtrless}(11')$. The latter ones contain all the dynamical and statistical information about the system. Therefore, relaxation phenomena, transport and thermodynamic properties of dense plasmas can be obtained by determination of these two-time functions. At this point, we remember again the most important properties of the Kadanoff–Baym equations:

(i) The Kadanoff–Baym equations include various quantum effects, i.e., quantum diffraction contributions, exchange and degeneracy (Pauli blocking).
(ii) They are formulated without any restriction with respect to the time and allow for the inclusion of arbitrary binary correlations at the initial time t_0.
(iii) We get the conservation laws of an interacting many-particle system.
(iv) The Kadanoff–Baym equations are completely determined by the self-energies Σ_a^{in} and $\Sigma_a^{\gtrless}$. The latter include the interaction in the system, and have to be used in appropriate approximations.

Therefore, progress in the non-equilibrium theory of strongly correlated plasmas can be achieved by the solution of these general equations. As the

Kadanoff–Baym equations are valid without any restriction with respect to the time, we are able to consider the temporal evolution of a plasma on short time scales ($t < \tau_{corr}$) starting from the initial time t_0. Therefore, it is possible to describe ultra-fast relaxation processes in the initial stage. In this stage, the correlations are being built up. As already mentioned, these processes cannot be described by conventional kinetic equations. In order to demonstrate the efficiency of the approach, let us consider the relaxation of the two-time correlation function $g_a^<(\boldsymbol{p}_1, t_1 t_1')$ and the Wigner distribution $f_a(\boldsymbol{p}_1, t)$ for spatially homogeneous systems. Furthermore, the influence of quantum and correlation effects on the temporal evolution of the energy is investigated. See also the monograph on quantum kinetic theory by Bonitz (1998).

Due to the complicated structure of the equations, only numerical evaluations are possible. Such calculations were performed, up to now, only for simple approximations for the self-energy, for example in Born approximation (Köhler 1995; Bonitz et al. 1996). In this approximation, the self-energy reads

$$\Sigma_a^{\gtrless}(\boldsymbol{p}_1, t_1 t_1') = (2s_a + 1)\hbar^2 \sum_b \int \frac{d\boldsymbol{p}_2}{(2\pi\hbar)^3} \frac{d\bar{\boldsymbol{p}}_1}{(2\pi\hbar)^3} \frac{d\bar{\boldsymbol{p}}_2}{(2\pi\hbar)^3} |V_{ab}(\boldsymbol{p}_1 - \bar{\boldsymbol{p}}_1)|^2$$
$$\times (2\pi\hbar)^3 \delta(\boldsymbol{p}_1 + \boldsymbol{p}_2 - \bar{\boldsymbol{p}}_1 - \bar{\boldsymbol{p}}_2) g_a^{\gtrless}(\bar{\boldsymbol{p}}_1, t_1 t_1') g_b^{\gtrless}(\bar{\boldsymbol{p}}_2, t_1 t_1') g_b^{\lessgtr}(\boldsymbol{p}_2, t_1' t_1) \,, \quad (7.3)$$

where the sum runs over the species, and V_{ab} denotes the two-body interaction potential. For plasmas, $V_{ab}(\boldsymbol{q})$ is taken to be the statically screened Coulomb potential

$$V_{ab}(\boldsymbol{q}) = V_{ab}^s(\boldsymbol{q}) = \frac{4\pi\hbar^2 e_a e_b}{q^2 + \hbar^2\kappa^2} \,; \quad (7.4)$$

here κ is the non-equilibrium inverse screening length given by

$$\kappa^2(t) = \frac{4}{\pi\hbar^3} \sum_a m_a e_a^2 \int_0^\infty dp f_a(p, t) \,. \quad (7.5)$$

In order to demonstrate specific features of the behavior of the two-time functions, the Kadanoff Baym equations were solved first for an ideal electron gas in a uniform background, see Semkat (2001), Semkat et al. (2003), and Bonitz and Semkat (2005). For fixed momentum $p_1 = \hbar k_1$, the results for the correlation function are plotted in the t_1', t_1-plane, Fig. 7.1. We see the constant distribution function along the diagonal $t_1' = t_1 = t$, and, in perpendicular direction, an oscillatory behavior with a frequency determined by the undamped single-particle excitations. If the interaction according to (7.3) and (7.4) is included, there is a temporal evolution of the distribution function along the diagonal in the t_1', t_1-plane as a result of the collisions. In perpendicular direction, we again observe oscillations, but now with a frequency modified by the interaction between the particles described in Born approximation. It turns out that the interaction produces a damping of the oscillations which is connected with the life time of the single-particle excitations.

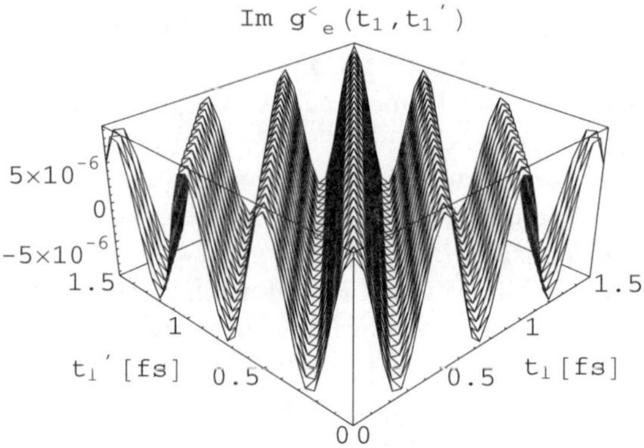

Fig. 7.1. Temporal evolution of the imaginary part of the single-particle correlation function $g^<$ for an ideal electron gas for a fixed wave number $k_1 = 0.8/a_B$. The initial state is a Fermi distribution function with $T = 10^4$ K and $n = 10^{21}$ cm^{-3}

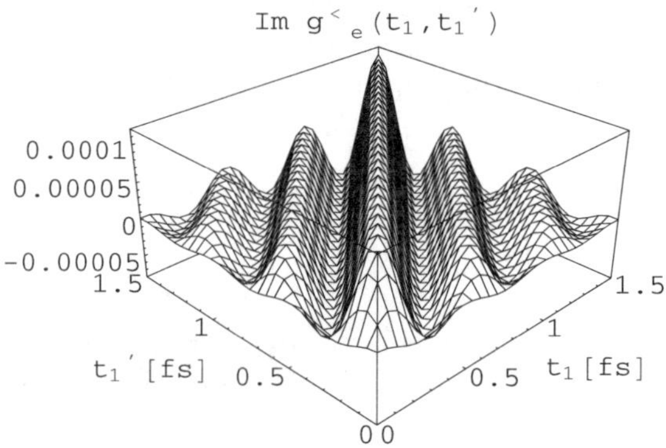

Fig. 7.2. Temporal evolution of the imaginary part of the single-particle correlation function $g^<$ for an interacting electron gas. The interaction is included via the self-energy in Born approximation given by (7.3). Momentum and initial state like in Fig. 7.1

With the knowledge of the correlation function $g_a^<(t_1, t_1')$, we are able to investigate the time relaxation of the macroscopic observables. Using the relations (3.221) and

$$\langle T \rangle = \sum_a \int \frac{d\boldsymbol{p}}{(2\pi\hbar)^3} \frac{p^2}{2m_a} (\pm i\hbar) g_a^<(p, tt')|_{t'=t} \,,$$

we determine, as an example, the temporal evolution of the mean kinetic and the mean potential energies of a quantum electron gas. Two cases are considered: i) The initial state is chosen to be uncorrelated, and ii) the initial state is correlated, described by

$$c(\boldsymbol{p}_1, \boldsymbol{p}_2; \boldsymbol{p}_1 + \boldsymbol{q}, \boldsymbol{p}_2 - \boldsymbol{q}; t_0)$$

$$= -\frac{1}{(i\hbar)^2} \frac{V_{ee}^s(\boldsymbol{q})}{k_B T} f(\boldsymbol{p}_1) f(\boldsymbol{p}_2) [1 - f(\boldsymbol{p}_1 + \boldsymbol{q})][1 - f(\boldsymbol{p}_2 - \boldsymbol{q})]|_{t_0}$$

with $V_{ee}^s(\boldsymbol{q})$ being the statically screened Coulomb potential.

The results in Born approximation are shown in Fig. 7.3. If the initial state of the plasma is considered to be uncorrelated, the mean value of the

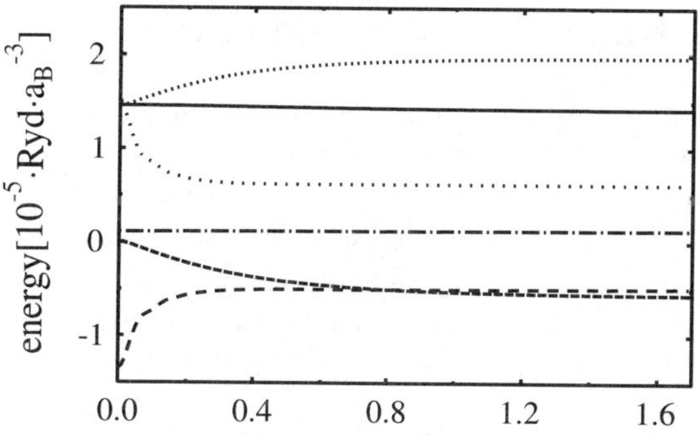

Fig. 7.3. Relaxation of the energy of a quantum electron gas. For the case without initial correlations, the kinetic energy is given by the *narrowly dotted line*, the potential energy by the *narrowly dashed* one. For the case with initial correlations the kinetic energy is given by a *wider dotted line*, the potential energy by the *wider dashed* one. In both cases the total energy is conserved (*full line* or *dash-dotted line*, respectively). Initial conditions like in Fig. 7.1

potential energy is zero. As time proceeds, correlations build up, leading to a finite negative potential energy. As a consequence of conservation of the total energy, this leads to an increase of the kinetic energy. Let us now consider the situation where the plasma is initially correlated. Then we have a finite negative value of the mean potential energy at $t_1 = t_1' = t_0$. In our calculations, we assumed an initial state with a large potential energy describing an extreme non-equilibrium situation. To reach the equilibrium balance of potential and kinetic energies, the modulus of the potential energy must decrease. Because of energy conservation, the kinetic energy has to decrease, too. Therefore, we observe a cooling effect corresponding to a decreasing temperature. Consequently, we find the interesting result that there is a lowering of the mean kinetic energy in the relaxation process if the system is initially stronger correlated than in the respective final equilibrium state. The result confirms that the energy balance in strongly coupled plasmas is indeed treated in accurate manner starting from the Kadanoff–Baym kinetic equations.

The Born approximation discussed here is rather simple and does not describe a number of important physical phenomena such as dynamical screening, bound states, and multiple scattering. To take these effects into account, more complex approximations for the self-energy are necessary. However, rigorous numerical solutions of the Kadanoff–Baym equations for such types of approximations are not available up to now. Therefore, it is of advantage to simplify the Kadanoff–Baym equations by taking the time–diagonal limit, i.e., $t_1 = t_1' = t$, $\tau = 0$. This level of description will be considered in the next sections.

7.2 The Time-Diagonal Kadanoff–Baym Equation

The time-diagonal Kadanoff–Baym equation we are going to derive now, is an equation for the Wigner function. Dynamical information such as the single-particle dispersion will be lost. We start from (7.2) and subtract the corresponding adjoint equation. The time-diagonal equation now follows for $t_1 = t_1' = t$. Next we introduce the substitution (3.172) and take into account the relation (3.173). Then the Fourier transform with respect to $\boldsymbol{r}$ of a time-diagonal equation just mentioned reads

$$
\begin{aligned}
&\left(\frac{\partial}{\partial t} + \frac{\boldsymbol{p}_1}{m_a}\nabla_{\boldsymbol{R}}\right) f_a(\boldsymbol{p}_1\boldsymbol{R}t) - \frac{1}{i\hbar}\int d\bar{\boldsymbol{r}}d\boldsymbol{r}e^{-\frac{i}{\hbar}\boldsymbol{p}\boldsymbol{r}} \\
&\times\Big[\Big(U_a(\boldsymbol{R}+\frac{\boldsymbol{r}}{2},t) + \Sigma_a^{\mathrm{HF}}(\boldsymbol{r}-\bar{\boldsymbol{r}},\boldsymbol{R}+\frac{\bar{\boldsymbol{r}}}{2},t)\Big)g_a^{<}(\bar{\boldsymbol{r}},\boldsymbol{R}+\frac{\bar{\boldsymbol{r}}-\boldsymbol{r}}{2},t) \\
&-\left(U_a(\boldsymbol{R}-\frac{\boldsymbol{r}}{2},t) - \Sigma_a^{\mathrm{HF}}(\boldsymbol{r}-\bar{\boldsymbol{r}},\boldsymbol{R}-\frac{\bar{\boldsymbol{r}}}{2},t)\right) g_a^{<}(\bar{\boldsymbol{r}},\boldsymbol{R}-\frac{\bar{\boldsymbol{r}}-\boldsymbol{r}}{2},t)\Big] \\
&= I_a^{\mathrm{in}}(\boldsymbol{p},\boldsymbol{R}t) - \frac{1}{\hbar}\int d\boldsymbol{r}\, e^{-\frac{i}{\hbar}\boldsymbol{p}_1\boldsymbol{r}}\int\limits_{t_0-t}^{0} d\bar{x}\{\Sigma_a^{>}\, g_a^{<} - \Sigma_a^{<}\, g_a^{>}\}_{x-\bar{x},X+\frac{\bar{x}}{2};\,\bar{x},X+\frac{\bar{x}-x}{2}} \\
&+\frac{1}{\hbar}\int d\boldsymbol{r}\, e^{-\frac{i}{\hbar}\boldsymbol{p}_1\boldsymbol{r}}\int\limits_{t_0-t}^{0} d\bar{x}\{g_a^{<}\, \Sigma_a^{>} - g_a^{>}\, \Sigma_a^{<}\}_{x-\bar{x},X+\frac{\bar{x}}{2};\,\bar{x},X+\frac{\bar{x}-x}{2}}\,.
\end{aligned}
\tag{7.6}
$$

Here, the subscript attached to the curly brackets means the set of arguments of the first factor, followed by that of the second factor in each summand. This equation is still a very general quantum kinetic equation. Like in the case of equations for the two-time correlation functions, we retained (i) the influence of initial correlations, (ii) retardation (memory effects), and (iii) the validity of conservation laws for nonideal systems. From (7.6) one gets all known kinetic equations by respective approximations for the self-energies and by appropriate assumptions with respect to the time behavior.

The r.h.s. of (7.6) represents a rather far-reaching non-Markovian generalization of collision integrals of Boltzmann-type kinetic equations. To get explicit expressions for the collision integral, two problems have to be solved:

(i) We have to find appropriate approximations for the self-energies Σ_a^{in} used in I^{in}, and $\Sigma_a^{\gtrless}$.
(ii) The two-time correlation functions have to be expressed as a functional of the Wigner function in order to find a closed equation for the latter quantity.

For the determination of the self-energies $\Sigma_a^{\gtrless}(t_1t_1')$, there exist, in dependence on the physical situation, standard approximations discussed in Chaps. 4 and 5. The simplest one including collisions is the Born approximation (7.3) used in the previous section. It leads to a generalized quantum

Landau equation as will be shown in the next section. A more general type of approximation including the higher order two-particle ladder diagrams is the binary collision approximation with self-energies given by

$$\Sigma_a^{\gtrless}(11') = \pm i\hbar \sum_b \int d\bar{\boldsymbol{r}}_2\, d\boldsymbol{r}_2\, \langle \boldsymbol{r}_1\boldsymbol{r}_2 | T_{ab}^{\gtrless}(t_1t_1') | \boldsymbol{r}_1'\bar{\boldsymbol{r}}_2\rangle g_b^{\lessgtr}(\bar{\boldsymbol{r}}_2t_1', \boldsymbol{r}_2t_1)\,. \quad (7.7)$$

This approximation allows us to account for bound states and multiple scattering processes. A detailed discussion of the binary collision approximation was given in Chap. 5. Introducing (7.7) into equation (7.6), we get a generalization of the Boltzmann equation (see also 7.6)).

A further physically relevant approximation for plasmas is the random phase approximation (RPA). It describes the charged particles' interaction including dynamical screening effects by the RPA dielectric function. A discussion of this approximation scheme to describe the dielectric and thermodynamic properties of dense plasmas was given in Chaps. 4 and 5. The RPA self-energies were found to be

$$\Sigma_a^{\gtrless}(11') = -i\hbar\, V_{aa}^{\gtrless}(11') g_a^{\gtrless}(11')\,. \quad (7.8)$$

Inserting this expression into (7.6), a generalized version of the Lenard–Balescu kinetic equation can be derived.

Let us now proceed with the second problem. In order to get a closed kinetic equation, one has to reconstruct the two-time correlation function from the single-time single-particle density matrix. This is the so called *reconstruction problem*, which was addressed first by Lipavský, Špička, and Velický (1986).

The first idea to do this is the Kadanoff–Baym ansatz (KBA).

$$\begin{aligned} \pm i g_a^{<}(t_1,t_1') &= f_a^{<}(t)\left\{ i g_a^R(\tau,t) - i g_a^A(\tau,t)\right\}\,, \\ i g_a^{>}(t_1,t_1') &= f_a^{>}(t)\left\{ i g_a^R(\tau,t) - i g_a^A(\tau,t)\right\}\,, \end{aligned} \quad (7.9)$$

where $\tau = t_1 - t_1'$ denotes the difference time, and $t = \frac{1}{2}(t_1 + t_1')$. As before, the upper sign refers to Bose particles and the lower one to Fermi particles. To get more compact expressions, we introduced the notation

$$f_a^{<}(t) = f_a(t) \quad , \quad f_a^{>}(t) = 1 \pm f_a(t)\,.$$

An extensively used approximation to determine the correlation functions is given in the framework of the quasi-particle picture discussed in Sect. 3.4.2. After Fourier transformation of $g_a^{<}$, we arrive at

$$\pm i g_a^{<}(\boldsymbol{p}\omega, \boldsymbol{R}t) = 2\pi\delta[\hbar\omega - E_a(\boldsymbol{p},\boldsymbol{R}t)] f_a(\boldsymbol{p},\boldsymbol{R}t) \quad (7.10)$$

with $E_a(\boldsymbol{p},\boldsymbol{R}t)$ being the quasi-particle energy determined by (3.193). We know that this approximation can be applied only for weakly damped quasi-particles as discussed in Sect. 3.4.1.

The KBA has the essential shortcoming that retardation and correlation effects are not described. A possibility to include retardation effects is given by the generalized Kadanoff–Baym ansatz (GKBA). This ansatz is due to Lipavský et al. (1986). In this paper, an equation for the reconstruction of the two-time correlation functions $g_a^{\gtrless}$ from their time-diagonal elements was derived. The simplest solution to that equation can be written as

$$\begin{aligned}
\pm i g_a^{<}(t_1,t_1') &= i g_a^R(t_1,t_1')\, f_a^{<}(t_1') - f_a^{<}(t_1)\, i g_a^A(t_1,t_1')\,, \\
i g_a^{>}(t_1,t_1') &= i g_a^R(t_1,t_1')\, f_a^{>}(t_1') - f_a^{>}(t_1)\, i g_a^A(t_1,t_1')\,. \qquad (7.11)
\end{aligned}$$

This ansatz is valid only in quasiparticle approximation. The KBA and the GKBA are approximations which represent restrictive assumptions as compared to the two-time equations.

Let as mention, therefore, finally a third approximative solution for the reconstruction problem derived by Bornath et al. (1996). This approximation goes beyond the quasiparticle approximation. It takes into account retardation in first order gradient expansion and correlation effects. Explicitly it was found

$$\begin{aligned}
\pm i g^{<}(\omega,t) &= 2\pi\delta(\omega-E)\, f(t) - \frac{\mathcal{P}'}{\omega-E}\frac{\partial}{\partial t} f(t) \\
-\frac{\mathcal{P}'}{(\omega-E)}(\pm i)\Sigma^{<}(\omega,t) &+ 2\pi\delta(\omega-E)\int\frac{d\bar\omega}{2\pi}\frac{\mathcal{P}'}{(\bar\omega-E)}(\pm i)\Sigma^{<}(\bar\omega,t)\,.
\end{aligned} \qquad (7.12)$$

This relation is clearly a generalization of (7.10). The second contribution of (7.12) describes retardation effects in first order gradient expansion. The third and the fourth terms describe correlations between quasiparticles. Equation (7.12) is, therefore, the non-equilibrium extended quasiparticle approximation.

After the discussion concerning the approximations for the self-energy and the reconstruction problem, we are now able to evaluate the collision integral on the r.h.s of (7.6). This is the aim of the following sections. In deriving (7.6), we have obtained a general (non-Markovian) kinetic equation which is nonlocal in time and space and which is valid at arbitrary space and time scales. An essential question is how the usual Boltzmann equation being local in space and time turns out to be an approximation of our nonlocal kinetic equation (Bornath et al. 1996).

In order to consider this problem, we have to evaluate integrals of the type (3.173). The way to do this is straightforward. We expand the collision integral about the macroscopic variables following the scheme of Sect. 3.4.2.

Let us first determine the expansion with respect to the *spatial retardation*. Using relation (3.174) only for the spatial retardation, the integrals (3.173) for $t_1 - t_1' = t = 0$ are given by

$$
\begin{aligned}
I(x,X) = \int d\bar{x}\, f\left(x-\bar{x}, X+\frac{\bar{x}}{2}\right) u\left(\bar{x}, X+\frac{\bar{x}-x}{2}\right) \\
\approx f(-\bar{t}, t+\frac{\bar{t}}{2}, \boldsymbol{p}, \boldsymbol{R}) u(\bar{t}, t+\frac{\bar{t}}{2}, \boldsymbol{p}, \boldsymbol{R}) \\
+\frac{i}{2}\Big(\nabla_{\boldsymbol{p}} f(-\bar{t}, t+\frac{\bar{t}}{2}, \boldsymbol{p}, \boldsymbol{R}) \cdot \nabla_{\boldsymbol{R}} u(\bar{t}, t+\frac{\bar{t}}{2}, \boldsymbol{p}, \boldsymbol{R}) \\
+\nabla_{\boldsymbol{R}} f(-\bar{t}, t+\frac{\bar{t}}{2}, \boldsymbol{p}, \boldsymbol{R}) \cdot \nabla_{\boldsymbol{p}} u(\bar{t}, t+\frac{\bar{t}}{2}, \boldsymbol{p}, \boldsymbol{R})\Big)\,. \qquad (7.13)
\end{aligned}
$$

We apply the expansion (7.13) to all terms of the r.h.s. of (7.6) and consider first the spatially local term

$$
\begin{aligned}
I_a &= \int_{-\infty}^{0} d\bar{t}\Big\{\Sigma_a^>(\bar{t}, t+\frac{\bar{t}}{2})\, g_a^<(-\bar{t}, t+\frac{\bar{t}}{2}) - \Sigma_a^<(\bar{t}, t+\frac{\bar{t}}{2})\, g_a^>(-\bar{t}, t+\frac{\bar{t}}{2}) \\
&\quad +g_a^<(\bar{t}, t+\frac{\bar{t}}{2})\, \Sigma_a^>(-\bar{t}, t+\frac{\bar{t}}{2}) - g_a^>(\bar{t}, t+\frac{\bar{t}}{2})\, \Sigma_a^<(-\bar{t}, t+\frac{\bar{t}}{2})\Big\}\,. \qquad (7.14)
\end{aligned}
$$

Now we expand this contribution with respect to the temporal retardation $\bar{t}/2$. Further we express all quantities $g_a^<(-\bar{t}, T)$, etc., by their Fourier transforms $g_a^<(-\omega, T)$. Then (7.14) is simply replaced by

$$
\begin{aligned}
I_a = \sum_{n=0}^{\infty} \frac{1}{n!} \int \frac{d\omega\, d\bar{\omega}}{(2\pi)^2} \int_{-\infty}^{0} \{e^{i(\omega-\bar{\omega})\bar{t}} + e^{-i(\omega-\bar{\omega})\bar{t}}\} \left(\frac{\bar{t}}{2}\right)^n \\
\times \frac{d}{dt}\{\Sigma_a^>(\omega, t)\, g_a^<(\bar{\omega}, t) - \Sigma_a^<(\omega, t)\, g_a^>(\bar{\omega}, t)\} = \sum_{n=0}^{\infty} I_{an}\,.
\end{aligned}
$$

Using well-known representations of the Dirac δ-function and of the principle value, we find for the terms $n = 0, 1$ of the expansion

$$
\begin{aligned}
I_{a0} &= (\pm i) \int \frac{d\omega}{2\pi} \{i\Sigma_a^>(\omega, t)\, g_a^<(\omega, t) - \Sigma_a^<(\omega, t)\, i g_a^>(\omega, t)\} \\
I_{a1} &= (\pm i) \frac{d}{dt} \int \frac{d\omega d\bar{\omega}}{(2\pi)^2} \frac{d}{d\omega} \frac{P}{\omega-\bar{\omega}} \{i\Sigma_a^>(\omega, t)\, g_a^<(\bar{\omega}, t) - \Sigma_a^<(\omega, t)\, i g_a^>(\bar{\omega}, t)\}\,. \qquad (7.15)
\end{aligned}
$$

Here, I_{a0} is the local approximation of the r.h.s. of (7.6), and I_{a1} is the first order correction with respect to the temporal retardation.

We now proceed in the same way with the other contributions of (7.13) and with the left hand side of (7.6). The result is a kinetic equation in first order gradient expansion (Bornath et al. 1996; Lipavský et al. 2001)

$$
\begin{aligned}
\left(\frac{\partial}{\partial t} + \nabla_{\boldsymbol{p}_1} E_a^{\mathrm{HF}}(\boldsymbol{p}_1, \boldsymbol{R}t) \cdot \nabla_{\boldsymbol{R}} - \nabla_{\boldsymbol{R}} E_a^{\mathrm{HF}}(\boldsymbol{p}_1, \boldsymbol{R}t) \cdot \nabla_{\boldsymbol{p}_1}\right) f_a(\boldsymbol{p}_1, \boldsymbol{R}t) \\
= I_{a0}(\boldsymbol{p}_1, \boldsymbol{R}t) + \sum_{k=1}^{3} I_a^k(\boldsymbol{p}_1, \boldsymbol{R}t)\,. \qquad (7.16)
\end{aligned}
$$

The l.h.s. is the drift term of a Boltzmann-like kinetic equation in which the kinetic energy and the external potential are replaced by the energy $E(\boldsymbol{p}_1, \boldsymbol{R}T)$ including Hartree–Fock contribution, i.e.,

$$E_a^{\mathrm{HF}}(\boldsymbol{p}, \boldsymbol{R}t) = \frac{p^2}{2m_a} + \Sigma_a^{\mathrm{HF}}(\boldsymbol{p}, \boldsymbol{R}t) + U_a(\boldsymbol{R}t)\,. \tag{7.17}$$

For simplicity, in $g_a^{\gtrless}$ and $\Sigma_a^{\gtrless}$ only the time and frequency variables are shown explicitly.

The second and the third terms come from the spatial retardation and are of importance for spatially inhomogeneous systems. They can be written as

$$\begin{aligned} I_a^2 &= \hbar\nabla_{\boldsymbol{R}} \int \frac{d\omega}{2\pi}\frac{d\bar{\omega}}{2\pi}\frac{(\mp)\mathcal{P}}{\bar{\omega}-\omega}\left[g_a^{<}(\omega,t)\nabla_{\boldsymbol{p}_1}\Sigma_a^{>}(\bar{\omega},t) - [\nabla_{\boldsymbol{p}_1} g_a^{>}(\bar{\omega},t)]\Sigma_a^{<}(\omega,t)\right], \\ I_a^3 &= \hbar\nabla_{\boldsymbol{p}_1} \int \frac{d\omega}{2\pi}\frac{d\bar{\omega}}{2\pi}\frac{(\mp)\mathcal{P}}{\omega-\bar{\omega}}\left[g_a^{<}(\omega,t)[\nabla_{\boldsymbol{R}}\Sigma_a^{>}(\bar{\omega},t) - \nabla_{\boldsymbol{R}} g_a^{>}(\bar{\omega},t)]\Sigma_a^{<}(\omega,t)\right]. \end{aligned} \tag{7.18}$$

The contributions I_a^1, I_a^2 and I_a^3 can be considered as re-normalization contributions to the drift term of the kinetic equation. In the next sections, we will show that the well-known Boltzmann-type collision integrals follow from the local contribution (7.15). At this point, we remember that the conservation laws and, therefore, the hydrodynamics following from Boltzmann-type kinetic equations describe only ideal systems. To go beyond this approximation, the gradient terms have to be accounted for, leading to nonideality corrections to the ideal balance equations (Bärwinkel 1969b; Klimontovich 1982).

It is interesting to consider the connection of the kinetic equation (7.16) with the generalized Landau–Silin equation (3.208) derived in Sect. 3.4.1. By simple rearrangements (add and subtract terms like $g^{<}\Sigma^{>}$ in the expressions for I_a^1, I_a^2 and I_a^3), and using the dispersion relations for $\mathrm{Re}g^R$ and $\mathrm{Re}\Sigma^R$, we can write, e.g.,

$$I_a^1 = (\pm i)\frac{d}{dT}\int\frac{d\omega}{2\pi}\left[g^{<}\frac{\partial Re\Sigma^R}{\partial\omega} - \Sigma^{<}\frac{\partial Re g^R}{\partial\omega}\right]. \tag{7.19}$$

Corresponding transformations are possible for I_a^2 and for I_a^3. Let us neglect now the second contribution (off-pole term). Further we apply the KB-ansatz (7.10). Then the kinetic equation (7.16) reduces to

$$\begin{aligned} &\left(\frac{\partial}{\partial t} + \nabla_{\boldsymbol{p}} E_a(\boldsymbol{p}, \boldsymbol{R}t)\cdot\nabla_{\boldsymbol{R}} - \nabla_{\boldsymbol{R}} E_a(\boldsymbol{p}, \boldsymbol{R}t)\cdot\nabla_{\boldsymbol{p}}\right) f_a(\boldsymbol{p}, \boldsymbol{R}t) \\ &= Z(\boldsymbol{p}, \boldsymbol{R}t)\Big[(\pm)i\Sigma_a^{<}(\boldsymbol{p}\omega, \boldsymbol{R}t) f_a^{>}(\boldsymbol{p}, \boldsymbol{R}t) - i\Sigma_a^{>}(\boldsymbol{p}\omega, \boldsymbol{R}t) f_a^{<}(\boldsymbol{p}, \boldsymbol{R}t)\Big]\Big|_{\omega=E_a}\,. \end{aligned} \tag{7.20}$$

Here we used

$$Z^{-1} = 1 - \frac{\partial}{\partial\omega}\Sigma_a^R(\boldsymbol{p}\omega, \boldsymbol{R}t)_{\omega=E_a(\boldsymbol{p},\boldsymbol{R}t)}\,, \tag{7.21}$$

where $E_a(\boldsymbol{p}, \boldsymbol{R}t)$ is the full quasiparticle energy determined from the dispersion relation

$$E_a(\boldsymbol{p}, \boldsymbol{R}t) = \frac{p^2}{2m_a} + Re\Sigma_a^R(\boldsymbol{p}\omega, \boldsymbol{R}t)_{\omega=E_a(\boldsymbol{p},\boldsymbol{R}t)} \,. \tag{7.22}$$

The collision integral at the r.h.s. of (7.20) is given in terms of the self-energy functions $\Sigma_a^{\gtrless}$ which can be interpreted as scattering rates with $\omega = E_a(\boldsymbol{p}, \boldsymbol{R}t)$.

Let us mention that the Landau–Silin equation has many shortcomings in comparison to the kinetic equation (7.16). In particular, only the quasiparticle energy is conserved. An extensive discussion of these problems may be found in the paper by Bornath et al. (1996).

Let us come back to the general equation (7.6). In many cases, we have to deal with weakly inhomogeneous systems. Then the collision integral may be considered in the local approximation with respect to the space variables, however, the retardation in time is fully retained. Fourier transformations can be carried out, and, for the time diagonal kinetic equation, we have

$$\begin{aligned}
&\left(\frac{\partial}{\partial t} + \frac{\boldsymbol{p}_1}{m_a}\nabla_{\boldsymbol{R}} - \nabla_{\boldsymbol{R}} U_a^{\text{eff}}(\boldsymbol{R}, t)\nabla_{\boldsymbol{p}_1}\right) f_a(\boldsymbol{p}_1, \boldsymbol{R}, t) = I_a^{\text{in}}(\boldsymbol{p}_1 \boldsymbol{R}, t) \\
&-2\text{Re}\int_{t_0}^{t} d\bar{t}\Big(g_a^<(\boldsymbol{p}_1\boldsymbol{R}, t\bar{t})\Sigma_a^>(\boldsymbol{p}_1\boldsymbol{R}, \bar{t}t) - g_a^>(\boldsymbol{p}_1\boldsymbol{R}, t\bar{t})\Sigma_a^<(\boldsymbol{p}_1\boldsymbol{R}, \bar{t}t)\Big) \,. \qquad (7.23)
\end{aligned}$$

We will consider this general equation for special expressions of the self-energy in the next sections.

7.3 The Quantum Landau Equation

The simplest time-diagonal kinetic equations is achieved if the self-energy $\Sigma_a^{\gtrless}$ is used in Born approximation given by (5.142). This approximation is meaningful for weakly correlated many-particle systems. The same structure follows if we replace, in the RPA self-energy (4.31), the dynamically screened potential by the statically screened one. Thus, the Born approximation (7.3) with a statically screened potential is a well defined approximation for plasmas.

We mention that this type of approximation retains all non-Markovian features of the time dependencies discussed above. In principle, in this approximation, the Kadanoff–Baym equations can be solved numerically. This was shown in Sect. 7.1. Nevertheless, it is worthwhile to discuss such an approximation in the time diagonal equation, too, as it leads to an interesting simplified non-Markovian kinetic equation.

We take the kinetic equation for weakly inhomogeneous systems, that means we start from equation (7.23). Furthermore, using the GKBA according to (7.11), the following generalized version of the quantum Landau kinetic equation is obtained

$$\begin{aligned}
&\left(\frac{\partial}{\partial t}+\frac{\boldsymbol{p}_1}{m_a}\nabla_{\boldsymbol{R}}-\nabla_{\boldsymbol{R}}U_a^{\text{eff}}(\boldsymbol{R},t)\nabla_{\boldsymbol{p}_1}\right) f_a(\boldsymbol{p}_1,\boldsymbol{R},t)=I_a^{\text{in}}(\boldsymbol{p}_{1,}t)\\
&-2\hbar^2\sum_b\int_{t_0}^{t}d\bar{t}\int\frac{d\boldsymbol{p}_2 d\bar{\boldsymbol{p}}_1 d\bar{\boldsymbol{p}}_2}{(2\pi\hbar)^6}\left|V_{ab}\left(\boldsymbol{p}_1-\bar{\boldsymbol{p}}_1\right)\right|^2\delta\left(\boldsymbol{p}_1+\boldsymbol{p}_2-\bar{\boldsymbol{p}}_1-\bar{\boldsymbol{p}}_2\right)\\
&\times\text{Re}\bar{g}_{ab}^{0R}(t,\bar{t})g_{ab}^{0A}(\bar{t},t)\left[\bar{f}_a\,\bar{f}_b(1-f_a)(1-f_b)-f_af_b(1-\bar{f}_a)(1-\bar{f}_b)\right]_{\bar{t}}\,.
\end{aligned}\tag{7.24}$$

Here, and in the following, we drop the macroscopic space variable $\boldsymbol{R}$ in the collision integrals. Only in special cases the $\boldsymbol{R}$-dependence will be written explicitly. In addition, we use the short-hand notations $f_a = f_a(\boldsymbol{p}_1,\boldsymbol{R},t)$, $\bar{f}_a = f_a(\bar{\boldsymbol{p}}_1,\boldsymbol{R},t)$ and $\bar{g}_{ab}^{0R} = g_a^R(\bar{\boldsymbol{p}}_1\boldsymbol{R},t\bar{t})g_b^R(\bar{\boldsymbol{p}}_2\boldsymbol{R},t\bar{t})$. The first term on the r.h.s. of (7.24) accounts for initial correlations. It is given by

$$\begin{aligned}
I_a^{\text{in}}(\boldsymbol{p}_{1,}t) &= 2\hbar^5\mathcal{V}\sum_b\int\frac{d\boldsymbol{p}_2 d\bar{\boldsymbol{p}}_1 d\bar{\boldsymbol{p}}_2}{(2\pi\hbar)^6}V_{ab}(\boldsymbol{p}_1-\bar{\boldsymbol{p}}_1)\delta(\boldsymbol{p}_1+\boldsymbol{p}_2-\bar{\boldsymbol{p}}_1-\bar{\boldsymbol{p}}_2)\\
&\quad\times\text{Im}\Big\{g_{ab}^{0R}(t,t_0)\,g_{ab}^{0A}(t_0,t)g_{ab}^{<}(t_0)\Big\},
\end{aligned}\tag{7.25}$$

where $g_{ab}^{<}(t_0) = g_{ab}^{<}(\boldsymbol{p}_1\boldsymbol{p}_2,\bar{\boldsymbol{p}}_1\bar{\boldsymbol{p}}_2;t_0)$ is the two-particle correlation function at the initial time t_0.

Equation (7.24) represents a generalization of the Markovian Landau kinetic equation which is often used in plasma and solid state physics. We still retained all the properties of the general time-diagonal equation. The kinetic equation is essentially an extension of the conventional Landau equation for times smaller than the correlation time and enables us to study the influence of initial correlations with an appropriate simple collision integral. Furthermore, we retain the full propagators $g_{ab}^{0R}(t,t') = g_a^R(t,t')g_b^R(t,t'$, where g_a^R has to be determined from the equation of motion for the retarded (advanced) single-particle Green's functions (3.148) used in Born approximation.

An essential simplification of the generalized Landau equation follows if single-particle propagators g_a^R for damped quasiparticles according to (3.185) are used. Introducing $\tau = t-\bar{t}$, we get

$$\begin{aligned}
I_a(\boldsymbol{p}_{1,}t) &= I_a^{\text{in}}(\boldsymbol{p}_{1,}t)+\frac{2}{\hbar^2}\sum_b\int_0^{t-t_0}d\tau\int\frac{d\boldsymbol{p}_2 d\bar{\boldsymbol{p}}_1 d\bar{\boldsymbol{p}}_2}{(2\pi\hbar)^6}\delta\left(\boldsymbol{p}_1+\boldsymbol{p}_2-\bar{\boldsymbol{p}}_1-\bar{\boldsymbol{p}}_2\right)\\
&\quad\times\left|V_{ab}\left(\boldsymbol{p}_1-\bar{\boldsymbol{p}}_1\right)\right|^2\exp\left[-\frac{\bar{\Gamma}_{ab}+\Gamma_{ab}}{2\hbar}\tau\right]\cos\left[\frac{1}{\hbar}(\bar{E}_{ab}-E_{ab})\tau\right]\\
&\quad\times\left[\bar{f}_a\,\bar{f}_b(1-f_a)(1-f_b)-f_af_b(1-\bar{f}_a)(1-\bar{f}_b)\right]_{t-\tau}\,.
\end{aligned}\tag{7.26}$$

Here, $E_{ab} = E_a + E_b$ is the two-particle energy with $E_a = p_1^2/2m_a + \mathrm{Re}\Sigma_a^R(\boldsymbol{p}_1)$. Further, $\Gamma_{ab} = \Gamma_a + \Gamma_b$ denotes the damping with $\Gamma_a = -2\mathrm{Im}\Sigma_a^R(\boldsymbol{p}_1)$. The non-Markovian equation (7.26) with self-energy corrections and initial correlations neglected, was derived, among others, by Klimontovich (1982).

The approximation of damped quasi-particles is not free of problems. According to its derivation, the GKBA is a consistent approximation for free particles or for Hartree–Fock self-energies only. Further, the adoption of a simple exponential shape of the propagators like in (7.26) is possible only under restrictive conditions as explained in Sect. 3.4. Moreover, the exponential damping causes serious problems in the relaxation behavior of the energy; see below.

The non-Markovian Landau equation has interesting properties. The equation is non-local with respect to the time. Therefore, it follows that the collision integral depends on the distribution function for all previous times, i.e., we have memory effects. Taking into account the properties of the retarded Green's function, the depth of the memory is restricted by the imaginary part of the poles of the Green's function in the complex energy plane, i.e., by the damping of the single-particle states. In the case that the memory depth is small, the damping may be neglected, and the Markovian limit may be taken by an expansion with respect to the retardation in the distribution functions. Introducing the substitution $\bar{E}_{12} - E_{12} = \omega$, we have to consider the expression

$$\int\limits_0^{t-t_0} d\tau \cos\omega\tau \left[\{\bar{F}^<_{ab}F^>_{ab} - F^<_{ab}\bar{F}^>_{ab}\} - \frac{\tau}{2}\frac{d}{dt}\{\bar{F}^<_{ab}F^>_{ab} - F^<_{ab}\bar{F}^>_{ab}\} \right]|_t \qquad (7.27)$$

with the short notation $F^{\gtrless}_{ab} = f^{\gtrless}_a(t)f^{\gtrless}_b(t)$. Now one can take the distribution functions out of the time integral, and the τ-integration can be performed. For the retardation expansion of the collision integral we arrive at

$$I_a(\boldsymbol{p}_1,t) = I_a^{\mathrm{in}}(\boldsymbol{p}_1,t) + I_a^0(\boldsymbol{p}_1,t) + I_a^1(\boldsymbol{p}_1,t)\,. \qquad (7.28)$$

Here, the local contribution is given by

$$I_a^0(\boldsymbol{p}_1,t) = \frac{2}{\hbar}\sum_b \int \frac{d\boldsymbol{p}_2 d\bar{\boldsymbol{p}}_1 d\bar{\boldsymbol{p}}_2}{(2\pi\hbar)^6} |V_{ab}(\boldsymbol{p}_1 - \bar{\boldsymbol{p}}_1)|^2 \delta(\boldsymbol{p}_1 + \boldsymbol{p}_2 - \bar{\boldsymbol{p}}_1 - \bar{\boldsymbol{p}}_2)$$
$$\times \frac{\sin\left[\frac{1}{\hbar}(\bar{E}_{ab} - E_{ab})(t - t_0)\right]}{\bar{E}_{ab} - E_{ab}} \{\bar{F}^<_{ab}(t)F^>_{ab}(t) - F^<_{ab}(t)\bar{F}^>_{ab}(t)\}\,, \qquad (7.29)$$

and, for the first order gradient contribution, we find

$$I_a^1(\boldsymbol{p}_1,t) = 2\sum_b \int \frac{d\boldsymbol{p}_2 d\bar{\boldsymbol{p}}_1 d\bar{\boldsymbol{p}}_2}{(2\pi\hbar)^6} |V_{ab}(\boldsymbol{p}_1 - \bar{\boldsymbol{p}}_1)|^2 \delta(\boldsymbol{p}_1 + \boldsymbol{p}_2 - \bar{\boldsymbol{p}}_1 - \bar{\boldsymbol{p}}_2)$$

$$\times \frac{d}{dE_{ab}} \frac{\cos\left[\frac{1}{\hbar}(E_{ab}-\bar{E}_{ab})\,(t-t_0)\right]-1}{E_{ab}-\bar{E}_{ab}} \frac{d}{dt} \left\{\bar{F}^{<}_{ab}(t)F^{>}_{ab}(t) - F^{<}_{ab}(t)\bar{F}^{>}_{ab}(t)\right\} . \tag{7.30}$$

In order to get (7.30), we used the relation

$$\int_0^{t-t_0} d\tau\tau\cos\omega\tau = -\frac{d}{d\omega}\frac{\cos[\omega(t-t_0)]-1}{\omega} .$$

The collision integrals considered above contain a typical spectral kernel produced by the propagators $g^{0R/A}_{ab}$ instead of the energy conserving δ-function in the Boltzmann-type kinetic equations. This spectral kernel is connected with a collisional broadening of the energy. Explicitly, this effect is accounted for in (7.26) by the cosine term, and in (7.29) by the the sine term. The latter one corresponds to the well-known slit function of time dependent perturbation theory. Because of the collisional broadening, the time evolution is not restricted to the energy shell. This leads to the correct energy conservation law for nonideal systems in Born approximation. The generalized Landau kinetic equation with the collision terms (7.28) conserves, therefore, density, momentum, and total energy.

Let us consider now the Boltzmann limit, i.e., $t_0 \to -\infty$. Initial correlations are being neglected. Then, using $\lim\limits_{\tau\to\infty} \frac{\sin(x\tau)}{x} = \pi\delta(x)$ and $\lim\limits_{\tau\to\infty} \frac{\cos(a\tau)-1}{a} = \frac{\mathrm{P}}{x}$, we get for the kinetic equation with the collision term I^0_a only

$$\begin{aligned} I^0_a &= \frac{1}{\hbar}\sum_b \int \frac{d\boldsymbol{p}_2 d\bar{\boldsymbol{p}}_1 d\bar{\boldsymbol{p}}_2}{(2\pi\hbar)^6} |V_{ab}(\boldsymbol{p}_1-\bar{\boldsymbol{p}}_1)|^2 \delta(\boldsymbol{p}_1+\boldsymbol{p}_2-\bar{\boldsymbol{p}}_1-\bar{\boldsymbol{p}}_2) \\ &\quad \times 2\pi\delta(\bar{E}_{12}-E_{12}) \left\{\bar{f}_a\,\bar{f}_b(1-f_a)(1-f_b) - f_a f_b(1-\bar{f}_a)(1-\bar{f}_b)\right\}_t . \end{aligned} \tag{7.31}$$

This is the well-known Markovian Landau kinetic equation which conserves the mean kinetic energy. If the first order gradient correction is accounted for taking $t_0 \to -\infty$, we have to add

$$\begin{aligned} I^1_a(\boldsymbol{p}_1,t) &= \sum_b \int \frac{d\boldsymbol{p}_2 d\bar{\boldsymbol{p}}_1 d\bar{\boldsymbol{p}}_2}{(2\pi\hbar)^6} \left|V_{ab}(\boldsymbol{p}_1-\bar{\boldsymbol{p}}_1)\right|^2 \delta(\boldsymbol{p}_1+\boldsymbol{p}_2-\bar{\boldsymbol{p}}_1-\bar{\boldsymbol{p}}_2) \\ &\times \frac{d}{dE_{ab}} \frac{P}{E_{ab}-\bar{E}_{ab}} \frac{d}{dt} \left\{\bar{f}_a\,\bar{f}_b(1-f_a)(1-f_b) - f_a f_b(1-\bar{f}_a)(1-\bar{f}_b)\right\}_t . \end{aligned} \tag{7.32}$$

This additional term gives a contribution which ensures the full energy conservation in the long time limit on the level of the Born approximation (Klimontovich 1982). If we introduce $I^0_a(\boldsymbol{p}_1, t; \varepsilon)$ by

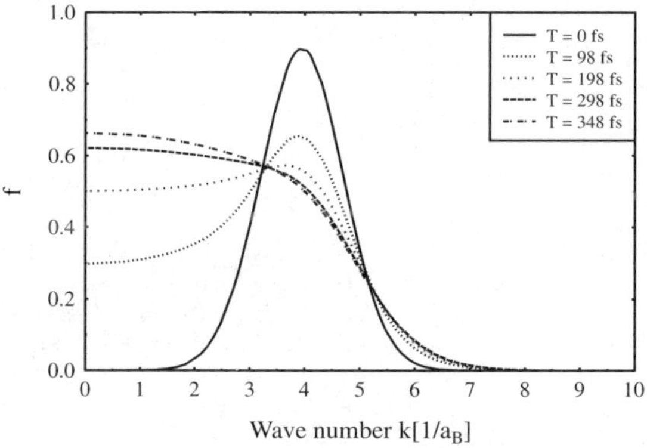

Fig. 7.4. Time evolution of the distribution function for the Landau collision integral using (7.35)

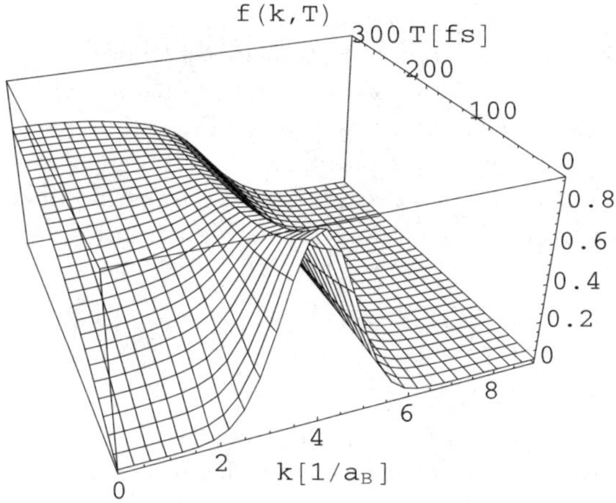

Fig. 7.5. Distribution function over time and momentum. The situation is that of Fig. 7.4; see text

$$I_a^0(\boldsymbol{p}_1, t; \varepsilon) = \frac{2}{\hbar} \sum_b \int \frac{d\boldsymbol{p}_2 d\bar{\boldsymbol{p}}_1 d\bar{\boldsymbol{p}}_2}{(2\pi\hbar)^6} \left|V_{ab}\left(\boldsymbol{p}_1 - \bar{\boldsymbol{p}}_1\right)\right|^2 \delta\left(\boldsymbol{p}_1 + \boldsymbol{p}_2 - \bar{\boldsymbol{p}}_1 - \bar{\boldsymbol{p}}_2\right)$$
$$\times \frac{\varepsilon}{(E_{ab} - \bar{E}_{ab})^2 + \varepsilon^2} \left\{\bar{f}_a\, \bar{f}_b(1 - f_a)(1 - f_b) - f_a f_b(1 - \bar{f}_a)(1 - \bar{f}_b)\right\}_t , \tag{7.33}$$

it is possible to write the collision integral $I_a = I_a^0 + I_a^1$ in the more compact and interesting form

$$I_a(\boldsymbol{p}_1, t) = \left(1 + \frac{\hbar}{2}\frac{d}{dt}\frac{d}{d\varepsilon}\right) I_a^0(\boldsymbol{p}_1, t; \varepsilon)|_{\varepsilon \to 0} . \tag{7.34}$$

The kinetic equations considered here are simple enough to allow for a numerical solution. Here we follow the calculations of Bonitz (1998), of Bonitz et al. (1999), Semkat and Bonitz (2000), Semkat (2001), and Bonitz and Semkat (2005). In Figs. 7.4 and 7.5, we show the irreversible time evolution of the

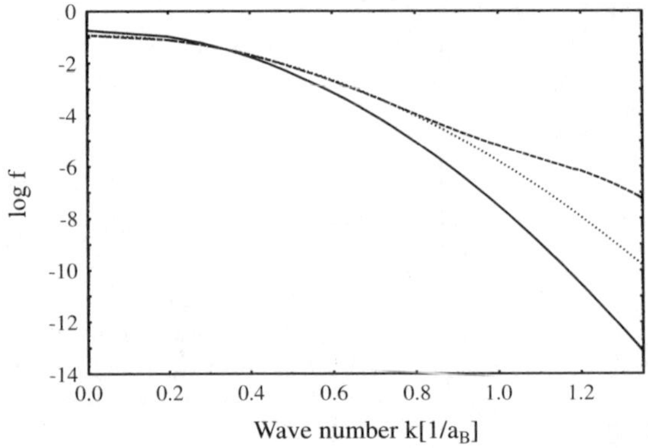

Fig. 7.6. Tail behavior of the distribution function. *Solid line*: Initial distribution (Fermi); *dashes*: Final distribution; *dots*: Fermi distribution with T calculated from $E_{kin,end}$

distribution function which follows from a numerical solution of the two-time KB equations, starting from a Gaussian distribution for the momenta ending up with a Fermi-like distribution. However, there is an essential difference to the usual Boltzmann equation which leads to a Fermi function for ideal systems. The final stationary momentum distribution function is now modified by the interaction. We obtain another tail behavior. This is shown in Fig. 7.6. The Gaussian was taken to be

$$f_e(k) = A \exp\left[\frac{(k-k_0)^2}{b^2}\right] \tag{7.35}$$

with $b = 1.06\ a_B^{-1}$, $k_0 = 3.95\ a_B^{-1}$, $A = 0.9$. Such data correspond to $n = 10^{18}\,\mathrm{cm}^{-3}$, $T = 290\,\mathrm{K}$ which is relevant for electron–hole plasmas.

The relaxation behavior of the kinetic, the potential, and of the total energies is shown in Fig. 7.7. In this figure, the temporal evolution of the energy is given comparing the results of the two-time Kadanoff–Baym equations with different approximations of the Landau equation (see Bonitz et al. (1999), Semkat and Bonitz (2000), Semkat (2001), and Bonitz and Semkat (2005)). Of course, the energy conservation should be fulfilled for the equation (7.24), in good agreement with the two-time equation. Problems arise if approximations for the propagators g_{ab}^{0R} are adopted. We see that the behavior of the potential energy is scarcely influenced by the approximations. On the other hand, there is good agreement between single- and two-time calculations, if the damping in (7.26) is neglected. With the damping taken into account, the kinetic energy does not achieve a saturation but increases permanently. Of course, this leads to a violation of the conservation of the total energy. For this simple approximation, the correct balance between memory and damping is destroyed. Therefore, this collision integral does not represent a conserving approximation. If memory and damping are neglected like in the collision integral (7.29), proper energy conservation is given. Let us summarize the information about the time evolution described on the basis of the Landau

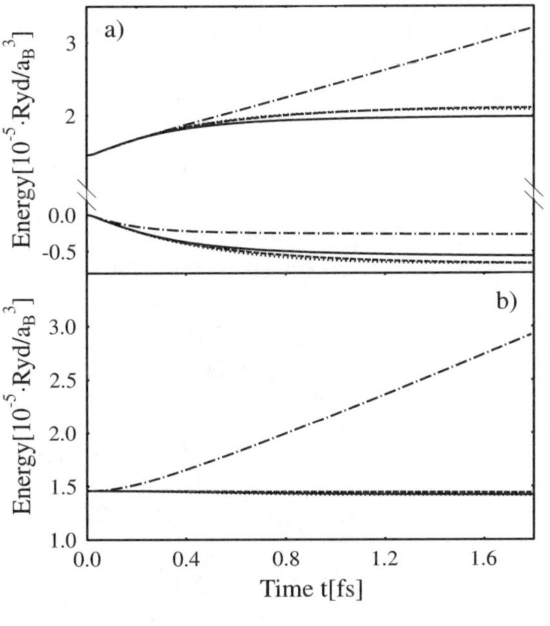

Fig. 7.7. Time evolution of kinetic (**upper curves**) and potential energies (**lower curves**) (**a**) and of the total energy (**b**) determined from the Landau equation. *Full lines*: two-time KB eqs., *dotted*: free GKBA, *dashed*: slit function, *dash-dot*: Lorentz with damping; after Bonitz et al. (1999), Semkat (2001)

kinetic equation. The most general expression for the Landau equation (7.24) describes the relaxation from the initial time up to the equilibrium state. Details of the kinetics follow from the approximation (7.28). A series expansion at the initial time t_0 shows that $I_a^0 \sim t$ and $I_a^1 \sim t^2$. Consequently, the short time behavior is, for the present, determined by I_a^0 and I_a^{in}. In this initial stage, especially correlations are being built up. Depending on the initial correlations, we observe an increase (heating) or a decrease (cooling) of the kinetic energy. When time goes on, the collision integrals degenerate to the long time behavior (7.31) and (7.32), respectively. In the long time limit, the temporal evolution is "on-shell", i.e., the kinetic and the potential energies are constant and, therefore,

$$\frac{d}{dt}\left\{\langle T\rangle + \langle V\rangle\right\}_{\text{Born}} = 0. \tag{7.36}$$

7.4 Dynamical Screening, Generalized Lenard–Balescu Equation

In the preceding section, we applied the Landau kinetic equation with a statically screened potential to describe the ultrashort relaxation behavior of nonideal plasmas. In general, however, the dynamic character of screening has to be taken into account. In order to do this, we will use the self-energy in RPA given by (4.244). As discussed in Chap. 4, this approximation allows us to study the properties of dense weakly correlated plasmas including screening of the long range Coulomb forces and collective excitations. After Fourier transformation with respect to the differencc variables in space, we get

$$\Sigma_a^{\gtrless}(\boldsymbol{pR},tt') = i\hbar \int \frac{d\boldsymbol{p}'d\boldsymbol{p}''}{(2\pi\hbar)^3} V_{aa}^{s\,\gtrless}(\boldsymbol{p}'\boldsymbol{R},tt')g_a^{\gtrless}(\boldsymbol{p}''\boldsymbol{R},tt')\delta(\boldsymbol{p}-\boldsymbol{p}'-\boldsymbol{p}'') . \quad (7.37)$$

For the derivation of a kinetic equation in this approximation, again, we focus on weakly inhomogeneous plasmas and start from the time diagonal equation (7.23). Inserting the self-energies (7.37), we find

$$\left(\frac{\partial}{\partial t} + \frac{\boldsymbol{p}_1}{m_a}\nabla_{\boldsymbol{R}} - \nabla_R U_a^{\text{eff}}(\boldsymbol{R},t)\nabla_{\boldsymbol{p}_1}\right) f_a(\boldsymbol{p}_1,t) = I_a(\boldsymbol{p}_1,t) \quad (7.38)$$

with the r.h.s. given by

$$I_a(\boldsymbol{p}_1,t) = I_a^{in}(\boldsymbol{p}_1,t) - 2\,\text{Re}\int_{t_0}^{t} d\bar{t} \int \frac{d\bar{\boldsymbol{p}}_1 d\boldsymbol{q}}{(2\pi\hbar)^3} i\hbar\delta(\bar{\boldsymbol{p}}_1 - \boldsymbol{p}_1 + \boldsymbol{q})$$
$$\times[V_{aa}^{>}(\boldsymbol{q},t\bar{t})g_a^{>}(\bar{\boldsymbol{p}}_1,t\bar{t})\, g_a^{<}(\boldsymbol{p}_1,\bar{t}t) - V_{aa}^{<}(\boldsymbol{q},t\bar{t})\, g_a^{<}(\bar{\boldsymbol{p}}_1,t\bar{t})\, g_a^{>}(\boldsymbol{p}_1,\bar{t}t)] . \quad (7.39)$$

As before, we dropped the variable $\boldsymbol{R}$ in the Wigner distribution function and in the collision integral. Together with the equation for the correlation functions $V_{ab}^{\gtrless}$ of the dynamically screened potential

$$\Delta_1 V_{ab}^{\gtrless}(12) = -4\pi \sum_{cd} e_a e_c \int d3\, [\Pi_{cd}^{R}(13)V_{db}^{\gtrless}(32) + \Pi_{cd}^{\gtrless}(13)V_{db}^{A}(32)] \quad (7.40)$$

which was derived already in Sect. 4.3, we get a closed system of equations for the description of a plasma coupled to the longitudinal electromagnetic field. These general equations are, in particular, nonlocal in time. We will first consider the local approximation which is essentially simpler. Then, macroscopic and microscopic dynamics are decoupled, and all quantities may be replaced by Fourier transforms with respect to the difference times. The local approximation is a very restrictive assumption. Especially, equation (7.40) reduces to the condition

$$\sum_c (V_{ac}^{<}(\boldsymbol{p}\omega,t)\Pi_{cb}^{>}(\boldsymbol{p}\omega,t) - V_{ac}^{>}(\boldsymbol{p}\omega,t)\Pi_{cb}^{<}(\boldsymbol{p}\omega,t)) = 0 . \quad (7.41)$$

Again, we dropped the macroscopic space variable $\boldsymbol{R}$. For the polarization function in RPA, we have $\Pi_{ab}^{\gtrless} = \delta_{ab}\Pi_{aa}^{\gtrless}$ with the explicit expression given by (4.108). This expression can be interpreted as being proportional to the rates of absorption and emission of a plasmon of momentum $\boldsymbol{q}$ and energy ω by a particle scattered from an initial state $\boldsymbol{p}$ to a state $\boldsymbol{p}+\boldsymbol{q}$. Therefore, (7.41) describes the local detailed balance between emission and absorption of plasmons, and it is equal to the fluctuation-dissipation theorem given by (4.248). Moreover, in the local approximation, the Bogolyubov condition of the weakening of initial correlations has to be adopted, i.e., $t_0 \to -\infty$ and $I_a^{\text{in}}(t) = 0$. Then we get

$$\begin{aligned} I_a^0(\boldsymbol{p}_1, t) &= -i\hbar \int \frac{d\bar{\boldsymbol{p}}_1 d\boldsymbol{q}}{(2\pi\hbar)^3} \frac{d\omega}{2\pi} \frac{d\bar{\omega}}{2\pi} \delta(\bar{\boldsymbol{p}}_1 - \boldsymbol{p}_1 + \boldsymbol{q}) \\ &\times \{ V_{aa}^{>}(\boldsymbol{q}, \omega, t)\, g_a^{>}(\bar{\boldsymbol{p}}_1, \omega - \bar{\omega}, t)\, g_a^{<}(\boldsymbol{p}_1, \omega, t) \\ &- V_{aa}^{<}(\boldsymbol{q}, \bar{\omega}, t)\, g_a^{<}(\bar{\boldsymbol{p}}_1, \omega - \bar{\omega}, t)\, g_a^{>}(\boldsymbol{p}_1, \omega, t) \} . \qquad (7.42) \end{aligned}$$

The further procedure is straightforward:

(i) With application of the local fluctuation-dissipation theorem (4.248), the correlation functions $V_{aa}^{\gtrless}$ are replaced by the retarded and advanced screened potentials $V_{aa}^{R/A}$, respectively.
(ii) The correlation functions $g_a^{\gtrless}$ are expressed in terms of the Wigner distributions using the KB ansatz (7.10).

Then we immediately get the kinetic equation with the collision integral I_a^0 in local approximation

$$\begin{aligned} &\left(\frac{\partial}{\partial t} + \frac{\boldsymbol{p}_1}{m_a} \nabla_{\boldsymbol{R}} - \nabla_{\boldsymbol{R}} U_a^{\text{eff}}(\boldsymbol{R}, t) \nabla_{\boldsymbol{p}_1} \right) f_a(\boldsymbol{p}_1, t) \\ &= \frac{1}{\hbar} \sum_b \int \frac{d\boldsymbol{p}_2 d\bar{\boldsymbol{p}}_1 d\bar{\boldsymbol{p}}_2}{(2\pi\hbar)^6} \left| V_{ab}^R(\boldsymbol{p}_1 - \bar{\boldsymbol{p}}_1, E_a - \bar{E}_a) \right|^2 \delta(\boldsymbol{p}_1 + \boldsymbol{p}_2 - \bar{\boldsymbol{p}}_1 - \bar{\boldsymbol{p}}_2) \\ &\times 2\pi\delta(E_{ab} - \bar{E}_{ab}) \{ \bar{f}_a \bar{f}_b (1 - f_a)(1 - f_b) - f_a f_b (1 - \bar{f}_a)(1 - \bar{f}_b) \}_t . \quad (7.43) \end{aligned}$$

As before, we used the notation $f_a = f_a(\boldsymbol{p}_1, \boldsymbol{R}t)$, $f_b = f_b(\boldsymbol{p}_2, \boldsymbol{R}t)$, etc., for the Wigner functions, and $E_{ab} = E_a + E_b$ with $E_a = E_a(\boldsymbol{p}_1, \boldsymbol{R}t)$, $E_b = E_b(\boldsymbol{p}_2, \boldsymbol{R}t)$ for the quasiparticle energies. The expression above has the typical form of a quantum Boltzmann-like collision integral. Here, the transition probability on the r.h.s of (7.43) is given in dynamically screened Born approximation with V_{ab}^R being local in space and time. The classical limit of the collision integral follows for small momentum transfer $\boldsymbol{q} = \boldsymbol{p}_1 - \bar{\boldsymbol{p}}_1$, and it was first derived by Lenard (1960) and by Balescu (1960). See also Guernsey (1960) and Guernsey (1962). The r.h.s. of (7.43) is called the Lenard–Balescu collision integral. There exist many derivations of the quantum Lenard–Balescu collision term, see Wyld and Pines (1962), Balescu (1963), Silin (1961). A field theoretical derivation was given by DuBois (1968). Equation (7.43), together with the screening equation (4.100) in RPA, provides a self-consistent system of equations to determine the particle distribution functions and the dielectric properties of weakly correlated quantum plasmas. If the variable $\boldsymbol{R}$ is dropped, the equation for the retarded screened potential reads

$$V_{ab}^R(\boldsymbol{q}\omega,t) = V_{ab}(\boldsymbol{q}) + \sum_{cd} V_{ac}^R(\boldsymbol{q}\omega,t)\Pi_{cd}^R(\boldsymbol{q}\omega,t)V_{db}(\boldsymbol{q}) = V_{ab}(\boldsymbol{q}) \,/\, \varepsilon^R(\boldsymbol{q}\omega,t)\,. \tag{7.44}$$

Expressions and properties of the dielectric function $\varepsilon^R(\boldsymbol{q}\omega,t)$ in RPA were given in detail in Chap. 4. At this point, it should be mentioned that the Lenard–Balescu collision integral (7.43) is a Markovian one, i.e., there are the shortcomings discussed already in Sect. 7.1.

In order to avoid the defects of Markovian equations, we have to go beyond the local approximation. Especially, retardation in time has to be taken into account. For this purpose, we return to (7.39) and generalize the above procedure leading to (7.43) by inclusion of the retardation in time. That means, the correlation functions $V_{aa}^{\gtrless}$ are replaced by the retarded or advanced screened potentials $V_{aa}^{R/A}$ using the general nonlocal form of the fluctuation-dissipation theorem (4.38), and the particle correlation functions $g_a^{\gtrless}$ are replaced by Wigner functions using the GKB ansatz (7.11). Neglecting initial correlations, we obtain, after a straightforward calculation

$$\begin{aligned} I_a(\boldsymbol{p}_1,t) = &-\frac{2}{\hbar^2}\sum_b \mathrm{Re}\int \frac{d\boldsymbol{p}_2 d\bar{\boldsymbol{p}}_1 d\bar{\boldsymbol{p}}_2}{(2\pi\hbar)^6}\,\delta(\boldsymbol{p}_1+\boldsymbol{p}_2-\bar{\boldsymbol{p}}_1-\bar{\boldsymbol{p}}_2)\int_{t_0}^{t} d\bar{t} \\ &\times\Bigg\{ e^{-\frac{i}{\hbar}[(\bar{E}_a-E_a)(t-\bar{t})]}\, f_a^>(\bar{\boldsymbol{p}}_1,\bar{t})\, f_a^<(\boldsymbol{p}_1,\bar{t}) \int_{t_0}^{t} dt' \int_{t_0}^{\bar{t}} dt'' V_{ab}^R(\boldsymbol{p}_1-\bar{\boldsymbol{p}}_1,tt') \\ &\times V_{ba}^A(\boldsymbol{p}_1-\bar{\boldsymbol{p}}_1,t''\bar{t}) \Big[\Theta(t'-t'')\, e^{-\frac{i}{\hbar}[(\bar{E}_b-E_b)(t'-t'')]}\, f_b^>(\bar{\boldsymbol{p}}_2,t'')\, f_b^<(\boldsymbol{p}_2,t'') \\ &+\Theta(t''-t')\, e^{-\frac{i}{\hbar}[(\bar{E}_b-E_b)(t'-t'')]} f_b^>(\bar{\boldsymbol{p}}_2,t')\, f_b^<(\boldsymbol{p}_2,t')\Big] - (\,>\longleftrightarrow<\,)\Bigg\}. \end{aligned} \tag{7.45}$$

This expression represents the non-Markovian generalization of the Lenard–Balescu collision integral. It was first given starting from (7.39) by Kuznetsov (1991) and then by Haug and Ell (1992). Together with the (nonlocal) screening equation (4.35), we have a self-consistent system for the particle and screening dynamics which generalizes the Markovian Lenard–Balescu equation given above. Similar to the Landau equation, one can find a gradient expansion of the Lenard–Balescu kinetic equation written as

$$\left(\frac{\partial}{\partial t}+\frac{\boldsymbol{p}_1}{m_a}\nabla_R-\nabla_R U_a^{\mathrm{eff}}(\boldsymbol{R},t)\nabla_{p_1}\right) f_a(\boldsymbol{p}_1,t) = \left(1+\frac{\hbar}{2}\frac{d}{dt}\frac{d}{d\varepsilon}\right) I_a^0(\boldsymbol{p}_1,t;\varepsilon)|_{\varepsilon\to 0}\,. \tag{7.46}$$

Here, the collision term $I_a^0(\boldsymbol{p}_1,t;\varepsilon)$ reads

$$I_a^0(\boldsymbol{p}_1, t; \varepsilon) = \sum_b \int \frac{d\boldsymbol{p}_2 d\bar{\boldsymbol{p}}_1 d\bar{\boldsymbol{p}}_2}{(2\pi\hbar)^6} \int \frac{d\omega}{2\pi} \left|V_{ab}^R(\boldsymbol{p}_1 - \bar{\boldsymbol{p}}_1, \omega)\right|^2$$
$$\times \delta(\boldsymbol{p}_1 + \boldsymbol{p}_2 - \bar{\boldsymbol{p}}_1 - \bar{\boldsymbol{p}}_2) \frac{2\varepsilon}{(E_{ab} - \hbar\omega)^2 + \varepsilon^2} \frac{2\varepsilon}{(\bar{E}_{ab} - \hbar\omega)^2 + \varepsilon^2}$$
$$\times \{\bar{F}_{ab}^<(t) F_{ab}^>(t) - F_{ab}^<(t) \bar{F}_{ab}^>(t)\} . \tag{7.47}$$

The notation $F_{ab}^{\gtrless}$ is used as in formula (7.27). In contrast to the Markovian Lenard–Balescu equation, the non-Markovian one gives the correct conservation laws for nonideal plasmas in the approximation considered. Let us find the energy conservation law following from (7.46). For this purpose, we multiply the kinetic equation with $p_1^2/2m_a$ and take the integral over $\boldsymbol{p}_1$. Performing the sum over the species, in the spatially homogeneous case we get

$$\frac{d}{dt} \{\langle T \rangle + \langle V \rangle\}_{\mathrm{RPA}} = 0. \tag{7.48}$$

Here, $\langle V \rangle$ is the mean value of the potential energy given in RPA, i.e., expressed in terms of the RPA dynamically screened potential (7.44),

$$\langle V \rangle = \langle V \rangle^{\mathrm{HF}} + \frac{1}{2} \sum_{ab} Tr_{12} \left[|V_{ab}^R(\bar{E})|^2 \frac{P}{E_{ab} - \bar{E}_{ab}} \{F_{ab}^< \bar{F}_{ab}^> - \bar{F}_{ab}^< F_{ab}^>\} \right] . \tag{7.49}$$

As before, the kinetic equation in first order gradient expansion ensures the conservation of the total energy in the respective approximation.

7.5 Particle Kinetics and Field Fluctuations. Plasmon Kinetics

The kinetic equation (7.43) is valid if the detailed balance (7.41) is fulfilled. In this case, the longitudinal field fluctuations $V_{aa}^{\gtrless}$ may be eliminated from the kinetics of particles. However, there are important and interesting situations far from equilibrium in which this condition is violated. Such systems appear under the influence of external fields or other external actions, which lead to plasma instabilities and turbulence. Generally speaking, the condition of detailed balance is violated, if gradients with respect to $\boldsymbol{R}$ and t cannot be neglected. In this case, it is necessary to formulate a system of equations both for the particle distribution functions, and for the field fluctuations as well. We start from the general equations (7.39) and (7.40) and apply the identity $ab - cd = 1/2(a-c)(b+d) + 1/2(a+c)(b-d)$. Using the definitions (4.179) for the correlation functions of the longitudinal field fluctuation and the relation $1/2(V^> - V^<) = i\mathrm{Im}V^R = iV\mathrm{Im}\varepsilon^{R^{-1}}$, we can write the generalized (non-Markovian) Lenard–Balescu collision integral in the form

$$
\begin{aligned}
I_a(\boldsymbol{p}_1,t) = &-2\mathrm{Re}\int_{t_o}^{t} d\bar{t}\int \frac{d\bar{\boldsymbol{p}}_1 d\boldsymbol{q}}{(2\pi\hbar)^3}\delta(\bar{\boldsymbol{p}}_1-\boldsymbol{p}_1+\boldsymbol{q})\\
&\times\left\{\frac{\hbar^2 e_a^2}{q^2}\langle\overline{\delta\boldsymbol{E}\delta\boldsymbol{E}}\rangle_{\boldsymbol{q},t\bar{t}}[g_a^>(\bar{\boldsymbol{p}}_1,t\bar{t})\,g_a^<(\boldsymbol{p}_1,\bar{t}t)-g_a^<(\bar{\boldsymbol{p}}_1,t\bar{t})g_a^>(\boldsymbol{p}_1,\bar{t}t)]\right.\\
&\left.-\hbar V_{aa}(q)\mathrm{Im}\varepsilon^{R-1}(\boldsymbol{q},t\bar{t})[g_a^>(\bar{\boldsymbol{p}}_1,t\bar{t})g_a^<(\boldsymbol{p}_1,\bar{t}t)+g_a^<(\bar{\boldsymbol{p}}_1,t\bar{t})g_a^>(\boldsymbol{p}_1,\bar{t}t)]\right\}.
\end{aligned}
\tag{7.50}
$$

We introduce the Wigner function using the GKB ansatz and finally substitute $\bar{t}=t-\tau$. Then we immediately get

$$
\begin{aligned}
I_a(\boldsymbol{p}_1,t) = &-\frac{2}{\hbar^2}\mathrm{Re}\int_0^{t-t_o} d\tau\int \frac{d\bar{\boldsymbol{p}}_1 d\boldsymbol{q}}{(2\pi\hbar)^3}\delta(\bar{\boldsymbol{p}}_1-\boldsymbol{p}_1+\boldsymbol{q})\\
&\times e^{-\frac{i}{\hbar}(E_a(\bar{\boldsymbol{p}}_1)-E_a(\boldsymbol{p}_1))\tau}\left\{\frac{\hbar^2 e_a^2}{q^2}\langle\overline{\delta\boldsymbol{E}\delta\boldsymbol{E}}\rangle_{\boldsymbol{q},\tau,t-\frac{\tau}{2}}[f_a^<(\boldsymbol{p}_1)-f_a^<(\bar{\boldsymbol{p}}_1)]_{t-\tau}\right.\\
&\left.-\hbar V_{aa}(q)\mathrm{Im}\varepsilon^{R-1}(\boldsymbol{q},\tau,t-\frac{\tau}{2})[f_a^>(\bar{\boldsymbol{p}}_1)\,f_a^<(\boldsymbol{p}_1)+\,f_a^<(\bar{\boldsymbol{p}}_1)\,f_a^>(\boldsymbol{p}_1)]_{t-\tau}\right\}.
\end{aligned}
\tag{7.51}
$$

This collision integral splits up into two parts. The first one is determined by the correlation function of the longitudinal electric field fluctuations (4.186) (micro-field fluctuations), i.e., by a fluctuation quantity. The second term contains the imaginary part of the inverse dielectric function and is, therefore, given by a dissipative quantity. We know from the dissipation–fluctuation relations (4.58) that these quantities are not independent.

Of course, (7.51) requires an additional equation for the correlation function of the field fluctuations. Taking into account (4.172) and (4.179), we have

$$
\begin{aligned}
e_a e_b\langle\overline{\delta\boldsymbol{E}(1)\delta\boldsymbol{E}(1')}\rangle &= \boldsymbol{\nabla}_1\boldsymbol{\nabla}_{1'}J_{ab}(11')\,,\\
J_{ab}(11') &= \frac{i\hbar}{2}(V_{ab}^>(11')+V_{ab}^<(11'))\,.
\end{aligned}
\tag{7.52}
$$

An equation for the symmetrized correlation function $J_{ab}(11')$ may be derived easily from (4.37). We get

$$
\begin{aligned}
\Delta_1 J_{ab}(11') \quad + \quad & 4\pi\sum_{cd} e_a e_c\int d\bar{1}\;\Pi_{cd}^R(1\bar{1})J_{db}(\bar{1}1')\\
&= -4\pi\sum_{cd} e_a e_c\int d\bar{1}\;Q_{cd}(1\bar{1})V_{db}^A(\bar{1}1')
\end{aligned}
\tag{7.53}
$$

with the source function $Q_{ab}(11')$ defined by

$$
Q_{ab}(11') = \frac{i\hbar}{2}(\Pi_{ab}^>(11')+\Pi_{ab}^<(11'))\,.
\tag{7.54}
$$

As shown in Sect. 4.3, the retarded/advanced screened potential can be calculated from

$$\Delta_1 V_{ab}^{R/A}(11') + 4\pi \sum_{cd} e_a e_c \int d\bar{1} \Pi_{cd}^{R/A}(1\bar{1}) V_{db}^{R/A}(\bar{1}, 1') = -4\pi e_a e_b \delta(1-1') \,. \tag{7.55}$$

Using this system of equations, we find the very general form of the fluctuation–dissipation theorem between the source of fluctuations and the field correlation function. It reads in terms of the response functions $V_{ab}^{R/A}$

$$J_{ab}(11') = J_{ab}^0(11') + \sum_{cd} \int d2d3 V_{ac}^R(12) Q_{cd}(23) V_{db}^A(3,1') \,. \tag{7.56}$$

In order to find the Markovian limit of the system of equations let us follow the scheme of gradient expansions like in Chap. 6 and in Sect. 7.3. We start with the kinetic equation for particles and expand the collision integral (7.51) with respect to τ. For the zeroth order term we arrive at

$$\begin{aligned} & I_a^0(\boldsymbol{p}_1, t) \\ & = -2\mathrm{Re} \int \frac{d\bar{\boldsymbol{p}}_1 d\boldsymbol{q}}{(2\pi\hbar)^3} \frac{d\omega}{2\pi} \frac{\sin\left[\frac{1}{\hbar}(\hbar\omega - E_a + \bar{E}_a)(t-t_0)\right]}{\hbar\omega - E_a + \bar{E}_a} \delta(\bar{\boldsymbol{p}}_1 - \boldsymbol{p}_1 + \boldsymbol{q}) \\ & \times \left\{ \frac{\hbar^2 e_a^2}{q^2} \langle \delta \boldsymbol{E} \delta \boldsymbol{E} \rangle_{\boldsymbol{q}\omega, t} [f_a^< - \bar{f}_a^<] - \hbar \mathrm{Im} \frac{V_{aa}(q)}{\varepsilon^R(\boldsymbol{q}\omega, t)} [\bar{f}_a^> f_a^< + \bar{f}_a^< f_a^>] \right\} . \end{aligned} \tag{7.57}$$

Similar to the Landau equation, we have in the collision integral (7.57), instead of the energy conserving δ-function, the slit function, this means an energy broadening. Therefore, the conservation of the full energy may be expected. In the limit $t_0 \to -\infty$, the slit function goes over into the kinetic energy conserving δ-function, and it follows the Markovian on-shell collision integral

$$\begin{aligned} I_a^0(\boldsymbol{p}_1, t) = & \int \frac{d\bar{\boldsymbol{p}}_1 d\boldsymbol{q}}{(2\pi\hbar)^3} \int \frac{d\omega}{2\pi} 2\pi \delta(\hbar\omega - E_a + \bar{E}_a) \delta(\bar{\boldsymbol{p}}_1 - \boldsymbol{p}_1 + \boldsymbol{q}) \\ & \times \left\{ \frac{\hbar^2 e_a^2}{q^2} \langle \overline{\delta \boldsymbol{E} \delta \boldsymbol{E}} \rangle_{\boldsymbol{q}\omega, t} [\bar{f}_a^< - f_a^<] + \hbar \mathrm{Im} \frac{V_{aa}(q)}{\varepsilon^R(\boldsymbol{q}\omega, t)} [\bar{f}_a^> f_a^< + \bar{f}_a^< f_a^>] \right\} . \end{aligned} \tag{7.58}$$

With the local approximation (4.186) for the field fluctuation and for the imaginary part of the dielectric function (4.111), the equivalence between (7.58) and (7.43) can be shown. In general we have to take into account that $\langle \bar{\delta} \boldsymbol{E} \delta \boldsymbol{E} \rangle_{\boldsymbol{q}\omega, \boldsymbol{R}t}$ and $\mathrm{Im} \varepsilon^{R^{-1}}(\boldsymbol{q}\omega, \boldsymbol{R}t)$ have gradient expansions, too.

Now we proceed with the scheme of Sect. 7.3 and determine the first order retardation contribution. The result is

$$I_a^1(\boldsymbol{p}_1,t) = 2\int \frac{d\bar{\boldsymbol{p}}_1 d\boldsymbol{q}}{(2\pi\hbar)^3}\int \frac{d\omega}{2\pi}\frac{d}{d\omega}\frac{P}{\hbar\omega - E_a + \bar{E}_a}\delta(\bar{\boldsymbol{p}}_1 - \boldsymbol{p}_1 + \boldsymbol{q})$$
$$\times\left\{[\frac{\hbar^2 e_a^2}{2q^2}\frac{d}{dt}\langle\overline{\delta\boldsymbol{E}\delta\boldsymbol{E}}\rangle_{\boldsymbol{q}\omega,t}]\,[\bar{f}_a^< - f_a^<] + \frac{e_a^2}{q^2}\langle\overline{\delta\boldsymbol{E}\delta\boldsymbol{E}}\rangle_{\boldsymbol{q}\omega,t}\frac{d}{dt}\,[\bar{f}_a^< - f_a^<]\right.$$
$$\left.+\hbar V_{aa}(q)\mathrm{Im}\frac{1}{\varepsilon^R(\boldsymbol{q}\omega,t)}\frac{1}{2}\frac{d}{dt}[\bar{f}_a^> f_a^< + \bar{f}_a^< f_a^>]\right\}. \tag{7.59}$$

Like in our previous considerations, we dropped the macroscopic space variable $\boldsymbol{R}$ for simplicity. It turns out that (7.59) does not give the complete first order correction. Additionally we have to take into account the gradient correction to $\langle\overline{\delta\boldsymbol{E}\delta\boldsymbol{E}}\rangle$ in I_a^0.

In addition to the particle kinetics, it is, therefore, necessary to find an equation for the field fluctuations. We start from the non-Markovian equation (7.40). It is convenient to transform this equation into

$$\sum_c \int d3\, V_{ac}^{R-1}(13)V_{ca}^{\gtrless}(32) = \sum_c \int d3\, \Pi_{ac}^{\gtrless}(13)V_{ca}^{A}(32)\,. \tag{7.60}$$

Here, $V_{ac}^{R-1} = V_{ac}^{-1} - \Pi_{ac}^R$. Again we consider the regime where all functions are slowly varying functions of the macroscopic variables $\boldsymbol{R}$ and t as compared to the dependence on the microscopic ones, $\boldsymbol{r}$ and τ. Following the scheme developed in Chap. 3 for the particle kinetics, for the first order gradient expansion of equation (7.60) we get

$$\sum_c[\mathrm{Re}V_{ac}^{R^{-1}}, iV_{ca}^{\gtrless}] + \sum_c[\mathrm{Re}V_{ac}^R, i\Pi_{ca}^{\gtrless}] = \sum_c(V_{ac}^<\Pi_{ca}^> - V_{ac}^>\Pi_{ca}^<)\,, \tag{7.61}$$

where the generalized Poisson bracket is given by (3.176). Now, using the definition (4.179) for the field fluctuations and (4.55) for the dielectric function, we find the kinetic equation

$$[\mathrm{Re}\varepsilon^R, \langle\overline{\delta\boldsymbol{E}\delta\boldsymbol{E}}\rangle] + \sum_{cd}[\mathrm{Re}V_{dc}^R, 4\pi Q_{cd}] = -2\mathrm{Im}\varepsilon^R\,\langle\overline{\delta\boldsymbol{E}\delta\boldsymbol{E}}\rangle - 8\pi\sum_{cd}\mathrm{Im}V_{dc}^R Q_{cd}\,. \tag{7.62}$$

In RPA, the real and imaginary parts of the dielectric function are given by the formulae (4.110) and (4.111), and for the source function, we have $Q_{ab} = Q_{aa}\delta_{ab}$. An explicit expression for Q_{aa} may be obtained in RPA using (4.106) in the definition (7.54):

$$Q_{aa}(\boldsymbol{q}\omega) = \hbar\int\frac{d\boldsymbol{p}}{(2\pi\hbar)^3}2\pi\delta(\hbar\omega - E_a(\boldsymbol{p}+\boldsymbol{q}) + E_a(\boldsymbol{p}))$$
$$\times[f_a^>(\boldsymbol{p}+\boldsymbol{q})f_a^<(\boldsymbol{p}) + f_a^<(\boldsymbol{p}+\boldsymbol{q})f_a^>(\boldsymbol{p})]\,. \tag{7.63}$$

Let us discuss the equation (7.62) in more detail. The simplest approximation is the local approximation, i.e., we neglect all gradient terms on the left hand side. Then we find the relation

$$\langle \overline{\delta \boldsymbol{E} \delta \boldsymbol{E}} \rangle_{\boldsymbol{q}\omega,t} = \sum_a \frac{4\pi V_{aa}(q) Q_{aa}(\boldsymbol{q}\omega,t)}{|\varepsilon^R(\boldsymbol{q}\omega,t)|^2} = \frac{\langle \overline{\delta \boldsymbol{E} \delta \boldsymbol{E}} \rangle^{\text{source}}_{\boldsymbol{q}\omega,t}}{|\varepsilon^R(\boldsymbol{q}\omega,t)|^2}, \tag{7.64}$$

which is just the fluctuation–dissipation theorem (4.186) for the correlation function of the longitudinal field fluctuations. It should be noticed that we have to account for the additional factor $1/4\pi\varepsilon_0$ on the r.h.s. if SI-units are used. This follows from the relation (7.52) and from the expression for the Coulomb potential $V_{ab}(q) = \hbar^2 e_a e_b/\varepsilon_0 q^2$ in this system.

However, as we discussed already, the fluctuation–dissipation relation is not valid in every case. Therefore, we must go back to (7.62) which is valid for more rapid variations of t and $\boldsymbol{R}$. A more appropriate form of that equation follows if we carry out the operation of the Poisson brackets at the left hand side. After some simple transformations, we obtain

$$\begin{aligned} \frac{\partial}{\partial t} \langle \overline{\delta \boldsymbol{E} \delta \boldsymbol{E}} \rangle &= -2Z\text{Im}\varepsilon^R \left[\langle \overline{\delta \boldsymbol{E} \delta \boldsymbol{E}} \rangle - 4\pi \sum_a \frac{V_{aa}\, Q_{aa}}{|\,\varepsilon^R\,|^2} \right] \\ &\quad -4\pi \sum_a [\text{Re}V^R_{aa}, Q_{aa}] + Z \frac{\partial \text{Re}\varepsilon^R}{\partial t} \frac{\partial}{\partial \omega} \langle \overline{\delta \boldsymbol{E} \delta \boldsymbol{E}} \rangle . \end{aligned} \tag{7.65}$$

This equation exhibits the shape of a kinetic equation for the time evolution of field fluctuations. The re-normalization function Z is defined by

$$Z^{-1}(\boldsymbol{q}\omega,t) = \frac{\partial \text{Re}\varepsilon^R(\boldsymbol{q}\omega,t)}{\partial \omega}. \tag{7.66}$$

Together with the kinetic equations for particles using the collision integrals (7.58) and (7.59), we have now a complete kinetic theory of particles and field fluctuations in first order gradient expansion.

The equations just given are rather general equations to describe the kinetics of field fluctuations. But usually they are considered only for plasma excitations $z_K(\boldsymbol{q},t) = \omega_K(\boldsymbol{q},t) - i\gamma_K(\boldsymbol{q},t)$ which are weakly damped, i.e., $\gamma_K/\omega_K \ll 1$. The damping γ_K and the frequencies ω_K are given by the relations (4.160) and (4.161). Then the spectral function $\text{Im}1/\epsilon^R(\boldsymbol{q}\omega)$ can be considered in the plasmon pole approximation given by (4.149). Further, we assume, in analogy to the particle kinetics, a Kadanoff–Baym ansatz of the form

$$iV^{\gtrless}_{ab}(\boldsymbol{q}\omega,t) = -N^{\gtrless}_K(\boldsymbol{q},t)\; 2\text{Im}V^R_{ab}(\boldsymbol{q}\omega,t)\,. \tag{7.67}$$

Here, the quantities $N^{\gtrless}_K$ have the meaning of the plasmon occupation numbers. In the approximation of weakly damped excitations, the expression (7.67) can be simplified according to

$$iV^{\gtrless}_{ab}(\boldsymbol{q}\omega,t) = V_{ab}(q) N^{\gtrless}_K(\boldsymbol{q},t) Z_K(\boldsymbol{q},t) 2\pi\delta(\omega - \omega_K(\boldsymbol{q},t))\,, \tag{7.68}$$

where $Z_K = Z(\omega)|_{\omega=\omega_K}$. In an electron–ion plasma, two kinds of modes $\omega_K(t)$ are of interest. We get a weakly damped excitation if we assume

$\hbar\omega/q \gg v_{T_e} \gg v_{T_i}$. Under this condition, only the electrons are involved in the oscillations. The excitations $\omega_K = \omega_L$ are plasmons which describe electron plasma waves (Langmuir waves). According to our considerations in Sects.4.5.2 and 4.6, we get for the corresponding dispersion relation

$$\omega_L(q) = \omega_{\rm pl}\left[1 + \frac{1}{2}\frac{q^2}{\hbar^2 m_e^2}\frac{\langle p^2\rangle_e}{\omega_{\rm pl}^2}\right] \tag{7.69}$$

with $\omega_{pl} = \omega_e = (4\pi e^2 n_e/m_e)^{1/2}$ being the electron plasma frequency. Apart from the plasmons, there are weakly damped excitations in which both electrons and ions are involved. They exist if the temperature of the electrons T_e is much bigger than the temperature of the ions T_i, i.e., if $T_e \gg T_i$, and if the phase velocity of the waves obeys the inequality $v_{T_e} \ll \hbar\omega/q \gg v_{T_i}$. Such excitations $\omega_K = \omega_i$ are known to be ion-acoustic waves. The corresponding dispersion relation reads

$$\omega_i(q) = c_s\frac{q}{\hbar}\ , c_s = \sqrt{\frac{kT_e}{m_i}}\,.$$

According to (7.68), the field fluctuations are completely determined by the occupation numbers $N_K^{\gtrless}$. From the definition of $\langle\overline{\delta\boldsymbol{E}\delta\boldsymbol{E}}\rangle$, we have

$$\begin{aligned}\langle\overline{\delta\boldsymbol{E}\delta\boldsymbol{E}}\rangle_{\boldsymbol{q}\omega,t} &= 4\pi^2\hbar(1+2N_K^{<}(\boldsymbol{q},t))Z_K(\boldsymbol{q},t)\delta(\omega-\omega_K(\boldsymbol{q},t))\\ &= \langle\overline{\delta\boldsymbol{E}\delta\boldsymbol{E}}\rangle_{\boldsymbol{q},t}2\pi\delta(\omega-\omega_K(\boldsymbol{q},t))\,.\end{aligned} \tag{7.70}$$

It is therefore sufficient to find an equation for $\langle\overline{\delta\boldsymbol{E}\delta\boldsymbol{E}}\rangle_{\boldsymbol{q}\omega,t}$. Such an equation may be derived immediately from (7.65) and (7.70). We get

$$\frac{\partial}{\partial t}\langle\overline{\delta\boldsymbol{E}\delta\boldsymbol{E}}\rangle_{\boldsymbol{q},t} = -2\gamma_K(\boldsymbol{q},t)\langle\overline{\delta\boldsymbol{E}\delta\boldsymbol{E}}\rangle_{\boldsymbol{q},t} - 4\pi\sum_a Z_K^2(\boldsymbol{q},t)V_{aa}(q)Q_{aa}(\boldsymbol{q}\omega_K)\,. \tag{7.71}$$

The first contribution on the right-hand side is determined by the Landau damping

$$\gamma_K(\boldsymbol{q},t) = Z_K(\boldsymbol{q},t){\rm Im}\varepsilon^R(\boldsymbol{q}\omega,t)|_{\omega=\omega_K}\,, \tag{7.72}$$

where the imaginary part of the dielectric function in RPA is given by formula (4.111) and describes, as well known, the balance between collision-less absorption and induced emission of plasmons.

The second contribution is determined by the source function Q_{aa}, which is given by (7.63). It describes the spontaneous emission of plasmons under the condition $\hbar\omega(\boldsymbol{q},t) = \boldsymbol{p}\cdot\boldsymbol{q}/m_a + q^2/2m_a$. In the classical limit, i.e., if $q = \hbar k \to 0$, for this contribution we get

$$\begin{aligned}&\sum_a V_{aa}(q)Q_{aa}(\omega_K,\boldsymbol{q})\\ &= \hbar\sum_a V_{aa}(q)\int\frac{d\boldsymbol{p}}{(2\pi\hbar)^3}2\pi\delta\left(\hbar\omega_K - \frac{\boldsymbol{p}\cdot\boldsymbol{q}}{m_a}\right)f_a(\boldsymbol{p},t)\,.\end{aligned} \tag{7.73}$$

This is related to the classical rate for the Čerenkov emission. For the Landau damping, in the classical limit we arrive at

$$\gamma_K(\boldsymbol{q},t) = Z_K(\boldsymbol{q},t)\sum_a V_{aa}(q)\int\frac{d\boldsymbol{p}}{(2\pi\hbar)^3}2\pi\delta\left(\hbar\omega_K - \frac{\boldsymbol{p}\cdot\boldsymbol{q}}{m_a}\right)\boldsymbol{q}\cdot\nabla_{\boldsymbol{p}}f_a(\boldsymbol{p},t)\,. \tag{7.74}$$

To complete this scheme we need the corresponding equation for the single-particle distribution function. Of course, this equation follows from the Lenard–Balescu kinetic equation (7.58). In addition, we have to apply the plasmon pole approximation. The equations (7.73), and (7.74) presented here are the basic equations of the quantum mechanical theory of unstable plasmas and turbulence, and they were for the first time elaborated by DuBois (1968).

In order to perform the classical limit of these equations, one has to expand the collision integral up to the second order with respect to the momentum transfer $\boldsymbol{q} = \hbar\boldsymbol{k}$. The resulting expression then takes the form of a Fokker–Planck collision term, i.e.,

$$I_a\,(\boldsymbol{p},t) = \frac{\partial}{\partial\boldsymbol{p}}\overset{\leftrightarrow}{\boldsymbol{D}}_a\,(\boldsymbol{p},t)\frac{\partial}{\partial\boldsymbol{p}}f_a(\boldsymbol{p},t) + \frac{\partial}{\partial\boldsymbol{p}}\boldsymbol{A}_a(\boldsymbol{p},t)f_a(\boldsymbol{p},t)\,. \tag{7.75}$$

The first term describes the diffusion in momentum space with the diffusion tensor $\overset{\leftrightarrow}{\boldsymbol{D}}_a$ and with the friction vector $\boldsymbol{A}_a$ defined by the equations

$$\overset{\leftrightarrow}{\boldsymbol{D}}_a\,(\boldsymbol{p},t) = \frac{e_a^2}{2}\int\frac{d\omega}{2\pi}\frac{d\boldsymbol{k}}{(2\pi)^3}2\pi\delta\left(\omega - \frac{\boldsymbol{k}\cdot\boldsymbol{p}}{m_a}\right)\frac{\boldsymbol{k}\otimes\boldsymbol{k}}{k^2}\overline{\langle\delta\boldsymbol{E}\delta\boldsymbol{E}\rangle}_{\boldsymbol{k}\omega,t}\,, \tag{7.76}$$

$$\boldsymbol{A}_a(\boldsymbol{p},t) = 4\pi e_a^2\int\frac{d\omega}{2\pi}\frac{d\boldsymbol{k}}{(2\pi)^3}2\pi\delta\left(\omega - \frac{\boldsymbol{k}\cdot\boldsymbol{p}}{m_a}\right)\frac{\boldsymbol{k}}{k^2}\frac{\mathrm{Im}\varepsilon^R(\boldsymbol{k}\omega,t)}{|\varepsilon^R(\boldsymbol{k}\omega,t)|^2}\,. \tag{7.77}$$

The classical limit of the equations (7.71), (7.73), and (7.74), together with the Fokker–Planck equation, is known as the quasi-linear theory of plasma fluctuations and of turbulence. It was established by Vedenov et al. (1961), and, independently, also by Drummond and Pines (1962). This theory is of great importance for the investigation of numerous phenomena in the physics of plasma waves. We want to refer to textbooks and review articles (Ichimaru 1992; Krall and Trivelpiece 1973; Kadomtsev 1965; Galeev and Sudan 1984).

An interesting problem is to find a kinetic equation for the plasmon occupation numbers. One can expect to derive such an equation by introduction of the ansatz (7.68) into (7.61). We neglect the second Poisson bracket on the left hand side of (7.61) and arrive at (DuBois 1968)

$$\left[\frac{\partial}{\partial t} + \hbar\frac{\partial\omega_K(\boldsymbol{p},\boldsymbol{R}t)}{\partial\boldsymbol{p}}\frac{\partial}{\partial\boldsymbol{R}} - \hbar\frac{\partial\omega_K(\boldsymbol{p},\boldsymbol{R}t)}{\partial\boldsymbol{R}}\frac{\partial}{\partial\boldsymbol{p}}\right]N_K^<(\boldsymbol{p},\boldsymbol{R}t)$$
$$= -2\gamma_K(\boldsymbol{p},\boldsymbol{R}t)N_K(\boldsymbol{p},\boldsymbol{R}t) + iZ_K(\boldsymbol{p},\boldsymbol{R}t)\sum_a V_{aa}(p)\Pi_{aa}^<(\boldsymbol{p}\omega_K,\boldsymbol{R}t)\,. \tag{7.78}$$

The physical contents of this kinetic equation for the occupation numbers of the plasmons is essentially equivalent to that of equation (7.71). But in addition, this equation is valid for inhomogeneous systems, too. With the knowledge of the occupation number $N_K^<(\boldsymbol{p}, \boldsymbol{R}t)$, the field fluctuations may be determined from (7.70).

Finally, let us come back to the collision integrals (7.58) and (7.59). In order to collect all first order contributions, we have to account for the first order gradient expansion for the correlation function of the field fluctuations in (7.58). We find this expansion by formal solution of (7.52) with respect to $\langle \overline{\delta \boldsymbol{E} \delta \boldsymbol{E}} \rangle$:

$$\langle \overline{\delta \boldsymbol{E} \delta \boldsymbol{E}} \rangle_\omega = -\frac{1}{2\mathrm{Im}\varepsilon^R(\omega)} \Bigg\{ 8\pi \sum_a \mathrm{Im} V_{aa}^R(\omega) Q_{aa}(\omega) + \frac{\partial}{\partial t} P' \int \frac{d\bar{\omega}}{2\pi} \frac{\mathrm{Im}\varepsilon^R(\bar{\omega})}{\bar{\omega} - \omega} \langle \overline{\delta \boldsymbol{E} \delta \boldsymbol{E}} \rangle_\omega + \frac{\partial}{\partial t} P' \sum_a \int \frac{d\bar{\omega}}{2\pi} \frac{\mathrm{Im} V_{aa}^R(\bar{\omega})}{\bar{\omega} - \omega} Q_{aa}(\omega) - \frac{d}{d\omega} \left[\left(\sum_a \frac{\partial \mathrm{Re} V_{aa}^R(\omega)}{\partial t} Q_{aa}(\omega) \right) + \frac{\partial \mathrm{Re}\varepsilon^R(\omega)}{\partial t} \langle \overline{\delta \boldsymbol{E} \delta \boldsymbol{E}} \rangle_\omega \right] \Bigg\} . \tag{7.79}$$

Here we used the dispersion relations for $\mathrm{Re}\varepsilon^R(\omega)$ and for $\mathrm{Im} V^R(\omega)$, and the representation (3.176) for the brackets. With (7.79), it is possible to find additional first order contributions to I_a^1. We will not carry out this idea explicitly here.

7.6 Kinetic Equation in Ladder Approximation. Boltzmann Equation

In the preceding sections, we considered kinetic equations which are based on the Born approximation, or on the screened Born approximation, respectively, for the self-energy. Born approximations are valid if the mean value of the potential energy is small as compared to that of the kinetic energy. On behalf of its perturbation theoretical character, the Born approximation fails for strong coupling. Effects based on strong coupling, e.g., the formation of bound states, cannot be described by perturbative approximations.

An approximation for the self-energy going beyond the Born approximation is the binary collision approximation, or the ladder approximation. Such an approximation describes all processes between two particles without any approximation with respect to the coupling parameter. It is the simplest approximation which allows for the formation of bound states.

The ladder approximation was discussed in detail in Chap. 5 and was used in Sect. 6.5 for the determination of thermodynamic properties. We are going to apply the ladder approximation in the time diagonal Kadanoff–Baym kinetic equation. Like in the other cases, we consider weakly inhomogeneous systems, i.e., we start from (7.23).

In Sect. 5.7, we found the following expression for the Fourier transform of the self-energy in ladder approximation

$$\Sigma_a^{\gtrless}(\boldsymbol{p}_1 t, t') = (\pm i\hbar) \sum_b \int \frac{d\boldsymbol{p}_2}{(2\pi\hbar)^3} \langle \boldsymbol{p}_1 \boldsymbol{p}_2 \,|\, \mathrm{T}_{ab}^{\gtrless}(t,t') \,|\, \boldsymbol{p}_2 \boldsymbol{p}_1 \rangle \, g_b^{\lessgtr}(\boldsymbol{p}_2 t, t') \,. \tag{7.80}$$

Here, the self-energy is given in terms of the T-matrix. The T-matrix $T_{ab}^{\gtrless}$ may be determined from (5.94) or from the optical theorem (5.102),

$$T^{\gtrless}(t,t') = \int_{-\infty}^{\infty} d\bar{t}\, d\tilde{t}\, T^R(t,\tilde{t}) \mathcal{G}_{12}^{\gtrless}(\tilde{t},\bar{t}) T^A(\bar{t},t') \,, \tag{7.81}$$

$$\mathcal{G}_{12}^{\gtrless}(t,t') = i g_1^{\gtrless}(t,t') g_2^{\gtrless}(t,t') \,,$$

and using the Lippmann–Schwinger equation for the retarded and advanced T-matrices (5.97). Here, a definition for $\mathcal{G}^{\gtrless}$ is used which is slightly different from that of (5.95).

Let us now introduce $\Sigma_a^{\gtrless}$ determined so far into the collision integral (7.23). We then get the kinetic equation (t=t')

$$\left(\frac{\partial}{\partial t} + \frac{\boldsymbol{p}_1}{m_a} \nabla_{\boldsymbol{R}} - \nabla_R U_a^{\mathrm{eff}}(\boldsymbol{R},t) \nabla_{\boldsymbol{p}_1} \right) f_a(\boldsymbol{p}_1, \boldsymbol{R}t) = I_a(\boldsymbol{p}_1, t) \tag{7.82}$$

with the collision integral

$$I_a(t) = I_a^{in}(t) + 2\, i \mathrm{Re}\, \mathrm{Tr}_2 \sum_b \int_{t_0}^{t} d\bar{t} [T_{ab}^{>}(t\bar{t}) g_a^{<}(\bar{t}t)\, g_b^{<}(\bar{t}t)$$
$$- T_{ab}^{<}(t\bar{t})\, g_a^{>}(\bar{t}t)\, g_b^{>}(\bar{t}t)] \,. \tag{7.83}$$

Applying the optical theorem (7.81) and the GKBA (7.11), from (7.83), we can derive the non-Markovian generalization of the Boltzmann equation. This was carried out in several papers (Bornath et al. 1996; Kremp et al. 1997; Bonitz 1998).

Here we will first consider the essentially simpler local approximation of the collision integral (7.83). Then, the macroscopic time t and the microscopic one τ are decoupled, the Bogolyubov condition may be adopted, and (7.83) reduces to (7.15), and, with (7.80), we get the collision integral

$$I_a^0(t) = \mathrm{Tr}_2 \sum_b \int \frac{d\omega}{2\pi} \left\{ i T_{ab}^{<}(\omega,t) i \mathcal{G}_{ab}^{>}(\omega,t) - i T_{ab}^{>}(\omega,t) i \mathcal{G}_{ab}^{<}(\omega,t) \right\} , \tag{7.84}$$

$$\mathcal{G}_{ab}^{\gtrless}(\omega,t) = i \int \frac{d\bar{\omega}}{2\pi} g_a^{\gtrless}(\omega-\bar{\omega},t) g_b^{\gtrless}(\bar{\omega},t) \,. \tag{7.85}$$

The further procedure is clear.

1. The correlation functions $g_b^{\gtrless}$ are replaced by Wigner distributions using the KBA (7.9).
2. With the application of the optical theorem in local approximation (5.106), the T-matrices $T_{ab}^{\gtrless}(\omega, t)$ are replaced by the retarded or advanced T-matrices.

We then immediately get the well-known form of the Boltzmann collision integral

$$I_a^B(\boldsymbol{p}_1, t) = \frac{1}{\hbar V} \sum_b \int \frac{d\boldsymbol{p}_2 d\bar{\boldsymbol{p}}_1 d\bar{\boldsymbol{p}}_2}{(2\pi\hbar)^6} \left| \langle \boldsymbol{p}_1 \boldsymbol{p}_2 \,|\, \mathrm{T}_{ab}^{\mathrm{R}}(E) \,|\, \bar{\boldsymbol{p}}_2 \bar{\boldsymbol{p}}_1 \rangle \right|^2$$
$$\times 2\pi\delta(\bar{E}_{12} - E_{12}) \left\{ \bar{f}_a \, \bar{f}_b (1 - f_a)(1 - f_b) - f_a f_b (1 - \bar{f}_a)(1 - \bar{f}_b) \right\}_t \,. \quad (7.86)$$

This expression again has the typical form of a Markovian collision integral with all shortcomings discussed earlier. The transition probability for the two-particle collisions is now given by the on-shell two-particle T-matrix, and bound states are therefore excluded. The occupation of the initial and final states in the collision process is determined by Fermi functions and Pauli blocking factors. Because of the argument of the delta-function, we have conservation of the kinetic energy only.

To obtain nonideality corrections to the quantum Boltzmann kinetic equations, the first order retardation terms must be taken into account. Retardation contributions to the Boltzmann equation were considered in papers by Bärwinkel (1969b), by Klimontovich (1982), and by Bornath et al. (1996). In our scheme, we must consider the retardation term I^1, together with the corresponding gradient expansion correction of Σ and $\mathcal{G}_{ab}^{\gtrless}$ in I^0. In terms of the T-matrix, we get for I^1

$$I_a^1 = \frac{\partial}{\partial t} \mathrm{Tr}_2 \sum_{\mathrm{b}} \int_{-\infty}^{+\infty} \frac{d\omega_1 d\omega_2}{(2\pi)^2} \frac{\mathcal{P}'}{\omega_1 - \omega_2}$$
$$\times \left\{ iT_{ab}^{>}(\omega_1, t) i\mathcal{G}_{ab}^{<}(\omega_2, t) - iT_{ab}^{<}(\omega_1, t) i\mathcal{G}_{ab}^{>}(\omega_2, t) \right\} . \quad (7.87)$$

Our aim is to find the complete first order corrections to the local approximation of the collision integral I_a^B. In this connection, we have to take into consideration that the collision integral I^0 still contains first-order gradient contributions in $\mathcal{G}_{12}^{<}$ and in the optical theorem for $T^{\gtrless}$ as well. Therefore, we need the gradient expansions of these two quantities.

The gradient expansion for $\mathcal{G}_{12}^{\gtrless}$ follows from the reconstruction formula (7.11), if we neglect the self-energy contributions

$$i\mathcal{G}_{12}^{\gtrless}(\omega, t) = 2\pi\delta(\hbar\omega - E_{12}) F_{12}^{\gtrless}(t) + \hbar \frac{d}{d\omega} \frac{P}{\hbar\omega - E_{12}} \frac{d}{dt} F_{12}^{\gtrless}(t) \,. \quad (7.88)$$

In order to find the gradient expansion of the optical theorem, we rewrite (7.81) to

$$\int d\bar{t}\, T^{R^{-1}}(t,\bar{t})T^{\gtrless}(\bar{t},t) = \int d\bar{t}\, \mathcal{G}_{12}^{\gtrless}(t,\bar{t})T^{A}(\bar{t},t)\,.$$

Then we may apply the scheme developed in Sect. 3.4.2. Further let us neglect contributions of the kind $\partial T^{R/A}/\partial t$. In this way, the optical theorem in first order gradient expansion may be obtained

$$\begin{aligned} iT^{\gtrless}(\omega,t) &= T^R(\omega)2\pi\delta\left(\hbar\omega - \bar{E}_{12}\right)\bar{F}_{12}^{\gtrless}T^A(\omega) - T^R(\omega)\frac{P'}{\omega - \bar{E}_{12}}\frac{\partial \bar{F}_{12}^{\gtrless}}{\partial t}T^A(\omega) \\ &- 2\pi\delta\left(\omega - \bar{E}_{12}\right)\mathrm{Im}\left(\frac{\partial T^R(\omega)}{\partial \omega}T^A(\omega)\right)\frac{\partial \bar{F}_{12}^{\gtrless}}{\partial t}\,. \end{aligned} \tag{7.89}$$

The first r.h.s. term of (7.89) is the well-known local version of the optical theorem. The next contributions are first order gradient corrections.

A further useful relation follows from the local contribution of (7.89). From this term, for the imaginary part of $T^R(\omega + i\varepsilon)$ we get

$$\mathrm{Im}T(\omega + i\epsilon, t) = T(\omega + i\epsilon, t)\mathrm{Im}\mathcal{G}(\omega + i\epsilon, t)T(\omega - i\epsilon, t)\,. \tag{7.90}$$

Taking the derivative with respect to ω and keeping in mind $\partial T(\omega+i\epsilon)/\partial\omega = \partial T(\omega + i\epsilon)/\partial i\varepsilon$, we get the so-called differentiated optical theorem

$$\frac{\partial}{\partial\omega}\mathrm{Re}T(\omega + i\epsilon) = T\frac{\partial \mathrm{Re}\mathcal{G}}{\partial\omega}T^* + i\frac{\partial T}{\partial\omega}\mathrm{Im}\mathcal{G}T^* - iT\mathrm{Im}\mathcal{G}\frac{\partial T^*}{\partial\omega}\,. \tag{7.91}$$

With the help of (7.88) and (7.89), the collision term I^0 is now easily obtained as

$$\begin{aligned} I_a^0 = I_a^B - \sum_b \mathrm{Tr}_b \Bigg\{ & T^R(E_{ab})\frac{\mathcal{P}'}{E_{ab} - \bar{E}_{ab}}T^A(E_{ab})\frac{d\bar{F}_{ab}^{<}}{dt} \\ -T^R(\bar{E}_{ab})\,\frac{P'}{E_{ab} - \bar{E}_{ab}}T^A(\bar{E}_{ab})\frac{dF_{ab}^{<}}{dt} &+ 2\pi\delta(E_{ab} - \bar{E}_{ab})\mathrm{Im}\left(\frac{\partial T^R}{\partial E}T_{ab}^A\right)\frac{d\bar{F}_{ab}^{<}}{dt}\Bigg\}, \end{aligned} \tag{7.92}$$

with I^B given by (7.86). The collision integral I^1 already contains a time derivative. Therefore, the quantities in it can be taken in the lowest order, and we get

$$I_a^1 = \frac{\partial}{\partial t}\sum_b \mathrm{Tr}_b\, T^R(\bar{E}_{ab})\frac{\mathcal{P}'}{E_{ab} - \bar{E}_{ab}}T^A(\bar{E}_{ab})\left[\bar{F}_{ab}^{>}F_{ab}^{<} - \bar{F}_{ab}^{<}F_{ab}^{>}\right]\,. \tag{7.93}$$

The additional terms to I^B are essentially determined by the "off-shell" T-matrix. For this quantity, we may derive a useful relation from the differentiated optical theorem. Using the dispersion relation for $\mathrm{Re}T(E)$, we arrive at the following equation

$$T^R(\bar{E}_{ab})\frac{\mathcal{P}'}{E_{ab}-\bar{E}_{ab}}\bar{N}_{ab}T^A(\bar{E}_{ab}) = T^R(E_{ab})\frac{\mathcal{P}'}{E_{ab}-\bar{E}_{ab}}\bar{N}_{ab}T^A(E_{ab}) \\ -2\pi\delta(E_{ab}-\bar{E}_{ab})\bar{N}_{ab}\mathrm{Im}(T^R T^{A\prime}) \qquad (7.94)$$

with $2\mathrm{Im}(T^R T^{A\prime}) = i\left(T^{R\prime}T^A - T^R T^{A\prime}\right)$.

We now denote all corrections beyond the Boltzmann collision integral by I^R, and write the kinetic equation in the shape

$$\left(\frac{\partial}{\partial t} + \frac{\boldsymbol{p}_1}{m_a}\nabla_{\boldsymbol{R}} - \nabla_R U_a^{\mathrm{eff}}(\boldsymbol{R},t)\nabla_{\boldsymbol{p}_1}\right) f_a(\boldsymbol{p}_1,\boldsymbol{R}t) = I_a^B(\boldsymbol{p}_1) + I_a^R(\boldsymbol{p}_1)\,. \qquad (7.95)$$

The first order retardation terms collected in $I^R(\boldsymbol{p}_1)$ are given by the following expression (Bornath et al. 1996; Kremp et al. 1997)

$$I_a^R(\boldsymbol{p}_1) = \sum_b \int \frac{d\boldsymbol{p}_2 d\bar{\boldsymbol{p}}_1 d\bar{\boldsymbol{p}}_2}{(2\pi\hbar)^9}\left(\frac{dF_{ab}^{\lessgtr}}{dt} - \frac{d\bar{F}_{ab}^{\lessgtr}}{dt}\right) \\ \times\left\{\left[\,|\langle \boldsymbol{p}_1\boldsymbol{p}_2|T_{ab}(\bar{E}_{ab})|\bar{\boldsymbol{p}}_1\bar{\boldsymbol{p}}_2\rangle\,|^2 + |\,\langle \boldsymbol{p}_1\boldsymbol{p}_2|T_{ab}(E_{ab})|\bar{\boldsymbol{p}}_1\bar{\boldsymbol{p}}_2\rangle\,|^2\right]\right. \\ \left.\times\frac{\mathcal{P}'}{E_{ab}-\bar{E}_{ab}} - 2\pi\delta(E_{ab}-\bar{E}_{ab})\mathrm{Im}(T_{ab}^R T_{ab}^{A\prime})\right\}. \qquad (7.96)$$

We show again that there is an interesting relation between I^B and I^R. For this reason, we define

$$I_a^B(\varepsilon) = \sum_b \int \frac{d\boldsymbol{p}_2 d\bar{\boldsymbol{p}}_1 d\bar{\boldsymbol{p}}_2}{(2\pi\hbar)^6}\int \frac{d\omega}{2\pi}\frac{2\varepsilon}{(E-\omega)^2+\varepsilon^2}\frac{2\varepsilon}{(\bar{E}-\omega)^2+\varepsilon^2} \\ \times\left|\langle \boldsymbol{p}_1\boldsymbol{p}_2|\,T_{ab}(\omega+i\varepsilon)\,|\bar{\boldsymbol{p}}_2\bar{\boldsymbol{p}}_1\rangle\right|^2\left(\bar{F}_{ab}^{\lessgtr} - F_{ab}^{\lessgtr}\right)\,. \qquad (7.97)$$

Then we easily see

$$I^R = \frac{1}{2}\frac{d}{dt}\frac{d}{d\varepsilon}I^B(\varepsilon)\bigg|_{\varepsilon\to 0}\,. \qquad (7.98)$$

The Boltzmann kinetic equation in first order retardation approximation has the compact shape (Bornath et al. 1996), (Kremp et al. 1997)

$$\left(\frac{\partial}{\partial t} + \frac{\boldsymbol{p}_1}{m_a}\nabla_{\boldsymbol{R}} - \nabla_R U_a^{\mathrm{eff}}(\boldsymbol{R},t)\nabla_{\boldsymbol{p}_1}\right) f_a(\boldsymbol{p}_1,\boldsymbol{R}t) \\ = \left(1 + \frac{1}{2}\frac{d}{dt}\frac{d}{d\varepsilon}\right) I_a^B(\varepsilon)\bigg|_{\varepsilon\to 0}\,. \qquad (7.99)$$

We now are able to investigate the problem of the conservation laws which follow from the Boltzmann equation in first order gradient approximation. In the paper by Bornath et al. (1996), we found that the kinetic equation (7.99) conserves the density, i.e., we have

$$\frac{d}{dt}\left(n^Q + n^{\mathrm{corr}}\right) = \frac{dn}{dt} = 0 \tag{7.100}$$

with the quasiparticle density contribution

$$n^Q(\boldsymbol{R},t) = \sum_a \int \frac{d\boldsymbol{p}}{(2\pi)^3} f_a^Q(\boldsymbol{p},\boldsymbol{R},t) \tag{7.101}$$

and the correlation part given by

$$n^{corr}(R,t) = \int \frac{d\boldsymbol{p}_1 d\boldsymbol{p}_2 d\bar{\boldsymbol{p}}_1 d\bar{\boldsymbol{p}}_2}{(2\pi)^9} i\pi\delta(E_{12} - \bar{E}_{12}) F_{12}^{<}$$
$$-2\mathrm{Im}\left(\frac{\partial}{\partial E}\left(\langle \boldsymbol{p}_1\boldsymbol{p}_2|T_{12}^R(E)|\bar{\boldsymbol{p}}_2\bar{\boldsymbol{p}}_1\rangle\right)\langle \bar{\boldsymbol{p}}_1\bar{\boldsymbol{p}}_2|T_{12}^A(E)|\boldsymbol{p}_2\boldsymbol{p}_1\rangle\right). \tag{7.102}$$

The conserved density consists, therefore, of two parts. In order to understand this result, let us consider, for a moment, the equilibrium limit. The first part then describes the connection between the density, the temperature, and the chemical potential for the quasiparticle approximation; the second part represents the influence of the interaction in binary collision approximation upon this connection. The conserved density is, therefore, the fugacity expansion of the density in generalized binary collision approximation. This was explained in detail in Sect. 6.5.

Furthermore, the kinetic equation leads to energy conservation. In order to show this we multiply (7.95) by the kinetic energy and take the trace over variable 1. Symmetrization with respect to 1 and 2, and to $\bar{1}$ and $\bar{2}$, respectively, leads to the simple result for the derivative of the kinetic energy T

$$\frac{\partial}{\partial t}\langle T\rangle = -\frac{\partial}{\partial t}\langle V\rangle\,, \tag{7.103}$$

where the average of the potential energy is

$$\langle V\rangle = \frac{1}{2}\mathrm{Tr}_{12}\left[\left|T^R(\bar{E})\right|^2 \frac{P}{E-\bar{E}}\left(F_{12}^{<} - \bar{F}_{12}^{<}\right)\right]. \tag{7.104}$$

This means, we have conservation of the total energy in binary collision approximation. The connection between conservation laws for nonideal many-particle systems and retardation was, for the first time, considered by Bärwinkel (1969b) for the binary collision approximation and later by Klimontovich (1982). However, Bärwinkel used the KBA for the reconstruction problem. Thus, the retardation was only partially considered and, therefore, the proper potential energy was not achieved.

7.7 Bound States in the Kinetic Theory

7.7.1 Bound States and Off-Shell Contributions

So far we ignored the role of bound states in the kinetic equations. Let us now consider a system with attractive interaction so that the formation of bound states is possible. We know from Chap. 2 and Sect. 6.5 that bound states essentially influence the equilibrium and non-equilibrium properties of many-component plasmas. On the time scale of the scattering processes, bound states are long living states. For this reason, we consider bound states as new composite particles. Therefore, a problem of special interest is the kinetic description of the formation and the decay of bound states, i.e., the ionization kinetics which was already considered from an elementary point of view in Chap. 2.

The investigation of bound states in the kinetic theory is a complicated problem. Bound states are known to be a result of strong coupling. The simplest approximation for the inclusion of bound states is, therefore, the binary collision approximation, developed in the preceding section. However, it is clear that the Markov approximation determined by the on-shell T-matrix does not describe bound states.

In Markov-like kinetic theories, bound states can be considered by using the condition of partial weakening of initial correlations (3.106). On this basis, a kinetic theory accounting for bound states was developed by Peletminski (1971), Klimontovich and Kremp (1981), McLennan (1982), Lagan and McLennan (1984), and in Klimontovich et al. (1987). In a more rigorous theory, formation of bound states takes place in the initial stage of the temporal evolution, which can be described only by non-Markovian kinetic equations or by the gradient correction terms to Markov equations as carried out above.

Contributions of bound states may only show up in I^1 given by (7.93); this quantity may be rewritten in the form

$$I^1 = \frac{\partial}{\partial t} \int \frac{d\boldsymbol{p}_2}{(2\pi)^3} \Bigg\{ \left(\frac{\partial}{\partial \omega} \mathrm{Re} \left\langle \boldsymbol{p}_1 \boldsymbol{p}_2 | T^R(\omega) | \boldsymbol{p}_2 \boldsymbol{p}_1 \right\rangle \right) \Bigg|_{\omega=\epsilon(p_1)+\epsilon(p_2)} f_1 f_2$$
$$- \int \frac{d\omega}{2\pi} \left\langle \boldsymbol{p}_1 \boldsymbol{p}_2 | T^<(\omega + \epsilon(p_2)) | \boldsymbol{p}_2 \boldsymbol{p}_1 \right\rangle \frac{\mathcal{P}'}{\omega - \epsilon(p_1)} (1 - f_1 - f_2) \Bigg\} . \qquad (7.105)$$

In the second contribution of the r.h.s. of (7.105), the T-matrix has to be taken off-shell. Inserting the bilinear expansion for the T-matrix derived in Sect. 5.5, (5.112), we can separate the bound state part and get (Bornath et al. 1996)

$$I^1 = I^1_{scatt} + \frac{\partial}{\partial t} \int \frac{d\boldsymbol{p}_2}{(2\pi)^3} \sum_{j\boldsymbol{P}} \langle \boldsymbol{p}_1 \boldsymbol{p}_2 | \Psi^{j\boldsymbol{P}} \rangle \langle \tilde{\Psi}^{j\boldsymbol{P}} | \boldsymbol{p}_2 \boldsymbol{p}_1 \rangle F_j(\boldsymbol{P}) \qquad (7.106)$$

with the bound states $|\Psi^{jP}\rangle$ and energy eigenvalues E_{jP} following from the effective Schrödinger equation (5.76). $F_j(\boldsymbol{P})$ is the distribution function of the bound states introduced by (5.88) where it was denoted by $N_j(\boldsymbol{P})$.

The unbound particles are represented by

$$f^F(\boldsymbol{p}_1) = f(\boldsymbol{p}_1) - \int \frac{d\boldsymbol{p}_2}{(2\pi)^3} \sum_{j\boldsymbol{P}} \langle \boldsymbol{p}_1 \boldsymbol{p}_2 | \Psi^{jP} \rangle \langle \tilde{\Psi}^{jP} | \boldsymbol{p}_2 \boldsymbol{p}_1 \rangle F_j(\boldsymbol{P}) \,. \qquad (7.107)$$

With the normalization

$$n^F = \int \frac{d\boldsymbol{p}}{(2\pi)^3} f^F(\boldsymbol{p}_1), \qquad n^b = \sum_j \int \frac{d\boldsymbol{P}}{(2\pi)^3} F_j(\boldsymbol{P}) \,, \qquad (7.108)$$

we obtain the necessary condition $N^F + 2N^b = N$. Here we took into account the indistinguishability of the particles leading to the Gibbs factor $\frac{1}{2!}$ in the unity relation

$$1 = \frac{1}{2!} \int \frac{d\boldsymbol{p}_1 d\boldsymbol{p}_2}{(2\pi)^3} |\boldsymbol{p}_1 \boldsymbol{p}_2\rangle^{-\,-} \langle \boldsymbol{p}_2 \boldsymbol{p}_1| \,. \qquad (7.109)$$

Now we return to the collision integral (7.106). Using the definition of f^F, we can eliminate the bound state part from equation (7.95) and obtain the Boltzmann equation for the free particles

$$\left(\frac{\partial}{\partial t} + \nabla_{\boldsymbol{p}_1} E_a^{\mathrm{HF}} \nabla_{\boldsymbol{R}} - \nabla_{\boldsymbol{R}} E_a^{\mathrm{HF}} \nabla_{\boldsymbol{p}_1} \right) f_a^F(\boldsymbol{p}_1, \boldsymbol{R}t) = I_a^B \; + \; I_a^R \,. \qquad (7.110)$$

For the complete description of the system, we need, of course, an equation for the distribution function of the bound states. Such an equation follows from the second equation of the Martin–Schwinger hierarchy or from the results of Sect. 3.3.3. In the binary collision approximation, the equation for the two-particle correlation function is collision-less, and, therefore, the time derivative of the distribution function of the atoms vanishes. Moreover, the kinetic equation for the free particles does not contain contributions which account for the formation and decay of bound states. In order to get a kinetic equation for changes of the bound state occupation, we therefore need three-particle or higher-order collisions. This problem will be dealt with subsequently.

7.7.2 Kinetic Equations in Three-Particle Collision Approximation

Let us first generalize the kinetic equation for the free particles; this means, we want to derive a kinetic equation for free particles which takes into account collisions between free and bound particles as well. Further we will deal again with a multi-component system. For this purpose, it is necessary to determine the self-energy Σ_a of species a at least in the three-particle approximation, which is given by the diagram expansion

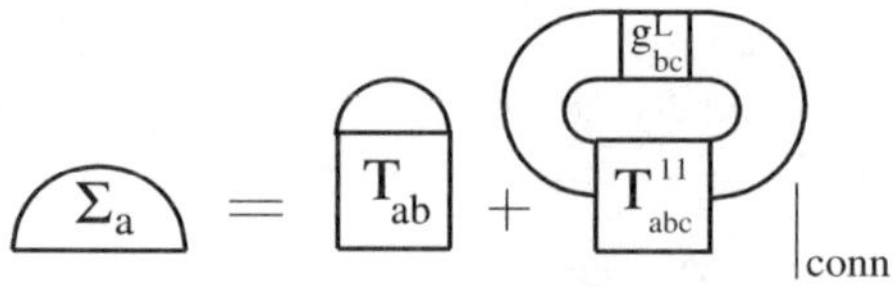

A corresponding cluster expansion for the two-particle density operator was given, e.g., in Klimontovich and Kremp (1981) and Klimontovich et al. (1987). The first contribution in the cluster expansion shown above represents the two-particle ladder sum and is expressed by the two-particle scattering matrix T_{ab}. In the second term, we have three-particle collision terms in ladder approximation. For non-degenerate systems, the latter contributions can be represented by the three-particle scattering matrix T^{11}_{abc} and the ladder approximation of the two-particle function, g^L_{bc}. Here, only the *connected* contribution has to be taken. Two- and three-particle exchange terms will be neglected for simplicity. The three-particle T-matrix T^{11}_{abc} defined on the Keldysh contour is given by

$$T^{11}_{abc} = V^1_{abc} + V^1_{abc} g_{abc} V^1_{abc} \,. \tag{7.111}$$

On the physical time axis, the diagrams lead to the following expression for the self-energy (Bornath 1987; Kremp et al. 1989)

$$\Sigma_a^{\gtrless} = \pm i \sum_b \mathrm{Tr}_b T_{ab}^{\gtrless} g_b^{\lessgtr} + \frac{(\pm i)^2}{2} \sum_{bc} \mathrm{Tr}_{bc} \left\{ T_{abc}^{11\gtrless} g_{bc}^{\lessgtr} - \sum_{k=2}^{3} t_{abc}^{k\gtrless} g_b^{\lessgtr} g_c^{\lessgtr} \right\} . \tag{7.112}$$

The self-energy above is now determined by the two- and three-particle scattering matrices. $t_{abc}^{\kappa\gtrless} = V_\kappa \mathcal{G}_{abc}^{\kappa\gtrless} V_\kappa$ are two-particle T-matrixes in the three-particle space (see Table 6.1). This contribution subtracts disconnected terms. For the further considerations, the local approximation is used.

Because of the existence of two-particle bound states, the three-particle scattering theory is rather complicated. The three-particle scattering states split up into channels which are classified by the asymptotic initial states. An excellent introduction into the multi-channel scattering theory may be found in the textbook by Taylor (1972). Without three-particle bound states, we have four channels with the asymptotic states $|\kappa, \alpha\rangle$. Here, κ is the channel number and α denotes the corresponding set of dynamical variables as explained in Table 7.1.

Each asymptotic state defines an eigen-state $|\kappa, \alpha+\rangle$ of the three-particle Hamiltonian. The asymptotic states are the eigen-states of the channel Hamiltonian H^κ_{abc} defined by $H_{abc} = H^\kappa_{abc} + V^\kappa_{abc}$ with V^κ_{abc} being the potential in the channel κ (see Table 7.1). It describes the interaction between free particles and sub-clusters. Further the channel-correlation functions $\mathcal{G}_{abc}^{\kappa\gtrless}$ are defined as $\mathcal{G}_{abc}^{0\gtrless} = g_a^{\gtrless} g_b^{\gtrless} g_c^{\gtrless}$, $\mathcal{G}_{abc}^{1\gtrless} = g_a^{\gtrless} g_{bc}^{\gtrless}$, etc.

Table 7.1. Classification of the scattering channels with the corresponding asymptotic states, energies, distribution functions, and channel potentials

κ	channel	$\|\kappa,\alpha\rangle$	E^{κ}_{abc}	f_κ	V^{κ}_{abc}
0	$a+\ b+\ c$	$\|p_a\rangle\,\|p_b\rangle\,\|p_c\rangle$	$E_a+E_b+E_b$	$f_af_bf_b$	$V_{ab}+V_{ac}+V_{bc}$
1	$a+(b+c)$	$\|p_a\rangle\,\|j_{bc}P_{bc}\rangle$	E_a+E_{bc}	f_aF_{bc}	$V_{ab}+V_{ac}$
2	$(a+c)+b$	$\|p_b\rangle\,\|j_{ac}P_{ac}\rangle$	E_b+E_{ac}	f_bF_{ac}	$V_{ab}+V_{bc}$
3	$(a+b)+c$	$\|p_c\rangle\,\|j_{ab}P_{ab}\rangle$	E_c+E_{ab}	f_cF_{ab}	$V_{ac}+V_{bc}$

Following the line of multi-channel theory, we find the T-matrix from the multi-channel optical theorem

$$T^{\kappa\kappa\gtrless}_{abc}(\omega,t)=i^2\sum_{\bar\kappa=0}^{3}T^{\kappa\bar\kappa}_{abc}(\omega+i\varepsilon,t)P^{\bar\kappa}\mathcal{G}^{\bar\kappa\gtrless}_{abc}(\omega,t)\tilde{T}^{\bar\kappa\kappa}_{abc}(\omega-i\varepsilon,t)\,,$$

where P^{κ} denotes the projection on the bound state parts of the two-particle correlation functions in the channels κ=1,2,3 (there is no action for κ=0). The retarded and advanced transition T-matrices have to be determined by the channel Lippmann–Schwinger equations

$$\begin{aligned}T^{\kappa\kappa'}_{abc}(z)&=V^{\kappa}_{abc}+i^2V^{\kappa}_{abc}\,g_{abc}(z)V^{\kappa'}_{abc}\,,\\ \tilde{T}^{\kappa\kappa'}_{abc}(z)&=V^{\kappa'}_{abc}+i^2V^{\kappa}_{abc}\,g_{abc}(z)V^{\kappa'}_{abc}\end{aligned}\qquad(7.113)$$

with the three-particle Green's function $i^2g_{abc}(z)=(z-H_{abc})^{-1}$. Now we are able to derive the kinetic equation with three-particle collisions. We will consider the three-particle part in the collision integral only in local approximation; this means that we start from equation (7.15). The three-particle contribution for the self-energy follows from (7.112) taking the local approximation and the Fourier transform with respect to the time difference $t-t'$. We will consider only the connected part. The disconnected terms are of relevance only for the scattering of three (free) particles described by the T-matrix T^{00}_{abc} Bornath (1987). Introducing this into equation (7.15), we obtain the three-particle collision term in compact operator notation

$$I^{(3)}_a=\frac{(\pm i)^2}{2}\mathrm{Tr}_{bc}\sum_{bc}\int\frac{d\omega}{2\pi}\left\{T^{11<}_{abc}(\omega,t)\mathcal{G}^{1>}_{abc}(\omega,t)-T^{11>}_{abc}(\omega,t)\mathcal{G}^{1<}_{abc}(\omega,t)\right\}\,.\qquad(7.114)$$

Here is

$$\mathcal{G}^{1\gtrless}_{abc}(\omega,t)=\int\frac{d\omega_1}{2\pi}\frac{d\omega_2}{2\pi}2\pi\delta(\omega-\omega_1-\omega_2)g^{\gtrless}_{a}(\omega_1,t)g^{\gtrless}_{bc}(\omega_2,t)\,.$$

With the help of the bilinear expansion of $g_{12}^{<}$, we may split up this expression into a bound and scattering part. Thus, we get

$$\mathcal{G}_{abc}^{1\gtrless}(\omega,t) = \int \frac{d\omega_1}{2\pi}\frac{d\omega_2}{2\pi} 2\pi\delta(\omega-\omega_1-\omega_2) g_a^{\gtrless}(\omega_1,t) g_{bc}^{b\gtrless}(\omega_2,t) + \Omega_1^R(\omega)\, \mathcal{G}_{abc}^{0\gtrless}(\omega,t)\, \Omega_1^A(\omega)\,, \tag{7.115}$$

where $\Omega_1^{R/A}(\omega)$ are two-particle Møller operators in the three-particle space with $\Omega_1^{R/A}(\omega) = \Omega_{bc}^{R/A}(\omega - \mathrm{H}_a)$. Let us first consider the two-particle scattering part of (7.114). We use the cyclic invariance of the trace and relations of the type $T_{abc}^{11^R}\Omega_1^R = T_{abc}^{00^R}$ (Bornath 1987; Bornath et al. 1988). We then obtain the following contribution to the three-particle collision integral

$$I_a^{(3sc)} = \frac{(\pm i)^2}{2}\mathrm{Tr}_{bc}\sum_{bc}\int\frac{d\omega}{2\pi}\left\{T_{abc}^{00^<}(\omega)\mathcal{G}_{abc}^{0>}(\omega,t) - T_{abc}^{00^>}(\omega)\mathcal{G}_{abc}^{0<}(\omega,t)\right\}. \tag{7.116}$$

We next consider the bound state contribution to the collision integral. We immediately get

$$I_a^{(3b)} = \frac{(\pm i)^2}{2}\mathrm{Tr}_{bc}\sum_{bc}\int\frac{d\omega}{2\pi}\frac{d\omega_1}{2\pi}\frac{d\omega_2}{2\pi}2\pi\delta(\omega-\omega_1-\omega_2)$$
$$\times\left\{T_{abc}^{11^<}(\omega)g_a^{>}(\omega_1,t)g_{bc}^{b>}(\omega_2,t) - T_{abc}^{11^>}(\omega)g_a^{<}(\omega_1,t)g_{bc}^{b<}(\omega_2,t)\right\}. \tag{7.117}$$

The further steps are clear. With the optical theorem, we may eliminate the $T^{\lessgtr}$ and express the collision integrals by the retarded and advanced T-matrices. We then get the operator form of the two contributions to the collision integral (7.114) (Schlanges 1985; Bornath 1987)

$$I_a^{(3sc)} = \frac{(\pm i)^2}{2}\mathrm{Tr}_{bc}\sum_{bc}\sum_{\kappa=0}^{3}\int\frac{d\omega}{2\pi}\left|T_{abc}^{0\kappa}(\omega+i\varepsilon)\right|^2$$
$$\times P^\kappa\left\{\mathcal{G}_{abc}^{\kappa<}(\omega,t)\mathcal{G}_{abc}^{0>}(\omega,t) - \mathcal{G}_{abc}^{\kappa>}(\omega,t)\mathcal{G}_{abc}^{0<}(\omega,t)\right\}, \tag{7.118}$$

$$I_a^{(3b)} = \frac{(\pm i)^2}{2}\mathrm{Tr}_{bc}\sum_{bc}\sum_{\kappa=0}^{3}\int\frac{d\omega}{2\pi}\left|T_{abc}^{1\kappa}(\omega+i\varepsilon)\right|^2$$
$$\times P^\kappa\left\{\mathcal{G}_{abc}^{\kappa<}(\omega,t)\mathcal{G}_{abc}^{1b>}(\omega,t) - \mathcal{G}_{abc}^{\kappa>}(\omega,t)\mathcal{G}_{abc}^{1b<}(\omega,t)\right\}. \tag{7.119}$$

Here, $\mathcal{G}_{abc}^{1b\gtrless}$ are the bound state parts of channel correlation functions given by (7.115). The last steps in deriving the three-particle collision integral are to carry out the trace and to introduce the distribution function by KB ansatzes.

In order to carry out the trace in the above equations, the collisional integral has to be considered in the space of the asymptotic states which is the direct sum of the channel subspaces $\mathcal{H}^\kappa$. In this space, the asymptotic channel states are orthogonal, and we have the following completeness relation

$$1 = \sum_\kappa \sum\!\!\!\!\!\!\int d(\kappa\alpha)|\kappa,\alpha\rangle\langle\alpha,\kappa|\,; \quad \langle\alpha,\kappa|\kappa',\alpha'\rangle = \delta_{\kappa\kappa'}\delta(\alpha,\alpha')\,. \tag{7.120}$$

where in the normalization a factor $(2\pi)^3$ has to be included for momentum states. Furthermore we have to determine the matrix elements of the correlation operators $\mathcal{G}^{\kappa<}_{abc}(\omega,t)$ in terms of the distribution function. We have, using the Kadanoff–Baym ansatz,

$$\begin{aligned}(\pm i)\langle \boldsymbol{p}\,|\,g^<_a(\omega,t)\,|\boldsymbol{p}'\rangle &= (2\pi)^3\delta(\boldsymbol{p}-\boldsymbol{p}')(\pm i)g^<_a(\boldsymbol{p},\omega,t)\\ &= (2\pi)^3\delta(\boldsymbol{p}-\boldsymbol{p}')\,2\pi\,\delta(\omega-E_a)f_a(\boldsymbol{p},t)\,.\end{aligned}$$

Let us further consider $\mathcal{G}^{1<}_{abc}$. This operator is diagonal with respect to $|j_{bc}P_{bc}\rangle|p_a\rangle$. In connection with the previous discussion we therefore get

$$\begin{aligned}&\langle \boldsymbol{p}_a|\langle j_{bc}\boldsymbol{P}_{bc}|\mathcal{G}^{1<}_{abc}(\omega,t)|j'_{bc}\boldsymbol{P}'_{bc}\rangle|\boldsymbol{p}'_a\rangle = (2\pi)^6\delta(\boldsymbol{p}_a-\boldsymbol{p}'_a)\,\delta(\boldsymbol{P}_{bc}-\boldsymbol{P}'_{bc})\delta_{j_{bc}j'_{bc}}\\ &\times 2\pi\delta\left(\omega-E_{j_{bc}\boldsymbol{P}_{bc}}-E_a(\boldsymbol{p}_a)\right)\,F_{j_{bc}}(\boldsymbol{P}_{bc},t)f_a(\boldsymbol{p}_a,t)\,.\end{aligned} \tag{7.121}$$

Similar formulae are valid for the other channels.

With all these relations, we find the following kinetic equation for the distribution function of the free particles.

$$\left(\frac{\partial}{\partial t}+\nabla_{\boldsymbol{p}_1}E^{\mathrm{HF}}_a\nabla_{\boldsymbol{R}}-\nabla_{\boldsymbol{R}}E^{\mathrm{HF}}_a\nabla_{\boldsymbol{p}_1}\right)\,f_a(\boldsymbol{p}_1,\boldsymbol{R}t)=\sum_b(I^B_{ab}+I^R_{ab})+\sum_{bc}I_{abc}\,. \tag{7.122}$$

Here I^B_{ab} and I^R_{ab} are well known from the preceding section and are given by (7.86) and (7.96). The three-particle contribution has the form (Klimontovich and Kremp 1981; Lagan and McLennan 1984; McLennan 1982; McLennan 1989; Schlanges 1985; Bornath 1987)

$$\begin{aligned}I_{abc} &= I^{3scatt}_{abc}+\frac{1}{2V}\sum_{\kappa=1}^{3}\int d(\kappa\alpha)\int\frac{d\boldsymbol{p}_b d\boldsymbol{p}_c}{(2\pi)^6}\left|\langle\boldsymbol{p}_a\boldsymbol{p}_b\boldsymbol{p}_c|T^{0\kappa}_{abc}|\kappa,\alpha\rangle\right|^2\\ &\times 2\pi\delta(E^0_{abc}-\bar{E}^\kappa_{abc})\{f_\kappa-f_af_bf_c\}\\ &+\frac{1}{2V}\sum_{\kappa=0}^{3}\int d(\kappa\alpha)\sum_{j_{bc}}\int\frac{d\boldsymbol{P}_{bc}}{(2\pi)^3}\left|\langle\boldsymbol{p}_a\boldsymbol{P}_{bc}j_{bc}|T^{1\kappa}_{abc}|\kappa,\alpha\rangle\right|^2\\ &\times 2\pi\delta(E^1_{abc}-\bar{E}^\kappa_{abc})\{f_\kappa-f_aF^j_{(bc)}\}\,.\end{aligned} \tag{7.123}$$

All quantities used in this equation are explained in Table 7.1.

With these equations, we achieved an interesting result. Applying the Boltzmann equation being generalized from binary to ternary collisions, we are now able to describe any of the elastic, inelastic and reactive three-particle processes. The first term of the r.h.s. of (7.123) describes the elastic scattering of three particles. This expression has the shape of a Boltzmann collision integral, i.e., we have

$$I_{abc}^{3\mathrm{scatt}} = \frac{1}{2V}\int \frac{d\boldsymbol{p}_b d\boldsymbol{p}_c d\bar{\boldsymbol{p}}_a d\bar{\boldsymbol{p}}_b d\bar{\boldsymbol{p}}_c}{(2\pi)^{12}} \left|\langle \boldsymbol{p}_a\boldsymbol{p}_b\boldsymbol{p}_c|T_{abc}^{00}(E^0+i\varepsilon)|\bar{\boldsymbol{p}}_a\bar{\boldsymbol{p}}_b\bar{\boldsymbol{p}}_c\rangle\right|_{\mathrm{conn}}^2$$

$$\times\; 2\pi\,\delta(E_{abc}^0-\bar{E}_{abc}^0)\left\{\bar{f}_a\bar{f}_b\bar{f}_c - f_af_bf_c\right\}\,. \tag{7.124}$$

In connection with this expression, the problem of secular divergencies occurs. It is well-known from the classical kinetic theory (Cohen 1968; Dorfmann and Cohen 1975; Weinstock 1963) that this expression contains terms which are divergent for long times. The origin of this behavior is the contribution of successive binary collisions, i.e., the contribution of three-particle processes of the kind

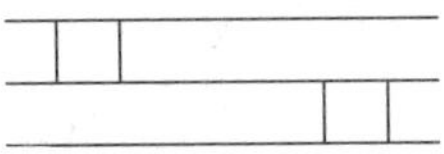

Peletminskij has shown that such processes contribute with terms like

$$\frac{1}{\varepsilon}\frac{\partial}{\partial t}I^B(\boldsymbol{p},\varepsilon) = \frac{1}{\varepsilon}\int d\boldsymbol{p}' I^B(\boldsymbol{p}')\frac{\delta}{\delta f(\boldsymbol{p}')}I^B(\boldsymbol{p},\varepsilon)\,;\;\varepsilon\to 0 \tag{7.125}$$

to the collision integral (7.124) (Akhiezer and Peletminskij 1980). There was shown furthermore that, in a correct cluster expansion of the quantum mechanical collision integral, these terms are compensated by additional terms. The re-normalized three-particle collision integral is, therefore, given by

$$I_{abc}^{3\mathrm{ren}} = I_{abc}^{3\mathrm{scatt}} - \frac{1}{\varepsilon}\frac{\partial}{\partial t}I^B(\boldsymbol{p},\varepsilon)\,. \tag{7.126}$$

In our approach, the compensating term arises if, for the correlation functions in the local contribution of the two-particle collision integral, the self-energy contributions are taken into account.

Let us consider now further terms of the collision integral (7.123). They describe scattering processes under participation of sub-clusters. These contributions do not have secular divergencies. The second part of the collision integral (7.123) describes reactive processes of the type

$$\begin{aligned} a+b+c &\rightleftarrows a'+(bc)', && \kappa=1;\\ a+b+c &\rightleftarrows (ac)'+b', && \kappa=2;\\ a+b+c &\rightleftarrows (ab)'+c', && \kappa=3. \end{aligned}$$

Of course, not all reactions are possible in certain systems. Let us consider, e.g., an H-plasma. In this case, the species a, b, c are electrons or protons. For $a=e_1$, $b=e_2$, $c=p_3$, we have

$$e_1 + e_2 + p_3 \rightleftarrows e_1 + (e_2 p_3), \qquad \kappa = 1;$$
$$e_1 + e_2 + p_3 \rightleftarrows e_2 + (e_1 p_3), \qquad \kappa = 2;$$
$$e_1 + e_2 + p_3 \rightleftarrows p_3 + (e_1 e_2), \qquad \kappa = 3\,.$$

The case $\kappa = 3$ is not possible. The third part of (7.123) contains, for $\kappa = 1$, the important processes of elastic and inelastic scattering of a bound pair (atom) on free particles (charged particles)

$$a + (bc) \rightleftarrows a' + (bc)'\,.$$

We furthermore find, for $\kappa = 2$ and $\kappa = 3$, the exchange reactions

$$a + (bc) \rightleftarrows b' + (ac)'\,, \qquad a + (bc) \rightleftarrows c' + (ab)'\,.$$

In order to discuss a general property of the collision integral (7.123), it is useful to split up the integral into two parts

$$I_a = \sum_{bc} I_{abc} = [I_a]_1 + [I_a]_2\,. \tag{7.127}$$

$[I_a]_1$ is determined by the terms $\kappa = 0, 1$ which means that this term describes collisions in which the particle a is always free. Therefore, the density n_a does not change, and we have

$$\int \frac{d\boldsymbol{p}}{(2\pi)^3}[I_a]_1 = 0\,. \tag{7.128}$$

The second part is given by contributions with $\kappa = 2, 3$ and describes the formation and decay of bound states involving particle a. Therefore, we now have

$$\int \frac{d\boldsymbol{p}}{(2\pi)^3}[I_a]_2 = W_a \neq 0\,. \tag{7.129}$$

Here, W_a is the source function which describes the change of the density n_a by chemical reactions.

The kinetic equation (7.122) for the distribution function of free particles f has to be completed by an equation for the distribution of the bound states, or the atoms, F_j. Here we will not give the derivation in detail. In principle, we have to start from an equation of the type (5.159), and the self-energy Σ_{ab} has to be determined in 3-particle collision approximation. We then get, from the time diagonal equation corresponding to (5.159), the following kinetic equation

$$\left(\frac{\partial}{\partial t} + \frac{\partial E_{j\boldsymbol{P}}}{\partial \boldsymbol{P}}\frac{\partial}{\partial \boldsymbol{R}} - \frac{\partial E_{j\boldsymbol{P}}}{\partial \boldsymbol{R}}\frac{\partial}{\partial \boldsymbol{P}}\right) F^j_{(ab)}(\boldsymbol{P}, \boldsymbol{R}, t) = \sum_c I^j_{(ab)c}(\boldsymbol{R}, \boldsymbol{P}, t) \tag{7.130}$$

for the distribution of the atoms. The collision integral in (7.130) has the shape (Klimontovich et al. 1987; Schlanges 1985; Bornath 1987)

$$I^j_{(ab)c}(\boldsymbol{P}, \boldsymbol{R}, t) = \frac{1}{V} \sum_{\kappa=0}^{3} \int d(\kappa\alpha) \int \frac{d\boldsymbol{p}_c}{(2\pi)^3} \left| \langle j \boldsymbol{P}_{ab} \boldsymbol{p}_c | T^{3\kappa}_{abc} | \kappa, \alpha \rangle \right|^2$$
$$\times\, 2\pi\delta(E^3_{abc} - \bar{E}^\kappa_{abc}) \left\{ f_\kappa - f_c F^j_{ab} \right\} . \tag{7.131}$$

This collision integral describes the following processes: (i) elastic scattering of atoms with free particles of species c, accounted for by T^{33}, (ii) formation and decay of bound states (ab) by collisions with particle c, described by T^{30}, and (iii) rearrangement reactions, represented by T^{32} and T^{31}.

7.7.3 The Weak Coupling Approximation. Lenard–Balescu Equation for Atoms

The collision integrals considered in the previous section are essentially given by the scattering operator $T^{\kappa\kappa'}$ determined by the set of equations (7.113). Under the condition that the interaction remains weak, these equations may be solved in first Born approximation taking $T^{\kappa\kappa'} = V^\kappa$. Then we get a significant simplification. Let us consider, e.g., T^{33} . In Born approximation, and for statically screened Coulomb interactions, we get (Klimontovich et al. 1987)

$$\langle j \boldsymbol{P}_{ab} \boldsymbol{p}_c | (V_{ac} + V_{bc}) | \bar{\boldsymbol{p}}_c \bar{\boldsymbol{P}}_{ab} \bar{j} \rangle = \frac{4\pi Z_c e^2\, \mathcal{P}_{j,\bar{j}}(\boldsymbol{q})}{\varepsilon(\boldsymbol{q}, 0)\, q^2} (2\pi)^3 \delta(\boldsymbol{q} + \boldsymbol{P}_{ab} - \bar{\boldsymbol{P}}_{ab}) , \tag{7.132}$$

where we used the notation

$$\mathcal{P}_{\alpha_{ab}, \bar{\alpha}_{ab}}(\boldsymbol{q}) = \int d\boldsymbol{r}\, \psi^*_{\alpha_{ab}}(\boldsymbol{r}) \psi_{\bar{\alpha}_{ab}}(\boldsymbol{r}) \left\{ Z_a e^{-i\frac{m_b}{M}\boldsymbol{q}\cdot\boldsymbol{r}} + Z_b e^{i\frac{m_a}{M}\boldsymbol{q}\cdot\boldsymbol{r}} \right\} \tag{7.133}$$

with $\alpha_{ab} = j$ for bound states and $\alpha_{ab} = p_{ab}$ (relative momentum) for scattering states. $\boldsymbol{P}_{ab} = \boldsymbol{p}_a + \boldsymbol{p}_b$ denotes the total momentum, and Z_a and Z_b are the charge numbers. The quantity (7.132) has a clear physical meaning. It may be interpreted as effective interaction between the free particle c and the bound pair (ab). The expression (7.133) characterizes the static charge distribution for bound states in the atom (ab). For large distances between the free particles and the atoms, or, equivalently, $\boldsymbol{q} \cdot \boldsymbol{r} < 1$, we may expand the exponentials in (7.133). Then (7.132) simplifies to the charge-dipole interaction between the charged particle and the atom given by

$$\langle |T^{33}| \rangle = i \frac{4\pi Z_c\, e}{\varepsilon(\boldsymbol{q}, 0)\, q^2}\, \boldsymbol{q} \cdot \boldsymbol{m}_{j\bar{j}}\, (2\pi)^3 \delta(\boldsymbol{q} + \boldsymbol{P}_{ab} - \bar{\boldsymbol{P}}_{ab}) , \tag{7.134}$$

where $\boldsymbol{m}_{j\bar{j}}$ is the dipole matrix element of the atom

$$\boldsymbol{m}_{j\bar{j}} = e \int d\boldsymbol{r}\, \psi^*_{\bar{j}}(\boldsymbol{r}) \left(Z_b \frac{m_a}{M} \boldsymbol{r} - Z_a \frac{m_b}{M} \boldsymbol{r} \right) \psi_{\bar{j}}(\boldsymbol{r}) . \tag{7.135}$$

The T-matrices of the other processes can also be considered in Born approximation. Additionally, we give the matrix element of the operator responsible for ionization, T^{30}

$$\begin{aligned}\langle j\boldsymbol{P}_{ab}\,\boldsymbol{p}_c|(V_{ac}+V_{bc})|\bar{\boldsymbol{p}}_c\bar{\boldsymbol{p}}_a\bar{\boldsymbol{p}}_b\rangle &= \frac{4\pi e^2 Z_c}{\varepsilon(\boldsymbol{q},0)\,q^2}\,\mathcal{P}_{j\boldsymbol{P}_{ab},\bar{\boldsymbol{p}}_a\bar{\boldsymbol{p}}_b}(\boldsymbol{q})\\ &= \frac{4\pi e^2 Z_c}{\varepsilon(\boldsymbol{q},0)\,q^2}\,\mathcal{P}_{j,\bar{p}_{ab}}(\boldsymbol{q})\;(2\pi)^3\delta(\boldsymbol{q}+\boldsymbol{P}_{ab}-\bar{\boldsymbol{P}}_{ab})\,.\end{aligned} \tag{7.136}$$

If we use the Born approximation considered here in the kinetic equation (7.122), it reduces to a Landau type equation for reacting systems.

In some interesting cases, the approximation of static screening is not justified because dynamical effects such as plasma oscillations, plasmons, and instabilities are of importance. The simplest way to introduce dynamical screening into the kinetic equation (7.122) is to replace the statically screened Coulomb potential by a dynamically screened one, i.e.,

$$\frac{1}{\varepsilon(\boldsymbol{q},0)\;q^2}\to\frac{1}{\varepsilon(\boldsymbol{q},\omega)\;q^2}\,,$$

where ω has to be chosen at energies which are determined by the arguments of the δ-functions of the collision integrals. So, in the expression (7.132), we have to take $\omega = E_{\bar{j}\bar{\boldsymbol{P}}}-E_{j\boldsymbol{P}}$.

Following this scheme, we get a Lenard–Balescu equation for bound states. We restrict ourselves to an electron–ion pair and, therefore, suppress the index (ab):

$$\frac{\partial}{\partial t}F_j(\boldsymbol{P},t) = I_j^{\text{scatt}}(\boldsymbol{P},t)+I_j^{\text{react}}(\boldsymbol{P},t). \tag{7.137}$$

Here, the first scattering integral accounts for the elastic and inelastic scattering processes; the explicit expression is

$$\begin{aligned}I_j^{\text{scatt}}(\boldsymbol{P},t) &= \frac{(2\pi)^3}{V}\sum_{\bar{j},c}\int\frac{d\bar{\boldsymbol{P}}d\boldsymbol{q}d\boldsymbol{p}}{(2\pi)^9}\,z_c^2\left|\mathcal{P}_{j\boldsymbol{P},\bar{j}\bar{\boldsymbol{P}}}(\boldsymbol{q})\right|^2\\ &\times\left|\frac{V(\boldsymbol{q})}{\varepsilon^R(\boldsymbol{q},\,E_{\bar{j}\bar{\boldsymbol{P}}}-E_{j\boldsymbol{P}})}\right|^2 2\pi\;\delta\left(E_{j\boldsymbol{P}}+E_c(\boldsymbol{p})-E_{\bar{j}\bar{\boldsymbol{P}}}-E_c(\boldsymbol{p}-\boldsymbol{q})\right)\\ &\times\left\{F_{\bar{j}}(\bar{\boldsymbol{P}})f_c(\boldsymbol{p}-\boldsymbol{q})\;-\;F_j(\boldsymbol{P})f_c(\boldsymbol{p})\right\}.\end{aligned} \tag{7.138}$$

with $V(q)=4\pi e^2/q^2$. The second contribution describes the formation and the decay of bound states in three-particle collisions and is given by

$$I_j^{\text{react}}(\boldsymbol{P},t) = \frac{(2\pi)^3}{V} \sum_{\bar{j},c} \int \frac{d\bar{\boldsymbol{p}}_a d\bar{\boldsymbol{p}}_b d\boldsymbol{q} d\boldsymbol{p}}{(2\pi)^{12}} z_c^2 \left|\mathcal{P}_{j\boldsymbol{P},\bar{\boldsymbol{p}}_a\bar{\boldsymbol{p}}_b}(\boldsymbol{q})\right|^2$$
$$\times \left| \frac{V(q)}{\varepsilon^R(\boldsymbol{q},\, \bar{E}_a + \bar{E}_b - E_{j\boldsymbol{P}})} \right|^2$$
$$\times 2\pi\delta\left(E_{j\boldsymbol{P}} + E_c(\boldsymbol{p}) - E_a(\bar{\boldsymbol{p}}_e) - E_b(\bar{\boldsymbol{p}}_p) - E_c(\boldsymbol{p}-\boldsymbol{q})\right)$$
$$\times \left\{ f_a(\bar{\boldsymbol{p}}_a) f_b(\bar{\boldsymbol{p}}_b) f_c(\boldsymbol{p}-\boldsymbol{q}) \; - \; F_{\bar{j}}(\boldsymbol{P}) f_c(\boldsymbol{p}) \right\}. \tag{7.139}$$

Rearrangement reactions are neglected in this equation. Corresponding equations may be obtained for distribution functions of the free particles. A more rigorous derivation of kinetic equations of this type was given by Klimontovich et al. (1987) and by Schlanges and Bornath (1997). The dielectric function may be used in RPA. The drawback of this approximation is that only free particles contribute to the screening. This approximation is not sufficient for partially ionized plasmas with a large fraction of neutrals. In such cases, the influence of correlated two-particle states and especially of that of bound states has to be taken into account. This is possible, if one derives a cluster expansion of the polarization function starting from the equations (4.21) for the correlation functions of density fluctuations. At this level, a dielectric function in thermodynamic equilibrium was derived (Röpke and Der 1979). A generalization to non-equilibrium systems was given by Schlanges and Bornath (1997). According to the work mentioned, the dielectric function of a partially ionized plasma is

$$\varepsilon^{R/A}(\boldsymbol{q},\omega,t) = 1 + \frac{4\pi}{q^2} \sum_a e_a^2 \int \frac{d\boldsymbol{p}}{(2\pi)^3} \frac{f_a(\boldsymbol{p}+\boldsymbol{q},t) - f_a(\boldsymbol{p},t)}{\omega \pm i0 + \epsilon_a(\boldsymbol{p}) - \epsilon_a(\boldsymbol{p}+\boldsymbol{q})}$$
$$+ \frac{4\pi}{q^2}\frac{1}{V} \left[\sum_K \sum_{K'} |\mathcal{P}_{K,K'}(\boldsymbol{q})|^2 \frac{F_{K'}(t) - F_K(t)}{\omega \pm i0 - E_{K'} + E_K} \right.$$
$$\left. - \int \frac{d\boldsymbol{p}_1 d\boldsymbol{p}_2 d\boldsymbol{p}_1' d\boldsymbol{p}_2'}{(2\pi)^{12}} \left|\mathcal{P}^0_{\boldsymbol{p}_1\boldsymbol{p}_2,\boldsymbol{p}_1'\boldsymbol{p}_2'}(\boldsymbol{q})\right|^2 \frac{f(\boldsymbol{p}_1',t)\, f(\boldsymbol{p}_2',t) - f(\boldsymbol{p}_1,t)\, f(\boldsymbol{p}_2,t)}{\omega \pm i0 - E(\boldsymbol{p}_1'\boldsymbol{p}_2') + E(\boldsymbol{p}_1\boldsymbol{p}_2)} \right]. \tag{7.140}$$

In this formula, the F_K are occupation numbers of a pair of free ($K = \boldsymbol{p}_1\boldsymbol{p}_2$) or of bound ($K = j, \boldsymbol{P}$) particles, respectively. The quantity $\mathcal{P}^0_{\boldsymbol{p}_1\boldsymbol{p}_2,\boldsymbol{p}_1'\boldsymbol{p}_2'}(\boldsymbol{q})$ follows from (7.133), if the two-particle scattering wave functions are replaced by plane waves. The expression (7.140) has the typical shape of a cluster expansion; beginning with the (ideal) RPA, we find, in the second term, the contributions of the two-particle bound and scattering correlations. The subtraction of uncorrelated terms compensates the spatially divergent scattering contributions. A detailed discussion of (7.140) for systems in thermodynamic equilibrium is given in Kraeft et al. (1986).

For the limiting case of vanishing ionization, the contribution of the two-particle correlations only remains, i.e., the dielectric function of the atomic gas. Such expression was, for the first time, given by (Klimontovich 1982). Using the principle of weakening of correlations, Klimontovich derived expressions for the partially ionized plasma from the atomic gas expression. However, there the volume divergencies were not correctly dealt with.

7.8 Hydrodynamic Equations

So far we have developed a quantum kinetic theory for nonideal plasmas starting from the Kadanoff–Baym equations for the two-time correlation functions. In this framework, generalizations of the well-known Landau, Lenard–Balescu, and Boltzmann equations were derived. Furthermore, the theory was generalized to systems with bound states to account for scattering processes with free and bound particles by corresponding three-particle collision integrals. In particular, the formation and decay of bound states in ionization and recombination reactions were included.

Now, we will use the kinetic equations to derive balance equations for macroscopic quantities such as number density, current, and energy density. We will do this on the level of the quasi-particle approximation, i.e., we start from the kinetic equations (3.208) and (7.20). Here, the collision integrals derived in the previous sections of this chapter are used with quasiparticle energies. For the free particles, this equation can be written as

$$\left[\frac{\partial}{\partial t} + \nabla_{\boldsymbol{p}} E_a(\boldsymbol{p}, \boldsymbol{R}t) \cdot \nabla_{\boldsymbol{R}} - \nabla_{\boldsymbol{R}} E_a(\boldsymbol{p}, \boldsymbol{R}t) \cdot \nabla_{\boldsymbol{p}}\right] f_a(\boldsymbol{p}, \boldsymbol{R}t) = I_a(\boldsymbol{p}, \boldsymbol{R}t) \tag{7.141}$$

where a labels the species of the free plasma particles ($a = e, i$) and bound states ($a = A_j$). The latter ones are described by the distribution function $f_{Aj} = F_j$ with j comprising the set of internal quantum numbers of the bound state level. The l.h.s of (7.141) has the known form of a drift term, but now generalized to weakly damped quasi-particles. Of course, the collision integral in (7.141) has to be specified using appropriate approximations for the self-energy functions as it was done in the previous sections. In this way, it allows to include different two- and three-particle scattering processes between the particles in the plasma as well as many-body effects.

As mentioned above, the l.h.s of (7.141) describes the drift of quasi-particles. The energy of free particles is

$$E_a(\boldsymbol{p}, \boldsymbol{R}t) = \frac{p^2}{2m_a} + \Delta_a(\boldsymbol{p}, \boldsymbol{R}t)\,, \tag{7.142}$$

and for the bound particles we have

$$E_{Aj}(\boldsymbol{P}, \boldsymbol{R}t) = \frac{P^2}{2M_A} + E_j^0 + \Delta_j(\boldsymbol{P}, \boldsymbol{R}t) \,. \tag{7.143}$$

Here, M_A denotes the mass, and E_j^0 is the binding energy of the isolated bound state. In comparison to ideal systems, the quasi-particle energies (7.142) and (7.143) include medium effects by the energy shifts Δ_a and Δ_j. Thus, they describe the modifications of the energies due to the influence of the surrounding plasma. In the following, bound states will be treated in the same manner like free particles.

The determination of the Wigner functions from the kinetic equations considered in this section would involve, in general, extensive numerical calculations, and provides a description of the evolution of the system on the kinetic stage. However, in many cases it is sufficient to focus on the hydrodynamic time scale where moments of the distribution function can be considered leading to balance equations for macroscopic quantities. To realize such a hydrodynamic level of description, we introduce the number density of quasi-particles of species a according to

$$n_a(\boldsymbol{R}, t) = \int \frac{d\boldsymbol{p}}{(2\pi\hbar)^3} f_a(\boldsymbol{p}, \boldsymbol{R}t) \,. \tag{7.144}$$

The current density is given by

$$\boldsymbol{j}_a(\boldsymbol{R}, t) = \int \frac{d\boldsymbol{p}}{(2\pi\hbar)^3} [\nabla_{\boldsymbol{p}} E_a(\boldsymbol{p}, \boldsymbol{R}t)] f_a(\boldsymbol{p}, \boldsymbol{R}t) \,, \tag{7.145}$$

where $\nabla_{\boldsymbol{p}} E_a$ is the velocity of quasi-particles with momentum $\boldsymbol{p}$. Further, it holds (Kadanoff and Baym 1962; Lifshits and Pitayevskij 1978)

$$\boldsymbol{j}_a(\boldsymbol{R}, t) = \int \frac{d\boldsymbol{p}}{(2\pi\hbar)^3} \frac{\boldsymbol{p}}{m_a} f_a(\boldsymbol{p}, \boldsymbol{R}t) = n_a(\boldsymbol{R}, t) \boldsymbol{u}_a(\boldsymbol{R}, t) \tag{7.146}$$

with $\boldsymbol{u}_a$ being the mean velocity. Integrating the kinetic equation (7.141) with respect to the momentum $\boldsymbol{p}$, we get the balance equation for the density

$$\frac{\partial}{\partial t} n_a(\boldsymbol{R}, t) + \nabla_{\boldsymbol{R}} \cdot \boldsymbol{j}_a(\boldsymbol{R}, t) = W_a(\boldsymbol{R}, t) \,. \tag{7.147}$$

A non-zero source function W_a follows from contributions in the collision term which do not conserve the number density of species a. Such contributions are included in the three-particle collision integrals and were discussed in Sect. 7.7.2. The source functions in the density balance equations for the free plasma particles result from terms due to ionization and recombination processes. Excitation and de-excitation processes are included additionally in the case of bound particles. Expressions for the source functions will be considered in the subsequent sections.

If spatially inhomogeneous plasmas are considered, diffusion processes in connection with chemical reactions can be of importance. The determination of the number densities from (7.147) then requires the knowledge of the current densities which are related to the mean velocities according to (7.146). The balance equation for $\boldsymbol{j}_a$ can be obtained by multiplying the kinetic equation (7.141) by the momentum $\boldsymbol{p}$, and subsequent integration over the latter one. If the system is assumed to be under the influence of an external potential U_a, the resulting equation of motion can be written as

$$\frac{\partial}{\partial t} m_a \boldsymbol{j}_a(\boldsymbol{R},t) + \nabla_{\boldsymbol{R}} \cdot \overset{\leftrightarrow}{\boldsymbol{\Pi}}_a (\boldsymbol{R},t) + n_a(\boldsymbol{R},t)\nabla_{\boldsymbol{R}} U_a(\boldsymbol{R},t) = \boldsymbol{X}_a(\boldsymbol{R},t) . \tag{7.148}$$

Here, the source function is given by

$$\boldsymbol{X}_a(\boldsymbol{R},t) = \int \frac{d\boldsymbol{p}}{(2\pi\hbar)^3} \boldsymbol{p} I_a(\boldsymbol{p},\boldsymbol{R}t) . \tag{7.149}$$

The second term on the l.h.s. of (7.148) gives the contribution of the pressure tensor $\overset{\leftrightarrow}{\boldsymbol{\Pi}}_a$. In the equation for the i-th component j_{ai} of the current it reads

$$(\nabla_{\boldsymbol{R}} \cdot \overset{\leftrightarrow}{\boldsymbol{\Pi}}_a)_i = \frac{\partial}{\partial R_j} \int \frac{d\boldsymbol{p}}{(2\pi\hbar)^3} p_i \frac{\partial E_a}{\partial p_j} f_a + \int \frac{d\boldsymbol{p}}{(2\pi\hbar)^3} \frac{\partial E_a}{\partial R_j} \delta_{ij} f_a . \tag{7.150}$$

Finally, let us consider the balance equation for the energy density. For this purpose, we take in mind that the total energy is not simply given by the sum of the quasi-particle energies. However, according to the Landau theory, the variation of the total energy density of the system due to an infinitesimal small change in the distribution function is given by (Lifshits and Pitayevskij 1978)

$$\delta\mathcal{E} = \sum_a \int \frac{d\boldsymbol{p}}{(2\pi\hbar)^3} E_a(\boldsymbol{p}) \delta f_a(\boldsymbol{p}) . \tag{7.151}$$

Then we get for the temporal change of the energy density of species a

$$\frac{\partial}{\partial t}\mathcal{E}_a(\boldsymbol{R},t) = \int \frac{d\boldsymbol{p}}{(2\pi\hbar)^3} E_a(\boldsymbol{p},\boldsymbol{R}t) \frac{\partial}{\partial t} f_a(\boldsymbol{p},\boldsymbol{R}t) . \tag{7.152}$$

Now, multiplying the kinetic equation (7.141) by the quasi-particle energy E_a and after integration over $\boldsymbol{p}$ we find

$$\frac{\partial}{\partial t}\mathcal{E}_a(\boldsymbol{R},t) + \nabla_{\boldsymbol{R}} \cdot \boldsymbol{j}_a^{\mathcal{E}}(\boldsymbol{R},t) + \nabla_{\boldsymbol{R}} U_a(\boldsymbol{R},t) \cdot \boldsymbol{j}_a(\boldsymbol{R},t) = S_a(\boldsymbol{R},t) . \tag{7.153}$$

This is just the balance equation for the energy density of quasi-particles with the energy current

$$\boldsymbol{j}_a^{\mathcal{E}} = \int \frac{d\boldsymbol{p}}{(2\pi\hbar)^3} E_a(\boldsymbol{p}) \nabla_{\boldsymbol{p}} E_a(\boldsymbol{p}) f_a(\boldsymbol{p}) , \tag{7.154}$$

and the source term

$$S_a = \int \frac{d\boldsymbol{p}}{(2\pi\hbar)^3} E_a(\boldsymbol{p}) I_a(\boldsymbol{p}) \,. \tag{7.155}$$

With (7.147), (7.148), and (7.153) we have found balance equations for quasi-particles in a nonideal plasma which, in general, may be partially ionized. Summation over the species gives the corresponding equations for the total number, current, and energy densities. Important quantities entering the equations for the species are the source functions W_a, $\boldsymbol{X}_a$, and S_a determined by the collision terms of the kinetic equation. They are given in terms of the Wigner distribution functions, and, therefore, additional assumptions are necessary for the latter ones to get a closed set of hydrodynamic equations. Alternatively, the transport quantities such as the current densities given by (7.146) and (7.154) can be calculated directly from their definitions. But this requires the determination of the distribution functions from (7.141) using appropriate approximations, too. Usually, one accounts for the fact that non-equilibrium processes are characterized by different time scales. In a relaxation towards thermodynamic equilibrium the system comes first into equilibrium with respect to the translational degrees of freedom, afterwards with respect to the internal ones. In the last stage, chemical equilibrium is established. If transport quantities under stationary conditions are considered, the assumption of weak deviation from thermodynamic equilibrium can be applied in many cases. Therefore, different approximations will be used in the next sections to describe non-equilibrium properties of dense non-ideal plasmas.

8. Transport and Relaxation Processes in Nonideal Plasmas

8.1 Rate Equations and Reaction Rates

The temporal evolution of the densities in a reacting plasma can be described by rate equations. Their source functions account for the excitation, deexcitation, ionization, and recombination processes, and they are usually expressed in terms of corresponding rate coefficients. The calculation of rate coefficients is one of the important problems of plasma physics. Most of the papers devoted to this problem are based on concepts which can be applied only to the ideal plasma state. For an overview we refer, e.g., to Biberman et al. (1987), Sobel'man et al. (1988). However, for dense nonideal plasmas, many-particle effects become important, and the question is how the rate coefficients are influenced by screening, self-energy effects, lowering of the ionization energy, and Pauli blocking.

There is a number of papers dealing with theoretical investigations of inelastic and reactive processes in nonideal plasmas. Cross sections and collision rate coefficients for inelastic electron–atom and electron–ion scattering assuming static screening of the Coulomb interaction were considered, e.g., by Whitten et al. (1984), Schlanges et al. (1988), Schlanges and Bornath (1993), Leonhardt and Ebeling (1993), Guttierrez and Diaz-Valdes (1994), and Jung (1995). In further papers, excitation processes and recombination rates were calculated accounting for collective effects due to dynamical screening (Vinogradov and Shevel'ko 1976; Weisheit 1988; Rasolt and Perrot 1989; Guttierrez and Girardeau 1990). Rate coefficients for hydrogen and hydrogen-like plasmas were investigated on the level of a dynamically screened Born approximation including effective ionization energies and atomic form factors in different approximations (Murillo and Weisheit 1993; Murillo and Weisheit 1998; Schlanges (a) et al. 1996; Bornath et al. 1997). The influence of correlation effects in the electron distribution function on the rate coefficients was considered by Starostin and Aleksandrov (1998). A formalism using effective statistical weights was described by Fisher and Maron (2003). Further approaches were developed on the basis of numerical simulation techniques (Klakow et al. 1996; Ebeling et al. 1996; Beule et al. 1996; Ebeling and Militzer 1997). In the following sections, we will investigate rate coefficients for nonideal plasmas in the framework of quantum kinetic theory.

8.1.1 *T*-Matrix Expressions for the Rate Coefficients

An introduction into the theory of rate coefficients for nonideal plasmas from an elementary point of view was given in Sect. 2.12. Now this problem will be treated in a more rigorous manner starting from the quantum kinetic equations for systems with bound states derived in Sect. 7.7. Such equations account for the formation and the decay of two-particle bound states in three-particle collisions as well as for many-particle effects. For simplicity, we consider a spatially homogeneous plasma consisting of electrons and singly charged ions with number densities n_e, n_i and bound electron–ion pairs (atoms) with densities n_A^j. Here, j denotes the set of internal quantum numbers. The plasma is assumed to be dense enough such that radiation processes can be neglected (Griem 1964; Fujimoto and McWhriter 1990).

We consider the balance equations for the densities obtained from the kinetic equations, and write them as rate equations. From (7.122), we get for the densities of the free particles ($a = e, i$)

$$\begin{aligned} \frac{\partial n_a}{\partial t} &= \int \frac{d\boldsymbol{p}_a}{(2\pi\hbar)^3} I_a(\boldsymbol{p}_a, t) \;=\; W_a(t) \\ &= \sum_{c=e,i} \sum_j (\alpha_c^j n_c n_A^j - \beta_c^j n_c n_e n_i)\,, \end{aligned} \tag{8.1}$$

and, in the same manner, we get from (7.130) for the bound particles

$$\begin{aligned} \frac{\partial n_A^j}{\partial t} &= \int \frac{d\boldsymbol{P}}{(2\pi\hbar)^3} I_A^j(\boldsymbol{P}, t) \;=\; W_A^j(t) \\ &= \sum_{c=e,i} \Big(\sum_{\bar{j}} (n_c n_A^{\bar{j}} K_c^{\bar{j}j} - n_c n_A^j K_c^{j\bar{j}}) + \beta_c^j n_c n_e n_j - \alpha_c^j n_c n_j \Big)\,. \end{aligned} \tag{8.2}$$

In this way, the source functions of the density balance equations are given in a form which is well-known from phenomenological theory, i.e., in (8.1) and (8.2) rate coefficients of inelastic and reactive processes are introduced. In the framework of the approach presented here, we are able to find quantum statistical expressions for these quantities. They can be easily derived from the respective three-particle collision terms of the kinetic equations given in Sect. 7.7.2 (Klimontovich 1975; Klimontovich and Kremp 1981). We want to mention again that for the further considerations the quasiparticle approximation is used with kinetic equations of the form (7.20) assuming the case $Z(\boldsymbol{p}, \boldsymbol{R}t) = 1$. Then, energies and distribution functions in the collision terms (7.123) and (7.131) are replaced by respective quasiparticle quantities (see Sect. 3.4.1). It is convenient to start from the kinetic equation (7.130) of the bound particles. The coefficient of (impact) ionization of an atom in the state j by collision with a free particle of species c then reads (Bornath and Schlanges 1993)

$$\alpha_c^j = \frac{1}{\hbar V}\int \frac{d\boldsymbol{P}}{(2\pi\hbar)^3}\frac{d\boldsymbol{p}_c}{(2\pi\hbar)^3}\frac{d\bar{\boldsymbol{P}}}{(2\pi\hbar)^3}\frac{d\bar{\boldsymbol{p}}}{(2\pi\hbar)^3}\frac{d\bar{\boldsymbol{p}}_c}{(2\pi\hbar)^3} 2\pi\delta\left(E^3_{(ei)c} - \bar{E}^0_{(ei)c}\right)$$
$$\times \left|\langle j\boldsymbol{P}|\langle \boldsymbol{p}_c|T^{33}_{eic}\left(E^3_{(ei)c} + i\varepsilon\right)|\bar{\boldsymbol{p}}_c\rangle|\bar{\boldsymbol{P}}\bar{\boldsymbol{p}}+\rangle\right|^2$$
$$\times \frac{f_c}{n_c}\frac{F_j}{n_A^j}\left(1-\bar{f}_c\right)\left(1-\bar{f}_i\right)\left(1-\bar{f}_e\right). \tag{8.3}$$

Here, we generalized the expression to account for phase space occupation effects with $f_c = f_c(\boldsymbol{p}_c, t)$ $(c = e, i)$ and $F_j = F_j(\boldsymbol{P}, t)$ being the distribution functions of the free and bound particles, respectively. Furthermore, we used the T-matrix relation

$$\langle j\boldsymbol{P}|\langle \boldsymbol{p}_c|T^{30}_{eic}|\bar{\boldsymbol{p}}_c\rangle|\bar{\boldsymbol{p}}_i\rangle|\bar{\boldsymbol{p}}_e\rangle = \langle j\boldsymbol{P}|\langle \boldsymbol{p}_c|T^{33}_{eic}|\bar{\boldsymbol{p}}_c\rangle|\bar{\boldsymbol{P}}\bar{\boldsymbol{p}}+\rangle, \tag{8.4}$$

where $|\bar{\boldsymbol{P}}\bar{\boldsymbol{p}}+\rangle$ denotes the scattering state of the electron–ion pair. The coefficient of recombination into an atomic state j by a three-body scattering process with a particle of species c is given by

$$\beta_c^j = \frac{1}{\hbar V}\int \frac{d\boldsymbol{P}}{(2\pi\hbar)^3}\frac{d\boldsymbol{p}_c}{(2\pi\hbar)^3}\frac{d\bar{\boldsymbol{P}}}{(2\pi\hbar)^3}\frac{d\bar{\boldsymbol{p}}}{(2\pi\hbar)^3}\frac{d\bar{\boldsymbol{p}}_c}{(2\pi\hbar)^3} 2\pi\delta\left(E^3_{(ei)c} - \bar{E}^0_{(ei)c}\right)$$
$$\times \left|\langle j\boldsymbol{P}|\langle \boldsymbol{p}_c|T^{33}_{eic}\left(E^3_{(ei)c} + i\varepsilon\right)|\bar{\boldsymbol{p}}_c\rangle|\bar{\boldsymbol{P}}\bar{\boldsymbol{p}}+\rangle\right|^2$$
$$\times \frac{\bar{f}_c}{n_c}\frac{\bar{f}_i}{n_i}\frac{\bar{f}_e}{n_e}\left(1-f_c\right)\left(1+F_j\right). \tag{8.5}$$

The $K_c^{\bar{j}j}$ are the coefficients of collisional excitation and deexcitation. They read

$$K_c^{j\bar{j}} = \frac{1}{\hbar V}\int \frac{d\boldsymbol{P}}{(2\pi\hbar)^3}\frac{d\boldsymbol{p}_c}{(2\pi\hbar)^3}\frac{d\bar{\boldsymbol{P}}}{(2\pi\hbar)^3}\frac{d\bar{\boldsymbol{p}}_c}{(2\pi\hbar)^3} 2\pi\delta\left(E^3_{(ei)c} - \bar{E}^3_{(ei)c}\right)$$
$$\times \left|\langle j\boldsymbol{P}|\langle \boldsymbol{p}_c|T^{33}_{eic}\left(E^3_{(ei)c} + i\varepsilon\right)|\bar{\boldsymbol{p}}_c\rangle|\bar{\boldsymbol{P}}\bar{j}\rangle\right|^2 \frac{f_c}{n_c}\frac{F_j}{n_A^j}(1-\bar{f}_c)(1+\bar{F}_j). \tag{8.6}$$

With (8.4), (8.5), and (8.6), expressions for the rate coefficients in terms of the three-particle T-matrix T^{33} are given. The latter is determined by the Lippmann–Schwinger equation

$$T^{kk}_{eic}(z) = V^k_{eic} + V^k_{eic}G_{eic}(z)V^k_{eic} \tag{8.7}$$

with $G_{eic}(z) = (z - H^{0\text{eff}}_{eic} + V^0_{eic})^{-1}$ being the resolvent of three interacting quasiparticles and V^k_{eic} being the effective scattering potential in the channel k. For $k = 3$, we have $V^3_{eic} = V_{ec} + V_{ic}$. Many-particle effects are also included in the scattering energies $E^3_{(ei)c} = E_j(\boldsymbol{P}, t) + E_c(\boldsymbol{p}_c, t)$ and $E^0_{(ei)c} = E_e(\boldsymbol{p}_e, t) + E_i(\boldsymbol{p}_i, t) + E_c(\boldsymbol{p}_c, t)$ with $E_c(\boldsymbol{p}_c, t)$ and $E_j(\boldsymbol{P}, t)$ being the energies of the free and bound quasiparticles.

8.1.2 Rate Coefficients and Cross Sections

Let us now deal with the calculation of the rate coefficients for nonideal plasmas. In particular, we consider hydrogen and hydrogen-like plasmas. Examples for the latter ones are highly ionized plasmas consisting of electrons, bare nuclei and hydrogen-like ions, but also alkali plasmas with electrons, singly charged ions and atoms. Further examples are electron–hole plasmas in semiconductors where the attractive Coulomb interaction leads to the formation of excitonic bound states. In all these systems, the inelastic and reactive collisions can be treated approximately as three-particle scattering processes. In most cases, the ionization and excitation rates are mainly determined by the contributions due to electron impact, i.e., the corresponding scattering processes with the heavy particles are negligible. This is not the case in electron–hole plasmas where the effective masses do not differ considerably from each other. However, electron–hole plasmas will not be considered here (for some results see Kremp et al. (1988)).

We start from the T-matrix expressions of the rate coefficients given by (8.3), (8.5), and (8.6) and assume the system to be in the stage of equilibrium with respect to the translational degrees of freedom. However, chemical equilibrium is not established, i.e., densities and temperatures are functions of time determined by the corresponding balance equations. For the electrons, the distribution function then takes the form $f_e = f_e^0$ with

$$f_e^0(p,t) = \frac{1}{\exp\{[p^2/2m_e + \Delta_e(p,t) - \mu_e]/k_B T\} + 1} . \tag{8.8}$$

To simplify the further treatment, we use the quasiparticle energies in the *rigid shift approximation* (Zimmermann 1988) (see Sect. 3.4.1). In this scheme, the self-energy corrections are replaced by thermally averaged shifts (3.204) which are momentum independent, and the energies of the free and bound particles are given by

$$E_a(p,t) = \frac{p^2}{2m_a} + \Delta_a(t), \qquad (a = e, i) \tag{8.9}$$

$$E_j(P,t) = \frac{P^2}{2M} + E_j^0 + \Delta_j(t) . \tag{8.10}$$

E_j^0 denotes the binding energy of the isolated atom. The shifts Δ_a and Δ_j are determined by formula (3.204). As discussed in Sect. 3.4.1, we have $\Delta_a = \mu_a^{\text{int}}$ and $\Delta_j = \mu_{Aj}^{\text{int}}$ with μ_a^{int} and μ_{Aj}^{int} being the interaction parts of the chemical potentials. In this way, it is possible to use the results of quantum statistical theory of plasmas in equilibrium. It should be noticed that the self-energy shifts in the distribution functions are compensated by the interaction parts of the chemical potentials.

In the approximations used here, a simple relation between the ionization and recombination coefficients can be derived. Accounting for the energy conserving delta function in the expressions (8.3) and (8.5), we arrive at (Schlanges and Bornath 1989)

$$\beta_c^j = \frac{n_A^j}{n_e n_i} \alpha_c^j \, e^{(\mu_e^{\rm id}+\mu_i^{\rm id}-\mu_{Aj}^{\rm id})/k_BT} \, e^{(\Delta_e+\Delta_i-\Delta_j)/k_BT} \,. \tag{8.11}$$

In comparison with the theory of ideal plasmas, we have now an additional contribution given in terms of the self-energy shifts.

In the further treatment, the adiabatic approximation is applied ($m_e \ll m_i$), and the heavy particles are considered to be non-degenerate. As we restrict ourselves to electron impact processes, the index c of the rate coefficients will be dropped. The coefficient of ionization by electron impact then reads

$$\begin{aligned} \alpha_j &= \frac{2\pi}{n_e\hbar} \int \frac{d\boldsymbol{p}_e}{(2\pi\hbar)^3} \frac{d\bar{\boldsymbol{p}}_e}{(2\pi\hbar)^3} d\bar{\boldsymbol{p}} \\ &\times \; \delta\left(E_j^0 + \frac{p_e^2}{2m_e} - \frac{\bar{p}^2}{2m_e} - \frac{\bar{p}_e^2}{2m_e} + \Delta_j - \Delta_e - \Delta_i\right) \\ &\times \; \left|\langle j|\langle \boldsymbol{p}_e|T_{eie}^{33}\left(E_{eie}^3 + i\varepsilon\right)|\bar{\boldsymbol{p}}_e\rangle|\bar{\boldsymbol{p}}+\rangle\right|^2 f_e^0 \left(1-\bar{f}_e^0\right)\left(1-\bar{f}_e^0\right) . \end{aligned} \tag{8.12}$$

It is appropriate to introduce an effective ionization energy for the bound state level j by

$$I_j^{\rm eff} = |E_j^0| + \Delta_e + \Delta_i - \Delta_j \,. \tag{8.13}$$

This quantity depends on density and temperature via the self-energy shifts. This leads to a lowering of the ionization energy. Furthermore, it is possible to introduce differential cross sections. For ionization of the bound particles in the state j by electron impact, we define

$$d\sigma_j^{\rm ion}(p_e,\bar{p}) = \frac{m_e^2\bar{p}_e}{(2\pi\hbar^2)^2 p_e} \int d\Omega_{\bar{p}_e} d\Omega_{\bar{p}} \left|\langle j|\langle \boldsymbol{p}_e|T_{eie}^{33}|\bar{\boldsymbol{p}}_e\rangle|\bar{\boldsymbol{p}}+\rangle\right|^2 \bar{p}^2 \,. \tag{8.14}$$

Here, $d\Omega$ is the element of the solid angle and $\bar{p}_e = \left(p_e^2 - 2m_e I_j^{\rm eff} - \bar{p}^2\right)^{1/2}$.

For non-degenerate systems, Maxwell distribution functions can be used and the phase space occupation terms (Pauli blocking) can be neglected. In this case, the ionization and excitation coefficients can be expressed in terms of total cross sections. The coefficient for electron impact ionization (8.12) then takes the form

$$\alpha_j = \frac{8\pi m_e}{(2\pi m_e k_B T)^{3/2}} \int_{I_j^{\rm eff}}^{\infty} d\epsilon \; \epsilon \; \sigma_j^{\rm ion}(\epsilon) \; e^{-\beta\epsilon} \,, \tag{8.15}$$

with the total cross section $\sigma_j^{\rm ion}(\epsilon)$ given by

$$\sigma_j^{\text{ion}}(\epsilon) = \int\limits_0^{\bar{p}_{\max}} d\bar{p}\, d\sigma_j^{\text{ion}}(\epsilon, \bar{p})\,. \tag{8.16}$$

Here, $\epsilon = p_e^2/2m_e$ denotes the kinetic energy of the incident electron. The upper limit of integration is $\bar{p}_{\max} = (p_e^2 - 2m_e I_j^{\text{eff}})^{1/2}$. In a similar manner, the excitation (deexcitation) coefficients can be written as

$$K_{j\bar{j}} = \frac{8\pi m_e}{(2\pi m_e k_B T)^{3/2}} \int\limits_{\Delta E_{j\bar{j}}^{\text{eff}}}^{\infty} d\epsilon\; \epsilon\; \sigma_{j\bar{j}}^{\text{exc}}(\epsilon)\; e^{-\beta\epsilon}\,. \tag{8.17}$$

with $\sigma_{j\bar{j}}^{\text{exc}}$ being the total excitation cross section. The effective excitation (deexcitation) energy is given by

$$\Delta E_{j\bar{j}}^{\text{eff}} = |E_j| - |E_{\bar{j}}| - \Delta_j + \Delta_{\bar{j}}\,. \tag{8.18}$$

We found that the coefficients of ionization and recombination are connected by the formula (8.11). In the non-degenerate case, it reduces to

$$\beta_j = \alpha_j\, \Lambda_e^3\, e^{I_j^{\text{eff}}/k_B T}\,. \tag{8.19}$$

Consequently, we find from the rate equation (8.1) in thermodynamic equilibrium

$$\frac{n_j}{n_e n_i} = \frac{\beta_j}{\alpha_j} = \Lambda_e^3 \exp\left(I_j^{\text{eff}}/k_B T\right). \tag{8.20}$$

This is just a mass action law for a nonideal plasma. It corresponds to the chemical equilibrium condition $\mu_e + \mu_i = \mu_{Aj}$.

With the approach developed so far, a quantum statistical theory of reaction rates for nonideal plasmas is given. Especially, it generalizes the elementary theory presented in Sect. 2.12. The influence of medium effects on the three-particle problem can be accounted for by solving effective Schrödinger and T-matrix equations. In this way, rate coefficients for strongly coupled plasmas can be calculated using the methods of many-body theory.

A simplified treatment was presented in Schlanges et al. (1988), Bornath (1987), and in Bornath et al. (1988). There, a modified Bethe-type cross section was used to account for the lowering of the ionization energy. It was already given in Sect. 2.11 by formula (2.240). Inserting the cross section into the expression (8.15), we get

$$\alpha_j = \alpha_j^{\text{ideal}} \exp[(\Delta_j - \Delta_e - \Delta_i)/k_B T]\,. \tag{8.21}$$

With formula (8.21), the ionization coefficient is expressed by the density independent (ideal) part

$$\alpha_j^{\text{ideal}} = \frac{10\pi a_B^2 |E_j|^{\frac{1}{2}}}{(2\pi m_e)^{\frac{1}{2}}} n^3\, h\left(\frac{|E_j|}{k_B T}\right), \quad h(x) = x^{\frac{1}{2}} \int_x^{\infty} \frac{e^{-t}}{t} dt\,, \tag{8.22}$$

and a nonideality contribution given in terms of self-energy shifts. Using the formula (8.21) in the relation (8.19), we get for the coefficient of three-body recombination

$$\beta_j = \alpha_j^{\mathrm{ideal}} \Lambda_e^3 \, e^{-E_j^0/k_B T} = \beta_j^{\mathrm{ideal}} \,. \tag{8.23}$$

Thus, the recombination coefficient in this simple approximation is equal to the one valid for ideal plasmas. This can be understood by the fact that only the lowering of the ionization energy was considered in formula (8.21).

In a more rigorous treatment, we have to determine the cross section from the three-particle T-matrix. This requires the solution of the corresponding Lippmann–Schwinger equation which is, in general, an extremely complicated problem. Instead, we will use a modified Born approximation which reflects already the main features of the influence of many-particle effects on the scattering processes in nonideal plasmas. In statically screened Born approximation, the total ionization cross section is given by

$$\sigma_j^{\mathrm{ion}}(\epsilon) = \frac{m_e^2}{2\pi\hbar^4 p_e^2} \int \frac{d\Omega_{p_e}}{4\pi} \int_0^{\bar{p}_{\max}} d\bar{p}\,\bar{p}^2 \int d\Omega_{\bar{p}} \int_{q_{\min}}^{q_{\max}} q\,dq \left| V_{ee}^{\mathrm{eff}}(\boldsymbol{q}) \mathcal{P}_{j\bar{\boldsymbol{p}}}(\boldsymbol{q}) \right|^2 . \tag{8.24}$$

$\mathcal{P}_{j\boldsymbol{p}}(\boldsymbol{q})$ is the atomic form factor considered below. For the excitation cross section there follows that

$$\sigma_{j\bar{j}}^{\mathrm{exc}} = \frac{m_e^2}{2\pi\hbar^4 p_e^2} \int \frac{d\Omega_{p_e}}{4\pi} \int_{q_{\min}}^{q_{\max}} q\,dq \left| V_{ee}^{\mathrm{eff}}(\boldsymbol{q}) \mathcal{P}_{j\bar{j}}(\boldsymbol{q}) \right|^2 . \tag{8.25}$$

The limits of integration are determined by the conservation of energy in the three-particle scattering process. They are

$$\bar{p}_{\max} = \left(p_e^2 - 2m_e I_j^{\mathrm{eff}}\right)^{1/2} \text{and} \quad q_{\min}^{\max} = p_e \pm \bar{p}_e \,. \tag{8.26}$$

In the case of ionization, we have $\bar{p}_e = \left(p_e^2 - 2m_e I_j^{\mathrm{eff}} - \bar{p}^2\right)^{1/2}$ and for excitation processes $\bar{p}_e = \left(p_e^2 - 2m_e \Delta E_{j\bar{j}}^{\mathrm{eff}}\right)^{1/2}$. The effective electron–electron potential is taken to be the statically screened Coulomb potential given by

$$V_{ab}^{\mathrm{eff}}(\boldsymbol{q}) = \frac{V_{ab}(q)}{\varepsilon(\boldsymbol{q},0)} \,, \qquad \varepsilon(\boldsymbol{q},0) = \frac{q^2 + \hbar^2\kappa^2}{q^2} \,, \tag{8.27}$$

with $V_{ab}(q) = 4\pi\hbar^2 e_a e_b / q^2$ being the Fourier transform of the Coulomb potential and κ is the inverse screening length. In adiabatic approximation, the form factor in the ionization cross section reads

$$\mathcal{P}_{j\bar{\boldsymbol{p}}}(\boldsymbol{q}) = \int d\boldsymbol{r}\, \Psi_j^*(\boldsymbol{r}) \Psi_{\bar{\boldsymbol{p}}}^+(\boldsymbol{r}) e^{\frac{i}{\hbar}\boldsymbol{q}\cdot\boldsymbol{r}} \,. \tag{8.28}$$

Here, $\Psi_j(\boldsymbol{r}) = \langle \boldsymbol{r}|j\rangle$ and $\Psi_{\bar{\boldsymbol{p}}}^+(\boldsymbol{r}) = \langle \boldsymbol{r}|\bar{\boldsymbol{p}}+\rangle$ denote the bound and scattering wave functions.

8.1.3 Two-Particle States, Atomic Form Factor

The wave functions in the expressions for the form factors have to be determined by an effective Schrödinger equation taking into account the influence of the surrounding plasma on the two-particle properties. In Chap. 5, such equations were discussed in different approximations. In the dynamically screened ladder approximation, we derived the effective wave equation (5.192) where many-particle effects are condensed in the dynamical self-energy correction and in the effective dynamically screened potential.

The expressions for the rate coefficients considered up to now were derived assuming statically screened interactions. Therefore, the wave functions in the atomic form factors are determined by (5.192) in the limiting case of statical screening. We get for non-degenerate plasmas

$$(\mathrm{H}^0_{ab} + \Delta_{ab} + \mathrm{V}^{\mathrm{eff}}_{ab})|\boldsymbol{P}\alpha\rangle = E_{\alpha\boldsymbol{P}}|\boldsymbol{P}\alpha\rangle \tag{8.29}$$

where $\alpha = j$ for bound states and $\alpha = \bar{\boldsymbol{p}}$ for scattering states. The effective potential is the Debye potential given by (8.27). The self-energy correction of the free particles is in lowest order

$$\Delta_{ab} = \Delta_a + \Delta_b \qquad \text{with} \quad \Delta_a = -\frac{1}{2}\kappa e_a^2\,, \tag{8.30}$$

which is just the Debye-shift. The bound state energy $E_{j\boldsymbol{P}}$ which includes the self-energy correction Δ_j follows from the solution of (8.29). In spite of the approximation involved, the statically screened Schrödinger equation includes important many-particle effects such as the lowering of the continuum edge and screening in the bound and scattering wave functions. It is convenient to solve (8.29) in position representation using partial wave expansion. The scattering wave function of relative motion is

$$\Psi^+_{\bar{\boldsymbol{p}}}(\boldsymbol{r}) = \frac{1}{(2\pi\hbar)^{3/2}} \sum_{\ell=0}^{\infty} \sqrt{\frac{\pi}{2}} \frac{R_{\bar{p}\ell}(r)}{\bar{p}} e^{i\left(\delta_\ell(\bar{p}) + \frac{\pi\ell}{2}\right)} (2\ell+1) P_\ell\left(\frac{\bar{\boldsymbol{p}}\cdot\boldsymbol{r}}{\bar{p}r}\right), \tag{8.31}$$

where δ_ℓ denotes the scattering phase shift. The wave function (8.31) is normalized such that the asymptotic behavior is given by

$$\Psi^+_{\bar{\boldsymbol{p}}}(\boldsymbol{r}) \to \frac{1}{(2\pi\hbar)^{3/2}} \left\{ e^{\frac{i}{\hbar}\bar{\boldsymbol{p}}\cdot\boldsymbol{r}} + \frac{e^{\frac{i}{\hbar}\bar{p}r}}{r} f_{\bar{p}}\left(\frac{\bar{\boldsymbol{p}}\cdot\boldsymbol{r}}{\bar{p}r}\right)\right\}$$

with the scattering amplitude $f_{\bar{p}}$ for an outgoing wave.

The radial parts of the wave functions are determined by

$$\left[\frac{\hbar^2}{2m_{ab}} \frac{1}{r^2} \frac{d}{dr}\left(r^2 \frac{d}{dr}\right) - \Delta_{ab} - \frac{\hbar^2 \ell(\ell+1)}{2m_{ab} r^2} - V^{\mathrm{eff}}_{ab}(r) + E\right] R_{\rho\ell}(r) = 0\,. \tag{8.32}$$

We have $\rho = n$ for bound states and $\rho = k$ for scattering states. For the continuous spectrum, the wave number is $k = \bar{p}/\hbar = \sqrt{\frac{2m_{ab}}{\hbar^2}(E - \Delta_{ab})}$ where m_{ab} is the reduced mass.

Now it is useful to perform a partial wave expansion of the form factor (8.28). It turns out that there is no dependence on the magnetic quantum number m. Thus, the label m will be omitted in the subsequent expressions. We consider the auxiliary quantity

$$F_{n\ell}(k,q) = \int \frac{d\Omega_{\boldsymbol{p}_e}}{4\pi} \int d\Omega_{\boldsymbol{k}} \left|\mathcal{P}_{(n\ell),\boldsymbol{k}}(\boldsymbol{q})\right|^2 , \tag{8.33}$$

which is needed in the calculations of cross sections. The partial wave expansion yields (Bornath and Schlanges 1993)

$$F_{n\ell}(k,q) = \frac{1}{k^2} \sum_{\ell'=0}^{\infty} \sum_{\ell''=0}^{\infty} (2\ell'+1)(2\ell''+1) \begin{pmatrix} \ell' & \ell'' & \ell \\ 0 & 0 & 0 \end{pmatrix} \begin{pmatrix} \ell' & \ell'' & \ell \\ 0 & 0 & 0 \end{pmatrix} I_{n,\ell}^{\ell',\ell''}(k,q) , \tag{8.34}$$

with $\begin{pmatrix} \ell' & \ell'' & \ell \\ 0 & 0 & 0 \end{pmatrix}$ being a special Wigner $3j$-symbol (Landau and Lifschitz 1988). Furthermore, we introduced the abbreviation

$$I_{n,\ell}^{\ell',\ell''}(k,q) = \left[\int_0^\infty dr r^2 R_{n\ell}(r) R_{k\ell'}(r) j_{\ell''}(qr/\hbar) \right]^2 , \tag{8.35}$$

where the spherical Bessel functions $j_\ell(qr/\hbar)$ result from the expansion of the factor $\exp(i\boldsymbol{q}\cdot\boldsymbol{r}/\hbar)$. Since there are selection rules with respect to the angular momentum quantum numbers, the expression (8.34) can be simplified in dependence of the quantum number ℓ. For example, we have for the ionization from s-states ($\ell = 0$)

$$F_{n0}(k,q) = \frac{1}{k^2} \sum_{\ell'=0}^{\infty} (2\ell'+1) I_{n,0}^{\ell',\ell'}(k,q) , \tag{8.36}$$

and for ionization processes from p-states, we get

$$F_{n1}(k,q) = \frac{1}{k^2} \sum_{\ell'=0}^{\infty} \left[\ell' I_{n,1}^{\ell',\ell'-1}(k,q) + (\ell'+1) I_{n,1}^{\ell',\ell'+1}(k,q) \right] . \tag{8.37}$$

The angle averaged form factors (8.33) are given in terms of the radial bound state and scattering state wave functions which have to be calculated by numerical solution of the effective Schrödinger equation (8.32).

8.1.4 Density Effects in the Cross Sections

Before we discuss results for the rate coefficients, it is of interest to investigate the influence of many-particle effects on the cross sections for ionization determined by expression (8.24) which represents a modified Born approximation. It is known that the Born approximation overestimates the cross section near the threshold. Accordingly, more sophisticated methods of quantum scattering theory have to be applied. However, this is a difficult problem because medium effects have to be taken into account. For this reason, we use the statically screened Born approximation which allows to study, in a simple manner, important features of the influence of many-particle effects on ionization cross sections: (i) there is a density and temperature dependent threshold energy (8.13). (ii) The wave functions in the expressions for the atomic form factor are determined by an effective Schrödinger equation (8.32). (iii) The scattering potential between the incident electron and the bound particles is a statically screened one.

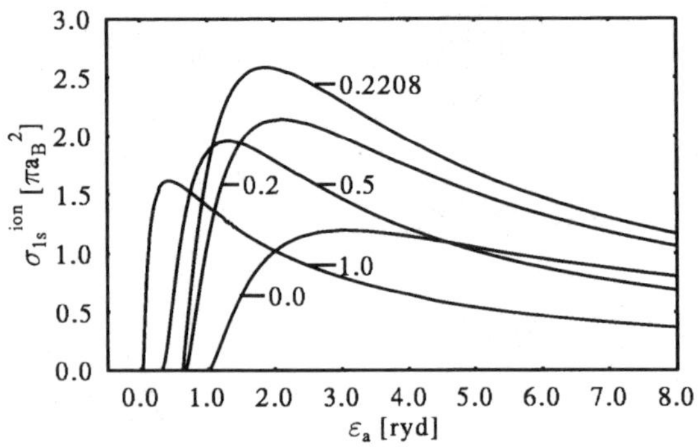

Fig. 8.1. Total cross section $\sigma_{1s}^{\mathrm{ion}}$ of electron impact ionization of the 1s atomic state in a hydrogen plasma versus impact energy for different screening parameters $\lambda_D = \kappa a_B$

We will discuss some results for ionization cross sections obtained for nonideal hydrogen plasmas (Schlanges and Bornath 1993; Bornath and Schlanges 1993). Calculations for electron–hole and for alkali plasmas are discussed in Kremp et al. (1988), Bornath et al. (1994). The wave functions in the form factor were calculated numerically from (8.32) using the Numerov method. In Fig. 8.1, the cross section of ionization from the atomic ground state is shown as a function of the electron impact energy for different values of the screening parameter $\lambda_D = \kappa a_B$. With growing plasma density (λ_D increases), the threshold energy decreases, which corresponds to the lowering of the ionization energy up to the Mott point.

Near the threshold, the cross section first shows an increase with density whereas the maximum lowers at higher values of λ_D. This is due to a competition between screening effects in the scattering potential and in the form factor. The increasing behavior results from the influence of plasma effects on the two-particle states. At certain densities, the higher lying bound states reach their respective Mott points. They merge into the continuum where they contribute as resonance states enhancing the cross section. A clear demonstration of this effect is possible if one considers the differential

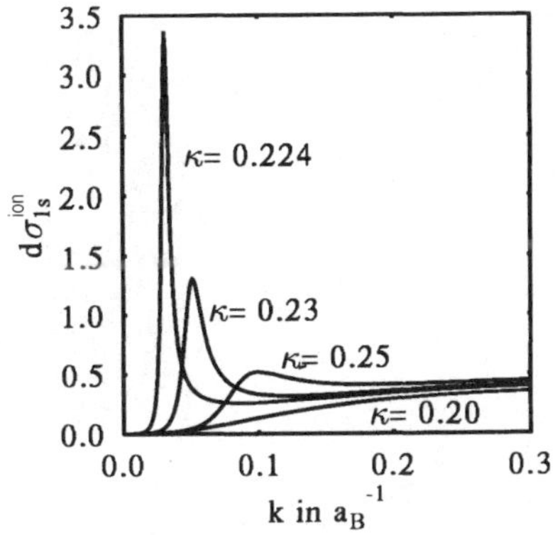

Fig. 8.2. Differential cross section for the 1s atomic state in a hydrogen plasma versus wave number of the ejected electron for different screening parameters λ_D. The electron impact energy is $\epsilon = 1Ryd$

cross section (Bornath et al. 1994; Rietz 1996). In Fig. 8.2, results are shown for screening parameters in the range $0.2 \leq \lambda_D \leq 0.25$. The differential cross section for $\lambda_D = 0.2$ is a smooth function of the energy of the ejected electron. In this case, the 2p-bound states still exist. However, for $\lambda = 0.224$, their respective Mott point was already reached, i.e., these bound states merged into the continuum. There they appear as resonances producing sharp peaks in the low energy range of the cross section.

8.1.5 Rate Coefficients for Hydrogen and Hydrogen-Like Plasmas

Let us now discuss the results following for the rate coefficients in the case of static screening (Schlanges et al. 1992; Schlanges and Bornath 1993; Bornath and Schlanges 1993; Prenzel (a) et al. 1996).

In Fig. 8.3, the coefficient of electron impact ionization of the atomic ground state in a hydrogen plasma as a function of the free electron density is shown. The coefficients were calculated from expression (8.15) using the numerical results for the cross sections discussed above. In order to demonstrate the influence of many-particle effects, the ionization coefficients were related to their values $\alpha_{1s}^{\text{ideal}}$ valid for the ideal plasma state. The latter follow from (8.15) with (8.24) in the low density limit ($\lambda_D = 0$) where many-particle effects are negligible. They are given in Table 8.1 for different temperatures. At low densities, the nonideality of the plasma is small. The rate coefficients are nearly density independent, and they merge into the values valid for the ideal plasma state ($\alpha_j/\alpha_j^{\text{ideal}} \to 1$). However, we observe a strong increase of the ionization coefficients with increasing density, due to the influence of the many-particle effects. The curves end at the Mott densities where the corresponding bound state vanishes. In addition, results are shown which follow from the analytical formula (8.21) where the only plasma effect is the lowering of the ionization energy. One can see that the results obtained in statically screened Born approximation show a similar exponential behavior as given by the analytical formula. Therefore, the density dependence

Table 8.1. Ionization coefficient α^{ideal} and recombination coefficient β^{ideal} for several temperatures

Temperature/K	$\alpha_{1s}^{\mathrm{ideal}}/cm^3s^{-1}$	$\beta_{1s}^{\mathrm{ideal}}/cm^6s^{-1}$
10000	8.331E-16	2.466E-30
15000	2.309E-13	1.932E-30
20000	4.104E-12	1.606E-30

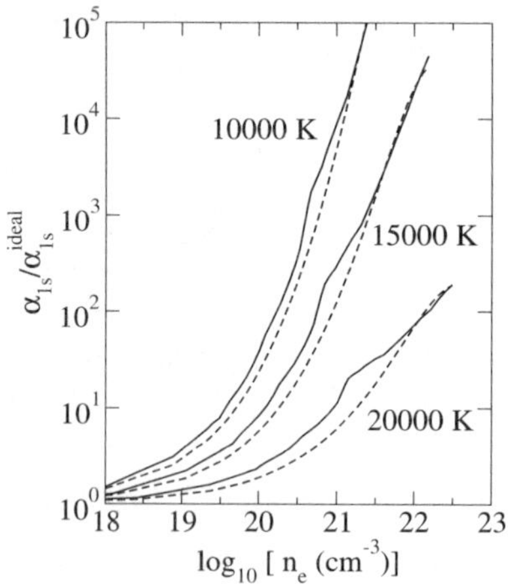

Fig. 8.3. Impact ionization coefficients for the atomic ground state in a hydrogen plasma as a function of the free electron density for several temperatures. For comparison, the results following from the analytical formula (8.21) are shown (*dashed*)

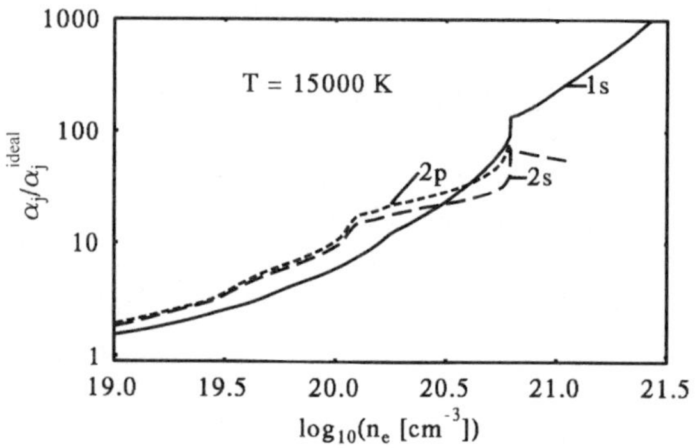

Fig. 8.4. Impact ionization coefficient for the 1s, 2s, 2p atomic states in hydrogen plasma versus free electron density for a temperature of $T = 15000\,\mathrm{K}$

of the ionization coefficient can mainly be attributed to the lowering of the ionization energy.

At higher densities, we observe a weak step-like behavior. This can be seen more clearly in Fig. 8.4 where the ionization coefficients are shown for the atomic ground state and for the lowest excited states. It turns out that the step-like behavior is connected with the density dependence of the form

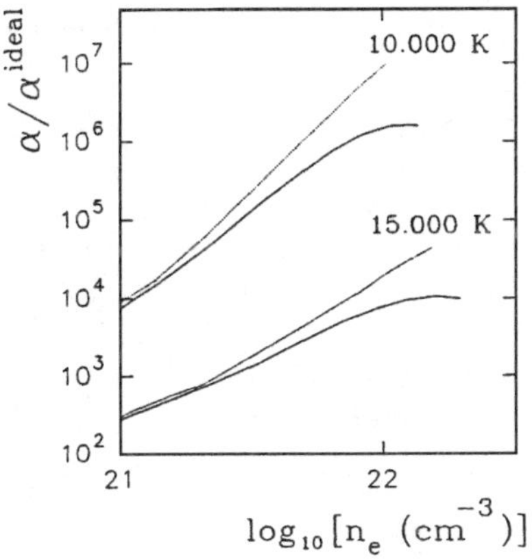

Fig. 8.5. Coefficients of ionization for the ground state of hydrogen atoms versus free electron density for different temperatures including degeneracy effects (**lower curves**). Results without degeneracy effects are given by the **upper curves**

factor of the bound-free transitions. In particular, it is due to the merging of higher lying bound states into the continuum as discussed above. Iglesias and Lee (1997) could reproduce this behavior approximately in a simple model accounting for the lowering of the continuum edge and the excitation coefficients for the higher-lying bound states.

In order to study effects of degeneracy, we start from expression (8.12) with the T-matrix used in Born approximation. Degeneracy is here taken into account by Fermi distribution functions and Pauli blocking factors. Furthermore, the screening length is calculated from (4.141). Results for the coefficient of ionization for the $1s$ atomic state are shown in Fig. 8.5. As expected, Pauli blocking reduces the ionization rates at high densities. However, this is of importance only for the ground state. According to the Mott criterion, the excited states decay at densities before degeneracy has to be taken into account.

With the calculated ionization coefficients α_j, it is rather simple to get the recombination coefficients β_j using the relations (8.11) and (8.19). The latter can be applied for the non-degenerate case. As shown in the previous section, β_j is equal to its ideal value ($\beta_j/\beta_j^{\rm ideal} = 1$) if α_j is calculated from formula (8.21). Thus, the recombination coefficient remains density independent in this approximation. The situation changes if the rate coefficients are calculated using the more rigorous theory developed in the previous section. Now many-particle effects lead to a density dependence of the recombination coefficient, too.

In Fig. 8.6, three-body recombination coefficients for several atomic states are shown for a hydrogen plasma. Like in the case of the ionization coefficients, β_j is related to its ideal part. On the left, the coefficient for the atomic ground state at different temperatures is presented. The right part of the figure shows the three-body recombination coefficients for the 1s, 2s, and 3s atomic states. The influence of many-particle effects on the recombination

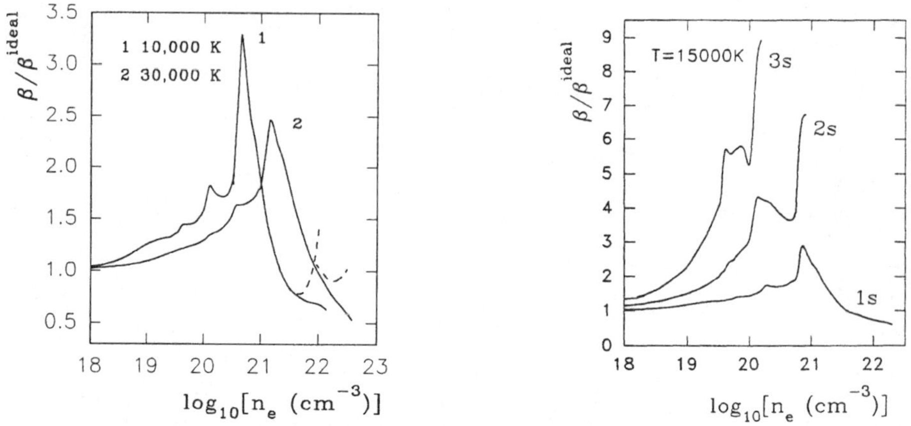

Fig. 8.6. Three-body recombination coefficients of different atomic states for a hydrogen plasma versus free electron density. **Left**: 1s atomic state for different temperatures. The results where degeneracy effects are neglected are given by the dashed curves. **Right**: 1s, 2s and 3s atomic states for a temperature $T = 15000\,\mathrm{K}$

coefficient is smaller compared to the ionization coefficient. Here, it should be noticed that recombination is a process without an effective threshold energy. However, the competition between screening of the scattering potential and the enhancement of continuum contributions in the atomic form factor leads to an interesting density dependence. We see an increase with increasing density and some maxima connected with the merging of higher lying bound states into the continuum. The influence of the nonideality effects of the plasma increases with increasing quantum number. In contrast to the excited states, the ground state recombination coefficient is a decreasing function at high densities due to plasma and degeneracy effects.

Finally, let us discuss results for the coefficients of excitation and deexcitation processes which were calculated from the expressions (8.17) and (8.25). Similar to the ionization and recombination coefficients, the following relation can be derived

$$K_{j'j} = K_{jj'} \exp\left(\Delta E^{\mathrm{eff}}_{jj'}/k_B T\right). \tag{8.38}$$

Instead of hydrogen, we consider a carbon plasma with hydrogen-like bound states. To treat the screening, the plasma is assumed to be (almost) fully ionized. In Fig. 8.7, coefficients of excitation and deexcitation are shown as a function of free electron density. We observe a relatively small influence of plasma effects. The reason is the weak density dependence of the effective excitation energy (8.18). There is a slight maximum behavior for certain excitation coefficients whereas the deexcitation coefficients are decreasing functions with increasing density.

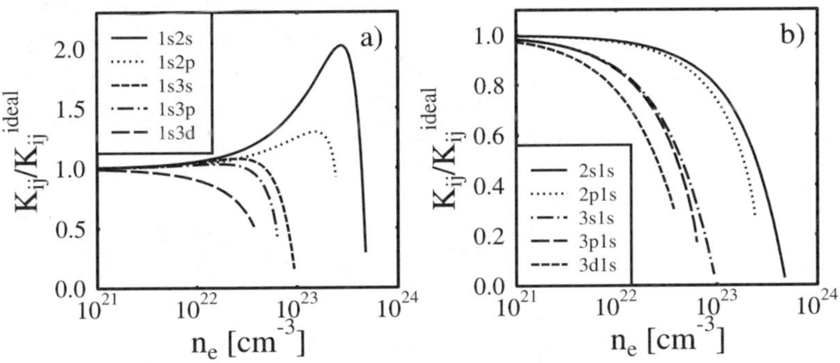

Fig. 8.7. Coefficients of excitation (**a**) and deexcitation (**b**) for hydrogen-like carbon ions versus free electron density. The values are related to their values for the ideal plasma. The temperature is $T = 50\,\mathrm{eV}$ ($T = 5.8 \times 10^5\,\mathrm{K}$)

In this section, we analyzed in an exemplary manner the consequences of nonideality effects on the rate coefficients for hydrogen and hydrogen-like carbon plasmas assuming statical screening. Further results from calculations for nonideal carbon, alkali, and electron–hole plasmas can be found in Kremp et al. (1988), Bornath et al. (1994), and Prenzel (a) et al. (1996).

Finally, we want to mention the interesting problem of the influence of an external static electric field on the ionization kinetics. The electric field modifies the ionization rates at lower densities, because the mean free path length is large, and the external field accelerates the electrons to energies which are sufficient for impact ionization. The competition between acceleration at low densities and lowering of ionization energy produces a minimum behavior of the ionization coefficient. This was shown by Kremp (a) et al. (1993).

8.1.6 Dynamical Screening

So far we calculated the rate coefficients using statically screened Coulomb potentials. Collective excitations of the plasma are not described in this approximation. In order to include such effects in the rate coefficients, dynamic screening has to be taken into account.

An appropriate starting point is the kinetic equation for two-particle bound states given by (7.137). It takes the form of a Lenard–Balescu-type kinetic equation where the collision terms of elastic, inelastic, and reactive processes are given in Born approximation with respect to the dynamically screened potential. The rate equations for the population densities n_j can be obtained integrating the kinetic equation with respect to the momenta. We will focus on the ionization and recombination coefficients. Their expressions follow from the collision integral (7.139). To include degeneracy effects, we use here the generalized collision integrals given in Schlanges and Bornath (1997), Bornath et al. (1997). The adiabatic approximation ($m_e \ll m_i$) is applied, and the heavy particles are assumed to be non-degenerate. Like in Sect. 8.1.2, the *rigid shift approximation* is used for the self-energy corrections. The coefficient of ionization for the bound state j by electron impact is then given by

$$\begin{aligned}\alpha_j &= \frac{1}{\hbar n_e}\int \frac{d\bar{\boldsymbol{p}}}{(2\pi\hbar)^3}\frac{d\boldsymbol{p}}{(2\pi\hbar)^3}d\boldsymbol{q}\,|\mathcal{P}_{j,\bar{\boldsymbol{p}}}(\boldsymbol{q})|^2\left|\frac{V_{ee}(q)}{\varepsilon^R(q,I_j^{\text{eff}}+\frac{\bar{p}^2}{2m_{ei}})}\right|^2\\ &\quad\times 2\pi\delta\Big(I_j^{\text{eff}}+\frac{\bar{p}^2}{2m_{ei}}-E_e(\boldsymbol{p})+E_e(\boldsymbol{p}-\boldsymbol{q})\Big)\\ &\quad\times f_e(\boldsymbol{p})\,[1-f_e(\boldsymbol{p}-\boldsymbol{q})]\,[1-f_e(\bar{\boldsymbol{p}})]\,,\end{aligned} \tag{8.39}$$

and we get for the three-body recombination coefficient

$$\begin{aligned}\beta_j &= \frac{1}{\hbar n_e^2}\int \frac{d\bar{\boldsymbol{p}}}{(2\pi\hbar)^3}\frac{d\boldsymbol{p}}{(2\pi\hbar)^3}d\boldsymbol{q}\,|\mathcal{P}_{j,\bar{\boldsymbol{p}}}(\boldsymbol{q})|^2\left|\frac{V_{ee}(q)}{\varepsilon^R(q,I_j^{\text{eff}}+\frac{\bar{p}^2}{2m_{ei}})}\right|^2\\ &\quad\times 2\pi\delta\Big(I_j^{\text{eff}}+\frac{\bar{p}^2}{2m_{ei}}-E_e(\boldsymbol{p})+E_e(\boldsymbol{p}-\boldsymbol{q})\Big)\\ &\quad\times f_e(\bar{\boldsymbol{p}})f_e(\boldsymbol{p}-\boldsymbol{q})\,[1-f_e(\boldsymbol{p})]\,.\end{aligned} \tag{8.40}$$

Here, E_e is the quasiparticle energy of the electrons according to (8.9). Furthermore, m_{ei} is the reduced mass of the bound electron–ion pair. The rate coefficients are given in terms of the dynamically screened potential with $\varepsilon^R(q,\omega)$ being the dielectric function.

In the following, we will assume that screening is mainly determined by the free plasma particles. Then the RPA expression (4.109) for the dielectric function is used. The effective ionization energy I_j^{eff} and the atomic form factor $\mathcal{P}_{j,\bar{\boldsymbol{p}}}$ are given by (8.13) and (8.28), respectively. They will be treated like in the previous sections. In particular, the wave functions will be calculated from the Schrödinger equation with a statically screened potential (8.32). Again, we consider the stage of local equilibrium. For non-degenerate systems, we have Boltzmann distribution functions, and the Pauli blocking terms can be neglected, i.e., $(1-f_e)\approx 1$. Then the ionization coefficient can be written in a form given by (8.15) with a cross section defined by (Schlanges (a) et al. 1996; Bornath et al. 1997)

$$\begin{aligned}\sigma_j^{\text{ion}} &= \frac{m_e^2}{2\pi\hbar^4 p_e^2}\int\frac{d\Omega_{p_e}}{4\pi}\int_0^{\bar{p}_{\max}} d\bar{p}\,\bar{p}^2\int d\Omega_{\bar{p}}\\ &\quad\times\int_{q_{\min}}^{q_{\max}} q\,dq\left|\frac{V_{ee}(q)}{\varepsilon(q,I_j^{\text{eff}}+\frac{\bar{p}^2}{2m_{ei}})}\right|^2|\mathcal{P}_{j\bar{\boldsymbol{p}}}(\boldsymbol{q})|^2\,.\end{aligned} \tag{8.41}$$

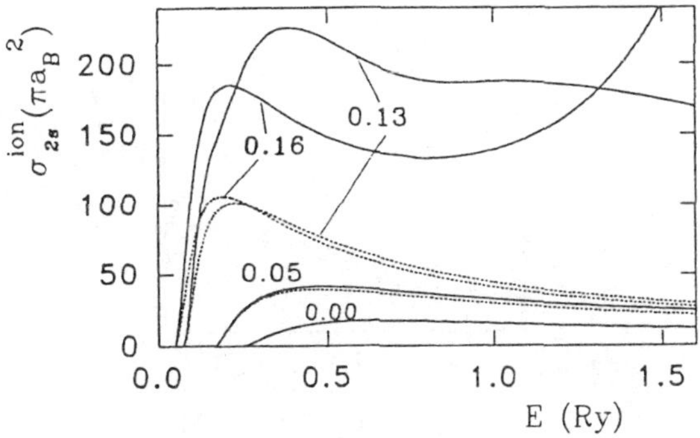

Fig. 8.8. Total ionization cross section σ_{2s}^{ion} versus electron (impact) energy. Results in the dynamic (*solid lines*) and in the static (*dotted lines*) approximations for the effective scattering potential are shown for different screening parameters $\lambda_D = \kappa a_B$

This expression is similar to (8.24). However, instead of a statically screened scattering potential, a dynamically screened one is used with an energy argument including the effective ionization energy. Many-particle effects are accounted for via the effective ionization energy, the screened atomic form factor, and the dynamically screened scattering potential. Numerical results for the cross section of ionization from the $2s$-state of a hydrogen atom are shown in Fig. 8.8.

The cross sections with dynamic screening (solid lines) are larger than the one with static screening (dotted lines). The reason is that static screening overestimates plasma density effects on the cross section. Since the same effective ionization energies and form factors are used, the behavior near the threshold is similar in both approximations. However, at higher plasma densities, the cross sections with dynamic screening show an irregular behavior for high impact energies: Instead of the expected decrease they grow with increasing energies. This indicates that the picture of scattering of a single electron on the bound state (atom) becomes inadequate. Indeed, with decreasing effective ionization energy, the energy argument of the dielectric function in (8.41) can have values near the plasma frequency. This is demonstrated in Fig. 8.10. Furthermore, at high energies the momentum transfer q is small, and one enters the region of the plasmon dispersion curve.

It turns out that ionization and recombination processes in dense plasmas are connected with the collective effects of absorption and emission of plasmons. In this case, an electron impact ionization cross section is not the appropriate quantity (Murillo and Weisheit 1993; Schlanges (a) et al. 1996). Consequently, it seems to be more appropriate to start from more general expressions of the rate coefficients. Using the fluctuation–dissipation theorem (4.64), it follows for the ionization coefficient (Schlanges (a) et al. 1996; Bornath et al. 1997)

$$\begin{aligned}\alpha_j &= \frac{1}{\hbar n_e}\int \frac{d\bar{\boldsymbol{p}}}{(2\pi\hbar)^3}d\boldsymbol{q}\,\left|\mathcal{P}_{j,\bar{\boldsymbol{p}}}(\boldsymbol{q})\right|^2 V_{ee}(q)\,2\mathrm{Im}\,\varepsilon^{-1}\left(\boldsymbol{q}, I_j^{\text{eff}}+\frac{\bar{p}^2}{2m_{ei}}\right)\\ &\quad\times n_B\left(I_j^{\text{eff}}+\frac{\bar{p}^2}{2m_{ei}}\right)\left(1-f_e^0(\bar{p})\right), \qquad (8.42)\end{aligned}$$

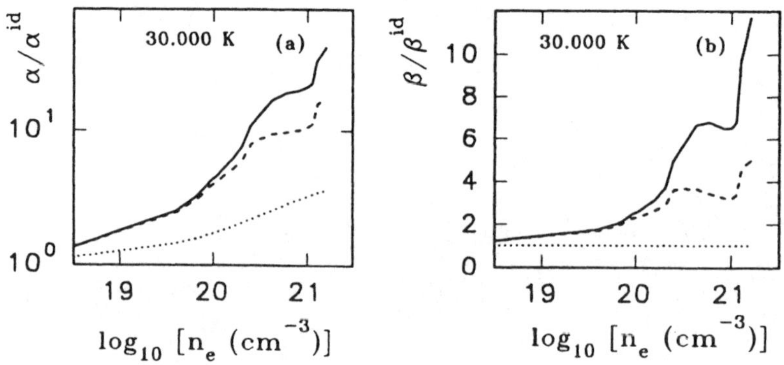

Fig. 8.9. Ionization coefficient (**a**) and recombination coefficient (**b**) versus free electron density for the 2s atomic state in a hydrogen plasma in different approximations: dynamical screened potential (*solid lines*), statical screening (*dashed lines*) and analytical formula (8.21) (*dotted lines*). The ratio of the rate coefficients to their ideal values is shown

with f_e^0 being the Fermi distribution function and $n_B(\omega) = (\exp(\beta\omega) - 1)^{-1}$. Similar expressions are found for the other rate coefficients.

Let us discuss some numerical results for the rate coefficients. Fig. 8.9 shows the coefficients of ionization and recombination for the atomic 2s state in a hydrogen plasma as a function of the free electron density in three different approximations. We observe a strong increase of the rates with increasing density due to the influence of many-particle effects. As discussed in the previous section, the reasons of this behavior arise mainly from the lowering of the ionization energy and from the enhancement of the atomic form factor due to the merging of higher-lying bound states into the continuum (Mott effect). Static screening in the effective potential tends to lower the rate coefficients whereas this effect is smaller for dynamic screening. Therefore, the inclusion of dynamic screening in RPA leads to higher rate coefficients. Furthermore, there is an additional contribution to the rates due to collective excitations (plasmons) at high densities.

Similar calculations to study ionization and recombination coefficients for hydrogen-like carbon plasmas were performed by Bornath et al. (1997) and Schlanges and Bornath (1998). Like in the case of hydrogen, the inclusion of dynamic screening leads to higher rates compared to static screening. Plasmon emission and absorption processes become especially important at higher densities. This is more drastic for excited states.

It was shown in the previous section that the influence of static screening gives small effects in the excitation and deexcitation coefficients for low-lying bound states. There is no remarkable change if dynamic screening is included. However, collective effects could become important for high-lying excited states with a small excitation energy.

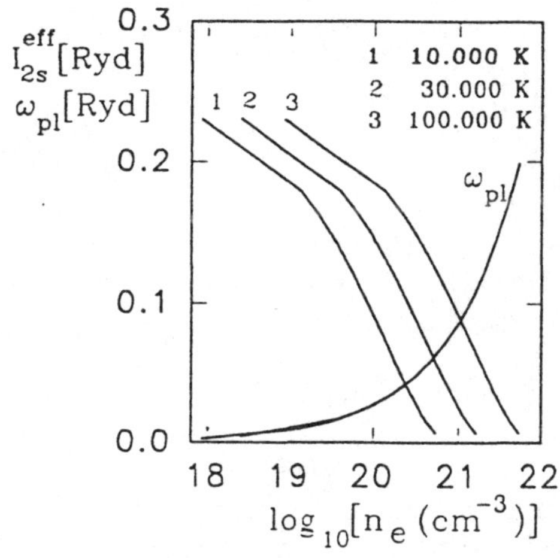

Fig. 8.10. Effective ionization energy I_{2s}^{eff} and plasma frequency versus free electron density

Of course, further improvements are necessary to get more accurate results for reaction rates of dense strongly coupled plasmas. Contributions beyond the Born approximation should be included, and dynamic screening has to be considered beyond the simple RPA scheme. Here, the influence of strong correlations on the damping of the plasmon excitations should be of importance. Furthermore, damping effects should enter the rates via the spectral functions of the one- and two-particle correlation functions.

8.2 Relaxation Processes

In the last years it has become possible to create and diagnose dense nonideal plasmas in experiments including laser-matter interaction, shock heating, exploding wires, capillary discharges, and others. In many of these examples, the plasma is in non-equilibrium, and phenomena such as ionization, recombination, and energy relaxation are of particular interest. They play an important role to understand dense plasmas under different laboratory and natural conditions.

In the following, relaxation processes in nonideal plasmas are considered using quantum kinetic theory. Simple approximations are used applicable to the regime of weak coupling. Like in the previous section, we assume the stage where the plasma particles can be described by local equilibrium distribution functions. This allows to describe the relaxation processes by the temporal evolution of macroscopic quantities such as densities and temperatures.

8.2.1 Population Kinetics in Hydrogen and Hydrogen-Like Plasmas

The occupation of bound state levels in dense nonideal plasmas can be described by rate equations discussed in the previous section. A non-degenerate spatially homogeneous plasma consisting of free electrons with number density n_e, bare nuclei of charge Z with density n_Z and hydrogen-like ions $(Z-1)$ with density n_{Z-1} will be considered. The latter are two-particle bound states with internal degrees of freedom characterized by the set of quantum numbers denoted by j.

Introducing the respective occupation densities, one has $n_{Z-1} = \sum_j n^j_{Z-1}$. Electro-neutrality will be assumed, i.e.. $n_e = Zn_Z + (Z-1)n_{Z-1}$. The system is considered to be dense enough in order to neglect radiation processes (Griem 1964; Fujimoto and McWhriter 1990). Furthermore, we only include the contributions due to electron impact and the respective inverse processes. Then, the rate equations can be written as

$$\frac{\partial n_e}{\partial t} = \sum_j (n_e n^j_{Z-1}\alpha_j - n_e^2 n_Z \beta_j), \tag{8.43}$$

$$\frac{\partial n^j_{Z-1}}{\partial t} = n_e^2 n_Z \beta_j - n_e n^j_{Z-1}\alpha_j + \sum_{\bar{j}} (n_e n^{\bar{j}}_{Z-1} K_{\bar{j}j} - n_e n^j_{Z-1} K_{j\bar{j}}). \tag{8.44}$$

Quantum statistical expressions for the coefficients of indexionizationionization α_j, three-body recombination β_j, and excitation (deexcitation) $K_{j\bar{j}}$ were given in Sect. 8.1. Many-body effects such as screening, self energies, and lowering of ionization energies are accounted for and lead to an additional dependence of the rate coefficients on density and temperature. An important result is the strong density dependence of the ionization coefficients mainly determined by the lowering of the ionization energies. The influence of many-body effects on the recombination, excitation and deexcitation coefficients was found to be weaker.

For the purpose of solving the rate equations, it is desirable to have analytic results for the rate coefficients. For that reason we will use here the rather simple approximation given by (8.21). In the following, j is understood to be the principal quantum number. The coefficient of ionization by electron impact then takes the form

$$\alpha_j = \alpha_j^{\text{ideal}} \exp[-(\Delta_e + \Delta_Z - \Delta^j_{Z-1})/k_B T]\,, \tag{8.45}$$

with the density independent ideal part α_j^{ideal} determined by formula (8.22). The nonideality correction is given in terms of momentum independent energy shifts determined in *rigid shift approximation* (see Sect. 3.4.1). We use for free particles $\Delta_a = -\kappa e_a^2/2, [a = e, Z]$ with $\kappa^2 = 4\pi e^2 n_e/k_B T$. Only screening by the electrons is taken into account to model a correct qualitative behavior for the range of larger coupling. The energy shifts for the hydrogen-like bound states are approximated by that of a charged particle with charge number $Z-1$, i.e., $\Delta^j_{Z-1} = -\kappa e^2(Z-1)^2/2$. In this approximation, the shift vanishes in the case of a hydrogen atom. The effective ionization energy is then

$$I_j^{\text{eff}} = |E_j| - Z\kappa e^2 \,. \tag{8.46}$$

The lowering of the ionization energy in the exponent of (8.45) gives rise to an increasing ionization coefficient with increasing plasma density. The recombination coefficient follows from

$$\beta_j = g_j \Lambda_e^3 \alpha_j \exp(I_j^{\text{eff}}/k_B T) \tag{8.47}$$

with g_j being the statistical weight of the j-th level. As already discussed in Sect. 8.1.2 (see also Sect. 2.12), in this simple approximation, β_j does not depend on the density. The same holds for the excitation and deexcitation coefficients which are, therefore, taken from Biberman et al. (1987).

Expressions of the form (8.45) were also used in Ebeling and Leike (1991) solving rate equations for nonideal hydrogen. A simple but interesting model was applied in Iglesias and Lee (1997) to study density effects on collision rates and population kinetics. A self-consistent approach based on the characterization of different electron states, and on the formalism of effective statistical weights for the bound states was given in Fisher and Maron (2003). Investigations based on a stochastic scheme and on simulations were performed in Beule et al. (1996), Ebeling et al. (1996), and in Ebeling and Militzer (1997).

A problem one is faced while solving rate equations for nonideal plasmas is the Mott effect. During the relaxation process, the screening length changes which affects the number of possible bound states. In the simple model, we used here, bound states vanish or appear abruptly at their respective Mott points, i.e., if $I_j^{\text{eff}} = 0$. This leads to unphysical jumps in the densities. That is why, as usual, the statistical weights are modified in order to allow for a soft transition at the Mott points (More 1982). In the following, we will treat this problem in a similar manner using a so-called Planck–Larkin redefinition of the statistical weights (Rogers 1986; Prenzel (b) et al. 1996; Bornath et al. 1998). For this purpose, we recall the discussion of the properties of thermodynamic quantities as a function of the coupling parameter given in Sect. 6.5.2. There it was shown that the total second virial coefficient is continuous at the Mott densities whereas the bound state and scattering state contributions show jumps which compensate each other. Applying Levinson's theorems, the second virial coefficient can be expressed by the Planck–Larkin sum of states. The latter is continuous at the Mott densities if effective ionization energies are used. Following this idea, we replace the statistical weights g_j by (Prenzel (b) et al. 1996)

$$g_j^{\rm eff} = g_j \left(e^{\beta I_j^{\rm eff}} - 1 - \beta I_j^{\rm eff} \right) / e^{\beta I_j^{\rm eff}} \, . \tag{8.48}$$

Now the rate equations give a vanishing population of bound state levels at their Mott points. For an interpretation of the densities $\tilde{n}_{Z-1}^j$ determined by these modified rate equations, let us consider the case of thermodynamic equilibrium. We then get

$$\frac{\tilde{n}_{Z-1}^j}{n_e n_Z} = g_j \Lambda_e^3 \left(e^{\beta I_j^{\rm eff}} - 1 - \beta I_j^{\rm eff} \right) . \tag{8.49}$$

For the total bound state density $\tilde{n}_{Z-1} = \sum_j \tilde{n}_{Z-1}^j$, this gives just the Saha equation with the Planck–Larkin sum of states. Following the discussion given by Rogers (1986), the Planck–Larkin scheme is appropriate to calculate thermodynamic functions. However, the corresponding densities are not the actual occupation numbers, which are relevant for the determination of line intensities. In the latter case, the actual occupation numbers n_{Z-1}^j have to be recalculated from the Planck–Larkin densities $\tilde{n}_{Z-1}^j$ using the formula

$$n_{Z-1}^j = \tilde{n}_{Z-1}^j \frac{e^{\beta I_j^{\rm eff}}}{e^{\beta I_j^{\rm eff}} - 1 - \beta I_j^{\rm eff}} \, . \tag{8.50}$$

The total ion density is given by $n_Z^{\rm tot} = n_Z + n_Z^{\rm corr}$, where the correlated part $n_Z^{\rm corr}$ which includes bound and scattering states is approximated by the Planck–Larkin densities $n_Z^{\rm corr} = \sum_j \tilde{n}_{Z-1}^j$. It turns out that this scheme can be applied only to low densities which corresponds to the restricted validity of an expansion up to the second virial coefficient (Bornath 1998).

Of course, the concept of bound states used here becomes questionable at high densities because the particles in the plasma are strongly correlated. The influence of the many-body effects leads to a damping of the two-particle states. There is an increasing delocalization of the wave functions and the spectral densities of the bound particles, respectively. To model such high density effects on level populations, Rogers introduced the following reduced statistical weight (Rogers 1990)

$$g_{nl}^{\rm eff}(\lambda_a) = (2l+1) \left(1 - e^{-(1-\lambda_c^{nl}/\lambda_a)^2} \right) . \tag{8.51}$$

Here, λ_a is the screening length following from the activity expansion method, and λ_c^{nl} is the critical screening length where the bound electron is delocalized and its wave function is extended over the entire system. Formula (8.51) does not follow rigorously from the many-body theory presented here. It gives a simple approach to account for correlation effects in dense plasmas. In particular, it avoids jumps at the Mott points and it gives a lower population of high lying excited levels as compared to the Planck–Larkin scheme (Prenzel 1999). In Fig. 8.11, numerical results for the temporal evolution of the

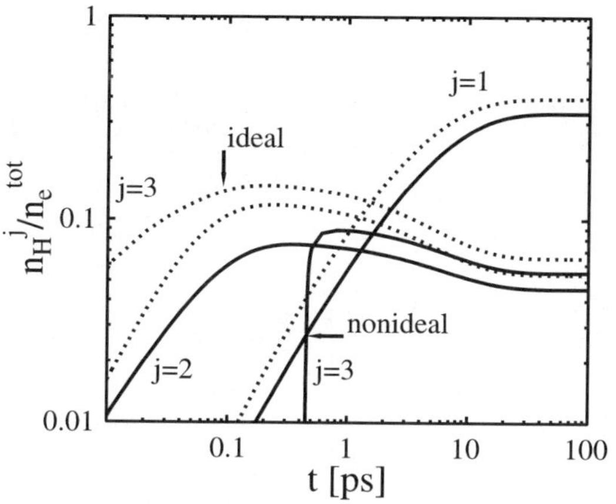

Fig. 8.11. Temporal evolution of the occupation of bound states (j denotes the principal quantum number) in an isothermal hydrogen plasma with $T = 3 \times 10^4$ K and $n_e^{tot} = 1.9 \times 10^{20}\,\mathrm{cm}^{-3}$. In the initial state, the plasma is assumed to be fully ionized

densities of atomic states in an isothermal hydrogen plasma are shown. The latter is assumed to be fully ionized in the initial state. The Planck–Larkin scheme is applied to avoid jumps at the Mott densities. In the first stage only the lowest two levels can exist. During recombination the free electron density decreases, and after some time the existence of the level with the principal quantum number $j = 3$ is possible according to the Mott criterion. For comparison, calculations were done with rate coefficients for the ideal plasma model accounting for the lowest three bound states (dotted lines). Nonideality effects lead to a lowering of the level population during the relaxation process. In the ideal as well as in the nonideal cases, the population is cascade like: the highest excited level is populated by the process of three-body recombination, the other levels mainly by deexcitation. After less than 1 *ps*, the excited levels are in partial equilibrium with the ground state, and finally, thermodynamic equilibrium is established.

8.2.2 Two-Temperature Plasmas

Equations for the Temperatures. A quite common situation during the relaxation of non-equilibrium plasmas is that each plasma component is described by its own local equilibrium distribution function characterized by a specific temperature. As the energy transfer between electrons and heavy particles (atoms, ions) is slow compared to other processes, the two-temperature model can be applied in many cases.

In this section, the theory developed so far is used to study density and temperature relaxation in nonideal partially ionized plasmas. Now in addition to the rate equations for the densities, the energy balance equations have to be taken into account. We use them in the quasiparticle approximation given by (7.153) starting from kinetic equations of the rather general form (7.20) with $Z(\boldsymbol{p}, \boldsymbol{R}t) = 1$. As for the calculations in Sect. 8.1, the energies and distribution functions in the collision integrals (7.123) and (7.131) are considered on the quasiparticle level (see Sect. 3.4.1).

We consider a spatially homogeneous plasma consisting of electrons, bare nuclei, and hydrogen-like ions. Using the V^S-approximation, the expression for total mean energy density was obtained to be (Ohde et al. 1996; Bornath 1998)

$$\mathcal{E}^{\text{tot}} = \sum_{a=e,Z,Z-1} \left(\frac{3}{2} n_a k_B T_a - \frac{n_a e_a^2}{4} \kappa \right) + \sum_j E^j_{Z-1} n^j_{Z-1} \, . \tag{8.52}$$

A nonideality correction is given by the second contribution on the r.h.s in terms of the inverse Debye screening length. At this point, we want to mention again the problem concerning the potential energy within the quasiparticle approximation (see also Sect. 6.1). For instance, for a fully ionized plasma consisting of electrons and singly charged ions, the nonideality correction in (8.52) is only the half of the correct classical value derived in Sect. 2.6. To reach a consistent description on that level, the kinetic theory should be considered within an extended quasiparticle approximation (see Sect. 3.4.1).

Using the quasiparticle approximation with (8.52), the following equations for the temperatures of electrons T_e and heavy particles T_h ($T_h = T_Z = T_{Z-1}$) were derived (Ohde et al. 1996),

$$\begin{aligned} \frac{\partial T_e}{\partial t} &= \frac{1}{k_1 k_4 - k_2 k_3} \Bigg\{ k_1 \Bigg[\sum_j \left(\frac{3}{2} k_B T_e + Y_1 + |E_j| \right) \frac{\partial n^j_{Z-1}}{\partial t} \\ &\quad + Z_{ei} + Z_{e(ei)} \Bigg] - k_3 \left[Y_2 \frac{\partial n_{Z-1}}{\partial t} - Z_{ei} - Z_{e(ei)} \right] \Bigg\} , \end{aligned} \tag{8.53}$$

$$\begin{aligned} \frac{\partial T_h}{\partial t} &= \frac{1}{k_1 k_4 - k_2 k_3} \Bigg\{ - k_2 \Bigg[\sum_j \left(\frac{3}{2} k_B T_e + Y_1 + |E_j| \right) \frac{\partial n^j_{Z-1}}{\partial t} \\ &\quad + Z_{ei} + Z_{e(ei)} \Bigg] + k_4 \left[Y_2 \frac{\partial n_{Z-1}}{\partial t} - Z_{ei} - Z_{e(ei)} \right] \Bigg\} . \end{aligned} \tag{8.54}$$

For the special case of a hydrogen plasma ($Z = 1$), the abbreviations are $Y_1 = -7\kappa e^2/8$ and $Y_2 = \kappa e^2/8$. Furthermore, we have

$$k_1 = \frac{3}{2} k_B (n_e + n_{(ei)}) + \frac{k_B}{32\pi\kappa} \kappa_i^4 , \qquad k_2 = \frac{k_B}{32\pi\kappa} \frac{T_h}{T_e} \kappa_e^2 \kappa_i^2$$

$$k_3 = \frac{k_B}{32\pi\kappa} \frac{T_e}{T_h} \kappa_e^2 \kappa_i^2 , \qquad k_4 = \frac{3}{2} n_e k_B + \frac{k_B}{32\pi\kappa} \kappa_e^4 \, .$$

Here, the inverse screening length is determined by

$$\kappa^2 = \sum_a \kappa_a^2 \qquad \kappa_a^2 = 4\pi \frac{e_a^2 n_a}{k_B T_a}, \quad a = e, i.$$

The abbreviations for the more general case of hydrogen-like plasmas are given in Bornath et al. (1998).

The equations (8.53) and (8.54) describe the two-temperature relaxation of a weakly coupled plasma accounting for the following contributions: (i) the energy loss and gain due to reaction processes determined by the rate equations (8.43) and (8.44), (ii) the energy exchange between the electrons and the heavy particles which is described here by the collision terms Z_{ei} and $Z_{e(ei)}$ of elastic scattering with screened interaction potentials, (iii) nonideality corrections on the level of the quasiparticle approximation.

For comparison, we give the corresponding balance equations for ideal plasmas, i.e., the nonideality corrections are neglected. We obtain

$$\frac{\partial T_e}{\partial t} = \frac{\sum_j \left(\frac{3}{2}k_B T_e + |E_j|\right) \frac{\partial n_{Z-1}^j}{\partial t} + Z_{ei} + Z_{e(ei)}}{\frac{3}{2} n_e k_B}, \tag{8.55}$$

$$\frac{\partial T_h}{\partial t} = \frac{Z_{ie} + Z_{(ei)e}}{\frac{3}{2}(n_Z + n_{Z-1})k_B}. \tag{8.56}$$

Detailed considerations concerning ionization and population kinetics for ideal plasmas are given, e.g., in Biberman et al. (1987).

Energy Transfer Rates. As mentioned above, the quantities Z_{eb} ($b = i, (ei)$) stand for the energy transfer rates due to elastic scattering between electrons and heavy particles ($Z_{eb} = -Z_{be}$). They are determined by the respective two- and three-particle collision integrals in the quantum kinetic equation (7.122). From the two-particle collision term (7.86), we get the T-matrix expression

$$\begin{aligned} Z_{ei} &= \frac{1}{\hbar V} \int \frac{d\boldsymbol{p}_e}{(2\pi\hbar)^3} \frac{d\boldsymbol{p}_i}{(2\pi\hbar)^3} \frac{d\bar{\boldsymbol{p}}_e}{(2\pi\hbar)^3} \frac{d\bar{\boldsymbol{p}}_i}{(2\pi\hbar)^3} \frac{p_e^2}{2m_e} 2\pi\delta(E_{ei} - \bar{E}_{ei}) \\ &\times |\langle \boldsymbol{p}_e \boldsymbol{p}_i | T_{ei} | \bar{\boldsymbol{p}}_i \bar{\boldsymbol{p}}_e \rangle|^2 \Big\{ \bar{f}_e \bar{f}_i (1 - f_e)(1 - f_i) - f_e f_i (1 - \bar{f}_e)(1 - \bar{f}_i) \Big\}. \end{aligned} \tag{8.57}$$

Similarly, $Z_{e(ei)}$ is expressed in terms of the three-particle T-matrix $T_{e(ei)}^{11}$. For non-degenerate plasmas and using local equilibrium distribution functions for all species, the expressions for the energy transfer rates can be simplified to give (Zhdanov 1982)

$$Z_{ab} = \frac{8 k_B n_a n_b \mu_{ab}}{m_a + m_b} \sqrt{\frac{2}{\pi \phi_{ab}}}\, Q_{ab} \left(T_b - T_a\right). \tag{8.58}$$

Here, μ_{ab} is the reduced mass, $\phi_{ab} = \phi_a \phi_b / (\phi_a + \phi_b)$, and $\phi_a = m_a / k_B T_a$. The quantity Q_{ab} can be written as

$$Q_{ab} = \int_0^\infty dz z^5 \exp(-z^2)\, Q_{ab}^T(z), \qquad z^2 = \phi_{ab}\hbar^2 k^2/2\mu_{ab}^2 \tag{8.59}$$

with Q_{ab}^T being the transport cross section. The latter is defined by (5.128) and can be calculated from the scattering phase shifts according to formula (5.135).

Energy transfer rates in T-matrix approximation were calculated, e.g., in Ohde et al. (1996) and Gericke et al. (2002). The phase shifts for the electron–ion scattering were determined by numerical solution of the radial Schrödinger equation with the statically screened Coulomb (Debye) potential. For the case of a hydrogen plasma, the elastic electron–atom scattering was treated on the basis of the close coupling equations of quantum scattering theory. Within perturbation theory and neglecting exchange effects, this scheme reduces to a Schrödinger equation which describes electron scattering in an effective atomic potential determined by a static and a polarization contribution (see Sect. 9.2.4). Let us focus on the rates Z_{ei} for the energy transfer between electrons and ions in a nonideal plasma. In particular, the behavior of the T-matrix approximation used here will be demonstrated by comparing the latter one with the results obtained in the framework of the well-known theories by Landau (1936) and Spitzer (1967). The simple formula due to Spitzer can be derived using the Fokker–Planck kinetic equation (Ichimaru 1992). We then have

$$Z_{ei} = \frac{3}{2} n_e k_B (T_i - T_e)/\tau_{ei}^{\mathrm{LS}}\,, \tag{8.60}$$

with τ_{ei}^{LS} being the respective relaxation time

$$\tau_{ei}^{\mathrm{LS}} = \frac{3 m_e m_i}{8\sqrt{2\pi}\, n_i Z_i^2 e^4 \ln \Lambda} \left(\frac{k_B T_e}{m_e} + \frac{k_B T_i}{m_i} \right)^{3/2} . \tag{8.61}$$

Here, the Coulomb logarithm is defined as $\ln \Lambda = \ln(r_D/r_{min})$ where $r_D = (k_B T_e/4\pi e^2 n_e)^{1/2}$ denotes the electron Debye length, $r_{min} = (\Lambda_e^2 + \rho_\perp^2)^{1/2}$ which is an interpolation between the de Broglie wavelength $\Lambda_e = \hbar/(2\pi m_e k_B T_e)^{1/2}$ and the distance of closest approach $\rho_\perp = Z_i e^2/k_B T_e$. Here, Z_i denotes the ion charge number. Obviously, the Landau–Spitzer approach fails for $\Lambda < 1$, i.e., it applies only to ideal or weakly coupled plasmas. This can be seen clearly in Fig. 8.12 where the electron–ion energy transfer rate is shown for a hydrogen plasma ($Z_i = 1$) as a function of the electron temperature for the case $T_i \ll T_e$. There is a good agreement between the T-matrix calculation and the Spitzer result at high temperatures whereas increasing deviations occur at lower temperatures where the plasma becomes strongly coupled.

There are several attempts to extend the Spitzer theory using an effective Coulomb logarithm which gives convergent results in the range of large

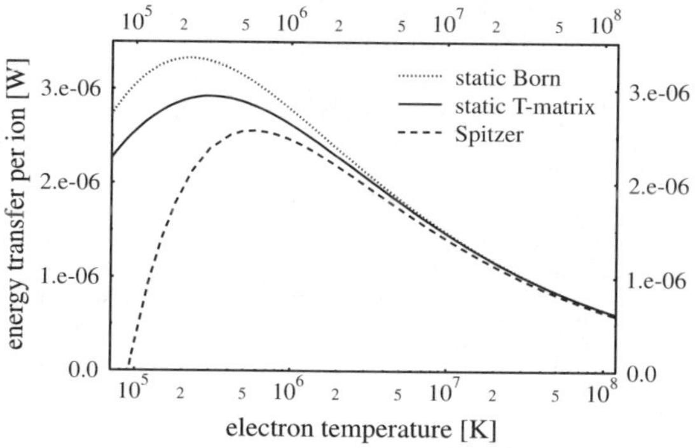

Fig. 8.12. Electron–ion transfer rate versus temperature for a hydrogen plasma with an electron density of $n_e = 1 \times 10^{22}\,\mathrm{cm}^{-3}$

coupling parameters. A brief discussion of such Spitzer-type theories is given in Gericke et al. (2002). Furthermore, degeneracy effects were taken into account (Brysk 1974) and numerical simulations were performed (Hansen and McDonald 1983; Reimann and Toepffer 1990). Based on quantum many-body theory, Dharma-wardana and Perrot (1998), Dharma-wardana and Perrot (2001) developed a model which allows to describe energy transfer for strong electron–ion coupling by coupled modes going beyond the Fermi golden rule approach. In Hazak et al. (2001), it was shown that, under certain circumstances, the Fermi golden role approach reduces to the Landau–Spitzer result, if weak electron–ion coupling is assumed.

Density–Temperature Relaxation. The balance equations (8.53) and (8.54) describe temperature relaxation in connection with population kinetics governed by the rate equations (8.43) and (8.44). On the other hand, the rate coefficients depend on the electron temperature. Besides the coupling which is due to reactions, there is also a coupling due to nonideality corrections. It should be mentioned again that many-particle effects are accounted for in simple and crude approximations assuming Debye screening in the ionization rates and in the nonideality corrections of the energy balance equations. Therefore, only a qualitative description of the behavior is possible. Improvements can be obtained using quantum kinetic equations for nonideal plasmas beyond the quasiparticle approximation. Such equations are given, e.g., by (7.122) including the retardation term and accounting for the potential energy in a proper way (see Sect. 7.6).

Let us discuss results obtained by a numerical solution of the coupled set of balance equations (8.43), (8.44), (8.53), and (8.54). In Fig. 8.13, density–temperature relaxation of a partially ionized hydrogen plasma is shown for three situations with different initial values for the electron temperature. The atoms are considered to be in the ground state, i.e., contributions of excited levels which are due to inelastic processes are neglected. Further details of the approach are given in Ohde et al. (1996). For comparison, results neglecting nonideality corrections (dotted curves) are shown. The three examples pre-

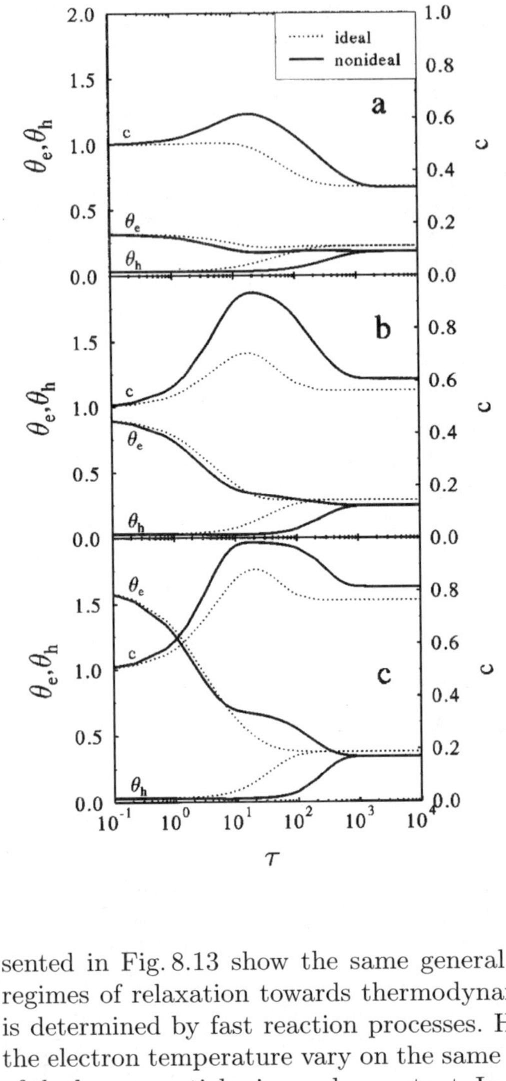

Fig. 8.13. Density and temperature relaxation of a nonideal hydrogen plasma (solid lines) for three initial values of the electron temperature $\theta_e = k_B T_e/|E_1|$ ($E_1 =$ 1 Ryd). The same initial values are chosen for the temperature of heavy particles $\theta_h = k_B T_h/|E_1|$ and for the degree of ionization $c = n_e/n_e^{tot}$. The total electron density is $n_e^{tot} = 10^{21}\,\mathrm{cm}^{-3}$, and for the unit of time we use $1\,\tau = 1.84 \times 10^{14}\,\mathrm{s}$. For comparison results are shown neglecting nonideality corrections (*dotted lines*)

sented in Fig. 8.13 show the same general behavior. There are mainly two regimes of relaxation towards thermodynamic equilibrium. The first regime is determined by fast reaction processes. Here, the degree of ionization and the electron temperature vary on the same time scale, while the temperature of the heavy particles is nearly constant. In the following regime, temperature equilibration occurs which is due to elastic scattering processes. Finally, the thermodynamic equilibrium state is established. The influence of nonideality effects leads to a reduction of the degree of ionization during relaxation which is a result of the lowering of the ionization energy.

The relaxation of a dense hydrogen-like carbon plasma accounting for the population of excited bound state levels is considered in Fig. 8.14. For simplicity, the energy transfer rates were calculated here in Born approximation using the Debye potential. The initial temperature of heavy particles was cho-

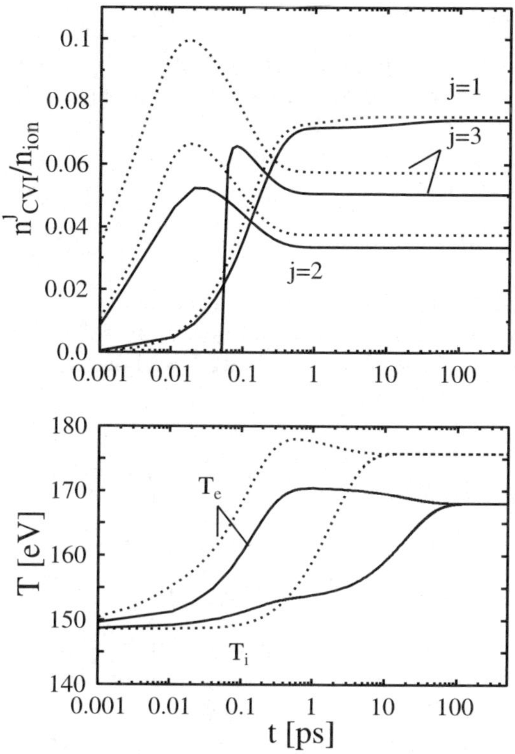

Fig. 8.14. Occupation numbers and temperatures of electrons and heavy particles versus time for a nonideal hydrogen-like carbon plasma. The initial state is a fully ionized plasma with $n_e^{tot} = 3.5 \times 10^{23}\,\mathrm{cm}^{-3}$, $T_e(0) = 150\,\mathrm{eV}$, and $T_i(0) = 149\,\mathrm{eV}$. For comparison, the ideal case is shown by *dotted lines*

sen to be $T_i = 149\,\mathrm{eV}$, i.e., nearly equal to that of the electrons. In Fig. 8.14, the temporal evolution of the occupation numbers of the hydrogen-like ions and that of the temperatures is shown. The energy release caused by the recombination leads to a sharp increase of the electron temperature in the beginning of the relaxation. Together with the decrease of the free electron density, this results in a weakening of the screening and in the occurrence of transient effects. After less than 0.1 ps, the third energy level is allowed to exist according to the Mott criterion. After 1 ps the chemical relaxation process is almost finished, whereas the equilibration of the temperatures needs a time considerably longer.

The initial electron temperature chosen in Fig. 8.14 is a typical temperature for laser-produced plasmas (Theobald et al. 1996). Usually, the initial temperature of the ions is lower, and is difficult to be estimated in experiments. That is why different situations are considered in model calculations (Bornath et al. 1998). On the other hand, there are non-equilibrium situations in shock experiments with electron temperatures less than the ion temperature (Celliers et al. 1992; Ng et al. 1995).

8.2.3 Adiabatically Expanding Plasmas

In many experiments, especially with laser produced plasmas, there are conditions closer to adiabatic ones. A simple model for the relaxation of a hydrogen-like plasma under such conditions is presented in this section. For simplicity, the case of a common temperature for the species is considered, and the plasma is assumed to be spatially homogeneous. A plasma under such conditions is characterized by the relation

$$dQ = T\,dS = dU + p\,dV = 0\,. \tag{8.62}$$

Since we are going to consider the range of weakly coupled plasmas, nonideality corrections in the pressure and in the internal energy are taken into account in lowest order on the level of the Debye limiting law. For the temperature equation, we get (Bornath et al. 1998)

$$\begin{aligned}\frac{\partial T}{\partial t} &= \sum_j \left\{ \frac{\frac{3}{2}k_B T - \frac{3}{4}\kappa e^2[1+Z^2-(Z-1)^2] + |E_j|}{\frac{3}{2}n_\sigma k_B + \frac{\kappa e^2}{4T}[n_e + Z^2 n_Z + (Z-1)^2 n_{Z-1}]} \right\} \frac{\partial n^j_{Z-1}}{\partial t} \\ &\quad - \left\{ \frac{p/V}{\frac{3}{2}n_\sigma k_B + \frac{\kappa e^2}{4T}[n_e + Z^2 n_Z + (Z-1)^2 n_{Z-1}]} \right\} \frac{\partial V}{\partial t}\,, \end{aligned} \tag{8.63}$$

with $n_\sigma = n_e + n_Z + n_{Z-1}$. In comparison with the ideal gas approximation, the relaxation of the temperature is changed by the inclusion of reactions and by nonideality effects. Equation (8.63) has to be solved together with the rate equations (8.43), (8.44), where nonideality effects are also taken into account.

In the following, we investigate an adiabatically expanding carbon plasma which is assumed to be fully ionized in the initial state. In particular, we apply our model to describe results of temporally resolved measurements of the electron density of dense plasmas (Theobald et al. 1996; Theobald et al. 1999). In the experiments, an intense sub-picosecond laser pulse creates a dense plasma in a thin Lexan-foil which is then irradiated by temporally correlated high harmonics. The basic idea is to deduce the electron density from the transmitted high harmonics accounting for the fact that light can only propagate in the plasma with electron densities $n_e < n_c$ where $n_c = \pi m_e c^2/(e^2\lambda^2)$ is the critical density.

In the calculations presented, the parameters given by the experimental conditions (Theobald et al. 1997) were used: total electron density $n_e^{tot} = 4 \times 10^{23}\,\mathrm{cm}^{-3}$, initial temperature $T(0) = 150\,\mathrm{eV}$, and thickness of the irradiated Lexan-foil $d_0 = 70\,\mathrm{nm}$. A one-dimensional expansion was considered assuming a constant expansion velocity u_0 equal to the ion sound velocity $u_0 = u_{ion} = (<Z_i> T_e(0)k_B/m_i)^{1/2}$. In the case considered, we have $u_0 = 8.5 \times 10^6\,\mathrm{cm/s}$, and one can write $\frac{1}{V}\frac{\partial}{\partial t}V = u_0/d$ with $d = d_0 + u_0\,t$. Results for the temporal evolution of the temperature and the free electron

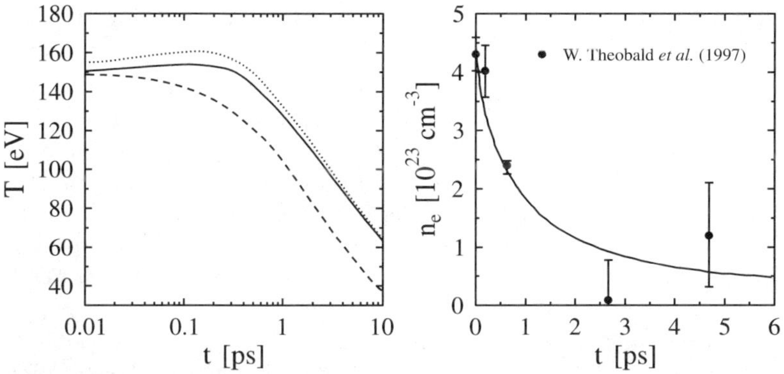

Fig. 8.15. Temporal evolution of an adiabatically expanding hydrogen-like carbon plasma with total electron density $n_e^{tot} = 4.3 \times 10^{23}\,\text{cm}^{-3}$. The initial conditions and parameters are given in the text. Temperature vs. time (**left**) is given for the nonideal plasma according to (8.63) (*solid line*). For comparison, results are given for an ideal plasma (*dotted*), and for an ideal plasma without reactions (*dashed*). The **right part** of the figure shows the free electron density versus time as compared to data from measurements (Theobald et al. 1997)

density are shown in Fig. 8.15. For the purpose of comparison, there are also shown results for the temperature for an ideal recombining plasma and for an ideal plasma without reactions. In the beginning, the temperature increases slightly due to the energy released in the recombination processes. The nonideality corrections in (8.63) have two effects: First, due to the lowering of the continuum edge, the heating by recombination is smaller than in the ideal case. Second, because the pressure is lower than the ideal one, the temperature decrease is less steep in the nonideal case. The free electron density as a function of time is shown in the right part of Fig. 8.15, where also a comparison with experimental data obtained by Theobald et al. (1997) is given.

Results for the population kinetics show transient effects due to screening, and a strong change in the composition in favor of the ground state population of hydrogen-like ions determined by the expansion with decreasing temperature and density (Bornath et al. 1998). It turns out that after 1 ps the level population has reached local thermodynamic equilibrium (LTE) conditions and can be calculated from nonideal Saha equations with parametric time dependence. For further investigations concerning optical properties of laser produced plasmas, we refer, e.g., to Theobald et al. (1999), Fehr et al. (1999), and to Röpke and Wierling (1998).

It should be mentioned again that a rather simple model was used here to describe a laser produced plasma in the expansion and recombination stage. Improvements are necessary in many directions. Simulations performed with the one-dimensional hydro-code MEDUSA showed for instance deviations from the adiabatic behavior of the expansion for longer delay times (Kingham et al. 1999).

8.3 Quantum Kinetic Theory of the Stopping Power

Beam-matter interaction is one of the key tools for creating, heating, and diagnosing dense plasmas (see, e.g., Golubev et al. (1998), Tahir et al. (2003)). In this connection, the stopping power is a central quantity for the characterization of the energy deposition by means of particle beams into a plasma. Especially, for investigations concerning topics related to inertial confinement fusion, the stopping power is being studied intensely (see Fusion-Symposia). Electron cooling of heavy ion beams is an other important application (Wolf et al. 1994; Winkler et al. 1997).

From the theoretical point of view, there exist several schemes to determine the stopping power. A standard approach is based on the Bethe stopping power expression and the corresponding generalizations (Bethe 1930; Bloch 1933; Deutsch et al. 1989; Ziegler 1999). Another widely used scheme to study the energy deposition by particle beams into plasmas applies the dielectric formalism (Lindhard 1954; Ichimaru 1992; Peter and Meyer-ter-Vehn 1991a; Boine-Frankenheim 1996). Correlation effects were considered using perturbation theory (Ashley et al. 1972; Pitarke et al. 1993), density functional theory (Nagy et al. 1989; Zaremba et al. 1995) and kinetic models (Sigmund 1982; Li and Petrasso 1993). A stopping power expression in terms of the force autocorrelation function can be found in a paper by Dufty and Berkovsky (1995). Simulation techniques such as molecular dynamics (MD) and particle in cell (PIC) simulations were developed to model the stopping power of plasmas for large coupling parameters (Zwicknagel et al. 1999). Magnetized plasmas were considered by Ortner et al. (2001), and some model plasmas were considered by Tkachenko et al. (2002).

A rigorous approach to investigate the stopping power of dense nonideal plasmas is quantum kinetic theory. Here, the beam-plasma interaction processes can be treated by appropriate collision terms (Kraeft and Strege 1988; Gericke et al. 1996; Gericke et al. 2002). The aim of this section is to present results for the stopping power using the quantum kinetic equations derived in Chap. 7.

8.3.1 Expressions for the Stopping Power of Fully Ionized Plasmas

The energy deposition by a particle beam into a plasma is usually described by the change of the kinetic energy of the projectile particles. The corresponding rate is referred to as stopping power. In the framework of kinetic theory, it is defined as

$$\frac{\partial}{\partial t}\langle E\rangle = \frac{1}{n_{\mathrm{b}}}\int \frac{d\boldsymbol{p}_{\mathrm{b}}}{(2\pi\hbar)^3}\,\frac{p_{\mathrm{b}}^2}{2m_{\mathrm{b}}}\,\frac{\partial}{\partial t} f_{\mathrm{b}}(\boldsymbol{p}_{\mathrm{b}},t)\;, \tag{8.64}$$

where the index b labels the beam particle quantities: n_{b} is the density, m_{b} the mass, and f_{b} is the distribution function. In many other cases, the force in direction of the beam particle velocity (x-direction) is called stopping power, too. With the latter definition we have

$$\frac{\partial}{\partial x}\langle E\rangle = \frac{1}{n_{\rm b}}\int \frac{d\boldsymbol{p}_{\rm b}}{(2\pi\hbar)^3}\,\frac{\boldsymbol{p}_{\rm b}\cdot\boldsymbol{v}}{v}\,\frac{\partial}{\partial t} f_{\rm b}(\boldsymbol{p}_{\rm b},t)\,, \tag{8.65}$$

which describes the change of kinetic energy per unit length. To calculate these quantities from their definitions (8.64) and (8.65), the beam particle distribution function $f_{\rm b}(\boldsymbol{p},t)$ has to be determined from an appropriate kinetic equation. In Chap. 7, different approximation schemes were applied to determine such equations for quantum plasmas. First, we will use kinetic equations for fully ionized plasmas in the Markovian form neglecting retardation effects. For weakly interacting particles, the RPA may be applied leading to the Lenard–Balescu kinetic equation (7.43). On the other hand, the quantum Boltzmann collision integral (7.86) follows in ladder approximation which allows for the inclusion of strong binary correlations. As the collision terms in the kinetic equations include the distribution functions, one has, in general, to solve a system of coupled kinetic equations for the beam and plasma particles. For a spatially homogeneous target plasma without external fields, this system can be written as

$$\frac{\partial}{\partial t} f_{\rm b}(\boldsymbol{p}_{\rm b},t) \;=\; \sum_c I_{{\rm b}c}(\boldsymbol{p}_{\rm b},t) \tag{8.66}$$

$$\frac{\partial}{\partial t} f_a(\boldsymbol{p}_a,t) \;=\; \sum_c I_{ac}(\boldsymbol{p}_a,t) \qquad (a=e,i)\,. \tag{8.67}$$

In the following, the indication of the spin sum is dropped for simplicity. The collision terms on the r.h.s. describe the different interaction processes where the sum runs over the beam and plasma particles ($c = {\rm b}, e, i$). The beam particles are characterized by a fixed charge number $Z_{\rm b}$, i.e., its change due to ionization and capture processes in the target plasma is not considered in our approach.

To simplify the problem, we make the following assumptions: (i) we consider a short initial time interval in which the target plasma is assumed to be in equilibrium, and the beam distribution does not change; (ii) the density of the beam should be small enough such that interactions between the beam particles can be neglected; (iii) furthermore, we assume a sharply peaked beam particle distribution function in the collision integral

$$f_{\rm b}(\boldsymbol{p}_{\rm b},t) = (2\pi\hbar)^3 n_{\rm b}\delta\left(\boldsymbol{p}_{\rm b} - m_{\rm b}\boldsymbol{v}(t)\right)\,, \tag{8.68}$$

where $\boldsymbol{v}$ is the velocity of the beam particles.

Inserting (8.66) into (8.64) and (8.65), one gets closed expressions for the stopping power determined by the collision terms $I_{{\rm b}c}$ where the beam particles are described by (8.68), and the plasma particles by equilibrium distribution functions. As mentioned above, kinetic equations in different approximations were considered in detail in Chap. 7. In order to derive explicit expressions for the stopping power, let us start here from a kinetic equation in the more general form (7.20). For the spatially homogeneous case and for single particle energies $E = p^2/2m$, we have

$$\begin{aligned} \frac{\partial}{\partial t} f_{\mathrm{b}}(\boldsymbol{p}_{\mathrm{b}}, t) &= -f_{\mathrm{b}}(\boldsymbol{p}_{\mathrm{b}}, t)\, i\Sigma_{\mathrm{b}}^{>}(\boldsymbol{p}_{\mathrm{b}}, \omega, t)|_{\omega=p^2/2m_{\mathrm{b}}} \\ &\quad \pm [1 \pm f_{\mathrm{b}}(\boldsymbol{p}_{\mathrm{b}}, t)]\, i\Sigma_{\mathrm{b}}^{<}(\boldsymbol{p}_{\mathrm{b}}, \omega, t)|_{\omega=p^2/2m_{\mathrm{b}}} . \end{aligned} \tag{8.69}$$

Let us first determine the stopping power in RPA what corresponds to a dynamically screened Born approximation on the level of the Lenard–Balescu kinetic equation. In this approximation, the self-energy functions are given by the expressions (4.245). For our purpose, we can there apply the spectral representation (4.64) valid for equilibrium situations. Then, we find from (8.64), neglecting terms proportional to the square of the beam-particle density,

$$\begin{aligned} \frac{\partial}{\partial t}\langle E\rangle &= \frac{2}{\hbar}\int \frac{d\boldsymbol{p}\, d\bar{\boldsymbol{p}}}{(2\pi\hbar)^6}\frac{p^2}{2m_{\mathrm{b}}} V_{\mathrm{bb}}(\boldsymbol{p}-\bar{\boldsymbol{p}})\, \mathrm{Im}\varepsilon^{-1}(\boldsymbol{p}-\bar{\boldsymbol{p}}, E_{\mathrm{b}}-\bar{E}_{\mathrm{b}}) \\ &\quad \times \left\{ n_B(E_{\mathrm{b}}-\bar{E}_{\mathrm{b}})\, f_{\mathrm{b}}(\bar{\boldsymbol{p}}) - \left[1 + n_B(E_{\mathrm{b}}-\bar{E}_{\mathrm{b}})\right] f_{\mathrm{b}}(\boldsymbol{p}) \right\} , \end{aligned} \tag{8.70}$$

where $V_{\mathrm{bb}}(q) = 4\pi\hbar^2 Z_{\mathrm{b}}^2 e^2/q^2$ and $\mathrm{n_B}(\omega) = [\exp(\hbar\omega/k_{\mathrm{B}}T) - 1]^{-1}$. Using (8.68), we get after a straightforward calculation (Kraeft and Strege 1988; Gericke et al. 1996)

$$\frac{\partial}{\partial t}\langle E\rangle = \frac{2\, Z_{\mathrm{b}}^2 e^2}{\pi v}\int_0^\infty \frac{dk}{k}\int_{\frac{\hbar k^2}{2m_{\mathrm{b}}}-kv}^{\frac{\hbar k^2}{2m_{\mathrm{b}}}+kv} d\omega\, \omega\, \mathrm{Im}\varepsilon^{-1}(k,\omega)\, \mathrm{n_B}(\omega) . \tag{8.71}$$

For the stopping power defined by (8.65), a similar calculation leads to (Gericke and Schlanges 1999)

$$\frac{\partial}{\partial x}\langle E\rangle = \frac{2\, Z_{\mathrm{b}}^2 e^2}{\pi v^2}\int_0^\infty \frac{dk}{k}\int_{\frac{\hbar k^2}{2m_{\mathrm{b}}}-kv}^{\frac{\hbar k^2}{2m_{\mathrm{b}}}+kv} d\omega \left[\omega - \frac{\hbar k^2}{2\, m_{\mathrm{b}}}\right] \mathrm{Im}\varepsilon^{-1}(k,\omega)\, \mathrm{n_B}(\omega) . \tag{8.72}$$

Expressions of this type were also obtained by Arista and Brandt (1981) and by Morawetz and Röpke (1996). Both formulae are given in terms of the imaginary part of the inverse dielectric function which can be calculated for arbitrary degeneracy, and by the Bose function $\mathrm{n_B}(\omega)$. In a more general treatment, the dielectric function beyond the RPA scheme can be used including local field corrections (Yan et al. 1985; Deutsch and Maynard

2000). It should be noticed that formulae (8.71) and (8.72) represent quantum statistical generalizations of results derived in the framework of classical dielectric theory. Indeed, if we take the high temperature expansion of the Bose function $n_B(\omega)$, neglect the terms $\hbar k^2/2m_b$ in (8.71) and (8.72), and use the dielectric function in the classical limit, we get

$$\frac{\partial}{\partial x}\langle E\rangle = -\frac{Z_b^2 e^2}{\pi v^2}\int_0^{k_{\max}}\frac{dk}{k}\int_{-kv}^{kv} d\omega\,\omega\,\mathrm{Im}\varepsilon^{-1}(k,\omega)\,. \qquad (8.73)$$

In this limit, we can write $\partial\langle E\rangle/\partial t = v\partial\langle E\rangle/\partial x$. The result (8.73) shows the well-known problem of divergencies for large k-values that occurs in the classical theory. This problem is usually solved by introducing an upper integration limit at $k = k_{\max}$ (Ichimaru 1992) (see also Sect. 2.5). In contrast, convergency is achieved in the quantum statistical expressions given above, i.e., cutoff procedures are avoided automatically. Modified expressions were used in investigations to consider magnetized plasmas (Seele et al. 1998; Nerisisyan 1998; Ortner et al. 2001) and the stopping of clusters (Deutsch 1995; Zwicknagel and Deutsch 1997).

The formulas (8.71) and (8.72) include dynamical screening by the RPA dielectric function. In this way, the influence of collective effects on the stopping power is described. However, the RPA scheme can only be applied to weakly interacting particles, i.e., strong correlations such as multiple scattering are not taken into account. Such effects can be described in the framework of the ladder approximation leading to the quantum Boltzmann kinetic equation. Assuming static screening of the long range Coulomb forces, this scheme allows to include strong binary correlations by a T-matrix treatment with the Debye potential. The Boltzmann collision integral (7.86) follows if the self-energies $\Sigma_b^{\gtrless}$ are used in ladder approximation (7.80). For the further treatment, we apply the ladder approximation to non-degenerate target plasmas. Inserting the Boltzmann collision term 7.86 into (8.64), we then obtain

$$\begin{aligned}\frac{\partial}{\partial t}\langle E\rangle &= \frac{1}{V\hbar}\sum_c\int\frac{d\boldsymbol{p}'\,d\boldsymbol{p}''\,d\bar{\boldsymbol{p}}'\,d\bar{\boldsymbol{p}}''}{(2\pi\hbar)^{12}}\,\frac{p'^2}{2m_b}\,\left|\langle\boldsymbol{p}'\boldsymbol{p}''|\,T^R_{bc}\,|\bar{\boldsymbol{p}}''\bar{\boldsymbol{p}}'\rangle\right|^2\\ &\quad\times 2\pi\delta(E_{bc}-\bar{E}_{bc})\left\{f_b(\bar{\boldsymbol{p}}')f_c(\bar{\boldsymbol{p}}'')-f_b(\boldsymbol{p}')f_c(\boldsymbol{p}'')\right\}\,. \qquad (8.74)\end{aligned}$$

It should be noticed again that b labels the beam particles, while c runs here over the different target species only. According to our assumptions, f_b is given by (8.68), and the plasma particles are described by Maxwellian distribution functions. Using relative and center of mass variables, some integrations can easily be performed. Introducing the transport cross section $Q^{(1)} = Q^T$ defined by (5.128), we find in binary collision approximation (Kraeft and Strege 1988; Gericke et al. 1996)

$$\frac{\partial}{\partial t}\langle E\rangle = \frac{1}{(2\pi)^2\hbar^3}\sum_c \frac{m_c^2}{m_{\rm bc}^3 m_{\rm b}}\frac{n_c\Lambda_c^3}{v}k_{\rm B}T\int\limits_0^\infty dp\,p^3\,Q_{\rm bc}^T(p)$$
$$\times\left\{k_-\exp\left(-\frac{m_c v_-^2}{2k_BT}\right)-k_+\exp\left(-\frac{m_c v_+^2}{2k_BT}\right)\right\}. \tag{8.75}$$

For the stopping power defined by (8.65), a similar calculation gives the following formula (Gericke and Schlanges 1999).

$$\frac{\partial}{\partial x}\langle E\rangle = -\frac{1}{(2\pi)^2\hbar^3}\sum_c \frac{m_c^2}{m_{\rm bc}^3}\frac{n_c\Lambda_c^3}{v}k_{\rm B}T\int\limits_0^\infty dp\,p^3\,Q_{\rm bc}^T(p)$$
$$\times\left\{p_-\exp\left(-\frac{m_c v_-^2}{2k_BT}\right)-p_+\exp\left(-\frac{m_c v_+^2}{2k_BT}\right)\right\}. \tag{8.76}$$

Here, $m_{\rm bc}$ is the reduced mass, and $\Lambda_c=(2\pi\hbar^2/(m_ck_BT)^{1/2}$ is the thermal wave length. Furthermore, we introduced the abbreviations $k_\pm = p\pm m_{\rm b}v + m_{\rm b}m_{\rm bc}k_BT/m_cp$, $p_\pm = 1\pm(m_{\rm bc}k_{\rm B}T)/(m_cp\,v)$, and $v_\pm = p/m_{\rm bc}\pm v$. The transport cross section

$$Q_{\rm bc}^T(p) = 2\pi\int_{-1}^{1} d\cos\theta(1-\cos\theta)\frac{d\sigma_{\rm bc}}{d\Omega}$$

is the central quantity in (8.75) and (8.76). It is connected with the T-matrix by the known relation (5.126).

The stopping power on the level of a quantum Landau equation follows using the cross section in statically screened Born approximation . Cross sections of two-particle scattering processes in plasmas were discussed in Sect. 5.6.

While the RPA accounts for dynamical screening, but is valid only in the weak coupling limit, the binary collision approximation accounts for multiple scattering contributions represented by higher order ladder terms of the T-matrix. An important ingredient of the RPA expressions (8.71) and (8.72) is the sharply peaked inverse dielectric function which describes collective excitations (plasmons). This contribution is not included in the T-matrix expressions (8.75)and (8.76), if the statically screened Coulomb potential is used. To include both dynamic screening effects and strong binary correlations, an ansatz according to Gould and DeWitt (1967) can be used. For the stopping power, this scheme reads (Gericke et al. 1996; Morawetz and Röpke 1996)

$$\frac{\partial}{\partial x}\langle E\rangle = \frac{\partial}{\partial x}\langle E\rangle^{\rm static}_{T\text{-matrix}} + \frac{\partial}{\partial x}\langle E\rangle^{\rm dynamic}_{\rm Born} - \frac{\partial}{\partial x}\langle E\rangle^{\rm static}_{\rm Born}\,. \tag{8.77}$$

In this combined model, the stopping power is given by the sum of the T-matrix expression (8.76) and of the dynamic RPA formula (8.72). The statically screened Born approximation has to be subtracted to avoid double counting. Consequently, we get a quantum T-matrix approach for the stopping power with a dynamically screened first Born approximation and statically screened higher order ladder terms. It should be noticed that screening is described in linear approximation which corresponds, in the classical case, to the level of the linearized Vlasov equation. However, in contrast to the linear response scheme, strong binary correlations are taken into account in the T-matrix contributions which go beyond the Born approximation. In this way, nonlinear coupling effects are included in the stopping power.

Let us now discuss the asymptotic behavior of the stopping power for large beam velocities. This limit is of special importance because in many beam-particle matter experiments high beam energies are used. We will restrict ourselves to results for the stopping power for ion beams in an electron gas. From the RPA expression (8.72), we get the well-known Bethe-type asymptotic expression (Brouwer et al. 1990)

$$\lim_{v\to\infty} \frac{\partial}{\partial x}\langle E\rangle = -\frac{Z_{\mathrm{b}}^2 e^2 \omega_{pl}^2}{v^2} \ln\left(\frac{2m_{\mathrm{be}}v^2}{\hbar\,\omega_{\mathrm{pl}}}\right), \tag{8.78}$$

with $\omega_{pl} = \omega_e = (4\pi e^2 n_e/m_e)^{1/2}$ being the electron plasma frequency. Like in the classical case, we find the relation $\partial\langle E\rangle/\partial t = v\partial\langle E\rangle/\partial x$. Many investigations have shown that formula (8.78) can be successfully used to describe experimental results for the stopping of ion beams at high velocities (see, e.g., Hoffmann et al. (1988), Jacobi et al. (1995)). It is interesting to note that the asymptotic formula following from the statically screened T-matrix expression (8.76) is smaller than (8.78) by a factor of 2 (Gericke et al. 1996). The reason is that the contribution of collective excitations is missing in the case of static screening, i.e., only one of the two elementary excitations is accounted for.

The combined model (8.77) leads to the same asymptotic behavior as given by (8.78) because the statically screened T-matrix and Born terms coincide at high beam velocities and, therefore, they cancel each other.

In the low velocity limit, the rate of energy transfer $\partial\langle E\rangle/\partial t$ is finite and positive at $v \to 0$ (Kraeft and Strege 1988; Gericke et al. 1996; Gericke et al. 1997). Here, an energy gain of the beam particles or a cooling of the target plasma is described. For the change of energy per unit length, we have in both cases $\lim_{v\to 0}\frac{\partial}{\partial x}\langle E\rangle \sim v$ (Gericke et al. 2001).

8.3.2 T-Matrix Approximation and Dynamical Screening

Let us now discuss numerical results for the stopping power of dense plasmas which were obtained from the different approximation schemes given in the previous section. In most cases, the plasma ions give small contributions to the stopping power; therefore, the free plasma electrons are considered to be the only target species.

In dynamically screened Born approximation, the stopping power was calculated from (8.71) and (8.72) using the dielectric function in RPA given by (4.109). To carry out the integration over ω, a sum rule was applied to include the contribution of the plasmon peak (Brouwer et al. 1990; Gericke 2000). The stopping power in T-matrix approximation was calculated from (8.76) and from the corresponding formula for $\partial\langle E\rangle/\partial t$ using the scattering phase shift representation (5.135) of the transport cross section. In statically screened Born approximation, the cross section given by (5.130) was used.

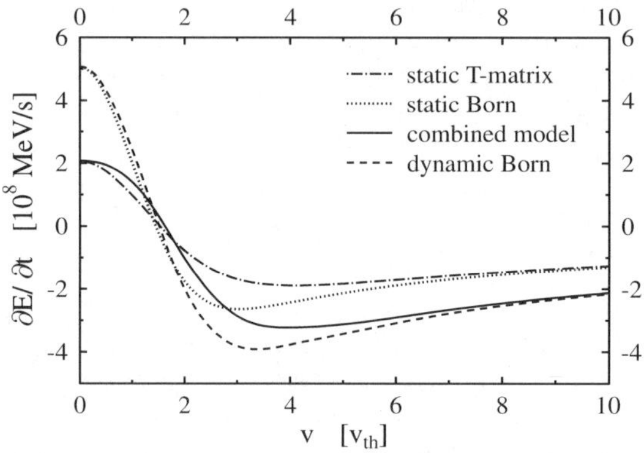

Fig. 8.16. Energy transfer of an electron beam into an electron gas in different approximations versus beam velocity ($v_{th} = \sqrt{k_B T/m_e}$). An electron gas target is considered with a temperature of $T = 3.75 \times 10^4$ K and a number density of $n_e = 5 \times 10^{19}\,\mathrm{cm}^{-3}$

In Fig. 8.16, results for the change of energy per unit time are shown for an electron beam. An energy gain is described at low beam velocities as long as the beam particle energy is smaller than the thermal energy of the electron gas. After passing zero, the curves show a maximum energy loss, and at high velocities, they approach the asymptotic behavior $\partial\langle E\rangle/\partial t \sim \ln v^2/v$. For low beam velocities, the Born approximations overestimate the change of kinetic energy of the beam particles, and dynamic screening effects are small for the plasma parameters considered. The inclusion of multiple scattering by the T-matrix reduces the stopping power for low and intermediate beam velocities.

The same qualitative behavior can be found for an ion beam travelling through an electron gas. However, the change of sign is located at much lower beam velocities, and the energy gain in the low velocity range is much smaller than that for an electron beam. Furthermore, the maximum energy loss is higher for heavy beam particles.

Results for the energy loss of a proton beam are shown in Fig. 8.17. All curves which correspond to different approximations show the same general behavior: At low velocities, there is a linear increase, and after passing the maximum energy loss, we observe a decrease according to $\partial\langle E\rangle/\partial x \sim \ln v^2/v^2$ for high beam velocities. Strong binary correlations, represented by the T-matrix approximation, reduce the stopping power. Furthermore, the statically and dynamically screened Born approximations coincide at low beam

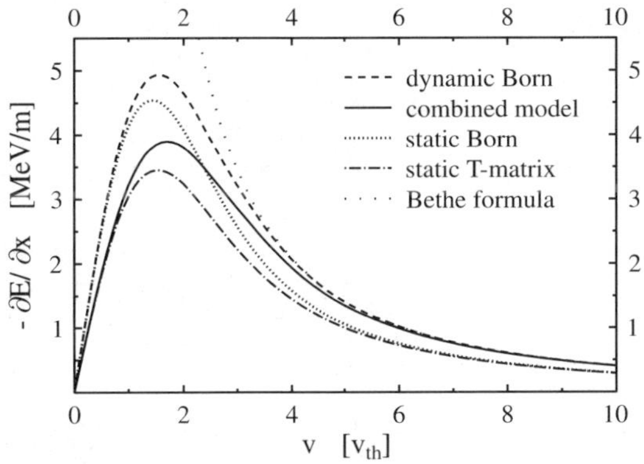

Fig. 8.17. Energy loss of a proton beam in an electron gas versus beam velocity ($v_{th} = \sqrt{k_B T/m_e}$). The temperature of the electron gas target is $T = 3.75 \times 10^4$ K, and the densities are $n_e = 5 \times 10^{17}\,\text{cm}^{-3}$

velocities. Here, the static T-matrix approximation dominates in the combined model. On the other hand, the weak coupling approximation is sufficient for high beam velocities. In this case, the behavior of the combined model is determined by the RPA asymptotic result given by the Bethe formula (8.78). Consequently, the combined model interpolates between the two limiting cases: the statically screened T-matrix approximation for low beam velocities, and the dynamic RPA for high velocities.

8.3.3 Strong Beam–Plasma Correlations. Z Dependence

Strong beam–plasma correlations are of special importance if the energy transfer from highly charged ions to a dense plasma is investigated. Here, the dependence of the stopping power on the beam ion charge number Z_b is an essential point to understand the beam-plasma interaction processes. In the case of slow ions, the coupling strength between the beam ions and the target plasma can be described by the parameter $Z_\text{b}\Gamma^{3/2}$ with Γ being the classical nonideality parameter of the plasma electrons.

Standard approaches to calculate the stopping power for plasmas, e.g., the dielectric (linear response) formalism and the simple Bethe type formula (8.78) predict an increase of the stopping power according to a Z_b^2 scaling law. The Born approximations discussed in the previous sections show this behavior, too. The reason is that these theories consider the beam–plasma interaction in the weak coupling limit which gives correct results only for sufficiently low ion charges and/or high beam velocities, and for high temperature plasmas. This behavior changes for situations where the coupling between the beam particles and the target plasma is strong. Deviations from the Z_b^2 scaling were first found experimentally by Barkas et al. (1956) and Barkas et al. (1963). Further investigations were made, e.g., in stopping power measurements by Andersen (a) et al. (1977) and Andersen (b) et al. (1977).

There exist several theoretical approaches to interpret these results (see, e.g., Ziegler (1999), Zwicknagel et al. (1999)). In this section, the quantum kinetic theory of the stopping power is used to consider the problem. It was found that the T-matrix expression (8.76) accounts for strong binary correlations between beam and plasma particles due to the higher order ladder terms beyond the Born approximation. For strong beam–plasma coupling, the inclusion of these terms leads to deviations from the Z_b^2 dependence. In this connection, it should be mentioned that we used the Debye potential in the calculation of the higher order ladder contributions. That means, the deviations from the Z_b^2 scaling obtained in our approach are not due to nonlinear screening but an effect of multiple scattering introduced by the T-matrix and, therefore, an effect of a more exact treatment of the two-particle scattering as compared to the Born approximation.

Let us consider, in more detail, the Z_b-dependence for low velocities where strong correlations are most important. For this purpose, the normalized stopping power $\partial\langle E\rangle/\partial x \,/\, Z_\mathrm{b}^2$ versus ion charge number is shown in Fig. 8.18 for a beam velocity $v = 0.2 \cdot v_{th}$. One observes again that the higher order ladder terms accounted for in the T-matrix approach cause a reduction of the stopping power. This effect increases with increasing coupling strength, i.e., for higher ion charge numbers and higher plasma densities. In order to give an estimate of strong beam–plasma coupling effects, let us consider the exponent of a Z_b^γ scaling ($\partial\langle E\rangle/\partial x|_{Z_\mathrm{b}} = Z_\mathrm{b}^\gamma \cdot \partial\langle E\rangle/\partial x|_{Z_\mathrm{b}=1}$) valid for the low velocity range. For very high plasma temperatures and small Z_b, the scaling is close to Z_b^2. But if the beam–plasma coupling increases, significant deviations from this scaling can be found (Gericke and Schlanges 1999). In fact, the scaling exponent varies in the range of $\gamma = 1.4 \ldots 2$. It is larger for higher temperatures and decreases with increasing plasma density and beam charge number. A considerable reduction of the Z_b scaling was also found in MD and PIC simulations (Zwicknagel et al. 1999; Zwicknagel et al. 1993; Zwicknagel et al. 1996) and solutions of the nonlinear Vlasov–Poisson equations (Peter

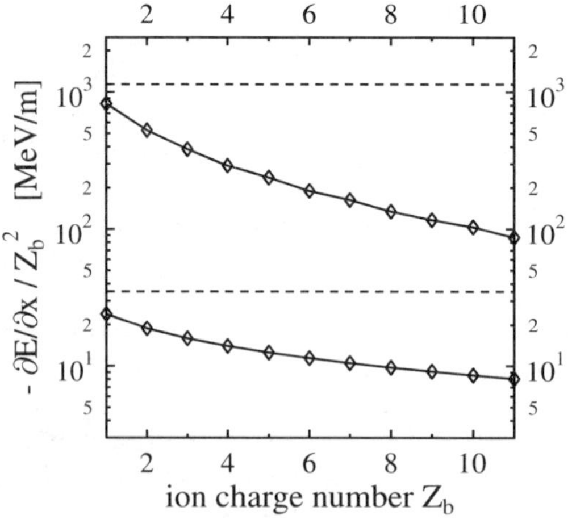

Fig. 8.18. Normalized stopping power for an ion beam versus ion charge number in Born (*dashed*) and T-matrix approximations (*solid*). An electron gas target is considered with a temperature $T = 5 \times 10^4\,\mathrm{K}$. The densities are $n_e = 3.4 \times 10^{19}\,\mathrm{cm}^{-3}$ (**lower lines**) and $n_e = 3.4 \times 10^{21}\,\mathrm{cm}^{-3}$ (**upper lines**). The beam velocity is $v = 0.2 v_{th}$

and Meyer-ter-Vehn 1991a; Boine-Frankenheim 1996). Similar scaling was found experimentally in electron cooling devices (Wolf et al. 1994; Winkler et al. 1997).

8.3.4 Comparison with Numerical Simulations

In this section, the results for the stopping power calculated from quantum kinetic equations are compared with data obtained by simulations. For this purpose data from PIC and MD simulations are used which were performed by Zwicknagel et al. (1993, 1996, 1999).

First, we focus on low beam velocities where strong beam–plasma correlations are of special importance. In this limit, the stopping power increases linearly with the beam velocity and, therefore, it is determined by the friction coefficient $(\partial\langle E\rangle/\partial x)/v$. As dynamic screening is of minor importance in the region of low velocities and not to high coupling parameters, we can consider the stopping power in statically screened T-matrix approximation. Expanding the exponential terms in (8.76) up to the order v^3, we find

$$\lim_{v\to 0}\frac{\frac{\partial\langle E\rangle}{\partial x}}{v} = -\frac{1}{6\pi^2\hbar^3}\sum_c \frac{m_c^4}{m_{\mathrm{bc}}^5}\frac{n_c\Lambda_c^3}{k_BT}\int\limits_0^\infty dp\,p^5\,Q^T_{\mathrm{bc}}(p)\exp\left(-\frac{m_cp^2}{2m_{\mathrm{bc}}^2k_BT}\right). \tag{8.79}$$

In Fig. 8.19, results are shown for the friction coefficient in T-matrix and in Born approximations. A good agreement can be observed between the T-matrix results and the simulation data whereas the Born approximation overestimates the friction coefficient. At high beam–plasma coupling ($Z_\mathrm{b}\Gamma^{3/2} > 5$), dynamic screening becomes important (Gericke 2002). Furthermore, the behavior of the T-matrix results is determined essentially by a dependence on the parameter $Z_\mathrm{b}\Gamma^{3/2}$. In contrast, the Born approximation depends additionally on the temperature.

Now we will investigate higher beam velocities. In Fig. 8.20, results for the stopping power calculated in different approximations are shown as a function of the beam velocity. A coupling parameter is considered which corresponds to a weakly coupled electron gas target. It should be noticed that essential contributions coming from quantum corrections are not expected for the parameters considered in this figure. At low beam velocities $v < v_\mathrm{th}$, both the statically screened T-matrix approximation and the T-matrix approach including dynamic screening (combined model) agree well with the PIC simulation data. On the other hand, the weak coupling theories (static and dynamic first Born approximations) deviate considerably. For beam velocities higher than the electron thermal velocity and up to $v \leq 3v_\mathrm{th}$, only the combined model leads to a reasonable agreement with the simulation data. This confirms the expected result that both strong collisions and dynamic screening effects determine the stopping power for moderate velocities. As already

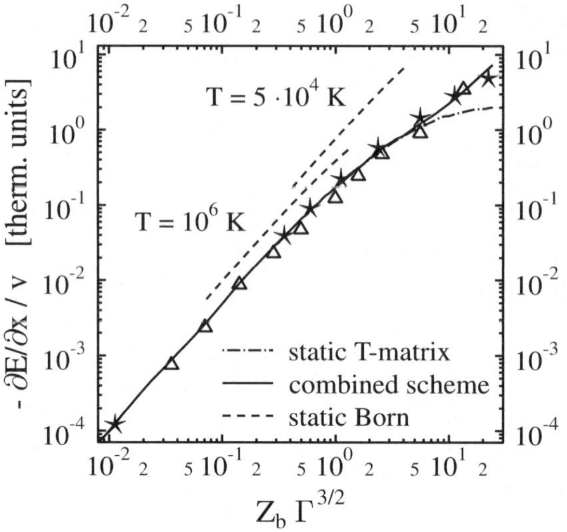

Fig. 8.19. Friction coefficient for an ion beam with $Z_\mathrm{b} = 10$ in T-matrix (*solid line*) and Born approximations (*dashed lines*). For comparison, data obtained by MD (*triangles*) and PIC (*asterisks*) simulations are given. The units for the stopping power and the beam velocity are $3k_BT/l$ (with $l = e^2/k_BT$) and $v_{th} = \sqrt{k_BT/m_e}$ (thermal units), respectively

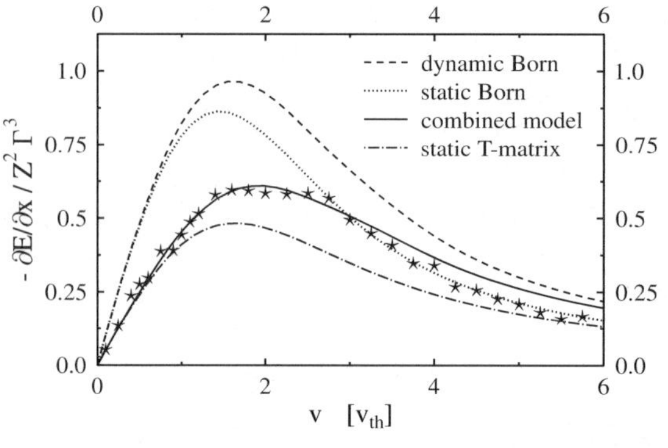

Fig. 8.20. Comparison of different approximations for the stopping power to data from PIC simulations (*asterisks*). Results are shown for an ion beam with a charge number $Z_\mathrm{b} = 5$ moving in an electron gas. Density and temperature are $n_e = 1.1 \times 10^{20}\,\mathrm{cm}^{-3}$ and $T = 1.6 \times 10^5\,\mathrm{K}$ ($Z_\mathrm{b}\Gamma^{3/2} = 0.12$) Thermal units are used (see Fig. 8.19)

outlined in Sect. 8.3.1, the combined model (8.77) approaches the well-known asymptotes given by formula (8.78) at high beam velocities. This behavior is not reproduced by the simulations which tend to give slightly smaller results at high beam velocities. At larger values of the coupling parameter $Z_\mathrm{b}\Gamma^{3/2}$, increasing deviations between the combined model and the simulation data can be observed (Gericke and Schlanges 1999). Here, one has to remind the fact that dynamical screening in the combined model (8.77) is accounted for in the first Born term and at the RPA level only. The nonlinear coupling effects found in the combined model are due to statically screened higher order ladder diagrams of the T-matrix. However, for moderate beam velocities and large coupling parameters, dynamic screening effects are expected to be significant in these terms, too. A more rigorous treatment requires an approach based on the dynamically screened ladder approximation. This is an extremely difficult problem and, up to now, a solution is not available.

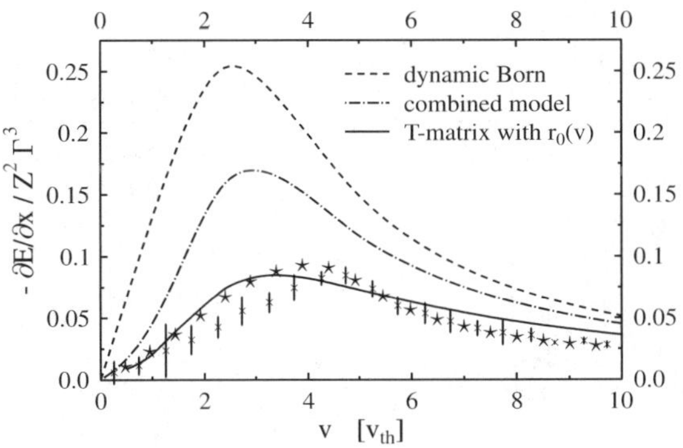

Fig. 8.21. Comparison of different approximation schemes for the stopping power (see text) with data from PIC (*asterisks*) and MD (*crosses*) simulations. The stopping of beam ions with the charge number $Z_\mathrm{b} = 10$ in an electron gas with $n_e = 9.9 \times 10^{22}\,\mathrm{cm}^{-3}$ and $T = 1.15 \times 10^5\,\mathrm{K}$ ($Z_\mathrm{b}\Gamma^{3/2} = 11.22$) is considered. Thermal units are used (see Fig. 8.19)

A phenomenological and simple approach to include dynamic screening effects in higher order ladder terms can be given by using the modified Debye potential

$$V_{ab}^S(r,v) = \frac{e_a e_b}{r} \exp\left(-r/r_0(v)\right) , \tag{8.80}$$

where dynamical screening is modelled by a velocity dependent screening length $r_0(v)$. Now the idea is to adjust the effective screening length in such a way that the correct asymptotic results for the stopping power are obtained. Zwicknagel et al. (1999) proposed the following screening length:

$$r_0(v) = r_D^e \left(1 + \frac{v^2}{v_{th}^2}\right)^{1/2} . \tag{8.81}$$

The Debye screening length of the electron gas $r_D^e = v_{th}/\omega_{pl}$ is obtained in the limit $v \to 0$, while $r_0(v) = v/\omega_{pl}$ follows for high beam velocities. Another possibility is to use $r_0(v)$ as a free parameter and to fit the stopping power in statically screened Born approximation to the dynamic RPA result (Gericke 2000). Both treatments lead to the correct asymptotic behavior described by the Bethe-type result (8.78) in the limit $v \to \infty$.

Results for the stopping of an ion beam in an electron gas target for the case of a large coupling parameter are shown in Fig. 8.22. As expected, the dynamically screened Born approximation (RPA) overestimates the stopping power considerably. The inclusion of strong binary correlations on the T-matrix level reduces the stopping power. Evidently, the T-matrix calculation using the modified Debye potential leads to a larger reduction than the combined model (8.77). Moreover, the model with the velocity dependent screening length agrees well with the simulation data at large beam–plasma coupling parameters (Gericke and Schlanges 2003). Further investigations show that the T-matrix approach with an effective screening length (8.81) coincides with the combined scheme (8.77) for weak and moderate beam–

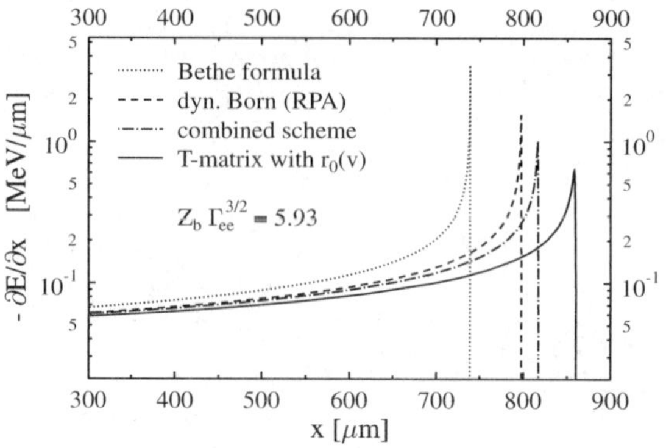

Fig. 8.22. Energy deposition by a beam ion versus distance considering different approximations for the stopping power. The beam particle is a $^{12}C^{+6}$ ion with an initial beam energy of $E = 6\,\text{MeV}$ per nucleon. The target is an electron gas with $n_e = 5 \times 10^{22}\,\text{cm}^{-3}$ and $T = 1 \times 10^5\,\text{K}$

plasma coupling ($Z_{\text{b}}\Gamma^{3/2} < 0.3$). Agreement with the Born approximation is only achieved for very hot plasmas.

It should be pointed out that the T-matrix approach with the velocity dependent Debye potential is not based on a rigorous quantum many-body theory. However, it provides a possibility to treat both strong binary collisions and dynamic screening with a standard approach to calculate the stopping power in terms of a cross section.

8.3.5 Energy Deposition in the Target Plasma

Now we will give a brief discussion of the influence of strong beam–plasma correlations on the slowing down process of a beam ion (Gericke and Schlanges 2003). The temporal evolution of the beam particle energy and position is determined by the set of equations

$$\dot{x} = v(t) \quad \text{and} \quad \dot{v} = \frac{1}{m_{\text{b}}} \frac{\partial}{\partial x} \langle E(v) \rangle \,, \tag{8.82}$$

with the initial conditions $v(0) = v_0 = (2E(v_0)/m_{\text{b}})^{1/2}$ and $x(0) = 0$. The quantity x denotes the distance, the a beam ion has moved in the plasma. We will consider here the stopping of light beam ions which can be assumed to be fully ionized except for the low velocities at the end of the stopping range. Figure 8.9 shows results for the energy loss versus penetration depth for fully ionized carbon ions with an initial beam velocity $v_0 \approx 28 v_{th}$. For such an initial condition, the different approximations show the same well-known behavior: first the energy transfer increases slowly and is then sharply peaked close to the point where the particle is stopped (Bragg peak). As the inclusion of strong binary collisions results in a smaller stopping force, the T-matrix schemes for the stopping power show a larger penetration depth compared to the RPA approach, and especially, also a larger one than the calculations based on the Bethe formula (8.78). It should be noticed that, in comparison to the total stopping range, there are considerable differences of the Bragg peak positions. Furthermore, the Bragg peak is more pronounced both in the Bethe and in the RPA calculations (Gericke and Schlanges 2003). Only for very hot plasmas, the results of the RPA and the T-matrix approximation coincide.

In the case of heavy ion stopping, the charge of the ions has to be determined self-consistently as a function of the beam velocity (Peter and Meyer-ter-Vehn 1991b). Rigorously, this requires a many-body approach to describe the kinetics of electron capture and loss processes accounting for the influence of the plasma medium. For simplification, semi-empirical formulas such as the Betz formula can be used (Betz 1983). This leads to a broadening of the Bragg peak (Gericke and Schlanges 2003).

8.3.6 Partially Ionized Plasmas

So far, we considered the energy loss of charged particle beams in fully ionized plasmas. However, the target plasma is partially ionized in many experiments and applications. In such situations, one has to account for the existence of bound states connected with processes such as ionization and excitation of plasma particles.

The first basic concepts to calculate the stopping of particle beams by bound target electrons were given in the pioneering papers of Bohr (1915), Bethe (1930), and Bloch (1933). Since that time, a lot of work has been done to further develop this theoretical approach (for reviews see, e.g., Ziegler (1999), Inokuti (1971)). Most of the papers considering the contributions of bound particles in plasma targets are based on the early concepts and result in modified Bethe formulas (Basko 1984; Peter and Meyer-ter-Vehn 1991b; Peter and Kärcher 1991; Couillaud et al. 1994).

In this section, we consider the problem in the framework of quantum kinetic theory (Gericke et al. 2002). As before, we assume beam ions with a fixed charge number Z_{b}, i.e., the evolution of the beam particle charge is not considered. We utilize the kinetic equations for systems with bound states derived in Sect. 7.7.2. This approach is particularly advantageous because it allows for the inclusion of the relevant two- and three-particle scattering processes as well as many-particle effects such as screening, lowering of the ionization energy and the Mott effect. We start from the kinetic equation (7.122) neglecting the retardation terms. Considering a spatially homogeneous target plasma, the stopping power is given by

$$\begin{aligned} \frac{\partial}{\partial x}\langle E\rangle &= \frac{1}{n_{\mathrm{b}}}\sum_{c}\int\frac{d\boldsymbol{p}_{\mathrm{b}}}{(2\pi\hbar)^3}\,\frac{(\boldsymbol{p}_{\mathrm{b}}\cdot\boldsymbol{v})}{v}\,I_{\mathrm{b}c}(\boldsymbol{p}_{\mathrm{b}}) \\ &+ \frac{1}{n_{\mathrm{b}}}\sum_{cd}\int\frac{d\boldsymbol{p}_{\mathrm{b}}}{(2\pi\hbar)^3}\,\frac{(\boldsymbol{p}_{\mathrm{b}}\cdot\boldsymbol{v})}{v}\,I_{\mathrm{b}cd}(\boldsymbol{p}_{\mathrm{b}})\,. \end{aligned} \tag{8.83}$$

The first term on the r.h.s. describes the two-particle scattering processes between the beam and the free plasma particles. Different approximations for this contribution to the stopping power have been discussed in the previous sections. The contribution of scattering processes with bound particles is determined by the three-particle collision integral I_{bcd} in the second term of (8.83). Here, we consider only reactions that keep the charge number of the beam ions constant.

The ionization of bound electron–ion pairs by beam particle impact is described by two terms in the three-particle collision integral (7.123). Neglecting the beam particle assisted recombination and using the relation (8.4) which can also be applied to the T-matrices T^{10} and T^{11}, we obtain for the ionization contribution of the stopping power (Schlanges et al. 1998; Gericke et al. 2002)

$$\begin{aligned}\frac{\partial}{\partial x}\langle E\rangle^{\text{ion}} &= \frac{1}{n_{\text{b}} V\hbar}\sum_j \int \frac{d\boldsymbol{P}}{(2\pi\hbar)^3}\frac{d\boldsymbol{p}_{\text{b}}}{(2\pi\hbar)^3}\frac{d\bar{\boldsymbol{P}}}{(2\pi\hbar)^3}\frac{d\bar{\boldsymbol{p}}_{\text{b}}}{(2\pi\hbar)^3}\frac{d\boldsymbol{p}_{ei}}{(2\pi\hbar)^3}\frac{(\boldsymbol{p}_{\text{b}}\cdot\boldsymbol{v})}{v} \\ &\times \Big\{\delta(E^0_{\text{b}(ei)} - \bar{E}^1_{\text{b}(ei)})\left|\langle\boldsymbol{p}_{\text{b}}|\langle+\boldsymbol{p}_{ei}\boldsymbol{P}|\,T^{11}_{\text{b}(ei)}\,|\bar{\boldsymbol{P}}j\rangle|\bar{\boldsymbol{p}}_{\text{b}}\rangle\right|^2 \bar{f}_{\text{b}}\,\bar{F}_j \\ &- \delta(E^1_{\text{b}(ei)} - \bar{E}^0_{\text{b}(ei)})\left|\langle\boldsymbol{p}_{\text{b}}|\langle j\boldsymbol{P}|\,T^{11}_{\text{b}(ei)}\,|\bar{\boldsymbol{P}}\boldsymbol{p}_{ei}+\rangle|\bar{\boldsymbol{p}}_{\text{b}}\rangle\right|^2 f_{\text{b}}\,F_j\Big\}, \end{aligned} \tag{8.84}$$

where $F_j = F_j(\boldsymbol{P})$ is the distribution function of the bound states with the total momentum $\boldsymbol{P}$, and $f_{\text{b}} = f_{\text{b}}(\boldsymbol{p}_{\text{b}})$ is the beam particle distribution function given by (8.68). The T-matrix T^{11} is determined by the Lippmann–Schwinger equation (8.7) and describes the transition of the electron–ion pair from the bound state $|j\rangle$ to the scattering state $|\boldsymbol{p}_{ei}+\rangle$ by beam–particle impact. The effective two-body interaction potential is used to be the statically screened Coulomb potential. The many-particle effects included in the scattering energies $E^1_{\text{b}(ei)}$ and $E^0_{\text{b}(ei)}$ are treated on the level of the *rigid shift approximation* as it was done for the rate coefficients in Sect. 8.1.2. The energies of the free and of the bound quasiparticles are then given by (8.9) with momentum independent self-energy shifts.

Since we want to consider arbitrary mass ratios of plasma and beam particles, it is appropriate to transform the momenta in (8.84) into Jacobi variables. As we consider a non-degenerate target plasma in local thermal equilibrium, the distribution of bound plasma particles is given by the Boltzmann distribution. Performing some of the integrations, the final result for the ionization contribution of the stopping power reads (Gericke et al. 2002)

$$\begin{aligned}\frac{\partial}{\partial x}\langle E\rangle^{\text{ion}} &= -\sum_j \frac{M_{ei}^2}{\mu_{\text{b}}^3}\frac{n_j \Lambda_{ei}^3}{(2\pi)^2\hbar^3}\frac{k_B T}{v}\int_0^\infty dk\, k^3\, Q_j^{\text{ion}}(k) \\ &\times\left\{p_-\,\exp\left(-\frac{M_{ei}v_-^2}{2k_BT}\right) - p_+\,\exp\left(-\frac{M_{ei}v_+^2}{2k_BT}\right)\right\}. \end{aligned} \tag{8.85}$$

Here we used $p_\pm = 1 \pm (\mu_{\text{b}} k_B T)/(M_{ei} k v)$ and $v_\pm = k/\mu_{\text{b}} \pm v$. k denotes the relative momentum between the beam particle and the electron–ion pair, and $\Lambda_{ei}=(2\pi\hbar^2/M_{ei}k_BT)^{1/2}$ is the thermal wavelength with $M_{ei}=m_e+m_i$. Furthermore, we introduced a transport cross section of ionization defined by

$$Q_j^{\rm ion}(k) = \frac{\mu_{\rm b}^2 g_b}{(2\pi)^2\hbar^4 k}\int_0^\infty dp_{ei}\, p_{ei}^2 \int d\Omega_{p_{ei}} \int_{-1}^{1} d\cos\vartheta$$
$$\times\left(1-\frac{g_b}{k}\cos\vartheta\right)\left|\langle\boldsymbol{k}|\langle j|\, T^{11}_{{\rm b}(ei)}\, |\boldsymbol{p}_{ei}+\rangle|\bar{\boldsymbol{k}}\rangle\right|^2 \qquad (8.86)$$

with the quantity $g_b=(k^2-\mu_{\rm b}p_{ei}^2/m_{ei}-2\mu_{\rm b}\, I_j^{\rm eff})^{1/2}$ and the reduced mass $\mu_{\rm b}=m_{\rm b}M_{ei}/(m_{\rm b}+M_{ei})$. ϑ is here the angle between the momenta $\boldsymbol{k}$ and $\bar{\boldsymbol{k}}$. The effective ionization energy is given by (8.13), i.e.,

$$I_j^{\rm eff} = |E_j^0| + \Delta_e + \Delta_i - \Delta_j\,.$$

A lowering of the ionization energy is described by the self-energy shifts Δ_a $(a=e,i)$ and Δ_j of the free and bound plasma particles, respectively.

It should be noticed that (8.85) has the same structure as the expression (8.76) for the free particle contribution to the stopping power assuming statically screened interactions. The different scattering processes are reflected by the different types of transport cross sections.

Excitation and deexcitation processes as well as elastic collisions of beam particles with composite plasma particles are described by collision integrals which are characterized by a bound state in the input and output channels. The derivation of corresponding expressions for the stopping power is similar to the one performed for the ionization contribution.

The solution of the effective three-particle problem described by the T-matrix T^{11} is a difficult task. A simple approximation is given by a modified Born approximation which was applied already in Sect. 8.1.2 to calculate rate coefficients for dense plasmas. Here, nonideality effects are taken into account by the statically screened Coulomb potential (8.27), by the medium dependent atomic form factor (8.28), and by the effective ionization energy $I_j^{\rm eff}$.

Let us consider some simplifications which are possible in the important limit of very high beam velocities. In this case, we can assume $\cos\vartheta\approx 1$ and $k-g_b\cos\vartheta\approx k^{-1}\mu_{\rm b}I_j^{\rm eff}$. Furthermore, the second term in the curly brackets of (8.85) is negligible, and in the first term, only momenta with $k\approx\mu_{\rm b}v$ contribute to the k-integral. The remaining integration can be performed analytically. Then we get for the ionization contribution (Gericke et al. 2002)

$$\frac{\partial}{\partial x}\langle E\rangle^{\rm ion} = -\sum_j I_j^{\rm eff}\, n_j\, \sigma_j^{\rm ion}(\mu_{\rm b}v)\,. \qquad (8.87)$$

In this limit, the stopping power is proportional to the effective ionization energy $I_j^{\rm eff}$, to the number density n_j of the bound states, and to the total ionization cross section $\sigma_j^{\rm ion}$. A similar formula can be found for processes which describe excitations of bound plasma particles.

To find an explicit expression for the stopping power in the high velocity limit, we need an analytic expression for the total ionization cross section for large impact energies. For the ionization of hydrogen-like bound states by electron impact, the modified Bethe cross section (2.240) was used by Schlanges et al. (1998). Here, we consider ionization by ion impact and neglect the energy shifts in the logarithm of the cross section because they are negligible for large impact energies. Then we obtain for ionization from the ground state

$$\frac{\partial}{\partial x}\langle E\rangle^{\text{ion}} = -16\pi a_B^2\, Z_{\text{b}}^2 \frac{n_1 I_1^{\text{eff}}|E_1^0|}{m_e v^2} \ln\left(\frac{2m_e v^2}{|E_1^0|}\right). \tag{8.88}$$

This represents a generalized Bethe formula for the stopping power of a target plasma with hydrogen-like bound states. The nonideality effect in the high velocity limit is condensed in the effective ionization energy I_1^{eff}. As there is a lowering of the ionization energy in dense plasmas, the ionization contribution to the stopping power is reduced as compared to that of an ideal plasma.

For target atoms or ions having more than one bound electron, we assume, for high impact energies, the cross section to be proportional to the number of bound electrons. We get for the stopping power of ideal target plasmas

$$\frac{\partial}{\partial x}\langle E\rangle^{\text{ion}} = -\frac{4\pi Z_{\text{b}}^2 e^4}{m_e v^2} \sum_{Z=0}^{Z_c} (Z_c - Z)\, n_Z \ln\left(\frac{2m_e v^2}{|E_Z|}\right). \tag{8.89}$$

Here, Z_c denotes the nuclear charge of the considered target species, n_Z is the number density, and E_Z is the ionization energy of an isolated Z-fold charged ion. Such formula was also found in Peter and Kärcher (1991). It should be mentioned that the expression (8.89) was successfully used to describe the energy loss of ions in weakly coupled, partially ionized plasmas with free electron densities $n_e < 10^{19}\, cm^{-3}$ (Stöckl et al. 1996; Wetzler et al. 1997; Golubev et al. 1998). However, deviations have been found for high plasma densities (Roth et al. 2000).

Let us now consider the total stopping power which is given by the sum of the contributions of the free and the bound plasma particles. If ionization is accounted for only in the bound particle contribution, we have

$$\frac{\partial}{\partial x}\langle E\rangle = \frac{\partial}{\partial x}\langle E\rangle^{\text{free}} + \frac{\partial}{\partial x}\langle E\rangle^{\text{ion}}. \tag{8.90}$$

The next task to calculate the stopping power of partially ionized plasmas is to determine the plasma composition. As the target plasma is assumed to be in equilibrium, the number densities of free and bound particles can be calculated from mass action laws. The derivation of mass action laws for nonideal plasmas and their applications were considered in the Chaps. 2 and 6. In the following, we will restrict ourselves to some special features of

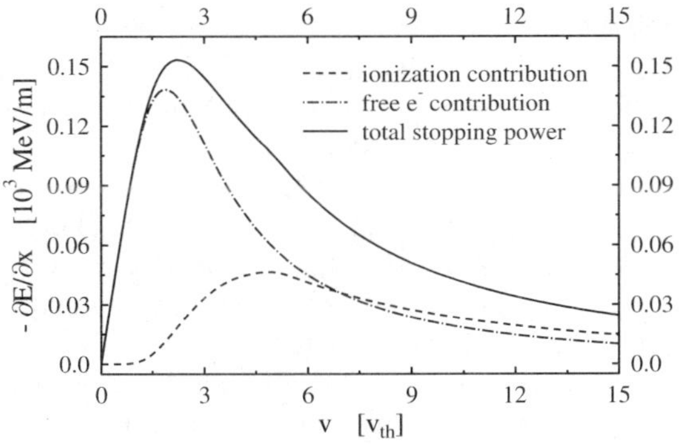

Fig. 8.23. Energy loss of a proton beam in a partially ionized hydrogen plasma vs. beam velocity ($v_{th} = \sqrt{k_B T/m_e}$). Temperature and total electron density of the plasma are $T = 20000\,\mathrm{K}$ and $n_e^{tot} = 10^{20}\,\mathrm{cm}^{-3}$, respectively. The degree of ionization is $\alpha = 0.19$

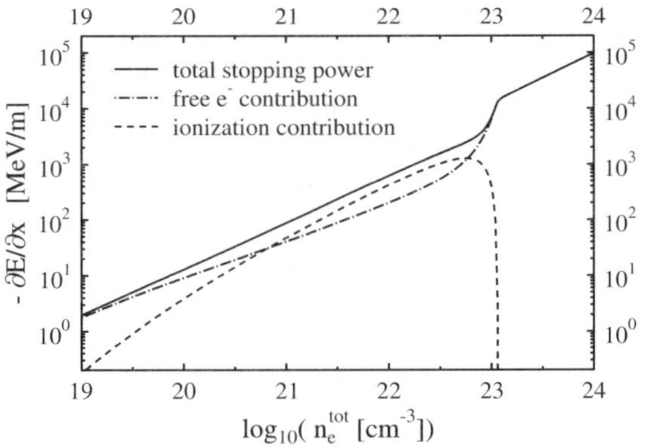

Fig. 8.24. Stopping power of a partially ionized hydrogen plasma with a temperature of $T = 25000\,\mathrm{K}$ as a function of the total electron density. A proton beam with an energy of 1 MeV per particle is considered

the stopping power of partially ionized plasmas. For this purpose, we consider hydrogen and use results for the plasma composition obtained from an ionization–dissociation model presented by Schlanges et al. (1995). This model accounts for the formation of atoms and molecules, and it describes pressure ionization (Mott transition) at high target densities.

Results for the energy loss of a proton beam in hydrogen with a degree of ionization $\alpha = 0.19$ are given in Fig. 8.23. The fraction of molecules is below 1%; they were treated as two (independent) atoms. The contribution of the free plasma electrons, we applied the T-matrix approach given by the combined scheme (8.77). In the bound particle contribution only atomic ionization from the ground state was accounted for. The latter was calculated using the expressions (8.85) and (8.86) with the three-particle T-matrix in statically screened Born approximation (Gericke et al. 2002).

Evidently, the free electron contribution is dominant at very low beam velocities. It also gives the major contribution for beam energies where the maximum of the total stopping power occurs. At higher beam velocities, the ionization contribution becomes relatively larger; it finally exceeds that of the free electrons for the considered plasma parameters. However, the contribution per bound electron is still smaller than that per free electron.

This fact leads to the observed enhancement of the stopping power of plasmas as compared to cold gases (Young et al. 1982; Jacobi et al. 1995).

The density dependence of the total stopping power and its contributions is demonstrated in Fig. 8.24 (Gericke et al. 2002). The beam consists of protons with 1MeV energy. For this case, the asymptotic formulas (8.78) and (8.88) could be used. To be consistent, the effective ionization energy in (8.88) was considered on the same approximation level as in the calculation of the plasma composition. Fig. 8.24 shows that with increasing density, there is a stronger increase of the ionization contribution in comparison to that of the free electrons due to the higher fraction of atoms. IIowever, we observe a sharp decrease of the stopping power due to ionization at densities of $n_e^{\mathrm{tot}} \approx 10^{23} cm^{-3}$ whereas the free electron contribution increases in this range. This behavior results from the lowering of the ionization energy which affects the stopping power directly and indirectly over the plasma composition. As the free electrons give a higher contribution per particle, the total stopping shows a strong increase, too. Here, the transition to the fully ionized plasma due to pressure ionization in dense hydrogen (Mott transition) is described.

9. Dense Plasmas in External Fields

9.1 Plasmas in Electromagnetic Fields

9.1.1 Kadanoff–Baym Equations

In this chapter, we will investigate the kinetic theory of a plasma being in interaction with an electromagnetic field. This is, in general, a complicated problem. Up to now, the interaction of the plasma with the electromagnetic field was taken into account only via the static Coulomb interaction between the plasma particles (longitudinal field). However, the motion of the plasma particles produces current and charge densities which lead to electromagnetic fields. Consequently, a complete description of the plasma requires a self-consistent application of the quantum-mechanical equations of motion of the plasma particles together with the quantized equations of electrodynamics. This means, the quantum electrodynamics of non-equilibrium plasmas has to be developed. A formulation of such theory was given by DuBois (1968) and by Bezzerides and DuBois (1972).

Moreover, the interaction of the plasma with external electromagnetic fields is of special interest. External fields induce macroscopic currents and give rise to transport processes. In the case of weak external fields, well-developed methods of linear response theory can be applied.

Of current interest, however, is also the behavior of plasmas under the influence of strong high-frequency fields. The rapid development of the short-pulse laser technology provides the possibility to produce fields of extremely high intensity such that the quiver velocity $v_0 = eE/m_e\omega$ can be large compared to the thermal velocity $v_{\rm th} = \sqrt{k_B T_e/m_e}$, and interesting effects of laser-matter interaction have to be expected.

The evolution of the correlation functions $g^{\lessgtr}$ of the non-relativistic plasma particles in an electromagnetic field is determined by the following Kadanoff–Baym equations

$$\left[i\hbar\frac{\partial}{\partial t_1} - \frac{1}{2m_a}\left(\frac{\hbar}{i}\nabla_1 - \frac{e_a}{c}\boldsymbol{A}(1)\right)^2 - e_a\phi(1)\right] g_a^{\gtrless}(1,1')$$
$$= \int d\bar{\boldsymbol{r}}_1\, \Sigma_a^{\rm HF}(1,\bar{\boldsymbol{r}}_1 t_1) g_a^{\gtrless}(\bar{\boldsymbol{r}}_1 t_1, 1') + \int_{t_0}^{\infty} d\bar{1}\, \Sigma_a^{\rm R}(1,\bar{1})\, g_a^{\gtrless}(\bar{1},1')$$

$$+ \int_{t_0}^{\infty} d\bar{1}\, \Sigma_a^{\gtrless}(1,\bar{1})\, g_a^{\mathrm{A}}(\bar{1},1') \, . \tag{9.1}$$

In addition, the equation of motion for the retarded and advanced Green's functions reads

$$\left[i\hbar \frac{\partial}{\partial t_1} - \frac{1}{2m_a} \left(\frac{\hbar}{i} \nabla_1 - \frac{e_a}{c} \boldsymbol{A}(1) \right)^2 - e_a \phi(1) \right] g_a^{R/A}(1,1')$$
$$- \int d2\, \Sigma_a^{R/A}(1,2) g_a^{R/A}(2,1') = \delta(1-1') \, . \tag{9.2}$$

In these equations, the electromagnetic field is characterized in a well-known manner by the electrodynamic potentials $\boldsymbol{A}$ and ϕ. The connection of the electrodynamic potentials with the field strengths $\boldsymbol{E}$ and $\boldsymbol{B}$ is given by the relations

$$\boldsymbol{E} = -\frac{1}{c} \frac{\partial}{\partial t} \boldsymbol{A} - \nabla \phi \;\; ; \quad \boldsymbol{B} = \nabla \times \boldsymbol{A} \, . \tag{9.3}$$

In the following, we sometimes will use the co-variant 4-vector notation with $A = (c\phi, \boldsymbol{A})$, $x = (c\tau, \boldsymbol{r})$, $X = (ct, \boldsymbol{R})$, $a_\mu b^\mu = a_0 b_0 - \boldsymbol{ab}$, etc.

In addition to the Kadanoff–Baym equations, we also need equations of motion for the electromagnetic fields, i.e., the Maxwell equations. In terms of A_μ, these familiar equations are given by

$$\Box A_\mu - \frac{\partial}{\partial x_\mu} \frac{\partial A_\nu}{\partial x_\nu} = -\frac{4\pi}{c} \left(j_\mu + j^{ext} \right) \tag{9.4}$$

with the four-current $j = (c\rho, \boldsymbol{j})$. The electrodynamic potentials are, however, not uniquely defined by (9.3). The field strengths are clearly not affected by a gauge-transformation

$$A'_\mu(x) = A_\mu(x) - \partial_\mu \chi(x) \, . \tag{9.5}$$

The gauge-arbitrariness is a basic but also useful principle of electrodynamics. Special choices of the gauge lead to interesting simplifications. For instance, in the Coulomb or transverse gauge, div $\boldsymbol{A} = 0$, no time derivatives appear in the equation for $A_0 = c\phi$. We get, e.g.,

$$\triangle \phi = -4\pi \left(\rho + \rho^{ext} \right) , \qquad \phi(\boldsymbol{r},t) = \int d\boldsymbol{r}' \, \frac{\rho(\boldsymbol{r}') + \rho^{ext}(\boldsymbol{r}')}{|\boldsymbol{r} - \boldsymbol{r}'|} \, , \tag{9.6}$$

$$\Box \boldsymbol{A} = -\frac{4\pi}{c} \left(\boldsymbol{j} + \boldsymbol{j}^{ext} \right) + \frac{1}{c} \nabla \frac{\partial \phi}{\partial t} = -\frac{4\pi}{c} \left(\boldsymbol{j}_T + \boldsymbol{j}_T^{ext} \right) . \tag{9.7}$$

Here, $\boldsymbol{j}_T$ is the transverse current density.

According to the equations above, we see that the transverse contributions were completely neglected in our considerations in previous chapters, and only the long-range longitudinal (electrostatic) field was considered. In the following, we will first neglect the self-consistent transversal field and consider only the influence of the external electromagnetic field on the plasma. Moreover, the external field will be dealt with classically.

The Kadanoff–Baym equations (9.1) remain covariant under gauge transformations, i.e., under the following common transformations of the four-potential and of the field operators ψ

$$A'_\mu(x) = A_\mu(x) - \partial_\mu \chi(x), \qquad \psi'_a(x) = e^{\frac{i}{\hbar}\frac{e_a}{c}\chi(x)}\psi_a(x). \tag{9.8}$$

The corresponding gauge transform of the Green's functions may be written as

$$g'_a(x, X) = e^{\frac{i}{\hbar}\frac{e_a}{c}\left[\chi(X+\frac{x}{2})-\chi(X-\frac{x}{2})\right]} g_a(x, X)\,. \tag{9.9}$$

With the help of the two-time correlation functions, we can determine mean values of physical quantities of the plasma in dependence of the electromagnetic field. In order to avoid the difficulty to work with correlation functions depending on the gauge, we follow an idea of Fujita (1966) and write up the theory in terms of correlation functions $\widetilde{g}(k, X)$ which are taken to be explicitly gauge-invariant. The gauge invariant correlation function is given by the modified Fourier transform

$$\widetilde{g}_a(k, X) = \int \frac{d^4x}{(2\pi)^4} \exp\left\{ i \int\limits_{-\frac{1}{2}}^{\frac{1}{2}} d\lambda\, x_\mu \left[k^\mu + \frac{e_a}{c} A^\mu(X + \lambda x)\right] \right\} g_a(x, X)\,. \tag{9.10}$$

Indeed, using the identity

$$\begin{aligned}\chi\left(X + \frac{x}{2}\right) - \chi\left(X - \frac{x}{2}\right) &= \int_{-\frac{1}{2}}^{\frac{1}{2}} d\lambda\, \frac{d}{d\lambda}\chi\left(X + \lambda x\right) \\ &= x_\mu \partial^\mu \int_{-\frac{1}{2}}^{\frac{1}{2}} d\lambda\, \chi\left(X + \lambda x\right)\,,\end{aligned}$$

one readily confirms that the phase factors cancel under any gauge transformation (9.9), and we have $g'(k, X) \equiv g(k, X)$ (Haug and Jauho 1996).

In the following, we focus on spatially homogeneous electric fields and use the vector potential gauge

$$A_0 = \phi = 0; \qquad \boldsymbol{A} = -c \int_{-\infty}^{t} d\bar{t}\, \boldsymbol{E}(\bar{t}). \tag{9.11}$$

In this case, relation (9.10) simplifies to

$$\widetilde{g}_a(\boldsymbol{k},\omega;\boldsymbol{R},t) = \int d\tau d\boldsymbol{r} \exp\left[i\,\omega\tau - \frac{i}{\hbar}\boldsymbol{r}\cdot\left(\boldsymbol{k} + \frac{e_a}{c}\int\limits_{t-\frac{\tau}{2}}^{t+\frac{\tau}{2}} \frac{dt'}{\tau}\boldsymbol{A}(t')\right)\right] g_a(\boldsymbol{r},\tau;\boldsymbol{R},t)\,. \tag{9.12}$$

The gauge-invariant Green's function $\widetilde{g}(\boldsymbol{k},\tau,t)$ is connected to the Wigner transformed quantity by

$$\widetilde{g}(k,t,t') = g\left(\boldsymbol{k} + \frac{e_a}{c}\int\limits_{t-\frac{\tau}{2}}^{t+\frac{\tau}{2}} d\bar{t}\,\frac{\boldsymbol{A}(\bar{t})}{\tau}, t, t'\right)\,. \tag{9.13}$$

Therefore, $\widetilde{g}$ follows from the Wigner transformed function $g_a(\boldsymbol{p},\tau,t)$ by replacing the canonical momentum $\boldsymbol{p}$ according to

$$\boldsymbol{p} = \boldsymbol{k} + \frac{e_a}{c}\int\limits_{t-\frac{\tau}{2}}^{t+\frac{\tau}{2}} dt'\,\frac{\boldsymbol{A}(t')}{\tau}\,. \tag{9.14}$$

In particular, for a harmonic electric field,

$$\boldsymbol{E}(t) = \boldsymbol{E}_0\cos\Omega t\,, \qquad \boldsymbol{A}(t) = -\frac{c\boldsymbol{E}_0}{\Omega}\sin\Omega t, \tag{9.15}$$

the substitution for the momentum takes the form

$$\boldsymbol{p} = \boldsymbol{k} + \frac{2\,\boldsymbol{E}_0}{\tau\,\Omega^2}\sin\Omega t\,\sin\frac{\Omega\tau}{2}. \tag{9.16}$$

For the Wigner-function (time diagonal correlation function), the relation 9.13 simplifies to

$$\widetilde{f}(k,t) = f(\boldsymbol{k} + \frac{e_a}{c}\boldsymbol{A}(t),\ t) = f(\boldsymbol{k} - e_a\int_{-\infty}^{t} d\bar{t}\,\boldsymbol{E}(\bar{t}),\ t)\,. \tag{9.17}$$

For the further considerations, we need the retarded and advanced Green's functions for free particles in an electromagnetic field. In this simple case, (9.2) is solved immediately by

$$g_a^R(\boldsymbol{p};\tau,t) = -\frac{i}{\hbar}\Theta(\tau)\exp\left[-\frac{i}{\hbar}\int\limits_{t-\frac{\tau}{2}}^{t+\frac{\tau}{2}} dt'\,[\boldsymbol{p} - \frac{e_a}{c}\boldsymbol{A}(t')]^2/2m_a\right]\,, \tag{9.18}$$

and g_a^A is obtained from the symmetry relation $g_a^A(\boldsymbol{p};\tau,t) = [g_a^R(\boldsymbol{p};-\tau,t)]^*$. From this result, we can calculate the spectral function $a(t,t')$

$$a_a(\boldsymbol{p};\tau,t) = \exp\left[-\frac{i}{\hbar}\int\limits_{t-\frac{\tau}{2}}^{t+\frac{\tau}{2}} dt'\,[\boldsymbol{p} - \frac{e_a}{c}\boldsymbol{A}(t')]^2/2m_a\right]. \qquad (9.19)$$

Obviously, the results (9.18) and (9.19) are gauge-dependent since $g_a^{R/A}$ and a_a are functions of the canonical momentum $\boldsymbol{p}$. However, one easily can obtain the corresponding gauge-invariant results by applying the transform (9.14), with the result

$$\widetilde{g}_a^R(\boldsymbol{k};\tau,t) = -\frac{i}{\hbar}\Theta(\tau)\,e^{-\frac{i}{\hbar}\left[\frac{k^2}{2m_a}\tau + S_a(\boldsymbol{A};\tau,t)\right]}, \qquad (9.20)$$

where

$$S_a(\boldsymbol{A};\tau,t) = \frac{e_a^2}{2m_a c^2}\left[\int_{t-\frac{\tau}{2}}^{t+\frac{\tau}{2}} dt'\,A^2(t') + \frac{1}{\tau}\left(\int_{t-\frac{\tau}{2}}^{t+\frac{\tau}{2}} dt'\,\boldsymbol{A}(t')\right)^2\right].$$

For a free particle without field, the spectral function shows free undamped oscillations along the τ-axis, (i.e., perpendicular to the diagonal in the t-t'-plane) with the single-particle energy $\epsilon_a(k) = k^2/2m_a$, and its Fourier transform is

$$a_a^{\rm free}(\boldsymbol{k};\omega,t) = 2\pi\delta[\hbar\omega - \epsilon_a(k)]\,. \qquad (9.21)$$

On the other hand, the result (9.18) reflects the influence of an electromagnetic field on the quasiparticle spectrum, while correlation effects were neglected. Equation (9.18) shows that the field causes a time-dependent shift of the single-particle energy which obviously reflects the well-known fact that the proper eigenstates of the system contain the electromagnetic field and are given by Volkov states (Volkov 1934). The spectrum may even contain additional peaks. This behavior becomes particularly apparent in the limiting case of a harmonic time dependence (9.15). The time integrations in S can be performed, and simple trigonometric relations lead to (Jauho and Johnson 1996)

$$S_a(\boldsymbol{A};\tau,t) = \varepsilon_a^{\rm pond}\,\tau\left[1 - \frac{\sin\Omega\tau\cos 2\Omega t}{\Omega\tau} + \frac{8\sin^2\Omega t\sin^2\frac{\Omega\tau}{2}}{(\Omega\tau)^2}\right], \qquad (9.22)$$

where $\varepsilon_a^{\rm pond}$ is the ponderomotive potential , i.e., the cycle-averaged ($T = 2\pi/\Omega$) kinetic energy-gain of a charged particle in an E-field

$$\varepsilon_a^{\rm pond} = \frac{e_a^2 E_0^2}{4m_a\Omega^2}\,.$$

The first term in the brackets leads to a shift of the single-particle energy, the average kinetic energy of the particles increases by $\varepsilon_a^{\rm pond}$. The remaining terms modify the spectrum qualitatively and give rise to additional peaks which are related to *photon sidebands* (Jauho and Johnson 1996). Now it is simple to generalize the GKBA for the reconstruction of the two-time correlation function from the Wigner distribution (Kremp et al. 1999). If we start from (7.11) and carry out the transition to the gauge-invariant quantities, we find

$$\pm\widetilde{g}_a^{\gtrless}(\boldsymbol{k};t_1,t_1') = \widetilde{g}_a^R(\boldsymbol{k};t_1,t_1')\,\widetilde{f}_a^{\gtrless}\left[\boldsymbol{k}-\boldsymbol{K}_a^A(t_1',t_1);t_1'\right]$$

$$-\widetilde{f}_a^{\gtrless}\left[\boldsymbol{k}-\boldsymbol{K}_a^A(t_1,t_1');t_1\right]\widetilde{g}_a^A(\boldsymbol{k};t_1,t_1'), \qquad (9.23)$$

with the notation

$$\boldsymbol{K}_a^A(t,t') \equiv \frac{e_a}{c}\boldsymbol{A}(t) - \frac{e_a}{c}\int_{t'}^{t} dt''\,\frac{\boldsymbol{A}(t'')}{t-t'}\,. \qquad (9.24)$$

9.1.2 Kinetic Equation for Plasmas in External Electromagnetic Fields

The description of plasmas with two-time correlation functions considered in the previous section is very general and allows for the calculation of statistical and dynamical properties of plasmas in electromagnetic fields. If we are only interested in the statistical properties, it is, as we know from Chap. 7, easier to consider the time-diagonal Kadanoff–Baym equation for the Wigner function. We get, like in Sect. 7.2, the general kinetic equation for the spatially homogeneous system

$$\frac{\partial}{\partial t}f_a(\boldsymbol{p},t) = -2\,\mathrm{Re}\int_{t_0}^{t} d\bar{t}\Big\{\Sigma_a^{>}(\boldsymbol{p};t,\bar{t})\;g_a^{<}(\boldsymbol{p};\bar{t},t) - \Sigma_a^{<}(\boldsymbol{p};t,\bar{t})\;g_a^{>}(\boldsymbol{p};\bar{t},t)\Big\}\,.$$

Note that there is no explicit dependence of the equation on the field.

It is more appropriate to transform equation (9.1) to an equation for the gauge-invariant Wigner function (9.14)

$$\frac{\partial}{\partial t}\widetilde{f}_a(\boldsymbol{k}_a,t) + e_a\boldsymbol{E}(t)\cdot\nabla_{\boldsymbol{k}}\widetilde{f}_a(\boldsymbol{k}_a,t) = -2\mathrm{Re}\int_{t_0}^{t} d\bar{t}\Big\{\widetilde{\Sigma}_a^{>}\widetilde{g}_a^{<} - \widetilde{\Sigma}_a^{<}\widetilde{g}_a^{>}\Big\}\,, \qquad (9.25)$$

where the arguments in the collision integral are given explicitly by

$$\Sigma_a^{\gtrless}g_a^{\lessgtr} \equiv \Sigma_a^{\gtrless}\left[\boldsymbol{k}_a + \boldsymbol{K}_a^A(t,\bar{t});t,\bar{t}\right]\,g_a^{\lessgtr}\left[\boldsymbol{k}_a + \boldsymbol{K}_a^A(t,\bar{t});\bar{t},t\right]\,.$$

This kinetic equation is still very general. The collision integral is a functional of two-time correlations functions. The further procedure to get a closed kinetic equation, however, is well known from Chap. 7:

1. The self-energy has to be specified in a certain approximation.

2. The two-time correlation functions have to be eliminated with the gauge-invariant GKBA (9.23).

In the following, we drop the tilde which denotes gauge invariant functions.

Let us first consider the self-energy in first Born approximation. A lengthy but straightforward calculation (Kremp et al. 1999) leads to the following kinetic equation

$$\left\{ \frac{\partial}{\partial t} + e_a \boldsymbol{E}(t) \cdot \nabla_{k_a} \right\} f_a(\boldsymbol{k}_a, t) = \sum_b I_{ab}(\boldsymbol{k}_a, t) \tag{9.26}$$

with the collision integral

$$\begin{aligned} I_{ab}(\boldsymbol{k}_a, t) &= 2 \int \frac{d\boldsymbol{k}_b d\bar{\boldsymbol{k}}_a d\bar{\boldsymbol{k}}_b}{(2\pi\hbar)^6} \delta(\boldsymbol{k}_a + \boldsymbol{k}_b - \bar{\boldsymbol{k}}_a - \bar{\boldsymbol{k}}_b) \frac{1}{\hbar^2} \left| V_{ab}(\boldsymbol{k}_a - \bar{\boldsymbol{k}}_a) \right|^2 \\ &\times \int_{t_0}^{t} d\bar{t} \, \mathrm{Re} \, e^{\left\{ \frac{i}{\hbar} \left[(\epsilon_{ab} - \bar{\epsilon}_{ab})(t - \bar{t}) - (\boldsymbol{k}_a - \bar{\boldsymbol{k}}_a) \cdot \boldsymbol{R}_{ab}(t, \bar{t}) \right] \right\}} \\ &\times \left\{ \bar{f}_a \bar{f}_b \left[1 - f_a\right] \left[1 - f_b\right] - f_a f_b \left[1 - \bar{f}_a\right] \left[1 - \bar{f}_b\right] \right\} \Big|_{\bar{t}} , \end{aligned} \tag{9.27}$$

where we used the notation $\epsilon_{ab} = \epsilon_a + \epsilon_b$, $\epsilon_a = p_a^2/2m_a$ and $f_a = f_a[\boldsymbol{k}_a + \boldsymbol{Q}_a(t, \bar{t}), \bar{t}]$. The quantities $\boldsymbol{Q}_a$ and $\boldsymbol{R}_{ab}$ are defined by (9.28) and (9.29), see below.

Equation (9.26) is a rather general non-Markovian kinetic equation which describes two-particle collisions in a weakly coupled quantum plasma in the presence of a spatially homogeneous time-dependent field. It generalizes previous results obtained for classical plasmas (Silin 1960; Klimontovich 1975). The time-dependent field modifies the collision integral in several ways:

1. The momentum arguments of the distribution functions are

$$\boldsymbol{k}_a + \boldsymbol{Q}_a(t, \bar{t}), \quad \text{with} \quad \boldsymbol{Q}_a(t, \bar{t}) \equiv -e_a \int_{\bar{t}}^{t} dt' \, \boldsymbol{E}(t') \, , \tag{9.28}$$

   i.e., they contain an additional retardation, the intra-collisional field contribution to the momentum. It describes the gain of momentum in the time interval $t - \bar{t}$ due to the field. In the case of a harmonic field given by (9.15), we have $\boldsymbol{Q}_a(t, \bar{t}) = -e_a \boldsymbol{E}_0 / \Omega \, (\sin \Omega t - \sin \Omega \bar{t})$. The result for a static field (Morawetz and Kremp 1993) is readily recovered by taking the limit $\Omega \to 0$, i.e., $\boldsymbol{Q}_a^{\mathrm{st}} = -e_a \boldsymbol{E}_0 (t - \bar{t})$.
2. Another modification occurs in the exponent under the time integral which essentially governs the energy balance in a two-particle collision: In addition to the usual collisional energy broadening (which has the form $\cos\{[\epsilon_{ab} - \bar{\epsilon}_{ab}](t - \bar{t})/\hbar\}$), there appears a field-dependent broadening. This effect is determined by the change of the distance between particles a and b due to the field given by

$$\boldsymbol{R}_{ab}(t,\bar{t}) = \left(\frac{e_a}{m_a} - \frac{e_b}{m_b}\right) \int_{\bar{t}}^{t} dt' \int_{t'}^{t} dt'' \boldsymbol{E}(t''). \tag{9.29}$$

We get for harmonic fields

$$\boldsymbol{R}_{ab}(t,\bar{t}) = \left(\frac{e_a}{m_a} - \frac{e_b}{m_b}\right) \left[\frac{\boldsymbol{E}_0 \cdot (t-\bar{t})}{\Omega} \sin \Omega t + \frac{\boldsymbol{E}_0}{\Omega^2}\Big(\cos \Omega t - \cos \Omega \bar{t}\Big)\right].$$

It is clear that the field has no effect on the scattering of identical particles, $\boldsymbol{R}_{aa} \equiv 0$. (Of course, the scattering rates (collision frequency) of identical particles will be modified by the field indirectly via the distribution function.) For static fields, we have $\boldsymbol{R}^{\rm st}_{ab}(t,\bar{t}) = (\frac{e_a}{m_a} - \frac{e_b}{m_b})\frac{\boldsymbol{E}_0}{2}(t-\bar{t})^2$.

The important nonlinear (exponential) dependence of the collision term (9.27) on the field strength will be discussed below more in detail.

As mentioned above, the kinetic equation (9.26) represents a generalized version of kinetic equations used in dense plasma physics. The neglection of quantum effects leads to the well-known classical kinetic equation derived by Silin (1960). On the other hand, for static fields, (9.26) reduces to the kinetic equation given by Morawetz and Kremp (1993). Finally, in the zero-field case, we get the non-Markovian form of the well-known quantum Landau equation (Klimontovich 1975).

Let us now consider the consequences of the nonlinear field dependence in the kinetic equation (9.26). This nonlinearity gives rise to interesting physical processes including generation of higher field harmonics and emission and/or absorption of multiple photons in the two-particle scattering in the case of high field strengths. To show this, we expand the spectral kernel of the collision integrals I_{ab} (second line in (9.27)) into a Fourier series making use of the Jacobi-Anger expansion

$$e^{iz\cos\theta} = \sum_{n=-\infty}^{\infty} (i)^n J_n(z) e^{in\theta}\,, \tag{9.30}$$

where J_n denotes the Bessel function of first kind. As a result, the collision integral in the kinetic equation (9.27) is transformed to

$$\begin{aligned} I_{ab}(\boldsymbol{k}_a,t) \;=\; & 2\mathrm{Re} \int \frac{d\boldsymbol{k}_b d\bar{\boldsymbol{k}}_a d\bar{\boldsymbol{k}}_b}{(2\pi\hbar)^6} \frac{1}{\hbar^2} \Big|V_{ab}(\boldsymbol{k}_a - \bar{\boldsymbol{k}}_a)\Big|^2 \delta(\boldsymbol{k}_a + \boldsymbol{k}_b - \bar{\boldsymbol{k}}_a - \bar{\boldsymbol{k}}_b) \\ & \times \sum_{n=-\infty}^{\infty} \sum_{l=-\infty}^{\infty} (i)^l J_n\left(\frac{\boldsymbol{q}\cdot\boldsymbol{w}^0_{ab}}{\hbar\Omega}\right) J_{n+l}\left(\frac{\boldsymbol{q}\cdot\boldsymbol{w}^0_{ab}}{\hbar\Omega}\right) \\ & \times \Big[\cos(l\Omega t) - i\sin(l\Omega t)\Big] \int_{t_0}^{t} d\bar{t}\; e^{\frac{i}{\hbar}\left(\epsilon_{ab} - \bar{\epsilon}_{ab} - \boldsymbol{q}\cdot\boldsymbol{w}_{ab}(t) + n\hbar\Omega\right)(t-\bar{t})} \\ & \times \left\{\bar{f}_a\bar{f}_b[1-f_a][1-f_b] - f_a f_b[1-\bar{f}_a][1-\bar{f}_b]\right\}\Big|_{\bar{t}}\,, \end{aligned} \tag{9.31}$$

where we used the notation

$$\boldsymbol{q} \equiv \boldsymbol{k}_a - \bar{\boldsymbol{k}}_a = \bar{\boldsymbol{k}}_b - \boldsymbol{k}_b \,; \quad \boldsymbol{w}_{ab}(t) \equiv \boldsymbol{v}_a(t) - \boldsymbol{v}_b(t);$$
$$\boldsymbol{w}^0_{ab} \equiv \boldsymbol{v}^0_a - \boldsymbol{v}^0_b; \quad \boldsymbol{v}_a(t) = \boldsymbol{v}^0_a \sin \Omega t; \quad \boldsymbol{v}^0_a \equiv \frac{e_a \boldsymbol{E}_0}{m_a \Omega}\,.$$

Obviously, $\boldsymbol{v}_a$ is the velocity of a classical particle of charge e_a in the periodic field.

This representation of the collision integral allows for a clear physical interpretation of the binary scattering process:

1. In a strong periodic field, Coulomb collisions with a shift $\boldsymbol{q}$ of the momentum give rise to the generation of higher harmonics of the field (sum over l). This has already been shown by Silin for classical plasmas. These terms are important on very short time scales, whereas they do not contribute to transport quantities which are averaged over times larger than the period of the field $2\pi/\Omega$.
2. Furthermore, it is obvious that collisions in strong harmonic fields are accompanied by emission and absorption of multiple photons, cf. the sum over n. Indeed, if the retardation in the distribution functions is omitted, and the initial time is shifted to minus infinity, the time integration in the collision term can be performed, giving rise to an energy delta function

$$\delta\left[\epsilon_{ab} - \bar{\epsilon}_{ab} + \boldsymbol{q} \cdot \boldsymbol{w}_{ab}(t) - n\hbar\Omega\right]\,.$$

This function describes the two-particle scattering process in the presence of a periodic electric field, which leads to multi-photon emission and absorption.

For high frequencies Ω, the collision integral (9.31) can be simplified by averaging over a period of the oscillating field

$$I_{ab}(\boldsymbol{k}_a, t) = 2 \int \frac{d\boldsymbol{k}_b d\bar{\boldsymbol{k}}_a d\bar{\boldsymbol{k}}_b}{(2\pi\hbar)^6} \frac{1}{\hbar^2} \left|V_{ab}(\boldsymbol{k}_a - \bar{\boldsymbol{k}}_a)\right|^2 \delta(\boldsymbol{k}_a + \boldsymbol{k}_b - \bar{\boldsymbol{k}}_a - \bar{\boldsymbol{k}}_b)$$
$$\times \sum_n J_n^2 \left(\frac{\boldsymbol{q} \cdot \boldsymbol{w}^0_{ab}}{\hbar\Omega}\right) \int d\bar{t} \cos\left[\frac{1}{\hbar}(\epsilon_{ab} - \bar{\epsilon}_{ab} - \boldsymbol{q} \cdot \boldsymbol{w}_{ab}(t) + n\hbar\Omega)(t - \bar{t})\right]$$
$$\times \left\{\bar{f}_a \bar{f}_b \left[1 - f_a\right]\left[1 - f_b\right] - f_a f_b \left[1 - \bar{f}_a\right]\left[1 - \bar{f}_b\right]\right\}\Big|_{\bar{t}}\,. \tag{9.32}$$

For classical plasmas, such an equation was given by Klimontovich (1975). Solutions of the kinetic equation (9.26) with the non-Markovian collision integral (9.32) cannot be determined analytically. Perturbative investigations are given in the subsequent sections. A numerical solution of this equation and the energy relaxation connected therewith was given in Haberland et al. (2001).

With the Born approximation, we used a simple approximation in order to demonstrate the essential modifications and physical aspects of the kinetic theory in strong and high-frequency fields. In order to proceed further, we take into account dynamical screening considering the V^s-approximation for the self-energy (Bonitz et al. 1999; Bornath et al. 2001). This approximation was intensely discussed in Chap. 4. It reads for the gauge-invariant Fourier transform

$$\Sigma_a^{\gtrless}(\boldsymbol{k};t,t') = i\hbar \int \frac{d\boldsymbol{q}}{(2\pi\hbar)^3} g_a^{\gtrless}(\boldsymbol{k}-\boldsymbol{q};t,t')\, V_{aa}^{s\gtrless}(\boldsymbol{q};t,t') \,. \tag{9.33}$$

The key quantities of this approximation are the correlation functions of the dynamically screened potential, $V^{s>}$ and $V^{s<}$, which are related to the correlation functions of the longitudinal field fluctuations. Within the random phase approximation (RPA), we have

$$\begin{aligned} V_{ab}^{s\gtrless}(\boldsymbol{q};t_1,t_2) &= \sum_c \int_{t_0}^{\infty} d\bar{t}\, V_{ac}(\boldsymbol{q})[\Pi_{cc}^R(\boldsymbol{q};t_1,\bar{t})V_{cb}^{s\gtrless}(\boldsymbol{q};\bar{t},t_2) \\ &\quad + \Pi_{cc}^{\gtrless}(\boldsymbol{q};t_1,\bar{t})V_{cb}^{s\,A}(\boldsymbol{q};\bar{t},t_2)] \,. \end{aligned} \tag{9.34}$$

Furthermore, we need the retarded and advanced screened potentials determined by the equation

$$\begin{aligned} V_{ab}^{s\,R/A}(\boldsymbol{q};t_1,t_2) &= V_{ab}(\boldsymbol{q})\delta(t_1-t_2) \\ &+ \sum_c \int_{t_0}^{\infty} d\bar{t}\, V_{ac}(\boldsymbol{q})\, \Pi_{cc}^{R/A}(\boldsymbol{q};t_1,\bar{t})\, V_{cb}^{s\,R/A}(\boldsymbol{q};\bar{t},t_2) \,. \end{aligned} \tag{9.35}$$

After insertion of this expression into (9.25), the collision integral can be written as

$$\begin{aligned} I_a(\boldsymbol{k}_a,t) &= 2\,\mathrm{Re} \int_{t_0}^{t} d\bar{t} \int \frac{d\boldsymbol{q}}{(2\pi\hbar)^3} \Big\{ V_{aa}^{s>}(\boldsymbol{q};t,\bar{t})\widetilde{\Pi}_{aa}^{<}\Big(\boldsymbol{k}_a+\boldsymbol{K}_a^A(t,\bar{t}),\boldsymbol{q}\,;\bar{t},t\Big) \\ &\quad - V_{aa}^{s<}(\boldsymbol{q};t,\bar{t})\,\widetilde{\Pi}_{aa}^{>}\Big(\boldsymbol{k}_a+\boldsymbol{K}_a^A(t,\bar{t}),\boldsymbol{q}\,;\bar{t},t\Big)\Big\} \,. \end{aligned} \tag{9.36}$$

Here, we introduced the auxiliary function

$$\widetilde{\Pi}_{aa}^{\lessgtr}(\boldsymbol{k}_a,\boldsymbol{q};\bar{t},t) = -i\hbar\, g_a^{\gtrless}(\boldsymbol{k}_a-\boldsymbol{q};t,\bar{t})\, g_a^{\lessgtr}(\boldsymbol{k}_a;\bar{t},t) \,.$$

There are different possibilities to transform the collision integral into another form. Usually the fluctuation–dissipation theorem for $V_{ab}^{s\gtrless}$ is used (like in Sect. 7.4). However, it is more convenient to use the screening equations (9.34) and (9.35) for the purpose of deriving source terms in the balance equations. Then we immediately get $I_a(\boldsymbol{k}_a,t) = \sum_b I_{ab}(\boldsymbol{k}_a,t)$ with

$$
\begin{aligned}
I_{ab}(\boldsymbol{k}_a,t) \;=\;& 2\,\mathrm{Re}\int \frac{d\boldsymbol{q}}{(2\pi\hbar)^3}\,V_{ab}(\boldsymbol{q})\int_{t_0}^{t} d\bar{t}_1\int_{t_0}^{t} d\bar{t}_2 \\
&\times\Big\{\Pi_{bb}^{R}(\boldsymbol{q};t,\bar{t}_2)\,V_{ba}^{s\,R}(\boldsymbol{q};\bar{t}_2,\bar{t}_1)\,\widetilde{\Pi}_{aa}^{<}\Big(\boldsymbol{k}_a+\boldsymbol{K}_a^{A},\boldsymbol{q}\,;\bar{t}_1,t\Big) \\
&+\Big[\Pi_{bb}^{R}(\boldsymbol{q};t,\bar{t}_2)V_{ba}^{s\,<}(\boldsymbol{q};\bar{t}_2,\bar{t}_1)+\Pi_{bb}^{<}(\boldsymbol{q};t,\bar{t}_2)V_{ba}^{s\,A}(\boldsymbol{q};\bar{t}_2,\bar{t}_1)\Big] \\
&\times\widetilde{\Pi}_{aa}^{A}\Big(\boldsymbol{k}_a+\boldsymbol{K}_a^{A},\boldsymbol{q};\bar{t}_1,t\Big)\Big]\Big\}\,. \qquad (9.37)
\end{aligned}
$$

So far, the collision integral (9.37) is given as a functional of the two-time correlation functions $g_a^{\gtrless}(t,t')$. In order to get an expression in terms of one-time distribution functions, we apply the gauge-invariant GKBA (9.23) in (9.37), and we finally obtain the collision integral (Bornath et al. 2001)

$$
\begin{aligned}
I_{ab}(\boldsymbol{k}_a,t) \;=\;& -\frac{2}{\hbar^2}\mathrm{Re}\int_{t_0}^{t} d\bar{t}_1\int_{t_0}^{t} d\bar{t}_2\int\frac{d\boldsymbol{q}}{(2\pi\hbar)^3}\int\frac{d\boldsymbol{k}_b}{(2\pi\hbar)^3}\,V_{ab}(\boldsymbol{q}) \\
&\times e^{-\frac{i}{\hbar}\{[\epsilon^a_{\boldsymbol{k}_a-\boldsymbol{q}}-\epsilon^a_{\boldsymbol{k}_a}](t-\bar{t}_1)-\boldsymbol{q}\cdot\boldsymbol{R}_a(t,\bar{t}_1)\}}\,e^{-\frac{i}{\hbar}\{[\epsilon^b_{\boldsymbol{k}_b+\boldsymbol{q}}-\epsilon^b_{\boldsymbol{k}_b}](t-\bar{t}_2)+\boldsymbol{q}\cdot\boldsymbol{R}_b(t,\bar{t}_2)\}} \\
&\times\Big\{\Big[f_b(\bar{t}_2)-\bar{f}_b(\bar{t}_2)\Big]\,V_{ba}^{s\,R}(\boldsymbol{q};\bar{t}_2,\bar{t}_1)\,f_a(\bar{t}_1)\Big[1-\bar{f}_a(\bar{t}_1)\Big] \\
&+\Big[f_b(\bar{t}_2)-\bar{f}_b(\bar{t}_2)\Big]V_{ba}^{s\,<}(\boldsymbol{q};\bar{t}_2,\bar{t}_1)\Big[f_a(\bar{t}_1)-\bar{f}_a(\bar{t}_1)\Big] \\
&+\bar{f}_b(\bar{t}_2)\Big[1-f_b(\bar{t}_2)\Big]V_{ba}^{s\,A}(\boldsymbol{q};\bar{t}_2,\bar{t}_1)\Big[f_a(\bar{t}_1)-\bar{f}_a(\bar{t}_1)\Big]\Big\}\,. \qquad (9.38)
\end{aligned}
$$

For a shorter notation, the momentum arguments of the distribution functions were dropped, i.e., it is $f_c(\bar{t}) = f_c[\boldsymbol{k}_c+\boldsymbol{Q}_c(t,\bar{t});\bar{t}]$ with $c=a,b$. For the functions $\bar{f}_a$ and $\bar{f}_b$, one has to replace $\boldsymbol{k}_c$ by $\boldsymbol{k}_a-\boldsymbol{q}$ or by $\boldsymbol{k}_b+\boldsymbol{q}$, respectively.

Again, the collision integral is essentially modified by the electromagnetic field. (i) The momentum arguments of the distribution functions are shifted by $\boldsymbol{Q}_c$, i.e., we observe the intra-collisional field effect. (ii) A further modification occurs in the exponential functions which essentially govern the energy balance. Additional terms appear leading to a field dependent broadening determined by the function $\boldsymbol{R}_{ab}$. (iii) The collision integral depends on the field in a nonlinear way.

9.1.3 Balance Equations. Electrical Current and Energy Exchange

One central quantity of the transport theory of plasmas in external electromagnetic fields is the electrical current defined by

$$
\boldsymbol{j}(t)=\sum_a e_a\int\frac{d\boldsymbol{k}_a}{(2\pi\hbar)^3}\,\frac{\boldsymbol{k}_a}{2m_a}f_a(k_a,t)=\sum_a e_a n_a\boldsymbol{u}_a(t)\,. \qquad (9.39)
$$

The balance equation for this quantity follows in well-known manner from the kinetic equation (9.25). We obtain

$$\frac{d\boldsymbol{j}}{dt} - \sum_a \frac{n_a e_a^2}{m_a}\boldsymbol{E} = \sum_{ab} \int \frac{d\boldsymbol{k}_a}{(2\pi\hbar)^3} \frac{e_a \boldsymbol{k}_a}{m_a} I_{ab}(\boldsymbol{k}_a)\,. \tag{9.40}$$

If collisions could be neglected completely, one would find

$$\boldsymbol{j}^{(0)} = \sum_a \frac{e_a^2}{m_a} n_a \int_{t_0}^{t} dt' \boldsymbol{E}(t')\,. \tag{9.41}$$

Defining $\boldsymbol{j} \equiv \boldsymbol{j}^{(0)} + \boldsymbol{j}^{(1)}$, we have

$$\frac{d\boldsymbol{j}^{(1)}}{dt} = \sum_{ab} \int \frac{d\boldsymbol{k}_a}{(2\pi\hbar)^3} \frac{e_a \boldsymbol{k}_a}{m_a} I_{ab}(\boldsymbol{k}_a)\,.$$

Now we substitute the collision integral (9.37), into the the r.h.s. and symmetrize with respect to a,b. Then the equation for the current $\boldsymbol{j}^{(1)}$ is given by

$$\begin{aligned}
\frac{d\boldsymbol{j}^{(1)}}{dt} &= \mathrm{Re} \int_{t_0}^{\infty} d\bar{t}_1 \int_{t_0}^{\infty} d\bar{t}_2 \int \frac{d\boldsymbol{q}}{(2\pi\hbar)^3} \frac{1}{2} \sum_{ab} \left(\frac{e_a}{m_a} - \frac{e_b}{m_b} \right) \\
&\times \boldsymbol{q}\, V_{ab}(q) \Big\{ \Pi^R_{bb}(\boldsymbol{q}; t, \bar{t}_2) V^{s\,R}_{ba}(\boldsymbol{q}; \bar{t}_2, \bar{t}_1) \Pi^<_{aa}(\boldsymbol{q}; \bar{t}_1, t) \\
&+ \Pi^R_{bb}(\boldsymbol{q}; t, \bar{t}_2) V^{s\,<}_{ba}(\boldsymbol{q}; \bar{t}_2, \bar{t}_1) \Pi^A_{aa}(\boldsymbol{q}; \bar{t}_1, t) \\
&+ \Pi^<_{bb}(\boldsymbol{q}; t, \bar{t}_2) V^{s\,A}_{ba}(\boldsymbol{q}; \bar{t}_2, \bar{t}_1) \Pi^A_{aa}(\boldsymbol{q}; \bar{t}_1, t) \Big\}\,,
\end{aligned} \tag{9.42}$$

where the upper limits of the time integrations are determined by the Heaviside functions contained in the advanced and retarded functions, respectively. It is easy to see that only collision terms I_{ab} with $a \neq b$ contribute to the balance of the current.

In the following, we will consider a two-component plasma consisting of electrons and ions ($m_i \gg m_e$). The adiabatic approximation leads to $\Pi^{R/A}_{ii}(\boldsymbol{q}; t, t') \approx 0$ and $i\hbar\, \Pi^<_{ii}(\boldsymbol{q}; t, t') \approx n_i$ with n_i being the ion density. Furthermore, $V_{ab}(q) = e_a e_b \hat{V}(q)$ is the Coulomb potential with $\hat{V}(q) = 1/(\varepsilon_0 q^2/\hbar^2)$ (in the remaining part of Sect .9.1 we use SI units). With these assumptions, expression (9.42) is simplified considerably

$$\begin{aligned}
\frac{d\boldsymbol{j}^{(1)}}{dt} &= n_i\, \mathrm{Re} \int_{t_0}^{t} d\bar{t}_1 \int_{\bar{t}_1}^{t} d\bar{t}_2 \int \frac{d\boldsymbol{q}}{(2\pi\hbar)^3} \\
&\times \frac{e_e e_i^2}{m_e i\hbar} \boldsymbol{q}\, \hat{V}(q)\, e_e^2\, \Pi^R_{ee}(\boldsymbol{q}; t, \bar{t}_2) \hat{V}^{s\,R}(\boldsymbol{q}; \bar{t}_2, \bar{t}_1)\,,
\end{aligned} \tag{9.43}$$

where only the electrons contribute to the dynamically screened potential $\hat{V}^{s\,R} = \hat{V} + \hat{V}\, e^2 \Pi^R_{ee}\, \hat{V}^{s\,R}$. Now it is useful to define the following quantity

$$\varepsilon^{-1}(\boldsymbol{q}; t, t') = \delta(t - t') + \int_{t'}^{t} d\bar{t}\, e_e^2\, \Pi^R_{ee}(\boldsymbol{q}; t, \bar{t}) \hat{V}^{s\,R}(\boldsymbol{q}; \bar{t}, t') \,. \tag{9.44}$$

This function represents a generalization of the dielectric function. It is a functional of the electron Green's functions, and it includes the full memory. The Wigner distribution functions can be introduced in an approximation by the generalized Kadanoff–Baym ansatz (GKBA) in its gauge invariant form (cf. Sect. 9.1.2).

With this definition, the balance equation (9.40) gets the interesting form (Bornath et al. 2001)

$$\begin{aligned} \frac{d\boldsymbol{j}}{dt} - \sum_a \frac{n_a e_a^2}{m_a} \boldsymbol{E} &= \frac{e_e n_i e_i^2}{m_e \hbar} \operatorname{Re} \int_{t_0}^{t} d\bar{t}_1 \int \frac{d\boldsymbol{q}}{(2\pi\hbar)^3} \\ &\quad \times \boldsymbol{q}\, \hat{V}(q)\, \frac{1}{i} \left[\varepsilon^{-1}(\boldsymbol{q}; t, \bar{t}_1) - \delta(t - \bar{t}_1) \right] . \end{aligned} \tag{9.45}$$

Let us now consider the energy balance, resulting from the second moment of the kinetic equation (9.25)

$$\frac{dW^{\mathrm{kin}}}{dt} - \boldsymbol{j} \cdot \boldsymbol{E} = \sum_{a,b} \int \frac{d\boldsymbol{k}_a}{(2\pi\hbar)^3} \frac{k_a^2}{2m_a} I_{ab}(\boldsymbol{k}_a) \,. \tag{9.46}$$

Here, W^{kin} is the mean kinetic energy defined by

$$W^{\mathrm{kin}}(t) = \sum_a \int \frac{d\boldsymbol{k}_a}{(2\pi\hbar)^3} \frac{k_a^2}{2m_a} f_a(k_a, t) \,. \tag{9.47}$$

It is possible to show (cf. Bornath et al. (2001)) that the right hand side of (9.46) is just $-(\mathrm{d}/\mathrm{d}t)W^{\mathrm{pot}}$ with the mean potential energy density given by

$$\begin{aligned} W^{\mathrm{pot}} &= \frac{1}{2}(i\hbar) \sum_{a,b} \int_{t_0}^{\infty} d\bar{t}_1 \int_{t_0}^{\infty} d\bar{t}_2 \int \frac{d\boldsymbol{q}}{(2\pi\hbar)^3} V_{ab}(q) \\ &\times \Big[\Pi^R_{bb}(\boldsymbol{q}; t, \bar{t}_2)\, V^{s\,R}_{ba}(\boldsymbol{q}; \bar{t}_2, \bar{t}_1)\, \Pi^<_{aa}(\boldsymbol{q}; \bar{t}_1, t) \\ &+ \Pi^R_{bb}(\boldsymbol{q}; t, \bar{t}_2)\, V^{s\,<}_{ba}(\boldsymbol{q}; \bar{t}_2, \bar{t}_1)\, \Pi^A_{aa}(\boldsymbol{q}; \bar{t}_1, t) \\ &+ \Pi^<_{bb}(\boldsymbol{q}; t, \bar{t}_2)\, V^{s\,A}_{ba}(\boldsymbol{q}; \bar{t}_2, \bar{t}_1)\, \Pi^A_{aa}(\boldsymbol{q}; \bar{t}_1, t) \Big] . \end{aligned} \tag{9.48}$$

Thus, the energy balance (9.46) reads now

$$\frac{dW^{\mathrm{kin}}}{dt} + \frac{dW^{\mathrm{pot}}}{dt} = \boldsymbol{j} \cdot \boldsymbol{E} \,, \tag{9.49}$$

i.e., the change of the total energy of the system of particles is equal to $\boldsymbol{j} \cdot \boldsymbol{E}$ which is in turn the energy loss of the electromagnetic field due to Poynting's theorem. Both the mean kinetic energy and the potential energy are functionals of the actual distribution functions which follow from the kinetic equation. It is an important feature of (9.49) that the total energy occurs on the left hand side. This means that a nonideal system is described by the underlying non-Markovian kinetic equation.

Let us come back to the equation of motion for the current (9.45). So far, this equation is not closed because the r.h.s. is a functional of the unknown electron distribution function. In the case of strong high-frequency fields, one can use an approximation which is due to Silin (Bornath et al. 2001). In the decomposition $f(\boldsymbol{k},t) = f^0(\boldsymbol{k},t) + f^1(\boldsymbol{k},t)$, the contribution $f^0(\boldsymbol{k},t)$ fulfills the equation

$$\left\{\frac{\partial}{\partial t} - e_e \frac{\partial \boldsymbol{A}}{\partial t} \cdot \nabla_{\boldsymbol{k}}\right\} f^0(\boldsymbol{k},t) = 0\,. \tag{9.50}$$

The term f^1 is due to collisions and is considered to be a small correction which can be neglected on the r.h.s. of (9.45). Equation (9.50) has the solution

$$\begin{aligned} f^0(\boldsymbol{k},t) &= f_0\left(\boldsymbol{k} + e_e \boldsymbol{A}(t)\right) \\ &= f_0\left(\boldsymbol{k} - m_e \boldsymbol{u}_e^0(t)\right) \end{aligned} \tag{9.51}$$

with $\boldsymbol{u}_e^0(t) = -(e_e/m_e)\boldsymbol{A}(t)$ being the velocity of a free electron in a homogeneous electric field, and f_0 being an arbitrary function depending on the initial conditions. Usually it is assumed to be a local equilibrium function.

We want to generalize this concept, in order to describe the case of weak or intermediate field strengths, too. In the following we will assume that the electron subsystem is in local thermodynamic equilibrium (with slowly varying temperature $T_e(t)$) with respect to a coordinate frame moving with the mean velocity $\boldsymbol{u}_e(t)$. This can be considered to be a generalization of the quasi-hydrodynamic approximation. The transformation from such coordinate system to a system at rest is given by

$$\tilde{\boldsymbol{r}} = \boldsymbol{r} - \int_{t_0}^{t} \mathrm{d}\bar{t}\, \boldsymbol{u}_e(\bar{t})\,, \tag{9.52}$$

where we assumed that the two systems coincide at time t_0. Now we consider

$$\varepsilon(\boldsymbol{q}, t_1 t_2) = \int d(\boldsymbol{r}_1 - \boldsymbol{r}_2) e^{-\frac{i}{\hbar}\boldsymbol{q}\cdot(\boldsymbol{r}_1 - \boldsymbol{r}_2)} \varepsilon(\boldsymbol{r}_1 t_1, \boldsymbol{r}_2 t_2)\,. \tag{9.53}$$

In the spatially homogeneous system considered, the function depends only on the difference $(\boldsymbol{r}_1 - \boldsymbol{r}_2)$. On the r.h.s., the transformation into the moving system results in

$$\begin{aligned}
\varepsilon(\boldsymbol{q},t_1t_2) &= \int d(\tilde{\boldsymbol{r}}_1-\tilde{\boldsymbol{r}}_2)e^{-\frac{i}{\hbar}\boldsymbol{q}\cdot\left[\tilde{\boldsymbol{r}}_1+\int_{t_0}^{t_1}\mathrm{d}\bar{t}\,\boldsymbol{u}_e(\bar{t})-\tilde{\boldsymbol{r}}_2-\int_{t_0}^{t_2}\mathrm{d}\bar{t}\,\boldsymbol{u}_e(\bar{t})\right]}\,\tilde{\varepsilon}(\tilde{\boldsymbol{r}}_1t_1,\tilde{\boldsymbol{r}}_2t_2) \\
&= e^{-\frac{i}{\hbar}\boldsymbol{q}\cdot\int_{t_2}^{t_1}\mathrm{d}\bar{t}\,\boldsymbol{u}_e(\bar{t})}\int d(\tilde{\boldsymbol{r}}_1-\tilde{\boldsymbol{r}}_2)e^{-\frac{i}{\hbar}\boldsymbol{q}\cdot(\tilde{\boldsymbol{r}}_1-\tilde{\boldsymbol{r}}_2)}\,\tilde{\varepsilon}(\tilde{\boldsymbol{r}}_1t_1,\tilde{\boldsymbol{r}}_2t_2) \\
&= e^{-\frac{i}{\hbar}\boldsymbol{q}\cdot\int_{t_2}^{t_1}\mathrm{d}\bar{t}\,\boldsymbol{v}_a(\bar{t})}\,\tilde{\varepsilon}(\boldsymbol{q},t_1,t_2)\,. \qquad (9.54)
\end{aligned}$$

As stated above, we assume

$$\tilde{\varepsilon}(\boldsymbol{q},t_1,t_2)\approx\varepsilon_{\mathrm{RPA}}(\boldsymbol{q},t_1-t_2)\,, \qquad (9.55)$$

where $\varepsilon_{\mathrm{RPA}}$ denotes the local equilibrium function. Therefore, we have

$$\varepsilon(\boldsymbol{q},t_1t_2) = e^{-\frac{i}{\hbar}\boldsymbol{q}\cdot\int_{t_2}^{t_1}\mathrm{d}\bar{t}\,\boldsymbol{u}_e(\bar{t})}\,\varepsilon_{\mathrm{RPA}}(\boldsymbol{q},t_1-t_2)\,. \qquad (9.56)$$

With $\boldsymbol{j}\approx e_en_eu_e$, the equation of motion for the current (9.45) now reads

$$\begin{aligned}
\frac{d\boldsymbol{j}(t)}{dt}-\sum_a\frac{n_ae_a^2}{m_a}\boldsymbol{E}(t) &= \frac{e_en_ie_i^2}{m_e\hbar}\,\mathrm{Re}\frac{1}{i}\int_{t_0}^{t}d\bar{t}_1\int\frac{d\boldsymbol{q}}{(2\pi\hbar)^3}\,\boldsymbol{q}\,\hat{V}(q) \\
\times\exp\left[\frac{i}{\hbar e_en_e}\boldsymbol{q}\int_{\bar{t}_1}^{t}\boldsymbol{j}_a(\bar{t})\,d\,\bar{t}\right] &\left[\varepsilon_{\mathrm{RPA}}^{-1}(\boldsymbol{q};t-\bar{t}_1)\;-\delta(t-\bar{t}_1)\right]\,. \qquad (9.57)
\end{aligned}$$

Equation (9.57) is a nonlinear non-Markovian equation for the current. The most interesting feature of this important relation is the non-perturbative character with respect to the external field. As a consequence, we find a nonlinear connection between plasma current and laser field. In the following, the relation (9.57) will serve as the basis for the investigation of typical effects of the field-plasma interaction, like higher harmonics in the current and inverse bremsstrahlung absorption and emission in a dense quantum plasma.

Equation (9.57) may be considered in two special cases. In the case of weak fields, it is convenient to linearize (9.57) with respect to $\boldsymbol{j}$ what leads to a linear response theory of plasmas in time-dependent fields. In the case of strong fields, such a simplification is not possible. For strong fields, a good approximation is to replace the full current $\boldsymbol{j}$ by $\boldsymbol{j}^{(0)}$, what corresponds to the Silin ansatz. We will consider these two cases in the next sections.

9.1.4 Plasmas in Weak Laser Fields. Generalized Drude Formula

Here, we will consider the weak field case, i.e., we assume a situation in which the thermal velocity v_{th} is larger than the quiver velocity v_0. Furthermore, we consider the electron–ion system in adiabatic approximation ($m_e\ll m_i$). Then we find the following simplified equation of motion for the current $\boldsymbol{j}(t)$ of the electrons

$$\frac{d\boldsymbol{j}(t)}{dt} - \omega_{pl}^2 \varepsilon_0 \boldsymbol{E}(t) = \frac{e_e n_i e_i^2}{m_e \hbar} \operatorname{Re} \frac{1}{\hbar} \int_{t_0}^{t} d\bar{t}_1 \left[\int_{\bar{t}_1}^{t} \frac{\boldsymbol{j}(\bar{t})}{e_e n_e} \, d\bar{t} \right]$$
$$\times \int \frac{d\boldsymbol{q}}{(2\pi\hbar)^3} \, \boldsymbol{q} \otimes \boldsymbol{q} \, \hat{V}(q) \left(\varepsilon_{\mathrm{RPA}}^{-1}(\boldsymbol{q}; t - \bar{t}_1) - \delta(t - \bar{t}_1) \right) . \tag{9.58}$$

For the further considerations, we adopt a harmonic time dependence of the field again, i.e., we have $\boldsymbol{E}(t) = E_0(e^{i\omega t} + e^{-i\omega t})/2$, and, therefore, the current is given by the ansatz $\boldsymbol{j}(t) = \boldsymbol{j}(\omega)e^{-i\omega t} + \boldsymbol{j}^*(\omega)e^{i\omega t}$. Inserting this ansatz into (9.58), we find equations for $\boldsymbol{j}(\omega)$ and $\boldsymbol{j}^*(\omega)$. It is sufficient to consider the equation for $\boldsymbol{j}(\omega)$ only. It follows

$$-i\omega \, \boldsymbol{j}(\omega) - \omega_{pl}^2 \, \varepsilon_0 \frac{\boldsymbol{E}_0}{2} = -\nu_{ei}(\omega)\boldsymbol{j}(\omega) , \tag{9.59}$$

where the complex dynamical electron–ion collision frequency $\nu_{ei}(\omega)$ was introduced by

$$\nu_{ei}(\omega) = i \frac{n_i e_i^2}{6\pi^2 m_e n_e \hbar^3} \frac{1}{\omega} \int_0^\infty dq \, q^4 \, V(q) \left[\varepsilon_{\mathrm{RPA}}^{-1}(\boldsymbol{q}; \omega) - \varepsilon_{\mathrm{RPA}}^{-1}(\boldsymbol{q}; 0) \right] . \tag{9.60}$$

This is an important result. If we solve the equation for the current (9.59), we obtain a generalization of the well-known Drude formula

$$\boldsymbol{j}(\omega) = \frac{\omega_{pl}^2}{-i\omega + \nu_{ei}(\omega)} \frac{\varepsilon_0 \boldsymbol{E}_0}{2} . \tag{9.61}$$

In contrast to the usual Drude formula, we now have a frequency dependent complex electron–ion collision frequency given in RPA by formula (9.60). From the generalized Drude formula (9.61), we immediately obtain some other important quantities. The conductivity is given by

$$\sigma(\omega) = \frac{\omega_{pl}^2 \, \varepsilon_0}{-i\omega + \nu_{ei}(\omega)} , \tag{9.62}$$

and the polarization $\boldsymbol{P}$ is given by

$$\boldsymbol{P} = \frac{\boldsymbol{j}}{-i\omega} = \frac{\omega_{pl}^2 \, \varepsilon_0}{-i\omega(-i\omega + \nu_{ei}(\omega))} \frac{\boldsymbol{E}_0}{2} . \tag{9.63}$$

With $\boldsymbol{D} = \varepsilon_0 \boldsymbol{E} + \boldsymbol{P} = \varepsilon_0 \varepsilon \boldsymbol{E}$, we then obtain for the dielectric function in the long wave limit

$$\varepsilon(\omega) = 1 - \frac{\omega_{pl}^2}{\omega(\omega + i\nu_{ei}(\omega))} . \tag{9.64}$$

It is interesting to consider the asymptotic behavior of such relations for high and low frequencies. For the conductivity, e.g., we get the asymptotic relations

$$\sigma(\omega) = \frac{\omega_{pl}^2 \varepsilon_0}{\omega}\left(i + \frac{\nu_{ei}(\omega)}{\omega}\right), \qquad \omega \to \infty, \tag{9.65}$$

$$\sigma(\omega) = \frac{\omega_{pl}^2 \varepsilon_0}{\nu_{ei}(0)}, \qquad \omega \to 0. \tag{9.66}$$

The high-frequency conductivity (9.65) with the classical limit of the collision frequency (9.60) was given in papers by Dawson and Oberman (1962), Oberman et al. (1962), and Dawson and Oberman (1963), and is also discussed in the monograph by Ecker (1972). Recently, a classical theory was given taking into account strong correlations between electrons and ions which leads to the Dawson–Oberman result in the weak coupling case (Felderhof 1998; Felderhof and Vehns 2004; Felderhof and Vehns 2005).

9.1.5 Absorption and Emission of Radiation in Weak Laser Fields

Important properties of the interaction between the plasma and the electromagnetic field can be described on the basis of Drude formulas. For example, the energy transfer between field and plasma (9.49) is given by $\boldsymbol{j} \cdot \boldsymbol{E}$. The energy which is transferred to the plasma is usually characterized by the average over one period of oscillation, i.e., by

$$\langle \boldsymbol{j} \cdot \boldsymbol{E} \rangle = \frac{1}{T}\int_0^T dt\, \boldsymbol{j}(t) \cdot \boldsymbol{E}(t) = \boldsymbol{E}_0 \cdot \mathrm{Re}\boldsymbol{j}(\omega). \tag{9.67}$$

Using the generalized Drude formula (9.61), we get with $\langle \boldsymbol{E}^2 \rangle = \boldsymbol{E}_0^2/2$

$$\frac{\langle \boldsymbol{j} \cdot \boldsymbol{E} \rangle}{\langle \varepsilon_0 \boldsymbol{E}^2 \rangle} = \frac{\omega_{pl}^2}{[\omega - \mathrm{Im}\nu_{ei}(\omega)]^2 + [\mathrm{Re}\nu_{ei}(\omega)]^2}\mathrm{Re}\nu_{ei}(\omega). \tag{9.68}$$

In the high-frequency limit, this simplifies to

$$\frac{\langle \boldsymbol{j} \cdot \boldsymbol{E} \rangle}{\langle \varepsilon_0 \boldsymbol{E}^2 \rangle} = \frac{\omega_{pl}^2}{\omega^2}\mathrm{Re}\nu_{ei}(\omega). \tag{9.69}$$

The energy transfer between field and plasma can be considered as emission and absorption of photons. In a strongly ionized plasma, emission of bremsstrahlung and absorption (inverse bremsstrahlung) are the dominating radiative processes. Let us first show, how the absorption coefficient of the inverse bremsstrahlung can be determined from the collision frequency (9.60). For this aim, we remember that the phase velocity of an electromagnetic wave in a plasma with $\mu = 1$ is given by $v = c/n = c/\sqrt{\varepsilon}$. For the complex refraction index $\sqrt{\varepsilon} = n' + in''$, one gets easily

$$n''(\omega) = \frac{\mathrm{Im}\varepsilon(\omega)}{2n'(\omega)}, \quad n'(\omega) = \frac{1}{\sqrt{2}}\sqrt{\mathrm{Re}\varepsilon + \sqrt{(\mathrm{Re}\varepsilon)^2 + (\mathrm{Im}\varepsilon)^2}}\,. \tag{9.70}$$

Using this relation in the formula for the propagation of a plane wave along the z axis,

$$e^{-i(\omega t - k_0 n z)} = e^{-k_0 n'' z} e^{-i(\omega t - k_0 n' z)}\,,$$

we see that $k_0 n''(\omega) = \alpha(\omega)$ has the meaning of an attenuation of the amplitude along the z-direction. Therefore, it is obvious to define

$$\alpha(\omega) = \frac{\omega}{2n'(\omega)c}\,\mathrm{Im}\varepsilon(\omega) \tag{9.71}$$

as the absorption coefficient. With relation (9.64), it follows finally that

$$\alpha(\omega) = \frac{\omega_{pl}^2}{2n'(\omega)c}\,\frac{\mathrm{Re}\nu_{ei}(\omega)}{|\omega + i\nu_{ei}(\omega)|^2}\,. \tag{9.72}$$

In the high frequency regime $\omega \gg \omega_{pl}$, we have $\mathrm{Re}\varepsilon \approx 1$, and $n'(\omega) \approx 1$. Then we may write

$$\alpha(\omega) = \frac{1}{c}\frac{\omega_{pl}^2}{\omega^2}\,\mathrm{Re}\nu_{ei}(\omega) = \frac{1}{c}\frac{\langle \boldsymbol{j}E\rangle}{\varepsilon_0\left\langle \boldsymbol{E}^2\right\rangle}\,. \tag{9.73}$$

With the knowledge of the inverse bremsstrahlung absorption, we are able to compute the emission of bremsstrahlung using Kirchhoff's law. We consider an equilibrium plasma. Then the emission can be found in terms of the absorption and the black body energy intensity by

$$\alpha(\omega)\frac{\hbar\omega^3}{4\pi^3 c^3(\exp[\frac{\hbar\omega}{k_BT}] - 1)} = I(\omega)\,, \tag{9.74}$$

where $I(\omega)$ and $\alpha(\omega)$ are the emission and the absorbtion coefficients, respectively. The emission coefficient determines the rate of radiation energy per unit volume, frequency, and solid angle

$$I(\omega) = \frac{\hbar\omega}{4\pi}\,\overline{\sigma(\omega, v)}\,, \tag{9.75}$$

with $\overline{\sigma(\omega, v)}$ being the averaged radiation cross section. The average has to be carried out over the velocities using a Maxwell distribution.

The formulas just given demonstrate clearly that the dynamical collision frequency (9.60) turns out to be a central quantity for the description of physical processes of plasmas in electromagnetic fields. Therefore, we investigate further properties of this quantity. For simplification, we restrict ourselves to the non-degenerate version of the expression (9.60), and using the results of Chap. 4, especially formula (4.115), we get

$$\mathrm{Re}\nu_{ei} = \frac{4}{3}\frac{\sqrt{2\pi}}{m_e^2}\frac{Ze^4 n_i}{v_{th}^3}\frac{\sinh x}{x}\int_0^\infty dq\,\frac{\exp\{-[x^2\frac{2}{(q\lambda_e)^2}+\frac{(q\lambda_e)^2}{8}]\}}{q\,\left|\varepsilon(q,\omega)\right|^2} \tag{9.76}$$

with $x = \hbar\omega/(2k_BT)$. This expression is a generalization of the expression given by Perel' and Eliashberg (1962), and by Klimontovich (1975). The analytical evaluation of the q-integration is possible under some assumptions with respect to the frequency as demonstrated below.

Let us first consider the situation $\omega \gg \omega_{pl}$. In this case, the polarization can be neglected. The convergence of the integral is ensured by the exponential factor. Using the substitution $(q\lambda_e)^2/8 = t$, it is now possible to transform the integral into the integral representation of the modified Bessel function $K_0(x)$. In this way, we find the following form for the high-frequency electron–ion collision frequency

$$\begin{aligned}\mathrm{Re}\nu_{ei} &= \frac{4}{3}\frac{\sqrt{2\pi}}{m_e^2}\frac{Ze^4 n_i}{v_{th}^3}\frac{\sinh x}{x}K_0(x)\\ &= \frac{4}{3}\frac{\sqrt{2\pi}}{m_e^2}\frac{Ze^4 n_i}{v_{th}^3}\frac{\exp[\frac{\hbar\omega}{k_BT}]-1}{\hbar\omega/k_BT}\exp\left[-\frac{\hbar\omega}{2k_BT}\right]K_0(x).\end{aligned} \tag{9.77}$$

This result was obtained without any arbitrary cut-off in the q-integration. The classical limit follows with $x \to 0$. Then the asymptotic relation $K_0(x) \to -\ln\frac{x}{2}$ can be applied, and we find

$$\mathrm{Re}\nu_{ei} = -\frac{4}{3}\frac{\sqrt{2\pi}}{m_e^2}\frac{Ze^4 n_i}{v_{th}^3}\ln\hbar\omega/4k_BT = \frac{4}{3}\frac{\sqrt{2\pi}}{m_e^2}\frac{Ze^4 n_i}{v_{th}^3}\ln\frac{q_{\max}}{q_{\min}}. \tag{9.78}$$

In order to illustrate the connection to the well-known classical formulas, we have rewritten the relation by introduction of the usual cut-off parameters which are given here by $q_{\max} = \frac{4}{\lambda_e}$, and $q_{\min} = \omega/v_{th}$. Expression (9.78) was derived by many authors (for instance by Silin (1964)) with small modifications of the so-called Coulomb logarithm.

Next we consider (9.77) for large $x = \hbar\omega/2k_BT$. In this case, the asymptotic representation $K_0(x) \to \sqrt{\pi/2x}\exp(-x)$ can be used, and we find

$$\mathrm{Re}\nu_{ei} = \frac{4}{3}\frac{\sqrt{2\pi}}{m_e^2}\frac{Z^2e^4 n_i}{v_{th}^3}\sqrt{\frac{\pi}{\frac{\hbar\omega}{k_BT}}}\frac{\exp\left(\frac{\hbar\omega}{k_BT}\right)-1}{\hbar\omega/k_BT}\exp\left(-\frac{\hbar\omega}{k_BT}\right). \tag{9.79}$$

With the knowledge of the collision frequency, we are able to determine the bremsstrahlung emission and absorption spectrum in the plasma. According to (9.73), we find approximations for the inverse bremsstrahlung or collisional absorption rate.

Now Kirchhoff's law can be used to determine the emission rate of the bremsstrahlung. Using, for example, the relation (9.77), the thermally averaged emission coefficient is

$$I(\omega) = \frac{16}{3} \frac{1}{\pi c^3} \frac{\sqrt{2\pi}}{m_e^2} \frac{Z^2 e^6 n_i n_e}{v_{th}} \exp\left(-\frac{\hbar\omega}{2k_B T}\right) K_0(x). \tag{9.80}$$

This is a well-known result of the theory of plasma radiation (Shkarofsky, Johnston, and Bachinski 1966). It was for the first time given by Green (1959) and DeWitt (1959). For further considerations, we refer to monographs on plasma radiation (Bekefi 1966).

A generalization of the expressions for the rates of the inverse bremsstrahlung and of the bremsstrahlung was recently given by Röpke (1988a), Reinholz et al. (2000), and by Wierling et al. (2001). In the latter, the effective electron–ion collision frequency was determined for strong collisions from the half-on-shell T-matrix. Furthermore, a combination of the T-matrix approximation with the dynamically screened expression (9.60) according to the Gould–DeWitt scheme was discussed.

The interesting problem of the interaction of lasers with semiconductors was dealt with, e.g., in Henneberger and Haug (1988) and in Henneberger et al. (2000).

Let us finally consider the limit $\omega \to 0$. In this case, it is not possible to neglect the screening. In statical screening approximation, we get from (9.76)

$$\mathrm{Re}\nu_{ei} = \frac{4}{3} \frac{\sqrt{2\pi}}{m_e^2} \frac{Ze^4 n_i}{v_{th}^3} \int_0^\infty dq \, \frac{\exp\left[-\frac{(q\lambda_e)^2}{8}\right]}{q\, |\varepsilon(q,0)|^2} = \frac{4}{3} \frac{\sqrt{2\pi}}{m_e^2} \frac{Ze^4 n_i}{v_{th}^3} L. \tag{9.81}$$

Then the static conductivity follows immediately from the relation (9.66). However, this procedure gives not the well-known Spitzer formulas (2.222) and (2.223) presented in Chap. 2. We get, instead of the pre-factor 1.0159, the value $3/(4\sqrt{2\pi}) = 0.299$. The reason for this discrepancy is that only five moments are considered to solve the underlying kinetic equation. If one considers higher moments or, equivalently, higher orders in the Sonine polynomial expansion, there is a rapid convergence to the exact Spitzer value within the adiabatic approximation. A better theory of the static conductivity is the topic of Sect. 9.2.

9.1.6 Plasmas in Strong Laser Fields. Higher Harmonics

Let us now consider the plasma under the influence of a strong high-frequency laser field. We have, therefore, a situation where the quiver velocity is larger than the thermal velocity, and the field frequency is larger than the collision frequency. For the first time, such a situation was discussed by Silin (1964) in the framework of the classical kinetic theory with a modified Landau collision integral, later on by Klimontovich (1975) in dynamically screened approximation, and more recently by Decker et al. (1994) in terms of the dielectric model. A quantum mechanical version of the dielectric model was presented by Kull and Plagne (2001). A ballistic model was used by Mulser and Saemann (1997) and by Mulser et al. (2000). Quantum mechanical treatments of the problem were given by Rand (1964), by Schlessinger and Wright (1979) and by Silin and Uryupin (1981). Effects of non-Maxwellian distributions were considered by Langdon (1980), and investigated in more detail in several subsequent papers (see, e.g., Jones and Lee (1982), Albritton (1983), Chichkov et al. (1992), and Haberland et al. (2001)).

Here, we start from the general equation (9.57). For a strong high-frequency laser field, a linearization is not possible. We follow Silin (1964) and assume that the influence of the collisions may be considered as a small perturbation compared to the external field in the case of a strong high-frequency laser field. We decompose the current according to $\boldsymbol{j} = \boldsymbol{j}^{(0)} + \boldsymbol{j}^{(1)}$ and assume $\boldsymbol{j}^{(0)} \gg \boldsymbol{j}^{(1)}$. Then equation (9.57) may be linearized with respect to $\boldsymbol{j}^{(1)}$, and we obtain

$$\begin{aligned}\frac{d\boldsymbol{j}^{(1)}}{dt} &= \frac{e_e n_i e_i^2}{m_e \hbar} \mathrm{Re} \frac{1}{i} \int_{t_0}^{t} d\bar{t}_1 \int \frac{d\boldsymbol{q}}{(2\pi\hbar)^3}\, \boldsymbol{q}\, \hat{V}(q) \exp\left[\frac{i}{\hbar}\frac{e_e \boldsymbol{q}}{m_e} \int_{\bar{t}_1}^{t} d\bar{t}\, \boldsymbol{A}(\bar{t})\right] \\ &\times \left[1 + \frac{q}{\hbar}\int_{\bar{t}_1}^{t} d\bar{t}\, \frac{\boldsymbol{j}^{(1)}(\bar{t})}{n_e e}\right] \left[\varepsilon^{-1}_{\mathrm{RPA}}(\boldsymbol{q}; t - \bar{t}_1) - \delta(t - \bar{t}_1)\right] . \end{aligned} \tag{9.82}$$

The external field is treated here in a non-perturbative manner. As a consequence, we find a nonlinear connection between the plasma current and the laser field which gives rise to non-linear effects, such as higher harmonics in the current and non-linear inverse Bremsstrahlung absorption in a dense quantum plasma.

In a first approximation, we neglect the contribution on the r.h.s. of (9.82) containing $\boldsymbol{j}^{(1)}$. Of course, we could not obtain Drude-like relations for the current in this case. For a harmonic field, $\boldsymbol{E} = \boldsymbol{E}_0 \cos\omega t$, one can expand (9.82) into a Fourier series using the relation (9.30),

$$\begin{aligned}\frac{d\boldsymbol{j}^{(1)}}{dt} &= \frac{e_e n_i e_i^2}{m_e \hbar} \mathrm{Re} \int \frac{d\boldsymbol{q}}{(2\pi\hbar)^3}\, \boldsymbol{q}\, \hat{V}(q) \sum_m \sum_n (-i)^{m+1}\, e^{im\omega t} \\ &\times J_n\left(\frac{\boldsymbol{q}\cdot\boldsymbol{v}_e^0}{\hbar\omega}\right) J_{n-m}\left(\frac{\boldsymbol{q}\cdot\boldsymbol{v}_e^0}{\hbar\omega}\right) \int_0^{t-t_0} d\tau\, e^{-\frac{i}{\hbar}n\omega\tau} \left[\varepsilon^{-1}_{\mathrm{RPA}}(\boldsymbol{q};\tau) - \delta(\tau)\right] , \end{aligned} \tag{9.83}$$

with $\boldsymbol{v}_e^0 = e_e \boldsymbol{E}_0 / m_e \omega$. For times $t \gg t_0$, which will be considered in the following, the integral in the second line is just the Fourier transform of the inverse dielectric function.

Integrating the above equation, one has

$$\begin{aligned} \boldsymbol{j}^{(1)}(t) &= \frac{e_e n_i e_i^2}{m_e \hbar} \operatorname{Im} \int \frac{d\boldsymbol{q}}{(2\pi\hbar)^3} \boldsymbol{q} \hat{V}(q) \sum_m \sum_n (-i)^{m+2} \frac{e^{im\omega t}}{m\omega} \\ &\times J_n\left(\frac{\boldsymbol{q}\cdot\boldsymbol{v}_e^0}{\hbar\omega}\right) J_{n-m}\left(\frac{\boldsymbol{q}\cdot\boldsymbol{v}_e^0}{\hbar\omega}\right) \left[\frac{1}{\varepsilon_{\mathrm{RPA}}(\boldsymbol{q}; -n\omega)} - 1\right], \quad (9.84) \end{aligned}$$

which is clearly a Fourier expansion of the current in terms of all harmonics,

$$\boldsymbol{j}(t) = \sum_{m=-\infty}^{\infty} \boldsymbol{j}_m(\omega)\, e^{-im\omega t}. \quad (9.85)$$

The Fourier coefficients $\boldsymbol{j}_m$ of the current $\boldsymbol{j}(t)$ (with $j_m = j^*_{-m}$) can be easily identified from (9.84). One can show that only the odd harmonics are allowed due to the symmetry of the interaction which is characterized by $\operatorname{Re}\varepsilon^{-1}(\boldsymbol{q},\omega) = \operatorname{Re}\varepsilon^{-1}(-\boldsymbol{q},-\omega)$ and $\operatorname{Im}\varepsilon^{-1}(\boldsymbol{q},\omega) = -\operatorname{Im}\varepsilon^{-1}(-\boldsymbol{q},-\omega)$. Using the properties of the Fourier coefficients, we find the expansion

$$\boldsymbol{j}(t) = 2\sum_{l=0}^{\infty}\{\mathrm{Re} j_{2l+1}(\omega)\,\cos[(2l+1)\omega t] + \mathrm{Im} j_{2l+1}(\omega)\,\sin[(2l+1)\omega t]\}. \quad (9.86)$$

In particular, we get for the real parts ($l = 0, 1, 2\ldots$)

$$\begin{aligned} \operatorname{Re}\boldsymbol{j}_{2l+1}(\omega) &= \frac{(-1)^l}{(2l+1)} n_i e_i^2 \int \frac{d\boldsymbol{q}}{(2\pi\hbar)^3} \frac{e_e}{m_e\hbar\omega} \boldsymbol{q}\hat{V}(q) \sum_{n=0}^{\infty} J_n\left(\tfrac{\boldsymbol{q}\cdot\boldsymbol{v}_e^0}{\hbar\omega}\right) \\ &\times \left[J_{n-(2l+1)}\left(\tfrac{\boldsymbol{q}\cdot\boldsymbol{v}_e^0}{\hbar\omega}\right) + J_{n+(2l+1)}\left(\tfrac{\boldsymbol{q}\cdot\boldsymbol{v}_e^0}{\hbar\omega}\right)\right] \operatorname{Im}\varepsilon^{-1}_{\mathrm{RPA}}(\boldsymbol{q}; -n\omega), \quad (9.87) \end{aligned}$$

whereas the imaginary parts follow to be

$$\begin{aligned} \operatorname{Im}\boldsymbol{j}_{2l+1}(\omega) &= \delta_{l,0}\frac{n_e e^2}{m_e}\frac{\boldsymbol{E}_0}{2} + \frac{(-1)^l}{(2l+1)} n_i e_i^2 \int \frac{d\boldsymbol{q}}{(2\pi\hbar)^3} \frac{e_e}{m_e\hbar\omega} \boldsymbol{q}\hat{V}(q) \\ &\times \left\{J_0\left(\tfrac{\boldsymbol{q}\cdot\boldsymbol{v}_e^0}{\hbar\omega}\right) J_{-(2l+1)}\left(\tfrac{\boldsymbol{q}\cdot\boldsymbol{v}_e^0}{\hbar\omega}\right) \left[\operatorname{Re}\varepsilon^{-1}_{\mathrm{RPA}}(\boldsymbol{q};0) - 1\right]\right. \\ &+ \sum_{n=1}^{\infty} J_n\left(\tfrac{\boldsymbol{q}\cdot\boldsymbol{v}_e^0}{\hbar\omega}\right) \left[J_{n-(2l+1)}\left(\tfrac{\boldsymbol{q}\cdot\boldsymbol{v}_e^0}{\hbar\omega}\right) - J_{n+(2l+1)}\left(\tfrac{\boldsymbol{q}\cdot\boldsymbol{v}_e^0}{\hbar\omega}\right)\right] \\ &\times \left.\left[\operatorname{Re}\varepsilon^{-1}_{\mathrm{RPA}}(\boldsymbol{q}; -n\omega) - 1\right]\right\}. \quad (9.88) \end{aligned}$$

Evaluation of these expressions allows to investigate the spectrum of the time dependent electrical current as a function of the electrical field strength, the frequency, and the plasma temperature and density.

The appearance of higher harmonics in a strong laser field (Silin 1999) is a very interesting effect. The higher harmonics in the current density can be calculated according to (9.87) and (9.88). In Fig. 9.1, the amplitudes of the

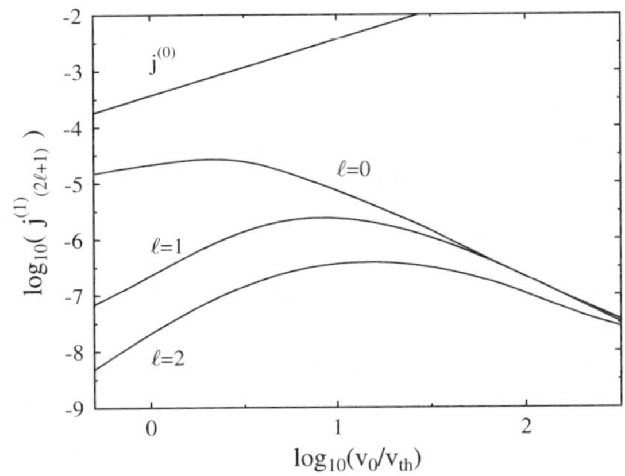

Fig. 9.1. Amplitudes $2[(\mathrm{Re} j_m)^2 + (\mathrm{Im} j_m)^2]^{1/2}$ of the $m = (2\ell + 1)$th harmonics of the current vs. field strength give by v_0/v_{th}. The parameters are $n_e = 4 \times 10^{21}\,\mathrm{cm}^{-3}$ and $T = 1.2 \times 10^5$ K

different harmonics are given as a function of the field strength. The harmonics have amplitudes increasing with the field strength up to maxima at certain values and decreasing afterwards. For high fields, the differences between the higher harmonics decrease. The ratio to $j^{(0)}$, however, becomes very small.

9.1.7 Collisional Absorption Rate in Strong Fields

Now we consider the energy transfer $\boldsymbol{j} \cdot \boldsymbol{E}$ between the laser field and the dense plasma. Let us start with the Fourier expansion (9.86). For the laser field, we assume $\boldsymbol{E} = \boldsymbol{E}_0 \cos \omega t$. Then we obtain after a simple calculation

$$\begin{aligned} \boldsymbol{j}(t) \cdot \boldsymbol{E}(t) &= \boldsymbol{E}_0 \cdot \Big\{ \mathrm{Re}\, \boldsymbol{j}_1(\omega) + \sum_{l=1}^{\infty} \Big[\mathrm{Re}\Big(\boldsymbol{j}_{2l+1}(\omega) + \boldsymbol{j}_{2l-1}(\omega) \Big) \cos(2l\omega t) \\ &\quad + \mathrm{Im}\Big(\boldsymbol{j}_{2l+1}(\omega) + \boldsymbol{j}_{2l-1}(\omega) \Big) \sin(2l\omega t) \Big] \Big\} . \end{aligned} \tag{9.89}$$

Besides the constant term, we have even harmonics only.

The dissipation of energy is determined by averaging over one cycle of oscillation:

$$\langle \boldsymbol{j} \cdot \boldsymbol{E} \rangle \equiv \frac{1}{T} \int_{t-T}^{t} dt'\, \boldsymbol{j}(t') \cdot \boldsymbol{E}(t') = E_0 \cdot \mathrm{Re}\, \boldsymbol{j}_1(\omega) . \tag{9.90}$$

The current $\boldsymbol{j}^{(0)}$ does not give a contribution here. We want to mention that, in our approximation, $\langle \boldsymbol{j} \cdot \boldsymbol{E} \rangle = \big\langle dW^{\mathrm{kin}}/dt \big\rangle$ holds. That means, the potential energy averaged over one oscillation cycle is constant

In the expression for $\mathrm{Re}\, \boldsymbol{j}_1$ (9.87), one can use the recursion formula for the Bessel functions

$$J_{n-1}(z) + J_{n+1}(z) = \frac{2n}{z} J_n(z) \, ,$$

which leads to

$$\langle \boldsymbol{j} \cdot \boldsymbol{E} \rangle = n_i e_i^2 \, 2 \int \frac{d\boldsymbol{q}}{(2\pi\hbar)^3} \, \hat{V}(q) \sum_{n=1}^{\infty} n\omega \, J_n^2 \left(\frac{\boldsymbol{q} \cdot \boldsymbol{v}_e^0}{\hbar\omega} \right) \mathrm{Im} \, \frac{1}{\varepsilon_{\mathrm{RPA}}(\boldsymbol{q}; -n\omega)} \, . \tag{9.91}$$

This result has a similar form as that of the nonlinear Dawson–Oberman model (Decker et al. 1994). We want to stress the fact that here the dielectric function is given by the quantum Lindhard form, whereas the dielectric theory by Decker et al. leads to the classical Vlasov dielectric function. Finally, using $\mathrm{Im}\,\varepsilon^{-1} = -\mathrm{Im}\,\varepsilon/|\varepsilon|^2$, we get

$$\langle \boldsymbol{j} \cdot \boldsymbol{E} \rangle = n_i e_i^2 \int \frac{d\boldsymbol{q}}{(2\pi\hbar)^3} \, \hat{V}(q) \sum_{n=1} n\omega \, J_n^2 \left(\frac{\boldsymbol{q} \cdot \boldsymbol{v}_e^0}{\hbar\omega} \right) \frac{2 \, \mathrm{Im} \, \varepsilon_{\mathrm{RPA}}(\boldsymbol{q}; n\omega)}{|\varepsilon_{\mathrm{RPA}}(\boldsymbol{q}; n\omega)|^2} \, . \tag{9.92}$$

The Lindhard dielectric function has to be calculated numerically, what can be done for arbitrary degeneracy. For a detailed discussion of the different quantum effects, however, it is advantageous to consider especially the non-degenerate case, in which some integrations can be done analytically. For the case of a Maxwellian electron distribution function, we get (Schlanges et al. 2000; Bornath et al. 2001)

$$\begin{aligned} \langle \boldsymbol{j} \cdot \boldsymbol{E} \rangle &= \frac{8\sqrt{2\pi} Z^2 e^4 n_e n_i \sqrt{m_e}}{(4\pi\varepsilon_0)^2 (k_B T)^{3/2}} \, \omega^2 \sum_{n=1}^{\infty} n^2 \int_0^{\infty} \frac{dq}{q^3} \frac{1}{|\varepsilon_{\mathrm{RPA}}(\boldsymbol{q}, n\omega)|^2} \\ &\times e^{-\frac{n^2 m_e \omega^2}{2k_B T q^2}} \, e^{-\frac{\hbar^2 q^2}{8 m_e k_B T}} \, \frac{\sinh \frac{n\hbar\omega}{2k_B T}}{\frac{n\hbar\omega}{2k_B T}} \int_0^1 dz \, J_n^2 \left(\frac{e \boldsymbol{E}_0 \boldsymbol{q}}{m_e \omega^2} \, z \right) . \end{aligned} \tag{9.93}$$

Formula (9.93) determines many interesting quantities of the laser–plasma interaction. Using the high-frequency limit, cf. the previous section, we may obtain the effective electron–ion collision frequency ν_{ei}, the electrical conductivity σ, and the energy absorption coefficient by inserting (9.93) into

$$\mathrm{Re}\nu_{ei} = \frac{\omega^2}{\omega_{pl}^2} \frac{\langle \boldsymbol{j} \cdot \boldsymbol{E} \rangle}{\langle \varepsilon_0 \boldsymbol{E}^2 \rangle}; \; \mathrm{Re}\sigma = \frac{\langle \boldsymbol{j} \cdot \boldsymbol{E} \rangle}{\langle \varepsilon_0 \boldsymbol{E}^2 \rangle}; \; \alpha(\omega) = \frac{1}{c} \frac{\langle \boldsymbol{j} \cdot \boldsymbol{E} \rangle}{\langle \varepsilon_0 \boldsymbol{E}^2 \rangle} \, . \tag{9.94}$$

In the classical limit $\hbar \to 0$, the collision frequency $\mathrm{Re}\nu_{ei}$ and the conductivity $\mathrm{Re}\sigma$ are well-known. They were derived for the first time by Klimontovich (1975). Later Decker et al. (1994) got such expressions in the framework of the nonlinear Dawson–Oberman model. The result for $\mathrm{Re}\nu_{ei}$ is

$$\mathrm{Re}\nu_{ei} = \frac{4Z}{\sqrt{2\pi}^3} \frac{\omega_{pl}}{n_e r_d^3} \left(\frac{v_{th}}{v_0}\right)^2 \left(\frac{\omega}{\omega_{pl}}\right)^2 r_d^2 \sum_{n=1}^{\infty} n^2 \int_0^{\infty} \frac{dq}{q^3} \frac{1}{|\varepsilon_{\mathrm{RPA}}(\boldsymbol{q}, n\omega)|^2}$$

$$\times \quad e^{-\frac{n^2 m_e \omega^2}{2k_B T q^2}} \int_0^1 dz\, J_n^2 \left(\frac{e\boldsymbol{E}_0 \boldsymbol{q}}{m_e \omega^2} z\right) . \tag{9.95}$$

The classical formulae exhibit the well-known problem of a divergency at large k which has to be overcome by some cut-off procedure. In contrast, in our quantum approach, no divergencies exist.

Let us now discuss the expression (9.93) more in detail. Quantum effects, marked by $\hbar$, occur there at different places. The first place is one of the exponential functions in (9.93) describing the quantum diffraction effect at large momenta q. This exponential function ensures the convergence of the integral. The second place is the term with the sinh function which is connected with the Bose statistics of multiple photon emission and absorption. Finally, quantum effects enter via the calculation of $|\varepsilon(q, n\omega)|^2$ itself. These effects will be discussed in detail in the next section.

Unfortunately, a further analytical simplification is only possible in limiting cases. Therefore, we will first consider the weak field and the strong field limits, respectively, following Kull and Plagne (2001). Afterwards we present a numerical evaluation of the general expression (9.93).

The asymptotic behavior of (9.93) is completely determined by the asymptote of the integral

$$I_n(a) = \int_0^1 dz\, J_n^2\,(a\, z)\,, \qquad a = \frac{e\boldsymbol{E}_0 \cdot \boldsymbol{q}}{m_e \omega^2} = \frac{\boldsymbol{v}_0 \cdot \boldsymbol{q}}{\omega} . \tag{9.96}$$

For small arguments, i.e., $az \to 0$, one easily gets from the series expansion of the Bessel function

$$I_n(a) = \frac{1}{2n+1} \frac{a^n}{2^n n!} . \tag{9.97}$$

In the opposite case of large arguments, i.e., for $az \to \infty$, one can use – under the restriction $az > n$ – the asymptotic representation for the Bessel function (see Abramowitz and Stegun (1984), p.110)

$$J_n^2(az) \to \frac{2\cos^2\left[\frac{\pi}{4} - \sqrt{(az)^2 - n^2} + n \arccos\left(\frac{n}{a\,z}\right)\right]}{\pi\, \sqrt{(az)^2 - n^2}} . \tag{9.98}$$

This function oscillates around the function (mean value of the integrand)

$$\frac{1}{\pi\, \sqrt{(az)^2 - n^2}} . \tag{9.99}$$

The amplitudes of these oscillations can be neglected for large az. Consequently, the integrand of (9.96) may be replaced for large az by (9.99). On behalf of the inequality $az > n$, one has to restrict the range of integration in (9.96) to $z < n/a$. Now it is possible to carry out the integration with the result

$$I_n(a) = \frac{1}{\pi a}\operatorname{arcosh}\left(\frac{a}{n}\right), \qquad a \geqq n\,. \tag{9.100}$$

For the further discussion, let us consider the collision frequency ν_{ei} which determines all interesting quantities. With the definition (9.94) and the relation (9.93), we get

$$\begin{aligned}\mathrm{Re}\nu_{ei} &= \frac{16\sqrt{2\pi}}{\boldsymbol{E}_0^2}\frac{Ze^2 n_i}{v_{th}^3}\omega^4\sum_{n=1}^{\infty} n^2\frac{\sin\, n\, x}{n\, x}\int_0^{\infty} dq\,\frac{1}{q}\frac{1}{|\varepsilon(q,n\omega)|^2} \\ &\times\ \exp\left\{-\left[(nx)^2\frac{2}{(q/q_{dB})^2}+\frac{(q/q_{dB})^2}{8}\right]\right\}\int_0^1 dz\, J_n^2\,(a\,z)\,, \end{aligned} \tag{9.101}$$

where $x = \hbar\omega/2k_BT$ and $\hbar q_{dB} = m_e v_{th} = \sqrt{m_e k_B T}\,$.

First we will consider the expression (9.101) in the weak field case, i.e., we assume $v_{th} > v_0$. Then ν_{ei} is determined by the asymptotic formula (9.97) and we get just the formula (9.76) resulting from linear response theory (9.100).

The asymptotic strong field behavior is more complicated. For the high-frequency field under consideration, we can approximate $|\varepsilon(q,n\omega)|^2 = 1$. We insert the asymptotic formula (9.100) in (9.101), introduce the new variable $\mu = a/n = \boldsymbol{v}_0\boldsymbol{q}/n\omega_0$, and take into account the cut-off due to $a/n \geqq 1$. We arrive at

$$\begin{aligned}\mathrm{Re}\nu_{ei} &= \frac{\omega_{pl}^4}{\pi^2}\frac{1}{n_e v_0^3}\sum_{n=1}^{\infty}\frac{1}{n}\frac{\sinh nx}{nx}\left(\frac{v_0}{v_{th}}\right)^3\sqrt{\frac{2}{\pi}} \\ &\times \int_1^{\infty} d\mu\,\frac{\operatorname{arcosh}(\mu)}{\mu^4}\exp\left\{-\frac{1}{2}\left[(nx)^2\frac{\mu^2 v_{th}^2}{v_0^2}+\frac{v_0^2}{\mu^2 v_{th}^2}\right]\right\}. \end{aligned} \tag{9.102}$$

Now we can use the fact that $\operatorname{arcosh}(\mu)$ is a slowly varying function. That means, it can be replaced by the constant $\operatorname{arcosh}(\mu_m)$ where μ_m is the maximum position of the remaining integrand, i.e.,

$$\mu_m = \frac{v_0}{v_{th}}\frac{1}{(nx)^2}\sqrt{2+\sqrt{4+n^2x^2}}\,. \tag{9.103}$$

We have to take into account the condition $\mu_m(nx) \geqq 1$. In the remaining integral, however, the lower integration limit may be extended to zero. The result is

$$\mathrm{Re}\nu_{ei} = \frac{\omega_{pl}^4}{\pi^2} \frac{1}{n_e v_0^3} \sum_{n=1}^{n_m} \frac{1}{n} \frac{\sinh nx}{nx} (\mu_m)(nx)^{\frac{3}{2}} \sqrt{\frac{2}{\pi}} K_{\frac{3}{2}}(nx) \tag{9.104}$$

with the Bessel function $K_{3/2}$. Here, the maximum order of multi-photon processes n_m is determined by the cut-off condition

$$n_m = \frac{1}{x} \left(\frac{v_0}{v_{th}} \right)^2 = \frac{8U_p}{\hbar\omega}, \tag{9.105}$$

where U_p is the ponderomotive potential. This interesting result was first derived by Kull and Plagne (2001). They also gave a further discussion of the asymptotic behavior of the series in (9.104). In the classical limit ($nx \ll 1$), one gets

$$\mathrm{Re}\nu_{ei} = \frac{Z}{\pi^2} \frac{\omega_{pl}}{n_e r_D^3} \left(\frac{v_{th}}{v_0} \right)^3 \ln\left(\frac{v_0}{v_{th}} \right) \ln\left(\frac{\sqrt{2}\, q_{dB}}{\omega / v_{th}} \right), \tag{9.106}$$

and in the quantum case, one has

$$\mathrm{Re}\nu_{ei} = \frac{Z}{\pi^2} \frac{\omega_{pl}}{n_e\, r_D^3} \left(\frac{v_{th}}{v_0} \right)^3 \ln\left(\frac{v_0}{v_{th}} \right) \ln\left(\frac{\sqrt{2}\, q_{dB}}{\omega / v_{th}} \sqrt{\frac{v_0}{v_{th}}} \right). \tag{9.107}$$

Again all cut-off parameters follow automatically from the quantum mechanical consideration.

The asymptotic strong field behavior was also considered in the paper by Silin (1964) in the framework of classical kinetic theory. In the Silin paper, the following asymptotic expression was proposed

$$\mathrm{Re}\nu_{ei} = \frac{Z}{\pi^2} \frac{\omega_{pl}}{n_e r_D^3} \left(\frac{v_{th}}{v_0} \right)^3 \left[\ln\left(\frac{v_0}{v_{th}} \right) + 1 \right] \ln\left(\frac{q_{\max}}{q_{\min}} \right). \tag{9.108}$$

In this pure classical consideration, the cut-off parameter $q_{\max}$ is the inverse Landau length $q_{\max} = 4\pi\varepsilon_0 kT/(Ze^2)$ and $q_{\min}$ is, as usual for high-frequency fields, given by $q_{\min} = \omega/v_{th}$.

9.1.8 Results for the Collision Frequency

In this section, we will present numerical results for the the electron–ion collision frequency in high-frequency laser fields based on the formulae derived in the preceding Sect. 9.1.7 (cf. Bornath et al. (2001)). Our emphasis will be to show the importance of a quantum approach. In the following, a hydrogen plasma which is assumed to be fully ionized is considered.

In Fig. 9.2, the collision frequency $\mathrm{Re}\nu_{ei}(\omega)$ in such a plasma is shown as a function of the quiver velocity. Here and in the subsequent figures, the real part is simply denoted by ν_{ei}. For comparison, there are curves given (dashed

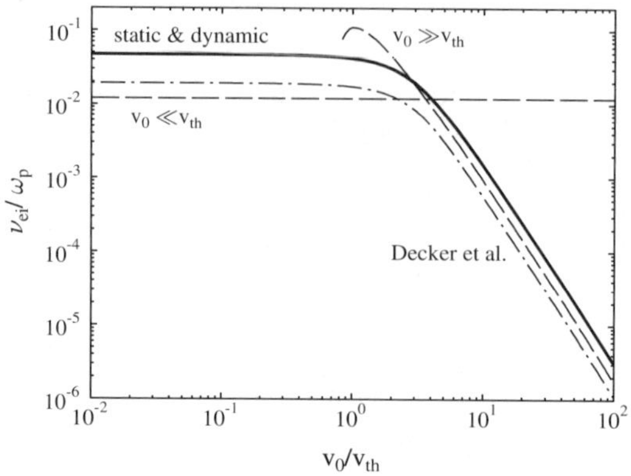

Fig. 9.2. Real part of the electron–ion collision frequency as a function of the quiver velocity $v_0 = eE_0/\omega m_e$ for an H-plasma in a laser field ($Z = 1$; $n_e = 10^{22}\,\mathrm{cm}^{-3}$; $T = 3 \times 10^5\,\mathrm{K}$; $\omega/\omega_p = 5$). For comparison, results from the theory of Decker et al. (*dash-dotted line*) and from the asymptotic formulas (9.78) and (9.108) of Silin (*dashed*) are given

line) that follow from the asymptotic formulas of Silin for the cases of small (9.78) and of large quiver velocities (9.108), respectively. Furthermore, the classical expression of Decker et al. (1994) is evaluated (dash-dotted line) [unfortunately, in their Eq. (22) a factor $2/(n_e\lambda_D^3)$ is missing]. Our quantum expression (9.101), was evaluated with the fully dynamical quantum Lindhard dielectric function and, for comparison, with static screening in the denominator of (9.92). These results are shown in Fig. 9.2 using solid lines. In the given logarithmic plot, there is almost no difference between these two cases. The static approximation for $1/|\varepsilon|^2$ slightly overestimates the effect of screening.

The qualitative behavior of the results from the classical dielectric theory of Decker et al. and from our quantum approach is very similar. The collision frequency is nearly constant for small field strengths up to $v_0/v_{\mathrm{th}} = 1$ and decreases then rapidly for higher fields. This is in agreement with the asymptotic formulae of Silin, too. There are, however, quantitative differences. These can be attributed to the use of Coulomb logarithms in the classical approaches which correspond to cutting procedures in the integral over momentum.

Now the dependence of the collision frequency on the coupling parameter $\Gamma = (e^2/4\pi\varepsilon_0)/dk_BT$ with $d = (4\pi n_i/3)^{-1/3}$ will be considered. First, in Fig. 9.3, the collision frequency is shown as a function of coupling parameter Γ for a small quiver velocity, i.e., a small field strength. Results of the evaluation of (9.92) and (9.94) are given by the upper solid line. The static screening results are the lower solid curve. Again the asymptotic formula of Silin for small quiver velocities and the classical expression of Decker et al. were evaluated. Furthermore, numerical results of Cauble and Rozmus (1985) are plotted. They considered small field strengths and used a memory function kinetic approach which allows to consider plasmas up to strong coupling. The data points in Fig. 9.3 correspond to their so-called Debye–Hückel mean field approximation. Finally, we compare also with numerical simulation re-

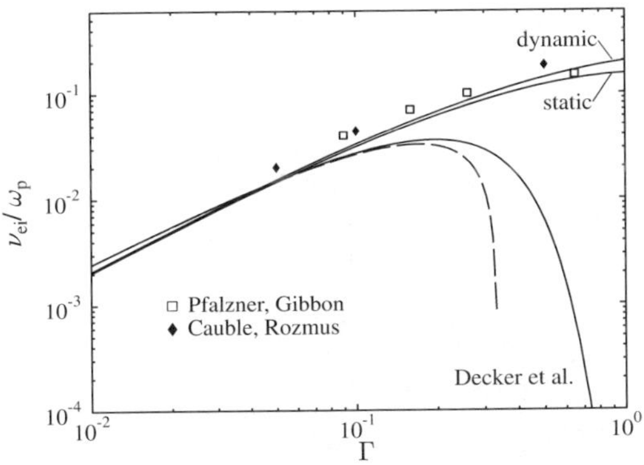

Fig. 9.3. Electron–ion collision frequency as a function of the coupling parameter Γ for an H-plasma in a laser field ($Z = 1$; $v_0/v_{th} = 0.2$; $n_e = 10^{22}\,\mathrm{cm}^{-3}$; $\omega/\omega_p = 3$). A comparison with the theory of Decker et al. and with the asymptotic formula (9.78) of Silin (*dashed line*) is given. Furthermore, results of Cauble and Rozmus, and simulation data of Pfalzner and Gibbon are shown

sults by Pfalzner and Gibbon (1998). They applied a tree code method to classical molecular dynamics simulations using a soft Coulomb potential.

According to Fig. 9.3, the collision frequency first increases with increasing coupling Γ. For small Γ, the dielectric theory and our theory give almost the same results. The values of the asymptotic formula of Silin are slightly larger. With increasing Γ, this asymptotic formula as well as the dielectric approach reach a maximum around $\Gamma \sim 0.2$ and sharply drop down afterwards. This behavior is governed by the Coulomb logarithm used in these approaches. It results from a cut-off procedure at large momenta k. Such a cut-off, inherent in many classical approaches, is avoided in our approach because the k integration is automatically convergent, cf. the second exponential function in (9.101). Therefore, the range of applicability of our approach is extended to higher values of Γ.

The agreement with the results of Cauble and Rozmus (1985) and of Pfalzner and Gibbon (1998) is rather good with the values of the present theory being slightly smaller. One has to take into account, however, that our approximation is a weak coupling theory whereas the approaches we compare with include the correlations in higher approximations. A generalization of the quantum kinetic approach used here will be considered in the next section. We want to mention that our results do not depend solely on Γ, but depend on temperature and on density (as well as the results of Cauble et al. do since the application of a modified potential takes into account short-range quantum effects).

Before we are going to elucidate the quantum effects, we want to present the behavior of the collision frequency versus Γ for a higher value of the electrical field strength ($v_0/v_{\rm th} = 10$). For this case, we compare our results with those from the classical dielectric theory (Decker et al. 1994) and with the asymptotic formula (9.108) for large quiver velocities (Silin 1964). The qualitative behavior is similar to that for $v_0/v_{\rm th} = 0.2$.

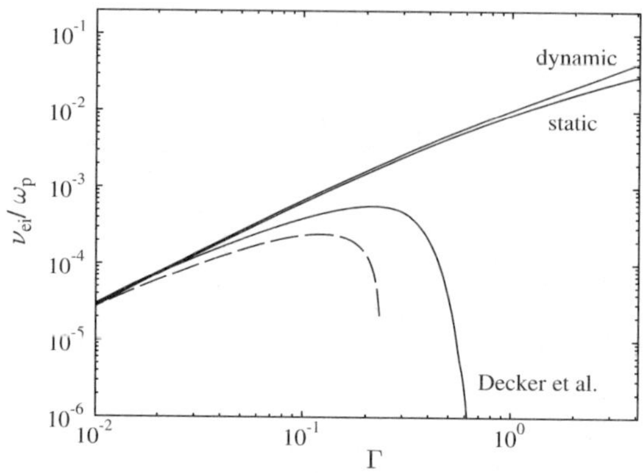

Fig. 9.4. Electron–ion collision frequency as a function of the coupling parameter Γ for an H-plasma in a laser field ($Z = 1$; $v_0/v_{th} = 10$; $n_e = 10^{22}\,\mathrm{cm}^{-3}$; $\omega/\omega_p = 5$). Comparison is given with the classical dielectric theory of Decker et al. and with the asymptotic formula (9.108) given by Silin (*dashed line*)

Now the consequences of the quantum approach in contrast to the classical dielectric theory will be investigated. We consider such parameters that the plasma can be assumed to be non-degenerate, i.e., (9.101) can be used. In this case, a direct comparison with the classical dielectric theory of Decker et al. is possible. Quantum effects indicated by $\hbar$ occur in (9.101) at two places. One is the quantum diffraction effect ensuring the convergence of the integral at large k (cf. the second exponential function). The other is the factor $\sinh x/x$ with $x = n\hbar\omega/(2k_BT)$. The classical theory uses, instead, a cut-off at $k_{\max} = mv_{\mathrm{th}}^2 4\pi\varepsilon_0/Ze^2$, and the sinh factor is missing.

An important feature of the expressions (9.92)–(9.101) is the sum over n which can be interpreted as a sum over the different multi-photon processes, i.e., the emission and absorption of energies $n\hbar\omega$. The different contributions ν_n in the sum $\nu_{ei} = \sum_n \nu_n$ depend on the field strength. It is obvious that, with increasing field, the number of terms contributing essentially to the sum is also increasing. The following two figures, Fig. 9.5 and Fig. 9.6, showing ν_n vs. n for two different field strengths illustrate this issue (full solution of (9.101) – solid line). Moreover, we compare with the classical dielectric theory (dotted line) and the case in which the factor $\sinh x/x$ in (9.101) is set to unity (dashed line). One observes that the differences between these three cases grow with increasing photon number n. The reason for the faster decreasing contributions in the classical approach is the hard cut-off in the k integration while the maximum of the integrand is shifted to higher k due to the exponential factor $\exp[-(n^2\omega^2 m_e)/(2k_BTk^2)]$. Thus the relative error increases with n.

The other quantum effect is connected with the $\sinh x$ factor which behaves for large $x(n)$ like $e^x/2x$. Therefore, this factor becomes more important for large n, the processes involving large numbers of photons are enhanced. This can be seen in Fig. 9.6 which considers the case $v_0/v_{\mathrm{th}} = 10$. The solid

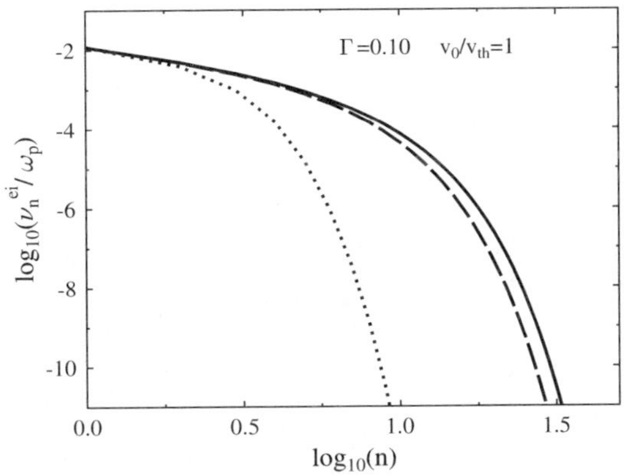

Fig. 9.5. Contributions ν_n vs. photon number n in a hydrogen plasma ($n_e = 10^{22}\,\mathrm{cm}^{-3}$; $\omega/\omega_p = 5$; $\Gamma = 0.1$) for $v_0/v_{th} = 1.0$. Present approach (*solid line*), sinh term neglected (*dashed line*), classical dielectric theory (*dotted line*)

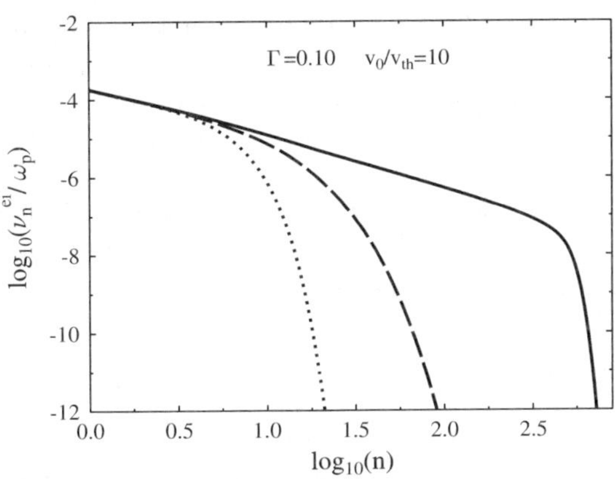

Fig. 9.6. Contributions ν_n vs. photon number n in a hydrogen plasma ($n_e = 10^{22}\,\mathrm{cm}^{-3}$; $\omega/\omega_p = 5$; $\Gamma = 0.1$) for $v_0/v_{th} = 10$. Present approach (*solid line*), sinh term neglected (*dashed line*); classical dielectric theory (*dotted line*)

curve corresponding to the full solution extends to much higher n values than that curve which results from a neglect of the sinh term. An interesting feature is the plateau-like behavior up to $n \sim 350$ with the subsequent sharp drop down. We can conclude at this point that, especially in the strong field case where multi-photon processes play an increasing role, it is important to treat the problem on the basis of quantum mechanics. This problem was also discussed in Kull and Plagne (2001) basing on an asymptotic solution of expression (9.101) for strong fields. In order to complete the discussion of the collision frequency, the dependence on the laser frequency is considered. This is shown in Fig. 9.7 for two different field strengths. The full dynamic solution is compared with the static screening approximation for $1/|\varepsilon|^2$. For large frequencies, the differences between the two approximations decrease. Collective effects in the dielectric function play a role only in the vicinity of

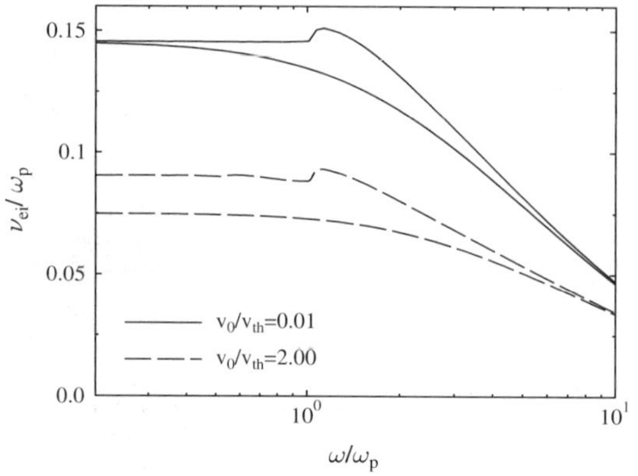

Fig. 9.7. Electron-ion collision frequency as a function of the laser frequency for a hydrogen plasma ($n_e = 10^{21}\,\mathrm{cm}^{-3}$; $T = 10^5\,\mathrm{K}$) for two different field strengths. The upper curve of each pair corresponds to the full dynamical screening, the lower one to the static screening approximation

the plasma frequency. This behavior, considered already in the frame of the classical theory by Dawson and Oberman (1962), can be seen also for higher field strengths. In the high field case, the static screening approximation deviates from the dynamical one also for the lower frequencies. This is caused by collective effects in the terms with higher n in the respective sum in (9.101) for arguments of the dielectric function $n\omega \approx \omega_{\mathrm{p}}$.

The discussion of the behavior of $\nu_{ei}(\omega)$ for laser frequencies around and below the plasma frequency has to be treated, of course, with some care because the underlying theory is a high-frequency approximation.

9.1.9 Effects of Strong Correlations

Let us come back to the dependence of the collisional absorption on the coupling parameter Γ. Up to now we considered the collision frequency in weak coupling approximation. A generalization of the results above for the case of a stronger coupling was given in the papers (Bornath et al. 2003; Schlanges et al. 2003; bsh) where formula (9.91) was generalized to

$$\langle \boldsymbol{j} \cdot \boldsymbol{E} \rangle = 2 \int \frac{d\boldsymbol{q}}{(2\pi\hbar)^3} \sum_{m=1}^{\infty} m\omega\, J_m^2(z)\, V_{ei}^2(q) n_i\, \mathcal{S}_{ii}(\boldsymbol{q})\, \mathrm{Im}\, \mathcal{L}_{ee}^R(\boldsymbol{q}; -m\omega)\,. \tag{9.109}$$

With the definition of a dielectric function of the electron subsystem,

$$\mathrm{Im}\, \varepsilon_{ee}^{-1}(\boldsymbol{q}; -m\omega) = V_{ee}(q) \mathrm{Im}\, \mathcal{L}_{ee}(\boldsymbol{q}; -m\omega)\,, \tag{9.110}$$

we arrive at

$$\langle \boldsymbol{j} \cdot \boldsymbol{E} \rangle = 2n_i \int \frac{d\boldsymbol{q}}{(2\pi\hbar)^3} \sum_{m=1}^{\infty} m\omega\, J_m^2(z)\, V_{ii}(q)\, \mathcal{S}_{ii}(\boldsymbol{q})\, \mathrm{Im}\, \varepsilon_{ee}^{-1}(\boldsymbol{q}; -m\omega)\,. \tag{9.111}$$

There are two generalizations. The first one is the occurrence of the static ion–ion structure factor

$$\mathcal{S}_{ii}(\boldsymbol{q}) = 1 + n_i \int d^3r \, [g_{ii}(\boldsymbol{r}) - 1] \, e^{-\frac{i}{\hbar}\boldsymbol{q}\cdot\boldsymbol{r}} \,, \qquad (9.112)$$

where $g_{ii}(r)$ is the pair correlation function. This enables us to consider a correlated ion subsystem.

Furthermore, the function $\mathcal{L}^R_{ee}$ is the exact density response function of the electron subsystem, not only the RPA function as in Bornath et al. (2001). The electron–electron interactions can be included, therefore, on a higher level. Appropriate approximations can be expressed via local field corrections (LFC), see, e.g., Hubbard (1958) and Ichimaru and Utsumi (1981),

$$\mathcal{L}^R_{ee}(\boldsymbol{q},\omega) = \frac{\chi^0_e(\boldsymbol{q},\omega)}{1 - V_{ee}(q)G(q)\chi^0_e(\boldsymbol{q},\omega)} \qquad (9.113)$$

with χ^0_e being the usual free-electron Lindhard polarizability, and the local field factor G is given in our notation by

$$G(\boldsymbol{q}) = -n_e^{-1} \int \frac{d\boldsymbol{q}'}{(2\pi\hbar)^3} \, \frac{\boldsymbol{q}\cdot\boldsymbol{q}'}{q'^2} [S_{ee}(\boldsymbol{q}-\boldsymbol{q}') - 1] \,. \qquad (9.114)$$

Figure 9.8 shows the influence of the LFC (the structure factor is calculated in hyper-netted chain (HNC) approximation). The collision frequency is given as a function of the coupling parameter Γ. For weak and moderate electric fields ($v_0/v_{\rm th} = 0.2$ and 3), deviations occur in the region $\Gamma > 1$ which increase with increasing coupling. For rather strong fields, the LFC has no influence up to a coupling of about $\Gamma = 10$. Furthermore, one can see that, for strong coupling, the influence of the field strength becomes smaller.

Let us now consider the influence of the ion–ion correlations. Effects of such correlations via static structure factors were already discussed by Dawson and Oberman (1963) starting from the linearized Vlasov equation and

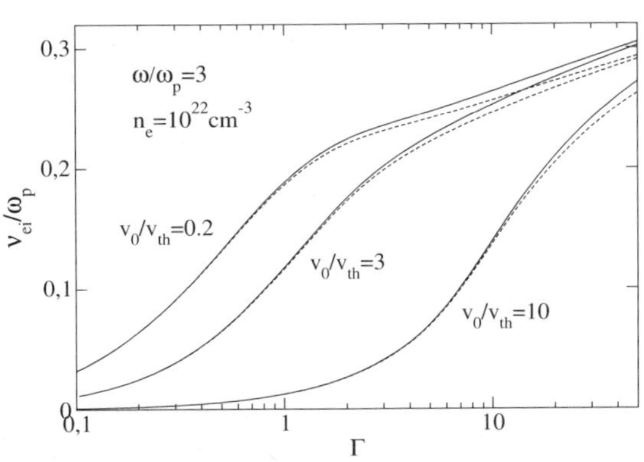

Fig. 9.8. Electron–ion collision frequency as a function of the coupling parameter Γ for different values of the quiver velocity. LFC in accordance with Ichimaru and Utsumi (1981) (solid), without LFC (dashed). S_{ii} is calculated in HNC approximation. The quiver velocity is defined as $v_0 = eE_0/m_e\omega$, $v_{\rm th} = (k_BT_e/m_e)^{1/2}$

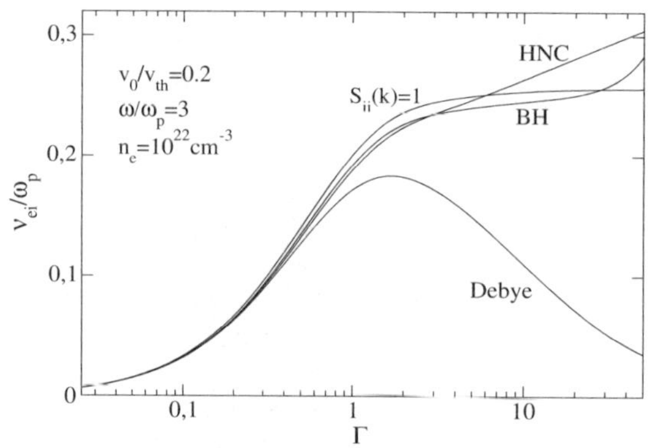

Fig. 9.9. Electron–ion collision frequency as a function of the coupling parameter Γ. Ion structure factor in different approaches: HNC, Debye, BH (Baus–Hansen formula)

recently in Hazak et al. (2002) using the quantum BBGKY hierarchy. In Fig. 9.9, the collision frequency is shown using different approximations for the static ion–ion structure factor S_{ii} (here the LFC is calculated as in Ichimaru and Utsumi (1981)). The structure factor was calculated numerically using a HNC code. For comparison, we used a semi-analytical formula given by Baus and Hansen (1979) which reads

$$S(k) = \frac{1}{1 - c(k)} = \frac{1}{1 + (k_D^2/k^2)h_p(kr_0)}, \tag{9.115}$$

where $k_D = (4\pi e^2 n/k_B T)^{1/2}$ is the Debye wave number and $h_p(y) = d_p(y)(2p-1)!!/y^p$, $d_p(y)$ being the spherical Bessel function of order p. Here, r_0 is an effective radius which can serve as a fit parameter, in a rough approximation it is the mean ion-sphere radius. For the details, we refer to Baus and Hansen (1979). In our calculations, we used the formula with $p = 1$. For $r_0 \to 0$, we get the Debye–Hückel result.

The inclusion of the structure factor decreases the collision frequency for small and moderate coupling up to a value of $\Gamma \approx 5$. For high values of the coupling parameter, there is a strong increase of the collision frequency. This increase is even stronger in the HNC calculation than in the semi-analytical formula of Baus and Hansen. As expected, the Debye approximation can be applied only for weak coupling. Finally, it should be mentioned that (9.111) can be applied to describe correlation effects on the collisional absorption in two-temperature plasmas. Investigations of such plasmas including MD simulations are given in a paper by Hilse et al. (2005).

9.2 The Static Electrical Conductivity

Transport properties like electrical and thermal conductivities are very important for the understanding of the behavior of plasmas. The theoretical investigation of such quantities is of considerable importance for astrophysical problems and inertial confinement fusion related experiments. The electrical conductivity is one of the most indicative and easily observed properties and, therefore, its investigation is of special importance to study nonideality effects in dense plasmas. For an overview about various theoretical calculations and experiments concerning the electrical conductivity of nonideal plasmas, we refer to, e.g., Gryaznov et al. (1980), Ebeling et al. (1983), Günther and Radtke (1984), Kraeft et al. (1986), Fortov and Yakubov (1999), Ebeling et al. (1991), Redmer (1997), Hensel and Warren (1999).

In Chap. 2, we have already shown in a more simple consideration that the electrical conductivity of dense plasmas is strongly influenced by nonideality effects like self energy, screening, bound states and, in connection with the chemical composition, by the lowering of the ionization energy. It was especially shown that the nonideality effects produce a minimum of the conductivity as a function of the density with a subsequent strong increase at high densities what can be interpreted as an insulator–metal transition. First theoretical investigations of this behavior of hydrogen plasmas were performed by Ebeling et al. (1977), Kremp (b) et al. (1983), Kremp (b) et al. (1984), and by Höhne et al. (1984).

Similar experimental observations in electron–hole plasmas were given by Meyer and Glicksman (1978). Due to considerable progress in experimental techniques, this effect was recently observed in a series of experiments measuring the electrical conductivity of different materials. This fact offered new and interesting possibilities to compare various transport theories for dense plasmas with the experimental results.

In this chapter, we want to consider the statical conductivity of strongly coupled plasmas from a more general point of view as in Chap. 2. The first subsection is devoted to the relaxation effect on the electrical conductivity in the linear response regime. In the second subsection, the Lorentz model is applied to the fully ionized plasma accounting for dynamical screening. Then the Chapman–Enskog approach is developed for dense partially ionized plasmas starting from quantum kinetic equations. Finally, the theory is applied to different materials in order to discuss the influence of nonideality on the electrical conductivity.

9.2.1 The Relaxation Effect

In Chap. 2, we considered the electrical conductivity under the following simple assumptions: i) The Lorentz model was used; i.e., the electron-electron interaction was not taken into account. ii) The collision integral was taken in Born approximation (quantum mechanical Landau equation). iii) The collision integral was independent of the electromagnetic field. iv) The lowering of the ionization energy was approximated by the Debye shift.

For a more general theory, one may start from quantum kinetic equations or from the linear response theory developed by Kubo (1957), Kubo (1965),

Kubo (1966), and by Zubarev et al. (1996). Following the line of the present book, the basis of our representation is the kinetic theory.

Let us first consider the influence of the field dependence of the collision integral on the static conductivity in the linear response regime. This problem was considered for the first time by Kadomtsev (1958). A consequent investigation on the basis of kinetic equations was given in the monograph by Klimontovich (1975) and earlier in papers by Klimontovich and Ebeling (1962) and Klimontovich and Ebeling (1973). Further considerations on this problem are found in Ebeling and Röpke (1979), Ebeling et al. (1983), Morawetz and Kremp (1993), and in Esser and Röpke (1998).

We start from the kinetic equation (9.26) given in Sect. 9.1.2 and linearize the collision integral with respect to the electric field. Then we get

$$\begin{aligned} \frac{\partial f_a(\boldsymbol{p}_a,t)}{\partial t} &+ e_a \boldsymbol{E}\cdot\nabla_{\boldsymbol{p}_a} f_a(\boldsymbol{p}_a,t) = \sum_b I_{ab}(\boldsymbol{p}_a,t) \\ &= \sum_b \left(I^0_{ab}(\boldsymbol{p}_a,t) + \tilde{I}_{ab}(\boldsymbol{p}_a,\boldsymbol{E},t)\right) . \end{aligned} \tag{9.116}$$

In this equation, I^0_{ab} is the field-independent collision integral in Born approximation and $\tilde{I}_{ab}$ collects the contributions of the collision integral linear with respect to the field. These linear terms represent the intra-collisional field effect and the collisional broadening. The field dependent contribution to the collision integral $\tilde{I}_{ab}$ reads

$$\tilde{I}_{ab} = I^{ICF}_{ab} + I^{CB}_{ab} . \tag{9.117}$$

Here, the intra-collisional field contribution is

$$\begin{aligned} I^{ICF}_{ab} &= 2\int \frac{d\boldsymbol{p}_b d\boldsymbol{q}}{(2\pi\hbar)^6}\frac{1}{\hbar^2}\left|V^s_{ab}(\boldsymbol{q})\right|^2 \, 2\,\boldsymbol{E}\frac{\mathcal{P}}{a^2}\left\{e_a\nabla_{p_a}+e_b\nabla_{p_b}\right\} \\ &\times \left\{\bar{f}_a\bar{f}_b\left[1-f_a\right]\left[1-f_b\right]-f_af_b\left[1-\bar{f}_a\right]\left[1-\bar{f}_b\right]\right\} , \end{aligned} \tag{9.118}$$

while the collisional broadening is represented by

$$\begin{aligned} I^{CB}_{ab} &= 2\int \frac{d\boldsymbol{p}_b d\boldsymbol{q}}{(2\pi\hbar)^6}\frac{1}{\hbar^2}\left|V^s_{ab}(\boldsymbol{q})\right|^2 \boldsymbol{E}\left[\left\{e_a\nabla_{p_a}+e_b\nabla_{p_b}\right\}\frac{\mathcal{P}}{a^2}\right] \\ &\times \left\{\bar{f}_a\bar{f}_b\left[1-f_a\right]\left[1-f_b\right]-f_af_b\left[1-\bar{f}_a\right]\left[1-\bar{f}_b\right]\right\} . \end{aligned} \tag{9.119}$$

The following abbreviation was used

$$\frac{\mathcal{P}}{a} = \frac{\mathcal{P}}{\boldsymbol{q}\cdot\left(\frac{\boldsymbol{p}_a}{m_a}-\frac{\boldsymbol{p}_b}{mb}\right)-\frac{q^2}{2m_{ab}}} , \tag{9.120}$$

with $\boldsymbol{q} = \boldsymbol{p} - \bar{\boldsymbol{p}}$ being the momentum transfer. Using the equations (9.116)–(9.119), the transport properties of a plasma can be completely determined in

Born approximation and linear with respect to the external electric field. The inclusion of the electric field into the collision integral leads to the additional terms I_{ab}^{ICF} and I_{ab}^{CB} which will subsequently be named relaxation contributions. In the framework of the linear response approximation, we can use equilibrium distribution functions in order to determine these contributions.

Let us now consider the influence of the relaxation terms on the electrical transport properties. For this purpose, it is useful to start from the balance equation for the current

$$\frac{d\boldsymbol{j}}{dt} - \sum_a \frac{n_a e_a^2}{m_a} \boldsymbol{E} = \sum_{ab} \int \frac{d\boldsymbol{p}_a}{(2\pi\hbar)^3} \frac{e_a \boldsymbol{p}_a}{m_a} I_{ab}(\boldsymbol{p}_a) \tag{9.121}$$

and use the collision integral from (9.26). Furthermore, we symmetrize the right hand side with respect to $\boldsymbol{p}_a and \bar{\boldsymbol{p}}_a$, make the substitution $\boldsymbol{p}_a - e_a \boldsymbol{E} \to \boldsymbol{p}_a$, and use $\boldsymbol{p}_a - \bar{\boldsymbol{p}}_a = \boldsymbol{q}$. Then we get

$$\begin{aligned}
\sum_{ab} \int \frac{d\boldsymbol{p}_a}{(2\pi\hbar)^3} \frac{e_a \boldsymbol{p}_a}{m_a} I_{ab}(\boldsymbol{p}_a) &= \sum_{ab} \frac{e_a}{m_a} \int \frac{d\boldsymbol{p}_a d\boldsymbol{p}_b d\boldsymbol{q}}{(2\pi\hbar)^9} \\
&\times \left\{ \frac{\boldsymbol{q}}{\hbar} \left|V_{ab}^s(q)\right|^2 \int_0^\infty \cos\left[\frac{1}{\hbar}\left(\boldsymbol{q}\cdot(\boldsymbol{v}_a - \boldsymbol{v}_b)\tau + \boldsymbol{q}\cdot\boldsymbol{R}_{ab}\right)\right] d\tau \right. \\
&\times \left. \left[f_a\left(\boldsymbol{p}_a - \frac{\boldsymbol{q}}{2}\right) f_b\left(\boldsymbol{p}_b + \frac{\boldsymbol{q}}{2}\right) - f_a\left(\boldsymbol{p}_a + \frac{\boldsymbol{q}}{2}\right) f_b\left(\boldsymbol{p}_b - \frac{\boldsymbol{q}}{2}\right) \right] \right\} .
\end{aligned} \tag{9.122}$$

As a consequence of the substitutions just indicated, the distribution functions do not depend on the electric field any longer. We further mention that, as a consequence of the substitutions, the sign in front of $\boldsymbol{q}\cdot\boldsymbol{R}_{ab}$ has changed. Intra-collisional field and collisional broadening are no longer separable. Since we are interested only in the linear response approximation, we may linearize (9.122) with respect to the electric field. We now take into account the relation

$$\int_0^\infty \tau^2 \sin[a\tau]\, d\tau = -\frac{d^2}{da^2} \int_0^\infty \sin[a\tau]\, d\tau = -\frac{d^2}{da^2} \frac{\mathcal{P}}{a} .$$

Here, $\mathcal{P}$ denotes the principal value. With the help of this relation, we may carry out the time integration in the linearized expression (9.122)

$$\begin{aligned}
\sum_{ab} \int \frac{d\boldsymbol{p}_a}{(2\pi\hbar)^3} \frac{e_a \boldsymbol{p}_a}{m_a} I_{ab}(\boldsymbol{p}_a) &= \sum_{ab} \int \frac{d\boldsymbol{p}_a}{(2\pi\hbar)^3} \frac{e_a \boldsymbol{p}_a}{m_a} I_{ab}^0(\boldsymbol{p}_a) \\
&+ \sum_{ab} \left(\frac{e_a}{m_a} - \frac{e_b}{m_b}\right) \frac{e_a}{m_a} \int \frac{d\boldsymbol{p}_a d\boldsymbol{p}_b d\boldsymbol{q}}{(2\pi\hbar)^6} \left\{ \left|V_{ab}^s(q)\right|^2 \frac{\mathcal{P}}{[\boldsymbol{q}\cdot(\boldsymbol{v}_a - \boldsymbol{v}_b)/\hbar]^3} \right. \\
&\times \left. \boldsymbol{q} \frac{\boldsymbol{q}\cdot\boldsymbol{E}}{\hbar^2} \left[f_a^0\left(\boldsymbol{p}_a - \frac{\boldsymbol{q}}{2}\right) f_b^0\left(\boldsymbol{p}_b + \frac{\boldsymbol{q}}{2}\right) - f_a^0\left(\boldsymbol{p}_a + \frac{\boldsymbol{q}}{2}\right) f_b^0\left(\boldsymbol{p}_b - \frac{\boldsymbol{q}}{2}\right) \right] \right\} ,
\end{aligned} \tag{9.123}$$

where f_a^0 are Maxwell distribution functions. Using the linearized form of the collision integral, we may collect all field dependent contributions, i.e., the relaxation terms, on the left hand side of the equation (9.121). Then the relaxation terms produce a re-normalization of the external field. This effect is well known from the Debye–Onsager theory of electrolyte solutions. The balance equation for the current now takes the form

$$\frac{d\boldsymbol{j}}{dt} - \sum_a \frac{n_a e_a^2}{m_a} \boldsymbol{E} \left(1 + \frac{\delta \boldsymbol{E}_a}{\boldsymbol{E}}\right) = \sum_{ab} \int \frac{d\boldsymbol{p}_a}{(2\pi\hbar)^3} \frac{e_a \boldsymbol{p}_a}{m_a} I_{ab}^0(\boldsymbol{p}_a)\,. \tag{9.124}$$

Here, $\delta \boldsymbol{E}_a$ is the relaxation field. It is given by

$$\delta \boldsymbol{E}_a = \frac{1}{e_a n_a} \sum_b \left(\frac{e_a}{m_a} - \frac{e_b}{m_b}\right) \int \frac{d\boldsymbol{p}_a d\boldsymbol{p}_b d\boldsymbol{q}}{(2\pi\hbar)^9} \left|\frac{V_{ab}(\boldsymbol{q})}{\varepsilon(q,0)}\right|^2 \frac{\mathcal{P}}{[\boldsymbol{q}\cdot(\boldsymbol{v}_a - \boldsymbol{v}_b)/\hbar]^3}$$
$$\times \boldsymbol{q}\,(\boldsymbol{q}\cdot\boldsymbol{E}) \left[f_a^0\left(\boldsymbol{p}_a - \frac{\boldsymbol{q}}{2}\right) f_b^0\left(\boldsymbol{p}_b + \frac{\boldsymbol{q}}{2}\right) - f_a^0\left(\boldsymbol{p}_a + \frac{\boldsymbol{q}}{2}\right) f_b^0\left(\boldsymbol{p}_b - \frac{\boldsymbol{q}}{2}\right)\right]. \tag{9.125}$$

For the further evaluation of the expression (9.125), let us introduce relative and center-of-mass momenta. The integration can be carried out immediately over the center-of-mass momentum. Furthermore, it is useful to introduce the function $h_{ab}(k)$ by

$$h_{ab}(k) = \frac{1}{2\sqrt{\pi}} \frac{n_a n_b}{(k_B T)^2} q_{ab}\,(e_a - e_b)\, \frac{\boldsymbol{k}\cdot\boldsymbol{E}}{k^2} \frac{V_{ab}(k)}{|\varepsilon(k,0)|^2} \exp\left(-\frac{k^2\lambda_{ab}^2}{4}\right)$$
$$\times \int_0^\infty \frac{dk'}{(2\pi)^3} \int_{-1}^1 du\, \exp\left(-k'^2\lambda_{ab}^2\right) \frac{\exp\left(-k\lambda_{ab}^2 k' u\right) - \exp\left(k\lambda_{ab}^2 k' u\right)}{k\lambda_{ab} k' u^3}, \tag{9.126}$$

with $\lambda_{ab} = \hbar/\sqrt{2m_{ab}k_B T}$ being the thermal wavelength and

$$q_{ab} = \frac{\left(\frac{e_a}{m_a} - \frac{e_b}{m_b}\right)}{\left(\frac{1}{m_a} + \frac{1}{m_b}\right)(e_a - e_b)}\,.$$

The function $h_{ab}(k)$ with $k = q/\hbar$ can be interpreted as the non-equilibrium part of the correlation function. With (9.126), we write the relaxation field (9.125) in the form

$$\delta \boldsymbol{E}_a = -\frac{1}{e_a n_a} \sum_b \int \frac{d\boldsymbol{k}}{(2\pi)^3} \boldsymbol{k}\, V_{ab}(k)\, h_{ab}(k)\,. \tag{9.127}$$

The expression (9.127) explains the physical meaning of $\delta \boldsymbol{E}_a$ to be the gradient of the averaged two-particle potential. It is similar to the definition of the relaxation field in the theory of electrolytes. Of course, $\delta \boldsymbol{E}_a$ vanishes for spherically symmetric correlation functions. Therefore, the relaxation field arises from the deformation of the correlation function by the electric field.

Let us now consider the correlation function in more detail. First, it is easy to see that this quantity is an antisymmetric function, i.e., $h_{ab} = -h_{ba}$. In order to carry out the integrations in (9.126), we use the relation

$$\int_0^\infty \frac{e^{-bx^2}}{x}(e^{ax} - e^{-ax})\, dx = \pi \,\mathrm{erfi}\left(\frac{a}{2\sqrt{b}}\right)$$

for the integration over p and then

$$\int \frac{\mathrm{erfi}(ax)}{x^3} dx = -\frac{x\, e^{a^2x^2}}{a\sqrt{\pi}} - \frac{\mathrm{erfi}(ax)}{2\, a^2} + x^2 \mathrm{erfi}(ax)$$

for the u integration. The error-function with imaginary argument $\mathrm{erfi}(x)$ is defined as $\mathrm{erfi}(x) = 2\pi^{-1/2}\int_0^x \exp(t^2)dt$. After some algebra, we obtain the quantum mechanical expression

$$h_{ab}(k) = \frac{n_a n_b}{(k_B T)^2}\, q_{ab}(e_a - e_b)\, \frac{\boldsymbol{k}\cdot\boldsymbol{E}}{k^2}\, \frac{V_{ab}(k)}{|\varepsilon(k,0)|^2}$$
$$\times \left\{1 - \frac{1}{2} + \frac{\sqrt{\pi}}{2}\left(\frac{1}{k\lambda_{ab}} - \frac{1}{2}k\lambda_{ab}\right)\exp\left(-\frac{1}{4}k^2\lambda_{ab}^2\right)\mathrm{erfi}\left(\frac{k\lambda_{ab}}{2}\right)\right\}. \tag{9.128}$$

Here, the first term is the Fourier transform of the classical contribution to the correlation function. With the knowledge of $h_{ab}(k)$, we are able to determine the relaxation field. In the following, we will consider a two-component plasma consisting of electrons and singly charged ions. Then we have $\delta E_a = \delta E_b = \delta E$. Introducing (9.128) into (9.127) and applying the integral representation for $\mathrm{erfi}(x)$

$$\mathrm{erfi}(x) = \frac{2}{\sqrt{\pi}}\, x\, e^{x^2}\, {}_1F_1(1, \frac{3}{2}; -x^2) = \frac{2}{\sqrt{\pi}}\, e^{x^2}\int_0^\infty e^{-y^2}\sin(2xy)\, dy\,,$$

we get for the relaxation field

$$\frac{\delta E}{E} = -\frac{1}{12}\frac{\kappa e^2}{k_B T}\left\{1 + \frac{\kappa^2\lambda_{ei}^2}{4} - \exp\left(\frac{\kappa^2\lambda_{ei}^2}{4}\right)\right.$$
$$\left.\times\left[\sqrt{\pi}\,\frac{\kappa\lambda_{ei}}{2}\left(\frac{3}{2} + \frac{\kappa^2\lambda_{ei}^2}{4}\right)\left(1 - \Phi\left(\frac{\kappa\lambda_{ei}}{2}\right)\right)\right]\right\}. \tag{9.129}$$

Here, $\kappa = (4\pi\sum_a n_a e_a^2/k_B T)^{1/2}$ is the inverse Debye length, and $\Phi(x)$ is the error integral. Under the condition $\kappa\lambda_{ei} < 1$, the expression can be expanded with respect to $\kappa\lambda_{ei}$. The result is

$$\frac{\delta E}{E} = -\frac{1}{12}\frac{\kappa e^2}{k_B T}\left\{1 - \frac{3\sqrt{\pi}}{4}\kappa\lambda_{ei} + (\kappa\lambda_{ei})^2 + \cdots\right\} . \tag{9.130}$$

The first term in (9.129) and (9.130) is the classical relaxation field strength. The corresponding contribution in the theory of electrolyte solutions was for the first time determined by Debye and Hückel and was later improved by Onsager (see also Klimontovich and Ebeling (1962)).

In connection with the result (9.129), some remarks should be made. Equation (9.129) follows if we start our calculations from the field-dependent Landau collision integral with statical screening, i.e., the more general dynamical screening in the Lenard–Balescu equation, $|V^s_{ei}(\boldsymbol{k}, \boldsymbol{v}\cdot\boldsymbol{k})|^2$, is replaced by $|V^s_{ei}(\boldsymbol{k}, 0)|^2$. By Klimontovich (1975), another possibility was proposed to simplify the dynamical screening by introducing a velocity-averaged potential. The consequence of such averaging is the replacement of $|V^s_{ei}(\boldsymbol{k}, \boldsymbol{v}\cdot\boldsymbol{k})|^2$ by $V_{ei}(k)V^s_{ei}(k, 0)$. On the basis of this idea only one potential was taken to be screened in the paper by Klimontovich and Ebeling (1973), and the result reads

$$\begin{aligned}\frac{\delta E}{E} &= -\frac{1}{6}\frac{\kappa e^2}{k_B T}\left\{1 + \frac{\sqrt{\pi}}{2\kappa\lambda_{ei}} - \frac{\sqrt{\pi}\,\kappa\lambda_{ei}}{4}\exp\left(\frac{\kappa^2\lambda_{ei}^2}{4}\right)\right.\\ &\qquad\left.\times\left(1 + \frac{2}{(\kappa\lambda_{ei})^2}\right)\left(1 - \Phi\left(\frac{\kappa\lambda_{ei}}{2}\right)\right) - \frac{1}{2}\right\}\\ &= -\frac{1}{6}\frac{\kappa e^2}{k_B T}\left\{1 - \frac{3\sqrt{\pi}}{8}\kappa\lambda_{ei} + \cdots\right\} .\end{aligned} \tag{9.131}$$

We remark that the term proportional to κ^2 in (9.131) coincides with the corresponding term of (9.130).

It is interesting to compare the different classical approximations for the correlation function h_{ab} with our quantum mechanical expression (9.128). The simplest approximation is the Debye–Hückel correlation function. The Fourier transform of this function reads

$$h_{ab}(k) = \frac{n_a n_b}{(k_B T)^2}\, q_{ab}\,(e_a - e_b)(\boldsymbol{k}\cdot\boldsymbol{E})\,\frac{4\pi e_a e_b}{k^2(k^2 + \kappa^2)} . \tag{9.132}$$

In the Debye–Hückel theory, the screening is accounted for only in the equilibrium correlation function, which enters (9.132) as the Debye function.

In the Onsager theory, the screening refers also to the non-equilibrium contribution (Falkenhagen et al. 1971; Kremp 1966; Kremp et al. 1966). Then we get for the Fourier transform of the correlation function (Kremp 1965)

$$h_{ab}(k) = \frac{n_a n_b}{(k_B T)^2}\, q_{ab}\,(e_a - e_b)\ \boldsymbol{k}\cdot\boldsymbol{E}\,\frac{4\pi e_a e_b}{(k^2 + q_{ab}\,\kappa^2)\ (k^2 + \kappa^2)} . \tag{9.133}$$

We want to discuss the consequences of the approximations (9.132) and (9.133) on the classical relaxation field $\delta E/E$ according to (9.127). We get with the Debye–Hückel distribution

$$\frac{\delta E}{E} = -\frac{1}{6} \frac{\kappa e^2}{k_B T} \tag{9.134}$$

in agreement with the classical limiting value of the paper by Klimontovich and Ebeling (1973), while the correct classical value according to Onsager reads

$$\frac{\delta E}{E} = -\frac{1}{6\,(1+\sqrt{q_{ei}}\,)} \frac{\kappa e^2}{k_B T} \,. \tag{9.135}$$

Obviously, the expression (9.129) is closer to the correct Onsager formula (9.135) than the expression (9.134) which is usually applied in plasma physics and was originally introduced by Kadomtsev (1958). A reproduction of the Onsager result from the kinetic theory, obviously demands a kinetic equation going beyond a Landau equation with external field. Possibly, a non-Markovian Balescu equation generalized to external fields gives the desired result.

Furthermore, it is interesting to mention that there are generalizations of the Onsager result taking into account short range forces. For the rigid sphere model for the ions, having the radius a, we get, e.g., the expansion (Kremp 1966; Kremp et al. 1966; Falkenhagen et al. 1971)

$$\frac{\delta E}{E} = -\frac{1}{6} \frac{\kappa a b}{1+\sqrt{q_{ei}}} \left\{ 1 - (\frac{1}{2b} + 1)\, \kappa a \, (1+\sqrt{q_{ei}}\,) + \cdots \right\} .$$

Here, $b = e^2/(k_B T a)$ is the Bjerrum parameter. We mention that the quantum expression $\kappa \lambda_{ab}$ corresponds, in the classical case, to κa.

9.2.2 Lorentz Model with Dynamic Screening, Structure Factor

In Sect. 2.11, the conductivity of the Lorentz model was considered on the basis of the Landau equation with static screening. Let us now treat the influence of the dynamical screening on the conductivity. For simplicity, we consider the Lorentz plasma again, i.e., electron–electron interaction is completely neglected.

In order to include the dynamical screening, we now have to start from the Lenard–Balescu equation (7.43) derived and discussed in Sect. 7.4. Then we have, instead of (2.207), the following linearized Lenard–Balescu collision integral

$$\begin{aligned} I_{ei}(\boldsymbol{p}) = I_{ei}[f_e^1, f_e^0] &= \frac{2\pi}{\hbar} \int_{-\infty}^{\infty} d\omega \int \frac{d\boldsymbol{p}'}{(2\pi\hbar)^3} \frac{d\boldsymbol{k}}{(2\pi)^3} \left| \frac{V_{ei}(k)}{\varepsilon(k,\, \omega)} \right|^2 \\ &\times f_i^0(p') \left[\frac{\bar{p}_z}{\bar{p}} f_e^1(\bar{p}) - \frac{p_z}{p} f_e^1(p) \right] \\ &\times \delta\left(\hbar\omega - E_e(\boldsymbol{p}) + E_e(\boldsymbol{p} - \hbar\boldsymbol{k})\right) \delta\left(\hbar\omega + E_i(\boldsymbol{p}') - E_i(\boldsymbol{p}' + \hbar\boldsymbol{k})\right) . \end{aligned} \tag{9.136}$$

In deriving this expression, another notation of the particle momenta was used, i.e., the momentum transfer is given by $\hbar \boldsymbol{k} = \boldsymbol{p} - \bar{\boldsymbol{p}}$. Now we carry out the following steps. We introduce $z = \omega/k$, $x = \cos\vartheta$ and $x' = \cos\vartheta'$. Furthermore, we make some rearrangements in the arguments of the δ-functions. Then (9.136) can be written as

$$I_{ei}[f_e^1, f_e^0] = \frac{2\pi}{\hbar} \int_{-\infty}^{\infty} dz \int \frac{d\boldsymbol{k}}{(2\pi)^3} \left| \frac{V_{ei}(k)}{\varepsilon(k,\ kz)} \right|^2$$
$$\times \frac{m_e}{\hbar k p} \delta\left(\frac{z m_e}{p} - x + \frac{\hbar k}{2p} \right) \left[\frac{\bar{p}_z}{\bar{p}} f_e^1(\bar{p}) - \frac{p_z}{p} f_e^1(p) \right]$$
$$\times 2\pi \int_0^{\infty} \frac{dp'}{(2\pi\hbar)^3} \left[p'^2 \int_{-1}^{1} dx' \frac{m_i}{p'} \delta\left(\frac{z m_i}{p'} - x' - \frac{\hbar k}{2p'} \right) f_i^0(p') \right] . \qquad (9.137)$$

We are now able to carry out the p' integration. For this purpose, we assume $f_i^0(p')$ to be a Maxwell distribution. Furthermore, we neglect the term $\hbar k/2p'$ in the δ-function according to an adiabatic approximation. Then the expression in the third line of (9.137) gives

$$2\pi \int_0^{\infty} \frac{dp'}{(2\pi\hbar)^3} [\ldots] = 2\pi m_i^2 kT \int_{m_i z}^{\infty} \frac{dp'}{(2\pi\hbar)^3} \frac{p'}{m_i k_B T} f_i^0(p')$$
$$= \frac{2\pi m_i^2 k_B T n_i \Lambda_i^3}{h^3} \exp[-m_i^2 z^2 / 2 m_i k_B T] . \qquad (9.138)$$

If we once more change the variables to $Y = z/\sqrt{2k_BT/m_i}$ and introduce (9.138) into (9.137), the collision integral takes the form

$$I_{ei}[f_e^1, f_e^0] = \frac{(2\pi)}{\hbar} \int_{-\infty}^{\infty} dY \int_0^{\infty} \frac{dk}{(2\pi)^3} k^2 \int_{-1}^{1} dx \int_0^{2\pi} d\phi \left| \frac{V_{ei}(k)}{\varepsilon(k,\ \sqrt{2k_BT/m_i}\, kY\ )} \right|^2$$
$$\times \frac{n_i e^{-Y^2}}{\sqrt{\pi}} \left[\frac{\bar{p}_z}{\bar{p}} f_e^1(\bar{p}) - \frac{p_z}{p} f_e^1(p) \right] \frac{m_e}{\hbar k p} \delta\left(x - \sqrt{\frac{2k_BT}{m_i}} \frac{m_e}{p} Y - \frac{\hbar k}{2p} \right) .$$
$$(9.139)$$

We can approximately ignore the second term in the δ-function, i.e., we have $m_e/p\hbar k\, \delta(x - \hbar k/2p) = \delta(\hbar \boldsymbol{p}\cdot\boldsymbol{k}/m_e - \hbar^2 k^2/2m_e)$. Consequently, we may write $\bar{p} = |\boldsymbol{p} - \hbar\boldsymbol{k}| = p$. We proceed now like in Sect. 2.11. Especially, we apply the relation (2.203) to $k_z = \boldsymbol{e}\cdot\boldsymbol{k}$ using a coordinate system where the z-direction coincides with the direction of the momentum $\boldsymbol{p}$. Performing the integration over the azimuthal angle, the contribution following from the second term of this relation vanishes. The first term of this relation gives for the expression in the brackets of (9.139)

$$\left[\frac{p_z - \hbar k_z}{\bar{p}} f_e^1(\bar{p}) - \frac{p_z}{p} f_e^1(p) \right] = -\frac{\hbar k_z}{p} f_e^1(p) = -\frac{\hbar^2 k^2}{2p^2} f_e^1(p) \frac{p_z}{p} . \qquad (9.140)$$

In the last step, we made use of the δ-function in (9.139) neglecting the second term in its argument.

Let us next introduce the averaged quantity (Boercker et al. 1982)

$$\langle|\varepsilon^{-1}|^2\rangle = \frac{1}{\sqrt{\pi}}\int_{-\infty}^{\infty} dY \frac{e^{-Y^2}}{\left|\varepsilon(k, \sqrt{\frac{2k_BT}{m_i}}\, kY)\right|^2} . \tag{9.141}$$

With (9.141) and (9.140), equation (9.139) finally leads to the following expression for the collision integral

$$\begin{aligned} I_{ei}[f_e^1, f_e^0] &= -\frac{f_e^1(p)}{\tau_{ei}(p)}\frac{p_z}{p} , \\ \frac{1}{\tau_{ei}(p)} &= \frac{n_i m_e}{2p^3}\int_0^{\frac{2p}{\hbar}} \frac{dk}{2\pi} k^3 V_{ei}^2(k) \langle|\varepsilon^{-1}|^2\rangle . \end{aligned} \tag{9.142}$$

With this expression, we have found the solution of the kinetic equation in the approximation considered given in terms of the relaxation time τ_{ei} . Following the lines of Sect. 2.11, the conductivity is given by (2.224).

It is interesting to consider the relaxation time in more detail. Especially, let us first consider the averaged square modulus of the inverse dielectric function. For this purpose we remember the relations of Chap. 4. In RPA, the dielectric function $\varepsilon(q, \omega = \sqrt{\frac{2k_BT}{m_i}} kY)$ has to be determined from

$$\varepsilon(k, \sqrt{\frac{2k_BT}{m_i}} kY) = 1 - V_{ee}\Pi_{ee}(k, \sqrt{\frac{2k_BT}{m_i}} kY) - V_{ii}\Pi_{ii}(k, \sqrt{\frac{2k_BT}{m_i}} kY) ,$$

where Π_{aa} is given by (4.107). On behalf of the large differences of their masses, electrons and ions contribute in different manner to the dielectric function. In the electron polarization function, the term $\sqrt{\frac{2k_BT}{m_i}}\, kY$ can be ignored compared to the kinetic energy in the denominator of (4.107), i.e., the static limit is sufficient, whereas we have to take into account the full dynamic expression in the ion polarization function. However, it is now sufficient to consider the classical limit of Π_{ii} according to (4.130). These approximations are justified in lowest order with respect to the mass ratio m_e/m_i. We now refer to (4.131) and write it in terms of Y, i.e., $\mathrm{Im}\varepsilon_i = \sqrt{\pi}\,\frac{\kappa_i^2}{k^2} Y \exp[-Y^2]$, with $\kappa_i^2 = 4\pi n_i e_i^2/k_BT$. We can then transform (9.141) to

$$\begin{aligned} \langle|\varepsilon^{-1}|^2\rangle &= \frac{k^2}{\kappa_i^2} P \int_{-\infty}^{\infty} \frac{dY}{\pi}\frac{1}{Y} \frac{\mathrm{Im}\varepsilon_i}{\left|1 - V_{ee}\Pi_{ee}(k,0) - V_{ii}\Pi_{ii}(k, \sqrt{\frac{2k_BT}{m_i}} kY)\right|^2} \\ &= \frac{1}{1 - V_{ee}\Pi_{ee}(k,0)}\frac{k^2}{\kappa_i^2} \mathcal{P}\int_{-\infty}^{\infty} \frac{dY}{\pi}\frac{\mathrm{Im}\varepsilon_i^{\mathrm{eff}-1}}{Y} . \end{aligned} \tag{9.143}$$

Here, $\varepsilon_i^{\text{eff}}$ is the dielectric function for the ion subsystem with an effective ion–ion interaction screened by the electrons

$$\varepsilon_i^{\text{eff}} = 1 - \frac{V_{ii}\Pi_{ii}(Y)}{1 - V_{ee}\Pi_{ee}(k,0)} = 1 - V_{ii}^{\text{eff}}(q)\Pi_{ii}(Y)\,. \tag{9.144}$$

The effective interaction given by

$$V_{ii}^{\text{eff}}(q) = \frac{V_{ii}(k)}{1 - V_{ee}\Pi_{ee}(k,0)}\,,$$

is an important quantity in the theory of dense plasmas and liquid metals (Kovalenko et al. 1990).

Now the integral in (9.143) can easily be carried out. The integral is just the dispersion relation (4.72) for $\omega' = 0$. Then (9.143) becomes

$$\left\langle |\varepsilon^{-1}|^2 \right\rangle = \frac{1}{1 - V_{ee}\Pi_{ee}(k,0)} \frac{k^2}{\kappa_i^2} \left[\text{Re}\varepsilon_i^{\text{eff}-1}(k,0) - 1 \right]\,.$$

After some simple transformations, we finally get the result (Boercker et al. 1982)

$$\begin{aligned} \left\langle |\varepsilon^{-1}|^2 \right\rangle &= \frac{1}{\varepsilon_e(k,0)\varepsilon(k,0)} = \frac{1}{|\varepsilon_e(k,0)|^2} S_{ii}(k)\,, \\ S_{ii}(k) &= \frac{1}{\varepsilon_e(k,0)^{-1}\varepsilon(k,0)} = \frac{\varepsilon_e(k,0)}{1 - V_{ee}(k)\Pi_{ee}(k,0) - V_{ii}(k)\Pi_{ii}(k,0)}\,, \end{aligned} \tag{9.145}$$

where $S_{ii}(k)$ is the static structure factor of the ions given by (4.204). Here the ion–ion interaction is screened by the dielectric function of the electrons.

Let us come back to the relaxation time and the conductivity. Equation (9.145) can be used in (9.142). Then the conductivity can be calculated by the equations

$$\begin{aligned} \tau_{ei}^{-1}(p) &= \frac{n_i m_e}{2p^3} \int_0^{\frac{2p}{\hbar}} \frac{dk}{2\pi} k^3 \left| \frac{V_{ei}(k)}{\varepsilon_e(k,0)} \right|^2 S_{ii}(k)\,, \\ \sigma &= (2s+1)\frac{4\pi}{3} e^2 \int_0^\infty \frac{dp}{(2\pi\hbar)^3} \frac{p^3}{m_e} \tau_{ei}(p) \frac{\partial}{\partial p} f_e^0(p)\,. \end{aligned} \tag{9.146}$$

This exact result for the Lorentz gas (9.146) differs from the Ziman formula for finite temperatures (2.227). Only in the limit $T \to 0$, the formulas (2.227) and (9.146) lead to the identical result if the same level of approximation is used for the dielectric function

$$\sigma = \frac{e^2}{m_e} \frac{12\pi^3\hbar^3}{m_e} \left[\int_0^{2k_{\text{F}}} \frac{dk}{2\pi} k^3 \left| \frac{V_{ei}(k)}{\varepsilon_e(k,0)} \right|^2 S_{ii}(k) \right]^{-1}\,.$$

In many situations, the Debye limit of the relaxation time is of interest. In this case, we have

$$\tau_{ei}^{-1}(p) = \frac{n_i m_e}{2p^3} \int_0^{\frac{2p}{\hbar}} \frac{dk}{2\pi} k^3 \left| \frac{4\pi e^2}{k^2+\kappa_e^2} \right|^2 \frac{k^2+\kappa_e^2}{k^2+\kappa_e^2+\kappa_i^2} \, .$$

Let us finally remark that the structure factor S_{ii} reduces to the expression (4.205) if $\varepsilon_e \simeq 1$.

9.2.3 Chapman–Enskog Approach to the Conductivity

We now consider the electrical conductivity of a partially ionized plasma consisting of electrons e, ions i, and atoms $A = (ei)$. The basic kinetic equations including all elastic and inelastic scattering integrals such as excitation, ionization, and recombination were derived in Sect. 7.7.2. We start, therefore, from (7.122) but neglect, for a first consideration, the inelastic processes. Then the electron kinetic equation can be written in the following form

$$\frac{\partial f_e(\boldsymbol{p}_e,t)}{\partial t} + e_e \boldsymbol{E}\left\{1+\frac{\delta \boldsymbol{E}}{\boldsymbol{E}}\right\}\cdot\nabla_{\boldsymbol{p}_e} f_e(\boldsymbol{p}_e,t) = \sum_{b=e,i} I_{eb}(\boldsymbol{p}_e,t) + I_{eA}(\boldsymbol{p}_e,t) \, . \quad (9.147)$$

This equation couples to other kinetic equations for the ions and atoms as the summation index runs over all species, i.e., electrons, ions, and atoms. Furthermore, the complicated momentum dependent relaxation contribution ΔI_a on the left hand side of (9.116) is replaced by the relaxation field, and we get in such a way just the balance equation (9.124) from (9.147).

The collision integrals on the r.h.s. of (9.147) were used here in T-matrix approximation as given in Sects. 7.6 and 7.7. The subscripts at the momenta which label the species will not be given in the following. Furthermore, the spin sum is dropped for simplicity like in previous sections. Then the collision integrals for the interaction between the charged particles can be written as

$$\begin{aligned} I_{eb}(\boldsymbol{p},t) &= \frac{1}{\hbar V} \int \frac{d\boldsymbol{p}' d\bar{\boldsymbol{p}} d\bar{\boldsymbol{p}}'}{(2\pi\hbar)^9} \left| \langle \boldsymbol{p}\,\boldsymbol{p}' \,|\, \mathrm{T}^R_{eb}(E_{eb}) \,| \bar{\boldsymbol{p}}\,\bar{\boldsymbol{p}}'\rangle \right|^2 \\ &\times 2\pi\delta(E_{eb} - \bar{E}_{eb}) \left\{ \bar{f}_e\, \bar{f}'_b (1-f_e)(1-f'_b) - f_e f'_b (1-\bar{f}_e)(1-\bar{f}'_b) \right\}_t \, . \end{aligned} \quad (9.148)$$

The collisions of electrons with atoms is taken to be the elastic scattering on the ground state and, consequently, we get from (7.123) the following collision integral neglecting the degeneracy of atoms,

$$\begin{aligned} I_{eA}(\boldsymbol{p},t) &= \frac{1}{\hbar V} \int \frac{d\boldsymbol{P}\, d\bar{\boldsymbol{p}}\, d\bar{\boldsymbol{P}}}{(2\pi\hbar)^9} \left| \langle \boldsymbol{p}\,\boldsymbol{P}\, | T^R_{eA}(E_{eA}) |\, \bar{\boldsymbol{P}}\,\bar{\boldsymbol{p}}\rangle \right|^2 \\ &\times 2\pi\delta(E_{eA} - \bar{E}_{eA}) \left\{ \bar{f}_e \bar{f}_A (1-f_e) - f_e f_A (1-\bar{f}_e) \right\}_t \, . \end{aligned} \quad (9.149)$$

For the distribution functions, we used $f_e = f_e(\boldsymbol{p},t)$, $f_A = f_A(\boldsymbol{P},t)$, etc.

Since all these equations contain the distribution functions of the atoms, electrons, and ions, we additionally need the plasma composition. In general, the time evolution of the plasma composition follows from the more general kinetic equation (7.122) or from the corresponding rate equations (8.1) and (8.2) developed in Sect. 8.1. In the stationary case, it is sufficient to use a mass action law (MAL). The MAL is essentially influenced by nonideality effects, especially by the lowering of the ionization energy. We know that the lowering of the ionization energy produces the nonmetal–metal transition.

In the following, we will extend the simple explanation of this effect discussed already in Chap. 2 by more rigorous considerations. Especially, we treat the scattering processes on the T-matrix level, and we include electron–electron collisions. Therefore, we cannot apply the adiabatic approximation like in Sect. 2.11 what led to a drastic simplification of the kinetic equation (Lorentz model). The main feature of the kinetic equation in that approximation was its linearity with respect to the distribution function. The complication in the present situation is due to the nonlinearity in the electron–electron collision integral.

In order to manage the nonlinearity, we take into account the fact that, due to the axial symmetry, the distribution function has the structure $f_e(\boldsymbol{p}) = f_e(p, \vartheta)$, where ϑ is the angle between the electric field and the momentum $\boldsymbol{p}$. For the further considerations, we define a coordinate system where the z-direction is $\hat{\boldsymbol{e}} = \boldsymbol{E}/E$. Thus an expansion of the distribution function in terms of Legendre polynomials can be used for the situation given, i.e.,

$$f_e(\boldsymbol{p}) = \sum_n f_e^n(p) P_n(\cos\vartheta) \approx f_e^0(p) + f_e^1(p)\cos\vartheta \tag{9.150}$$

with $\cos\vartheta = p_z/p$. The first term on the r.h.s of (9.150) is the isotropic part of the distribution function, and the second one describes the deviation due to the anisotropy. The functions $f_e^0(p)$ and $f_e^1(p)$ fulfill the relations (2.206). If the expansion (9.150) is inserted into the kinetic equation (9.147) and the orthogonality relations of the Legendre polynomials are used, the resulting expression can be broken down into two equations. These equations are coupled with each other and, especially, the isotropic part f_e^0 depends on the field, too. That means the equations are very useful to determine the isotropic distribution for plasmas in strong electric fields (see Davidov (1936), Golant et al. (1980)) and the influence of f_e^0 on quantities such as rate coefficients (Kremp (a) et al. 1993; Morawetz et al. 1993; Schlanges (b) et al. 1996). In the following, we restrict ourselves to the simpler case of an isotropic stationary plasma in linear response approximation. Then f_e^0 is the Maxwell or the Fermi distribution function and, therefore, only one of the equations is relevant. This equation, we consider here in the form

$$e_e E_z^{\mathrm{eff}} \frac{p_z}{p} \frac{\partial}{\partial p} f_e^0 = \sum_{b=e,i} I_{eb}[f^0 f^1] + I_{eA}[f^0 f^1]\,. \tag{9.151}$$

In order to solve this equation, we apply, like in section 2.11, the Chapman–Enskog perturbation theory (Chapman and Cowling 1952). For the further considerations, we use the ansatz

$$f_e(p) = f_e^0(p) + f_e^1(p)\cos\vartheta = f_e^0(p) + f_e^0(p)\left[1 - f_e^0(p)\right]\Phi_e(p)p_z \qquad (9.152)$$

for the distribution function and linearize the collision integrals with respect to Φ_e. Furthermore, we use the representation (5.116) for the T-matrix and introduce the cross section corresponding to (2.201). Then we get, e.g., for the electron–electron collision integral

$$I_{ee} = \frac{1}{m_e}\int \frac{d\boldsymbol{p}'}{(2\pi\hbar)^3} d\Omega\, |\boldsymbol{p}-\boldsymbol{p}'|\,\frac{d\sigma_{ee}}{d\Omega}\; f_e^0(p)\; f_e^0(p')[1 - f_e^0(\bar{p})]\,[1 - f_e^0(\bar{p}')]$$
$$\times\, [\bar{p}_z\Phi_e(\bar{p}) + \bar{p}'_z\Phi_e(\bar{p}') - p_z\Phi_e(p) - p'_z\Phi_e(p')]\,. \qquad (9.153)$$

Here, momentum and energy conservation was used, i.e., $\boldsymbol{p}+\boldsymbol{p}' = \bar{\boldsymbol{p}}+\bar{\boldsymbol{p}}'$ and $|\boldsymbol{p}-\boldsymbol{p}'| = |\bar{\boldsymbol{p}}-\bar{\boldsymbol{p}}'|$. Now we apply the relation (2.203) to every term in (9.153) with an appropriate choice of the z-direction for the integration over $\boldsymbol{p}'$. The integral over the respective azimuthal angles then vanishes, and we get

$$I_{ee} = \frac{1}{m_e}\int \frac{d\boldsymbol{p}'}{(2\pi\hbar)^3} d\Omega\; |\boldsymbol{p}-\boldsymbol{p}'|\,\frac{d\sigma_{ee}}{d\Omega}\; f_e^0(p)\; f_e^0(p')\;[1 - f_e^0(\bar{p})][1 - f_a^0(\bar{p}')]$$
$$\times\, \frac{p_z}{p^2}\,[(\boldsymbol{p}\cdot\bar{\boldsymbol{p}})\,\Phi_e(\bar{p}) + (\boldsymbol{p}\cdot\bar{\boldsymbol{p}}')\,\Phi_e(\bar{p}') - (\boldsymbol{p}\cdot\boldsymbol{p})\,\Phi_e(p) - (\boldsymbol{p}\cdot\boldsymbol{p}')\,\Phi_e(p')]\,. \qquad (9.154)$$

The collision integrals for the interaction between the electrons and the heavy particles (atoms and ions) can easily be linearized using the adiabatic approximation. Assuming Boltzmann statistics for the heavy particles we find like in Sect. 2.11,

$$I_{eb} = \frac{n_b}{m_e^2}\int d\bar{\boldsymbol{p}}\;\frac{d\sigma_{eb}}{d\Omega}\;\delta(E_p - E_{\bar{p}})\, f_e^0(p)[1 - f_e^0(\bar{p})]$$
$$\times\; \frac{p_z}{p^2}\,[(\boldsymbol{p}\cdot\bar{\boldsymbol{p}})\,\Phi_e(\bar{p}) - (\boldsymbol{p}\cdot\boldsymbol{p})\,\Phi_e(p)]\,. \qquad (9.155)$$

Now we insert the collision integrals into (9.151), which turns out to be an inhomogeneous linear integral equation for the function Φ_e. The method of solution is standard. The unknown function Φ_e is expanded into a series of properly chosen orthogonal polynomials. As a result, the integral equation is transformed into an infinite set of algebraic equations for the coefficients of the expansion (Chapman and Cowling 1952). Different approximate solutions of this set are obtained by truncation at different levels.

The appropriate polynomials that should be used here for non-degenerate plasmas are the Sonine polynomials $S_{3/2}^{(\nu)}$ (Schirmer 1955; Schirmer and Friedrich 1958). They are defined as

$$S^{(\nu)}_{3/2}(x^2) = \sum_{k=0}^{\nu}(-x^2)^k \frac{\Gamma(\nu+\frac{5}{2})}{k!\,(\nu-1)!\,\Gamma(k+\frac{5}{2})}\,. \tag{9.156}$$

The Sonine polynomials obey the following orthogonality relations

$$\int_0^\infty dx\; e^{-x^2} x^4\, S^{(\mu)}_{3/2}(x^2)\, S^{(\nu)}_{3/2}(x^2) \;=\; \delta_{\mu\nu}\frac{\Gamma(\nu+\frac{5}{2})}{2\;\nu!}\,. \tag{9.157}$$

Now we assume Φ_e to be expanded as

$$\Phi_e(p) = \sum_{\nu=0}^{\infty} a_\nu\, S^{(\nu)}_{3/2}(x^2)\,, \tag{9.158}$$

where a_ν are constants still to be determined, and $x^2 = p^2/(2m_e k_B T)$. We insert this expansion into (9.151), multiply by $x^2\, S^{(\mu)}_{3/2}(x^2)$, and integrate over $\boldsymbol{p}$. Taking into account (9.157), we obtain the following infinite set of linear algebraic equations (Devoto 1966) for the coefficients a_ν:

$$\frac{3\sqrt{\pi}}{8}\frac{eE^{\mathrm{eff}}}{k_B T}\sqrt{\frac{m_e}{2k_B T}}\,\delta_{\mu,0} = \sum_b \sum_{\nu=0}^{\infty} a_\nu I^{\mu\nu}_{eb}\,. \tag{9.159}$$

The quantities $I^{\mu\nu}_{eb}$ are determined by the collision integrals and have the following form for the electron-heavy particle collisions ($b = i, A$)

$$I^{\mu\nu}_{eb} = n_b \int_0^\infty dz\, z^5\, Q^{(1)}_{eb}(z)\, S^{(\mu)}_{3/2}(z^2)\; S^{(\nu)}_{3/2}(z^2)\, e^{-z^2}\,. \tag{9.160}$$

For the electron–electron scattering, we get:

$$I^{\mu\nu}_{ee} = n_e\sqrt{2}\sum_{s=1}^{\nu+\mu-1}\int_0^\infty dz\, z^{2s+5}\, Q^{(m)}_{ee}(z)\, A^{\mu\nu}_s\; e^{-2z^2} \tag{9.161}$$

with coefficients $A^{\mu\nu}_s$ which can be found, for instance, in the paper by Chin et al. (1988). In the expressions (9.160) and (9.161), the dimensionless quantity z is given by $z^2 = p^2_{eb}/2m_e k_B T$ where p_{eb} is the relative momentum of the respective scattering process. The transport cross sections are given by

$$Q^{(m)}_{eb}(p) = \int_0^{2\pi} d\phi \int_{-1}^{1} d\cos\theta(1-\cos^m\theta)\,\frac{d\sigma_{eb}}{d\Omega}\,, \tag{9.162}$$

with $m = 1$ for electron-heavy particle collisions ($b = i, A$). As in previous chapters, we will use the notation $Q^T_{eb} = Q^{(1)}_{eb}$. The electron–electron transport cross section ($b = e$) is given by (9.162) with $m = 2$ for $\nu + \mu < 6$.

With the distribution function thus known, the electrical current can be determined from

$$\boldsymbol{j}_e = e_e \int \frac{d\boldsymbol{p}}{(2\pi\hbar)^3} \frac{\boldsymbol{p}}{m_e} f_e(\boldsymbol{p}) = \sigma \boldsymbol{E} \,. \tag{9.163}$$

One can see rather easily that the electrical conductivity is determined by the coefficient a_0 of the expansion (9.158) only. We get

$$\sigma = \frac{e n_e}{m_e} \frac{k_B T}{E} a_0 \,. \tag{9.164}$$

To determine the coefficient a_0, we have to solve the infinite set of equations (9.159) for the coefficients a_ν. Of course, this can be done only approximately. We solve the system up to a finite number $\nu = n$ of terms in the expansion (9.158). This level is named an approximation of $(n+1)th$ order. Using the determinants

$$I^{(n)} = \begin{vmatrix} I^{00} & I^{10} & . & . & . & I^{n0} \\ I^{10} & I^{11} & . & . & . & I^{n1} \\ . & . & . & & & . \\ . & . & & . & & . \\ . & . & & & . & . \\ I^{n0} & I^{n1} & . & . & . & I^{nn} \end{vmatrix} \qquad I^{(n)}_{00} = \begin{vmatrix} 1 & 0 & . & . & . & 0 \\ 0 & I^{11} & . & . & . & I^{n1} \\ . & . & . & & & . \\ . & . & & . & & . \\ . & . & & & . & . \\ 0 & I^{n1} & . & . & . & I^{nn} \end{vmatrix} ,$$

we get the conductivity in the $(n+1)th$ order with the help of Cramer's rule

$$\sigma^{(n)} = \frac{3\sqrt{\pi}}{8} \frac{e^2 n_e}{\sqrt{2 m_e k T}} \frac{I^{(n)}_{00}}{I^{(n)}} \,. \tag{9.165}$$

The elements of the determinants are $I^{\mu\nu} = I^{\mu\nu}_{ee} + I^{\mu\nu}_{ei} + I^{\mu\nu}_{eA}$.

Equation (9.165) is the basic relation for the calculations of the plasma conductivity. There are papers where (9.165) was evaluated in the case of completely ionized plasmas. We mention the calculations by Williams and DeWitt (1969) and by Rogers et al. (1981). Further work was done by Morales et al. (1989) and Redmer et al. (1990). In these papers, the Gould–DeWitt collision integral was used and, therefore, dynamical screening and strong collisions were accounted for.

It is interesting to write the lowest order approximations for the electrical conductivity following from the expression (9.165). We easily get

$$\sigma^{(0)} = \frac{3\sqrt{\pi}}{8} \frac{e^2 n_e}{\sqrt{2 m_e k T}} \frac{1}{I^{00}} , \tag{9.166}$$

$$\sigma^{(1)} = \sigma^{(0)} \frac{I^{00} I^{11}}{I^{00} I^{11} - I^{10} I^{10}} , \tag{9.167}$$

$$\sigma^{(2)} = \sigma^{(0)} \left(1 - \frac{I^{10}(I^{10} I^{22} - I^{20} I^{21}) - I^{20}(I^{10} I^{21} - I^{20} I^{11})}{I^{00}(I^{11} I^{22} - I^{21} I^{21})} \right)^{-1} , \tag{9.168}$$

where we have used $I^{\nu\mu} = I^{\mu\nu}$. In the expressions given above, we dropped the contribution which is due to the relaxation effect. It gives an additional factor $(1+\delta E/E)$ in (9.166). In order to test the quality of the approximation, we may apply the equations (9.166)–(9.168) to the Lorentz plasma model and compare the result with the well-known exact solution (see Sect. 2.11). We find that, for practical purposes, it is sufficient to consider the approximation given by $\sigma^{(2)}$.

Let us now consider degenerate plasmas. The calculation for the non-degenerate quantum regime was simplified by the fact that the l.h.s. of the equations (9.159) is proportional to $\delta_{0\mu}$. For degenerate Fermi systems, we have to use Fermi distribution functions in equation (9.151), and Sonine polynomials are no longer appropriate. In papers by Lampe (1968a), Lampe (1968b), polynomials P_i were therefore constructed such that each P_i is orthogonal to P_j, where the functions P_i, P_j are orthogonal if

$$\int_0^\infty dp f^0(p)[1 - f^0(p)]\, p^4 P_i(p^2) P_j(p^2) = 0 \qquad i \neq j\,, \tag{9.169}$$

where $f^0(p)$ is the Fermi function. The first three polynomials read

$$\begin{aligned} P_0(x^2) &= 1 \\ P_1(x^2) &= x^2 + h_1 \\ P_2(x^2) &= x^4 + h_2 x^2 + h_3\,. \end{aligned} \tag{9.170}$$

The quantities h_i are given in terms of Fermi integrals (see Lampe (1968a), Lampe (1968b)). In the non-degenerate case, these polynomials reduce to Sonine polynomials. Now we expand Φ_e with respect to P_j and proceed in the same way as above. We consider a fully ionized plasma and, for simplification, the cross sections are taken in Born approximation. One can then find the first Chapman–Enskog polynomial approximations given by

$$\begin{aligned} \sigma^{(0)} &= \frac{3e^2 n_e^2 k_B T}{m_e^2} \frac{1}{a_{00}} \\ \sigma^{(1)} &= \sigma^{(0)} \left(1 - \frac{a_{10}^2}{a_{00}\, a_{11}}\right)^{-1}. \end{aligned} \tag{9.171}$$

The elements a_{ij} are determined by the electron–ion and by the electron–electron collision integrals according to

$$a_{ij} = \frac{2}{m_e^2} \int \frac{d\boldsymbol{p}}{(2\pi\hbar)^3} P_i(p^2) \frac{1}{p_z} \left[\frac{1}{(2s+1)} I_{ee}(P_j) + I_{ei}(P_j)\right]. \tag{9.172}$$

The zeroth approximation is determined by electron–ion collisions only. If we use the collision integral (9.155) in Born approximation in (9.172), we get after a simple calculation

$$\sigma^{(0)} = \frac{3e^2 n_e^2 k_B T}{2\pi\hbar\, n_i} \Big\{ \int \frac{d\boldsymbol{p}}{(2\pi\hbar)^3}\, d\boldsymbol{k}\, k^2 \left| \frac{V_{ei}(k)}{\varepsilon(k)} \right|^2 \times \delta(E_{\boldsymbol{p}} - E_{\boldsymbol{p}-\hbar\boldsymbol{k}})\, f_e^0(p) \left[1 - f_e^0(p)\right] \Big\}^{-1} \qquad (9.173)$$

with $V_{ei}(k) = 4\pi e_e e_i / k^2$ being the Coulomb potential, and $\varepsilon(k)$ is the static RPA dielectric function. Further reformulation of (9.173) gives

$$\sigma^{(0)} = \frac{12\pi^3 n_e^2 e^2 \hbar^3}{m_e^2 n_i} \left\{ \int_0^\infty dk\, k^3 f_e^0\left(\frac{k}{2}\right) \left| \frac{V_{ei}(k)}{\varepsilon(k)} \right|^2 \right\}^{-1} . \qquad (9.174)$$

In the limit $T \to 0$, the Fermi function becomes a unit step function, so that equation (9.174) reduces to

$$\sigma^{(0)} = \frac{12\pi^3 n_e^2 e^2 \hbar^3}{m_e^2 n_i} \left\{ \int_0^{2k_f} dk\, k^3 \left| \frac{V_{ei}(k)}{\varepsilon(k)} \right|^2 \right\}^{-1} . \qquad (9.175)$$

The approach developed by Lampe (1968a), Lampe (1968b) is of interest if electron–electron collisions have to be taken into account in the calculations of transport properties of degenerate quantum plasmas. In particular, the first order approximation for the electrical conductivity $\sigma^{(1)}$ given by (9.171) allows for first corrections due to electron–electron scattering. As already mentioned, these scattering processes are not included in the zeroth approximation $\sigma^{(0)}$ which follows in the lowest order of the polynomial expansion. The resulting expressions in Born approximation given by (9.174) and (9.175) are similar to the Ziman formula (Ziman 1972; Boercker, Rogers, and DeWitt 1982; Ichimaru and Tanaka 1985) with the following differences: In the formulas (9.174) and (9.175), the screening is realized by the full dielectric function $\varepsilon(k)$ of the electron–ion system corresponding to the derivation from the static Lenard–Balescu kinetic equation. The dielectric function in the Ziman formula includes only the electronic component. However, the Ziman formula accounts for the static ion–ion structure factor. As compared to the Lorentz expressions for the conductivity of degenerate plasmas, given by (2.217) for static screening and by (9.146) accounting for dynamical screening, the formulas (9.174) and (9.175) represent drastic approximations.

We return to (9.165) which will be used to calculate the static electrical conductivity for non-degenerate partially ionized plasmas. In order to determine the conductivity, one has to deal with two main problems. First, we have to calculate the transport cross section by solving the corresponding quantum mechanical few-body scattering problem. Second, the plasma composition, e.g., the density of the electrons, ions, and atoms, has to be determined from generalized mass action laws such as derived in Chap. 6. This program will be carry out for different materials in the following subsections.

9.2.4 Partially Ionized Hydrogen Plasma

Transport Cross Sections. A hydrogen plasma is considered to consist of electrons, protons, atoms, and molecules. We describe the electron–proton and electron–electron interactions in a simple manner by statically screened Coulomb (Debye) potentials and determine the transport cross sections from their phase shift representations (5.135) and (5.136) by numerical solution of the radial Schrödinger equation. In this way, the cross sections are given on the level of the statically screened T-matrix approximation (see Sect. 5.6).

The more complicated contribution of electron–atom scattering is included in the three-particle collision term (7.123) of the kinetic equation (7.122). As mentioned in the previous section, we will restrict ourselves to elastic scattering from the atomic ground state. Instead of solving the Lippmann–Schwinger equation for the three-particle T-matrix by perturbation theory (Kremp (b) et al. 1983), we apply here a close-coupling-type method starting from the Schrödinger equation. To give a more general formulation, we consider the scattering on a many-electron atom (ion) described by the full scattering state $|\alpha \boldsymbol{p}_e+\rangle$, where $\boldsymbol{p}_e$ is the momentum of the incident electron and α denotes the set of quantum numbers of the target. The scattering state is determined by

$$\mathrm{H}\,|\alpha\,\boldsymbol{p}_e+\rangle = E_{\alpha \boldsymbol{p}_e}|\alpha\,\boldsymbol{p}_e+\rangle\,. \tag{9.176}$$

Here, H is the full Hamiltonian which can be written in the form

$$\mathrm{H} = \mathrm{H}_e + \mathrm{H}_\mathrm{A} + \mathrm{V}_\mathrm{eA}\,, \quad \mathrm{V}_\mathrm{eA} = \mathrm{V}_{e0} + \sum_j \mathrm{V}_{ej}\,, \tag{9.177}$$

where H_e is the Hamiltonian of the incident electron, while H_A is the target (atomic) Hamiltonian. The quantity V_{eA} denotes the potential in the channel of elastic electron–atom scattering, where V_{e0} is the two-body potential of the interaction between the electron and the nucleus, and V_{ej} are the potentials between the incident electron and the target electrons. It is convenient to expand the full scattering wave function in terms of the target states $\varphi_\alpha(R) = \langle R|\alpha\rangle$,

$$\langle \boldsymbol{r}_1\,R|\alpha\,\boldsymbol{p}_e+\rangle = \sum_\alpha F_\alpha(\boldsymbol{r}_1)\varphi_\alpha(R)\,, \tag{9.178}$$

where $\boldsymbol{r}_1$ is the position vector of the incident electron, and R denotes the complete set of the target coordinates. The sum runs over the discrete and continuous states of the target including all channels. Inserting the expansion (9.178) into (9.177), one gets (Taylor 1972)

$$[\mathrm{H}_e - (E - E_\alpha)]\;F_\alpha(\boldsymbol{r}_1) \;=\; -\sum_{\alpha'} \mathrm{V}_{\alpha\alpha'}(\boldsymbol{r}_1)\,F_{\alpha'}(\boldsymbol{r}_1) \tag{9.179}$$

with the potential matrix

$$\mathrm{V}_{\alpha\alpha'}(\boldsymbol{r}_1) = \int dR\, \varphi^*_\alpha(R)\, \mathrm{V}_{\mathrm{eA}}(\boldsymbol{r}_1, R)\, \varphi_{\alpha'}(R)\,. \tag{9.180}$$

Exchange processes are not included here. The result (9.179) represents a system of coupled equations for the expansion coefficients $F_\alpha(\boldsymbol{r}_1)$. The latter can be interpreted as single-particle wave functions describing the multichannel scattering process of an electron hitting the target. The usefulness of this set of coupled equations is that it can serve as an appropriate starting point for approximate solutions to the problem. As already mentioned, we consider elastic scattering on a target in the ground state ($\alpha = 0$). For this purpose, we write the equation for F_0 in the form

$$[\mathrm{H}_e - (E - E_0) + V_{00}(\boldsymbol{r}_1)]\ F_0(\boldsymbol{r}_1) \ = -\sum_{\alpha'\neq 0} \mathrm{V}_{0\,\alpha'}(\boldsymbol{r}_1)\, F_{\alpha'}(\boldsymbol{r}_1)\,. \tag{9.181}$$

Here, E_0 is the atomic ground state energy. The simplest approximation which can be applied to this equation is the one-state approximation restricting to the (diagonal) ground state contribution $V_{00}(\boldsymbol{r}_1)$ and neglecting the r.h.s. completely. In this way, the task reduces to a Schrödinger equation describing the elastic scattering of the electron on the static charge distribution of the target. For a better description, the coupling terms on the r.h.s. of (9.179) have to be treated in improved approximations (Sobel'man et al. 1988). We determine the coupling term by solving the set of equations (9.179) within a perturbation theory. The result is a closed equation for F_0 including the static field and, additionally, the effect of the polarizability of the charge distribution of the atom produced by the incident electron.

For the elastic scattering of an electron on a hydrogen atom in the ground state, the resulting equation then can be written as

$$\left[\mathrm{H}_e - E + E_0 + V^{\mathrm{stat}}_{\mathrm{eH}}(\boldsymbol{r}_1) + V^{\mathrm{pol}}_{\mathrm{eH}}(\boldsymbol{r}_1)\right]\ F_0(\boldsymbol{r}_1) \ = \ 0\,. \tag{9.182}$$

On the l.h.s., the electron–atom interaction is described by a static and by a polarization contribution. The static electron–atom potential can easily be evaluated from (9.180) for $\alpha = \alpha' = 0$. The result is

$$V^{\mathrm{stat}}_{\mathrm{eH}}(r_1) = -e^2 \left(\frac{1}{r_1} + \frac{1}{a_B}\right) e^{-2r_1/a_B}\,. \tag{9.183}$$

The polarization potential reads

$$V^{\mathrm{pol}}_{\mathrm{eH}}(r_1) = -\sum_{\alpha\neq 0} \frac{|\int d\boldsymbol{r}_2 \varphi^*_0(\boldsymbol{r}_2) V_{e2}(\boldsymbol{r}_1 - \boldsymbol{r}_2)\varphi_\alpha(\boldsymbol{r}_2)|^2}{\mathrm{H}_e + E_\alpha - E_0}\,, \tag{9.184}$$

where $\varphi_0(\boldsymbol{r}_2)$ is the atomic ground state wave function, and V_{e2} denotes the electron–electron potential. With the latter, we account for plasma effects by using a statically screened Coulomb potential. In order to get a simple analytic expression for the polarization potential, the adiabatic approximation is used and the dipole approximation is applied. Introducing a radius r_0 which permits an interpolation to the behavior for small distances, we get from (9.184) (Kremp (b) et al. 1983; Redmer et al. 1987; Starzynski et al. 1996)

$$V_{\mathrm{eH}}^{\mathrm{pol}}(r_1) = -\frac{e^2 \alpha_{pol}}{2(r_1^2 + r_0^2)^2}(1 + \kappa r_1)^2 e^{-2\kappa r_1} \,. \tag{9.185}$$

Here, κ is the inverse screening length. For the atomic polarizability, we use $\alpha_{pol} = 4.5 a_B^3$, and the radius r_0 is chosen to be $r_0 = 1.456 a_B$.

The result is that the elastic scattering of the electron on the atomic ground state is approximately described by a Schrödinger equation with an effective potential given by the sum of (9.183) and (9.185). After partial wave expansion, we arrive at

$$\left[\frac{\hbar^2}{2m_e}\frac{d^2}{dr^2} - \frac{\hbar^2 l(l+1)}{2m_e r^2} - V_{\mathrm{eH}}^{\mathrm{stat}}(r) - V_{\mathrm{eH}}^{\mathrm{pol}}(r) + \frac{\hbar^2 k^2}{2m_e}\right] f_l(r) = 0 \tag{9.186}$$

with the radial part of the electron scattering wave function $f_l(r)$, and $k = \sqrt{2m_e(E - E_0)/\hbar^2}$ being the wave number. On the T-matrix level, the transport cross section for the elastic electron–atom scattering can then be calculated using the phase shift formula (5.135) with phase shifts determined by a numerical solution of the radial Schrödinger equation (9.186) (Kosse 2002). In order to include the elastic scattering of electrons on hydrogen molecules in a simple manner, we also start from a Schrödinger equation and model the electron–molecule interaction approximately by an effective potential given by (9.185) with the parameters $\alpha_{pol} = 5.4 a_B^3$ and $r_0 = 1.2 a_B$.

In Fig. 9.10, we present numerical results for the T-matrix cross sections. In the presented case of weak screening, the transport cross sections of electron–proton and electron–electron scattering are essentially larger in the

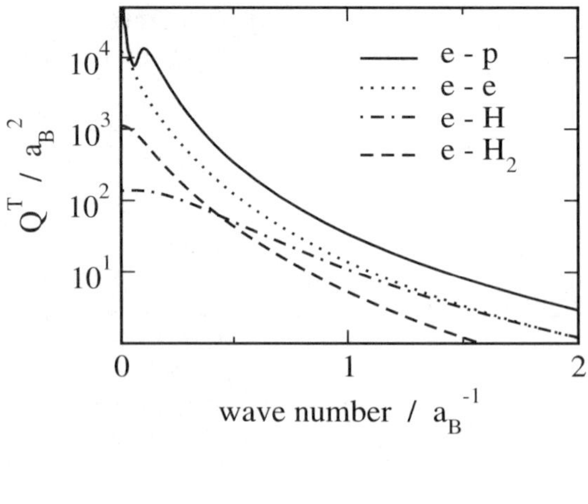

Fig. 9.10. Transport cross sections for different scattering processes in a hydrogen plasma with an inverse Debye screening length $\kappa = 0.05 a_B^{-1}$. Here, the notation Q^T is also used for the e–e cross section given by (5.136)

low energy range than those of elastic electron–atom and electron–molecule scattering. Furthermore, there is only a weak dependence of the electron–atom cross section on the screening parameter.

Results for the Electrical Conductivity. To proceed with the calculation of the electrical conductivity, we have to determine the plasma composition. This will be done using mass action laws. In the following, we will focus on the general behavior of the conductivity for partially ionized hydrogen. For this purpose, we use here results for the plasma composition obtained from a rather simple ionization–dissociation model presented by Schlanges et al. (1995). This model accounts for the formation of atoms and molecules and describes pressure ionization (Mott transition) at high densities. The plasma composition for hydrogen as a function of the total ion (proton) density is given in the left part of Fig. 9.11, where we show the change of the number fractions $\alpha_e = n_e/n_i^{\rm tot}$ (e), $\alpha_H = n_H/n_i^{\rm tot}$ (H) and $\alpha_{H_2} = 2n_{H_2}/n_i^{\rm tot}$ (H_2) for a temperature $T = 20000\,\mathrm{K}$. The differences in the composition at high densities in comparison to the results shown in Sect. 6.7 follow from different choices of hard sphere diameters.

In the right part of Fig. 9.11, results for the electrical conductivity calculated in different approximations are given. The solid curve shows the conductivity in third order Sonine polynomial expansion according to (9.168) accounting for electron–electron, electron–proton, electron–atom, and, additionally, electron–molecule scattering on the T-matrix level. The dash-dotted curve gives results neglecting electron-neutral scattering. It should be noted that degeneracy effects relevant at very high densities are not accounted for in

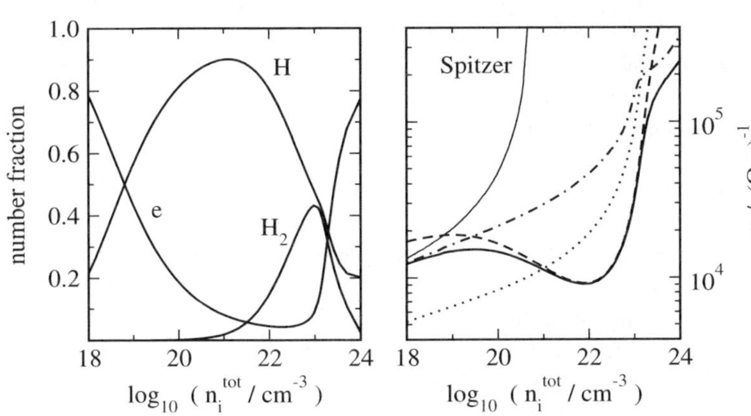

Fig. 9.11. Plasma composition (**left**) and electrical conductivity (**right**) of a hydrogen plasma as a function of the total ion (proton) density for a temperature $T = 20000\,\mathrm{K}$. Conductivity results using different approximations and models are given: All scattering processes included (*solid*); electron-neutral scattering processes neglected (*dash-dotted*); Lorentz approximation (*dashed*); Ziman-type formula (*dotted*)

the conductivity expression (9.168) used to calculate both curves. However, such effects are included in the dashed curve which was calculated from the Lorentz formula (2.217). There, electron–electron scattering is neglected, but all the processes including heavy particles (p, H, H_2) are taken into account. For comparison, the electrical conductivities calculated from the Ziman-type expression (9.174) and from the Spitzer formula (2.223) are shown, too. The latter can be applied in the low density region whereas it diverges with increasing density due to the Coulomb logarithm. On the other hand, the generalized Ziman formula (2.227) describes the limiting behavior of the electrical conductivity at very high densities where the quantum system is highly degenerate (Reinholz et al. 1995). In the calculations of the curves in Fig. 9.11, the inverse screening length $\kappa^2 = \kappa_e^2 = (8\pi\beta e^2/\Lambda_e^3)I_{-1/2}(\beta\mu_e^{\mathrm{id}})$ was used, i.e., only the contribution of the free electrons was taken into account.

Let us proceed with a brief discussion of the results for the conductivity presented in Fig. 9.11. At very low densities, the electrical conductivity is determined by the scattering processes between the free charged particles. Here, the inclusion of electron–electron scattering reduces the conductivity by about 40% compared to the Lorentz model. With increasing density, the degree of ionization decreases due to the formation of atoms and molecules (see left part of Fig. 9.11). The behavior of the conductivity changes drastically in this region if the contributions of electron-neutral scattering processes are taken into account. First, there is a decrease determined by the reduced fraction of the free electrons and by the scattering processes with the neutral particles. Furthermore, we observe the typical nonideality effect described by a minimum behavior of the conductivity with a subsequent strong increase at densities of about $10^{23}\,\mathrm{cm}^{-3}$ (Ebeling et al. 1977; Kremp (b) et al. 1983; Kremp (b) et al. 1984; Höhne et al. 1984; Reinholz et al. 1995). This increase results from the increase of the degree of ionization due to the lowering of the ionization and dissociation energies determined by strong correlation and quantum many-particle effects at high densities. Here, the Mott transition is described, i.e., pressure ionization of the bound particles lead to the transition into a dense, fully ionized plasma state (nonmetal–metal transition).

Isotherms of the electrical conductivity calculated from the expression (9.168) are shown in Fig. 9.12. At lower temperatures, the typical minimum behavior discussed above occurs in the density range of the partially ionized plasma. The minimum is deeper for lower the temperatures. The minimum behavior disappears with increasing temperature, since the formation of atoms and molecules is reduced. Here, the plasma becomes fully ionized over the entire density range.

The electrical conductivity and other transport coefficients of dense hydrogen plasmas were investigated intensely in the last 30 years. We mention calculations for the dense fully ionized plasma state accounting for strong correlations and electron degeneracy based on the Ziman formula and its generalizations (Minoo et al. 1976; Itoh et al. 1983; Ichimaru and Tanaka 1985;

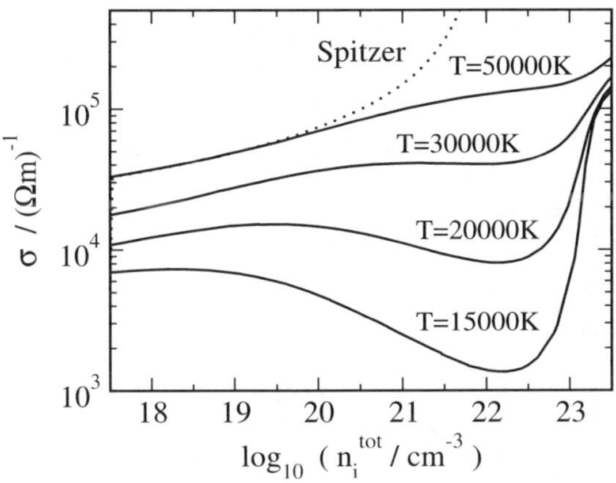

Fig. 9.12. Electrical conductivity for a partially ionized hydrogen plasma as a function of the total ion (proton) density for different temperatures

Boercker 1981; Boercker et al. 1982). Density-functional theory was applied by Perrot and Dharma-wardana (1984) and by Perrot and Dharma-wardana (1987). Furthermore, there are calculations using self-consistent field theories (Djuric et al. 1991; Tkachenko and de Cordoba 1998). T-matrix results for the electrical conductivity using a statically screened potential were presented by Meister and Röpke (1982). Based on the kinetic equation of Gould and DeWitt (1967), T-matrix calculations accounting for dynamical screening in the Born cross section were performed in papers by Williams and DeWitt (1969), Rogers et al. (1981), Morales et al. (1989), and Redmer et al. (1990). Furthermore, low-density expansions of the electrical conductivity and interpolation formulae were given (Röpke 1988b; Röpke and Redmer 1989; Essser et al. 2003). Results for the static conductivity of hot, dense hydrogen obtained by molecular dynamics simulations can be found in papers by Hansen and McDonald (1981) and by Kwon et al. (1996).

Transport coefficients for nonideal partially ionized hydrogen plasmas were investigated within an extension of the linear response theory by Zubarev (Röpke 1983; Höhne et al. 1984; Reinholz et al. 1995), and starting from generalized quantum kinetic equations (Kremp (b) et al. 1983; Kremp (b) et al. 1984; Schlanges et al. 1984; Schlanges et al. 1985). Transport cross sections were used in different approximations, and the plasma composition was determined by mass action laws accounting for nonideality effects such as the lowering of the ionization energy. In this way, one can describe the minimum behavior of the conductivity and the Mott transition discussed above. First experimental results for the electrical conductivity of nonideal hydrogen plasmas in the low density range were obtained by Radtke and Günther (1976). Later on, experiments were focused on fluid hydrogen at very high pressures (Weir et al. 1996; Nellis et al. 1999). In those shock-compression measurements of the electrical conductivity, metallization of hydrogen could be observed at pressures of 1.4 Mbar and temperatures around 3000 K. A sharp increase in the electrical conductivity of hydrogen and inert gases was reported in experiments using multiple shock compression in the megabar range as a consequence of pressure ionization (Fortov et al. 2003). This parameter range was theoretically studied for hydrogen, e.g., by MD simulations (Lenoski et al. 1997), by application of percolation theory (Likalter 1998), and by quantum statistical calculations considering the hopping conductivity (Redmer et al. 2001).

9.2.5 Nonideal Alkali Plasmas

Transport Cross Sections. In this section, we consider dense alkali plasmas consisting of free electrons, singly charged ions, and atoms. The ions and atoms are assumed to be in ground states. In order to calculate the transport cross sections for the scattering of an electron on the bound particle complexes in the plasma, we apply the approach presented in the previous section. In addition, exchange processes between the incident and the valence electrons are accounted for in the lowest order term of the expansion (9.178). The resulting effective Schrödinger equation for the wave function $F_0(\boldsymbol{r}_1)$ of the incident electron then takes the form (Stone and Reitz 1963)

$$\left[\mathrm{H}_e - E + E_0 + V_{\mathrm{eA}}^{\mathrm{stat}}(\boldsymbol{r}_1) + V_{\mathrm{eA}}^{\mathrm{pol}}(\boldsymbol{r}_1)\right] F_0(\boldsymbol{r}_1)$$
$$= \mp \int d\boldsymbol{r}_2 \varphi_0^*(\boldsymbol{r}_2)\,\left[2E_0 - E + V_{e2}(\boldsymbol{r}_1 - \boldsymbol{r}_2)\right]\, F_0(\boldsymbol{r}_2)\,\varphi_0(\boldsymbol{r}_1)\,, \quad (9.187)$$

which is an integro-differential equation including contributions of static interaction, polarizability and exchange. Here, φ_0 denotes the wave function of the valence electron in the atomic ground state, and E_0 is the corresponding energy eigenvalue. The expression on the r.h.s. represents the electron exchange contribution, where the upper sign refers to the singlet state and the lower one to the triplet state. On the l.h.s., the electron–atom interaction is described by a static and by a polarization contribution. Because of the special electronic configuration of alkali atoms, it is useful to distinguish between contributions resulting from the valence electron and those from the inner shells of the ions, and we write the static potential as

$$V_{\mathrm{eA}}^{\mathrm{stat}}(r_1) = V_{\mathrm{e2}}^{\mathrm{stat}}(r_1) + V_{\mathrm{e,ion}}^{stat}(r_1)\,. \quad (9.188)$$

The first term gives the averaged potential determined by the valence electron, and the second one gives the averaged potential seen by the incident electron in the static field of the ion. To describe the static electron–ion interaction, a pseudo-potential with two adjustable parameters is used, which was proposed in Green et al. (1969):

$$V_{\mathrm{e,ion}}^{\mathrm{stat}}(r) = -\frac{e^2}{r}\left[e^{-\kappa r} + \frac{Z-1}{H\left(e^{r/d}-1\right)+1}\right] \quad (9.189)$$

Table 9.1. Parameters of the pseudo-potentials

	ion		atom			
	H	$d[a_B]$	H	$d[a_B]$	$\alpha_{pol}[a_B^3]$	$r_0^2[a_B^2]$
Li	0.85	0.4335	2.62	1.635	162.9	9.46
Na	1.337	0.493	2.33	0.849	222.12	10.56
K	1.85	0.667	2.62	0.95	280.5	16.50
Rb	3.5	0.8	4.65	1.066	306.0	17.53
Cs	5.121	1.0	6.54	1.29	382.4	20.05

with static Debye screening for the long range part. Here, Z is the nuclear charge number and κ is the inverse Debye screening length. The parameters H and d were fitted to the energy spectrum of the valence electron of the alkali atom under consideration and are given in Table 9.1 (Starzynski et al. 1996).

The electron–electron static potential in (9.188) was determined numerically. Together with the static electron–ion potential, it gives the total static part of the electron–atom interaction. The latter part was also parametrized by a Green-type potential which is given by the second term on the right hand side of (9.189) replacing $(Z-1)$ by Z. The corresponding fit parameters can also be found in Table 9.1.

The polarization potential is considered in adiabatic approximation accounting only for the polarizability of the valence electron. Thus, it is approximated by the expression (9.184) and finally by (9.185) applied to alkali atoms. The atomic polarizibilities α_{pol} and the parameters r_0 are given in Table 9.1, too.

If a partial wave expansion is performed, the following integro-differential equations can be derived from (9.187) (Stone and Reitz 1963)

$$\left[\frac{d^2}{dr^2}+k^2-\frac{l(l+1)}{r^2}-V_{\rm eA}^{\rm stat}(r)-V_{\rm eA}^{\rm pol}(r)\right]f_l(r)$$
$$=\pm\varphi_0(r)\Big\{(E_0-k^2)\delta_{l,0}\int_0^\infty f_0\varphi_0 d\bar r+\frac{2}{2l+1}\Big[r^l\int_0^\infty f_l\varphi_0\bar r^{-(l+1)}d\bar r$$
$$+r^{-(l+1)}\int_0^r f_l\varphi_0\bar r^l d\bar r-r^l\int_0^r f_l\varphi_0\bar r^{-(l+1)}d\bar r\Big]\Big\}. \qquad (9.190)$$

In this equation Heaviside units were used ($e^2/2=2m_e=\hbar=1$).

The transport cross sections of elastic electron–atom scattering are given by $Q_{eA}^T=\frac{1}{4}Q_{eA}^{(s)}+\frac{3}{4}Q_{eA}^{(t)}$, where (s) refers to the singlet and (t) to the triplet state. These cross sections can be calculated using their phase shift representation given by (5.135). The phase shifts were obtained by numerical solution of (9.190) applying a method developed by Marriott (1958). For the electron–ion transport cross section, the phase shifts where determined from a Schrödinger equation with the pseudo-potential given by (9.189). The

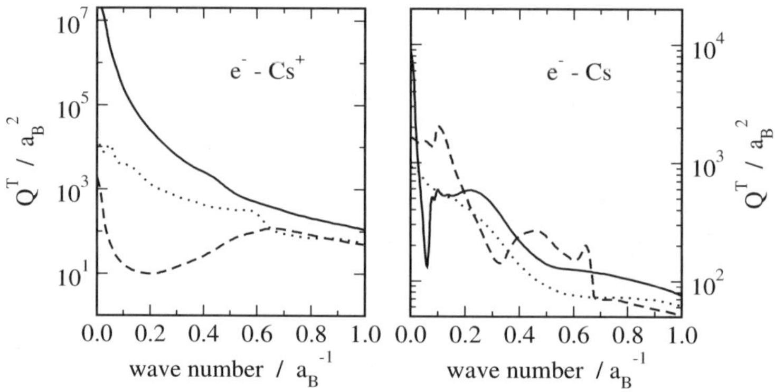

Fig. 9.13. Transport cross sections for the elastic scattering of an electron on the Cs^+ ion (**left**) and on the Cs atom (**right**) for different screening parameters: $\kappa a_B = 0.001$ (*solid*), $\kappa a_B = 0.1$ (*dotted*), $\kappa a_B = 0.6$ (*dashed*)

electron–electron scattering was treated like in the case of hydrogen using the model of the Debye potential.

In the following, some results for transport cross sections are given. In the left part of Fig. 9.13, the cross sections are shown for e–Cs^+ scattering as a function of the wave number for different screening parameters κa_B. The cross sections are lowered with increasing plasma screening. Especially, resonance states appear which lead to a characteristic structure in the low energy region. Calculation on the same level were performed for other alkali plasmas (Starzynski et al. 1996). A comparison of the results for different alkali ions shows the differences expected in the high-energy region due to the influence of the ion core electrons on the effective electron–ion interaction potential. Here, the cross section is higher for elements with higher nuclear charge Z which results from the more complex ionic core.

The transport cross sections for elastic scattering of electrons on Cs atoms in the ground state are displayed in the right part of Fig. 9.13. Like for the electron–ion scattering, we observe a resonance-like behavior in the low energy region. At higher scattering energies, the cross section decreases with increasing screening parameters.

Results for the cross section and the conductivity of Li plasmas are shown in Fig. 9.14.

It is interesting to consider the effects of the different interaction contributions on the transport cross sections. This is shown in the left part of Fig. 9.14 for the electron-Li atom scattering in the case of weak screening. The solid curve is the result including all contributions, i.e., the effective Schrödinger equation (9.190) was solved taking into account the static and polarization potentials as well as exchange. This is compared to the cross sections obtained without exchange (dotted), and neglecting the exchange term as well as the static potential (dashed). The latter results represent solutions

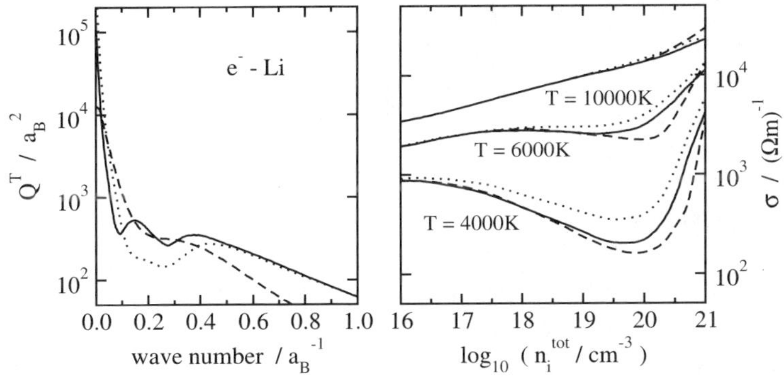

Fig. 9.14. Transport cross sections for the elastic scattering of electrons on Li atoms for $\kappa = 0.001a_B^{-1}$ in different approximations (**left**) and isotherms of the electrical conductivity for a partially ionized Li plasma (**right**): all contributions in the electron–atom scattering included (*solid*), static and polarization potential without exchange (*dotted*), polarization potential only (*dashed*)

of the Schrödinger equation with a screened polarization potential only. This simple treatment of electron–atom scattering has frequently been utilized to determine the effect of partial ionization on the electron mobility (Reinholz et al. 1989; Redmer et al. 1992; Arndt et al. 1990).

The results show a significant effect of exchange contributions in the low energy region. At higher energies $ka_B > 0.5$, the transport cross section which includes all contributions merges into that which accounts only for static and polarization interactions without exchange.

Results for the Electrical Conductivity. First we want to make some remarks concerning the plasma composition. A quantum statistical approach to calculate the plasma composition and the equation of state of partially ionized alkali plasmas was developed in several papers (for an overview, see Redmer (1997)). Some details of the model used here to calculate the electrical conductivity for Li, Na, K, Rb, and Cs plasmas is described in Starzynski et al. (1996). Results for the degree of ionization are given there for different temperatures. For example, Table 9.2 presents the results for a temperature of $T = 4000\,\mathrm{K}$. The electrical conductivity for a Li plasma, calculated from the expression (9.168), is shown in the right part of Fig. 9.14. At high temperatures, there is a monotonous increase with the density. This behavior changes drastically at lower temperatures ($T < 5000\,\mathrm{K}$). We observe the typical minimum behavior which corresponds to the partially ionized plasma state. For densities near $n_i > 10^{20}\,\mathrm{cm}^{-3}$, the conductivity is strongly increasing what indicates the occurrence of the Mott transition in the plasma. Like in the case of hydrogen, such transition can be interpreted as a nonmetal-to-metal transition between a state with a low degree of ionization (partially ionized plasma) and another one with a high degree of ionization (fully ionized plasma) because of pressure ionization in the dense plasma. The same

Table 9.2. Degree of ionization for alkali-atom plasmas at 4000 K

$\log\left(n_i^{\text{tot}}[\text{cm}^{-3}]\right)$	Li	Na	K	Rb	Cs
16.0	0.13	0.18	0.48	0.56	0.69
16.5	0.076	0.11	0.31	0.38	0.51
17.0	0.045	0.065	0.20	0.25	0.34
17.5	0.025	0.038	0.12	0.15	0.22
18.0	0.015	0.022	0.073	0.094	0.14
18.5	0.0088	0.013	0.045	0.060	0.092
19.0	0.0054	0.0082	0.030	0.040	0.065
19.5	0.0038	0.0055	0.024	0.033	0.056
20.0	0.0035	0.0050	0.032	0.045	0.075
20.5	0.0072	0.011	0.088	0.10	0.15
21.0	0.047	0.060	0.21	0.24	0.29

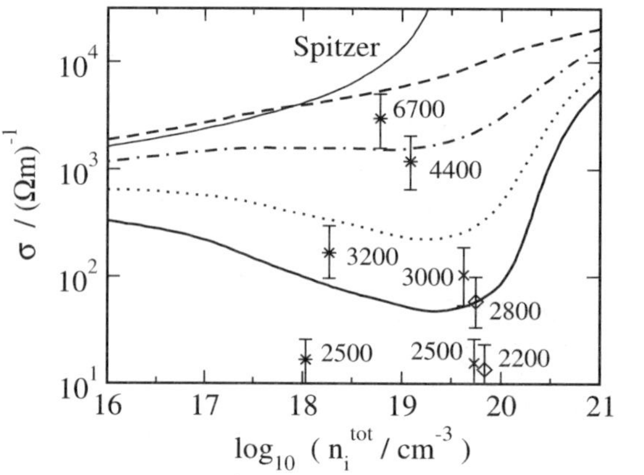

Fig. 9.15. Electrical conductivity for a Cs plasma as a function of the total ion density. Isotherms for $T = 2500\,\text{K}$ (*solid*), $T = 3000\,\text{K}$ (*dotted*), $T = 4000\,\text{K}$ (*dash-dotted*), $T = 6000\,\text{K}$ (*dashed*). The results are compared to experimental data given by Lomakin and Lopatin (1983) ($\times$), by Isakov et al. (1984) ($\diamond$), and by Borzhijewskii et al. (1988) ($*$)

general behavior is obtained for other alkali-atom plasmas (Starzynski et al. 1996). The different conductivity curves in Fig. 9.14 which refer to the same temperature demonstrate the influence of different approximations for the electron–atom transport cross sections on the electrical conductivity.

A comparison with experimental data for the electrical conductivity of dense Cs plasmas obtained by Lomakin and Lopatin (1983), Isakov et al. (1984), and Borzhijewskii et al. (1988) is given in Fig. 9.15. The experimental data show a characteristic scattering and, due to the complexity of the experiments, rather large error bars. However, the experimental values are systematically smaller than the present results. No data were found for the high density plasma region where the Mott effect leads to the sharp increase of the conductivity. Here, only data for the low temperature and high density Cs vapor are available (Hensel et al. 1991).

There are many papers dealing with the equilibrium and non-equilibrium properties of dense alkali plasmas and fluids. Electrical transport properties for fully ionized Cs plasma were studied in detail by Meister and Röpke (1982) and by Sigeneger et al. (1988). The calculations were based on the T-matrix approximation for the scattering cross sections using effective potentials. Calculations of the transport properties for partially ionized alkali plasmas were performed, e.g., by Chin et al. (1988), Reinholz et al. (1989), Bialas et al. (1989), Arndt et al. (1990), and Starzynski et al. (1996) utilizing different approximations for the electron–atom interaction. The existence of clusters was considered, e.g., by Gogoleva et al. (1984).

A lot of work was done to investigate the thermodynamic properties of dense alkali-atom plasmas (Alekseev and Iakubov 1983; Richert et al. 1984; Redmer and Röpke 1985; Hernandez 1986; Redmer et al. 1988; Redmer and Röpke 1989). Dense plasma effects were considered such as the formation of atoms and higher complexes as well as pressure ionization. At low temperatures $T < 2500K$ (for Na through Cs), the thermodynamic instability of the liquid-gas phase transition occurs and, simultaneously, a metal-nonmetal transition takes place near the critical point (see Hensel and Uchtmann (1989), and Hensel (1990)). Applying the model of a partially ionized vapor, the variation of the thermodynamic (Reinholz et al. 1989), transport (Redmer et al. 1992; Reinholz and Redmer 1993), and magnetic properties of Cs (Redmer and Warren 1993a; Redmer and Warren 1993b) along the coexistence line could be explained. For an overview over many of the theoretical and experimental investigations concerning alkali metal plasmas and fluids we refer to Redmer (1997), Fortov and Yakubov (1999), and to Hensel and Warren (1999).

9.2.6 Dense Metal Plasmas

Due to the extreme conditions typical for strongly coupled plasmas, experiments are difficult to perform in this regime. Considerable progress in experimental investigations of the electrical conductivity was achieved over the last 15 years in studying dense metal plasmas such as Al, Cu, W, Zn, and others. New techniques were developed to measure the conductivity for densities and temperatures corresponding to ion–ion coupling parameters of the order $\Gamma_{ii} \approx 1$ and higher. We mention the reflectivity probe experiments (Milchberg et al. 1988; Ng et al. 1994; Mostovych and Chan 1997), capillary discharges (Shepherd et al. 1988; Benage et al. 1994), and wire explosions (DaSilva and Kunze 1994; Kloss et al. 1996; DaSilva and Katsouros 1998; Krisch and Kunze 1998; Benage et al. 1999; Haun et al. 2002). Furthermore, experiments were done using an isochoric plasma in a confining vessel (Renaudin et al. 2002; Recoules et al. 2002). Most of the techniques were applied to create expanded fluid metals and dense metal plasmas. In particular, data for the electrical conductivity for aluminum, copper, zinc, tungsten and others are now available in the strongly coupled plasma regime.

In the following, we will present calculations of the dc electrical conductivity for dense metal plasmas within the quantum kinetic approach. Comparison with experimental results will be given with special attention to nonideality effects such as the minimum behavior of the conductivity in the range of lower temperatures.

As before, we consider the system to be a partially ionized plasma described in the chemical picture. The plasma species are free electrons, ions, and atoms with densities n_e, n_z and n_0. The ions in the different charge state z and the atoms are assumed to be in the ground state. Like in the previous subsections, the cross sections which enter the conductivity expression are determined by a simplified treatment of the different scattering processes, i.e., effective two-body potentials are used accounting for plasma screening in the static approximation. The electron–electron as well as the electron–ion interactions are described by Debye potentials with the screening length according to (2.73), including only the contribution of the free electrons. The electron–atom interaction is modelled by the screened polarization potential given by the analytic expression (9.185). To determine the parameter r_0, the relation $r_0^4 = \alpha_{\rm pol}\, a_B/2Z^{1/3}$ is used with Z being the nuclear charge number (Joachain 1975). Finally, the transport cross sections were calculated on the T-matrix level using their respective phase shift representations.

In order to determine the plasma composition, we extend the chemical picture introduced in Chap. 6 to partially ionized plasmas with ions in different charge states, i.e., we consider ionization and recombination processes according to

$$X_{z-1} \rightleftharpoons X_z + e \qquad (z = 1\ \ldots\ Z)\,. \tag{9.191}$$

Assuming chemical equilibrium with $\mu_{z-1} = \mu_z + \mu_e$, the composition of the plasma is then determined by the following system of coupled mass action laws for the particle number densities

$$n_{z-1} = \frac{g_{z-1}}{g_z} n_z \exp\left[\frac{1}{k_B T}(|E_z^0| + \mu_e^{\rm id} + \mu_e^{\rm corr} + \mu_z^{\rm corr} - \mu_{z-1}^{\rm corr})\right], \tag{9.192}$$

with $|E_z^0|$ being the ionization energy to get the ion in the charge state z, and g_z denotes the statistical weight. The ideal part of the electron chemical potential $\mu_e^{\rm id}$ is determined by (2.24) for arbitrary degeneracy. Nonideality is accounted for by the correlation parts of the chemical potentials $\mu_a^{\rm corr}$. Padé interpolation formulas for the Coulomb part of the free energy density f given by Ebeling et al. (1991) and Förster (1991) are used to calculate the contributions of the free-charged particle interaction to the chemical potentials according to $\mu_a = \partial f/\partial n_a$ (see also Sect. 6.3.4). The effect of the electron–atom interaction was included on the level of a second virial coefficient in Born approximation for the polarization potential. Using this scheme,

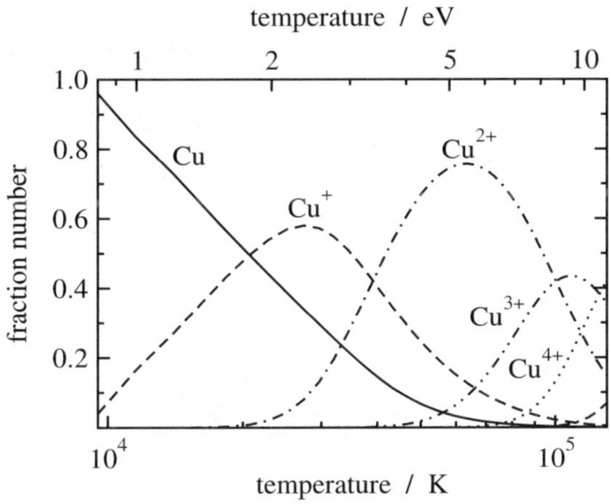

Fig. 9.16. Composition of a Cu plasma as a function of the temperature for a mass density $\rho = 0.2\,\text{g/cm}^3$

we follow the approach applied by Redmer (1999) to determine the plasma composition of metal plasmas. In this paper, the conductivity was calculated within the linear response theory.

As an example we consider results for copper. The respective plasma composition calculated from (9.192) is shown in Fig. 9.16 (Kosse 2002). Here, the number fractions $\alpha_z = n_z/n_i^{\text{tot}}$ are given as a function of the temperature for a mass density $\rho = 0.2\,\text{g/cm}^3$. The total ion density is $n_i^{\text{tot}} = n_0 + \sum_z n_z$. The system of mass action laws was solved including charge states up to $z = 10$. The data for the ionization energies and the statistical weights were taken from the NIST-Database (2002). At temperatures below $T = 20000\,\text{K}$, a partially ionized plasma is described with atoms, singly charged ions, and electrons to be the relevant species. As expected, higher ionic charge states become important at higher temperatures. A remarkable fraction of atoms is found up to $T = 80000\,\text{K}$.

Using the results obtained for the plasma composition and the transport cross sections, the electrical conductivity was finally calculated from (9.168) (Kosse 2002). In Fig. 9.17, the conductivity for copper is shown as a function of the total ion density. Theoretical results are shown for temperatures $T = 10000\,\text{K}$, $T = 20000\,\text{K}$, and $T = 30000\,\text{K}$, and they are compared to experimental data given by DaSilva and Katsouros (1998). These authors measured the electrical conductivity for copper and aluminum in this temperature range creating the dense plasma state by rapid vaporization of metal wires in a water bath. The plasma temperature was determined indirectly by making use of the SESAME equation of state tables (Lyon and Johnson 1987). Furthermore, the Thomas–Fermi model developed in (More 1991) was used to estimate the average ion charge state.

The results given in Fig. 9.17 show interesting features. For low temperatures ($T = 10000\,\text{K}$), a minimum behavior is seen from the theoretical curves

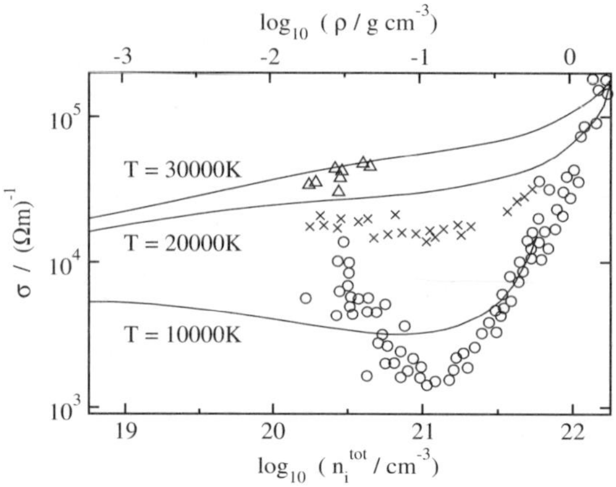

Fig. 9.17. Electrical conductivity of a Cu plasma as a function of the total ion density for different temperatures. The theoretical results (solid) are compared with experimental data given by DaSilva and Katsouros (1998): $T = 10000\,\mathrm{K}$ (o), $T = 20000\,\mathrm{K}$ (×), $T = 30000\,\mathrm{K}$ (△)

as well as from the experimental data. First the conductivity decreases with increasing density. Then the minimum appears with a subsequent strong increase at high densities. As mentioned already in the previous sections, this typical behavior of the electrical conductivity for partially ionized plasmas has been found by theoretical investigations many years ago. The minimum results from the reduced fraction of free electrons determined by the mass action law and by their reduced mobility due to the scattering on atoms. The sharp increase at high densities is a consequence of the lowering of the ionization energy and can be interpreted as a nonmetal–metal transition. At higher temperatures, the minimum behavior becomes less pronounced what is in qualitative agreement with the experimental data given in Fig. 9.17 for $T = 20000\,\mathrm{K}$ and $T = 30000\,\mathrm{K}$. The strong increase of the conductivity in Fig. 9.17 at low temperatures ($T = 10000\,\mathrm{K}$) is not completely reproduced by the theoretical results presented. In this range of low temperatures and very high densities, the plasma chemical potential has an instability which leads to an ambiguous behavior of the degree of ionization as a function of the total density. Therefore, only results up to densities where the stability condition is fulfilled were shown.

Similar results for the electrical conductivity of copper were obtained by Redmer (1999) and Kuhlbrodt and Redmer (2000) using linear response theory and mass action laws such as given by (9.192). In those papers, transport properties were determined for other metals too, and the results were compared to available experimental data.

As a second example, we consider dense zinc plasmas. In Fig. 9.18, the theoretical results are compared with experimental data presented by Haun et al. (2002) and Kosse (2002). In the experiment, the plasma was produced by vaporizing a wire in a glass capillary. In contrast to the experiment performed by DaSilva and Katsouros (1998), the plasma temperature was spectroscopically determined by fitting a Planck function to the measured spectrum.

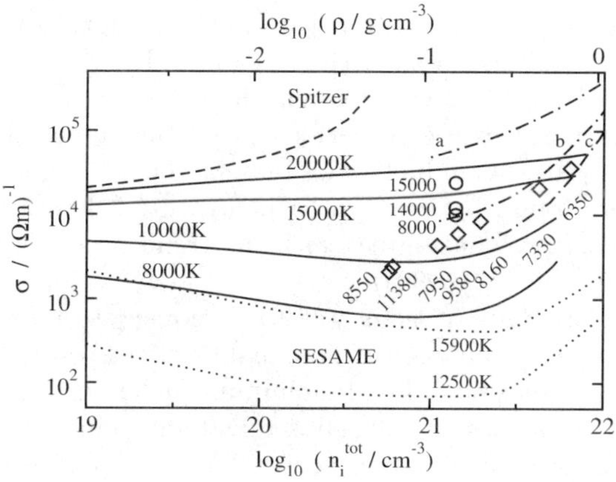

Fig. 9.18. Electrical conductivity of a Zn plasma as function of the total ion density for different temperatures. The theoretical results (*solid curves*) are compared with experimental data. The *dash-dotted curves* are the results obtained by Kuhlbrodt and Redmer (2000): $T = 8000\,\mathrm{K}$ (**a**), $T = 10000\,\mathrm{K}$ (**b**), $T = 20000\,\mathrm{K}$ (**c**)

The theoretical curves in Fig. 9.18 show the same general feature like in the case of copper. The experimental data were obtained in the density range of about $n_i^{\mathrm{tot}} = (1-10) \times 10^{21}\,\mathrm{cm}^{-3}$. Here, an increase of the conductivity connected with the Mott transition is observed for temperatures below $T = 10000\,\mathrm{K}$. Theoretical results for the conductivity of zinc were also given by Kuhlbrodt and Redmer (2000) for the high density region using linear response theory, and neglecting electron–atom scattering. For comparison, those results are given by the dash-dotted curves. Furthermore, SESAME data (Lyon and Johnson 1987) are shown. The latter differ considerably from the experimental data and from the theoretical results presented.

There exist various models which were applied to calculate transport properties of dense metal plasmas. Here we only mention the theoretical approaches performed by Tkachenko and de Cordoba (1998), Ebeling et al. (1991), Djuric et al. (1991), Perrot and Dharma-wardana (1987), Lee and More (1984), Ichimaru and Tanaka (1985), Rinker (1985), and Rinker (1988). Comparison of experimental data for the electrical conductivity with some of these models can be found in DaSilva and Katsouros (1998) and Redmer (1999).

Finally, we make some remarks concerning the electrical conductivity of aluminum which was measured in several experimental investigations. The theoretical results for the conductivity of aluminum obtained within the approach used here show the same qualitative behavior like those of copper and zinc presented in Figs. 9.17 and 9.18 (Kosse 2002). The deviations from the experimental data given by DaSilva and Katsouros (1998) are larger than those for copper. The theoretical results are systematically higher what is also the case in comparison with the data by Krisch and Kunze (1998). At this point, we want to mention that the theoretical results for lower temperatures sensitively depend on the plasma composition and the treatment of the electron–

atom scattering, respectively (Kuhlbrodt et al. 2001; Kosse 2002). Thus, considerable changes are possible by modifications of the model used here.

Measurements of the electrical conductivity for aluminum in the range of high densities and of high temperatures were performed by Benage et al. (1999) using an exploding wire z pinch. Conductivities by a factor 2–3 higher than the experimental data were found by the approach presented here for high temperatures and lower densities what agrees with the trend according to some other theories (see Benage et al. (1999)).

To conclude, the aim of the previous sections was to demonstrate more qualitatively the main features of the influence of nonideality effects on the electrical conductivity of dense plasmas such as the minimum behavior connected with the formation of bound states and the subsequent sharp increase as a consequence of the lowering of the ionization energy. For this purpose, we considered different materials and applied simple concepts based on the chemical model with mass action laws and effective two-body potentials. The latter were considered using rather simple approximations. In particular, the static (Debye) screening model was applied which becomes a poor approximation in the region of strong coupling.

Of course, a more rigorous treatment has to be applied to determine the transport and thermodynamic properties of dense strongly coupled plasmas in a self-consistent way. Here, we mention the approach developed by Perrot and Dharma-wardana (1995) and by Perrot and Dharma-wardana (1987). In particular, the warm dense matter regime is a challenge for further theoretical investigations. In this connection, the application and the development of *ab initio* molecular dynamics simulations such as performed by Silvestrini (1999), Recoules et al. (2002), and by Desjarlais et al. (2002) become of increasing importance.

References

Abramowitz, M. and B. Stegun (1984). *Handbook of Mathematical Functions.* Frankfurt/M: Harry Deutsch–Thun.

Abrikosov, A. A., L. P. Gor'kov, and N. N. Dzyalozhinski (1975). *Methods of Quantum Field Theory in Statistical Physics.* New York: Dover Publ.

ACTEX-OPAL (2000). *http://www.pat.llnl.gov/Research/OPAL/Download* .

Adamyan, V. M. and I. M. Tkachenko (1983). *Sov. Phys. High. Temp. Phys.* **21**, 307.

Akhiezer, A. I. and S. V. Peletminskij (1980). *Methods of Statistical Physics.* Moscow 1977; Elmsford, N.Y.: Nauka 1977; Pergamon.

Alastuey, A., F. Cornu, and A. Perez (1994a). *Phys. Rev.* **E49**, 1077.

Alastuey, A., F. Cornu, and A. Perez (1994b). *Phys. Rev.* **E51**, 1725.

Alastuey, A. and A. Perez (1996). *Phys. Rev.* **E53**, 5714.

Albritton, J. R. (1983). *Phys. Rev. Lett.* **50**, 2078.

Alekseev, V. A. and I. T. Iakubov (1983). *Phys. Rep.* **96**, 1.

Andersen (a), H. H., J. Bak, H. Knudsen, P. Møller-Petersen, and B. Nielsen (1977). *Nucl. Instrum. Methods* **140**, 537.

Andersen (b), H. H., J. Bak, H. Knudsen, and B. Nielsen (1977). *Phys. Rev.* **A16**, 1929.

Arista, N. R. and W. Brandt (1981). *Phys. Rev.* **A23**, 1898.

Arndt, S. (1993). *PhD thesis.* Greifswald: Greifswald University.

Arndt, S., W. D. Kraeft, and J. Seidel (1996). *phys. stat. sol.* **b194**, 601.

Arndt, S., F. Sigeneger, F. Bialas, W. D. Kraeft, M. Luft, T. Meyer, R. Redmer, G. Röpke, and M. Schlanges (1990). *Contrib. Plasma Phys.* **30**, 273–283.

Ashley, J. C., R. H. Ritchie, and W. Brandt (1972). *Phys. Rev.* **B5**, 2393.

Aviram, J., S. Goshen, and R. Thieberger (1975). *J. Chem. Phys.* **62**, 425.

Aviram, J., Y. Rosenfeld, and R. Thieberger (1976). *J. Chem Phys.* **64**, 4741.

Bagnier, S., P. Blottiau, and J. Clérouin (2000). *Phys. Rev.* **E63**, 015301(R).

Balescu, R. (1960). *Phys. Fluids* **3**, 52.

Balescu, R. (1963). *Statistical Mechanics of Charged Particles.* London: Interscience.

Barkas, W. H., W. Birnbaum, and F. M. Smith (1956). *Phys. Rev.* **101**, 778.

Barkas, W. H., N. J. Dyer, and H. H. Heckmann (1963). *Phys. Rev. Lett.* **11**, 26.

Bartsch, G. P. and W. Ebeling (1975). *Beitr. Plasmaphys.* **11,15**, 393,25.

Bärwinkel, K. (1969a). *Z. Naturforsch. A* **24**, 22.

Bärwinkel, K. (1969b). *Z. Naturforsch. A* **24**, 38.

Basko, M. M. (1984). *Fiz. Plasmy [Sov. J. Plasma Phys.]* **10**, 1195[689].

Baumgartl, B. J. (1967). *Z. Physik* **198**, 148.

Baus, M. and J. P. Hansen (1979). *J. Phys. C: Solid State Phys.* **12**, L55.

Baym, G. and L. P. Kadanoff (1961). *Phys. Rev.* **125**, 287.

Bekefi, G. (1966). *Radiation Processes in Plasmas.* New York: Wiley.

Belov, S. I., G. V. Boriskov, A. I. Bykov, I. Il'kaev, N. B. Luky'anov, A. Y. Matveev, O. L. Michailova, V. D. Selemir, G. V. Simakov, R. F. Trunin, I. P. Trusov, V. D. Urlin, V. E. Fortov, and A. N. Shuikin (2002). *Zh. Eksp. Teor. Fiz. Lett.* **76**, 433.

Benage, J. F., W. R. Shanahan, and M. Murillo (1999). *Phys. Rev. Lett.* **83**, 2953.

Benage, J. F., W. R. Shanahan, E. G. Sherwood, L. A. Jones, and R. J. Trainor (1994). *Phys. Rev. E* **49**, 4391.

Bertero, M. and G. A. Viano (1967). *Nuclear Phys.* **B1**, 317.

Beth, G. E. and E. Uhlenbeck (1936). *Physica* **3;4**, 729;914.

Bethe, H. (1930). *Ann. Phys. (N.Y.)* **5**, 325.

Betz, H. D. (1983). In H. S. W. Massey, E. W. McDaniel, and B. Bederson (Eds.), *Applied Atomic Collision Physics.* Orlando: Academic Press.

Beule, D., W. Ebeling, and A. Förster (1996). *Physica* **A228**, 140.

Beule, D., W. Ebeling, A. Förster, H. Juranek, R. Redmer, and G. Röpke (2001). *Phys. Rev.* **E63**, 060202.

Bezkrovniy (a), V., M. Schlanges, D. Kremp, and W. D. Kraeft (2004). *Phys. Rev.* **E69**, 061204.

Bezkrovniy (b), V., V. S. Filinov, D. Kremp, M. Bonitz, M. Schlanges, W. D. Kraeft, P. R. Levashov, and V. E. Fortov (2004). *Phys. Rev. E* **70**, 057401.

Bezzerides, B. and D. F. DuBois (1972). *Ann. Phys. (N.Y.)* **70**, 10.

Bialas, F., K. Schmidt, M. Schlanges, and R. Redmer (1989). *Contrib. Plasma Phys.* **29**, 413.

Biberman, L. M., V. S. Vorob'ev, and I. T. Iakubov (1987). *Kinetics of Nonequilibrium Low–Temperature Plasmas.* New York: Consultants Bureau.

Bloch, F. (1933). *Ann. Phys. (N.Y.)* **16**, 285.

Blomberg, C. and B. Bergersen (1972). *Can. J. Phys.* **50**, 2286.

Boercker, D. B. (1981). *Phys. Rev.* **A23**, 1969.

Boercker, D. B. and J. W. Dufty (1979). *Annals of Phys.* **119**, 43.

Boercker, D. B., F. J. Rogers, and H. E. DeWitt (1982). *Phys. Rev. A* **25**, 1623.

Bohm, D. and D. Pines (1953). *Phys. Rev.* **92**, 609.

Bohr, N. (1915). *Philos. Mag.* **30**, 581.

Boine-Frankenheim, O. (1996). *Phys. Plasmas* **3**, 792.

Bollé, D. (1979). *Ann. Phys. (N.Y.)* **121**, 131.

Bonch–Bruevich, V. L. and S. V. Tyablikov (1961). *Metod Funktii Grina v Statisticheskoi Mekhaniki (Green's function method in statistical mechanics).* Moscow: Fizmatgiz.

Bonev, S. A., B. Militzer, and G. Galli (2004). *Phys. Rev. B* **69**, 014101.

Bonitz, M. (1998). *Quantum Kinetic Theory.* Stuttgart/Leipzig: Teubner.

Bonitz, M., T. Bornath, D. Kremp, M. Schlanges, and W. D. Kraeft (1999). *Contrib. Plasma Phys.* **39**, 329.

Bonitz, M., D. Kremp, D. C. Scott, R. Binder, W. D. Kraeft, and H. Köhler (1996). *J. Phys.: Condens. Matter* **8**, 6057–6071.

Bonitz, M., N. H. Kwong, D. Semkat, and D. Kremp (1999). *Contrib. Plasma Phys.* **39**, 37.

Bonitz, M. and D. Semkat (2005). *Eds.: M. Bonitz and D. Semkat, Introduction to Computational Methods in Many–Body Physics.* Princeton: Rinton Press.

Bonitz, M., D. Semkat, and H. Haug (1999). *Eur. Phys. J.* **B9**, 309.

Bornath, T. (1987). *PhD thesis.* Rostock: Rostock University.

Bornath, T. (1998). *Habilitation (B–Dissertation).* Rostock: University of Rostock.

Bornath, T. (2004). *private communication* .

Bornath, T., D. Kremp, W. D. Kraeft, and M. Schlanges (1996). *Phys. Rev. E* **54**, 3274.

Bornath, T., D. Kremp, and M. Schlanges (1999). *Phys. Rev.* **E60**, 6382.

Bornath, T., T. Ohde, and M. Schlanges (1994). *Physica* **A211**, 344–358.
Bornath, T. and M. Schlanges (1993). *Physica* **A196**, 427–440.
Bornath, T., M. Schlanges, P. Hilse, and D. Kremp (2001). *Phys. Rev. E* **64**, 26414.
Bornath, T., M. Schlanges, P. Hilse, and D. Kremp (2003). *J. Phys. A: Math. Gen.* **36**, 5941.
Bornath, T., M. Schlanges, and D. Kremp (1988). *Contrib. Plasma Phys.* **28**, 57.
Bornath, T., M. Schlanges, F. Morales, and R. Prenzel (1997). *JQSRT* **58**, 501.
Bornath, T., M. Schlanges, and R. Prenzel (1998). (see also the erratum in: Phys. plasmas, **5**,3797(1998)). *Phys. Plasmas* **5**, 1485–1490.
Borzhijewskii, A. A., V. A. Sechenov, and V. I. Khorunzhenko (1988). *Teplofiz. Vyss. Temp.* **26**, 722.
Botermans, W. and R. Malfliet (1990). *Phys. Rep.* **198**, 115.
Brooks, H. (1951). *Phys. Rev.* **83**, 879.
Brouwer, H. H., W. D. Kraeft, M. Luft, T. Meyer, P. P. J. M. Schram, and B. Strege (1990). *Contrib. Plasma Phys.* **30**, 263.
Brown, L. S. and L. G. Yaffe (2001). *Phys. Rep.* **340**, 1–164.
Brysk, H. (1974). *Plasma Physics* **16**, 927.
Callen, H. B. and T. A. Welton (1951). *Phys. Rev.* **83**, 34.
Carr, W. J. and A. A. Maradudin (1964). *Phys. Rev.* **133**, 371.
Cauble, R. and W. Rozmus (1985). *Phys. Fluids* **28**, 3387.
Celliers, P., A. Ng, G. Xu, and A. Forsman (1992). *Phys. Rev. Lett.* **68**, 2305.
Chabrier, G. J. (1990). *J. Phys. (Paris)* **51**, 1607.
Chapman, S. and T. G. Cowling (1952). *The Mathematical Theory of Non–uniform Gases.* London: University Press.
Chichkov, B. N., S. A. Shumsky, and S. A. Uryupin (1992). *Phys. Rev.* **A45**, 7475.
Chin, Y. J., R. Radtke, and R. Zimmermann (1988). *Contrib. Plasma Phys.* **28**, 101.
Chow, P. C. (1972). *ApJ* **40**, 730.
Christensen-Dalsgaard, J. and W. Däppen (1992). *Astron. Astrophys. Rev.* **4**, 267.
Clérouin, J. and J. F. Dufreche (2001). *Phys. Rev.* **E64**, 066406.
Cohen, E. G. D. (1968). *The Kinetic Theory of Dense Gases.* Amsterdam: North Holland.
Collins, G. W., L. B. DaSilva, P. Celliers, D. M. Gold, M. E. Ford, R. J. Wallace, A. Ng, S. V. Weber, K. S. Budil, and R. Cauble (1998). *Science* **281**, 1178.
Collins, L. A., S. R. Bickham, J. D. Kress, S. Mazevet, T. J. Lenosky, N. J. Troullier, and W. Windl (2001). *Phys. Rev.* **B63**, 184110.
Couillaud, C., R. Deicas, P. Nardin, M. A. Beuve, J. M. Guihaumé, M. Renaud, M. Cukier, C. Deutsch, and G. Maynard (1994). *Phys. Rev.* **E49**, 1545.
Craig, R. A. (1968). *J. Math. Phys.* **9**, 605.
Danielewicz, P. (1984). *Ann. Phys. (N.Y.)* **152**, 239.
Danielewicz, P. (1990). *Ann. Phys. (N.Y.)* **197**, 154.
Däppen, W., D. Mihalas, D. G. Hummer, and B. W. Mihalas (1988). *ApJ* **332**, 261.
Dashen, R., S. Ma, and H. J. Bernstein (1969). *Phys. Rev.* **187**, 345.
DaSilva, A. W. and J. D. Katsouros (1998). *Phys. Rev. E* **57**, 5945.
DaSilva, A. W. and H.-J. Kunze (1994). *Phys. Rev. E* **49**, 4448.
DaSilva, L. B., P. Celliers, G. W. Collins, K. S. Budil, N. C. Holmes, T. W. Barbee, B. A. Hammel, J. D. Kilkenny, R. J. Wallace, M. Ross, R. Cauble, A. Ng, and G. Chiu (1997). *Phys. Rev. Lett.* **78**, 483.
Davidov, B. (1936). *Phys. Z. der Sowjetunion* **9**, 443.
Dawson, J. and C. Oberman (1962). *Phys. Fluids* **5**, 517.
Dawson, J. M. and C. Oberman (1963). *Phys. Fluids* **6**, 394.

Decker, C. D., W. B. Mori, J. M. Dawson, and T. Katsouleas (1994). *Phys. Plasmas* **1**, 4043.

Desjarlais, M. P., J. D. Kress, and L. A. Collins (2002). *Phys. Rev.* **E66**, 025401.

Deutsch, C. (1995). *Phys. Rev.* **E51**, 619.

Deutsch, C. and G. Maynard (2000). *Recent Res. Devel. Plasmas* **1**, 1–23.

Deutsch, C., G. Maynard, R. Bimbot, D. Gardes, S. Della-Negra, M. Dumail, B. Kubica, A. Richard, M. F. Rivet, A. Servajean, C. Fleurier, A. Sanba, D. H. H. Hoffmann, K. Weyrich, and H. Wahl (1989). *Nucl. Instr. and Meth. in Phys. Res. A* **278**, 38.

Devoto, R. S. (1966). *Phys. Fluids* **9**, 1266.

DeWitt, H. E. (1959). University of california radiation report. **5377**.

DeWitt, H. E. (1961). *J. Nucl. Energy* **C2**, 27.

DeWitt, H. E. (1962). *J. Math. Phys.* **3**, 1216.

DeWitt, H. E. (1966). *J. Math. Phys.* **7**, 616.

Dharma-wardana, M. W. C. and F. Perrot (1998). *Phys. Rev.* **E58**, 3705.

Dharma-wardana, M. W. C. and F. Perrot (2001). *Phys. Rev.* **E63**, 069901.

Djuric, Z., A. A. Mihajlov, V. A. Nastasyuk, M. Popovic, and I. M. Tkachenko (1991). *Phys. Letters* **A155**, 415.

Dorfmann, J. R. and E. G. D. Cohen (1975). *Phys. Rev.* **A 12**, 292.

Drawin, H. W. and F. Emard (1977). *Physica* **C85**, 333.

Drummond, W. E. and D. Pines (1962). *Nucl. Fusion, Suppl.* **3**, 1049.

DuBois, D. F. (1968). In W. E. Brittin, A. O. Barut, and M. Guenin (Eds.), *Lectures in Theoretical Physics, Vol. IX C.* New York.

Dufty, J. W. and M. Berkovsky (1995). *Nucl. Instr. and Meth. in Phys. Res.* **B96**, 629.

Ebeling, W. (1967). *Ann. Physik (Leipzig)* **19**, 104.

Ebeling, W. (1969a). *Physica* **43**, 293.

Ebeling, W. (1969b). *Ann. Physik (Leipzig)* **22**, 33,383,392.

Ebeling, W. (1971). *phys. stat. sol. (b)* **46**, 243.

Ebeling, W. (1993). *Contrib. Plasma Phys.* **33**, 492.

Ebeling, W. et al. (1983). *Transport Properties of Dense Plasmas.* Berlin, Basel: Akademie-Verlag, Birkhäuser Verlag.

Ebeling, W., A. Förster, V. E. Fortov, V. K. Gryaznov, and A. Y. Plolishchuk (1991). *Thermophysical Properties of Hot Dense Plasmas.* Stuttgart/Leipzig: Teubner Verlags-Gesellschaft.

Ebeling, W., A. Förster, D. Kremp, and M. Schlanges (1989). *Physica* **A159**, 285–300.

Ebeling, W., A. Förster, and V. Y. Podlipchuk (1996). *Phys. Rev. Lett.* **A218**, 297.

Ebeling, W. and M. Grigo (1980). *Ann. Physik (Leipzig)* **37**, 21.

Ebeling, W. and M. Grigo (1982). *J. Solution Chem.* **11**, 151.

Ebeling, W., H. J. Hoffmann, and G. Kelbg (1967). *Beitr. Plasmaphys.* **7**, 233.

Ebeling, W., Y. L. Klimontovich, W. D. Kraeft, and G. Röpke (1983). Kinetic equations and conductance theory of dense coulomb fluids. In W. Ebeling et al. (Ed.), *Transport Properties of Dense Plasmas.* Basel: Birkhäuser.

Ebeling, W., W. D. Kraeft, and D. Kremp (1977). In *Proc. of the XIIIth Int. Conf. on Phenomena in Ionized Gases*, pp. 73.

Ebeling, W. and I. Leike (1991). *Physica A* **170**, 682.

Ebeling, W. and B. Militzer (1997). *Phys. Letters* **A226**, 298.

Ebeling, W. and W. Richert (1985). *Phys. Letters A* **108**, 80.

Ebeling, W., W. Richert, W. D. Kraeft, and W. Stolzmann (1981). *phys. stat. sol. (b)* **104**, 193.

Ebeling, W. and G. Röpke (1979). *Ann. Physik (Leipzig)* **36**, 429.

Ebeling, W. and R. Sändig (1973). *Ann. Physik (Leipzig)* **28**, 289.

Ebeling (a), W., W. D. Kraeft, and D. Kremp (1976). *Theory of Bound States and Ionization Equilibrium in Plasmas and Solids.* Berlin, Moscow: Akademie-Verlag, Mir.

Ebeling (b), W., W. D. Kraeft, D. Kremp, and K. Kilimann (1976). *phys. stat. sol. (b)* **78**, 241–253.

Ecker, G. and W. Kröll (1966). *Z. Naturforsch.* **21a**, 2012, 2023.

Ecker, G. and W. Weizel (1956). *Ann. Physik (Leipzig)* **17**, 126.

Ecker, G. H. (1972). *Theory of Fully Ionized Plasmas.* New York and London: Academic Press.

Elton, R. C. (1990). *X–Ray Lasers.* San Diego: Academic Press.

Esser, A. and G. Röpke (1998). *Phys. Rev.* **E58**, 2446.

Essser, A., R. Redmer, and G. Röpke (2003). *Contrib. Plasma Phys.* **43**, 33.

Falkenhagen, H. (1971). *Theorie der Elektrolyte.* Leipzig: S. Hirzel.

Falkenhagen, H., W. Ebeling, and W. D. Kraeft (1971). Mass transport properties of ionized dilute electrolytes. In S. Petrucci (Ed.), *Ionic Interactions.* New York: Academic Press.

Fehr, R. (1997). *PhD thesis.* Greifswald: Greifswald University.

Fehr, R. and W. D. Kraeft (1994). *Phys. Rev.* **E40**, 463.

Fehr, R. and W. D. Kraeft (1995). *Contrib. Plasma Phys.* **35**, 463–479.

Fehr, R., M. Schlanges, and W. D. Kraeft (1999). *Contrib. Plasma Phys.* **39**, 81.

Felderhof, B. U. (1998). *J. Stat. Phys.* **93**, 307.

Felderhof, B. U. and T. Vehns (2004). *J. Chem. Phys.* **121**, 2536.

Felderhof, B. U. and T. Vehns (2005). In *Proceedings of the Workshop Kinetic Theory of Nonideal Plasmas. M. Bonitz, Ed., Kiel University.*

Fennel, W., W. D. Kraeft, and D. Kremp (1974). *Ann. Physik (Leipzig)* **31**, 171–81.

Fetter, A. L. and J. D. Walecka (1971). *Quantum theory of many particle systems.* New York: McGraw Hill.

Feynman, R. P. and A. R. Hibbs (1965). *Quantum Mechanics and Path Integrals.* New York: McGraw–Hill.

Filinov, V. S., M. Bonitz, and V. E. Fortov (2000). *Zh. Eksp. Teor. Fiz. Letters* **72**, 245.

Filinov, V. S., M. Bonitz, D. Kremp, W. D. Kraeft, and V. Fortov (2001). , unpublished.

Filinov, V. S., M. Bonitz, P. R. Levashov, V. E. Fortov, W. Ebeling, M. Schlanges, and S. W. Koch (2003). *J. Phys. A: Math. Gen.* **36**, 6069.

Fisher, D. V. and Y. Maron (2003). *JQSRT* **81**, 147.

Fisher, M. (1996). *J. Phys.: Condens. Matter* **8**, 9103.

Fisher, M. E. and Y. Levin (1993). *Phys. Rev. Lett.* **71**, 3826.

Förster, A. (1991). *PhD thesis.* Berlin: Humboldt-University.

Fortov, V. (2003). *Plasma Phys. Control. Fusion* **45**, A1–A16.

Fortov, V. E., V. Y. Ternovoi, M. V. Zhernokletov, M. a. Mochalov, A. L. Mikhailov, a. S. Filimonov, A. A. Pyalling, V. B. Mintsev, V. K. Gryaznov, and I. L. Iosilevskii (2003). *JETP* **97**, 259.

Fortov, V. E. and I. T. Yakubov (1999). *The Physics of Nonideal Plasma.* London: World Scientific.

Fujimoto, T. and R. W. P. McWhriter (1990). *Phys. Rev. A* **42**, 6588.

Fujita, S. (1966). *Introduction to Nonequilibrium Statistical Mechanics.* Philadelphia: Saunders.

Fujita, S. (1971). *Phys. Rev.* **A4**, 1114.

Fusion-Symposia. In *Proceedings of the International Symposium on Heavy Ion Inertial Fusion, Heidelberg 1998, Nucl. Instr. and Meth. in Phys. Res* **A415** (1998); *San Diego 2000, Nucl. Instr. and Meth. in Phys. Res* **A464** (2001).

Galeev, A. A. and R. N. Sudan (1984). *Basic Plasma Physics, Vol.II.* Amsterdam: North Holland.
Galitski, V. M. and A. B. Migdal (1958). *Zh. Eksp. Teor. Fiz.* **34**, 139.
Gau, R., M. Schlanges, and D. Kremp (1981). *Physica* **109A**, 531.
Gell-Mann, M. and K. A. Brueckner (1957). *Phys. Rev.* **106**, 364.
Gericke, D. O. (2000). *PhD thesis.* Greifswald: Greifswald University.
Gericke, D. O. (2002). *Laser and Particle Beams* **20**, 471 and 643.
Gericke, D. O., M. S. Murillo, and M. Schlanges (2002). *Phys. Rev.* **E65**, 036418.
Gericke, D. O. and M. Schlanges (1999). *Phys. Rev.* **E60**, 904.
Gericke, D. O. and M. Schlanges (2003). *Phys. Rev.* **E67**, 037401.
Gericke, D. O., M. Schlanges, and W. D. Kraeft (1996). *Phys. Letters* **A222**, 241–245.
Gericke, D. O., M. Schlanges, and W. D. Kraeft (1997). *Laser & Particle Beams* **15**, 523.
Gericke, D. O., M. Schlanges, T. Bornath, and W. D. Kraeft (2001). *Contrib. Plasma Phys.* **41**, 147.
Gericke, D. O., M. Schlanges, and T. Bornath (2002). *Phys. Rev.* **E65**, 036406.
Gibson, W. G. (1971). *Phys. Letters* **36A**, 403.
Glenzer (a), S. H., G. Grigori, F. J. Rogers, D. H. Froula, S. U. Pollaine, and R. C. Wallace (2003). *Phys. Plasmas* **10**, 2433.
Glenzer (b), S. H., G. Gregori, R. W. Lee, F. J. Rogers, S. W. Pollaine, and O. L. Landen (2003). *Phys. Rev. Lett.* **90**, 175002.
Gluck, P. (1971). *Nuovo Cimento* **38**, 67.
Gogoleva, V. V., V. Y. Zitserman, A. Y. Polishchuk, and I. T. Iakubov (1984). *Teplofiz. Vyss. Temp.* **21**, 170.
Golant, V. E., A. P. Zhilinski, and I. E. Sakharov (1980). *Fundamentals of Plasma Physics.* New York: Wiley.
Golden, K. and G. Kalman (1979). *Phys. Rev.* **A19**, 2112.
Golubev, A., M. Basko, A. Fertman, A. Kozodaev, N. Mesherayakov, B. Sharkov, A. Vishnevskiy, V. Fortov, M. Kulish, V. Gryaznov, V. Mintsev, E. Golubev, A. Pukhov, V. Smirnov, U. Funk, S. Stoeve, M. Stetter, H. P. Flierl, D. H. H. Hoffmann, J. Jacobi, and I. Iosilevski (1998). *Phys. Rev.* **E57**, 3363.
Gorobchenko, V. D., V. N. Kohn, and E. G. Maksimov (1989). The dielectric function of thr homogeneous electron gas. In L. V. Keldysh, D. A. Kirzhnitz, and A. A. Maradudin (Eds.), *The Dielectric Function of Condensed Systems.* Amsterdam: North–Holland.
Gould, H. A. and H. E. DeWitt (1967). *Phys. Rev.* **155**, 68.
Green, A. E. S., D. L. Sellin, and A. S. Zachor (1969). *Phys. Rev.* **181**, 1.
Green, J. (1959). *Astrophys. J.* **130**, 693.
Gregori, G., S. H. Glenzer, F. J. Rogers, S. M. Pollaine, O. L. Landen, C. Blancard, G. Faussurier, P. Renaudin, S. Kuhlbrodt, and R. Redmer (2004). *Phys. Plasmas* **11**, 2754.
Gregori, G., S. H. Glenzer, W. Rozmus, R. W. Lee, and O. L. Landen (2003). *Phys. Rev.* **E67**, 026412.
Griem, H. (1974). *Spectral Line Broadening by Plasmas.* New York: Academic Press.
Griem, H. R. (1964). *Plasma Spectroscopy.* New York: McGraw–Hill.
Griem, H. R. (1997). *Princples of Plasma Spectroscopy.* Cambridge: Cambridge University Press.
Groot, S. R. and P. Mazur (1962). *Non-Equilibrium Thermodynamics.* Amsterdam: North Holland Publ. Comp.

Gryaznov, V. K., I. L. Iosilevskii, Y. G. Krasnikov, N. N. Kuznetsova, V. I. Kucherenko, G. B. Lappo, B. N. Lomakin, G. A. Pavlov, E. E. Son, and V. E. Fortov (1980). *Thermophysical Properties of the Working Medium of Gas-Phase Nuclear Reactors (in Russian).* Moscow: Atomizdat.
Guernsey, R. L. (1960). *PhD thesis.* Michigan: University of Michigan.
Guernsey, R. L. (1962). *Phys. Fluids* **5**, 322.
Günter, S., L. Hitzschke, and G. Röpke (1991). *Phys. Rev.* **A44**, 6834.
Günther, K. and R. Radtke (1984). *Properties of Weakly Nonideal Plasmas.* Berlin: Akademie-Verlag.
Guttierrez, F. A. and J. Diaz-Valdes (1994). *J. Phys. B: At. Mol. Opt. Phys.* **27**, 593.
Guttierrez, F. A. and M. D. Girardeau (1990). *Phys. Rev* **A42**, 936.
Haberland, H., M. Bonitz, and D. Kremp (2001). *Phys. Rev.* **E64**, 026405.
Hall, A. G. (1975). *J. Phys. A: Math. Gen.* **8**, 214.
Hansen, J. P. and I. R. McDonald (1981). *Phys. Rev.* **A23**, 2041.
Hansen, J. P. and I. R. McDonald (1983). *Phys. Letters* **A97**, 42.
Hansen, J. P. and I. R. McDonald (1986). *Theory of Simple Liquids, 2nd edition.* New York: Academic.
Hansen, J. P., G. M. Torrie, and P. Vieillefosse (1977). *Phys. Rev.* **A16**, 2153.
Haronska, P., D. Kremp, and M. Schlanges (1987). *Wiss. Z. Univ. Rostock* **36**, 18.
Haug, H. and C. Ell (1992). *Phys. Rev. B* **44**, 8721.
Haug, H. and A. P. Jauho (1996). *Quantum Kinetics in Transport and Optics of Semiconductors.* Heidelberg-New York: Springer.
Haug, H. and S. W. Koch (1993). *Quantum Theory of Optical and Electronic Properties of Semiconductors.* Singapore: World Scientific.
Haug, H. and D. B. T. Thoai (1978). *phys. stat. sol. (b)* **85**, 561.
Haun, J., H.-J. Kunze, S. Kosse, M. Schlanges, and R. Redmer (2002). *Phys. Rev.* **E65**, 046407.
Hazak, G., N. Metzler, M. Klapisch, and J. Gardner (2002). *Phys. Plasmas* **9**, 345.
Hazak, G., Z. Zinamon, Y. Rosenfeld, and M. W. C. Dharma-wardana (2001). *Phys. Rev.* **E64**, 066441.
Henneberger, K. and H. Haug (1988). *Phys. Rev.* **B38**, 5759.
Henneberger, K., U. Moldzio, and H. Güldner (2000). In *Progress in Nonequilibrium Green's Functions; Ed. M. Bonitz, World Scientific, Singapore*, pp. 180–211.
Hensel, F. (1990). *J. Phys.: Condens. Matter* **2**, SA33.
Hensel, F., M. Stolz, G. Hohl, R. Winter, and W. Götzlaff (1991). *J. Phys. IV(Paris) Colloq. 1* **C-5**, 191.
Hensel, F. and H. Uchtmann (1989). *Ann. Rev. Phys. Chem.* **40**, 61.
Hensel, F. and W. W. Warren (1999). *The Liquid-Vapor Transition of Metals.* Princeton: Princeton University Press.
Hensel, J. C., T. G. Phillips, and G. A. Thomas (1977). The electron–hole liquid in semiconductors: Experimental aspects. In H. Ehrenreich, F. Seitz, and D. Turnbull (Eds.), *Solid State Physics, Vol. 32.* New York: Academic Press Inc.
Hernandez, J. P. (1986). *Phys. Rev. Lett.* **57**, 3183.
Herzberg, G. and L. L. Howe (1959). *Can. J. Phys.* **37**, 636.
Hill, T. L. (1956). *Statistical Mechanics.* New York: McGraw-Hill.
Hilse, P., M. Schlanges, T. Bornath, and D. Kremp (2005). *submitted.*
Hirschfelder, J. O., C. F. Curtiss, and R. B. Bird (1954). *Molecular theory of gases and liquids.* New York: Wiley & Sons.
Hoffmann, D. H. H. et al. (1988). *Zeitschr. f. Physik* **A30**, 339.
Hoffmann, H. J. and W. Ebeling (1968a). *Physica* **39**, 593.
Hoffmann, H. J. and W. Ebeling (1968b). *Contrib. Plasma Phys.* **8**, 43.
Höhne, F., R. Redmer, G. Röpke, and H. Wegner (1984). *Physica A* **128**, 643.

Höll, A., R. Redmer, G. Röpke, and H. Reinholz (2004). *Eur. Phys. J.* **D29**, 159.
Holmes, N. C., M. Ross, and W. J. Nellis (1995). *Phys. Rev.* **B52**, 15835.
Huang, K. (1963). *Statistical Mechanics.* New York and London: Wiley.
Hubbard, J. (1958). *Proc. Roy. Soc.* **A243**, 336.
Hummer, D. G. and D. Mihalas (1988). *ApJ* **331**, 794.
Ichimaru, S. (1973). *Basic Principles of Plasma Physics.* London: Benjamin.
Ichimaru, S. (1982). *Rev. Mod. Phys.* **54**, 1017.
Ichimaru, S. (1992). *Statistical Plasma Physics I.* Redwood City, CA: Addison-Wesley Publishing Company.
Ichimaru, S. and S. Tanaka (1985). *Phys. Rev. A* **32**, 1790.
Ichimaru, S. and K. Utsumi (1981). *Phys. Rev.* **B24**, 7385.
Iglesias, C. A. and R. W. Lee (1997). *JQSRT* **58**, 637.
Inokuti, M. (1971). *Rev. Mod. Phys.* **43**, 297.
Isakov, I. M., A. A. Likalter, B. N. Lomakin, A. D. Lopatin, and V. E. Fortov (1984). *Zh. Eksp. Teor. Fiz.* **87**, 832.
Itoh, N., S. Mitake, H. Iyetomi, and S. Ichimaru (1983). *ApJ* **273**, 774.
Izuyama, T. (1973). *Progr. Theor. Phys.* **50**, 841.
Jacobi, J., D. H. H. Hoffmann, W. Laux, R. W. Müller, H. Wahl, K. Weyrich, E. Bogasch, B. Heimrich, C. Stöckl, H. Wetzler, and S. Miyamoto (1995). *Phys. Rev. Lett.* **74**, 1550.
Jauho, A. P. and K. Johnson (1996). *Phys. Rev. Lett.* **76**, 4576.
Joachain, C. (1975). *Quantum Collision Theory.* Amsterdam: North Holland.
Johnson, J. K., A. Panagiotopoulos, and K. E. M. Gubbins (1994). *Mol. Phys.* **81**, 717.
Jones, R. D. and K. Lee (1982). *Phys. Fluids* **25**, 2307.
Jung, Y. D. (1995). *Phys. Plasmas* **2**, 332.
Juranek, H. and R. Redmer (2000). *J. Chem. Phys.* **112**, 3780.
Juranek, H., R. Redmer, and Y. Rosenfeld (2002). *J. Chem. Phys.* **117**, 1768.
Juranek, H., R. Redmer, and W. Stolzmann (2001). *Contrib. Plasma Phys.* **41**, 131.
Kadanoff, L. P. and G. Baym (1962). *Quantum Statistical Mechanics.* New York: Benjamin.
Kadomtsev, B. B. (1958). *Sov.Phys.-JETP* **6**, 117.
Kadomtsev, B. B. (1965). *Plasma Turbulence.* New York: Academic.
Kalman, G. and K. Golden (1990). *Phys. Rev.* **A41**, 5516.
Kelbg, G. (1963). *Ann. Physik (Leipzig)* **12**, 219.
Keldysh, V. L. (1964). *Zh. Eksp. Teor. Fiz. [Sov. Phys. JETP]* **47** [**20**], 1515 [235].
Kerley, G. I. (1983). In J. M. Haile and G. A. Mansoori (Eds.), *Molecular Based Study of Fluids.* Washington DC: ACS.
Kilimann, K., W. D. Kraeft, and D. Kremp (1977). *Phys. Letters* **61A**, 393–395.
Kilimann, K., D. Kremp, and G. Röpke (1983). *Teor. Mat. Fiz.* **55**, 448.
Kingham, R., W. Theobald, R. Häßner, R. Sauerbrey, R. Fehr, D. O. Gericke, M. Schlanges, W. D. Kraeft, and K. Ishikawa (1999). *Contrib. Plasma Phys.* **39**, 53.
Kirzhnitz, D. A. (1976). *Usp. Fiz. Nauk* **119**, 367.
Klakow, D., H. Matuszok, P.-G. Reinhard, and C. Toepffer (1996). In *Proceedings of the Int. Conf. on the Physics of Strongly Coupled Plasmas, Eds. W. D. Kraeft and M. Schlanges; World Scientific, Singapore*, pp. 37.
Kliewer, K. L. and R. Fuchs (1969). *Phys. Rev.* **181**, 552.
Klimontovich, Y. L. (1975). *Kinetic Theory of Nonideal Gases and Nonideal Plasmas.* Moscow: Nauka.
Klimontovich, Y. L. (1982). *Kinetic Theory of Nonideal Gases and Plasmas.* Oxford: Pergamon.

Klimontovich, Y. L. and W. Ebeling (1962). *Soviet Physics JETP* **43**, 146.

Klimontovich, Y. L. and W. Ebeling (1973). *Soviet Physics JETP* **36**, 476.

Klimontovich, Y. L. and W. D. Kraeft (1974). *High Temp. Physics (USSR)* **12**, 212–19.

Klimontovich, Y. L. and D. Kremp (1981). *Physica* **A109**, 517.

Klimontovich, Y. L., D. Kremp, and W. D. Kraeft (1987). *Adv. Chem. Phys.* **58**, 175–253.

Klimontovich, Y. L. and V. P. Silin (1952). *Zh. Eksp. Teor. Fiz.* **23**, 151.

Kloss, A., T. Motzke, R. Grossjohann, and H. Hess (1996). *Phys. Rev. E* **54**, 5851.

Knaup, M., P. G. Reinhard, and C. Toepffer (2001). *Contrib. Plasma Phys.* **41**, 159.

Knudson, M. D., D. L. Hamson, J. E. Bailey, C. A. Hall, J. R. Asay, and C. Deeney (2004). *Phys. Rev.* **B69**, 144209.

Knudson, M. D., D. H. Hanson, J. E. Bailey, C. A. Hall, and J. R. Asay (2001). *Phys. Rev. Lett.* **87**, 225501.

Knudson, M. D., D. L. Hanson, J. E. Bailey, C. A. Hall, and J. R. Asay (2003). *Phys. Rev. Lett.* **90**, 035505.

Köhler, H. S. (1995). *Phys. Rev. C* **51**, 3232.

Kolos, W. and L. Wolniewicz (1965). *J. Chem. Phys.* **43**, 2429.

Kosse, S. (2002). *PhD thesis.* Greifswald: Greifswald University.

Kovalenko, N. P., Y. P. Krasnii, and S. A. Trigger (1990). *Statiticheskaya teoriya zhidkikh metallov (Statistical Theory of Liquid Metals.* Moskva (Moscow): Nauka.

Kraeft, W. D., S. Arndt, W. Däppen, and A. Nayfonov (1999). *ApJ* **516**, 369–370.

Kraeft, W. D., W. Ebeling, and D. Kremp (1969). *Phys. Letters* **29A**, 466–467.

Kraeft, W. D., W. Ebeling, D. Kremp, and G. Röpke (1988). *Ann. Physik (Leipzig)* **45**, 429–442.

Kraeft, W. D. and R. Fehr (1997). *Contrib Plasma Phys* **37**, 173.

Kraeft, W. D. and P. Jakubowski (1978). *Ann. Physik* **35**, 293–302.

Kraeft, W. D., K. Kilimann, and D. Kremp (1975). *phys. stat. sol. (b)* **72**, 461.

Kraeft, W. D. and D. Kremp (1968). *Z. Physik* **208**, 475–485.

Kraeft, W. D., D. Kremp, W. Ebeling, and G. Röpke (1986). *Quantum Statistics of Charged Particle Systems.* Berlin and London/New York: Akademie–Verlag and Plenum.

Kraeft, W. D., D. Kremp, and K. Kilimann (1973). *Ann. Physik (Leipzig)* **29**, 177–189.

Kraeft, W. D. and T. Rother (1988). *Contrib. Plasma Phys.* **28**, 117–129.

Kraeft, W. D., M. Schlanges, and D. Kremp (1986). *J. Phys. A* **19**, 3251.

Kraeft, W. D., M. Schlanges, D. Kremp, J. Riemann, and H. E. DeWitt (1998). *Z. Phys. Chem.* **204**, 199.

Kraeft, W. D., M. Schlanges, J. Riemann, and H. E. DeWitt (2000). *J. Phys. IV* **10**, Pr5–167–170.

Kraeft, W. D., M. Schlanges, J. Vorberger, and H. E. DeWitt (2002). *Phys. Rev. E* **66**, 046405.

Kraeft, W. D., J. Seidel, and S. Arndt (1995). *Ukr. J. Phys.* **40**, 430–433.

Kraeft, W. D. and W. Stolzmann (1979). *Physica* **97A**, 306–318.

Kraeft, W. D. and B. Strege (1988). *Physica A* **149**, 313.

Kraeft (a), W. D., D. Kremp, K. Kilimann, and H. E. DeWitt (1990). *Phys. Rev. A* **42**, 2340–2345.

Kraeft (b), W. D., B. Strege, and M. Girardeau (1990). *Contrib. Plasma Phys.* **30**, 563.

Krall, N. A. and A. W. Trivelpiece (1973). *Principles of Plasma Physics.* New York: McGraw.

Kramers, H. C. (1927). *Atti. Congr. Intern. Fisica, Como* **2**, 545.

Kremp, D. (1965). *Wiss. Z. Univ. Rostock, MN Reihe* **14**, 281.
Kremp, D. (1966). *Ann. Physik (Leipzig)* **18**, 237.
Kremp, D. and W. Bezkrowniy (1996). *J. Chem. Phys.* **104**, 1.
Kremp, D., M. Bonitz, W. D. Kraeft, and M. Schlanges (1997). *Ann. Phys. (N.Y.)* **258**, 320–359.
Kremp, D., T. Bornath, M. Bonitz, and M. Schlanges (1996). *Physica B* **228**, 72.
Kremp, D., T. Bornath, M. Schlanges, and M. Bonitz (1999). *Phys. Rev. E* **60**, 4725.
Kremp, D., R. Gau, M. Schlanges, and W. D. Kraeft (1977). *Physica* **86A**, 613.
Kremp, D. and W. D. Kraeft (1968). *Ann. Physik (Leipzig)* **20**, 340–347.
Kremp, D. and W. D. Kraeft (1972). *Phys. Letters* **38A**, 167.
Kremp, D., W. D. Kraeft, and W. Ebeling (1966). *Ann. Physik (Leipzig)* **18**, 246.
Kremp, D., W. D. Kraeft, and W. Ebeling (1971). *Physica* **51**, 146–164.
Kremp, D., W. D. Kraeft, and W. Fennel (1972). *Physica* **A62**, 461–73.
Kremp, D., W. D. Kraeft, and M. Schlanges (1993). *Contrib. Plasma Phys.* **33**, 567.
Kremp, D., W. D. Kraeft, and M. Schlanges (1998). In *Proceedings of the Int. Conf. on Strongly Coupled Coulomb Systems; Eds. G. J. Kalman and J. M. Rommel and K. Blagoev, Plenum Press*, pp. 601.
Kremp, D., M. Schlanges, and T. Bornath (1985). *J. Stat. Phys.* **41**, 661.
Kremp, D., M. Schlanges, and T. Bornath (1986). *ZIE–preprint 86-3*, 33.
Kremp, D., M. Schlanges, and T. Bornath (1988). *phys. stat. sol. (b)* **147**, 747.
Kremp, D., M. Schlanges, and T. Bornath (1989). *The Dynamics of Systems with Chemical Reactions; Ed. J. Popielawski.* Singapore: World Scientific.
Kremp, D., D. Semkat, and M. Bonitz (2000). In *Progress in Nonequilibrium Green's Functions; Ed. M. Bonitz, World Scientific, Singapore*, pp. 34–44.
Kremp (a), D., W. Ebeling, H. Krienke, and R. Sändig (1983). *J. Stat. Phys.* **33**, 99.
Kremp (a), D., K. Kilimann, W. D. Kraeft, H. Stolz, and R. Zimmermann (1984). *Physica* **127A**, 646.
Kremp (a), D., K. Morawetz, M. Schlanges, and V. Rietz (1993). *Phys. Rev.* **E47**, 635.
Kremp (b), D., G. Röpke, and M. Schlanges (1983). The electrical conductivity of partially ionized hydrogen plasmas. In W. Ebeling et al. (Ed.), *Transport Properties of Dense Plasmas.* Berlin, Basel: Akademie-Verlag, Birkhäuser.
Kremp (b), D., M. Schlanges, M. Bonitz, and T. Bornath (1993). *Phys. Fluids B* **5**, 216–229.
Kremp (b), D., M. Schlanges, and K. Kilimann (1984). *Phys. Letters* **100A**, 149–152.
Kremp (c), D., W. D. Kraeft, and A. J. D. Lambert (1984). *Physica* **127A**, 72–86.
Krisch, I. and H.-J. Kunze (1998). *Phys. Rev. E* **58**, 6557.
Kronig, R. (1926). *J. Opt. Soc. Am.* **12**, 547.
Kubo, R. (1957). *J. Phys. Soc. Japan* **12**, 570.
Kubo, R. (1965). *Statistical Mechanics.* Amsterdam: North Holland.
Kubo, R. (1966). *Rep. Progr. Phys.* **29**, 255.
Kuhlbrodt, S. (2003). *PhD thesis.* Rostock: Rostock University.
Kuhlbrodt, S. and R. Redmer (2000). *Phys. Rev.* **E62**, 7191.
Kuhlbrodt, S., R. Redmer, A. Kemp, and J. Meyer-ter-Vehn (2001). *Contrib. Plasma Phys.* **41**, 3.
Kull, H. J. and L. Plagne (2001). *Phys. Plasmas* **8**, 5244.
Kuznetsov, A. V. (1991). *Phys. Rev. B* **44**, 8721.
Kwon, I., L. Collins, J. Kress, and N. Troullier (1996). *Phys. Rev.* **E54**, 2844.
Kwong, N. H. and M. Bonitz (2000). *Phys. Rev. Lett.* **84**, 1768.

Lagan, S. and J. A. McLennan (1984). *Physica* **128 A**, 178.
Lampe, M. (1968a). *Phys. Rev.* **170**, 306.
Lampe, M. (1968b). *Phys. Rev.* **174**, 276.
Landau, L. D. (1936). *Phys. Z. Sowjetunion* **10**, 154.
Landau, L. D. and E. M. Lifschitz (1988). *Lehrbuch der Theoretischen Physik III.* Berlin: Akademie-Verlag.
Landau, L. D. and E. M. Lifshits (1977). *Theoretische Physik, Vol.3.* Berlin: Akademie-Verlag.
Landau, L. D. and Y. B. Zeldovich (1943). *Acta Physicochim.* **18**, 1994.
Langdon, A. B. (1980). *Phys. Rev. Lett.* **44**, 575.
Langmuir, I. (1928). *Proc. Nat. Acad. Sci. U. S.* **14**, 627.
Lee, Y. T. and R. M. More (1984). *Phys. Fluids* **27**, 1273.
Lenard, A. (1960). *Ann. Phys. (N.Y.)* **3**, 390.
Lenoski, T. J., J. D. Kress, L. A. Collins, and I. Kwon (1997). *Phys. Rev. B* **55**, R11907.
Lenosky, T. J., S. R. Bickham, and J. D. Kress (2000). *Phys. Rev.* **B61**, 1.
Leonhardt, U. and W. Ebeling (1993). *Physica* **A192**, 249.
Li, C.-K. and R. D. Petrasso (1993). *Phys. Rev. Lett.* **70**, 3059.
Lifschitz, E. M. and L. M. Pitajewski (1983). *Physikalische Kinetik.* Berlin: Akademie-Verlag.
Lifshits, E. M. and L. P. Pitayevskij (1978). *Teoreticheskaya Fizika, vol. IX. Statisticheskaya Fizika.* Moscow: Nauka.
Likalter, A. A. (1998). *Zh. Eksp. Teor. Fiz. (JETP)* **113 (86)**, 1094 (598).
Lindhard, J. (1954). *Kgl. Danske Videnskab. Mat. Fys. Medd.* **28**, 8.
Lipavský, P., K. Morawetz, and V. Špička (2001). Kinetic equations for strongly interacting dense fermi systems. *Annales de Physique* **26**, 1–254.
Lipavský, P., V. Špička, and B. Velický (1986). *Phys. Rev. A* **34**, 6933.
Lomakin, B. N. and A. D. Lopatin (1983). *Teplofiz. Vyss. Temp.* **21**, 170.
Lotz, W. (1968). *Z. Phys.* **216**, 241.
Lundquist, B. J. (1967). *Phys. Cond. Matter* **6**, 206.
Lyon, S. P. and J. D. Johnson (1987). *SESAME: The Los Alamos National Laboratory Equation of State Database; Eds. S. P. Lyon and J. D. Johnson.* Los Alamos, NM: Los Alamos National Laboratory.
Macke, W. (1950). *Z. Naturforsch.* **5a**, 192.
Mahan, G. D. (1990). *Many Particle Physics, 2nd edition.* New York: Plenum.
Mansoori, G. A., N. F. Carnahan, K. E. Starling, and T. W. Leland (1971). *J. Chem. Phys.* **54**, 1523.
Manzke (a), G., Q. Y. Peng, K. Henneberger, U. Neukirch, K. Hauke, K. Wundke, J. Gutowski, and D. Hommel (1998). *Phys. Rev. Lett.* **80**, 4943.
Manzke (b), G., Q. Y. Peng, U. Moldzio, and K. Henneberger (1998). *phys. stat. sol. (b)* **206**, 37.
Marriott, R. (1958). *Proc. Phys. Soc.* **72**, 121.
Martin, P. C. (1967). *Phys. Rev.* **161**, 143.
Martin, P. C. and J. Schwinger (1959). *Phys. Rev.* **115**, 1342.
Mayer, J. E. (1950). *J. Chem Phys.* **18**, 1426.
McLennan, J. A. (1982). *J. Stat. Phys.* **28;57**, 521;887.
McLennan, J. A. (1989). *Introduction to Nonequilibrium Statistical Mechanics.* Englewood Cliffs, NJ: Prentice Hall.
Meister, C. V. and G. Röpke (1982). *Ann. Physik (Leipzig)* **39**, 133.
Mermin, N. D. (1970). *Phys. Rev.* **B1**, 2362.
Meyer, J. R. and M. Glicksman (1978). *Phys. Rev.* **B17**, 3227.
Mihalas, D., W. Däppen, and D. G. Hummer (1988). *ApJ* **331**, 815.

Mihara, N. and R. D. Puff (1968). *Phys. Rev.* **174**, 221.
Mikhailovskii, A. B. (1974). *Theory of Plasma Instabilities.* New York: Consultants Bureau.
Milchberg, H. M., R. R. Freeman, S. C. Davey, and R. M. More (1988). *Phys. Rev. Lett.* **61**, 2364.
Militzer, B. (2000). *PhD thesis.* Urbana Champaign: University of Illinois.
Militzer, B. and D. M. Ceperley (2000). *Phys. Rev. Lett.* **85**, 1890.
Militzer, B. and D. M. Ceperley (2001). *Phys. Rev.* **E63**, 066404.
Militzer, B., D. M. Ceperley, J. D. Kress, J. D. Johnson, L. A. Collins, and S. Mazevet (2001). *Phys. Rev. Lett.* **87**.
Minoo, H., C. Deutsch, and J. P. Hansen (1976). *Phys. Rev.* **A14**, 840.
Montroll, E. W. and J. C. Ward (1958). *Phys. Fluids* **1**, 55.
Morales, F., M. K. Kilimann, R. Redmer, M. Schlanges, and F. Bialas (1989). *Contrib. Plasma Phys.* **29**, 425–430.
Morawetz, K. and D. Kremp (1993). *Phys. Letters* **A173**, 317.
Morawetz, K. and G. Röpke (1996). *Phys. Rev.* **E54**, 4134.
Morawetz, K., M. Schlanges, and D. Kremp (1993). *Phys. Rev. E* **48**, 2980–2988.
More, R. (1988). *Phys. Fluids* **31**, 3059.
More, R. M. (1982). *JQSRT* **27**, 345.
More, R. M. (1991). Physics of laser plasmas, vol. 3. In M. N. Rosenbluth and R. Z. Sagdeev (Eds.), *Handbook of Plasma Physics.* Amsterdam: Elsevier.
Morita, T. (1959). *Progr. Theor. Phys. (Japan)* **22**, 757.
Morozov, V. G. and G. Röpke (1999). *Ann. Phys. (N.Y.)* **278**, 127.
Mostovych, A. N. and Y. Chan (1997). *Phys. Rev. Lett.* **79**, 5094.
Mostovych, A. N., Y. Chan, T. Lehecha, A. Schmitt, and J. D. Sethian (2000). *Phys. Rev. Lett.* **85**, 3870.
Mott, N. F. (1961). *Phil. Mag.* **6**, 287.
Mulser, P., F. Cornolti, E. Besuelle, and R. Schneider (2000). *Phys. Rev.* **E63**, 63.
Mulser, P. and A. Saemann (1997). *Contrib. Plasma Phys.* **37**, 211.
Münster, A. (1969). *Statistical Thermodynamics Vol.I.* Berlin: Springer.
Murillo, M. S. and J. C. Weisheit (1993). In *Strongly Coupled Plasma Physics, Eds. H. Van Horn and S. Ichimaru, University of Rochester Press*, pp. 233–236.
Murillo, M. S. and J. C. Weisheit (1998). *Phys. Rep.* **302**, 1.
Nagy, I., A. Arnau, and P. M. Echenique (1989). *Phys. Rev.* **A40**, 987.
Nellis, W. J., A. C. Mitchell, M. van Thiel, G. J. Devine, and R. J. Trainor (1983). *J. Chem. Phys.* **79**, 1480.
Nellis, W. J., S. T. Weir, and A. C. Mitchell (1999). *Phys. Rev.* **B59**, 3434.
Nerisisyan, H. B. (1998). *Phys. Rev.* **E58**, 3686.
Newton, R. G. (1982). *Scattering Theory of Waves and Particles.* New York: Springer.
Ng, A., P. Celliers, A. Forsman, R. M. More, Y. T. Lee, F. Perrot, M. W. C. Dharma-wardana, and G. A. Rinker (1994). *Phys. Rev Lett.* **72**, 3351.
Ng, A., P. Celliers, G. Xu, and A. Forsman (1995). *Phys. Rev.* **E52**, 4299.
NIST-Database (2002). Atomic spectra database. *http://physics.nist.gov/cgi-bin/AtData/main_asd* , April–04.
Norman, G. E. and A. N. Starostin (1968). *Teplofiz. Vyss. Temp.* **6**, 410.
Numerov, B. (1923). *Publ. Observ. Astrophys. Centre Russie* **11**.
Nussenzveig, H. M. (1973). *Acta Phys. Austr.* **38**, 130.
Nyquist, H. (1928). *Phys. Rev.* **32**, 110.
Oberman, C., A. Ron, and J. M. Dawson (1962). *Phys. Fluids* **5**, 1514.
Ohde, T., M. Bonitz, T. Bornath, D. Kremp, and M. Schlanges (1995). *Phys. Plasmas* **2**, 3214.

Ohde, T., M. Bonitz, T. Bornath, D. Kremp, and M. Schlanges (1996). *Phys. Plasmas* **3**, 1241.
Onsager, L., L. Mittag, and M. Stephen (1982). *Ann. Physik* **18**, 71.
Ortner, J., M. Steinberg, and I. M. Tkachenko (2001). *Contrib. Plasma Phys.* **41**, 293.
Peletminski, S. V. (1971). *Teor. Mat. Fiz.* **6**, 123.
Penrose, P. (1960). *Phys. Fluids* **3**, 258.
Perel', V. P. and G. M. Eliashberg (1962). *Sov. Phys. JETP* **14**, 633.
Perrot, F. and M. W. C. Dharma-wardana (1984). *Phys. Rev.* **A29**, 1378.
Perrot, F. and M. W. C. Dharma-wardana (1987). *Phys. Rev.* **A36**, 238.
Perrot, F. and M. W. C. Dharma-wardana (1995). *Phys. Rev.* **E52**, 5352.
Peter, T. and B. Kärcher (1991). *J. Appl. Phys.* **69**, 3835.
Peter, T. and J. Meyer-ter-Vehn (1991a). *Phys. Rev.* **A43**, 1998.
Peter, T. and J. Meyer-ter-Vehn (1991b). *Phys. Rev.* **A43**, 2015.
Petschek, A. G. (1971). *Phys. Letters* **34A**, 411.
Pfalzner, S. and P. Gibbon (1998). *Phys. Rev.* **E57**, 4698.
Pines, D. (1962). *The Many Body Problem.* New York: Benjamin.
Pines, D. and F. Nozieres (1958). *Theory of Quantum Fluids.* New York: Benjamin.
Pitarke, J. M., R. M. Ritchie, P. M. Echenique, and E. Zaremba (1993). *Europhys. Lett.* **24**, 613.
Prenzel, R. (1999). *PhD thesis.* Rostock: University of Rostock.
Prenzel (a), R., T. Bornath, and M. Schlanges (1996). *Phys. Letters* **A223**, 453.
Prenzel (b), R., M. Schlanges, T. Bornath, and D. Kremp (1996). In *Proc. of the Int. Conf. on the Physics of Strongly Coupled Plasmas, Binz 1995; Eds. W. D. Kraeft and M. Schlanges; World Scientific, Singapore*, pp. 189.
Prigogine, I. (1963). *Nonequilibrium Statistical Mechanics.* New York: Wiley Interscience.
Puff, R. D. (1965). *Phys. Rev. A* **137**, 406.
Quinn, J. J. and R. A. Ferrel (1958). *Phys. Rev.* **112**, 812.
Radtke, R. and K. Günther (1976). *J. Phys.* **D9**, 1131.
Rajagopal, A. K. and C. K. Majumdar (1970). *J. Math. and Phys. Sciences* **4**, 109.
Rand, S. (1964). *Phys. Rev.* **136**, B231.
Rasolt, M. and F. Perrot (1989). *Phys. Rev. Letters* **62**, 2273.
Recoules, V., P. Renaudin, C. J, P. Noiret, and G. Zérah (2002). *Phys. Rev.* **E66**, 056412.
Redmer, R. (1997). *Phys. Rep.* **282**, 35.
Redmer, R. (1999). *Phys. Rev.* **E59**, 1073.
Redmer, R., H. Reinholz, G. Röpke, R. Winter, F. Noll, and F. Hensel (1992). *J. Phys.: Condens. Matter* **4**, 1659.
Redmer, R. and G. Röpke (1985). *Physica* **A130**, 523.
Redmer, R. and G. Röpke (1989). *Contrib. Plasma Phys.* **29**, 343.
Redmer, R., G. Röpke, S. Kuhlbrodt, and H. Reinholz (2001). *Phys. Rev.* **B63**, 233104.
Redmer, R., G. Röpke, F. Morales, and K. Kilimann (1990). *Phys. Fluids B* **2**, 390.
Redmer, R., G. Röpke, and R. Zimmermann (1987). *J. Phys. B* **20**, 4069.
Redmer, R., T. Rother, K. Schmidt, W. D. Kraeft, and G. Röpke (1988). *Contrib. Plasma Phys.* **28**, 41–55.
Redmer, R. and W. W. Warren (1993a). *Phys. Rev.* **B48**, 14892.
Redmer, R. and W. W. Warren (1993b). *Contrib. Plasma Phys.* **33**, 374.
Redmer, R. and W. W. Warren Jr. (1993). *Phys. Rev.* **B48**, 14892.
Ree, F. H. (1988). In S. C. Schmidt and N. C. Holmes (Eds.), *Shock Waves in Condensed Matter.* Elsevier Science Publishers.
Reimann, U. and C. Toepffer (1990). *Laser and Particle Beams* **8**, 771.

Reinholz, H. and R. Redmer (1993). *J. Non-Cryst. Solids* **156-158**, 654.
Reinholz, H., R. Redmer, and S. Nagel (1995). *Phys. Rev.* **E52**, 5368.
Reinholz, H., R. Redmer, G. Röpke, and A. Wierling (2000). *Phys. Rev.* **E62**, 5648.
Reinholz, H., R. Redmer, and D. Tamme (1989). *Contrib. Plasma Phys.* **29**, 395.
Renaudin, P., C. Blancard, G. Faussurier, and P. Noiret (2002). *Phys. Rev. Lett.* **88**, 215001.
Resibois, P. (1965). *Physica* **31**, 645.
Rice, T. M. (1977). The electron–hole liquid in semiconductors: Theoretical aspects. In H. Ehrenreich, F. Seitz, and D. Turnbull (Eds.), *Solid State Physics, Vol. 32*. New York: Academic Press Inc.
Richert, W., S. A. Insepov, and W. Ebeling (1984). *Ann. Physik (Leipzig)* **41**, 139.
Riemann, J. (1997). *PhD thesis*. Greifswald: Greifswald University.
Riemann, J., M. Schlanges, H. E. DeWitt, and W. D. Kraeft (1995). *Physica* **A219**, 423.
Rietz, V. (1996). *PhD thesis*. Rostock: Universität Rostock.
Rinker, G. A. (1985). *Phys. Rev.* **B31**, 4207,4220.
Rinker, G. A. (1988). *Phys. Rev.* **A37**, 1284.
Robnik, M. and W. Kundt (1983). *Astron. Astrophys.* **120**, 227.
Rogers, F. J. (1971). *Phys. Rev.* **A4**, 1145.
Rogers, F. J. (1986). *ApJ* **320**, 723.
Rogers, F. J. (1990). *ApJ* **352**, 689.
Rogers, F. J. (2000). *Phys. Plasmas* **7**, 51.
Rogers, F. J. and H. E. DeWitt (1973). *Phys. Rev.* **A8**, 1061.
Rogers, F. J., H. E. DeWitt, and D. B. Boercker (1981). *Phys. Letters* **82A**, 331.
Rogers, F. J., H. C. Graboske, and H. E. DeWitt (1971). *Phys. Letters A* **34**, 127.
Rogers, F. J., H. C. Graboske, and D. J. Harwood (1970). *Phys. Rev. A* **1**, 1577.
Rogers, F. J. and D. A. Young (1997). *Phys. Rev. E* **56**, 5876.
Röpke, G. (1983). *Physica* **121A**, 92.
Röpke, G. (1988a). *Phys. Rev.* **E57**, 4673.
Röpke, G. (1988b). *Phys. Rev.* **A88**, 3001.
Röpke, G. (1998). *Phys. Rev.* **E57**, 4673.
Röpke, G. and R. Der (1979). *phys. stat. sol. (b)* **92**, 501.
Röpke, G., K. Kilimann, D. Kremp, and W. D. Kraeft (1978). *Phys. Letters* **68A**, 329–332.
Röpke, G. and R. Redmer (1989). *Phys. Rev.* **A39**, 907.
Röpke, G. and A. Wierling (1998). *Phys. Rev.* **E57**, 7075.
Röpke, G. and A. Wierling (1999). *Phys. Letters* **A260**, 365.
Ross, M. (1996). *Phys. Rev. B* **54**, R9589.
Ross, M. (1998). *Phys. Rev.* **B58**, 669.
Ross, M., F. H. Ree, and D. A. Young (1983). *J. Chem. Phys.* **79**, 1487.
Ross, M. and L. H. Yang (2001). *Phys. Rev.* **B64**, 174102.
Roth, M., C. Stöckl, W. Süss, O. Iwase, D. O. Gericke, R. Bock, and D. H. H. Hoffmann (2000). *Europhys. Lett.* **50**, 28–34.
Saumon, D. and G. Chabrier (1989). *Phys. Rev. Lett.* **62**, 2397.
Saumon, D. and G. Chabrier (1991). *Phys. Rev. A* **44**, 5122.
Saumon, D. and G. Chabrier (1992). *Phys. Rev. A* **46**, 2084.
Schäfer, W., R. Binder, and K. H. Schuldt (1986). *Z. Physik* **B63**, 407.
Schäfer, W. and J. Treusch (1986). *Z. Phys. B: Condensed Matter* **63**, 407.
Schepe, R. (2001). *PhD thesis*. Rostock: Rostock University.
Schepe, R., T. Schmielau, D. Tamme, and K. Henneberger (1998). *phys. stat. sol. (b)* **206**, 273.
Schirmer, H. (1955). *Z. Physik* **142**, 1.
Schirmer, H. and J. Friedrich (1958). *Z. Physik* **151**, 174.

Schlanges, M. (1985). *B–Dissertation (Habilitation).* Rostock: University of Rostock.
Schlanges, M., M. Bonitz, and A. Tschtschjan (1995). *Contrib. Plasma Phys.* **35**, 109.
Schlanges, M., M. Bonitz, and A. Tschttschjan (1993). In *Strongly Coupled Plasma Physics, Eds. H. Van Horn and S. Ichimaru, University of Rochester Press*, pp. 77–80.
Schlanges, M. and T. Bornath (1989). *Contrib. Plasma Phys.* **29**, 527–536.
Schlanges, M. and T. Bornath (1993). *Physica* **A192**, 262–279.
Schlanges, M. and T. Bornath (1997). *Contrib. Plasma Phys.* **37**, 239–249.
Schlanges, M. and T. Bornath (1998). In *Proceedings of the Int. Conf. on Strongly Coupled Coulomb Systems; Eds. G. J. Kalman and J. M. Rommel and K. Blagoev, Plenum Press*, pp. 95–102.
Schlanges, M., T. Bornath, P. Hilse, and D. Kremp (2003). *Contrib. Plasma Phys.* **43**, 360.
Schlanges, M., T. Bornath, and D. Kremp (1988). *Phys. Rev. A* **38**, 2174–2177.
Schlanges, M., T. Bornath, and D. Kremp (1992). In W. Ebeling, A. Förster, and R. Radtke (Eds.), *Physics of Non-ideal Plasmas.* Leipzig: Teubner.
Schlanges, M., T. Bornath, D. Kremp, M. Bonitz, and P. Hilse (2000). *J. Phys. IV France* **10**, Pr5–323.
Schlanges, M., D. O. Gericke, W. D. Kraeft, and T. Bornath (1998). *Nuclear Instr. A* **415**, 517–524.
Schlanges, M. and D. Kremp (1982). *Ann. Physik (Leipzig)* **39**, 69.
Schlanges, M., D. Kremp, and H. Keuer (1984). *Ann. Physik (Leipzig)* **41**, 54–66.
Schlanges, M., D. Kremp, and W. Kraeft (1985). *Beitr. Plasmaphysik (Contrib. Plasma Phys.)* **25**, 233–239.
Schlanges (a), M., T. Bornath, R. Prenzel, and D. Kremp (1996). In *Atomic Processes in Plasmas, Tenth Topical APS Conference, Eds. A. L. Osterheld and W. H. Goldstein, AIP Conference Proceedings no.381*, pp. 215.
Schlanges (b), M., T. Bornath, V. Rietz, and D. Kremp (1996). *Phys. Rev. E* **53**, 2751–2756.
Schlessinger, L. and J. Wright (1979). *Phys. Rev.* **A20**, 1934.
Schmielau, T. (2001). *PhD thesis.* Rostock: Rostock University.
Schmielau, T. (2003). In M. Bonitz and D. Semkat (Eds.), *Progress in Nonequilibrium Green's Functions II.* Singapore: World Scientific.
Schmielau, T., G. Manzke, D. Tamme, and K. Henneberger (2000). *phys. stat. sol. (b)* **221**, 215.
Schmielau, T., G. Manzke, D. Tamme, and K. Henneberger (2001). In *Proc. 25th Int. Conf. Phys. Semicond., Springer, Berlin*, pp. 81.
Schwinger, J. (1961). *J. Math. Phys.* **2**, 407.
Seele, C., G. Zwicknagel, C. Toepffer, and P.-G. Reinhard (1998). *Phys. Rev.* **E57**, 3368.
Seidel, J. (1977). *Z. Naturforsch.* **a32**, 1195.
Seidel, J., S. Arndt, and W. D. Kraeft (1995). *Phys. Rev.* **E52**, 5387.
Selchow, A., G. Röpke, A. Wierling, H. Reinholz, T. Pschiwul, and G. Zwicknagel (2001). *Phys. Rev.* **E64**, 056410.
Semkat, D. (2001). *PhD thesis.* Rostock: Rostock University.
Semkat, D. and M. Bonitz (2000). In *Progress in Nonequlibrium Green's Functions, Ed. M. Bonitz, World Scientific, Singapore*, pp. 504.
Semkat, D., M. Bonitz, K. D, M. S. Murillo, and D. O. Gericke (2003). In *Progress in Nonequilibrium Green's Functions II, M. Bonitz and D. Semkat, Eds., World Scientific Singapore*, pp. 83.
Semkat, D., D. Kremp, and M. Bonitz (1999). *Phys. Rev. E* **59**, 1557.

Semkat, D., D. Kremp, and M. Bonitz (2000). *J. Math. Phys.* **41**, 7458.
Shepherd, R. L., D. R. Kania, and L. A. Jones (1988). *Phys. Rev. Lett.* **61**, 1278.
Shindo, K. (1970). *J. Phys. Soc. Japan* **29**, 287.
Shkarofsky, I. P., T. W. Johnston, and M. P. Bachinski (1966). *Particle Kinetics of Plasmas.* London: Addison Weseley.
Sigeneger, F., S. Arndt, R. Redmer, M. Luft, D. Tamme, W. D. Kraeft, G. Röpke, and T. Meyer (1988). *Physica* **A152**, 365.
Sigmund, P. (1982). *Phys. Rev.* **A26**, 2497.
Silin, V. P. (1960). *Zh. Eksp. Teor. Fiz.* **38**, 1771.
Silin, V. P. (1961). *Zh. Eksp. Teor. Fiz.* **41**, 861.
Silin, V. P. (1964). *Zh. Eksp. Teor. Fiz.* **47**, 2254.
Silin, V. P. (1999). *Quantum Electron.* **29**, 11.
Silin, V. P. and S. A. Uryupin (1981). *Sov. Phys. JETP* **54**, 485.
Silvestrini, P. L. (1999). *Phys. Rev. B* **60**, 16382.
Simon, A. H., S. J. Kirch, and J. P. Wolfe (1992). *Phys. Rev.* **B46**, 10098.
Singwi, K. S., A. Sjolander, M. P. Tosi, and R. H. Land (1970). *Phys. Rev.* **B1**, 1044.
Singwi, K. S. and M. P. Tosi (1981). In F. Seitz, D. Turnbull, and H. Ehrenreich (Eds.), *Solid State Physics, Vol. 36, p. 177.* New York: Academic Press.
Singwi, K. S., M. P. Tosi, R. H. Land, and A. Sjolander (1968). *Phys. Rev.* **176**, 589.
Smith, L. M. and J. P. Wolfe (1986). *Phys. Rev. Letters* **57**, 2314.
Smith, W. R. and B. Triska (1994). *J. Chem. Phys.* **100**, 3019.
Sobel'man, I. I., L. A. Vainshtein, and E. A. Yukov (1988). *Excitation of Atoms and Broadening of Spectral Lines.* Berlin: Springer.
Spitzer, L. (1967). *Physics of Fully Ionized Gases.* New York: Interscience.
Spitzer, L. and R. Härm (1953). *Phys. Rev.* **89**, 977.
Starostin, A. N. and N. L. Aleksandrov (1998). *Phys. Plasmas* **5**, 2127.
Starzynski, J., R. Redmer, and M. Schlanges (1996). *Phys. Plasmas* **3**, 1591–1602.
Stöckl, C., O. Boine-Frankenheim, M. Roth, W. Süss, H. Wetzler, W. Seelig, M. Kulish, M. Dornik, W. Laux, P. Spiller, M. Stetter, S. Stoeve, J. Jacobi, and D. H. H. Hoffmann (1996). *Laser and Particle Beams* **14**, 561.
Stolz, H. (1974). *Einführung in die Vielelektronentheorie der Kristalle.* Berlin.
Stolz, H. and R. Zimmermann (1979). *phys stat sol (b)* **94**, 135.
Stolzmann, W. and T. Blöcker (1996). In *Proc. Int. Conf. on the Physics of Strongly Coupled Plasmas, Binz 1995; Singapore 1996, World Scientific, W. D. Kraeft and M. Schlanges, Eds.*, pp. 95.
Stolzmann, W. and W. D. Kraeft (1979). *Ann. Physik (Leipzig)* **36**, 388–398.
Stolzmann, W., W. D. Kraeft, and I. Fromhold-Treu (1989). *Contrib. Plasma Phys.* **29**, 377–388.
Stone, P. M. and J. R. Reitz (1963). *Phys. Rev.* **131**, 2101.
Tahir, N. A., A. R. Piriz, A. Shutov, D. Varentsov, S. Udrea, D. H. H. Hoffmann, H. Juranek, R. Redmer, R. F. Portugues, I. Lomonosov, and V. E. Fortov (2003). *J. Phys. A: Math. Gen.* **36**, 6129.
Taylor, J. R. (1972). *Scattering Theory.* New York: John Wiley & Sons.
Theobald, W., R. Hässner, R. Kingham, R. Sauerbrey, R. Fehr, D. O. Gericke, M. Schlanges, W. D. Kraeft, and K. Ishikawa (1999). *Phys. Rev.* **E59**, 3544.
Theobald, W., R. Hässner, C. Wülker, and R. Sauerbrey (1996). *Phys. Rev. Lett.* **77**, 298.
Theobald, W., R. Häßner, C. Wülker, and R. Sauerbrey (1997). *private communication .*
Tkachenko, I. M., J. Alcober, and J. L. Muñoz-Cobo (2002). *Contrib. Plasma Phys.* **42**, 467.

Tkachenko, I. M. and P. F. de Cordoba (1998). *Phys. Rev.* **E57**, 2222.
Tonks, L. and I. Langmuir (1929). *Phys. Rev.* **33**, 195.
Totsuji, H. and S. Ichimaru (1974). *Progr. Theor. Phys.* **52**, 42.
Trubnikov, B. A. and V. F. Elesin (1964). *Zh. Eksp. Teor. Fiz.* **47**, 1279.
van Kampen, N. G. and B. U. Felderhof (1967). *Theoretical Methods in Plasma Physics.* Amsterdam.
Van Regenmorter, H. (1962). *ApJ* **132**, 906.
Vedenov, A. A. and A. I. Larkin (1959). *Zh. Eksp. Teor. Fiz.* **36**, 1133.
Vedenov, A. A., E. P. Velikhov, and R. Z. Sagdeev (1961). *Nucl. Fusion* **1**, 82.
Vinogradov, A. V. and V. P. Shevel'ko (1976). *Zh. Eksp. Teor. Fiz.* **71**, 1037.
Vlasov, A. A. (1938). *Zh. Eksp. Teor. Fiz.* **8**, 291.
Volkov, D. M. (1934). *Z. Physik* **94**, 125.
Vorberger, J. (2005). *PhD thesis.* Greifswald: Greifswald University. *submitted.*
Vorberger, J., M. Schlanges, and W. D. Kraeft (2004). *Phys. Rev.* **E69**, 046407.
Waech, T. G. and R. B. Bernstein (1967). *J. Chem. Phys.* **46**, 4905.
Wagner, M. (1991). *Phys. Rev.* **B44**, 6104.
Weinstock, J. (1963). *Phys. Rev.* **132**, 454.
Weir, S. T., A. C. Mitchell, and W. J. Nellis (1996). *Phys. Rev. Lett.* **76**, 1860.
Weisheit, J. C. (1988). *Adv. Atomic and Molecular Phys.* **25**, 101.
Wetzler, H., W. Süss, C. Stöckl, A. Tauschwitz, and D. H. H. Hoffmann (1997). *Laser and Particle Beams* **15**, 449.
Whitten, L. B., N. F. Lane, and J. C. Weisheit (1984). *Phys. Rev.* **A29**, 945.
Wierling, A. (1997). *PhD thesis.* Rostock: Rostock University.
Wierling, A., T. Millat, G. Röpke, R. Redmer, and H. Reinholz (2001). *Phys. Plasmas* **8**, 3810.
Wigner, E. P. and H. B. Huntington (1935). *J. Chem. Phys.* **3**, 764.
Williams, R. H. and H. E. DeWitt (1969). *Phys. Fluids* **12**, 2326.
Winkler, T., K. Bechert, F. Bosch, H. Eickhoff, B. Franzke, F. Nolden, H. Reich, B. Schlitt, and M. Steck (1997). *Nucl. Instr. Meth. Phys. Res.* **A391**, 12.
Wolf, A., C. Ellert, M. Grieser, D. Habs, B. Hchadel, R. Repnow, and D. Schwalm (1994). In J. Bosser (Ed.), *Beam Cooling and Related Topics.* Genf: CERN.
Wolniewicz, L. (1966). *J. Chem. Phys.* **45**, 515.
Wyld, W. and D. Pines (1962). *Phys. Rev.* **127**, 1851.
Yan, X.-Z., S. Tanaka, S. Mitake, and S. Ichimaru (1985). *Phys. Rev.* **A32**, 1785.
Young, F. C., D. Mosher, S. J. Stephanakis, S. A. Goldstein, and T. A. Mehlhorn (1982). *Phys. Rev. Lett.* **49**, 549.
Zaremba, E., A. Arnau, and P. M. Echenique (1995). *Nucl. Instr. and Meth. in Phys. Res.* **B96**, 619.
Zhdanov, V. M. (1982). *Transport Processes in Multi-Component Plasmas (in Russian).* Moscow: Energoizdat.
Ziegler, J. F. (1999). *J. Appl. Phys.* **85**, 1249.
Ziman, J. M. (1961). *Phil. Mag.* **6**, 1013.
Ziman, J. M. (1972). *Principles of the Theory of Solids.* London: Cambridge University Press.
Zimmermann, R. (1976). *phys. stat. sol. (b)* **76**, 191.
Zimmermann, R. (1988). *Many Particle Theory of Highly Excited Semiconductors.* Leipzig: Teubner.
Zimmermann, R., K. Kilimann, W. D. Kraeft, D. Kremp, and G. Röpke (1978). *phys. stat. sol. (b)* **90**, 175–187.
Zubarev, D., V. Morozov, and G. Röpke (1996). *Statistical Mechanics of Nonequilibrium Processes.* Berlin: Akademie-Verlag.
Zubarev, D. N. (1960). *Uspekhi Fiz. Nauk [Soviet Phys. Uspekhi]* **71** [**3**], 71 [320].

Zubarev, D. N. (1974). *Nonequilibrium Staistical Thermodynamics.* New York: Consultants Bureau.
Zwanzig, R. (1960). *J. Chem. Phys.* **33**, 1338.
Zwicknagel, G. and C. Deutsch (1997). *Phys. Rev.* **E56**, 970.
Zwicknagel, G., D. Klakow, P.-G. Reinhard, and C. Toepffer (1993). *Contrib. Plasma Phys.* **33**, 395.
Zwicknagel, G., P.-G. Reinhard, C. Seele, and C. Toepffer (1996). *Fusion Eng. and Des.* **32-33**, 395.
Zwicknagel, G., C. Toepffer, and P.-G. Reinhard (1999). *Phys. Rep.* **309**, 117.

Index

Springer Series on

ATOMIC, OPTICAL, AND PLASMA PHYSICS

Springer Series on

ATOMS + PLASMAS

Editors: G. Ecker P. Lambropoulos I.I. Sobel'man H. Walter
Founding Editor: H.K.V. Lotsch

1 **Polarized Electrons**
2nd Edition
By J. Kessler

2 **Multiphoton Processes**
Editors: P. Lambropoulos and S.J. Smith

3 **Atomic Many-Body Theory**
2nd Edition
By I. Lindgren and J. Morrison

4 **Elementary Processes in Hydrogen-Helium Plasmas**
Cross Sections and Reaction Rate Coefficients
By R.K. Janev, W.D. Langer, K. Evans Jr., and D.E. Post Jr.

5 **Pulsed Electrical Discharge in Vacuum**
By G.A. Mesyats and D.I. Proskurovsky

6 **Atomic and Molecular Spectroscopy**
Basic Aspects and Practical Applications
3rd Edition
By S. Svanberg

7 **Interference of Atomic States**
By E.B. Alexandrov, M.P. Chaika, and G.I. Khvostenko

8 **Plasma Physics**
Basic Theory with Fusion Applications
3rd Edition
By K. Nishikawa and M. Wakatani

9 **Plasma Spectroscopy**
The Influence of Microwave and Laser Fields
By E. Oks

10 **Film Deposition by Plasma Techniques**
By M. Konuma

11 **Resonance Phenomena in Electron–Atom Collisions**
By V.I. Lengyel, V.T. Navrotsky, and E.P. Sabad

12 **Atomic Spectra and Radiative Transitions**
2nd Edition
By I.I. Sobel'man

13 **Multiphoton Processes in Atoms**
2nd Edition
By N.B. Delone and V.P. Krainov

14 **Atoms in Plasmas**
By V.S. Lisitsa

15 **Excitation of Atoms and Broadening of Spectral Lines**
2nd Edition, By I.I. Sobel'man, L. Vainshtein, and E. Yukov

16 **Reference Data on Multicharged Ions**
By V.G. Pal'chikov and V.P. Shevelko

17 **Lectures on Non-linear Plasma Kinetics**
By V.N. Tsytovich

18 **Atoms and Their Spectroscopic Properties**
By V.P. Shevelko

19 **X-Ray Radiation of Highly Charged Ions**
By H.F. Beyer, H.-J. Kluge, and V.P. Shevelko

20 **Electron Emission in Heavy Ion–Atom Collision**
By N. Stolterfoht, R.D. DuBois, and R.D. Rivarola

21 **Molecules and Their Spectroscopic Properties**
By S.V. Khristenko, A.I. Maslov, and V.P. Shevelko

22 **Physics of Highly Excited Atoms and Ions**
By V.S. Lebedev and I.L. Beigman

23 **Atomic Multielectron Processes**
By V.P. Shevelko and H. Tawara

24 **Guided-Wave-Produced Plasmas**
By Yu.M. Aliev, H. Schlüter, and A. Shivarova

25 **Quantum Statistics of Nonideal Plasmas**
By D. Kremp, M. Schlanges and W.-D. Kraeft

26 **Atomic Physics with Heavy Ions**
By H.F. Beyer and V.P. Shevelko